A Dictionary of
# Plant Sciences

**Michael Allaby** has written many books on environmental science and especially on climatology and meteorology. These include the *Encyclopedia of Weather and Climate*; the *Facts on File Weather and Climate Handbook*; and the *DK Guide to Weather*. He is the General Editor of the *Oxford Dictionary of Zoology*, and co-author of the *Dictionary of Earth Sciences*.

### Contributors and Advisers

Michael Allaby     Diana Sainsbury
Robin Allaby        T. C. Whitmore
Michael Kent

A Dictionary of

# Plant
# Sciences

*Edited by*
MICHAEL ALLABY

OXFORD
UNIVERSITY PRESS

# OXFORD

UNIVERSITY PRESS

Great Clarendon Street, Oxford OX2 6DP

Oxford University Press is a department of the University of Oxford
It furthers the University's objective of excellence in research, scholarship,
and education by publishing worldwide in

Oxford New York

Auckland Bangkok Buenos Aires Cape Town Chennai
Dar es Salaam Delhi Hong Kong Istanbul Karachi Kolkata
Kuala Lumpur Madrid Melbourne Mexico City Mumbai Nairobi
São Paulo Shanghai Taipei Tokyo Toronto

Oxford is a registered trade mark of Oxford University Press
in the UK and in certain other countries

Published in the United States
by Oxford University Press Inc., New York

First published as an Oxford University Press paperback 1992
and simultaneously in a hardback edition under the title *The Concise Oxford Dictionary of Botany*
Second edition 1998 retitled *A Dictionary of Plant Sciences*
Reissued with new covers 2004
Revised edition 2006

British Library Cataloguing in Publication Data

Data available

Library of Congress Cataloging in Publication Data

Data available

ISBN 978-019-860891-2

1

Typeset by SPI Publisher Services, Pondicherry, India
Printed by Clays Ltd, St Ives plc

# From the Preface to the First Edition

This dictionary aims to address the needs of students as well as those of anyone whose profession or interest exposes him or her to books, articles, or scientific and technical papers on environmental topics.

The dictionary is not a textbook and does not pretend to be one. The first decision the editor of a dictionary must make involves defining the word 'dictionary' itself. The book might contain a limited number of fairly long entries, making it rather like an encyclopaedia. Alternatively, it might contain a much larger number of much shorter entries consisting only of definitions. A third approach, which has been adopted here, amounts to a compromise between the two extremes. This dictionary contains many very short entries, sometimes of only two or three words. Other entries are longer, going beyond the simple definition to describe complex biological processes as succinctly as is practicable. The result of the compromise is that the book is able to compress into one volume explanations of well over 5000 terms. These are drawn from ecology, Earth sciences, Earth history, evolution, genetics, plant physiology, biochemistry, cytology, and biogeography. There are brief biographical notes on individuals who have made outstanding contributions to the disciplines relevant to botany, and almost one-third of the entries describe taxonomic groups of seed plants, ferns, algae, mosses, and liverworts, and also fungi, bacteria, and slime moulds.

Entries are fairly extensively cross-referenced to words or expressions that have entries of their own or sometimes, for simplicity, to another part of speech (an adjective cross-referenced to its noun, for example) where the entry is easily recognizable. The cross-references use the customary *see* and *compare* within or at the end of entries, and asterisks before expressions that have entries of their own (e.g. *glycolysis). The cross-references should carry the reader from one entry to another, but, of course, the system involved another compromise. Too many cross-references interrupt the flow of the text, making it more difficult to follow rather than easier. Terms are cross-referenced, therefore, only where the cross-reference may be helpful.

MICHAEL ALLABY
*Wadebridge, Cornwall*

# Preface to the Second Edition

Five years have now elapsed since the first edition of this dictionary was published. During that time, popular interest in the conservation of the countryside has remained intense. Indeed, it has increased to the extent that young people have taken to occupying tree-houses and tunnels in areas they seek to protect from development. Obviously, environmental activists have a keen awareness of plant life and will find explanations of botanical terms helpful.

In revising the dictionary, I have been helped greatly by Robin Allaby, who has scrutinized my revisions and made many of his own. This edition contains many new entries. Some of these are technical terms used by botanists. Others are from molecular biology, the discipline that is now doing so much to illuminate the evolutionary history of plants and their taxonomic relationships.

Each of the entries in the first edition has been examined afresh. Many have been modified, often in only small ways, in order to clarify them or bring them up to date. We have also taken the opportunity to change the title of the dictionary. We believe *A Dictionary of Plant Sciences* is a more accurate description of the book, reflecting the broad approach we have adopted to what is, in truth, a group of interrelated disciplines rather than a single self-contained one.

The revisions do not disguise the fact that the original compilation was the work of a team and would have been impossible without the enthusiastic support and hard work of Mike Kent, Diana Sainsbury, Tim Whitmore, and Robin Allaby. Their work is retained, and if mistakes or omissions have survived, the fault is mine, not theirs.

MICHAEL ALLABY
www.michaelallaby.com

# Contents

**Aapa mires**  *Fens, sometimes called 'string bogs', of circumpolar distribution, found in northern Scandinavia and *boreal Canada. They are *soligenous mires with ridges arranged roughly normal to the slope, along the contours of the terrain. Water occupies the linear depressions between the ridges.

**abacá (Manila hemp)**  The strongest of the hard, natural fibres. It is extracted from the *pseudostems of *Musa textilis* (*see* MUSA), a relative of the edible bananas. Grown mainly in the Philippines, and native to that region, it is used to make ropes that are resistant to humidity, salt water, and fresh water, and to make special, strong papers (e.g. for tea-bags).

**abaxial**  Of a surface, directed away from the axis. *Compare* ADAXIAL.

**Abies (fir; family *Pinaceae)**  A genus of coniferous trees in which the leaves are crowded on the twigs, often approximately in 2 rows; they are needle-like and single, leaving a round, flat scar when they fall, thus producing a smooth twig. The female cones are borne erect; they shatter at maturity, but the woody axis persists. Many fine ornamental species are now widely cultivated for their lofty, deeply pyramidal, *monopodial crowns; but they require a moist climate and are sensitive to atmospheric pollution. There are 40–50 species, widespread throughout the northern hemisphere, especially on mountains.

**abiogenesis**  The development of living organisms from non-living matter, as in the supposed origin of life on Earth, or in the concept of spontaneous generation which was once held to account for the origin of life but which modern understanding of evolutionary processes has rendered outdated.

**abiotic**  Non-living; devoid of life. *See* BIOTIC.

**abscisic acid (abscisin II, dormin)**  A terpenoid (*see* TERPENE) compound that is one of the five major *plant hormones. Although it is synthesized principally in the *chloroplasts, it occurs throughout the plant body and is particularly concentrated in the leaves, fruits, and seeds. It has powerful growth-inhibiting properties generally and also promotes leaf *abscission and the *senescence of plants and/or their organs, and induces the closing of *stomata and dormancy in seeds and buds. Its effect is antagonistic to the plant *growth hormones, and is thought to act by inhibiting the synthesis of *protein and *nucleic acids.

**abscisin II**  *See* ABSCISIC ACID.

**abscission**  The rejection of plant organs, e.g. of leaves in autumn. This occurs at an abscission zone, where *hydrolytic *enzymes reduce cell adhesion. The process can be promoted by *abscisic acid and inhibited by respiratory poisons, and is controlled in nature by the proportion and gradients of *auxin and *ethylene. Other *hormones may be involved.

**absent rings**  *See* TREE-RING.

**absolute pollen frequency  (APF)**  The expression of *pollen data from sediments in terms of the absolute numbers (for each species, genus, or family) per unit area of surface and, where deposition rates are known, per unit time. In certain circumstances this approach gives clearer information than does the traditional expression as *relative pollen frequencies (RPFs). APFs are particularly useful in site comparisons in which one or more high-pollen producers vary. For example, when trees first appear in the regional pollen rain their prolific pollen may, in an RPF method, give the impression of declining herbaceous species, whereas examination by an APF method will show constant values for herb species.

**absorption**  The uptake of substances, usually nutrients, water, or light, by cells or *tissues.

a

**absorption spectrum** A graph that shows the percentage of each wavelength of light absorbed by a pigment (e.g. *chlorophyll, which absorbs mainly in the red and blue parts of the spectrum).

**abstriction** The detachment and release of a *spore by the constriction of the tissue by which it is attached.

*Abutilon* (family *Malvaceae) A genus of trees and shrubs, native to tropical and warm temperate climates, whose members have hairy leaves and branches, and bright, lantern-like flowers, mainly orange or red. The fruit is a *schizocarp and the stigmas (*see* CARPEL) are usually apical. There are more than 100 species. The genus includes *A. theophrasti* (China jute).

*Acacia* (family *Fabaceae, subfamily *Mimosoideae) A genus of plants, most of which are trees (wattles), although a few are climbers. Typically the leaves are *bipinnate, with numerous tiny leaflets, or phyllodic (*see* PHYLLODY). Acacias are important for timber, fuel wood, tannin, gum arabic (especially *A. senegal*), perfumes (*A. farnesiana*), and florists' 'mimosa' (usually *A. dealbata*). There are about 1200 species, most of them in the seasonal tropics and subtropics.

*Acaena* (family *Rosaceae, subfamily Rosoideae, *tribe Sanguisorbeae) A genus of wind-pollinated, mat-forming plants that occur mainly in the subtropics and temperate regions. There are about 100 species; 90 species are known in S. America, 14 in New Zealand, 2 in Australia, and others in New Guinea, S. Africa, and California.

*Acalypha* (family *Euphorbiaceae) A genus of shrubs and small trees in which the flowers are unisexual and apetalous, males having elongate, twisted *antherlobes, and females much-branched stigmas (*see* CARPEL); some species bear *catkins (e.g. the ornamental *A. hispida*, 'cat's tail'). *A. wilkesiana* and others have colourful foliage. There are no close relatives in the family; the resemblance to *Urticaceae is only superficial. There are about 430 species, found in the tropics and subtropics.

**Acanthaceae** A family, mainly of shrubs, in which the leaves are simple, *opposite, and *decussate. The flowers are bisexual and *zygomorphic, and the *bracts are often showy. There are 4 or 5 fused *sepals and *petals; 2–4 *epipetalous *stamens; a *superior, *bilocular *ovary; and numerous, axile (*see* PLACENTATION) *ovules. The fruit is a *capsule. There are several ornamentals (e.g. *Acanthus*, *Aphelandra*, and *Thunbergia*). There are 357 genera, comprising about 4350 species, most of them tropical, but with temperate outliers.

*Acanthus* (family *Acanthaceae) A genus of shrubs and perennial *herbs, most of which are *xeromorphic and have spiny leaves. Some species are cultivated as ornamentals. The upper lip of the *corolla is lacking. The pattern for the decoration on Corinthian column capitals is supposedly based on the leaves of *A. spinosus*. There are 30 known species, found in southern Europe (*A. mollis* is bear's breech), Asia, and Africa (*A. illicifolius* grows in mangrove swamps).

**accessory cell** *See* SUBSIDIARY CELL.

**accessory pigments** In *photosynthesis, pigments that can absorb light energy and pass the electrons they emit to *primary pigments.

**accidental species** One of five classes of fidelity used by the *Braun–Blanquet school of phytosociology in the description and classification of plant communities. Accidentals are rare species in the community present either as chance invaders from another community or as *relics from a previous community. *Compare* EXCLUSIVE SPECIES; INDIFFERENT SPECIES; PREFERENTIAL SPECIES; and SELECTIVE SPECIES.

**acclimatization** (acclimation, hardening) The changes involving the synthesis of *proteins, *membranes, and *metabolites that occur in a plant in response to chilling or freezing temperatures, which protect *tissues or confer tolerance to the cold. The term may also be applied to a range of physiological adjustments which occur in a plant when it is subjected to unusual environmental conditions.

**accumulated temperature** The sum, counted in degrees, by which the actual air temperature rises above or falls below a datum level over a prolonged period. The datum level is usually set at a value relevant to an ecological study or to crop production (e.g. the critical temperature for sustained plant growth is 6°C). If the mean temperature on a particular day is $m°$ above or below $(-m°)$ the datum level and remains there for $n$ hours (= $n/24$ days), then the accumulated temperature for that day is $mn/24$ degree-days. Degree-days can be added to give the accumulated temperature over a week, month, season, or year.

**accumulator 1.** In plant *succession studies, a pioneer plant species whose activities are claimed to enrich the *abiotic environment with nutrients. **2.** A plant that accumulates certain elements, such as selenium.

**-aceae** A standardized suffix used to indicate a family of plants in the recognized codes of classification (e.g. *Rosaceae, the rose family).

**acellular slime moulds** *See* MYXO-MYCETES.

**acentric** Applied to a fragment of a *chromosome, formed during cell division, that lacks a *centromere. The fragment will be unable to follow the rest of the chromosomes in migration towards one or other pole as it has lost its point of attachment to the *spindle.

*Acer* (family *Aceraceae) A genus of trees in which the leaves are without *stipules, *opposite, and entire, or often *palmately lobed (but *pinnate in *A. negundo*, box elder). Regular, *pentamerous flowers are borne in *racemes, *corymbs, or *umbels. The fruit is a *samara. These trees are important for timber, many small Japanese species and varieties are ornamentals, and in N. America maple syrup is obtained mainly from *A. saccharum*. Field maple (*A. campestre*) is wild in Britain; sycamore (*A. pseudoplatanus*), from southern Europe, is naturalized in Britain. There are 111 known species, mainly northern temperate, and a few in tropical mountains. China has most native species.

**Aceraceae** A small family of 2 genera of trees, *Acer* and *Dipteronia* (which comprises 2 species found only in China). The family is related to *Hippocastanaceae and *Sapindaceae.

**acervulus** An asexual, *conidia-bearing structure that is formed by certain *fungi *parasitic in plants. It consists of a mat of fungal tissue which bears a layer of *conidiophores; initially formed within the plant *tissues, it later breaks through to the surface to release conidia.

*Acetabularia* **(mermaid's cup, mermaid's wineglass)** A genus of *dasycladalean algae in which the mature *thallus consists of a single *axis or 'stalk' with a whorl of *gametangial sacs at the top. Found in warm regions in shallow, sheltered seas, attached to rocks, shells, etc. Fossil *Acetabularia* dating from the *Tertiary have been found.

**acetic acid bacterium** A *bacterium that produces acetic acid from ethanol (ethyl alcohol).

*Acetobacter* (family *Acetobacteraceae) A genus of *Gram-negative, *aerobic *bacteria. Cells are ovoid or rod-shaped, motile or non-motile. Most strains can oxidize ethanol (ethyl alcohol) to carbon dioxide, forming acetic acid as an intermediate. Some species are used in the manufacture of vinegar. They are found on fruits and vegetables, in alcoholic beverages (in which they can cause spoilage), etc.

**Acetobacteraceae** A family of *aerobic, *chemo-organotrophic, *Gram-negative bacteria which, typically, can oxidize ethanol to acetic acid.

**achene** A small, usually single-*seeded, dry, *indehiscent *fruit.

**achira** *See* CANNA.

*Acholeplasma* *See* ACHOLEPLASMAT-ACEAE.

**Acholeplasmataceae (order Mycoplasmatales)** A family of *Gram-negative, *chemo-organotrophic bacteria that do not require the presence of sterols (*see* STEROID) for growth. Cells are spherical, *pleomorphic, or filamentous. They are found as *parasites in a variety of mammals and birds. There is 1 genus (*Acholeplasma*).

*Achras* (family *Sapotaceae) A genus of plants that includes *A. zapota* (*Manilkara zapota*) which yields the delicious fruit chiku, or sapodilla plum, for which it is now widely cultivated, and chicle, formerly the elastic component of chewing gum. There are 4 species, found in northern tropical America.

**A-chromosome**   *See* B-CHROMOSOME.

**acicular**   Pointed or needle-shaped.

**acid**   According to the Brønsted–Lowry theory, a substance that in solution liberates hydrogen ions or protons. The Lewis theory states that it is a substance that acts as an electron-pair acceptor. An acid reacts with a base to give a salt and water (neutralization), and has a *pH of less than 7.

**acid-fast bacteria**   Certain *bacteria of the order Actinomycetales, including *Mycobacterium species, which, after being treated with certain dyes, are not decolorized on subsequent treatment with a mineral acid.

**acidic dye**   A dye which consists of an organic *anion that combines with and stains positively charged *macromolecules. It is used particularly for staining *cytoplasm. *Compare* BASIC DYE.

**acidic grassland**   A grassland that occurs on *acid soils: it is usually derived from former woodland as a consequence of centuries of grazing and, to a lesser extent, burning. In Britain and much of north-western Europe the dominant grasses are species of *Agrostis* (bent) and *Festuca* (fescue) (*see* HEAVY METAL TOLERANCE). This type of vegetation is most extensive in upland areas, but the associated plant species tend to be different, and the name 'grass heath' is considered more appropriate. In North America broomsedge (*Andropogon virginicus*), Elliott's broomsedge (*A. elliotti*), supina bluegrass (*Poa supina*), and Canada bluegrass (*P. compressa*) are typical grasses of acidic parts of the prairie.

**acidophilic**   **1.** Refers to the propensity of a cell, its components, or its products to become stained by an *acidic dye. **2.** *See* ACIDOPHILOUS.

**acidophilous**   (acidophilic)   Applied to 'acid-loving' organisms, i.e. organisms that grow best in acidic habitats.

**acid soil**   A soil having a *pH less than 7.0. Degrees of soil acidity are recognized. Soil is regarded as 'very acid' when the reaction is less than pH 5.0. The *USDA lists five standard ranges of soil acidity (less than pH 4.5, extremely acid; 4.5–5.0, very strongly acid; 5.1–5.5, strongly acid; 5.6–6.0, medium acid; and 6.1–6.5, slightly acid). Surface soil horizons of acid brown earths have a reaction of pH 5.0 or less.

***Acoelorraphe*** (formerly *Paurotis*; family *Arecaceae*)   A *monotypic palm genus (*A. wrightii*) which is a low, clump-forming, *monoecious, fan palm with greyish leaves, native to the Florida Everglades, and widely cultivated.

***Acorus*** (family *Araceae*)   A genus of 2 species of warm temperate Asia and N. America. *A. calamus* (sweet flag), a marsh plant, is widely cultivated for its fragrant leaves and *rhizomes which have medicinal properties. The rhizomes are *sympodial, the flowers bisexual.

**acquired characteristics**   The characteristics that are acquired in the lifetime of an organism, according to early evolutionary theorists such as *Lamarck. Lamarck further suggested that traits acquired in one generation in response to environmental stimuli would be transmitted to the *gametes and inherited by the next generation. Thus, over several generations a particular type of organism would become better adapted to its environment. The kinds of acquisition envisaged by Lamarck and their heritability are now discredited, although there has been a recent revival of some aspects of the theory in modified form.

**acrasin**   A chemotactic substance involved in the aggregation response of *cellular slime moulds (order Acrasiales). It is now known to be cyclic AMP (adenosine monophosphate).

**Acrasiomycetes**   (division *Myxomycota)   A class of *slime moulds in which the feeding stage generally consists of individual amoeboid cells (*myxamoebae). *Fruiting bodies are formed by aggregation of myxamoebae and differentiation of the resulting *pseudoplasmodium. Acrasiomycetes are found in soil, decomposing plant material, etc.

***Acremonium*** (class *Hyphomycetes)   A *form-genus of *fungi which form *septate *mycelia and which produce *conidia in *phialids. Some species produce *antibiotics called cephalosporins. (The form genus includes fungi formerly called *Cephalosporium.*)

**Acrisols** *Acid soils that have an *argic horizon with a *cation-exchange capacity lower than 24 cmol$_c$/kg. Acrisols are a reference soil group in the FAO *soil classification.

**acrocarpous moss** A type of moss in which the stems are erect and in which the archegonia (i.e. female sex organs), and hence the *capsules, are borne at the tips of stems or branches. Acrocarpous mosses usually show little or no branching and typically grow in erect tufts. *Compare* PLEUROCARPOUS MOSS.

**acrocentric** Applied to a *chromosome with a *centromere located nearer one end than the other. During the *anaphase stage of cell division, movement of an acrocentric chromosome towards one pole results in the chromosome being shaped like a 'J', as opposed to the normal 'V' shape of a *metacentric chromosome (in which the centromere is in the middle).

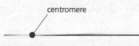

centromere

*Acrocentric*

**acropetal** Growing or developing upwards from the base or point of attachment, so that the oldest parts are at the base and the youngest are at the tip. *Compare* BASIPETAL.

**acropetal movement** The movement of substances within the plant toward its *root and *shoot apices. *Compare* BASIPETAL MOVEMENT.

**acrostichoid** *See* ACROSTICHUM.

**Acrostichum** (family *Pteridaceae) A genus of ferns in which the *rhizome is creeping, the leaves are *pinnate, and the upper pinnae are reduced and covered below by *sporangia (the 'acrostichoid' condition, which also evolved independently in other ferns). There are 3 species in mangrove forests throughout the tropics (*A. aureum* is the commonest). Its young leaves are edible.

**actino-** A prefix that means 'radiating', derived from the Greek *aktis-inos*, meaning 'ray'.

**Actinobacteria** A phylum of *Bacteria comprising organisms, many of which form mycelia (*see* MYCELIUM). Most live in soil and some fix nitrogen (*see* NITROGEN FIXATION), forming *root nodules on certain non-leguminous plants. Because they form mycelia they were formerly mistaken for fungi and known as actinomycetes.

**actinomorphic** Radially symmetrical, as is a *daisy flower.

**actinomycete** *See* ACTINOBACTERIA.

**actinorrhiza** *See* ROOT NODULE.

**actinostele** A *monostele type of *protostele in which the cross-section of the *xylem is star-shaped or lobed. *Compare* HYPOPHLOIC HAPLOSTELE, HAPLOSTELE, and SOLENOSTELE.

**Actinostrobus** (subdivision *Coniferophyta, family *Cupressaceae) A genus of 2 species, endemic to south-western Australia, of trees found as *emergents on heaths and sands in the south and west. *A. pyramidalis* (swamp cypress) is a pyramidal conifer up to 4 m tall, with dense, bright-green foliage reaching to the ground. It is cultivated.

**Actinotus** (family *Apiaceae) A genus that is thought to link the evolution of endemic Australian genera of Apiaceae to cosmopolitan genera. Known as 'flannel flowers', most members of the genus are *biennial herbs found in areas of low fertility and in some arid regions. There are 17 species found in Australia, with 1 in New Zealand.

**action spectrum** A graph of the efficiency of different wavelengths of light in promoting a given photoresponse, as in *photosynthesis or *phototropism.

**activation energy** (energy of activation) The energy that must be delivered to a system in order to increase the incidence within it of reactive molecules, thus initiating a reaction. It is an important feature of *enzymes that they greatly lower the activation energy of many metabolic reactions.

**activator** A metal ion that functions in conjunction with either an *enzyme or its *substrate in order to bring about a reaction.

**active chamaephyte** See CHAMAE-PHYTE.

**active dispersal** See DISPERSAL.

**active layer** A seasonally thawed surface soil layer, between a few centimetres and about 3 m thick, that lies above the permanently frozen *horizon in a periglacial environment. It may be subject to considerable expansion on freezing, especially if *silt-sized particles dominate, and during the melt it may become very mobile. See also MOLLISOLS.

**active pool** The part of a *biogeochemical cycle in which the nutrient element under consideration exchanges rapidly between the *biotic and *abiotic components. Usually it is smaller than the *reservoir pool, and is sometimes referred to as the 'exchange' or 'cycling' pool.

**active site** The part of an *enzyme molecule that binds it to the *substrate or substrates to form an enzyme–substrate complex. The conformation is not absolute and may alter according to reaction conditions.

**active transport** The transport of substances across a membrane against a concentration gradient. Such processes use energy, the source often being the hydrolysis of *ATP.

**actual evapotranspiration** (AE) The amount of water that evaporates from the surface and is transpired by plants if the total amount of water is limited. Compare POTENTIAL EVAPOTRANSPIRATION.

**aculeate** Prickly, pointed; the word is derived from the Latin *aculeatus*, meaning sting, from *acus*, needle.

**acuminate** Tapering to a point.

**Adanson, Michel** (1727–1806) A French botanist and plant collector, who worked as a clerk to a trading mission in Senegal, where he discovered many plants that were previously unknown. In the 1750s he returned to France with a large collection of plants and seeds. He was the first European to describe the baobab, which he observed in West Africa, although specimens were later found to be more widely distributed. He estimated the age of the tree he saw as about 5000 years; *radiocarbon dating has confirmed an age of 1000 years for some specimens and less precise methods have estimated greater ages for others. The baobab genus (*Adansonia*) is named for him.

**Adansonia** (family *Bombacaceae) A genus of trees, several species of which are pollinated by ants that inhabit modified spines. Baobab (*A. digitata*) is famous for its hugely swollen trunk, commonly 15 m in girth, and reaching 35 m in height. Other species swell, but less so. Baobab provides human and animal food and medicines. There are 9 species, occurring in the seasonal tropics of Africa, Madagascar, and north-western Australia.

**adaptation** That which fits an organism both generally and specifically to exploit a given environmental zone.

**adaptive breakthrough** An evolutionary change by the acquisition of a distinctive *adaptation that permits a population or *taxon to move from one adaptive zone to another. At the most extreme such moves might be from water to land, or from land to air.

**adaptive enzyme** See INDUCIBLE ENZYME.

**adaptive pathway** A series of small adaptive steps, rather than a single large one, which leads from one adaptive zone across an environmental and adaptive threshold into another adaptive zone. In effect, small changes accumulate so that the organism is virtually pre-adapted to enter the new zone.

**adaptive peaks and valleys** A symbolic contour map showing the *Darwinian fitness or *adaptive value of genotypic combinations will usually display adaptive peaks and valleys: these occur at points at which the fitness is respectively strong or weak. The distribution of the population of a given species will therefore coincide with that of the adaptive peaks.

**adaptive radiation 1.** A burst of *evolution, with rapid divergence from a single ancestral form, resulting in the exploitation of an array of *habitats. The term is applied at many taxonomic levels. **2.** A term used as a synonym for *cladogenesis by some authors.

**adaptive value (Darwinian fitness, fitness, selective value)** The balance of genetic advantages and disadvantages that determines the ability of an individual organism (or *genotype) to survive and reproduce in a given environment. The environment, and the competition or struggle for survival within it, determine which individuals are fittest to achieve this, the 'fittest' being the individual (or genotype) that produces the largest number of offspring that later reach reproductive maturity. Such *natural selection has been described as the 'survival of the fittest'. *See also* SELECTION.

**adaxial** Of a surface, directed towards the axis. *Compare* ABAXIAL.

**adder's tongue** *See* OPHIOGLOSSUM.

**additive genetic variance** *See* HERITABILITY.

**adelphoparasite** A parasite (*see* PARASITISM) that has as its host a species closely related to itself, often within the same family or genus.

**Adenanthera** (family *Fabaceae, subfamily *Mimosoideae) A genus of trees, several of them common as shade trees. *A. pavonina* has hard, scarlet seeds, used for centuries in India and Sri Lanka as units in one of the oldest weight systems known, for weighing gold, etc. There are 4 species, occurring from southern China to the Pacific islands.

**adenine** A *purine base which occurs in both *DNA and *RNA.

**adenosine** The *nucleoside formed when adenine is linked to ribose sugar.

**adenosine diphosphate (ADP)** High-energy phosphoric *ester (i.e. *nucleotide) of the *nucleoside adenosine. It can undergo *hydrolysis to adenosine monophosphate and inorganic phosphate, the reaction releasing 34 kJ/mol of energy at *pH 7.

**adenosine triphosphate (ATP)** High-energy phosphoric *ester (i.e. *nucleotide) of the *nucleoside adenosine, which functions as the principal energy-carrying compound in the cells of all living organisms. Its *hydrolysis to ADP (*adenosine diphosphate) and inorganic phosphate is accompanied by the release of a relatively large amount of free energy (34 kJ/mol at *pH 7) which is used to drive many metabolic functions.

**Adiantaceae** A family of ferns in which *sori are borne on the undersides of the veins of the *pinnules, without *indusia. Their distribution is cosmopolitan but they occur most abundantly in the moist tropics. There are about 56 genera and 1100 species.

***Adiantum*** (maidenhair ferns; family *Adiantaceae) A genus of ferns in which all the *fronds are alike, usually with black, glossy *petioles and wide, fan-shaped leaflets. *Rhizomes are scaly. *Sori are borne near the vein ends on the reflexed tips of the leaflets. Some are cultivated as ornamentals. There are about 200 species. Most occur in the moist tropics and warm temperate regions, as far as the southern British Isles.

**adnate 1.** Applied to the *gill of an *agaric that is attached to the *stipe by most or all of its width. **2.** Applied to *analogous close attachments of two organs in higher plants, e.g. of a leaf to a stem. *Compare* ADNEXED and CONNATE.

**adnexed 1.** Applied to the *gill of an *agaric which is attached to the *stipe by only a part of its width. **2.** Applied to *analogous loose attachments of two organs in higher plants. *Compare* ADNATE.

***Adoxa*** *See* ADOXACEAE.

**Adoxaceae** A family comprising 3 genera of obscure affinities; *monotypic *Adoxa* (moschatel), is of global distribution in northern-hemisphere, cool temperate woodlands. It is a herb 5–10 cm tall, with *ternate leaves and a terminal inflorescence of 4 outward- and 1 upward-facing flowers which are 5- or (in the terminal flower) 4-lobed and light green; the *ovary is *inferior. The fruit is a *drupe.

**ADP** *See* ADENOSINE DIPHOSPHATE.

**adpressed** Pressed close (e.g. as the leaves of heather (*Calluna vulgaris*) are pressed close to the stem of the plant).

**adsorption** The physical binding of a particle of a particular substance to the surface of another by adhesion or penetration. In soils, it is the attachment of an ion, molecule, or compound to the charged surface of a particle, usually of *clay or *humus, where replacement or exchange may take place. Ions carrying positive charges (e.g. those of calcium, magnesium, sodium, and potassium) become attached to, or adsorbed by, negatively charged surfaces (e.g. those of clay or humus).

**adsorption complex** Various materials of the soil, mainly *clay and *humus, and to a lesser degree other particles, that are capable of adsorbing ions and molecules.

**adventitious** Growing from an unusual position, e.g. *roots from a leaf or stem.

**adventive embryony** Embryonic development (*see* EMBRYO) without *fertilization from cells that are not egg cells.

**AE** *See* ACTUAL EVAPOTRANSPIRATION.

**-ae** A standardized suffix used to indicate a class of plants in the recognized codes of classification.

**Aechmea** (family *Bromeliaceae) A genus of *epiphytes, a few of which, with showy inflorescences, have now become house plants. *A. magdalenae* (pita), of Colombia, provides fibres for cloth and cordage. There are 172 species.

**aecidiosorus** *See* AECIUM.

**aecidiospore** *See* AECIOSPORE.

**aeciospore** (aecidiospore) A *spore produced by certain *fungi of the order *Uredinales. Chains of aeciospores are produced in an *aecium which develops in the *tissues of the host plant.

**aecium** (aecidiosorus) A structure formed by certain *rust fungi (order Uredinales) in the *tissues of a host plant. A mature aecium is often cup-shaped, with chains of *spores (*aeciospores) being produced from the base of the cup; in some rust fungi the aecium is a more diffuse structure.

**Aegiceras** (family *Myrsinaceae) A genus of small, shrubby trees. *A. corniculatum* grows in mangrove forest and is *viviparous. The bark contains a *saponin and is used as a fish poison. There are 2 species, found in the Old World tropics.

**Aegle** (family *Rutaceae) A genus of 3 species. Bael fruit (*A. marmelos*) grows wild and is extensively cultivated in India, from where it has spread to all of seasonal tropical Asia. It is highly tanniferous and is a valuable remedy for dysentery.

**aerenchyma** In a plant, *tissue containing large, intercellular air spaces.

**aerial mycelium** Part of a *mycelium held aloft from the *substrate.

**aerobe** An organism that can grow only in the presence of air, i.e. one that requires oxygen for growth.

**aerobic** **1.** Of an environment: one in which air (oxygen) is present. **2.** Of an organism: one requiring the presence of oxygen for growth, i.e. an aerobe. **3.** Of a process: one that occurs only in the presence of oxygen.

**aerobic fermentation** Cellular *respiration that requires oxygen. The process involves *glycolysis, the *citric acid cycle, and the *respiratory chain.

**aerochory** *See* ANEMOCHORY.

**aerodynamic method** A method for monitoring rates of *primary productivity; it is particularly useful in studies of woodland or forest communities. The carbon-dioxide concentration from ground level to above canopy height is measured at regular intervals, or continuously, by sensors attached to a mast. Differences in carbon-dioxide concentration reflect differences in rates of *photosynthesis and *respiration and hence in productivity. The chief advantages of this approach are that it is non-destructive and causes minimal disturbance to the environment of the vegetation being studied. Other non-destructive methods usually enclose the vegetation (e.g. in polythene or plexiglass chambers), which can make results unrepresentative of normal field conditions. *Compare* HARVEST METHOD; CHLOROPHYLL METHOD; and GAS-EXCHANGE METHOD; *see* FLUX STUDY.

**aerogenic** Applied to *Bacteria that can produce gas during the metabolism of certain types of substrate.

**aerole** In a leaf with *reticulate venation, an area surrounded by veins.

**aeropalynology** The study of *pollen grains and *spores in the atmosphere. This is important for allergy studies (e.g. of hay fever), for understanding the spread of diseases in man, other animals, and agricultural plants; and recently for *palaeoecologists attempting to improve the interpretation of microfossil data.

**aerophore** A root that grows upward out of water or waterlogged soil, thereby helping to aerate the root system.

**aerotaxis** A change in direction of locomotion, in a motile organism or cell, made in response to a change in the concentration of oxygen in its immediate environment.

**aerotolerant** Applied to organisms (usually bacteria) that, although normally *anaerobic, are not killed by the presence of air.

***Aeschynanthus*** (family *Gesneriaceae) A genus of *herbs, *bole climbers, or *epiphytes, many of which have fleshy leaves. The flowers are vivid red, strongly protandrous (*see* PROTANDRY), and bird-pollinated. There are about 100 species, found in forest undergrowth from warm temperate China to tropical South-east Asia.

***Aeschynomene*** (family *Fabaceae, subfamily *Papilionatae) A genus of 150 species of tropical and subtropical shrubs with sensitive leaves. The wood of *A. indica* (shola) was traditionally used to make pith helmets.

***Aesculus*** (horse chestnut, buckeye; family *Hippocastanaceae) A genus of deciduous trees that have large, sticky buds, opposite *palmate leaves, and showy terminal *panicles of white or red flowers. There are about 13 species, 1 found in deciduous forests of south-eastern Europe, 5 in India and eastern Asia, and 7 in N. America. They are much cultivated and the wood is sometimes used to make furniture.

**aestivation** (estivation) The arrangement of *sepals and *petals in a flower bud before opening.

**aethalium** In certain *myxomycetes, a large *fruiting body consisting of masses of *spores contained within a surface crust; it may be stalked or *sessile.

**Aextoxicaceae** A *monotypic family (*Aextoxicon punctatum*), which is a large tree of the warm temperate rain forests of Chile. The leaves are (sub) *opposite, and *lepidote. Flowers are borne in *axillary *racemes and are unisexual and *pentamerous. The fruit is a *drupe. Probably it is related to the *Monimiaceae; formerly it was placed incorrectly in the *Euphorbiaceae.

**affinity index** A measure of the relative similarity of composition of two samples, e.g. $A = c/\sqrt{a+b}$, where $a$ and $b$ are the numbers of species in one sample but not in the other and vice versa respectively, and $c$ is the number of species common to both. The reciprocal, $\sqrt{a+b}/c$, indicates the ecological and *phytosociological distance ($D$) between samples.

**afforestation** 1. The establishment of forest by natural *succession or by the planting of trees on land where they did not grow formerly. 2. Historically, the act of defining an area of land that henceforth would become subject to *forest laws.

**aflatoxin** One of a group of *toxins produced by fungi of the genus *Aspergillus*. In animals, the toxins have carcinogenic and other toxic properties. Feedstuffs contaminated with aflatoxins have caused severe outbreaks of disease among farm livestock.

**African bass** *See* PIASSAVA.

**a**

**African–Indian desert floral region** Part of the African subkingdom of R. Good's (*The Geography of the Flowering Plants*, 1974) palaeotropical kingdom. The flora is relatively poor and specialized, the number of endemic (*see* ENDEMISM) genera probably just exceeding 50. *Phoenix dactylifera* (the date palm) belongs to this region, but, not surprisingly perhaps, there are no garden plants. *See also* FLORAL PROVINCE and FLORISTIC REGION.

**African oak** *See* OCHNACEAE.

**African pencil cedar** *See JUNIPERUS*.

**African walnut** *(Coula edulis)* *See* OLACACEAE.

**Afro–alpine vegetation** On the highest mountains of Africa, above the dwarf or *elfin woodlands, are found shrublands and grasslands. These are alpine (*alpine zone) in character (hence the term 'Afro–alpine') and, besides containing plants found at lower altitudes in middle and high latitudes, they also include endemics (*see* ENDEMISM), e.g. the giant lobelias and groundsels on Ruwenzori and Kilimanjaro.

**afrormosia** An important, attractive hardwood from western tropical Africa, produced by *Pericopsis elata*.

**agamospermy** *See* APOMIXIS.

**agar** A complex *polysaccharide obtained from certain types of *seaweed. When heated with water and subsequently cooled to about 45°C, agar readily forms a gel (jelly); an agar gel supplemented with nutrients is used widely as a medium for the culture of *bacteria and other micro-organisms.

**agaric** **1.** Any *fungus that forms a *mushroom-like *fruit body (i.e. one that is umbrella-shaped and bears *gills; *compare* BOLETALES). **2.** Any fungus belonging to the order *Agaricales or, more specifically, to the family *Agaricaceae.

**Agaricaceae (order *Agaricales)** A family of *fungi in which the *spores in masses typically appear either dark brown or colourless, but not rust-coloured or cinnamon-brown. The *stipe typically bears a ring. They are *saprotrophic and found on the ground in grassland, woods,

etc. *Agaricus* (including the edible *mushroom) and *Lepiota* are typical genera.

**Agaricales (subclass *Holobasidiomycetidae)** An order of *fungi that typically form fleshy *fruit bodies which are typically mushroom-shaped and bear *gills. There are many families; the order includes many of the *'mushrooms' and *'toadstools'. Most members of the order are *saprotrophic, growing on wood, soil, dung, etc. They are important agents of decay.

**agaricide** A natural or synthetic chemical substance that is toxic to *agarics.

*Agaricus* **(family *Agaricaceae)** A genus of *fungi which includes the cultivated *mushroom. *A. brunnescens* (formerly *A. bisporus*), as well as several edible wild species; there are also some poisonous species. The *gills are typically pinkish when young, becoming dark brown to black as they mature. They are found on the ground in fields, woods, etc., and are cosmopolitan in distribution. *See also* FIELD MUSHROOM; HORSE MUSHROOM; and WOOD MUSHROOM.

**agarophyte** A seaweed from which *agar can be obtained.

*Agathis* **(kauri, kauri pines, Manila copal; family *Araucariaceae)** A genus of large, coniferous trees that have broad, elliptical leaves; valuable, pale, light timber; and resin (Manila copal) in the bark. There are 13 known species, found in tropical to warm temperate forests. They are native from Malesia to Queensland and New Zealand. *See also* DAMMAR.

**Agavaceae** A family of *monocotyledons, some of which are woody. Most are robust *rosette plants, but a few are trees or climbers. The leaves are crowded, straplike, many-spined, and succulent. Inflorescences are *racemose; the flowers usually bisexual; the *perianth united; the *ovary *superior or *inferior and *trilocular. The fruit is a *capsule or *berry. Several produce tough fibres. There are about 18 genera, with some 580 species; they are found in the tropics and subtropics, mainly in arid regions, and especially in America.

**Agave** (family *Agavaceae) A genus of *rosette plants that produce, after many years, a terminal *inflorescence and then die. *A. americana* (the century plant) provides the Mexican beverage pulque, and its juice may be distilled to produce mescal. Several *Agave* species provide fibres (e.g. sisal hemp and istle fibre). There are about 300 species, found in subtropical and tropical parts of arid America.

**age-and-area hypothesis** J. C. Willis suggested in 1922 that, all other things being equal, the area occupied by a *taxon is directly proportional to the age of that taxon. Thus in a *polytypic genus, the species with the smallest area of distribution would be the youngest in the genus. However, things are rarely ever equal, and Willis's ideas, though much discussed, have never gained acceptance as a law or rule.

**agglomerative method** A system of hierarchical classification that proceeds by grouping together the most similar individuals, and subsequently groups into progressively larger and more heterogeneous units. At each stage the groups or individuals linked are those giving the minimum increase in group heterogeneity.

**aggregate** A group of soil particles that adhere together in a cluster; the smallest structural unit, or ped, of soil. Aggregates join together to make up the major structural units.

**aggregation** A process in which soil particles coalesce and adhere to form soil *aggregates. The process is encouraged by the presence of bonding agents such as organic substances, *clay, iron oxides, and ions (e.g. calcium and magnesium).

**aggressin** A toxic substance that is produced by certain micro-organisms which are pathogenic in animals or humans. Aggressins inhibit the defence mechanisms of the host organism.

**Aglaia** (family *Meliaceae) A genus of trees, most of which are medium-sized. They have twigs with *stellate hairs, and leaves that are large and *pinnate (or, rarely, *simple). The flowers are tiny,

borne in *panicles, and often fragrant. The fruit is one- or few-seeded, the *pericarp leathery and *dehiscent, and the seed large, with a fleshy jacket. A few members of this genus are timber species (including the former genus *Amoora*). There are 100 species, occurring from China to the western Pacific.

**Agonomycetales** (Mycelia Sterilia) An order of imperfect *fungi in which there is apparently no specialized reproductive structure of any kind (i.e. no sexually produced *spores, *conidia, etc.).

**agric horizon** A *mineral-soil *diagnostic horizon formed from accumulation of *clay, *silt, and *humus, which has moved down from an overlying, cultivated soil layer. It is a soil *horizon created by agricultural management, and is identified by its near-surface position, and by *colloids accumulated in the *pores of the soil. The name is from the Latin *ager* meaning 'field'.

**Agrimonia** (family *Rosaceae) A genus of *perennial *herbs that have *pinnate leaves with small leaflets alternating with the larger pairs, and long *racemes of 5-petalled flowers with many *stamens; the *receptacle bears hooked spines that persist and become woody in the fruit, as does the deeply hollowed receptacle. There are about 15 species, occuring in the northern temperate zone.

**Agrobacterium** (family *Rhizobiaceae) A genus of *bacteria most of which are capable of causing the formation of *galls in plants. No species can fix atmospheric nitrogen. *A. tumefaciens* causes *crown gall in many types of plant and is a serious plant *pathogen in many parts of the world. These bacteria are found also in soil.

**agroclimatology** The scientific study of the ways climate affects agriculture.

**agro-ecosystem** An agricultural ecosystem, e.g. a cereal crop.

**agroforestry** A system of mixed arable farming and forestry that has been practised for millenniums in many parts of the tropics (e.g. in home gardens in Java), especially in rain-forest environments. Tree crops are interspersed with food-crop patches on a continuous basis, sometimes as hedges (so-called 'alley-cropping' or 'corridor farming'). The system ensures a continuous food supply, some continuous economic return, and the avoidance of soil degradation by exposure to tropical sunlight and rainfall during prolonged clearance; and it is suitable for small-scale family farmers. The trees act as 'nutrient pumps' and bring plant mineral nutrients from the depths up to the surface soil layers that are occupied by the roots of the crop plants.

**agrometeorology** The study of the relationship between conditions in the surface layers of the atmosphere and those in the surface of the Earth, as this affects agriculture.

**agronomy** A branch of agriculture concerned with the theoretical and practical production of crops, and with the management of soils.

***Agrostis tenuis*** See HEAVY METAL TOLERANCE.

**Aizoaceae** A family of *herbs or low shrubs that have fleshy, undivided leaves and showy flowers which are regular and have many petals. The plants occur in temperate southern continents including Australia, but the family is most developed in S. Africa. Hottentot figs (*Carpobrotus*) of S. Africa are often cultivated as ornamentals, and have edible fruits. There are about 115 genera and 2410 species.

***Ajuga*** (family *Lamiaceae) A genus of *herbs many of which have *radical leaves forming a *rosette. The leaves are *simple or lobed, sometimes toothed, without *stipules. The flowers are in at least 2- or many-flowered *whorls, often forming a terminal *inflorescence. They are bisexual, with 5 fused *sepals forming a toothed bell shape. The *corolla has a short upper lip, and a larger, 3-lobed lower lip with a ring of hairs in the corolla tube. The *stamens are attached to the corolla tube. The *ovary is *superior, with 2 *bilocular *carpels and 4 *ovules. The fruit is an oval *nutlet. There are approximately 50 species, found in the Old World in a wide variety of habitats.

**akinete** A thick-walled resting cell formed by certain *cyanobacteria. Akinetes

can survive unfavourable environmental conditions and can germinate to produce new vegetative growth.

**alan** See *SHOREA ALBIDA*.

**Alangiaceae** A family of 1 genus and 20 species of small trees, whose affinities are obscure. The leaves are *exstipulate, alternate, and *simple. The flowers are *cymose, regular, and bisexual. There are 4–40 *sepals, *petals, and *stamens. The petals are sometimes basally fused. The *ovary may be *superior or *inferior, and *unilocular, with 1 pendulous *ovule and a *disc. The fruit is a *drupe, and has a fleshy *endosperm. The family is confined to the Old World tropics.

**alar cell** In mosses (*Musci), 1 of the cells that occur at the junction of *phyllode and stem.

***Alaria*** (order *Laminariales) A genus of brown *algae, in which the *thallus consists of a branched, root-like *holdfast, a short *stipe, and a thin, leaf-like blade with a pronounced *midrib. During the summer, spores are produced in characteristic club-shaped 'leaflets' arising from the short stipe. There are several species; only *A. esculenta* (*dabberlocks) occurs in Britain. *Alaria* species are found attached to rocks at and below low-water mark, often on very exposed shores.

**Albeluvisols** Soils that have an *argic horizon with an irregular upper boundary. Albeluvisols are a reference soil group in the FAO *soil classification.

**Albian** A stage in the *Cretaceous, dated at about 112–99.6 Ma ago, underlain by the *Aptian and overlain by the *Cenomanian. It is known to contain a great variety of molluscs, with the gastropods in particular being useful zonal indicators between continents. The Gault and Speaton Clays of England are Albian.

**albic** Applied to an almost white soil in which there is little *clay or oxides coating the sandy or silty particles. The albic *horizon lies at or below the surface.

**albinism** In plants, a deficiency of *chromoplasts, *carotenoid-containing plastids that colour ripe *fruits and flowers.

**Albizia** (family *Fabaceae, subfamily *Mimosoideae) A genus of tropical plants, sometimes known (incorrectly) as *Albizzia*, that comprises shrubs and trees with feathery leaves and fluffy yellow flower heads, similar to *Acacia* (to which the genus is closely related). Several species are valuable timber trees. There are 150 species, found throughout the tropics and subtropics.

**Albuginaceae** (order *Peronosporales) A family of *fungi which are obligate parasites (*see* PARASITISM) of flowering plants, causing diseases known as *'white rust' or 'white blister'. *Sporangiophores develop beneath the host *epidermis, causing it to swell and burst. The white powdery *sporangia are then dispersed by wind and rain.

**Albugo** (family *Albuginaceae) A genus of *fungi; *A. candida* is a *parasite of *crucifers and is particularly common on *Capsella bursa-pastoris* (shepherd's purse).

**albuminous** Applied to seeds that contain *endosperm when mature. *Compare* EXALBUMINOUS.

**Alchemilla** (lady's mantle; family *Rosaceae) A genus of *herbs and low shrubs in which the *stipules are leaf-like; the leaves alternate, *palmately veined or lobed; and the flowers are green, small, with 4 *sepals, borne in *cymes, with no *petals. An *epicalyx of 4 lobes is present. There are 4 or 5 *stamens. The fruit is dry, and formed from the single *carpel. Many apomictic (*see* APOMIXIS) species occur in Europe, and several local endemics in E. Africa. The plants occur in open, grassy habitats and some are cultivated. There are about 250 species, found in the northern temperate zone and in the mountains of tropical Africa.

**alcohol** A hydrocarbon in which a hydrogen atom is replaced by a hydroxyl (OH) group. An alcohol is designated as primary, secondary, or tertiary, according to whether the carbon to which the hydroxyl group is attached is bound to one, two, or three other carbons. See ETHANOL.

**alcoholic fermentation** A process that occurs in plants that do not require oxygen for *respiration. *Glucose is broken down to *ethanol and carbon dioxide, with the release of energy that is used for *ATP production.

**alcrete** *See* DURICRUST.

**aldehyde** An organic compound that contains group –CHO.

**alder** *See* ALNUS.

**aldose** A monosaccharide or its derivative that contains an *aldehyde group.

**aldotriose** A 3-carbon monosaccharide that contains an aldehyde group.

**Alectoria** (order *Lecanorales) A genus of *lichens in which the *thallus is *fruticose, greenish-grey or dark-coloured, and lacks the central, whitish strand of *Usnea* species. *Spores (2–4 per *ascus) are brown when mature. Species are found on trees, rocks, etc. *See also* BEARD-LICHEN.

**-ales** A standardized suffix used to indicate an order of plants in the recognized codes of classification (e.g. *Filicales, the ferns).

**Aleuria** (order *Pezizales) A genus of *fungi which typically form large or small, saucer- or cup-shaped, red, orange, or yellow *apothecia on the ground in woods, fields, etc. *A. aurantia* (orange peel fungus) is the commonest British species, and is found on soil in woods, lawns, grass verges, etc. in autumn.

**aleuriospore** 1. A thick-walled, single-celled, asexual *fungal *spore formed terminally on an aerial *hypha. 2. An asexual spore that is liberated only on breakage of the hypha that bears it.

**Aleurites** (family *Euphorbiaceae) A genus of trees that produce large, round fruits with a thick, fleshy *pericarp, containing a few big, very hard, oily seeds. *A. moluccana* (candlenut) of Malesia provides oil for candles, soaps, paints, and varnishes; *A. fordii* and 2 relatives, of China, provide tung oil for varnishes and paints. There are 6 species, occurring from tropical Asia to the Pacific.

**aleurone grain** A seed cell storing protein.

**aleurone layer** A layer of cells below the *testa of some seeds (e.g. barley), which contains hydrolytic *enzymes (including *amylases and *proteases) for the digestion of the food stored in the *endosperm. The production of enzymes is activated by *gibberellins when the seed is soaked in water prior to germination.

**aleuroplast** A *leucoplast type of *plastid that is involved in the storage of proteins.

**Alfisols (grey-brown podzolics)** An order of *mineral soils that have *clay-enriched or argillic B horizons (see ARGIC HORIZON); are alkaline to intermediate in reaction, with the base saturation in the B horizon more than 35 per cent; are usually derived from base-rich parent materials; and are drier than −15 bars moisture potential for at least three months when plants could grow. Alfisols occur under deciduous woodland or grassland in humid areas of the world and are generally productive agricultural soils. *Brown earths are the equivalent in Britain.

**alga (pl. algae)** The common (non-taxonomic) name for a *protist resembling a relatively simple plant that is never differentiated into root, stem, and leaves. It contains chlorophyll *a* as the primary photosynthetic pigment, has no true vascular (water-conducting) system, and there is no sterile layer of cells surrounding the reproductive organs. There are three different types of algae: red (Rhodophyceae); brown (Phaeophyta); and several groups of green algae. Green plants have evolved from green algae. Algae range in form from single-celled *eukaryotes to plant-like organisms several metres long, for example seaweeds. Algae can be found in most habitats on Earth, although the majority occur in freshwater or marine environments.

**algal bloom** A sudden growth of *algae in an aquatic ecosystem. This can occur naturally in spring or early summer when *primary production exceeds consumption by aquatic *herbivores. Algal blooms may also be induced by nutrient enrichment of waters due to *pollution. They are a characteristic symptom of *eutrophication.

**algal layer** In *lichens, the tissue that contains the *phycobiont.

**algin** A salt (usually the sodium salt) of alginic acid, a complex polysaccharide present in the cell walls and between the cells in brown seaweeds (*Phaeophyta). When extracted from the seaweed and mixed with water, many algins make very thick, viscous solutions. They have a number of commercial uses (e.g. as stabilizers in ice-cream and other foods, and as suspending agents in paints and pharmaceutical preparations).

**algology** The study of algae (more usually known as phycology).

**Alisols** Soils that have an *argic horizon with a *cation-exchange capacity of more than 24 mol$_c$/kg clay, and a *base saturation of less than 50% within 100 cm of the surface. These are soils with high concentrations of aluminium. Alisols are a reference soil group in the FAO *soil classification.

**alkaline soil** Soil with a *pH greater than 7.0. Degrees of soil alkalinity are recognized. The *USDA lists soils with a pH 7.4–7.8 as mildly alkaline; 7.9–8.4 as moderately alkaline; 8.5–9.0 as strongly alkaline; and more than 9.0 as very strongly alkaline. Soil is not regarded as highly alkaline unless the reaction is between 8.0 and 10.0. The full range of the pH scale (0–14) is not used in soils, as the reaction of most soils is between pH 3.5 and pH 10.0. Base saturation of 100 per cent indicates a pH of about 7.0 or higher. Alkaline soils are usually rich in calcium ions. Plants characteristic of such soils are called *calcicole.

**alkaliphile** An *extremophile (domain *Archaea) that thrives in environments where the pH is above 9.0.

**alkaloid** One of a group of more than 1000 basic, nitrogenous, normally heterocyclic compounds of a complex nature that exist in combination with organic acids in plants belonging to approximately 1200 species. Alkaloids are *secondary plant compounds that in some species may confer a degree of protection against attacks by herbivores. Many alkaloids are very poisonous to humans and in small amounts are the active ingredients in many therapeutic drugs.

Pharmacologically powerful alkaloids derived from plants include caffeine, cocaine, morphine, nicotine, and strychnine.

**alkanet**   *See* ANCHUSA.

**Allamanda** (family *Apocynaceae) A neotropical genus of 12 species of plants, including *A. cathartica*, a *sarmentose shrub with showy yellow flowers, that is now widely cultivated.

**Allantospermum** (family Ixonanthaceae) A genus of biggish trees in which the leaves are *simple; the flowers are bisexual and *pentamerous; and the fruit is a *capsule. There is 1 species occurring in Borneo, Malaya, and Madagascar: a remarkable disjunction in range.

**allele**   The common shortening of the term 'allelomorph'. One of two or more forms of a *gene arising by *mutation and occupying the same relative position (*locus) on *homologous *chromosomes. When in the same cell, alleles may undergo pairing during *meiosis. They may be distinguished by their differing effects on the *phenotype. The existence of two forms of a gene may be termed 'diallelism', and that of many forms, 'multiple allelism'. The commonness of an allele in a population is termed the 'allele frequency'.

**allele frequency**   *See* ALLELE.

**allelomorph**   A term that is commonly shortened to *'allele'.

**allelopathy**   The release into the environment by an organism of a chemical substance that acts as a *germination or growth inhibitor to another organism. Typical substances include *alkaloids, terpenoids (*see* TERPENE), and phenolics (*see* PHENOL). The phenomenon was described originally for heath and scrub communities, notably the Californian *chaparral, but is now thought to be a widespread anti-competition mechanism in plants (e.g., barley inhibits competing weeds by means of root secretions). It is, however, extremely difficult to demonstrate in natural ecosystems. Allelopathy is also found in other organisms (e.g., antibiotics may be produced by fungi to inhibit competing bacteria, when the term 'antibiosis' may be used). It is a form of interference

competition, substituting space for a resource.

**Allerød**   A late glacial (i.e. late *Devensian) period marking a prolonged warmer oscillation or *interstadial during the general phase of ice retreat in north-western Europe. *Radiocarbon dating suggests it lasted from about 12 000 years BP to 10 800 years BP. Pollen records (*pollen zone) for the north-western European area indicate a cool temperate flora with birch (*Betula species) widespread, in marked contrast to the preceding and following, colder, *Dryas, phases.

**alley-cropping**   *See* AGROFORESTRY.

**alliance**   In *phytosociology, a grouping of closely related associations. *See also* PLANT ASSOCIATION.

**Allium** (onions, garlic; family *Liliaceae) A genus of bulbous herbs, all of which contain oils with a pungent onion or garlic odour. The leaves are undivided, parallel-veined, and all *radical. The flowers are borne in *umbels enclosed at first in a *spathe, which later splits. The *perianth consists of 6 free, equal segments, and 6 *stamens. *Bulbils often partly replace the flowers. The fruit is a 3-celled *capsule. Many species are cultivated as vegetables or for their flowers. There are about 700 species found in the north temperate zone, south to Mexico and Ethiopia.

**allochthonous**   Applied to material that did not originate in its present position (e.g. plant material in a deposit, such as lake sediment, which did not grow at that location but was introduced by some process). *Compare* AUTOCHTHONOUS.

**allogamy**   *Fertilization that involves *pollen and *ovules from different flowers. *See also* AUTOGAMY, GEITONOGAMY, and XENOGAMY.

**allogenic**   Applied to successional change arising from a change in *abiotic environmental conditions. *Compare* AUTOGENIC; *see also* SUCCESSION.

**allograft** (homograft) A graft of tissue from a donor of one *genotype to a host of a different genotype but of the same species. If the donor and recipient are not closely related, the graft is likely to be rejected. If the graft takes place from one part to another part of the same individual it is called an autograft.

**allometry** The growth of one part of an organism at a different rate from that of the whole organism.

***Allomyces*** **(order *Blastocladiales)** A genus of *fungi in which the *mycelium is branched and coenocytic (*see* COENOCYTE), attached to the substratum by branching *rhizoids.

**allopatric** Applied to species that occupy separate *habitats and that do not occur together in nature. *Compare* PARAPATRIC and SYMPATRIC.

**allopatric speciation** The formation of new species from the ancestral species as a result of the geographical separation or fragmentation of the breeding population. Separation may be due to climatic change, causing the gradual fragmentation of the population in a few surviving favourable areas (e.g. during glaciation or developing aridity), or may arise from the chance migration of individuals across a major *dispersal barrier. Genetic divergence in the newly isolated daughter populations ultimately leads to new species; divergence may be gradual or, according to punctuationist models (*see* PUNCTUATED EQUILIBRIUM), very rapid. The populations must evolve some sort of sexual or genetic isolating mechanism that prevents them from interbreeding should they come into contact again later.

**allopatry** The occurrence of species in different geographical regions. The differences between closely related species usually decrease (i.e. the characteristics converge) when species are separated, in a process called character displacement, which may be morphological or ecological. *Compare* SYMPATRY.

**allopolyploid** A polyploid formed from the union of genetically distinct *chromosome sets (usually from different species). Examples are cultivated wheats. Many new species of flowering plants are thought to have originated by allopolyploidy.

**allosteric effect** The binding of a *ligand to one site on a *protein molecule in such a way that the properties of another site on the same protein are affected. Some *enzymes are allosteric proteins, and their activity is regulated through the binding of an *effector to an allosteric site.

**allspice** *See* PIMENTA.

**alluvial** Applied to the environments, action, and products of rivers or streams. Alluvial deposits (alluvium) are (clastic, detrital materials transported by a stream or river and deposited over the river's flood plain. The term is also applied to surface flow, as in *alluvial fans, bajadas, etc.

**alluvial fan** A mass of sediment deposited at some point at which there is a sharp decrease in gradient, e.g. between a mountain range and a plain. Essentially, a fan is the land-bound equivalent of a river-delta formation.

**alluvium** *See* ALLUVIAL.

**almond** *See* AMYGDALUS.

***Alnus*** **(alder; family *Betulaceae)** A genus of deciduous trees which are common in wet places. The roots have nitrogen-fixing nodules. The leaves are *simple; the flowers are unisexual and borne in *catkins, the female flowers developing into a woody, persistent, cone-like structure. The trees are useful for timber and as ornamentals. There are 35 species, occurring in the north temperate zone, extending south to the Himalayas and Andes. *See also* FRANKIA.

***Alocasia*** **(family *Araceae)** A genus of rhizomatous (*see* RHIZOME) herbs. The stems and leaves of *A. macrorrhiza* (giant taro) are edible, but are grown nowadays mainly for pig food. There are 70 species, found in the eastern tropics.

**aloeswood** **(eaglewood)** *See* AQUILARIA.

***Alopecurus myosuroides*** **(black grass)** *See* HERBICIDE.

**alpha-amino acid** An *amino acid in which the *amino group is attached to the number two, or 'alpha', carbon, adjacent to the *carboxyl group. Compounds of this type represent the basic building blocks of *peptides and *proteins.

**alpha helix** The right-handed, or less commonly left-handed, coil-like configuration of a *polypeptide chain, which represents the secondary structure of some protein molecules, particularly the globular variety (*see* GLOBULAR PROTEIN). The configuration is maintained through intra-chain hydrogen bonding between >CO and >NH groups of *peptide bonds.

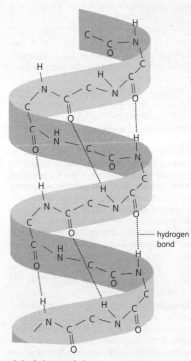

*Alpha helix or α-helix*

**alpha-mesohaline** *See* BRACKISH.

*Alphitonia* **(family *Rhamnaceae)** A genus of small trees and shrubs, several species of which are fast-growing pioneers of secondary forest. There are 6 species, occurring in Malesia, Australia, and the western Pacific islands.

**alpine zone** A region that occurs above the *tree-line and below the *snow-line on temperate and tropical mountains. The vegetation is characterized by an absence of trees, and varies greatly with aspect, the greatest contrasts being between the wet side and the dry, leeward side of the mountains concerned. The elevation of the lower limit of the zone increases from about 1000 m above sea level in Scotland to over 2000 m in the Swiss Alps, and to 3700 m in the western Himalaya.

*Alstonia* **(family *Apocynaceae)** A genus of small to big trees in which the crown is often pagoda-like, with distinct *whorls of branches. The *bole often has huge buttresses. The leaves are often whorled; the flowers are *cymose. The fruits are paired, woody *follicles containing many seeds and having a tuft of silky hairs at both ends. The timber is easily worked and is sometimes very light. The bark yields latex which is used in traditional medicine. There are 43 species, occurring in the Old World subtropics and tropics.

*Alternanthera* **(family *Amaranthaceae)** A genus of about 80 species of *herbs and shrubs whose single *ovule and *bilocular *stamens distinguish them from other members of the large family. Some species are grown for their ornamental leaves. The genus is cosmopolitan, mainly subtropical.

*Alternaria* **(class *Hyphomycetes)** A *form-genus of *fungi which form yellowish-brown *conidia which are divided by transverse and longitudinal *septa. There are many species, including a number of important plant *pathogens (e.g. *A. solani* causes early blight of potatoes; *A. radicina* causes black rot of carrot seedlings and stored carrots).

**alternation of generations** An alternate development of two types of individual in the *life cycle of an organism. Usually one type reproduces asexually and the other sexually. (The term is often restricted to organisms that have a *haploid generation alternating with a *diploid generation.) In *mosses, *vascular plants, many *algae, and some *fungi, for example, a *haploid phase, during which

a

*gametes are produced by *mitosis (*gametophyte phase), alternates with a *spore-producing *diploid (*sporophyte) phase. In an isomorphic alternation of generations (found in some algae, for example) the sporophyte and gametophyte are morphologically similar or identical; in a heteromorphic alternation of generations they are dissimilar (e.g. in mosses the gametophyte is the dominant and conspicuous generation, whereas in higher plants it is the sporophyte that forms the conspicuous plant).

***Althaea* (marsh mallow, hollyhock, etc.; family *Malvaceae)** A genus of *herbs that have showy, regular, 5-petalled flowers and an *epicalyx of 6–9 segments joined below into a cup-like *involucre hiding the *calyx. Several species (e.g. *A. rosea*, hollyhock) are cultivated for their flowers or (e.g. *A. officinalis*, marsh mallow) for the mucilage used in confectionery or medicine. There are about 12 species occurring in temperate Eurasia.

***Altingia* (family *Hamamelidaceae)** A genus of trees which produce valuable timber. The buds are scaly; the flowers unisexual, very reduced, and borne in small heads. The fruits forms as rounded, woody heads, and are *dehiscent, containing many winged seeds. There are 7 species, occurring from Assam and southern China to Java.

**altitudinal vegetation zones** With increased altitude conditions usually become cooler and damper, so that the vegetation of mountains of considerable elevation shows a corresponding zonation. In the tropics the zonation may extend from rain forest on the lower slopes, to alpine communities at heights above about 3500 m. On the other hand, in progressively higher latitudes the elevation of the *tree-line gradually descends, eventually to sea level.

***Alyxia* (family *Apocynaceae)** A genus of about 120 species of shrubs found in Asia, Australia, and the south-western Pacific, about 9 species being endemic to Australia. They have small, 5-petalled, white flowers, and fleshy, orange-red fruits. New Caledonian species accumulate nickel.

***Amanita* (family Amanitaceae)** A genus of *fungi which includes some of the most poisonous known (e.g. *Amanita phalloides*, see DEATH CAP; *Amanita virosa*, see DESTROY-ING ANGEL). A few species are edible. The fruit bodies are *mushroom-shaped with white *gills. Typically a *volva is present, and remnants of the *universal veil may adhere to the surface of the *pileus. Some species form *mycorrhizal association with higher plants. *See also* BLUSHER; CAESAR'S MUSHROOM; FALSE DEATH CAP; FLY AGRARIC; GRISETTE; and PANTHER CAP.

**Amanitaceae (order *Agaricales)** A family of *fungi in which the *fruit bodies are *mushroom-like; the *spores are white or pale-coloured. Both *universal and *partial veils are present in some species. They are found on the ground, chiefly in woodland.

**Amaranthaceae (cockscombs, celosias, etc.)** A family of mostly *annual herbs in which the leaves are *entire and without *stipules. The flowers are small, borne in dense, spike-like *cymes, and are mostly unisexual; they have 3–5 *scarious, green or brightly coloured *perianth segments; 3–5 *stamens; and a 1-celled *ovary which forms a small, dry fruit. *Celosia cristata* (cockscomb) is cultivated for its red flowers. There are 71 genera, with about 800 species, mostly tropical or warm temperate, and occurring as weeds in cooler regions.

**Amaryllidaceae (daffodils, narcissi, etc.)** A family of bulbous herbs which differ from *Liliaceae only in the *inferior *ovary in their flowers. The flowers are solitary or occur in *umbels. The inflorescence is enclosed at first in a *scarious *spathe. The flowers are regular and showy, with 6 similar *perianth segments in 2 *whorls, and sometimes with a trumpet-like *corona. There are 6 *stamens. Many species are cultivated for their flowers. There are about 85 genera, with some 1100 species, mainly of warm temperate regions but extending into the tropical and cool temperate zones.

**amatoxins** Poisonous substances present in the *fruit bodies of certain species of *Amanita. When eaten by humans, amatoxins cause severe gastrointestinal symptoms and degeneration of liver and kidneys. They may be lethal even in small quantities.

**Amazon floral region** Part of R. Good's (*The Geography of the Flowering Plants*, 1974) *neotropical kingdom, corresponding to one of the formations of the rain forest. Because of extensive flooding by the Amazon river, the vegetation can be divided into that which is above the flood-level, called 'igapo', and that which is below it, called 'ete'. The flora is one of the richest in the world, and includes about 100 endemic (*see* ENDEMISM) genera, with 3000 endemic species, e.g. *Hevea brasiliensis* (Para rubber) and *Theobroma cacao* (cocoa). *See also* FLORAL PROVINCE and FLORISTIC REGION.

**amber** *Fossil resin that was exuded from coniferous trees during the *Tertiary Period.

**amboyna** Timber from *Pterocarpus indicus*. *See* PTEROCARPUS.

**ambrosia fungi** Any of various types of *fungus which grow in the tunnels of wood-boring ambrosia beetles (e.g. *Xyleborus* species, family Scolytidae); the beetles use the fungi as food.

**ameiosis** *Meiosis in which the nucleus divides only once, so the number of *chromosomes is not reduced.

**Amelanchier** (June berries; family *Rosaceae) A genus of non-thorny, deciduous, small trees or shrubs with *simple leaves and flowers that are *pentamerous, with 2–5 *carpels united to the *receptacle below but free from it above, each carpel divided into 2 cells by a false septum. The flowers are also in *racemes, distinguishing the genus from *Crataegus* and *Cotoneaster*. Several species are cultivated for their flowers, often becoming naturalized. There are 6 species, occurring in the north temperate zone; most are N. American, but one occurs in Europe.

**amensalism** The interaction of species populations, in which one population is inhibited while the other is unaffected. *Compare* COMMENSALISM; COMPETITION; MUTUAL INHIBITION; MUTUALISM; NEUTRAL-

ISM; PARASITISM; PREDATION; and PROTOCO-OPERATION; *see also* SYMBIOSIS.

**American cherry** *See* PRUNUS SERO-TINA.

**American cranberry** *See* VACCINIUM.

**American red gum** *See* LIQUIDAMBAR.

**American sap gum** *See* LIQUIDAMBAR.

**American whitewood** *See* LIRIODEN-DRON.

**amethyst deceiver** *See* LACCARIA.

**Amherstia** (family Caesalpiniaceae) A *monotypic genus comprising *A. nobilis*, a small tree that, despite being difficult to grow, is prized in cultivation for its *racemes of spectacular, large, pink-red flowers and pinkish tassels of pendant young leaves. The *stamens are fused as a tube. The tree is native to Burma.

**amide** A compound derived from ammonia ($NH_3$) by the replacement of one or more of the hydrogens by organic acid groups. The resultant amide is designated primary, secondary, or tertiary according to the number of hydrogens replaced.

**amine** An organic base derived from ammonia ($NH_3$) by the replacement of one or more of the hydrogens by organic radical groups. The resultant amine is designated primary, secondary, or tertiary according to the number of hydrogens replaced.

**amino acid** An organic compound containing an acidic *carboxyl group and a basic amino ($NH_2$) group. The general formula for naturally occurring amino acids is $R–CH(NH_2)–COOH$, where R is a variable grouping of atoms (fundamentally, a carbon chain or ring). Amino acids constitute the basic building blocks of *peptides and *proteins and are classified as: (*a*) neutral, basic, or acidic; or (*b*) non-polar, polar, or charged (depending on the net electrical charge resulting from the configuration of the molecule).

**amino-acid sequence** The sequence of *amino acid residues in a *polypeptide chain that represents the primary structure of a *protein. This sequence is unique to each protein and influences the protein's secondary, tertiary, and quaternary structures.

**amino group** The radical group–$NH_2$.

**aminopeptidase** An *enzyme that catalyses the *hydrolysis of *amino acids in a *polypeptide chain by acting on the *peptide bond adjacent to the essential free *amino group.

**amino sugar** A monosaccharide in which an *amino group has been substituted for one or more *hydroxyl groups.

**aminotransferase** See TRANSAMINASE.

**amitosis** Nuclear division in which the cell is constricted into 2 without the *chromosomes becoming visible, the *nuclear envelope breaking down, or a *spindle forming.

**ammonium fixation** The adsorption of ammonium ions ($NH_4^+$) into inter-layer sites of the *clay minerals, similar to locations of potassium in hydrous mica, which renders them unavailable to plants.

**Amomum** (family *Zingiberaceae) A genus of 90 species of Old World tropical, rhizomatous (see RHIZOME), aromatic herbs, whose flowers are usually borne on leafless *scapes. *A. aromaticum*, of India, is cardamom.

**Amoora** See AGLAIA.

**Amorphophallus** (family *Araceae) A genus of giant herbs producing a *corm-like *rhizome that can weigh 50–70 kg and is cultivated for food in some species. The plants bear few leaves, which are several times *palmate, on a long erect stalk, and are sometimes huge, up to several metres across. The inflorescence has a powerful, foetid odour. It is sometimes massive (2–4 m tall in *A. titanum*) and the *spathe is blotched purple. There are about 90 species, native to Africa and Asia.

**amorphous** Applied to materials that have no regular atomic structure, or in which no extensive portions have a regular structure; but small units, irregularly aligned or stacked, may have a regular structure.

**amphi-Atlantic species** The plants that are found on both sides of the northern Atlantic, along the seaboards of eastern North America and western Europe. This disjunct distribution pattern is thought to be largely inherited from a single circum-North-Atlantic coastal plain flora. Typical representatives are pipewort (*Eriocaulon septangulare*), a slender aquatic herb, and Scots lovage (*Ligusticum scoticum*), an umbelliferous plant.

**amphibiotic** Applied to an organism that can be either *parasitic on or *symbiotic with a particular host organism.

**amphicribral** Applied to a *vascular bundle in which the *phloem surrounds the *xylem. *Compare* AMPHIVASAL.

**Amphidinium** See DINOPHYTA.

**amphidiploid** An *allopolyploid in which the genetic behaviour of the constitutive *genomes is *diploid, such that *bivalents between chromosomes originating from different genomes do not form during *meiosis.

**amphigastrium** In *Jungermanniales, a leaf that forms on the undersurface of the stem. Amphigastria occur in rows and are smaller than the leaves on the upper surface.

**amphimixis** Sexual reproduction. *Compare* APOMIXIS.

**amphiphloic** A *vascular bundle *morphology in which the *phloem occurs as concentric cylinders outside and inside the *xylem.

**amphiphloic siphonostele** A *mono-stele type of *siphonostele which appears in cross-section as 1 ring of *phloem around the outside of the *xylem and another around the inside of the xylem ring, but outside the *pith. *Compare* ECTOPHLOIC SIPHONOSTELE.

**amphithecium** In *Bryophyta, the outer layer of cells in the young *sporophyte.

**amphitrophic** Applied to an organism that can carry out photosynthesis in the presence of light and that can also grow *chemotrophically in the dark.

**amphitropical species** Species having disjunct distribution patterns, one part of the range being to the north of the equator, the other to the south, and geographically quite separate. These disjunctions probably arose in the *Pleistocene when the climatic belts were telescoped, and migration across the equator would have been easier.

**amphitropous** Applied to the position of an *ovule that is attached at its middle, its long axis lying parallel to the *placenta.

**amphivasal** Applied to a *vascular bundle in which the *xylem surrounds the *phloem. *Compare* AMPHICRIBRAL.

**amplexicaul** *See* STIPULE.

**ampulliform** Flask-shaped or bottle-shaped.

***Amyema*** (family *Loranthaceae) A genus of plants of the mistletoe family which are stem hemiparasites, usually with red or yellow flowers, and seeds spread by birds. The 92 species occur in South-east Asia, the Pacific, and (32 species) in Australia.

**amygdaliform** Almond-shaped.

***Amygdalus*** (almond, peach; family *Rosaceae) A genus of 40 species, from temperate Asia, but almond, peach, apricot, and nectarine have long been cultivated and are grown in many parts of the world. They are close to *Prunus and often not kept in a separate genus, but have *axillary buds in threes and fruits that are usually downy, with the stone pitted or wrinkled.

**amylase** A member of a group of *enzymes that hydrolyse (*see* HYDROLYSIS) starch or *glycogen by the splitting of glucosidic bonds (*see* GLUCOSIDE), so giving rise to the sugars *glucose, dextrin, or maltose. They are widely distributed in plants and animals, but occur particularly richly in germinating seeds e.g. those of barley, peas, and maize) in which the amylase mobilizes food reserves for the growth of the seedling.

**amyloid** Starch-like.

**amylolytic** Capable of digesting starch.

**amylopectin** A branched-chain *polysaccharide that is found in native starches composed of *glucose units joined by α–1,4 glycosidic bonds (*see* GLYCOSIDE) and at points of branching by α–1,6 bonds.

**amyloplast** A *plastid that synthesizes and stores *starch, to the exclusion of other activities.

**amylose** A long, unbranched-chain, *polysaccharide component of native *starch composed of *glucose units joined by α–1,4, glycosidic (*see* GLYCOSIDE) bonds.

***Anabaena*** A genus of filamentous *cyanobacteria (section IV) in which the filaments when viewed under the microscope resemble strings of beads. Thick-walled, clear *heterocysts can be seen at intervals along or at the ends of the filaments. *Akinetes may be formed. Species are capable of fixing atmospheric nitrogen. They are common in aquatic environments, where they are usually planktonic, buoyed up by gas vacuoles. *A. azollae* lives in *symbiotic association with the small floating fern, *Azolla*.

**anabolism** That part of cellular metabolism that encompasses those reactions which require energy and that result in the biosynthesis of needed compounds.

**Anacardiaceae** A family of resinous trees, shrubs, and climbers, in which the leaves are alternate, mostly *pinnate, and *exstipulate. The flowers are regular, and mostly bisexual; they have 5 fused *sepals, 5 free *petals, 5–10 *stamens, a *disc, and a *superior *ovary which is 1- to 5-locular, each *locule having a pendulous *ovule. The fruit is usually a *drupe. There are 73 genera, with about 850 species, which are found throughout the tropics, with a few northern outliers.

***Anacardium*** (family *Anacardiaceae) A genus of trees that have *simple leaves. The fruit is a kidney-shaped *nut with a resinous, irritant *pericarp surmounting a swollen, fleshy, edible *peduncle. *A. occidentale* (cashew nut) is a bushy tree, extensively cultivated. There are 8 species, of neotropical distribution.

***Anacystis*** A genus of *cyanobacteria; freshwater forms consist of free-floating masses of spherical cells, while terrestrial forms may be single-celled or colonial.

**anaerobe** An organism that can grow only in the absence of oxygen.

**anaerobic 1.** Of an environment: one in which air (oxygen) is absent. **2.** Of an organism: one able to grow only in the absence of oxygen, i.e. an anaerobe. **3.** Of a process: one that can occur only in the absence of oxygen.

**anagenesis** In the original sense, evolutionary advance. Now the term is often applied more widely, to virtually all sorts of evolutionary change, along a single, unbranching lineage.

**analogous variation** Those features with similar functions which have developed independently in unrelated taxonomic groups, in response to a similar way of life. For example, phyllodes (flattened petioles performing leaf functions) which are found on *Acacia* species are analagous to leaves. *Compare* HOMOLOGOUS and CONVERGENT EVOLUTION.

**Ananas** (family *Bromeliaceae*) A genus of plants in which the stem is short and leafy, trapping water in the overlapping leaf bases. *Inflorescences are terminal, forming a *compound structure in the fruit beyond which there develops a tuft of leaves. The pineapple is *A. comosus*. There are 8 species of terrestrial, tropical American habitats.

**anaphase** A stage that occurs once in *mitosis and twice in *meiosis and that involves the separation of chromosomal material to give two sets of *chromosomes which will eventually form part of new cell nuclei. The separation is controlled by the *spindle. In anaphase of mitosis and anaphase II of meiosis, the *centromere becomes functionally double and daughter chromosomes separate from the equator, moving towards the opposite poles of the spindle. The spindle then elongates and pushes the two groups of chromosomes further apart. In anaphase I of meiosis, the centromere does not divide.

**Anaplasmataceae** (order *Rickettsiales*) A family of *bacteria in which the cells are very small and obligately parasitic (*see* PARASITISM) in the blood of vertebrates. They are transmitted from one host to another by arthropod vectors, e.g. ticks. Some can cause disease. There are 4 genera.

**anastomosis** In woody plants, the linking of branches.

**anatomy** The details of the structure of an organism, as revealed by dissection. The term is sometimes used synonymously with *morphology.

**anatropous** Applied to the position of an *ovule where the *micropyle faces the *placenta and lies close to the base of the funicle. *Compare* CAMPYLOTROPOUS and ORTHORTROPOUS.

**Ancalomicrobium** A genus of *Gramnegative *bacteria not assigned to any taxonomic family. The cells are variable in shape, according to conditions, but typically have 2–8 stalk-like extensions. Reproduction is by *budding from one site on the cell body (not from stalks). These bacteria can grow under *aerobic or *anaerobic conditions, and are *chemoorganotrophs. They are found in aquatic environments.

**Anchusa** (alkanets, etc.; family *Boraginaceae*) A genus of usually bristly *herbs with flowers that are blue or purple, *pentamerous, with scales closing the throat of the *corolla-tube, and oblong leaves. Several species are cultivated for their flowers. There are about 35 species, found in Europe and Asia.

**Ancistrocladaceae** A family of woody climbers in which the twig tips are modified to form coiled hooks. The leaves are *simple and alternate, with *stipples absent or soon *caducous. Flowers are *cymose, regular, and bisexual. They have 5 fused *sepals and *petals (the former enlarging in the fruit (a *nut)), 10 (5) *stamens, the *ovary is *half-inferior with a single basal *ovule, and there are 3 *connate *styles. There is 1 genus, containing 12 species, with a disjunct distribution in W. Africa and South-east Asia.

**Andean floral region** Part of R. Good's (*The Geography of the Flowering Plants*, 1974) *neotropical kingdom. The region is elongated and extends from Colombia to southern Chile, spanning both tropical and temperate climates, but it also includes the Galápagos Islands. These islands are included as, although they have numerous endemic (*see* ENDEMISM) species, they have only two important endemic genera: *Leiocarpus* and *Scalesia*. The mainland flora is rich in endemic genera, however, and has supplied many economic and garden plants. *See also* FLORAL PROVINCE and FLORISTIC REGION.

**andic horizon** A *soil horizon formed by the moderate weathering of volcanic rocks. The name is from the Japanese *an* meaning 'dark', and *do* meaning 'soil'.

**Andira** (family *Fabaceae, subfamily *Papilionatae) A genus of trees some of which yield useful timber. The fruit is globose, woody, and *dehiscent. *A. inermis* (angelin) is widely planted as an ornamental. There are about 20 species, occurring in tropical W. Africa and America.

**Andosols** Soils that have developed from volcanic material and have either a *vitric horizon more than 30 cm below the surface, or that have weathered volcanic rocks within 25 cm of the surface. Andosols are a reference soil group in the FAO *soil classification.

**Andreaeidae** (class *Musci) A subclass of *acrocarpous mosses with several unique characteristics; they are sometimes included in a separate class (Andreaeopsida). The plants are small and characteristically black or dark brown. The *protonema is thalloid. The *capsules differ from those of other mosses in that they open by (usually 4) longitudinal slits. They form scattered patches on rocks in mountainous and Arctic regions.

**androchory** (anthropochory, brotochory) Dispersal of spores or seeds by humans.

**androdioecious** Applied to a *dioecious species in which male and hermaphrodite flowers occur on different plants. *Compare* ANDROMONOECIOUS, GYNODIOECIOUS, and GYNOMONOECIOUS.

**androecious** Applied to a plant that possesses only male flowers. *See also* STAMINATE.

**androecium** A collection of *stamens that form the male reproductive organs of a flowering plant. These may be borne together with the female reproductive organs (the *gynoecium, comprising *stigma, *style, and *ovary) within the same flower of an individual plant, in which case the plant is referred to as hermaphrodite; or they may be borne on the same individual but in different flowers, in which case the plant is termed *monoecious; or they may be borne separately on different individuals, in which case the plant is termed *dioecious.

**andromonoecious** Applied to a *monoecious species in which male and hermaphrodite flowers occur separately on the same plant. *Compare* ANDRODIOECIOUS, GYNODIOECIOUS, and GYNOMONOECIOUS.

**anemo-** A prefix, derived from the Greek *anemos*, meaning 'wind', that is used to associate the remainder of the word with the concept of wind.

**anemochory** (aerochory) Dispersal of spores or seeds by wind.

**Anemone** (family *Ranunculaceae) A genus of poisonous, *perennial herbs (or, rarely, low shrubs) that have *rhizomes, *palmate, lobed leaves some of which are *radical, the stem leaves occurring in a *whorl of 3 some way below the flowers, and solitary flowers (or occasionally 2 or 3 flowers together). They have a *perianth

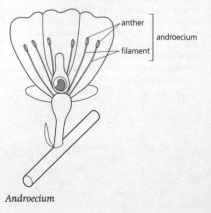

*Androecium*

of 1 whorl, which is petal-like, and many free *stamens and *carpels. The fruit is a head of unplumed *achenes. There are about 120 species, almost cosmopolitan and especially found in northern temperate regions.

**anemophily** Pollination of plants by wind.

**aneucentric** Applied to an aberrant *chromosome possessing more than one *centromere.

**aneuploid** A cell or organism whose *nuclei possess a *chromosome number that is greater by a small number than the normal chromosome number for that species. Instead of having an exact multiple of the *haploid number of chromosomes, one or more chromosomes are represented more times than the rest. An aneuploid typically results from non-*disjunction of one or more pairs of *homologous chromosomes.

**Angaraland** The name given by the Austrian geologist Eduard Suess (1831–1914) to a small shield in north central Siberia, where *Precambrian rocks are exposed, and which was considered to be the nucleus for subsequently developed structural features in Asia.

**Angelica** (family *Apiaceae) A genus of tall, *perennial herbs that have stout, hollow stems. They have 2 or 3 leaves, which are *pinnate with broad oval leaflets. The *umbels are *compound, with many rays, few *bracts, but many *bracteoles. The fruits are ovoid, and flattened dorsally, with 2 broad, marginal rays, and 3 dorsal ridges on each *carpel. *A. archangelica* is cultivated and the young stalks candied as a sweetmeat. There are some 50 species, occurring in the northern hemisphere and in New Zealand.

**angiocarpy** A type of development of a fungal *fruit body in which the *spore-bearing *tissue is enclosed for part of the period.

**angiosperm** A flowering plant, distinguished by producing seeds that are enclosed fully by fruits. Angiosperms are the most highly evolved of plants, and appear first in rocks of Lower *Cretaceous age. All of them are included in the division *Anthophyta.

**Angiospermae** Formerly, one of the 2 divisions of the Spermatophyta (seed plants), the other being the Gymnospermae. The term is now used only informally. (*See* ANGIOSPERM.)

**Angophora** (family *Myrtaceae, subfamily Leptospermoideae) A genus of trees and shrubs which are closely allied to eucalypts, but which have flower petals and no *operculum, and opposite leaves. They are found in or near rain forests in the east. *A. costata* (Sydney red gum or smooth-barked apple) is a tree growing to at least 14 m, with red bark and white flowers in spring and summer. It is the dominant tree of deep sands behind coastal dunes in New South Wales, and on alluvial sands. There are 8 species, all endemic to Australia.

**Angostura** (family *Rutaceae) A genus comprising 30 species, found in tropical America, in which the flowers are sympetalous and zygomorphic. The bark is the basis of Angostura bitters.

**angsana** Timber from *Pterocarpus indica*. *See* PTEROCARPUS.

**Anigozanthos** (kangaroo paw; family *Haemodoraceae) A genus of herbs that have unusually shaped spikes and *racemes of tubular, woolly flowers which may be orange, red, or green. They are now widely cultivated, and species hybrids are developed for increased vigour, greater flower numbers, and different colours. *A. manglesii* (red-and-green kangaroo paw) is the state emblem of W. Australia. There are 11 species, endemic to south-western Australia.

**anion** An *ion that carries a negative electrical charge.

**anion-exchange capacity** The total exchangeable *anions that a soil can adsorb, measured as moles per gram of soil.

**aniso-** Prefix meaning unequal, from the Greek *an* and *isos*, 'not' and 'equal'.

**anisogamy** The fusion of *gametes of different sizes. *Compare* ISOGAMY; *see also* OOGAMY.

***Anisoptera*** **(mersawa; family \*Diptero-carpaceae)** A genus of large, evergreen trees that yield valuable, usually \*siliceous, timber. They are native from Assam to New Guinea, and are found in lowland tropical rain forest. There are 11 species.

**annatto** *See* Bixa.

***Annona*** **(family \*Annonaceae)** A genus of shrubs or small trees, producing large fruits that are aggregate, individual \*berries being immersed in the fleshy \*receptacle. Many are edible—e.g. neotropical *A. muricata* (soursop), *A. reticulata* (bullock's heart or custard apple), *A. cherimola* (cherimoya or custard apple), and *A. squamosa* (sweetsop, custard apple, or sugar apple). There are about 100 species, occurring in the warm temperate to tropical regions of Africa and America.

**Annonaceae** A family of shrubs, \*monopodial trees, and climbers, that have aromatic tissues. The twigs have \*septate pith. The leaves are \*simple, alternate, and \*exstipulate. The flowers are bisexual or unisexual, often fragrant, and their parts are free. They have 3 \*sepals, 6 \*petals, and many \*stamens. The few to many \*carpels are \*superior, and there are 1 to many \*ovules. Fruits typically are clustered and fleshy. *Asimina triloba* (pawpaw), of N. America, is one of the few temperate species. There are 128 genera, with about 2000 species, almost entirely tropical, and found mainly in the Old World, especially in lowland rain forests.

**annual** Applied to a plant that completes its \*life cycle (from germination to flowering to seed production and the death of vegetative parts) within a single \*growing season. *Compare* BIENNIAL; EPHEMERAL; and PERENNIAL; *see also* THEROPHYTE.

**annual meadow grass** *See* POA.

**annual ring** *See* TREE RING.

**annular** In the shape of a series of rings.

**annulate lamellae** Flat, membranous \*cisternae, which bear regularly spaced pores. They are derived from the nuclear envelope and apparently represent an in-termediate stage between this and the \*endoplasmic reticulum.

**annulus** **1.** In the \*fruit bodies of certain \*agarics, a remnant of the \*partial veil that adheres as a ring of tissue around the \*stipe. **2.** A ring of cells around the \*sporangium of some \*ferns. The walls of these cells become progressively thicker around the circumference so inducing tension as they dehydrate, eventually causing the rupture of the sporangium. (The point of breaking is called the stomium.) The lid of the sporangium then curls back, releasing the tension induced by the annulus in a spring-like motion serving to aid \*spore dispersal. **3.** The thickened ring found around the pores of some pollen grains, e.g. the grasses (\*Poaceae).

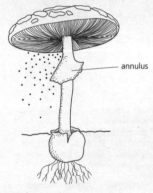

annulus

*Annulus*

**anode** A positive electrode, to which negative \*ions (anions) are attracted.

**anomocytic** Applied to a \*stoma that lacks morphologically differentiated \*subsidiary cells.

**Antarctic beech** *See* NOTHOFAGUS.

**anther** The terminal portion of a \*stamen of a flowering plant. The \*pollen sacs containing \*pollen are borne on the anther.

**antheridiophore** In some members of the \*Marchantiales, a structure that bears the antheridia (*see* ANTHERIDIUM). *Compare* ARCHEGONIOPHORE.

a

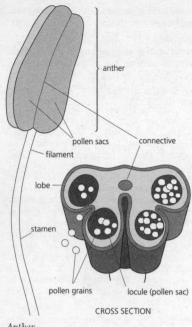

CROSS SECTION

*Anther*

**antheridium** The male sex organ or *gametangium, within which male *gametes are formed, in *algae, *fungi, *bryophytes (mosses, liverworts, etc), and *pteridophytes (ferns).

**antherocyte (spermatocyte)** A cell that develops into an *antherozoid without undergoing further division.

**antherozoid (spermatozoid)** The male *gamete produced in the (antheridia of *fungi, *algae, *bryophytes, *pteridophytes, and some *gymnosperms. The cell is *haploid and can move by means of *flagella (the number of which varies from species to species and can be used as a diagnostic feature). The majority of antherozoids have tape-like cells with a long slender nucleus which facilitates entry in to the neck of the *archegonia. Antherozoids usually develop in the antheridium (but in some gymnosperms, e.g. *Ginkgo, they develop from a cell in the *pollen tube of the *microspore).

**anthesis** **1.** The time of flowering in a plant. This appears to be a response to a combination of factors including day-length, temperature, and rainfall, but may also be initiated by the addition of *gibberellins, one of a group of growth-promoting substances. **2.** The opening of a flower bud.

***Anthocephalus*** **(family *Rubiaceae)** A genus of fast-growing, medium-sized, pioneer trees of lowland rain and seasonal forests, characteristically of moist alluvial soils. They yield useful timber and have been planted widely, but mainly unsuccessfully, throughout the humid tropics. There are 2 species, native to Nepal, and to the region from Yunnan to New Guinea.

**Anthocerotales** **(hornworts; class *Hepaticae)** An order of *thallose *liverworts which have a number of unique characteristics; they are some-times regarded as a separate class, Anthocerotae (Anthocerotopsida). The capsule is long and narrow and splits into 2 valves only (*compare* HEPATICAE); it remains green for a long period. The green cells of the *thallus each contain a single large *chloroplast. Two genera occur in Britain: *Anthoceros* and *Phaeoceros*. They are found e.g. on wet soil in fields, banks, ditches, etc.

**anthocyanins** Water-soluble, nitroge-nous pigments which contribute to the autumnal colours of the leaves of tem-perate climate plants.

**Anthophyta** The division that includes all flowering plants (*angiosperms).

**anthracnose** A general term for any of several plant diseases in which symptoms include the formation of dark, often sunken, spots on leaves, *fruit, etc. An example is bean anthracnose, a disease of dwarf and runner beans caused by *Colletotrichum lindemuthianum*. Grape an-thracnose, caused by *Elsinoe ampelina*, is a serious disease in parts of northern Europe; leaves develop irregular greyish spots, and portions of dead leaf *tissue may drop out.

***Anthriscus*** **(family *Apiaceae)** A genus of herbaceous plants that have fern-like,

2 or 3 times *pinnate leaves and *compound *umbels. Usually there are no *bracts but several *bracteoles. The flowers are small and white. The fruits are ovoid to oblong, smooth or spiny, and hardly ridged. *A. cereifolium* (chervil) has aniseed-flavoured leaves and is cultivated as a vegetable. There are some 12 species, found in temperate Eurasia and in N. Africa.

**anthropic horizon** The surface horizon of a *mineral soil that is produced by very long periods of cultivation and fertilization by humans. It is dark brown in colour, contains at least 1% organic carbon, is relatively deep (more than 50 cm), has a base saturation of more than 50%, and has more than 250 ppm $P_2O_5$ soluble in 1% citric acid. The name is from the Greek *anthropos* meaning 'human'.

**anthropochory** *See* ANDROCHORY.

**Anthropogene** The name that is sometimes given informally to the *Holocene, since this is the time during which human activity has significantly altered the natural environment.

**Anthrosols** Soils that have been strongly affected by human activities such as ploughing, irrigation, or the addition of manure. Anthrosols are a reference soil group in the FAO *soil classification.

**Antiaris** (family *Moraceae) A genus of trees that produce poisonous latex which is much sought as a constituent for arrow poisons. *A. toxicaria* (ipoh or upas tree) gained a false reputation in Europe for being excessively poisonous. There are 4 species, occurring in tropical Africa, Madagascar, and Indo-Malaysia.

**antibiosis** *See* ALLELOPATHY.

**antibiotic** An antimetabolite obtained from or produced by a living bacterium, fungus, or plant, which, in very small amounts, is toxic or lethal to other organisms (usually other bacteria or fungi). The term may also refer to chemical derivatives of naturally occurring antibiotics or to synthetic substances with similar properties. Under natural conditions the ability to produce an antibiotic presumably confers a competitive advan-

tage on the organism. Some antibiotics are important in the treatment of animal diseases caused by micro-organisms.

**anticlinal** At right angles to a surface. *Compare* PERICLINAL.

**anticodon** A triplet sequence of nucleotides in *transfer-RNA that during protein synthesis (*see* RIBOSOME) binds by base pairing to a complementary sequence, the so-called *codon, in *messenger-RNA attached to a ribosome. There are at least 20 different types of anticodon, each encoding for a specific *amino acid carried by the transfer-RNA.

**Antidesma** (family *Stilaginaceae) A genus of shrubs or small trees that produce spikes of minute, unisexual flowers with a prominent *disc. The fruit is a small, juicy *drupe, usually flattened. The genus is often placed in *Euphorbiaceae, but has no close relatives there, and perhaps is intermediate to *Icacinaceae. There are about 160 species, occurring in the Old World subtropics and tropics, especially in Asia.

**antigen** A molecule, normally of a *protein although sometimes of a *polysaccharide, usually found on the surface of a cell in an animal, whose shape causes the production in the invaded organism of *antibodies that will bind specifically to the antigen. The reaction is so highly specific that it can be used in the identification of viruses and to distinguish between proteins extracted from different plant species (*see* SEROTAXONOMY).

**anti-oxidant** A compound, usually organic in nature, that prevents or retards the oxidation by molecular oxygen (autoxidation) of materials such as food, rubber, and plastics. It acts by scavenging the free radicals generated in autoxidation chain reactions, and thus provides an alternative oxidation pathway. It does not act indefinitely as it is destroyed in the process.

**antipodal** Usually refers to three *haploid nuclei which are formed during megasporogenesis (*see* SPOROGENESIS) in plants; all are located opposite the micropylar end of an *ovule. *See also* MEGASPORANGIUM and MICROPYLE.

**a**

***Antirrhinum*** **(family \*Scrophulariaceae)** A genus of \*herbs in which the lower leaves are \*opposite, the upper leaves alternate, and all are \*simple. The flowers are borne in \*racemes and have strongly 2-lipped \*corollas, the lower 3-lobed lip having a projecting bulge or 'palate' which closes the mouth of the flower; the tube is unspurned, but with a swelling below. *A. majus* (snap-dragon) is cultivated in many coloured forms and has been naturalized beyond its southern European natural range. There are 42 species, most Mediterranean, but also occurring in western N. America.

***Apeiba*** **(family \*Tiliaceae)** A genus of fast-growing trees, with fibrous bark, some of which yield useful timber. There are 10 species, found in tropical S. America.

**aperturate** Applied to a structure having openings (apertures), especially to a \*pollen grain with regions where only \*nexine is present or \*exine is lacking completely.

**apex** The \*distal part of a leaf, shoot, or root.

**APF** *See* ABSOLUTE POLLEN FREQUENCY.

***Aphanizomenon*** A genus of filamentous \*cyanobacteria. The filaments aggregate into raft-like bundles and float at the surface of lakes, ponds, etc. Freshwater blooms of some strains of *A. flos-aquae* can lead to poisoning of animals and contamination of water supplies.

***Aphanocapsa*** *See* SYNECHOCYSTIS.

**Aphyllophorales** **(subclass \*Holobasidiomycetidae)** An order of \*fungi in which the development of the \*fruit body is \*gymnocarpic; the \*fruit body may be woody, corky, or leathery (but not fleshy), and the \*hymenium is typically borne on teeth or within tubes. There are many families. Most are \*saprotrophic, and found on wood, soil, plant debris, etc.

**aphyllous** Without leaves.

**Apiaceae** A family, formerly known as Umbelliferae, of dicotyledonous (*see* DICOTYLEDON) \*herbs, with a few shrubs, in which the alternate leaves are usually much divided and have sheathing stalks at the base. The flowers are characteristically borne in \*umbels, which are usually compound, and have a tiny \*calyx of 5 teeth, or no calyx. There are 5 free \*petals, which are often notched and sometimes very unequal in size; 5 \*stamens alternating with the petals; and an \*inferior \*ovary of 2 fused \*carpels, which ripens into 2 separating but \*indehiscent parts, which may remain suspended from the tip of the \*axis. The carpels each bear 5 (or sometimes 9) ridges, usually with 4 oil-canals (vittae) between the main ridges and 2 more on the inner faces. This large family contains many important food plants, e.g. *Daucus* (carrot), *Pastinaca* (parsnip), and *Apium* (celery); while others are very poisonous, e.g. *Oenanthe* (water dropwort), or are used medicinally. There are some 420 genera, with 3100 species, found throughout most of the world, but mainly in northern temperate regions.

**apical** Pertaining to the \*apex.

**apical dominance** A condition in plants where the stem apex prevents the development of lateral branches near the apex. It is controlled by \*auxins produced at the apical bud. Close to the bud, the auxin concentration is high and inhibits lateral growth; further back, the concentration is lower and branching occurs. Removal of the apical buds results in branching.

**apical meristem** *See* MERISTEM.

**apitong** *See* DIPTEROCARPUS.

***Apium*** **(family \*Apiaceae)** A genus of \*glabrous \*herbs that have mostly \*pinnate leaves and leaf-opposed \*umbels. The fruits are broad-ovoid to oblong, laterally compressed \*carpels with 5 low ridges. *A. graveolens* var. *dulce* (celery) is a valuable vegetable; var. *rapaceum* is celeriac, with edible, turnip-like roots. There are about 20 species, of cosmopolitan distribution.

**aplanate** Lying in a plane; leaves may be displayed on the twigs of a plant to give aplanate foliage.

**aplanetism** A condition in which no motile stage is formed.

**aplanogamete** A non-\*motile \*gamete.

**aplanospore** A non-\*motile, asexual \*spore, found in some algae and ferns.

**apo-** A prefix, derived from the Greek *apo*, meaning 'away' or 'from', that imparts a sense of 'away from' or 'separate'.

**apocarpous** With the *carpels free.

**Apocynaceae** A family of trees, climbers, or shrubs, in which the leaves are mostly *opposite or *whorled. The flowers are *cymose, regular, and bisexual; they have a 5-lobed *calyx and *corolla tubes, 5 *stamens, the *ovary *superior or half *inferior, and 2 *carpels which are free or fused. Fruits are paired, fleshy or woody, and *dehiscent. There are 215 genera, with about 2100 species, whose distribution is tropical with a few temperate outliers.

**apoenzyme** The portion of a conjugated *enzyme (i.e. one containing a non-protein component) that is a *protein.

**apogamy** *See* APOMIXIS.

**apogeotropism** The type of *tropism in which plant organs (e.g. shoots) grow against the force of gravity.

**apomict** A plant produced by *apomixis, a form of *asexual reproduction.

**apomixis** In plants, a form of asexual reproduction that gives the appearance of sexual reproduction. *Parthenogenesis is an example of apomixis, as is the seed production of dandelions (*Taraxacum officinale*). Since apomixis does not involve fertilization or *meiosis, the progeny are genetically identical to their parents.

**apomorph** An evolutionarily advanced ('derived') character state. The opposite of plesiomorph.

**apomorphic** Applied to features possessed by a group of biological organisms that distinguish those organisms from others descended from the same ancestor. The term means 'new-featured' and refers to 'derived' characters which have appeared during the course of evolution.

**apophysis** In the *sporophyte of mosses (*Musci), a swollen region between the *seta and *capsule. Its cells are rich in *chloroplasts and it contributes to the nourishment of the *sporangium.

**apoplast** Areas of a plant that lie outside the *plasmalemma, such as *cell walls and dead tissues of the *xylem. The apoplast may represent one of the main pathways of water movement through the plant.

**apospory** The development of a *diploid embryo sac in some plants by the somatic division of a *nucellus or integument cell without *meiosis. It is a form of agamospermy, in which a seed is produced without *fertilization.

**Apostasiaceae** A family of monocotyledonous *herbs that have *plicate leaves. The flowers are *racemose, with 6 free *perianths, 2–3 *stamens, the *filaments and *style united, the *ovary *inferior and *trilocular, and numerous axile (*see* PLACENTATION) ovules. The fruit is a *capsule. The family is primitive, and closely related to *Orchidaceae. There are 2 genera, with about 20 species, found from India to tropical Australia.

**apothecium** A roughly cup-shaped or dish-shaped *ascocarp, in which the *asci line the inner surface and are thus exposed to the atmosphere.

**apple** *See* MALUS.

**apple moss** The common name for the moss *Bartramia pomiformis* (order *Bryales). The name derives from the resemblance of the immature *capsules to miniature green apples; as they age the capsules become brown and wrinkled. The moss is found on sandy soil, rock ledges, etc.

**apple scab** A very common disease of apple trees (*Malus*) in which the most obvious symptom is the appearance of superficial, dark, corky scabs on the fruit. In general the disease is not serious in itself, although more serious secondary infections can occur. The causal agent of apple scab is a *fungus, *Venturia inaequalis*.

**appressorium** In certain parasitic *fungi, an attachment organ consisting of a flattened *hypha which presses closely to the *tissues of the host as a preliminary stage in the infection process.

**Aptian** One of the stages of the Lower *Cretaceous Period in Europe, dated at about 125–112 Ma ago.

**aquifer** A body of rock, unconsolidated gravel, or *sand stratum that is capable of storing significant quantities of water, that is underlain by impermeable material, and through which *ground water moves. An unconfined aquifer is one in which the *water table defines the upper water limit. A confined aquifer is sealed above and below by impermeable material. A perched aquifer is an unconfined ground-water body supported by a small impermeable or slowly permeable unit. *See* ARTESIAN WATER.

**Aquifoliaceae** A family of trees or shrubs that have leathery, alternate, evergreen, sometimes spiny, *simple, *exstipulate leaves. They are usually *dioecious. The flowers, borne in small *cymes, are regular, *tetra- or *pentamerous. The *petals and *stamens are usually free. The fruit is a small *drupe with 3 or more stones. *Ilex aquifolium* (holly), native in western and central Europe, is much grown for its foliage and for the scarlet fruits of the female trees. There are 4 genera, with more than 420 species, found throughout the world, except in Arctic regions, but centred on the tropics.

**Aquilaria** (family *Thymelaeaceae) A genus of trees whose timber is pale and soft, sometimes becoming diseased to form a hard, dense, resiniferous *heartwood (aloeswood or eaglewood) yielding a valuable oil for medicinal use and incense. There are 15 species, occurring in India, and from southern China to Malesia.

**Aquilegia** (columbines; family *Ranunculaceae) A genus of poisonous, *perennial herbs, which are usually erect, with alternate, 2 or 3 times *ternate, *compound leaves. The flowers are showy, regular, *pentamerous, with *petaloid *sepals and long-spurred *petals, and many free *stamens and *carpels which produce several-seeded *follicles in fruit. Several species and hybrids are cultivated for their flowers. There are about 70 species, found mostly in the northern temperate zone.

**Aquitanian** Defined first from the Aquitanian Basin, France, the Aquitanian Age marks the beginning of the *Miocene,

23.03 Ma ago. It also marks the start of the Upper *Cenozoic (*Neogene) Period of time. The Aquitanian is itself characterized by the appearance of the planktonic foraminiferid *Globigerinoides primordia*.

**arabinose** An *aldose that contains 5 carbon atoms.

**Arabis** (family *Brassicaceae) A genus of *calcicole *herbs, often forming mats, in which the stems are erect, and the leaves are usually oblong, sessile, and numerous up the stem. The flowers are 4-petalled, white, or pale yellow. The fruits are compressed, much elongated, and beakless (*see* BEAK). Many species are cultivated for their flowers. There are about 120 species, occurring throughout the northern temperate zone, and in the mountains of Africa.

**Araceae** A major family of monocotyledons comprising tiny to giant herbs and many *bole climbers, which produce sap with irritant crystals. The leaves commonly have net-veins. Flowers are borne in a spike (spadix) subtended by an often colourful *spathe; they are small, usually unisexual, the male apical (*see* RACEME), and usually lack a *perianth. The fruit is a *berry. There are 106 genera, and about 2950 species, with pantropical distribution (especially America), with temperate outliers.

**Arachis** (family *Fabaceae, subfamily *Papilionatae) A genus of *perennial *herbs which are *geocarpic (after flowering the *inflorescence axis elongates so that the pods open below ground). *A. hypogaea* (peanut, groundnut, monkey nut) is not known in the wild, but is widely cultivated for the edible seed and oil; it is also used for soap and lubrication; the residues make cattle feed, and the green haulms make excellent fodder. There are 12 species, occurring in America from the Amazon River south to Argentina.

**arachnoid** Resembling a spider's web (spiders belong to the class Arachnida).

**Araliaceae** A family of shrubs, climbers, and soft-stemmed *pachycaul trees, and a few herbs, which is the tropical counterpart to the *Umbelliferae. The leaves are often large, and usually *compound. The flowers are usually borne in compound *umbels and are small and regular, with 4 or 5 reduced *sepals, 5 or 3 free *petals and *stamens, a *disc, and the *ovary *inferior with 5 loculi (*see* LOCULE) and 1 pendulous *ovule. The fruit is a 5-seeded *drupe. There are 58 genera, with about 800 species, most of them tropical.

***Araucaria* (family *Araucariaceae)** A genus of tall, columnar, coniferous trees, in which the crown is *monopodial with radial limbs. The leaves are sharply pointed, crowded, and tough. The timber is valuable. *A. bidwillii* (bunya bunya) of Queensland produces edible seeds *A. angustifolia* (Paraña pine) is a major timber tree in Brazil. Norfolk Island pine is *A. heterophylla*. Monkey-puzzle (*A. araucana*), native to the southern Andes in Chile and Argentina, also produces edible seeds. There are 18 species, occurring in tropical to temperate forests in New Guinea, Queensland, Melanesia; 2 of the species are found in S. America, and 13 of them in New Caledonia. *See* DISJUNCT DISTRIBUTION.

**Araucariaceae** A family of conifers comprising large trees which produce valuable timber. Several species are cultivated. The leaves are broad to *lanceolate. The male cone is a *catkin; the female is *globose and often massive, disintegrating at maturity. There are 2 genera (*Agathis* and *Araucaria*), native to the southern hemisphere (but with fossils in the northern hemisphere), and found mainly in Malesia and Melanesia, but also in S. America, and extending north of the equator in Malaysia and the Philippines.

**arbor vitae** *See* THUJA.

**arbuscule** A tuft of branching *fungal *hyphae formed in certain types of *mycorrhiza.

***Arbutus* (family *Ericaceae)** A genus of evergreen shrubs or small trees that have dark green leaves and clusters of small, white or pink, bell-shaped flowers. The fruit is a *globose, warty *berry with mealy flesh, edible in *A. unedo* (strawberry tree, native in Ireland and south-western Europe). American *A. menziesii* (madroña laurel) has useful timber. There are 14 species occurring in N. and Central America, southern Europe, the Canary Islands, and western Asia.

**Archaea** Single-celled organisms including the *phenotypes: *methanogens; *sulphur-reducing organisms; and *extremophiles. **1.** In the widely used five-kingdom system of classification, the Archaea (also called Archaebacteria) is ranked as one of the two subkingdoms within the *kingdom *Bacteria (*see also* EUBACTERIA). **2.** In the three-domain system of classification, Archaea is one of the *domains, comprising organisms formerly known as the archaebacteria and placed in a kingdom of that name. The former kingdom Archaebacteria has been split into two kingdoms: *Crenarchaeota and *Euryarchaeota.

**Archaean** An eon of geologic time following the *Hadean and preceding the *Proterozoic eons, which together comprise what was formerly known as the *Precambrian. The Archaean eon lasted from 3800–2500 Ma and is divided into the Neoarchaean (2800–2500 Ma), Mesoarchaean (3200–2800 Ma), Palaeoarchaean (3600–3200 Ma), and Eoarchaean (3800–3600 Ma) eras. The Archaean was formerly known as the Archaeozoic or Azoic.

**Archaebacteria** *See* ARCHAEA.

***Archaeocalamites radiatus*** First described by *Brongniart, in 1828, one of the earliest recorded equisetaleans or horsetails, a group of plants that is noted for the jointed nature of the stem. At each joint there is a ring of short branches.

***Archaeopteris*** An early progymnosperm (*see* PROGYMNOSPERMOPSIDA) found in the *Frasnian stage of the *Devonian Period. It is known first by its fronds, and is the first representative of the Archaeopteridales.

**archegoniophore** In certain members of the *Marchantiales, a structure that bears the archegonia (*see* ARCHEGONIUM). *Compare* ANTHERIDIOPHORE.

**archegonium** A female sex organ of *liverworts, *mosses, *ferns, and most *gymnosperms. It is usually a flask-shaped organ, comprising a swollen base or venter containing a single y egg-cell and a slender elongated neck containing one or more layers of cells.

**archesporium** The tissue that gives rise to *spore mother cells.

***Archontophoenix*** (family *Arecaceae) A genus of palms in which the trunk is solitary and hoop-marked and the leaves are *pinnate, with a conspicuous crown-shaft, below which the much-branched pendulous inflorescences appear. There are 2 species, native to subtropical eastern Australia and widely planted as stately ornamentals. *A. alexandrae* (the king palm) is of very similar general appearance to *Dictyosperma album* (DICTYOSPERMA).

**Arctic-alpine species** A species that is found both in the Arctic and on the higher mountains in the temperate zone. Examples include *Salix herbacea* (least willow) and *Saxifraga oppositifolia* (opposite-leaved saxifrage). Although characteristic of the Arctic region and the Alps, Rockies, and Himalayas, they also occur in upland areas between the mountain ranges and the Arctic, and represent relics of the late *Pleistocene *tundra.

**Arctic and subarctic floristic region** Part of R. Good's (*The Geography of the Flowering Plants*, 1974) boreal kingdom, corresponding to the treeless wastes north of the *boreal conifer zone. There are virtually no endemic (*see* ENDEMISM) genera, which suggests that this is possibly the youngest flora in the world. The flora has two components: one Arctic, which includes the endemic species; the other *Arctic-alpine. *See also* FLORAL PROVINCE and FLORISTIC REGION.

**Arctic heath** Arctic heath is found in the low and middle Arctic belts of the *tundra. Normally it is dominated by members of the heath family (*Ericaceae) or by heath-like plants, e.g. *Vaccinium vitis-idaea* (crowberry). Arctic heaths tend to be restricted to relatively well-drained sites that are sheltered and snow-covered in winter.

**Arctic scrub** A scrub composed of plants that average about 60 cm high, and including dwarf willows (*see* SALIX), birches (*see* BETULA), or both. It occurs in damp hollows and along the edge of water

in the low Arctic belt of the *tundra. Northwards it becomes more stunted and limited in area, and more or less disappears beyond the centre of the middle Arctic belt.

**arcuate** Curved or arched.

**area cladistics** A technique that employs relationships between organisms to reconstruct past distributions and the positions of continents, independently of geological data. Genetic and morphological characteristics reveal patterns of relationships from which former geographic distributions can be inferred. Plotting past distribution patterns identifies the location of geographic barriers responsible for the isolation of organisms and the consequent divergence of species. This allows the geographic areas to be arranged hierarchically by adding the geographic information to *cladograms, producing diagrams called *areagrams.

**area-effect speciation** The speciation by increased differentiation of two *subspecies with incompatible *gene-complexes, so that selection is strongly against *hybrids.

**areagram** A diagram derived from a *cladogram but with the addition of data referring to the former geographic distribution of organisms. The areagram resembles a cladogram, but instead of showing the relationships among groups of organisms it shows geographic relationships and the way these have changed over time. An areagram shows the order in which geographic areas separated from one another.

***Areca*** (family *Arecaceae) A genus of palms which grow solitarily or in small clumps. The leaves are *pinnate with a conspicuous crown-shaft. Inflorescences occur *distally with only male flowers, and near the base triads a big female is flanked by 2 males. The fruits have a fleshy mesocarp (*see* PERICARP). The betel-nut comes from *A. catechu*. There are about 50 species, native to tropical Asia and the western Pacific.

**Arecaceae (palms)** A family, formerly known as Palmae, of solitary or clumped, small or tall, unbranched trees whose leaves are *palmate or *pinnate, the *leaflets being *plicate. The flowers are small, *trimerous, and unisexual or bisexual, the *inflorescence having various *bracts. Fruits are mostly 1-seeded *berries or *drupes, fleshy to dry, and *indehiscent, varying greatly in size from pea-like to the double coconut (*Lodoicea). Many species are ornamentals, and many yield valuable economic products, including thatch (e.g. *Nypa), starch (e.g. *Metroxylon), and oil (*Orbignya, *Cocos, and *Elaeis). There are about 207 genera and 2675 species, chiefly tropical, but some subtropical, and a few occurring as temperate outliers.

**arenaceous** Sandy, or *sand-like in appearance or texture. The term is applied to clastic sedimentary rocks with a grain size 0.0625–2.00 mm. Three main groups of arenaceous rocks are recognized: quartz sandstones (quartzites), which contain 95 per cent quartz; arkoses, which have greater than 25 per cent feldspar; and greywackes, which essentially are poorly sorted sediments with rock (lithic) fragments in a mud matrix. See also ARKOSE.

**Arenaria** (family *Caryophyllaceae) A genus of low *herbs or dwarf shrubs which have slender shoots, tiny, *opposite leaves, and small, usually white flowers with 5 undivided *petals, 10 *stamens, and 3 *styles (rarely more). The fruit is a tiny *capsule, usually 6-toothed on opening. Several matforming species are cultivated as rockgarden plants (sandworts); others are weeds. There are about 150 species, cosmopolitan in distribution, but most common in the northern temperate zone to the Arctic.

**Arenga** (family *Arecaceae) A genus of palms found in their native habitat ranging from treelets to stout trees. The leaves are *pinnate, with pinnate nervation, the *sheath coarsely fibrous. *Inflorescences often develop *basipetally, paired male flowers falling before the solitary female develops. The fruits have a soft, irritant mesocarp (see PERICARP) and 1–3 seeds. The sugar palm is A. pinnata. There are 17 species, native from the Himalayas to Hainan and New Guinea.

**areolate** Divided into small areas (areolae) by cracks or lines.

**argic horizon (argillic horizon)** A subsurface *soil horizon, at least one-tenth the thickness of the overlying horizon, that is identified by the illuvial (see ILLUVIATION) accumulation of silicate *clays. The amount of clay necessary is defined in comparison with the quantity in the overlying eluvial (see ELUVIATION) horizon, but it is at least 20% more. *Cutans may be used to identify an argic horizon. The name is from the Latin argilla meaning 'white clay'.

**argillaceous** Applied to rocks that are *silt- or *clay-sized sediments with grains less than 0.0625 mm in diameter. They account for more than 50 per cent of sedimentary rocks and most have a very high clay mineral content. Many contain a high percentage of organic material and can be regarded as potential source rocks for hydrocarbons.

**argillans** See CUTAN.

**argillic horizon** See ARGIC HORIZON.

**arginine** An aliphatic, basic, polar *amino acid, that contains the *guanido group.

**Aridisol** An order of *mineral soils that are found in arid environments. These soils have very little organic matter in their surface *horizons, but may contain calcium carbonate or gypsum, and/or soluble-salt accumulations. The order includes infertile *alkaline and *saline soils of deserts. Except for a few tolerant species, the vegetation is very sparse, making the soil prone to erosion.

**aridity index** An indication of moisture deficit. All climatic classifications include arid categories, defined either by quantitative or, more usually, by mainly subjective criteria. C. W. Thornthwaite first used the term 'aridity index' and calculated it as 100× the water deficit/the potential evaporation. The most frequently used index is that devised by the United Nations Environment Programme to express the degree of climatic dryness. It is calculated by dividing the mean annual precipitation by the mean annual potential evaporation.

**aril** A usually fleshy and often brightly coloured outgrowth from the *funicle or *hilum of a seed, a third *integument. Arils probably often aid *seed dispersal, by drawing attention to the seed after the *fruit has *dehisced, and by providing food as an attractant and reward to the disperser. The aril of the nutmeg (*see* MYRISTICA) produces the spice mace, and the seed itself is the nutmeg.

**arillate** Possessing an *aril, a diagnostic characteristic of the seeds of some flowering plants, especially tropical trees.

**aristogenesis** An outmoded theory holding that evolution proceeds along a determined path. The modern view is that *natural selection does not direct evolution towards any particular kind of organism or physiological attribute, nor is there any mysterious inner guiding force. *See also* ENTELECHY; NOMOGENESIS; and ORTHOGENESIS.

**Aristolochiaceae (birthworts)** A family of climbers and a few *herbs in which the flowers are usually *zygomorphic and often foetid. They have a *calyx that is often enlarged and highly coloured, no *corolla, 6 to many *stamens, and the *ovary *inferior and 4–6 locular (*see* LOCULE). The *capsule contains many small seeds. The roots are often medicinal. *Aristolochia* is Dutchman's pipe. There are 7 genera, with about 410 species, of warm temperate to tropical distribution.

**arkose** An *arenaceous rock that contains quartz and 25 per cent or more of feldspar. The feldspar is easily destroyed during transportation or chemical change, and the implication is that arkoses were deposited rapidly under fairly arid environmental conditions. Most were also deposited near to land and probably in close proximity to a granitic area.

**Armillaria (family *Tricholomataceae)** A genus of *fungi in which the *fruit bodies, which typically occur in clusters, are *mushroom-shaped but very variable in colour (ochre to brown) and appearance; the *spores are white. It includes *A. mellea* (honey fungus or honey tuft), a destructive *parasite of trees, shrubs, and other plants. *A. mellea* can spread from one host plant to another by means of thick, black *rhizomorphs ('boot-laces' or 'shoe-straps'). It can also live *saprotrophically, e.g. on dead tree-stumps, and forms a symbiotic *mycorrhizal association with certain orchids.

**Aromadendron (family *Magnoliaceae)** A genus of tropical rain-forest trees which are sometimes huge, growing to 60 m. The fruiting *carpels are *indehiscent, and *concrescent as a fleshy *syncarp. There are 4 species, occurring in Malaya, Sumatra, Java, and Borneo.

**arrow-grass (*Triglochin*)** *See* JUNCAGINACEAE.

**arrowroot** *See* MARANTA and TACCA.

**artefact (artifact)** **1.** A man-made object. **2.** Something observed that is not naturally present but has arisen as a result of the process of observation or investigation.

**Artemisia (family *Asteraceae)** A genus of *perennial (or, rarely, *annual) *herbs or low shrubs, often aromatic, with leaves that are alternate, and much divided *pinnately into narrow segments. The flower heads are tiny, usually numerous, often woolly, and gathered into *racemes or *panicles. The *receptacle is flat and naked, and all the *florets are tubular and surrounded by overlapping, *scariousedged *bracts. There is no *pappus. The genus includes mugworts and wormwoods. Wormwood (*A. absinthium*) is used for flavouring beers and other drinks; *A. tridentata* is sagebrush of the south-western USA. There are about 300 species, occurring mostly in dry grasslands of the northern temperate semi-arid and Arctic zones.

**artesian water** *Ground water that originally is confined in an *aquifer and that reaches the land surface owing to the high hydraulic pressure that may be developed in a confined aquifer when this has a synclinal form. The London Basin, England, provided artesian water during the nineteenth century from a chalk aquifer sealed by *clays. Many desert oases are created by the emergence of artesian water in springs or pools.

**Arthrobacter (order \*Actinobacteria)** A genus of \*bacteria. The cells undergo a change in form during culture in the laboratory, from roughly spherical to irregularly rod-shaped. They are \*chemoorganotrophic, and can grow only in the presence of air. They are found chiefly in soil.

**arthrospore** A type of fungal \*spore formed as a result of the fragmentation of a \*hypha.

**artifact** *See* ARTEFACT.

**artificial classification** The ordering of organisms into groups on the basis of non-evolutionary features (e.g. the grouping together of plants according to the number and situation of their stamens, styles, and stigmas rather than their evolutionary relationships). *Compare* NATURAL CLASSIFICATION.

**artificial selection** Selection by humans of individual plants from which to breed the next generation, because these possess the most marked development of the required attributes. Typically the process is repeated in successive generations until those attributes are fixed in the descendant offspring. Such artificial selection can result in dramatic changes, like those that took place in the domestication of plants from their wild forebears. *See* GENETIC ENGINEERING.

**Artocarpus (family \*Moraceae)** A genus of small to medium trees that yield latex from the bark and useful timber. The leaves have big \*stipules and are commonly divided. Male flowers are borne in \*catkins, female flowers in heads. The fruit is multiple, formed on the fleshy \*receptacle by \*concrescent \*achenes surrounded by a fleshy \*perianth, often \*cauliflorous, and gigantic, in *A. heterophyllus* (jak or jackfruit) attaining $1 \times 0.6$ m. The fruit of *A. altilis* (breadfruit) is starchy and bread-like when roasted. There are 31 species, occurring in tropical South-east Asia.

**Arundo (family \*Poaceae)** A genus of reeds that includes *A. donax* which may attain 7 m and is utilized for baskets, etc., and formerly for arrow shafts. It is also good for making paper. Gardeners' garters is a variegated horticultural variety. There are 3 species, found in the Mediterranean region and Taiwan.

**Ascencion and St Helena floral region** Part of R. Good's (*The Geography of the Flowering Plants*, 1974) \*palaeotropical kingdom, within the African subkingdom, comprising just two islands. This is probably the smallest floral region. There are 5 endemic (*see* ENDEMISM) genera, 3 of them being woody members of the \*Compositae. The proportion of endemism at the species level was very high, but much of this distinctive native flora has been exterminated by domesticated animals. *See also* FLORAL PROVINCE and FLORISTIC REGION.

**asci** The plural of \*ascus.

**ascidiate** *See* ASCIDIUM.

**ascidium (adj. asciidiate)** A plant organ shaped like a cup or pitcher.

**Asclepiadaceae** A family of erect or twining shrubs or \*perennial \*herbs that are sometimes fleshy and usually have white sap. Leaves are \*opposite or \*whorled. Inflorescences are \*cymose. The flowers are regular, and \*pentamerous, the \*petals contorted, the \*stamens often with two pollinia (*see* POLLINIUM), and the \*ovary \*half-inferior. The fruit consists of paired \*follicles. The seeds are tiny and plumed. There are 348 genera, with about 2900 species, mainly of the tropics and subtropics, especially of S. America.

**Asclepias (milkweeds; family \*Asclepiadaceae)** A genus comprising 120 species of herbs that occur in America, especially in the USA.

**ascocarp** The \*fruit body of an \*ascomycete within which \*asci and \*ascospores are formed.

**ascogonium** A female sex organ formed by certain \*fungi of the subdivision \*Ascomycotina.

**ascolichen** Any lichen in which the \*mycobiont is an \*ascomycete.

**ascolocular** Applied to \*ascocarps in which the \*asci are contained within cavities in a \*stroma.

**ascomycete** A fungus of the subdivision *Ascomycotina.

**Ascomycotina (division *Eumycota)** A subdivision of *fungi in which sexually produce *spores are contained within *asci and are called *ascospores. There are many thousands of species. They are *saprotrophic or *parasitic and are found in a wide variety of habitats. Some species can cause disease in animals or plants.

**Ascophyllum (family *Fucaceae, order *Fucales)** A genus of brown seaweeds which resemble *Fucus, but which can be distinguished by the absence of a *midrib. The *thallus is typically flattened and straplike, and air *bladders are usually present; it may reach several metres in length. *Spores develop in yellowish *receptacles, resembling sultanas, borne on short stalks arising from either side of the *fronds. The only British species is A. nodosum. See EGG WRACK.

**ascorbic acid** The water-soluble vitamin C that occurs in large quantities in fruit and vegetables, a deficiency of which causes scurvy in humans. In some plants, ascorbic acid plays an important part in aerobic respiration, replacing *cytochromes as an electron carrier.

**ascospore** A sexually produced, *haploid *spore formed within an *ascus by *fungi of the subdivision *Ascomycotina.

**ascostroma** A type of fungal *fruit body in which *asci develop in or within a *stroma.

**ascus** A minute, bag-like structure within which *ascospores develop in *fungi of the *Ascomycotina. An ascus often contains 8 ascospores which, in many species, are discharged explosively.

**asexual generation** See ALTERNATION OF GENERATIONS.

**asexual reproduction** A type of reproduction without the sexual processes of *gamete formation. In plants it consists of *vegetative reproduction, *apospory, and *apogamy. It occurs by *budding (*gemmation) in many *liverworts and *mosses. See also APOMIXIS.

**ash (Fraxinus)** See OLEACEAE.

**asoka** Saraca asoca, a sacred Indian tree. See SARACA.

**Asparagus (family *Liliaceae)** A genus of small shrubs or *perennial *herbs that have creeping, underground stems and green, needle-like branchlets replacing the leaves which are reduced to papery scales. A. officinalis is cultivated for its edible young shoots which are then forced and sometimes blanched. A. setaceus produces attractive, feathery foliage used in flower bouquets. There are about 100 species found throughout the Old World.

**aspartic acid** An aliphatic, acidic, polar *alpha-amino acid.

**aspen** See POPULUS.

**aspergillosis** A general term for any human or animal disease caused by a *fungus of the genus *Aspergillus. In humans the most common form of the disease is a lung infection which may lead to pneumonia; the disease is uncommon.

**Aspergillus (class *Hyphomycetes)** A *form-genus of *fungi that form a well developed, septate *mycelium and in which *conidia develop within *phialids. Some species have *perfect states classified in the Eurotiales. They are *saprotrophic, but some can also cause disease in humans (see ASPERGILLOSIS).

**Aspleniaceae** A family of ferns, distinguished among the advanced (*leptosporangiate) ferns by its oblong or linear superficial *sori, borne along one or both sides of the veins; the *indusium is often much elongated, sometimes double, or absent. There are 78 genera and about 2700 species, of cosmopolitan distribution.

**Asplenium (spleenworts; family *Aspleniaceae)** A cosmopolitan genus of 650 species of ferns with firm scales on the *rhizome, but leaves that are not densely scaly below. Leaf veins are usually free; the *sori oval to linear, usually separate and *simple, with the *indusium opening towards the midrib.

**asporogenous** Does not produce *spores.

**assai** See EUTERPE.

**assemblage** A collection of plants and/or animals characteristically associated with a particular environment. They can be used as an indicator of that environment (e.g. in geobotanical exploration). The term has a neutral connotation. Its use does not imply any specific relationship between the component organisms, whereas terms such as 'community' imply interactions.

**assemblage zone (coenozone, faunizone)** A stratigraphic unit or level of rock strata that is characterized by an assemblage of fossil plants and/or animals.

**assembly rules** The principles that underlie the aggregation of species to make up a community.

**assimilate** **1.** The portion of the food energy consumed by an organism that is metabolized by that organism. Some food, or in the case of a plant some light energy, may pass through the organism without being used. **2.** To engage in *assimilation.

**assimilation** The incorporation of new materials, acquired by the digestion of food or *photosynthesis, into the internal structure of an organism.

**association** *See* PLANT ASSOCIATION.

**association analysis** A hierarchical method of classification which is divisive and monothetic. Association analysis uses $\chi^2$ as a measure of association between pairs of species (attributes) found at a range of sample sites (individuals). The species or attribute with the highest overall sum of $\chi^2$ values (i.e. the strongest links) with all other species is selected as the basis for subdivision into two groups of sites or of individuals, either having or lacking that attribute. The process is then repeated for each new group until no further subdivision is required.

**association measure** Any measure of the link between two variables as shown by quantitative or qualitative data describing their characteristics. Links between continuous variables (e.g. mass and height or crop yield and rainfall) may be assessed statistically using *correlation and regression methods. Links between qualitative characteristics (e.g. aspect and slope form) may be assessed statistically using contingency tables and tests based on $\chi^2$ or an appropriate non-parametric test. Sometimes continuous variables can be more conveniently tested in qualitative form, e.g. height as short, medium, or tall, especially when the link with some other qualitative variable is wanted, e.g. aspect and tree height.

**associes** A phytosociological term used in the Clementsian (American) and British traditions and generally implying a sub-climax community in a *successional sequence.

**assortative mating** A sexual reproduction in which the pairing of male and female is not random, but involves a tendency for males of a particular kind to breed with females of a particular kind, or the converse. If the two partners in each pair tend to be more alike than is expected by chance, then it is referred to as positive assortative mating. Examples include positive assortative mating for height in wind-pollinated plants and for flower colour and time of development to sexual maturity in plants and in many insects with one generation per year.

*Aster* **(family *Asteraceae)** A genus of mostly *perennial *herbs in which the leaves are *simple and spirally arranged, the flower heads have *imbricate *bracts and flat, naked *receptacles; ray *florets, usually present, are white, blue, or red, but never yellow but the *disc florets are yellow. The *achenes are flattened and smooth with a *pappus of short, bristly hairs. Many N. American species are cultivated as Michaelmas daisies and have become naturalized in Europe and elsewhere. There are about 250 species, mostly northern temperate.

**Asteraceae** The largest family of flowering plants, formerly known as Compositae, and represented in every part of the world except for Antarctica. The flowers are individually small, but are clustered into a head (*capitulum) resembling a single flower, with *florets seated on *receptacles of varied form, and the head surrounded by an *involucre, or *bracts resembling a *calyx. Florets often have no calyx, but sometimes a hairy or scaly *pappus develops in the fruit. The *corolla is tubular or strap-like and 5-lobed. The 5 *stamens are joined into a tube. The *ovary is *inferior and one-celled, forming an *achene. Some have all the florets similar, either all tubular or all strap-shaped; in some (e.g. *Bellis*, daisy) the outer florets are strap-like and female or neuter, while the inner florets are tubular and usually hermaphrodite. The Asteraceae is closely related to the *Dipsacaceae (which, however, has long, usually free stamens, conspicuous calyx teeth, a cup-like epicalyx to each flower, and the ovule pendulous in the ovary, rather than basal). The Asteraceae includes the largest *angiosperm genus, *Senecio* (ragworts and groundsels) with more than 2000 species; and many valuable cultivated plants, e.g. *Lactuca sativa* (lettuce), *Chrysanthemum*, *Dahlia*, and *Helianthus annuus* (sunflower), as well as many common wild plants. The family comprises 1314 genera with about 21 000 species.

**Asticcacaulis** A genus of *Gram-negative *bacteria which is not assigned to any taxonomic family. It is similar to *Caulobacter* in its life cycle, but differs in that the stalk occurs eccentrically at one pole or on one side of the cell, and does not function in attachment. Cells adhere to a substrate by adhesive material near one end of the cell. It is found in *habitats similar to those of *Caulobacter*.

**atactostele** A *dictyostele type of *siphonostele, typical of *Monocotyledons, in which the *meristeles are randomly distributed *ectophloic siphonosteles. *Compare* EUSTELE, PLECTOSTELE, and POLYSTELE.

**atavism** The reappearance of a character after several generations, the character being the expression of a *recessive *gene or of *complementary genes. The character, or individual possessing this character, is sometimes referred to as a 'throw-back'.

**Athrotaxis** (family *Taxodiaceae) A genus of conifers that have small, often scale-like, *adpressed or spreading, spirally arranged leaves, and small (to 3 cm long) *globose cones. Each scale has a curved-out spine at its tip, and ripens orange in autumn. They are tall trees, cultivated in some temperate arboreta. There are 2 species, confined to Tasmania.

**Athyrium** (family *Aspleniaceae) A genus of ferns with large, soft *rhizome scales, and 1–3 *pinnate, soft, hairless *fronds; *sori are borne on *receptacles on a side branch of a vein; *indusia are hooked, or crescent-shaped, and toothed, or (rarely) absent. There are about 180 species. They are cosmopolitan, but most occur in temperate eastern Asia.

**Atlantic North American floral region** Part of R. Good's (*The Geography of the Flowering Plants*, 1974) boreal kingdom. It is a large and extensive flora with 100–200 endemic (*see* ENDEMISM) genera. *Robinia* (locust tree) is a well-known example, and there is also *Franklinia*, a garden tree that was obtained from a single plant of a single species, and which is now extinct in the wild. As a whole the flora strongly resembles those of temperate Eurasia, and of China and Japan. *See also* FLORAL PROVINCE and FLORISTIC REGION.

**Atlantic Period** A period in post-glacial times (i.e. post-*Devensian or *Flandrian) from about 7500 years BP to 5000 years BP which, according to pollen evidence (*see* POLLEN ZONE), was warmer than the present, and moist, with oceanic climatic conditions prevailing throughout north-western Europe. It corresponds to Pollen Zone VIIa, which throughout north-western Europe is characterized by the most *thermophilous species found in post-glacial pollen records. The *climatic optimum of the post-glacial, or current *Flandrian *interglacial, is dated to the early Atlantic period. *Compare* BOREAL PERIOD.

**atmometer** An instrument that is used for measuring the rate of evaporation of water into air. By using an atmometer alongside a *potometer, it is possible to compare the rate of *transpiration from a plant with evaporation from a purely physical system.

**ATP** *See* ADENOSINE TRIPHOSPHATE.

**Atrichum undulatum** *See* CATHERINE'S MOSS.

**Atriplex** (family *Chenopodiaceae) A genus of *herbs and small shrubs most of which are *halophytes. The stems are often striated which red or white, and the plant is often mealy. The leaves are alternate, toothed or lobed, linear or triangular, and without *stipules. The flowers are inconspicuous, unisexual, and held in small *cymes. The *perianth of the male flower is composed of 2–5 segments, usually green in colour, and there is a similar number of *stamens. The female flower is without any perianth, but is enclosed by 2 persistent *bracteoles, which can be entire or lobed. The *superior *ovary is composed of 3 fused *carpels, and a single *locule with one *ovule. The fruit is an *achene. There are about 100 species, found throughout the world, often as components of open communities or grassland, on saline soils (e.g. in coastal areas).

**atropous** *See* ORTHOTROPOUS.

**Atterberg limits** A series of thresholds that are observed when the water content of a soil is steadily changed. The 'contraction limit' occurs when sufficient water is added to a dry soil for contraction cracks to close. The addition of further water leads to plastic deformation at the 'plastic limit'. The 'liquid limit' occurs when just enough water is then added for the soil to behave like a liquid. Knowledge of these limits is important for understanding and predicting hill-slope failure. The limits were devised in 1911 by the Swedish soil physicist Albert Mauritz Atterberg (1846–1916).

***Aucoumea*** (family *Burseraceae) A *monotypic genus comprising the big West African rain-forest timber tree okoume or gaboon (*A. klaineana*), which also produces resin.

***Aucuba*** (family *Cornaceae) A genus of *dioecious, evergreen shrubs in which the leaves are *opposite, and the fruit is a *berry. *A. japonica* tolerates shade and is commonly planted. There are 3 or 4 species, occurring in temperate eastern Asia.

**auger** A tool used primarily for soil sampling, but also for sampling peats and other unconsolidated sediments. The simplest and most universal form has a screw head to bore the soil or sediment. Alternative auger heads are available for more specialized needs. Standard augers sample to one metre depth, but extension

rods can be attached enabling sampling at deeper levels.

**auricle** An appendage, which may resemble the ear of an animal, occurring at the base of a leaf-blade.

***Auricularia*** (order *Auriculariales) A genus of *fungi in which the *fruit bodies are gelatinous but firm, and may be cup-shaped, ear-shaped, discoid, etc. The *hymenium is borne on one side of the fruit body. *A. auricula* (also known as *Hirneola auricula-judae*) is the 'Jew's ear' fungus which is a common *parasite on elder (*Sambucus*) and other trees. *A. mesenterica* has a *resupinate *fruit body with a velvety surface and a jutting, shelf-like upper margin. It is common throughout the year on hardwood stumps and logs.

**Auriculariales** (subclass *Phragmobasidiomycetidae) An order of *fungi in which the *basidium is divided by transverse *septa. *Fruit bodies are typically gelatinous or waxy. They are *saprotrophic, or *parasitic on plants. Genera include *Auricularia.

**Auriscalpiaceae** (order *Aphyllophorales) A family of *fungi in which the *hymenium is borne on tooth-like spines, or on *gills. The family contains genera formerly included in the *Hydnaceae.

**austral** An adjective, from the Latin *australis*, that means 'southern' or 'of the south'.

**Australia cedar** *See* TOONA.

**Australian bluebell** *See* WAHLENBERGIA.

**Australian bower plant** *See* PANDOREA.

**Australian laurel** *See* PITTOSPORUM.

**Australian rosewood** *See* HETERODENDRUM.

**Australian sassafras** *See* DORYPHORA.

**Australian turpentine** *See* SYNCARPIA.

**aut-** A prefix, derived from the Greek *auto-* (meaning 'self') that means 'self' or 'individual' (e.g. 'autotroph', 'self-feeder').

**autapomorph** An *apomorph character state that is unique to a particular species or lineage in the group under consideration.

**autecology** The ecology of individual organisms and populations, including physiological ecology, animal behaviour, and population dynamics. Usually only one or two species are studied. *Compare* SYNECOLOGY. *See also* POPULATION ECOLOGY.

**authigenic** Applied to mineral particles formed at or close to the surface, or to soils containing such particles.

**autochory** Dispersal of spores or seeds by the parent organism.

**autochthonous** Applied to material which originated in its present position (e.g. the plant material in a deposit, such as *peat, which actually grew where it is found, rather than being brought in by outside influences). *Compare* ALLOCHTHONOUS.

**autoecious** Applied to a parasitic organism in which all the stages in its life cycle occur on the same host, e.g. certain rust fungi (*Urediniomycetes). *Compare* HETEROECIOUS.

**autogamy** In a flower, self-fertilization.

**autogenic** Applied to a *successional change resulting from modification of the environment by vegetation, e.g. by producing *humus or providing shade. *Compare* ALLOGENIC; *see also* SUCCESSION.

**autograft** *See* ALLOGRAFT and GRAFT.

**autolysis** The destruction of a cell or some of its components through the action of its own hydrolytic *enzymes. It is a process that is particularly marked in organisms undergoing metamorphosis.

**autophagic vacuole** A membrane-bound *vacuole, derived from within the cell, that contains material to be digested. *See also* LYSOSOME.

**autopolyploid** A *polyploid organism that originates by the multiplication of a single *genome (set of *chromosomes) such that all the chromosomes come from sets within one single *species. Autopolyploidy has been used commercially in breeding crop plants, including sugar beet and tomatoes, to improve their vigour and

growth, although the fertility of autopolyploids tends to be reduced.

**autoradiograph** A photographic print that is made by the action of a radioactive substance (e.g. carbon-14) upon a sensitive photographic plate.

**autosome** Any *chromosome in the cell *nucleus other than a *sex chromosome.

**autotetraploid** An *autopolyploid organism with four similar *genomes. They occur either naturally by the spontaneous accidental doubling of a 2*n* genome to 4*n*, or artificially, through the use of colchicine. They are present in many commercially important crop plants because, as with other *polyploids, they tend to be associated with increased size of the plant (through increased cell size, fruit size, stomata size, etc.).

**autotroph 1.** An organism that is capable of synthesizing complex organic materials from simple inorganic substrates. The term includes photosynthetic autotrophs that use sunlight as a source of energy and chemosynthetic autotrophs that obtain energy from inorganic reactions (e.g. iron oxidation). **2.** An organism that uses carbon dioxide as its main or sole source of carbon. *Compare* HETEROTROPH.

**autoxidation** *See* ANTI-OXIDANT.

**autumn crocus** *See* COLCHICUM.

**auxin** A *hormone that promotes longitudinal growth in the cells of higher plants. Typically, low concentrations of auxin promote growth whereas high concentrations inhibit it. Auxins are produced at the growing points of stems and roots, and promote growth by increasing the rate of cell elongation rather than that of cell division. They are involved in the curvature of parts of the plant towards light (phototropism) or gravity (geotropism), and the initiation of *cambium activity in association with *cytokinins; and they may control fruit growth or leaf fall, and inhibition of lateral-bud development in favour of apical buds. A natural example of an auxin is indole-3-acetic acid (IAA), which has been isolated from fungi and bacteria, and from the endosperm of corn, as well as from urine and saliva in humans. Auxins have also been synthesized, and are widely used to regulate growth in a variety of plants of agricultural and horticultural importance. Some of these may have differential effects on different plants. For example 2, 4-dichlorophenoxyacetic acid (2, 4-D) is toxic to dicotyledons but not to monocotyledons, and is used to control weeds in cereal crops and lawns.

**auxospore**   A spore produced by some diatoms (*Bacillariophyta) following a series of cell divisions in which daughter cells become progressively smaller. This sequence ends with the release of a naked auxospore which expands before its siliceous cell wall forms. Auxospores can also result from *apomixis, *parthenogenesis, or sexual reproduction.

**auxotroph**   An organism (usually a *bacterium) that, as a result of a *mutation, has lost the ability to synthesize a particular substance essential for its growth, e.g. an amino acid or a sugar. The organism must be able to obtain this component from its environment if it is to survive and grow. *Compare* PROTOTROPH.

**available nutrients**   Any elements or compounds in the soil solution that can be absorbed readily into plant roots, and that function as nutrients to growing plants. The available amount is usually much less than the total amount of that plant nutrient in the soil.

**available water**   In soil, the water that can be absorbed readily by plant roots. It is usually taken to be water held in the soil under a pressure of 0.3 to about 15 bars.

***Avena***   **(oat; family *Poaceae)** A genus of grasses that are cultivated for their grain, from which such foods as oatcakes and porridge are made. *A. sativa*, the cultivated species, is closely related to, and may be descended from, *Avena fatua* (wild oat) which is now a serious weed of cereal crops (*see* HERBICIDE). There are about 25 species, native to the Old World and cultivated in temperate regions throughout the world.

***Averrhoa***   **(family *Oxalidaceae)**   A genus of two species of small trees that are widely cultivated throughout the tropics

for their edible, fleshy, many-seeded fruits. *A. bilimbi* (bilimbing) has acid fruits like small cucumbers; *A. carambola* (carambola) has elongate 5-angled fleshy *berries, sweet or acid, and is also used for cleaning brassware.

***Avicennia***   **(family *Verbenaceae or Avicenniaceae)**   A genus of small trees which have greyish, *opposite leaves. The flowers are small and regular, the *calyx and *corolla *imbricate, the *ovary *superior. The fruit is a roundish, bivalved *capsule with 1 seed. There are 14 species, occurring on tropical coasts, mainly in mangrove forests at their ecological and geographical limits.

**avocado pear**   *See* PERSEA.

**awn**   In grasses (*Poaceae), the continuation of the central nerves of the *lemmas or *glumes, which often form a bristle-like structure projecting from the *spikelet.

**axenic**   Applied to a culture of an organism that consists of one type of organism only (i.e. that is free from any contaminating organisms).

**axes**   *See* AXIS.

**axial**   Pertaining to the *axis.

**axil**   In a plant, the upper angle where a small stem joins a larger one, or where a leaf stalk joins the stem.

**axile**   Applied to a plant structure that is attached to the central *axis.

**axillary**   Borne in an *axil. For example, an axillary bud lies between the twig and a subtending leaf.

**axis**   **(pl. axes)**   The main (or central) stem of a plant or infloresence.

**axoneme**   The '9 + 2' formation of *microtubules that forms the principal structural feature of flagella (*see* FLAGELLUM).

***Azadirachta***   **(family *Meliaceae)**   A genus of big trees which yield valuable timber. The fruit is a *capsule, with a few large seeds which have fleshy jackets. There are two species, occurring in Indo-Malaysian rain forests.

***Azalea***   *See* RHODODENDRON.

**Azollaceae** A specialized family of free-floating, aquatic ferns of uncertain affinities. The branched stems bear roots, and 2-lobed, overlapping leaves. *Sori bear either *microsporangia or *megasporangia. The megasporangia are each provided with floats; *microspores are numerous and grouped together into 'massulae' which bear barbed hairs (glochidia) that hook on to the *megaspores. The leaves have a basal cavity inhabited by a *cyanobacterium (*Anabaena azollae*) which fixes nitrogen. There is 1 genus, *Azolla*, with 6 species, of tropical and warm temperate distribution in America and Asia. They are not native in Europe, but are naturalized there.

**Azotobacter** (family *Azotobacteraceae) A genus of *bacteria characterized by the production of differentiated resting cells called *cysts. Vegetative cells are typically ovoid in shape and can carry out *nitrogen fixation. There are several species, found in soil and water.

**Azotobacteraceae** A family of *bacteria in which the cells are typically *Gram-negative and rod-shaped or variable in shape. They are *chemoorganotrophic, and can grow only in the presence of air. Species are capable of *nitrogen fixation. They are found in soil, water, plant material, etc. There are 2 genera.

**azygospore** A type of fungal *spore which resembles a *zygospore but which develops *parthenogenetically.

**B** *See* BORON.

**babaçu palms** *See* ORBIGNYA.

**Baccaurea** (family *Euphorbiaceae) A genus of small, *dioecious trees in which the flowers are borne in clustered spikes or *racemes, some *cauliflorous. The fruit has a firm, fleshy *pericarp which sometimes splits, and 1–6 large seeds, each with a juicy, sour jacket. Several species are edible, and some are cultivated. There are 80 species, occurring in lowland rain forest from Malesia to the Pacific islands.

**Bacillariophyta** A division of microscopic *algae (known as diatoms) which are mostly unicellular, but which may be colonial or filamentous. The cell wall (frustule) is composed of silica and consists of two halves, one of which overlaps the other like the lid on a box. The frustule is commonly delicately ornamented. Most diatoms are photosynthetic, but there are also species that lack *chlorophyll and live *heterotrophically among decaying marine algae. Some species are capable of a gliding motility when in contact with a surface. Cells with bilateral symmetry are said to be pennate, while those with radial symmetry are called centric. Pennate diatoms are found in both freshwater and marine habitats, either as plankton or attached to rocks, etc.; centric diatoms are predominantly marine and planktonic. The silica frustules are an important constituent of deep-sea deposits. There are more than 10 000 species of diatoms.

**Bacillus** A genus of *Bacteria in which the cells are rod-shaped, often *motile, and typically Gram-positive (*see* GRAM REACTION). *Endospores can be formed in the presence of air. *Bacillus* species are *chemoorganotrophic. Some can grow only in the presence of air; others can grow in either the presence or absence of air. There are many species, found in a wide range of *habitats. Some species can cause disease in vertebrate animals (e.g. anthrax), or in insects; insecticidal species, particularly *B. thuringiensis*, are used in the biological control of insect pests.

**bacillus** A bacterial cell that is rod-shaped (i.e. longer than it is wide).

**backcross** A cross of an $F_1$ *hybrid or *heterozygote with an individual of *geno-type identical to that of one or other of the two parental individuals. Matings involving a *hybrid genotype and a *recessive parental genotype are used in genetic analyses to determine *linkage and *cross-over values.

**back mutation** A reverse *mutation (i.e. reversion) in which a mutant *gene (called the non-wild-type form) reverts to the original standard form (the wild-type form).

**backshore** The part of a beach that is above the level of normal high spring tides. This zone is usually dry; only when exceptionally high tides or storms occur does wave action influence this part of a beach. Characteristic plants of this area include *halophytes and plants with *xeromorphic adaptations.

**backswamp** An area of low, ill-drained ground on a flood plain away from the main channel. It stands slightly lower than adjacent *alluvial fans extending from the valley sides, and is below natural levées that rise towards the main channel. It is a site of slow accumulation of *silts and *clays, usually inhabited by *marsh plants.

**Bacteria** In *taxonomy, a *kingdom comprising 11 main groups of *prokaryotes. These are: purple (photosynthetic); Gram positive (*see* GRAM REACTION); *cyanobacteria; green non-sulphur; spirochaetes; flavobacteria; green sulphur; Planctomyces; Chalmydiales; Deinococci; and Thermatogales. Most bacteria are single-celled and most have a rigid *cell wall. Cell division usually occurs by *binary fission; *mitosis never occurs. Bacteria are almost universal in distribution and may live as *saprotrophs, *parasites, *symbionts, pathogens, etc. They have many important roles in nature, e.g. as agents of decay and mineralization, and in the recycling of elements (such as nitrogen) in the *biosphere. Bacteria are also important to humans, e.g. as causal agents of certain diseases, as agents of spoilage of food and other commodities, and as useful agents in the industrial production of commodities such as vinegar, *antibiotics, and many types of dairy products. The oldest fossils known are of bacteria, from rocks in S. Africa that are apparently 3200 million years old. These must have been *heterotrophic bacteria, feeding off organic molecules dissolved in the oceans of that time. The first photosynthetic bacteria, of *anaerobic type, appeared a little later, about 3000 Ma ago. In the widely used five-kingdom system of classification, Bacteria is one of the kingdoms, and the only kingdom in the superkingdom Prokarya. The kingdom Bacteria comprises two subkingdoms: *Archaea and *Eubacteria. In the three-domain system of classification, the kingdom Bacteria is the only kingdom in the *domain Eubacteria.

**bactericidal** Applied to any agent that kills *bacteria.

**bacteriochlorophylls** Types of *chlorophyll occurring in small vesicular or tubular bodies attached to or continuous with the *cell membrane in the *anaerobic, anoxygenic, photosynthetic bacteria (*Rhodospirillales). Bacteriochlorophylls show minor structural differences from chlorophylls found in *cyanobacteria and higher plants.

**bacteriocin** A *protein, produced by a particular strain of *bacteria, that is lethal (or inhibitory) to other, often related, strains of bacteria.

**bacteriophage** (phage) A type of *virus which infects *bacteria. Infection with a bacteriophage may or may not lead to the death of the bacterium, depending on the phage and sometimes on conditions. A given bacteriophage usually infects only a single species or strain of bacterium. Phages can be found in most natural environments in which bacteria occur.

**bacteriorhodopsin** A *protein pigment found in certain species of *bacteria of the genus *Halobacterium. Bacteriorhodopsin is responsible for a unique method of harnessing the energy of sunlight for use by the bacteria.

**bacteriostatic** Applied to an agent that prevents the growth of at least some types of *bacteria without actually killing them.

**bacterium** 1. A single bacterial cell. 2. A particular prokaryotic organism (see PROKARYOTE).

**bacteroid** A modified bacterial cell, particularly of the type formed by species of *Rhizobium within the root nodules of leguminous plants.

**Bacteroidaceae** A family of (typically) *Gram-negative *bacteria in which the cells are rod-shaped and *motile (with many *flagella) or non-motile. They are *chemoorganotrophic, and can grow only in the absence of air. They are found in a wide range of habitats. There are many genera.

***Bacteroides*** (family *Bacteroidaceae) A genus of *Gram-negative *bacteria which can grow only in the absence of air. They are found, for example, in the mouth and intestinal tract of humans and other animals, in the rumen, and in sewage. Some species can act as *opportunist pathogens and have been isolated from various soft-tissue infections. There are many species.

**badlands** A term originally applied to the intricately eroded plateau country of southern Dakota, Nebraska, and northern Dakota, but now widened to refer to any barren terrain that has been similarly dissected. It is most common in areas of infrequent but intense rainfall and little vegetation cover.

**bael fruit** See AEGLE.

**baeocyte** A type of reproductive cell (previously called an endospore), many of which are formed within a cell in *cyanobacteria (section II).

***Baeomyces*** (order *Lecanorales) A genus of *lichens in which *apothecia are borne on short stalks, resembling miniature mushrooms. *Spores are multicellular. They are found mainly on acidic substrates, e.g. acid rocks or *peaty soil.

**bakanae disease** (foolish seedling disease) An important disease of rice caused by the *fungus *Gibberella fujikuroi*. Early symptoms include excessive growth of the stem, apparently caused by *gibberellins produced by the *pathogen.

**bakers' yeast** The strains of *Saccharomyces cerevisiae* used in bread-making. *Compare* BREWERS' YEAST.

**balanced load**    *See* GENETIC LOAD.

**balanced polymorphism** A genetic *polymorphism that is stable, and is maintained in a population by *natural selection, because the *heterozygotes for particular *alleles have a higher *adaptive value (fitness) than either *homozygote. This condition is also referred to as *overdominance, as opposed to underdominance, where the heterozygote has a lower fitness, giving rise to unstable equilibrium.

**Balanopaceae (order Fagales)** A small family of dicotyledonous (*see* DICOTYLEDON) trees and shrubs. The leaves are alternate or *whorled. The flowers are simple; male flowers are borne in *catkins on separate trees from the female flowers, which are solitary with scale-like *perianths. The fruit is an acorn-like *drupe. The family comprises 1 genus, with 9 species, native to Queensland and the south-western Pacific.

**Balanophoraceae (order Santalales)** A family of plants which are obligate parasites (*see* PARASITISM) attached to roots. A tuberous, subterranean, knobbly *rhizome with no chlorophyll bears stout aerial spikes on which are borne minute male and female flowers and then tiny *nuts, or *drupes. There are 18 genera and 44 species, occurring in rain forests throughout the tropics.

**balata** The latex of *Manilkara bidentata* (family *Sapotaceae), obtained from incising the trunk, which was formerly of considerable commercial importance.

**balau** The Malayan name for trees of the section *Shorea* (family *Dipterocarpaceae) which are valuable for their dark, dense timber.

**balk (bauk, baulk)** A roughly squared, large piece of timber used for beams, etc.

**ballistospore** A type of fungal *spore which is projected violently from its *sporophore at maturity.

**Balsaminaceae** A family of *herbs that have brittle, translucent stems, and *simple leaves. The flowers have 3–5 *sepals and 5 *petals, usually all coloured and strongly *zygomorphic, the lowest sepal being large, spurred, and petal-like. The fruits are *capsules, exploding if touched when ripe. Several are cultivated for their showy flowers. There are 2 genera, one (*Hydrocera*) *monotypic, occurring only in South-east Asia, the other (*Impatiens*) (balsams, or touch-me-nots) of about 850 species, mostly of tropical Asia and Africa, but with a few in northern temperate America and Eurasia, and in S. Africa.

**balsams** (*Impatiens*) *See* BALSAMINACEAE.

**bamboos** *See* BAMBUSACEAE.

**Bambusaceae (bamboos)** A family closely related to the grasses (*Poaceae) and sometimes considered to be a subfamily (Bambusoideae) of them. Nearly all have woody stems; a very few are herbs. The tallest grow to 40 m, the stoutest to 90 cm girth. Usually they are clump-forming, with a big underground *rhizome. They either fruit annually, or at long intervals and then die. The fruit is dry, or occasionally fleshy. Many are of major economic importance for building, scaffolding, or paper-making, or for the edible young shoot tip. Some whole cultures in Asia are based on bamboos. There are about 45 genera, mainly tropical and subtropical, and most diverse in S. America.

**Bambusoideae** *See* BAMBUSACEAE.

**banak** *See* VIROLA.

**banana** *See* MUSA.

**Bangiales (division *Rhodophyta)** An order of red algae which are fairly simple in their structural organization. Two common genera are *Bangia* and *Porphyra*; *Bangia* is filamentous and can live in freshwater and marine habitats, while *Porphyra* is membranaceous and is found in intertidal regions on temperate coasts. *See also* LAVER.

**Banks, Sir Joseph** (1743–1820) A British explorer (he sailed around the world with James Cook), who was a noted patron of science and did much to promote research and sent botanical expeditions to many countries. During his voyage on the *Endeavour* he named Botany Bay. His herbarium, now in the possession of the Natural History Museum (formerly the British Museum (Natural History)), and his books and manuscripts, now divided between the British Library and the Natural History Museum, are of major importance. He was honorary director of the Royal Botanic Gardens at Kew, became president of the Royal Society in 1778, and was knighted in 1795.

**Banksia** (family *Proteaceae*) A genus of shrubs and small trees, whose inflorescences are dense, stout, showy spikes of up to 1000 flowers, developing into woody cones of *follicles, plus *bracts and *bracteoles. There are 71 species, which are quintessentially Australian.

**banyans** Species of fig tree (*Ficus*) that have spreading crowns and descending prop roots. They commence life as *epiphytes; then descending anastomosing (*see* ANASTOMOSIS) roots strangle the trunk of the host tree. The Buddha attained enlightenment beneath a banyan (*F. religiosa*). The famous banyan in Calcutta Botanic Garden, *F. bengalensis*, dates from 1782 and is more than 300 m in circumference.

**baobab** *See* ADANSONIA.

**barachory** (**clitochory**) Dispersal of spores or seeds by their own weight.

**Baragwanathia longifolia** An early representative of the club mosses (*Lycopsida*) that is known from the Lower *Devonian. During the Devonian and *Carboniferous, this group reached the peak of its diversification.

**barberry** *See* BERBERIS.

**Barbula** (order *Pottiales*) A large genus of mosses, species of which are not always easy to distinguish. The plants are mostly small, forming tufts, cushions, or short turfs. Leaves may be blunt or pointed, but lack the hair-point typical of the related genus *Tortula*. *Capsules are cylindrical and erect, with an oblique, beaked lid; the *peristome has short, straight teeth in some species (e.g. *B. tophacea* and *B. recurvirostra*) and long, twisted teeth in others (e.g. *B. fallax*, *B. unguiculata*, and *B. convoluta*). There are many species, occurring mainly in temperate regions, and found on the ground, on walls, on garden paths, etc.

**Barclaya** (family *Nymphaeaceae*) A genus of water-lilies which produce large, solitary flowers, with an *inferior *ovary. There are 3 or 4 species, found in damp places on the floor of Indo-Malaysian tropical rain forests.

**bark** The outer skin of a tree trunk, outside the secondary, vascular, *cambium. It is composed of *phloem tissues, which occur as living inner and dead outer zones. The outer zone is penetrated by the *cork layers (or periderms) formed from cork cambia (or phelloderms), and is sometimes called the rhytidome. The bark surface is variously sculptured, e.g. smooth, scaly, or fissured; and these surface features are partly determined by the arrangement of the phloem tissues and periderms.

**barley** *See* HORDEUM.

**barley yellows** A virus disease of barley and other cereals in which infected plants are stunted and yellow in colour. The virus is transmitted by aphids.

**bar of Sanio** *See* CRASSULA.

**barren** A *community of few and scattered plants that occupy less than half the available ground area. Barrens occur within the Arctic *tundra, typically dominated by a single plant species e.g. *Dryas octopetala* (mountain avens); and often a single particle size predominates in the soil.

**barren lands** A term that was once used to describe the *tundra of northern

Canada, a region characterized by sparse vegetation, a harsh climate, and *permafrost.

**barrier island** A segmented barrier-bar complex found between two tidal inlets. Barrier-island systems have a lagoonal area on their landward side, and often have dunes and vegetation on the exposed barrier. There are three main hypotheses to explain the origin of barrier islands: (a) the building up of submarine bars; (b) spit progradation parallel to the coast and segmentation by inlets; and (c) submergence of subaerial coastal beach ridges by a rise in sea-level. Barrier islands are most common in areas of low tidal range.

***Barringtonia*** (family *Lecythidaceae) A genus of mostly small tropical trees whose flowers are nocturnal, sweet-scented, borne in spikes, and often *cauliflorous. They have many *stamens, and the *ovary is *inferior. The fruits are *indehiscent, with 1 seed. *B. asiatica* is a widespread species found along sandy coasts. There are 39 species, occurring in Africa, Madagascar, and from Indo-Malaysia to the Pacific.

**Barringtoniaceae** A group of a plants which includes the genus *Barringtonia* and which is often regarded as a separate family, but all of whose genera are now sometimes placed within the family *Lecythidaceae.

**Bartonellaceae** (order *Rickettsiales) A family of *bacteria that are *parasitic in vertebrate hosts, usually in or on the red blood cells. They can be cultivated in cell-free media in the laboratory. They can cause disease in humans. There are 2 genera.

***Bartramia pomiformis*** See APPLE MOSS.

**basal body** A structure located in the *cytoplasm of cells, from which flagella (*see* FLAGELLUM) project. Normally it is composed of 9 sets of *microtubules, each set arranged in triplets, embedded in a dense, granular matrix.

**base** **1.** According to the Brønsted–Lowry theory, a substance that in solution can bind and remove hydrogen ions or protons. The Lewis theory states that it is a substance that acts as an electron-pair donor. A base reacts with an acid to give a salt and water (a process called neutralization) and has a *pH greater than 7. **2.** *See* BASE PAIR.

**base analogue** A *purine or *pyrimidine base (*see* BASE PAIR) that differs slightly in structure from the normal base, but that because of its similarity to that base may act as a *mutagen when incorporated into *DNA. Once in place, these bases, which have pairing properties unlike those of the bases they replace, can produce mutations by causing insertions of incorrect *nucleotides opposite them during replication. Though the original base analogue exists only in a single strand, it can cause a nucleotide-pair substitution that is replicated in all DNA copies descended from that original strand. An example is 5-bromo-uracil (5BU), an analogue of *thymine that has bromine at the C-5 position in place of the $CH_3$ group found in thymine.

***Basella*** (family Basellaceae) A small genus of five species of tropical *herbs, related to *Chenopodiaceae, which have a swollen, perennial, underground stem annually producing twining shoots. *B. alba* (Indian spinach) is cultivated throughout the tropics for these shoots.

**base pair** **1.** Two *nucleotides on separate DNA strands that are connected through hydrogen bonds. **2.** A unit of measurement of a length of double-stranded DNA.

**base saturation** Expressed as a percentage of the total *cation-exchange capacity, the extent to which the exchange sites of the soil's adsorption complex are 'saturated' (or occupied) by exchangeable basic cations, or by cations other than hydrogen and aluminium.

**basic dye** A dye which consists of an organic *cation that combines with and stains negatively charged *macromolecules (e.g. *nucleic acids). It is used particularly for staining cell nuclei which contain nucleic acids. *Compare* ACIDIC DYE.

**basic grassland** A vegetation pattern that occurs on soils with an alkaline reaction, particularly those developed in chalk or limestone. In Britain, and NW Europe generally, basic grasslands include some of the oldest areas of rough grazing, dating from Neolithic times. Floristically they are comparatively diverse and there are numerous characteristic herbs other than grasses.

**basic soil** A soil with a *pH greater than 7.0. *See* ALKALINE SOIL.

**basidiocarp** The fruit body of a *basidiomycete.

**basidiolichen** Any *lichen in which the *mycobiont is a *basidiomycete.

**basidiomycete** A fungus that belongs to the subdivision *Basidiomycotina.

**Basidiomycotina (division *Eumycota)** A subdivision of *fungi in which the sexually produced *spores are *basidiospores. The subdivision includes most of the familiar macroscopic fungi, including *mushrooms, *toadstools, *bracket fungi, etc. Most are *saprotrophic, but some are *parasitic on plants.

**basidiospore** A sexually produced *spore, borne on a *basidium in a *fungus of the *Basidiomycotina.

**basidium** A microscopic structure upon which sexually produced *spores are formed externally. Basidia are characteristic of *fungi of the *Basidiomycotina.

**basifugal movement** *See* ACROPETAL MOVEMENT.

**basipetal** Growing or developing from apex to base, so that the oldest parts are nearest the apex and the youngest are nearest the base. *Compare* ACROPETAL.

**basipetal movement** Movement of substances towards the basal region of the plant from the root and shoot apices. *Compare* ACROPETAL MOVEMENT.

**basket fern** A fern that grows upright, rather than prostrate.

**basophilic** Applied to a cell, its components, or products that can be stained by a *basic dye.

**bass** *See* RAFFIA.

*Basidium*

*Bassia* **(family *Chenopodiaceae)** A genus of *perennial *herbs and small shrubs of warm temperate to hot regions, including Australia, where they are found mainly in the arid and littoral zones on lime-rich and saline soils (e.g. the Nullarbor plain). The plants are halophytic (*see* HALOPHYTE), with deeply penetrating roots and other *xerophytic characters. The flowers are small and borne in a *cymose *inflorescence, usually wind-pollinated. The seed has a ring-shaped or semicircular *embryo enclosing the *endosperm. There are 26 species in central Europe, Asia, Africa, and Australia.

**basswood** *See* TILIA.

**bast** *See* RAFFIA.

**Bathonian** A Middle *Jurassic stage, dated at about 167.7–164.7 Ma ago, commonly represented by carbonate sediments in many areas of Europe. It is known to contain an abundant fauna of invertebrates including the oyster *Catinula knorri*.

**Bauhin, Gaspard (1550–1624)** A French anatomist and herbalist, who, in 1582, was appointed professor of Greek, and in 1588 of anatomy and botany at Basle; eventually he became rector of the university and dean of his faculty. He wrote *Pinax theatri botanici* (1623), which is a concordance to earlier nomenclature (still used as such) and an attempt at a formal system of botanical classification. He also completed 3 of the 12 parts he had planned of *Theatrum botanicum*, although only one part was published, in 1658. His son, Jean Gaspard Bauhin (1606–85) also became professor of botany at Basle. *Bauhinia* is named after the Bauhin family.

**Bauhin, Jean (1541–1613)** A French physician and botanist, the elder brother of Gaspard *Bauhin, whose major work, *Historia plantarum universalis*, was unfinished when he died and was published in 1650–1. It describes more than 5000 plants and was a standard reference book for the next 100 years.

**Bauhinia** (family *Fabaceae, subfamily *Caesalpinoideae) A genus of plants, most of which are woody climbers with curiously flattened, or corrugated, stems. A few are small trees. The flowers are very showy. The leaves are simple, more or less split into 2 lobes, with many main veins. Because of the bilobed leaves, the genus is named in honour of the *Bauhin brothers, Jean and Gaspard. There are 250 species of *Bauhinia*, distributed through the tropics and subtropics.

**bauk** *See* BALK.

**baulk** *See* BALK.

**baumgrenze** *See* TREE-LINE.

**bay laurel** *See* *LAURUS*.

**bay rum** *See* *PIMENTA*.

**B-chromosome** A *chromosome that occurs in addition to the normal chromosome set (A-chromosomes) and to which it shows no *homology. B-chromosomes occur widely in flowering plants, are devoid of functional genes, and display an irregular, non-Mendelian, mode of inheritance.

**Bdellovibrio** A genus of *Gram-negative bacteria which is not assigned to any taxonomic family. The cells are small, curved rods, usually motile with a single *flagellum at one end. They are *aerobic. They parasitize other bacteria by penetrating them, and then growing and dividing within them. They are found in soil, in freshwater and marine habitats, and in sewage, etc.

**beach** An accumulation of *sand and gravel found at the landward margin of the sea or a lake. Upper and lower limits approximate to the position of highest and lowest tidal water levels. The angle of slope and the sedimentary structures of a beach are related to the grain size of the beach materials, and to the nature of wave activity and other sedimentary processes active in the area.

**beach cusp** One of a series of regularly spaced crescent-shaped structures forming local relief along a beach face. The horns or 'headlands' of the cusp are composed of coarse *sand or gravel, and point seaward down the beach. The intervening troughs or 'bays' are made up of finer sand. The height of beach cusps is usually in the order of several centimetres, although larger examples have been described. The size and spacing of cusps appears to be related to the nature of waves breaking on the beach. The 'headlands' and 'bays' form distinct *microhabitats for benthic *microflora.

**beach rock** A cemented beach *sand deposit that develops within the intertidal zone by the precipitation of needle-like crystals of aragonite in the *pore space between the grains. The cementation process is relatively rapid, taking as little as 10 years for a lithified rock to develop. The precipitation of the cement is favoured by a warm climate, and may be aided by algal or bacterial action.

**beak** 1. A rigid cylindrical or conical projection on the tip of a *fruit (as in *Geranium*). 2. A projection in Orchidaceae that separates the *anther from the stigmatic surface below it in the flower.

**bean gall** A red or pink, kidney-bean-shaped *gall, found on the leaves of *Salix species (willow) and caused by the small sawfly, *Pontania proxima* (family, Tenthredinidae). A second brood laid in the early autumn by the adults that emerged in summer may over-winter in the gall. Sometimes the term 'bean gall' is used more generally to describe any red or brown gall caused by any *Pontania* species. Many galls may occur on a single leaf and, like other gall-causing species, the insects support a large community of *parasites and hyperparasites.

**beans** *See* PHASEOLUS.

**beard-lichen** (**beard-moss**) The common name for the tufted, hair-like *thallus of a *lichen of the genera *Usnea, *Alectoria, or *Bryoria.

**beard-moss** *See* BEARD-LICHEN.

**bear's breech** *See* ACANTHUS.

**beat up** A foresters' phrase, meaning to replace dead trees with new ones, especially during the early years of the establishment or re-establishment of a plantation.

**Beaumontia** (family *Apocynaceae) A genus of woody climbers with handsome flowers, which are sometimes cultivated. There are 8 species, occurring from China to India and Java.

**Bedsonia** An obsolete genus of *bacteria now included in the genus *Chlamydia*.

**beech** *See* FAGUS.

**beech bark disease** A disease of beech trees (*Fagus) caused by a variety of *Nectria coccinea. Symptoms include yellowing of foliage and die-back; infected trees may be killed. The disease is spread from tree to tree by beech scale insects.

**beefsteak fungus** A common name for the *fruit body of *Fistulina hepatica*. The flesh is red-brown with a gelatinous upper surface, and yields a reddish juice; *spores are formed within tubes which open via pores on the pale underside of the fruit body. It is *parasitic on trees, particularly oak. It is fairly common in late summer and autumn. *See also* FISTULINA.

**beetroot** (*Beta*) *See* CHENOPODIACEAE.

**beet yellows** A virus disease of sugar beet, beetroot, etc. The leaves of infected plants become reddish-yellow, dry, and brittle. The virus is transmitted by aphids.

**Beggiatoa** (order *Cytophagales) A genus of gliding *bacteria which form colourless, non-sheathed filaments. Granules of sulphur are formed in the cells in the presence of sulphide. These bacteria are found in freshwater, estuarine, and marine habitats. They are sometimes regarded as colourless *cyanobacteria.

**Begoniaceae** A family of mostly succulent *herbs or undershrubs, which are almost entirely tropical, comprising 2 genera and about 900 species. The stems are pointed, with alternate, *simple, often unequal-sided leaves. They are *monoecious, and the flowers are usually irregular, borne in *cymes, and showy. There are 2–5 free *sepals and *petals, and many *stamens in the male flower; and the *ovary is usually *inferior. The fruit is a *capsule or *berry. *Begonia* (890 species) is much cultivated, but is not hardy in climates with frosty winters.

**Beilschmiedia** (family *Lauraceae) A genus of small to medium trees, some of which yield ornamental timber. There are 200 species, occurring in the tropics, Australia, and New Zealand.

**belian** (**Borneo ironwood**) *See* EUSID-EROXYLON.

**bellflowers** *See* CAMPANULACEAE.

**Belliolum** (family *Winteraceae) A genus of small trees in which the flowers are white, with numerous *petals and *stamens, and 1 or a few free *carpels. There are 8 species, occurring in the Solomon Islands and New Caledonia.

**bell morel** The common name for the *fruit body of the *ascomycetous fungus *Verpa conica* (order *Pezizales). It is rare in Britain.

**Beltian bodies** Sausage-shaped organs at the tips of the leaves of certain *Acacia* species found in the African savannah that secrete oils and proteins as food for ants living in nests they have hollowed out in leaf bases; the ants also feed on nectar. In return for the food, the ants defend the tree from attack by herbivores and cut away parts of adjacent plants that threaten to shade the acacia's leaves.

**Benedict's test** A procedure used to detect the presence of a reducing sugar in solution. It depends upon the fact that an alkaline solution of copper (II) sulphate is reduced to insoluble copper (I) oxide by reducing sugars, to give a red precipitate.

**Bennettitales** An extinct *gymnosperm (Gymnospermae) order ranging from the *Triassic into the *Cretaceous. They resembled cycads (*Cycadophyta), and possessed reproductive structures that must have looked more like flowers than cones.

**ben oil** *See* MORINGA.

**Bentham, George (1800–84)** A British botanist, the nephew of the utilitarian philosopher Jeremy Bentham for whom he worked as secretary (1826–32). He prepared the *Flora Hongkongensis* (1861), seven volumes of the *Flora Australiensis* (1863–78), and what is considered his most important work, three volumes of *Genera Plantarum*, in collaboration with J. D. *Hooker (1862–83). His *Handbook of the British Flora* was first published in 1858, and its seventh edition in 1924 (the fifth and sixth editions were prepared by Hooker).

**benthic** *See* BENTHOS.

**benthic zone** The lowermost region of a freshwater or marine profile in which the *benthos resides. In bodies of deep water where little light penetrates to the bottom the zone is referred to as the benthic abyssal region and productivity is relatively low. In shallower (e.g. coastal) regions, where the benthic zone is well lit, the zone is referred to as the benthic *littoral region and it supports some of the world's most productive ecosystems.

**benthos** (adj. **benthic**) In freshwater and marine ecosystems, the collection of organisms attached to or resting on the bottom sediments.

**Bentinckia** (family *Arecaceae) A genus of palms in which the trunk is solitary and hoop-marked. The leaves are *pinnate, with a long crown-shaft, below which appear the much-branched inflorescences. *B. nicobarica* is widely cultivated. There are 2 species, one native to India, the other to the Nicobar Islands.

**benzoin** *See* STYRAX.

**Berberidaceae** A family of *herbs and shrubs in whose flowers the *perianth is regular, *tri- or *dimerous, and composed of free parts in 2–7 *whorls, the whorls being either alike or different. The *ovary consists of a single *carpel, forming a *berry or *capsule. There are 15 genera and about 570 species, mainly confined to the northern temperate zone (but *Berberis* is almost cosmopolitan).

**Berberis** (barberry; family *Berberidaceae) A genus of spiny, *simple-leaved shrubs whose flowers are *trimerous, with spines in threes, the *perianth usually consisting of 5 yellow or orange *whorls. The fruit is an edible *berry. Barberries are much cultivated for ornament, and formerly were grown for their edible fruits. They are a secondary host of wheat rust, so are often eliminated from wheat-growing areas. There are about 450 species, of cosmopolitan distribution in temperate zones and in the mountains of tropical Africa.

**Beringia** An area comprising the Bering Strait and adjacent Siberia and Alaska. At various times in the late *Mesozoic and *Cenozoic, the Strait was dry land and so provided for plants and animals an important migration route between the Palaearctic and Nearctic biogeographical regions.

**Bering land bridge** A link (*see* LAND BRIDGE) between Siberia and Alaska that has existed intermittently during the *Cenozoic, permitting the migration of species between Eurasia and North America. It provided the means by which human invasion of the Americas from Asia took place at the close of the last glaciation. The Bering land bridge was the only route into North America once the direct link between Europe and North America had been broken by the opening of the Atlantic Ocean.

**Bermuda buttercup** *See* OXALIS.

**berry** A fleshy, *indehiscent, many-*seeded *fruit containing no hard parts except the seeds. A banana is a berry; others include tomato, grape, date, and gooseberry.

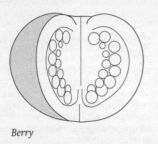

*Berry*

**Berrya** (family *Tiliaceae) A genus of trees which have fibrous bark and which yield valuable timber. There are 3–5 species, occurring in open places from Indo-Malaysia to the Pacific.

**Bertholletia** (family *Lecythidaceae) A *monotypic genus (*B. excelsa*, the Brazil-nut) of big, locally common, rain-forest trees of tropical America. The fruit is a hard, spherical, woody *capsule opening by a terminal plug and containing numerous very hard, woody seeds with copious, oily *endosperm.

**Beta** (beetroot, sugar beet) *See* CHENOPODIACEAE.

**beta-galactosidase** An *inducible enzyme that is responsible for the catalytic *hydrolysis of *beta-galactoside.

**beta-galactoside** The *glycoside formed from the β-stereoisomer of *galactose.

**beta-mesohaline** *See* BRACKISH.

**betel-nut** *See* ARECA.

**betle pepper** *See* PIPER.

**Betula** (birches; family *Betulaceae) A genus of fast-growing and light-demanding trees, some of which provide valuable, pale timber. Male flowers are borne in *catkins, female flowers in clusters; and there is no *perianth. The fruit is a small, winged *nut. There are about 60 species, occurring in the northern temperate and Arctic regions.

**Betulaceae** A family of deciduous shrubs and trees in which the leaves are *simple, with *stipules. The flowers are unisexual, the males borne in *catkins with 2–12 *stamens; the females clustered, the *ovary *inferior and composed of two fused *carpels. They are wind-pollinated in early spring. The fruit is a one-seeded *nut, often winged. The family includes the birches (*Betula*), alders (*Alnus*), hornbeams (*Carpinus*), and hazels (*Corylus*). There are 6 genera and about 150 species occurring in the northern hemisphere, extending to the tropical mountains.

**Beyeria** (family *Euphorbiaceae) A genus of 12 species of shrubs that are endemic (*see* ENDEMISM) to Australia where they are widespread.

**bicentric distribution** The occurrence of a species or other *taxon in two widely separated places, but nowhere in between. For example, the tulip tree genus (*Liriodendron*) has species in eastern North America and in China, but none in between, and southern beeches (*Nothofagus* species) occur in New Guinea and Australasia, and in Chile, on opposite sides of the Pacific Ocean, but nowhere else. A number of plants endemic (*see* ENDEMISM) to Scandinavia occur only in both of two widely separated montane areas, possibly reflecting the former existence of nunataks above the late *Pleistocene ice sheet. *Compare* UNICENTRIC DISTRIBUTION.

**bicollateral bundle** A *vascular bundle in which *phloem lies on both sides of the *xylem.

**biennial** Applied to a plant that lives for two years. During the first season food may be stored for use during flower and *seed production in the second year. If the plant is damaged, for example by insect attack, it may survive into a third year and flower again. This happens when the cinnabar moth (*Callimorpha jacobaeae*) attacks ragwort (*Senecio* spp).

**bifid** Split in two.

**Bifidobacterium** A genus of *Gram-positive *bacteria in which the cells are *pleomorphic and often Y- or V-shaped. They are *anaerobic, *fermenting glucose to produce acetic acid and lactic acid. They are found in the gut in a wide range of mammals, fowls, and bees; certain species are common in human infants' intestines. They are non-pathogenic.

**bifurcate**   Forked, with 2 branches.

**Bignoniaceae**   A family of mainly woody climbers in which the leaves are *pinnate and *opposite. The flowers are showy, borne in *cymose clusters, *zygomorphic, and *pentamerous, with fused *sepals and *petals, and with the *ovary *superior. The fruit is usually a *capsule with numerous small, winged seeds. There are 112 genera with 225 species, most occurring in tropical central and S. America.

**big tree**   See *SEQUOIADENDRON* and TAXODIACEAE.

**bilberry**   See *VACCINIUM*.

**bilimbing**   See *AVERRHOA*.

**Billardiera** (family   *Pittosporaceae) A genus of 8 species of *sarmentose sub-shrubs, endemic (*see* ENDEMISM) to Australia, which have bell-shaped flowers and fleshy fruits.

**billet**   A piece of wood that has been cut for fuel.

**bilocular**   Having two *locules.

**binary fission**   A division of one cell into two similar or identical cells: a common method of *asexual reproduction in single-celled organisms.

**binding hyphae**   Filaments that run transversely through the *prosenchyma tissue in the *medulla of brown seaweeds (*Phaeophyta).

**bindweed**   See *CALYSTEGIA*.

**binomial classification** (binomial classification) The systematic description of *species by means of 2 names, both in Latin. The first name, with an initial capital letter, is that of ,the genus into which the species is placed, the second that of the species itself. (For example, the name *Betula papyrifera* (paper birch) comprises the species *papyrifera* within the genus *Betula*.) This method of classification for plants was introduced in 1753 by Carolus *Linnaeus in his

*Species Plantarum.* The precise procedures and rules for naming newly discovered plants and *bacteria are laid down in the *International Code of Botanical Nomenclature and the International Code for Bacteriological Nomenclature.

**binominal classification**   See BINO-MIAL CLASSIFICATION.

**bintangor**   See *CALOPHYLLUM*.

**binuang**   See *OCTOMELES*.

**bioassay**   The use of living cells or organisms to make quantitative and/or qualitative measurements of the amounts or activity of substances.

**biochemical oxygen demand**   See BIOLOGICAL OXYGEN DEMAND.

**biochore**   A *biotic district, i.e. a distinctive *community and environment, and as such one of many precursors of the ecosystem concept. The term is also used by *Raunkiaer to describe precise biological boundaries between major climatically determined plant regions and their main subdivisions, as identified by his life-form recording system (i.e. percentages of *phanerophytes, *chamaephytes, etc.).

**biochrome**   See PIGMENT.

**biochronology**   The measurement of units of geological time by means of biological events. Biochronologists often derive their correlations from widespread and distinctive events in the biological history of the world, based on the first and last appearances of organisms.

**biocide**   A natural or synthetic substance toxic to living organisms. Some ecologists advocate use of this term instead of *'pesticides', since most pesticides are also toxic to species other than the target pest species. Indirectly, pesticides may also affect non-target organisms detrimentally in many other ways (e.g. by loss of food species or loss of shelter) so that the effects of pesticides may be felt throughout a whole ecosystem. The term 'biocide' indicates this property more clearly than 'pesticide'.

**bioclimatology**   The study of climate, with particular reference to the environments of living organisms, especially to those of agricultural plants and animals, and humans, together with the disease *vectors affecting humans and commercially important plants and animals.

**biocoenosis** (pl. **biocoenoses**) The living part of a *biogeocoenosis, comprising the phytocoenosis (the primary producers), the zoocoenosis (the secondary producers or consumers), and the microbiocoenosis (the decomposer organisms). It is equivalent to the *biotic component in an *ecosystem.

**biodiversity** A portmanteau term, which gained popularity in the late 1980s, that describes all aspects of biological diversity, especially including species richness, *ecosystem complexity, and genetic variation.

**bio-energetics** The study of energy transformations in living organisms, in particular the formation of *ATP in photosynthesis and by other mechanisms, and its subsequent use in metabolic processes.

**biogenesis** The principle that a living organism can arise only from another living organism, a principle contrasting with concepts such as that of the spontaneous generation of living from non-living matter. The term is currently more often used to refer to the formation from or by living organisms of any substance, e.g. coal, chalk, chemicals, etc.

**biogenic** Applied to the formation of rocks, traces (fossils), or structures as a result of the activities of living organisms.

**biogeochemical cycle** The movement of chemical elements from organism to physical environment to organism in a more or less circular pathway. They are termed 'nutrient cycles' if the elements concerned are essential to life. The form and quantity of an element varies through the cycle, with amounts in the inorganic reservoir pools usually greater than those in the active pools. Exchange between the system components is achieved by physical processes (e.g. weathering) and/or biological processes (e.g. protein synthesis and

decomposition). The latter form the vital negative-feedback mechanisms that regulate the cycles. Cycles may be described as varying from perfect to imperfect. A perfect cycle (e.g. the nitrogen cycle) has a readily accessible abiotic, usually gaseous, reservoir and many negative-feedback controls. By contrast, the phosphorus cycle, which has a sedimentary reservoir accessed only by slow-moving physical processes, has few biological feedback mechanisms; it is an imperfect cycle. Human activities can disrupt these cycles, leading to pollution. Theoretically, perfect cycles are more resilient than imperfect cycles.

**biogeochemical exploration** See GEO-BOTANICAL EXPLORATION.

**biogeocoenosis** A term equivalent to 'ecosystem', often used in Russian and Central European literature, and attributed to V. Sukachaev who is believed to have coined it in 1947. A biogeocoenosis comprises a *biocoenosis (the *biome, or living *community), together with its *habitat, which is usually termed an ecotope.

**biogeographical barrier** The various disjunctive geographical groupings of plants and animals are usually delimited by one or more barriers to migration, which prevent faunal and/or floral mixing. Such barriers may be climatic, involving temperature and the availability of water, or physical, involving for example mountain ranges or expanses of sea water.

**biogeographical province** A biological subdivision of the Earth's surface, usually on the basis of taxonomic rather than ecological criteria, that embraces both faunal and floral characteristics. The hierarchical status of such a unit, and the total number of such units, varies from one authority to another.

**biogeographical region** A biological subdivision of the Earth's surface, delineated on the same general principles as a *biogeographical province, but having superior taxonomic status. The provinces are grouped into regions, of which the following are generally recognized: Antarctic, Australasian,

Ethiopian, Nearctic, Neotropical, Oceanian, Oriental, and Palaearctic.

**biogeography**   A diverse subject, traditionally focusing on the distribution of plants and animals at different taxonomic levels, past and present. Modern biogeography, however, also lays great stress on the ecological character of the world vegetation types, and on the evolving relationship between humans and their environment.

**biological conservation**   Active management to ensure the survival of the maximum *diversity of species, and the maintenance of genetic variety within species. The term also implies the maintenance of *biosphere functions, e.g. *biogeochemical cycling, without which the basic resources for life would be lost. Biological conservation embraces the concept of long-term sustained resource use or sustained yield from the biosphere, which may conflict with species conservation in some circumstances. Conservation of species and biological processes is unlikely to succeed without simultaneous conservation of *abiotic resources.

**biological control**   In general, the control of the numbers of one organism as a result of natural predation by another or others. Specifically, the human use of natural predators for the control of pests, weeds, etc.

**biological oxygen demand**   (BOD, biochemical oxygen demand)   An indicator of the *polluting capacity of an effluent where pollution is caused by the take-up of dissolved oxygen by micro-organisms that decompose the organic material present in the effluent. It is measured as the weight (mg) of oxygen used by one litre of sample effluent stored in darkness at 20°C for five days.

**biological productivity**   The productivity of organisms and ecosystems, as defined by *primary, *secondary, and community productivities. *See also* PRIMARY PRODUCTIVITY.

**bioluminescence**   The production by living organisms of light without heat. Bioluminescence is a property of many types of organism, e.g. the *fruit bodies of certain *fungi, certain (mostly marine) *bacteria, dinoflagellates, etc.

**biomass (standing crop)**   The total weight of the living components (*producers, *consumers, and *decomposers) in an ecosystem at any moment, usually expressed as dry weight per unit area. Biomass is a quantity per unit area; *productivity is a rate of biomass gain per unit area.

**biome**   A biological subdivision of the Earth's surface that reflects the ecological and physiognomic character of the vegetation. Biomes are the largest geographical *biotic *communities that it is convenient to recognize. They broadly correspond with climatic regions, although other environmental controls are sometimes important. They are equivalent to the concept of major plant formations in plant ecology, but are defined in terms of all living organisms and of their interactions with the environment (and not only with the dominant vegetation type). Typically, distinctive biomes are recognized for all the major climatic regions of the world, emphasizing the *adaptation of living organisms to their environment, e.g. *tropical rain forest biome, temperate forest biome, *boreal forest (*taiga) biome, *desert biome, and *tundra biome. Wetlands and mountainous regions are often considered as biomes, because although they occur in all parts of the world their distinctive characters make them markedly different from the biomes surrounding them. Oceans are sometimes regarded as a single biome. Agricultural land is coming to be classed as a biome, as are urban areas.

**biometrics**   *See* BIOMETRY.

**biometry (biometrics)**   Quantitative biology, i.e. the application of mathematical and statistical concepts to the analysis of biological phenomena.

**bionomic strategy**   The characteristic features of an organism or population (i.e. size, longevity, fecundity, range, and migratory habit) that give maximum *fitness for the organism in its environment.

**biophage**   *See* CONSUMER ORGANISM.

**biospecies**   A group of interbreeding individuals that is isolated reproductively from all other groups.

**biosphere** The part of the Earth's environment in which living organisms are found, and with which they interact to produce a steady-state system, effectively a whole-planet ecosystem. Sometimes it is termed 'ecosphere' to emphasize the interconnection of the *biotic and *abiotic components.

**biosphere reserve** A type of conservation area designated by UNESCO in an attempt to establish an international network of protected areas encompassing examples of all the Earth's major vegetation and physiographic types. Biosphere reserves contain virgin vegetation, plus various kinds of cultural landscape, in the whole of which conservation is practised.

**biostratigraphy** A branch of stratigraphy that involves the use of fossil plants and animals in the dating and correlation of the stratigraphic sequences of rock in which they are discovered. A zone is the fundamental division recognized by biostratigraphers.

**biosynthesis** The formation of compounds by living organisms.

**biota** Plants and animals occupying a place together, e.g. marine biota, terrestrial biota.

**biotechnology** The exploitation of micro-organisms for industrial and other human purposes; the term usually refers particularly to techniques involving genetic manipulation in micro-organisms. *See* GENETIC ENGINEERING.

**biotic** Applied to the living components of the *biosphere or of an ecosystem, as distinct from the non-living, *abiotic, physical, and chemical components.

**biotic association** A community of plants and animals. *See also* COMMUNITY AND PLANT ASSOCIATION.

**biotic climax** *See* PLAGIOCLIMAX.

**biotic factor** The influence upon the environments of organisms resulting from the presence and activities of other organisms (e.g. the casting of shade and competition), as distinct from a physical, *abiotic, environmental factor.

**biotope** An environmental region characterized by certain conditions and populated by a characteristic *biota or *community.

**biotopographic unit** A small *habitat unit with distinctive topography formed by the activities of an organism, e.g. an ant hill, or a tussock. The term may also refer to small topographic units that, by their aspect, position, or other characteristics, generate distinctive micro-environments for living organisms. Examples include solar or shade slopes, windward slopes, and similar units in sand-dunes.

**biotroph** A parasitic organism that obtains its nutrients from the living tissues of its host organism.

**biotype 1.** A naturally occurring group of individuals with identical genomes. **2.** A physiological race, i.e. a group of individuals identical in structure but showing differences in physiological, biochemical, or *pathogenic characters.

**biozone** The total range of a given species defined within specific time limits.

**bipinnate** Of leaves, twice *pinnate.

**bipolar distribution** The distribution of an organism that is found in the high latitudes of both hemispheres, e.g. the plant *Koenigia islandica* is found both in the Arctic and in Tierra del Fuego at the southern tip of South America.

**birch** *See* BETULA.

**bird-of-paradise flower** *See* STRELITZIA.

**bird's-nest fungi** A common name for those *fungi of the *Nidulariales that form cup-shaped or funnel-shaped *fruit bodies within which occur the peridioles (spore cases), resembling eggs within a miniature bird's-nest.

**birthwort** *See* ARISTOLOCHIACEAE.

**bisaccate** Having 2 bladders or air sacs (e.g. certain *pollen grains).

**Bischofia** (family *Staphyleaceae) A *monotypic genus (*B. javanica*) of trees, formerly placed incorrectly in the Euphorbiaceae, that have *stipulate, spiral, *pinnate leaves composed of 3–5-serrate leaflets. They are *dioecious, the flowers being apetalous, and having 5 *sepals, 5 *stamens, and a trilocular (*see* LOCULE) *ovary with 2 pendulous *ovules. The fruits are leathery or fleshy, and *indehiscent. The trees yield useful timber. They are native in Georgia (Colchic forest), eastern to central China, Malesia, and the Pacific islands.

**biseriate**  Arranged in 2 rows.

**bisexual**  Applied either to a species comprising individuals of both sexes, or to a *hermaphrodite organism (in which an individual plant possesses both *stamens and *pistils in the flower).

**bitis**  *See PALAQUIUM.*

**bitunicate**  Applied to an *ascus in which the outer and inner layers of the ascus wall separate during *ascospore release.

**Biuret reaction**  A colorimetric reaction that serves as a basis for the qualitative and quantitative determination of *proteins. It depends upon the fact that in a strongly alkaline solution, compounds such as tripeptides (*see* PEPTIDE) and proteins, which have two *amide or *peptide bonds linked directly or through intermediate carbon atoms, react with copper sulphate to give a violet colour. Dipeptides and most *amino acids do not give this reaction.

**bivalent**  In genetics, applied to two *homologous *chromosomes when they are paired during the *prophase of the first meiotic division (*see* MEIOSIS). Bivalents do not occur during prophase II, because homologous chromosomes have segregated.

**Bixa**  The sole genus of the family Bixaceae, comprising 1 species of tropical American shrubs, *B. orellana* (annatto), which is extensively cultivated and contains a vivid orange pigment in the *testa, widely used as a food colourant.

**Bixaceae**  *See BIXA.*

**black bean**  *See CASTANOSPERMUM.*

**blackberry**  *See RUBUS.*

**black-boy**  *See XANTHORRHOEA.*

**black bryony**  (*Tamus communis*)  *See* DIOSCOREACEAE.

**black currant**  *See RIBES* and GROSSULARIACEAE.

**black grass**  (*Alopecurus myosuroides*)  *See* HERBICIDE.

**blackleg**  The name given to a number of plant diseases in which symptoms include blackening of the base of the stem, often followed by the collapse of the stem. The diseases may be caused by any of several *fungi or, less commonly, *bacteria.

**black mildew** (dark mildew) A plant disease caused by a *fungus belonging to the family Meliolaceae (order Dothideales). The name may also refer to the fungus itself. *Compare* SOOTY MOULD.

**black mustard**  *See BRASSICA.*

**black-stem rust**  A disease that affects a wide range of cereals and other grasses. Symptoms include the appearance on stems and leaves of patches of reddish-brown spores. The causal agent is a *fungus, *Puccinia graminis*; this organism requires two hosts to complete its life cycle, the other host being the barberry (*Berberis*).

**black walnut**  *See JUGLANS.*

**bladder**  A bag, formed from plant *tissue, in which metabolic products, or air, may be stored. In some plants, e.g. *Fucus vesiculosus* (wrack), air held in bladders causes parts of the plant to float. In others, e.g. *Utricularia* species (bladderworts), bladders borne on the leaves are

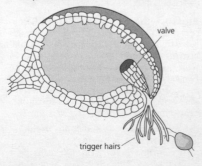

*Bladder*

traps that can be expanded rapidly in response to the stimulation of a trigger hair, drawing in small aquatic animals which are then digested.

**bladderwort** *See* LENTIBULARIACEAE.

**bladder wrack** The common name for the brown seaweed *Fucus vesiculosus*. The *thallus is variable; it is typically flattened and strap-like, repeatedly forked, and has prominent *midribs and (usually) paired air-*bladders which give buoyancy to the plant. Bladder wrack grows, often abundantly, on rocks, breakwaters, etc., between the high- and low-water marks. ('Bladder wrack' is sometimes used as a synonym for *'egg wrack'.)

**blade** The thin, flat part of a leaf.

**blanket bog** An *ombrogenous peat *bog community, typical of flat or moderately sloping areas in very wet, oceanic climates with high *humidity. In the British Isles, blanket bogs are widespread on the Pennine summits of England, in north-west Scotland, in Wales, and in parts of Ireland. In other parts of the world, this oceanic type of peatland is also found in Norway, Newfoundland, Tierra del Fuego, the Falkland Islands, and the Ruwenzori Mountains of East Africa.

**blanket weed** *See* CLADOPHORA.

**blast disease** A common disease of rice, caused by the *fungus *Pyricularia oryzae*. Lesions develop on the leaf-sheaths and on the stems, and the weakened stems are easily broken.

**blastochory** Dispersal of a plant by means of offshoots.

**Blastocladiales (class *Chytridiomycetes)** An order of *fungi in which the *thallus usually consists of *rhizoid-bearing *hyphae or *mycelium, and in which sexual reproduction involves the fusion of *motile *gametes which may be morphologically similar or different. The species are chiefly *saprotrophic, occurring in soil, fresh water, etc.

**Blastomycetes (subdivision *Deuteromycotina)** A class of *fungi that characteristically form single *yeast-like cells that reproduce by *budding; i.e., a new cell is formed by extrusion from the periphery of the parent cell. In some cases a poorly developed *mycelium may be formed. Species are found in a wide variety of habitats.

**blastospore** A fungal spore which is formed by *budding.

**Blechnaceae** A family of terrestrial ferns with scaly *rhizomes, to which the *fronds are not jointed. The *sori are short, or joined together in a continuous line, on a vein parallel to the midrib of the *pinna. The *indusium is outside the sorus, attached to the leaf surface or margin. *Spore-bearing fronds are often distinct from the vegetable ones. There are 10 genera, of cosmopolitan distribution, containing more than 260 species, of which 220 are in the genus *Blechnum*.

**bleeding** In plants, exudation of the contents of the *xylem stream at a cut surface owing to *root pressure.

**bleicherde** Ashy-grey soil that forms the leached layer in a *podzol soil.

**blending inheritance** An inheritance in which the characters of the parent appear to blend to form an intermediate state in the offspring, and in which there is no apparent segregation in later generations. The concept was proposed originally by biologists in the nineteenth century, including *Darwin, but later it was discredited as a model of inheritance after the results of *Mendel's experiments had been recognized.

**blepharoplast** A cylindrical body that is composed of parallel peripheral rods connected to the axial filaments of flagella (*see* FLAGELLUM).

**blewit** The common name for the *fruit body of *Lepista* species, e.g. wood blewit is *L. nuda* (*Tricholoma nudum*), field blewit is *L. saeva* (*Tricholoma saevum*). Both species are edible.

**blight** A non-specific term applied to any of a wide range of unrelated plant diseases. The causal agent is usually a *fungus. *See* (for example) POTATO BLIGHT.

**bloom** *See* ALGAL BLOOM.

**bluebell 1. (Australian)** *See* WAHLENBERGIA. **2. (European, *Endymion* species)**

*See* LILIACEAE. **3.** The bluebell of Scotland is *Campanula rotundifolia* of the family *Campanulaceae.

**blue-green algae**   *See* CYANOBACTERIA.

**blue mould**   **1.** The common name for the visible mat of *mycelium, bearing bluish *conidia, that is formed by species of *Penicillium*, e.g. on fruit. **2.** A disease of tobacco caused by *Peronospora tabacina*.

**blue pincushion**   *See* BRUNONIACEAE.

**blue-stain**   A common type of staining of freshly cut timber, caused by the growth of fungi (particularly species of *Ceratocystis*) in the sapwood.

**blusher**   The common name for the *fruit body of *Amanita rubescens*. It is *mushroom-shaped, with a reddish-brown or brown *cap bearing scattered whitish patches (remnants of the *universal veil). The flesh turns pink or reddish on breaking or bruising. The stem is white. It is common on ground in woodland during late summer and autumn.

**BOD**   *See* BIOLOGICAL OXYGEN DEMAND.

**bodh-tree (pipal-tree)** *See* BANYAN and *Ficus*.

**Boehmeria (family   *Urticaceae)**   A genus of trees and shrubs that includes *B. nivea* (China grass or rhea) which is cultivated in China for the long, tough, silky fibre of the inner bark, and in the tropics (as the variety *tenacissima*) for ramie fibre. There are about 50 species, occurring in the tropics and northern subtropics.

**bog**   A plant *community of acidic, wet areas. Decomposition rates in it are slow, favouring *peat development. In Britain and high northern latitudes typical plants include bog-mosses (*Sphagnum* species), sedges (e.g. *Eriophorum* (cottongrass) species), and heathers (e.g. *Calluna vulgaris* and *Erica tetralix*). Insectivorous plants (e.g. sundews, *Drosera* species) are especially characteristic; they compensate for low nutrient levels by trapping and digesting insects. Strictly the term 'bog' should be applied only to *ombrotrophic peatland, that is, those receiving water only from precipitation. These peatlands may be *raised bogs, *blanket bogs, or tropical bog forests, such as those found in the coastal regions of

Sarawak. Sometimes the term bog is applied to other types of peatland, such as 'valley bog' and 'string bog', but these are *rheotrophic *mires and should be regarded as *fens rather than bogs.

**bogbean**   *See* MENYANTHACEAE.

**bog peat**   *See* PEAT.

**bole**   The trunk of a tree.

**Boletales (subclass *Holobasidiomyce-tidae)**   An order of *fungi which typically form *mushroom-shaped, fleshy *fruit bodies in which the *hymenium lines the inner surface of tubes (pores) on the underside of the cap (as in *Boletus*) or forms a layer on the *gills (as in *Paxillus*). Genera include both edible and poisonous species. They are found on the ground under trees, with which they commonly form *mycorrhizal associations.

**bolete**   A *fungus of the order *Boletales. *See also* CEP.

**Boletus**   *See* BOLETALES and CEP.

**bolling**   The permanent trunk of a pollarded tree. *See* POLLARD.

**Bølling interstadial**   A relatively warm period that occurred towards the end of the last (*Weichselian) glaciation in north-western Europe, and which is named after the type site in Denmark. The event took place about 13 000–12 000 *radiocarbon years BP.

**bolochory**   Dispersal of spores or seeds by means of propulsive mechanisms.

**bolting**   Premature flowering and seed production.

**Bombacaceae**   A family of trees, closely related to *Malvaceae, in which the trunk is often stout, and the leaves often palmately *compound (*see* PALMATE) and usually scaly. The flowers are often large, and hermaphrodite. They have 5 *sepals and *petals, usually partly united, numerous free *stamens, and a *superior *ovary of 2–5 fused *carpels. The fruit is a big, woody *capsule, often spiny, or tessellated. There are 30 genera, with about 250 species, distributed in the tropics.

**Bombax (family   *Bombacaceae)**   A genus of deciduous trees that have thorny trunks and twigs. The inner fruit wall and seeds are covered with down. *B. ceiba* of India and Sri Lanka produces samul or silk cotton used in upholstery. There are 8 species, occurring in the Old World tropics.

**Bonnetiaceae** A family of shrubs and small trees, closely related to *Theaceae, in which the leaves are *simple, without *stipules. The flowers are *pentamerous, with many *stamens, and a *superior *ovary of 3 or 5 fused *carpels. The fruit is a *capsule. There are 3 genera, with 22 species, occurring in the tropics of Asia and America.

**boot-lace fungus** The common name for the *fungus *Armillaria mellea*, so called because of the resemblance of its black *rhizomorphs to boot-laces.

**bor** Within the closed-forest subzone of the *boreal-conifer belt in Eurasia, open woodlands dominated by *Pinus sylvestris (Scots pine) occur on dry sand-plains. They are known locally as bors, and similar vegetation types are known in Canada.

**Boraginaceae** A family of *herbs, or rarely shrubs or trees, most of which are rough or *hispid, usually with rounded stems, and with alternate, *exstipulate, entire leaves. The flowers are usually regular and borne in curved *cymes. The *calyx is 5-toothed, the *corolla 5-lobed, saucer-, funnel-, or bell-shaped, and is often pink in bud, changing later to blue. There are 5 *stamens, alternating with the corolla-lobes. The *ovary is 4-celled, or less often 2-celled, with the *style inserted between the 4 lobes, or terminal. The fruit is normally composed of 4 *nutlets. There are 156 genera, with about 2500 species, which are cosmopolitan in distribution, but centred in southern Europe and eastern Asia.

**Borassodendron** (family *Arecaceae) A genus of stout, solitary palm trees that have massive fan leaves and a smooth, knife-edged stalk. They are *dioecious, the little-branched, stout inflorescence forming among the leaves. The flowers occur in groups of 2–6, enclosed in *bracts. The fruits are massive, *globose, 12 cm in diameter, and in pendulous bunches, with a fibrous *pericarp. There are 3 seeds, with thick, stony walls protruding as flanges into the

*endosperm. There are 2 species, native to rain forest in southern Thailand, Malaya, and Borneo.

**Borassus** (lontar, palmyra palm, tal; family *Arecaceae) A genus of stout, solitary palm trees, with massive fan leaves which have short, spiny stalks split at the base, through which the sparsely branched, stout *inflorescence hangs. They are *dioecious, the flowers sunken in groups. The fruits are big, *globose, and 18 cm in diameter. The *pericarp is fibrous or fleshy, and is edible when ripe. There are 3 seeds, with thick, stony walls. The leaves provide writing material; the flower stalk is a source of toddy. The trees are much planted. There is 1 (or several *polymorphic) species, native to Africa, Madagascar, Sri Lanka, India, and seasonal areas from South-east Asia to Queensland.

**bordered pit** A *pit in which the pit cavity is partly covered by an overarching extension of the cell wall produced by *secondary thickening. A pit lacking such an overarched extension is known as a simple pit.

**Bordetella** A genus of *bacteria not assigned to a taxonomic family. The cells are small, ovoid, *Gram-negative, and non-motile, or motile with many *flagella. They are *chemo-organotrophic, and can grow only in the presence of air. They are found as *parasites and *pathogens of the respiratory tract in mammals. (*B. pertussis* is the causal agent of whooping cough.)

**boreal** Pertaining to the north (from Boreas, the Greek god of the north wind).

**boreal climate** The climate associated with the boreal (taiga) forest zone of Eurasia, where it extends to 65–70°N in the west and 50°N in the east, and N. America, where it extends from the fringe of the tundra southwards to 55°N in the east. Winters are long and cold, with temperatures below 6°C for 6–9 months, and with short summers with temperatures averaging more than 10°C. Precipitation, as snow in winter, typically amounts to 380–635 mm per annum.

**boreal forest** The circumpolar, sub-arctic forest of high northern latitudes that is dominated by conifers. To the north it is bounded by *tundra and to the south by temperate, broad-leaved, deciduous forest, steppe, or semi-*desert. The American and Asian parts of the forest are floristically more diverse than the European parts.

**Boreal Period** A period in post-glacial (i.e. post-*Devensian or *Flandrian) times from about 8800–7500 years BP, which preceded the ^climatic optimum of early *Atlantic times. Pollen records (*pollen zone) typically show an increasing abundance of *thermophilous tree species and also indicate the drier, more continental conditions that characterized the ensuing Atlantic period. The early Boreal corresponds to late Pollen Zone V; otherwise the Boreal is linked with Pollen Zone VI, which is sometimes subdivided to give Zones VIa, VIb, and VIc, according to the most abundant tree *pollen represented. For Britain, the Boreal period is significant as the last period in post-glacial times in which Britain was joined to mainland Europe by a land bridge across the Dover Strait.

**boreal zone** *See* CIRCUMBOREAL DISTRIBUTION.

**Borneo camphor** *See* DRYOBALANOPS.

**Borneo ironwood** (belian) *See* EU-SIDEROXYLON.

**boron** (B) An element that is commonly found as boric acid in the soil solution, and which is essential for healthy plant growth. It forms complexes with many organic molecules and is involved in phenolic metabolism (*see* PHENOL) and in membrane function, but its primary role is not yet understood.

**Boronia** (family *Rutaceae) A genus chiefly of shrubs less than a metre tall that have strongly aromatic leaves and 4-petalled, star-like flowers which are usually pink. They are found on heathland and in dry sclerophyll forests (*see* SCLEROPHYLLOUS VEGETATION) in shady places. There are 96 species, found mainly in temperate Australia, with a few also in New Caledonia.

**Borrelia** (family Spirochaetaceae, order *Spirochaetales) A genus of flexible, spirally shaped *bacteria, which are

*anaerobic *chemo-organotrophs. There are many species, found mainly on mucous membranes in humans, other mammals, and birds. Some are *pathogenic: certain species cause relapsing fever and are transmitted by ticks or lice.

**boscus** (subboscus) The wood or undergrowth produced by *coppice growth.

**Botrychium** (family *Ophioglossaceae) A genus of *eusporangiate ferns with both the sterile and fertile blades of the leaf *pinnately lobed or divided, the massive *sporangia *subsessile on the fertile pinnae, opening by a transverse slit without a true *annulus. There are 23 species, cosmopolitan in distribution, but most are northern temperate.

**Botrydium** (division *Xanthophyta) A genus of *siphonaceous algae in which the *thallus is a spherical or pear-shaped, coenocytic (*see* COENOCYTE) vesicle up to 2 mm in diameter. It grows on damp mud, soil, etc., to which it is anchored by colourless, branched *rhizoids.

**Botrytis** (class *Hyphomycetes) A *form-genus of *fungi which form septate *mycelia and produce dark-coloured, non-septate *conidia. They are *saprotrophic on plant debris, and also *parasitic in plants, causing a number of *important diseases. *B. cinerea* causes *grey mould on a wide range of plants and fruits; *B. aclada* (*B. allii*) causes neck-rot of onions. *Compare* NOBLE ROT.

**bottlebrush** *See* MELALEUCA.

**Bougainvillea** (family *Nyctaginaceae) A genus of spiny, *sarmentose shrubs or climbers that produce flowers in threes, surrounded by three coloured *bracts. They are extensively cultivated, with numerous varieties. There are 14 species, native to tropical America.

**bowstring hemp** *See* SANSEVIERIA.

**box elder** *See* ACER.

**box model** *See* COMPARTMENT MODEL.

**boxwood** *See* BUXUS.

**BP** Before the present (which is taken to be 1950). The initials should not be confused with BC. The year 1950 is taken as 'present' because the term BP is usually applied to radiocarbon dates (*see* RADIOCARBON

b

DATING) and the atmospheric testing of atomic weapons has greatly modified the radiocarbon content of all organisms since about 1950, so this is taken as the datum horizon for this method.

**Brachychiton** (family *Sterculiaceae) A genus of trees with swollen, bottle-like stems. They are deciduous only at flowering, which is irregular. Their leaves are entire or lobed, and the bell-shaped flowers are cream, pink, or red. There are about 30 species, confined to Australia. They occur widely in arid zones as well as in monsoon forest.

**Brachycome** (family *Asteraceae) A genus of small, *annual and *perennial *herbs which have *simple or divided leaves and flowers, about 3 cm in diameter, that are white, pink, blue, purple, or yellow. There are 66 species, mainly restricted to Australia, with 3 in New Zealand and 1 in New Guinea.

**Brachythecium** (order *Hypnobryales) A genus of *pleurocarpous mosses in which branching may be irregular or *pinnate. The leaves are imbricate to spreading, and each has a nerve extending at least half-way up the leaf. The *capsule is inclined to horizontal, ovoid to cylindrical, and borne on a deep-red *seta. *Brachythecium* is a cosmopolitan genus with many species, found in a wide variety of habitats. *B. rutabulum* is common in all but the most acid habitats, particularly in damp, shady situations. It grows on rocks, soil, wood, and often in lawns. Its shoots grow to 12 cm. Branching is irregular, with the branches erect, ascending, or *arcuate. The leaves are 2–3 mm long, broadly oval, and tapering to a sharp point. They are bright green to yellowish-green, and glossy.

**bracken** See *PTERIDIUM*.

**bracket fungus** Any one of a large group of fungi (order Aphyllophorales), many of which produce fruiting bodies that project from the trunks of trees. *Heterobasidion annosum* (*Fomes annosus*) is a parasite of coniferous trees that is the principal cause of decay in conifers.

**brackish** Applied to water that is *saline, but less so than sea water. According to the Venice system, brackish waters

are classified by the chlorine they contain and divided into zones. The zones, with their percentage chlorinity (mean values at limits) are: euhaline 1.65–2.2; polyhaline 1.0–1.65; meso-haline 0.3–1.0; alpha-mesohaline 0.55–1.0; beta-mesohaline 0.3–0.55; oligohaline 0.03–0.3; fresh water 0.03 or less.

**bract** A leaf, usually much reduced or modified, which subtends a flower or inflorescence in its *axis.

**bracteate** Bearing *bracts.

**bracteole** A little *bract borne on the flower stalk above the bract and below the *calyx.

**Bradyrhizobium** (family *Rhizobiaceae) A genus of *Gram-negative bacteria that was formerly included as 'slow-growing' strains in the genus *Rhizobium*. Some species induce the formation of *root nodules in certain legumes (*Fabaceae), particularly tropical legumes.

**brain fungus** The common name for the *fruit body of the fungus *Sparassis crispa*.

**brake** An area of *bracken (or other fern), scrub, or underwood.

**branch** A lateral stem that arises from another of the same form.

**brash** A foresters' term, meaning to cut off the lower branches of young trees, especially conifers, in plantations.

**Brassica** (family *Brassicaceae) A genus of *annual to *perennial *herbs which have tap roots and erect, branched stems. The flowers are usually yellow and are always four-petalled, the elongated fruit having a lower part (containing several seeds) that opens by 2 valves each with one strong vein, and an upper, *indehiscent part (the beak) with 0–3 seeds. Several are important vegetables (cabbage, turnip, etc.) or sources of useful oil (rape, black mustard, etc.). There are about 30 species, whose distribution is centred on the Mediterranean.

**Brassicaceae** A family, formerly known as Cruciferae (because the *petals are arranged in a cross), of mostly *annual or *perennial *herbs with alternate leaves and *racemes of 4-sepalled, 4-petalled flowers. There are usually 2 outer *stamens plus 4 larger, inner ones. The *ovary consists of 2 *carpels but produces dry fruits of various forms, either *capsules or *indehiscent. Brassicas provide many valuable vegetables, e.g. *Brassica oleracea*, cabbage, etc., and are sources of useful oils. No members of the family are poisonous. There are 390 genera with about 3000 species, cosmopolitan but mainly temperate.

**Braun-Blanquet, Josias (1884–1980)** A Swiss ecologist who devised the most widely used of European *phytosociological methods for the description of vegetation communities, and who was generally acknowledged as the foremost authority on such schemes. He developed the framework for his system in this doctoral study of the vegetation of the central Cévennes (1915), and the fully developed method was published in his authoritative textbook *Pflanzensoziologie* (Springer, Berlin, 1928) which he revised and updated most recently for a third, enlarged edition published in 1964. He worked at, and for many years was director of, the Station Internationale de Géobotanique, Méditerranéenne et Alpine (SIGMA) at Montpellier, France. *See also* ZURICH-MONT-PELLIER SCHOOL OF PHYTOSOCIOLOGY.

**braunerde**  *See* BROWN EARTH.

**Brazil-nut**  *See* BERTHOLLETIA.

**Brazil tulipwood**  *See* DALBERGIA.

**Brazil-wood**  *See* CAESALPINIA.

**breadfruit**  *See* ARTOCARPUS.

**bread mould**  Loosely, any mould found growing on bread; e.g., red bread mould is *Neurospora sitophila* and black bread mould is *Rhizopus nigricans*.

**breadnut**  *See* BROSIMUM.

**breakage and reunion**  In genetics, the established model of *crossing-over by the physical breakage and crosswise reunion of broken *chromatids during the pairing of *homologous *chromosomes at the *prophase I stage of *meiosis. The points at which chromosomes cross over are termed *chiasmata. The result

of crossing-over is the mutual exchange between chromatids of parts which contain corresponding *loci.

**breckland**  A distinctive locality in south-western Norfolk and north-western Suffolk, England, with a mosaic of grass heath and *Pteridium aquilinum* (bracken) and *Calluna vulgaris* (heather or ling) communities. Forest once grew over the breckland. Pollen analysis (*see* PALYNOLOGY) has demonstrated that the forest was cleared by Neolithic farmers.

**breed**  An artificial mating group derived (by humans) from a common ancestor, usually for agriculture (e.g. domesticated animals and crop plants), or for genetic analysis, or for pleasure (e.g. cats and dogs).

**breeding dispersal**  *See* PHILOPATRY.

**breeding size**  The number of individuals in a population involved in reproduction during a particular generation. It does not include the non-breeding element of the population.

**breeding true**  Producing offspring with *phenotypes for particular characters (i.e. organisms may breed true for some characters but not for others) that are identical to those of the parents. *Homozygous individuals necessarily breed true (unless *mutations arise), whereas *heterozygotes rarely do so.

**brewers' yeast**  The strains of *Saccharomyces cerevisiae* used in brewing beer. *Compare* BAKERS' YEAST.

**brezales**  *See* MEDITERRANEAN SCRUB.

**brigalow scrub**  The semi-arid scrub vegetation that occurs in parts of Australia; *Acacia* species provide the main vegetation.

**bristlecone pine**  Two pine species from California which are famous for their longevity and have been used to develop an exceptionally long, arid-site, tree-ring chronology. The oldest living specimens date back more than 4600 years, but cross-dating these with remnants of dead bristlecone pines has extended the arid-site chronology to more than 8200 BP. A 5500-year chronology has been developed for bristlecone pines at the upper tree limit. These pines are also used to calibrate the *radiocarbon-dating method to allow for fluctuations in atmospheric $^{14}C{:}^{12}C$ ratios as revealed by measuring the $^{14}C{:}^{12}C$ ratios of individual *tree rings in the long, absolutely dated, tree-ring series. *Pinus longaeva* is the Great Basin bristlecone pine and *P. aristata* is the mountain bristlecone pine.

**bristle-ferns** *See* TRICHOMANES.

**British ragwort** *See* SENECIO.

**Broad** In East Anglia, England, the name given to a freshwater lake, usually fringed by reeds, that is connected to a slow-flowing river near to its estuary.

**broad bean** *See* VICIA.

**broad-leaved** Applied to *angiosperm trees, in which leaves are broad, compared with the needle- or scale-like leaves of *gymnosperms.

**broad-leaved evergreen forest** North of the tropics, temperate, broad-leaved forests occur in the *formation types: (*a*) broad-leaved evergreen, and (*b*) broad-leaved *deciduous, the evergreen variants being restricted to the coastal plain of the Gulf of Mexico, S. Japan, and central China. South of the tropics, however, the broad-leaved forests are mainly *evergreen, the only exception occurring in Chilean Patagonia. This forest type requires plentiful rainfall, well distributed throughout the year.

**bromatia** The swellings that develop at the tips of *hyphae of *fungi cultivated by parasol ants; bromatia serve as food for the ants.

**Bromeliaceae** A distinctive family of monocotyledons (*see* MONOCOTYLEDON) with no close relatives, most of which are herbs and *epiphytes, a few being terrestrial. Most have leaves crowded together with an amplexicaul (*see* STIPULE) base forming a tank from which water is absorbed. The inflorescence is terminal, sometimes with showy *bracts. The flowers are regular and *trimerous. The fruit is a *berry or *capsule. There are 46 genera, with about 2110 species, entirely confined to the New World, except for one in W. Africa, and mostly tropical.

**Brongniart, Adolphe Théodore (1801–76)** A French botanist, who became a professor at the Muséum d'Histoire Naturelle, Paris, in 1831 and retained the position until his death. He was particularly noted for his work on the classification and distribution of fossil plants and their relationships with existing forms. In 1854 he founded the Société Botanique de France and was its first president.

***Brosimum*** (family *Moraceae) A genus of trees whose flowers are aggregated in heads of one sunken female and many males. The fruit is an *achene embedded in the fleshy *receptacle. The achene of *B. alicastrum* is the breadnut or rámon. *B. utile* is the cow-tree or milk-tree of Venezuela. It produces copious latex of similar appearance to cow's milk and that is used as a substitute for it. There are 13 species, occurring in tropical America.

**brotochory** *See* ANDROCHORY.

***Broussonetia*** (family *Moraceae) A genus of trees which are *dioecious, the female flowers being borne in heads. They produce multiple fruit. The inner bark of *B. papyrifera* contains fibre good for making paper (paper mulberry, Japan) and tapa cloth. There are 7 or 8 species, occurring from eastern Asia to Polynesia.

**Brown, Robert (1773–1858)** A British botanist who contributed greatly to the adoption of a natural system of plant taxonomy, but who is best known for his discovery of *Brownian motion. He was the first person to distinguish between *angiosperm and *gymnosperm, made a particular study of sexual processes in higher plants, originated the microscopical study of fossil plants, and strongly influenced the study of plant geography. In 1800–5 he took part in a botanical survey of the northern coast of Australia and returned to Britain with specimens of nearly 4000 species, many of them unclassified. He was appointed librarian of the Linnean Society of London, where he remained until 1822, and from 1810 was librarian and botanist to Sir Joseph *Banks, inheriting the use of his library and collections. These were transferred to the British Museum in 1827, with Brown as keeper, and when the botany department of the Museum was established in 1835, following the acquisition of the Sloane collection, Brown was appointed keeper, a position he held until his death.

**brown algae** The common name for seaweeds of the division *Phaeophyta, although not all members are strictly brown in colour: most are olive-brown when wet, becoming almost black on drying.

*Brownea* (family *Fabaceae, subfamily *Caesalpinoideae) A genus of small trees that are notable for their limp tassels of red or white young leaves. There are 30 species, occurring in tropical America.

**brown earth** A freely draining, and only slightly horizonated, soil-*profile type. It has a mull *humus in the surface *horizon and very little differentiation of horizons below. Brown earths are well-weathered and slightly leached soils, with a *cambic horizon in the middle part of the profile (also known as braunerde and now included in the *Inceptisols). Brown-earth types of soil are very productive and, although their natural *climax vegetation in humid, temperate latitudes is deciduous forest, they have been used extensively for agriculture. The soil type is common in southern England. *See also* ALFISOLS.

**brown forest soil** A little used soil-*profile term that has been applied to both acid *brown earths and *brown podzolics.

**Brownian motion** The random movement of small particles which are dispersed in a colloidal solution or suspension. *See* BROWN, ROBERT.

**brown oak** 1. The wood from an oak tree (*Quercus*) that has been infected with *Fistulina hepatica* (*beefsteak fungus). The wood has a rich, dark brown colour. Although mechanically weaker than wood from uninfected trees, it is nevertheless highly prized by cabinet-makers for decorative purposes. 2. A disease of oak trees caused by *Fistulina hepatica*.

**brown podzolic soil** A freely draining, leached soil *profile that has developed acid surface *horizons, a *mor surface *humus, and a clearly visible enrichment of translocated iron oxide

in the middle, or B, horizon. The profile has been leached to the early stage of *podzolization, identified by the movement down-profile of iron and aluminium compounds.

**brown rot** 1. A very common disease of fruit caused by *fungi of the genus *Sclerotinia* (usually *S. fructigena*). Fruits susceptible to it include apples, pears, cherries, plums, etc. Initially, soft brown patches appear on the fruit, often at a site of injury, and these gradually spread until the whole fruit is affected. Fungal *conidia are produced from small, pale-coloured, cottony patches which typically appear on the rotting fruit. 2. A type of timber decay in which the wood turns a reddish-brown colour and becomes cracked and eventually crumbly in texture. Fungi that cause brown rots are usually unable to break down the lignin component of wood.

**browse wood** Thin branches of trees that are fed to cattle and deer in winter.

*Brucella* A genus of *bacteria not assigned to a taxonomic family. The cells are ovoid or shortly rod-shaped, and are *Gram-negative. They are *chemo-organotrophic, and can grow only in the presence of air. There are several species, found as *parasites and *pathogens in a range of mammalian hosts. *B. abortus* can cause abortion in cattle and can also infect humans.

*Bruguiera* (family *Rhizophoraceae) One of the principal genera of trees of mangrove forests, having roots with erect *pneumatophores and a trunk without *stilt roots. They are *viviparous. There are 6 species, occurring in the tropics from E. Africa to Polynesia.

**Brunoniaceae** A *monotypic family. *Brunonia australis* (blue pincushion) is related to *Goodeniaceae, which is a small herb with a rosette of leaves, endemic (*see* ENDEMISM) to Australia and Tasmania and present throughout temperate regions of the country in open communities. The regular, *pentamerous flowers are borne terminally on long stems in densely clustered, hemispherical, bracteate heads. The 5 *sepals are united in a lobed tube, and the 5 *petals are narrow and not

**b**

winged. The *ovary is *superior and enclosed in the hollow *receptacle. The fruit is a small *nut.

**bruyère** See ERICA.

**Bryales (subclass *Bryidae)** An order of medium-sized to large *acrocarpous mosses. The leaves are oval to lance-shaped; each usually has a nerve and a border of distinct cells. The *capsule is spherical to cylindrical, inclined or drooping, with a double *peristome. There are many genera, found in a wide range of habitats.

**Bryidae (class *Musci)** A subclass of mosses sometimes known as 'true mosses' and sometimes elevated to the level of a class (Bryopsida). They form by far the largest group of mosses. The *protonema is typically filamentous, and from it arise buds which develop into *gametophytes. The gametophyte may be *acrocarpous or *pleurocarpous, and varies greatly in size and shape. The *capsule opens by a lid; a *peristome may or may not be present. They are found all over the world in most types of habitat.

**bryology** The study of bryophytes (Bryophyta).

**Bryophyta (bryophytes)** A division of plants which for some authors includes the mosses (*Musci) and liverworts (*Hepaticae), but is now often taken to include only the mosses; liverworts having been assigned divisional status as *Hepatophyta. Bryophytes differ from *algae in that the multicellular *gametangium is surrounded by a protective jacket of sterile cells; gametangia of algae are usually unicellular and never have a protective jacket of sterile cells. Although bryophytes lack differentiated water-conducting vessels, and rely largely or entirely on water absorbed from rain falling on the plants, or from a moist atmosphere, some larger species may have simple water-conducting cells. They lack true *roots, but possess root-like *rhizoids which anchor them to a substrate and which can absorb water and minerals. The plants all show a *heteromorphic *alternation of generations, with a green vegetative *gametophyte (the familiar moss or liverwort plant) and a *sporo-

phyte which typically takes the form of a (usually stalked) *capsule and which is partially or wholly *parasitic on the gametophyte. Most bryophytes are land plants and are found worldwide in a range of habitats. They are known from Devonian rocks, but there is no evidence to link them with either the green algae or the more advanced *pteridophytes.

**Bryopsida** A *taxon used by some authors as a synonym of *Musci, and by others as a class which includes mosses more usually included in the subclass *Bryidae, i.e. mosses excluding *Sphagnum and the *Andreaeidae.

**Bryoria (order *Lecanorales)** A genus of *lichens which resemble *Alectoria species but differ in forming colourless *spores (8 per *ascus). See also BEARD-LICHEN.

**Bryum (order *Bryales)** A genus of mosses in which the plants are usually densely tufted. The leaves are ovate or broadly oblong-lanceolate; the cells composing the leaf are uniform and are mainly hexagonal or rhomboid. The *capsule is *pyriform to sub-cylindrical, horizontal, or pendulous, but never erect; the *peristome is double. Bryum is a large cosmopolitan genus, with about 1050 species, some of which are difficult to distinguish. They are mainly *terricolous or *saxicolous. B. argenteum is easily recognizable on sight: it has a characteristic silvery-green colour, and grows abundantly in many places where little else can grow, e.g. on pavements in towns, roadsides, etc.

**Bubbia (family *Winteraceae)** A genus of small trees, closely related to Belliolum, comprising 30 species that occur in New Guinea, Australia, and the south-western Pacific.

**Buchanania (family *Anacardiaceae)** A genus of trees that are characteristic of regrowth forest, and yield useful timber. There are 25 species, occurring from India to Australia.

**buckbean** See MENYANTHACEAE.

**buckeye** See AESCULUS and HIPPOCASTANACEAE.

**buckler ferns** See DRYOPTERIS.

**buckminsterfullerene** See CARBON.

**buckthorn** See RHAMNACEAE.

**buckwheat** See POLYGONACEAE.

**buckwheat tree** (*Cliftonia mono-phylla*) See CYRILLACEAE.

**bud** An immature shoot, protected by tough scale leaves, from which the stem and leaves or flowers may develop.

**budding** 1. The formation of *buds as a result of cell division in a localized area of a shoot. In general, *budding is promoted by *cytokinins and is inhibited by *auxins. 2. The grafting of a bud on to a plant. 3. Gemmation, a form of *asexual reproduction in which a new individual develops within the body wall or cell membrane of the parent, causing a bud-like swelling, then detaches itself to commence an independent life. In certain single-celled micro-organisms a new cell is formed by extrusion or out-growth from an existing cell. See also ASEXUAL REPRODUCTION.

**Buddleja** (family *Loganiaceae) A genus of trees and shrubs plus a few herbs, many of them ornamentals, which are rich in *alkaloids. The flowers are clustered, attractive to butterflies, *tetramerous, and have a *superior *ovary. There are about 100 species, occurring in the tropics and subtropics, and especially in eastern Asia. The genus is sometimes separated into its own family, Buddlejaceae.

**bud-fission** A type of *budding that occurs in certain types of *fungi: the *'bud' or *daughter cell is separated from the parent cell by a *septum.

**bud scale** See CATAPHYLL.

**buffer** A solution of a weak acid and its conjugate weak base that resists changes in *pH that would otherwise result from the addition of an acid or base. The buffering properties of weak electrolytes in living organisms are of great importance, since most cells can survive only within narrow pH limits.

**building phase** A stage in the cyclical pattern of changes that is typical of many plant and *heathland communities, including forest grasslands and heaths. In grassland, the term refers to the accumulation of wind-borne particles to form a small hummock around any grass seedling that chances to invade a hollow, which in turn results from the erosion of a former hummock in the degenerate phase. In heathlands, the term describes the bushy phase of growth in individual *Calluna vulgaris (heather) plants, lasting from about 7 to 13 years of age. At this stage *net primary production in the new shoots is at a maximum and the dense, bushy plant allows little light to reach the ground; hence it tends to suppress other species commonly associated with *C. vulgaris* in the pioneer, mature, and degenerate phases. *Compare* HOLLOW PHASE; PIONEER PHASE; MATURE PHASE; and DEGENERATE PHASE; *see also* CYCLICAL VEGETATION CHANGE and MOSAIC.

**bujedales** See MEDITERRANEAN SCRUB.

**bulb** An underground storage organ, comprising a short, flattened stem with roots on its lower surface, and above it fleshy leaves or leaf bases, surrounded by protective scale leaves. It may provide the means for *vegetative reproduction, or for the survival of the plant from one season to the next.

**bulbil** Any small, *bulb-like structure, usually formed in a leaf *axil, that separates from the parent plant and functions in vegetative reproduction.

**Bulbine** (family *Liliaceae) A genus comprising about 30 species of small, geophytic (*see* GEOPHYTE) herbs, found in the tropics, especially in S. Africa, which form *tubers and have succulent, linear leaves and spikes of yellow, star-like flowers.

**Bulgaria** (order *Helotiales) A genus of *fungi which form gelatinous *apothecia within which the unicellular brown and/or colourless *ascospores are produced. *B. inquinans* ('Pope's buttons') is a common species, with black, globular to cup-shaped, rubbery fruit bodies, each 2–4 cm across, which grow in clusters on logs, especially those of oak.

**bulk density** The mass per unit volume of soil (sampled as a clod or core), dried to constant weight at 105°C. *Compare* PARTICLE DENSITY.

**bullate** Bearing blister-like swellings; blister-like in appearance.

**bulliform cell** One of the rows of large, epidermal cells found in the leaves of certain grasses and believed to be involved in the unrolling of young leaves and in the rolling and unrolling of leaves in response to their water status.

**bullock's heart** See ANNONA.

**bulrush** See SCIRPUS and TYPHACEAE.

**bunch grass prairie** See PALOUSE PRAIRIE.

**bundle cap** A layer of *sclerenchyma or thickened *parenchyma cells at the tip of a *vascular bundle.

**bundle sheath** A layer surrounding a *vascular bundle, usually consisting of *parenchyma cells but sometimes of *sclerenchyma. See C$_4$ PATHWAY.

**bunt (stinking smut)** Common bunt is a seed-borne disease of wheat (and other grasses) caused by *Tilletia caries or T. foetida (T. laevis). The grain in an infected ear of wheat is replaced by masses of fungal spores which have a characteristic fishy smell; the spores are released when the wheat crop is harvested. Dwarf bunt is a similar disease caused by T. contraversa (T. brevifaciens). Karnal bunt is caused by Neovossia indica (Tilletia indica); it occurs in northern India, Afghanistan, Iraq, and Pakistan.

**bunya bunya** See ARAUCARIA.

***Burckella* (family *Sapotaceae)** A genus of 11 species of timber trees of lowland rain forest that occur from the Moluccas to Samoa.

**Burdigalian** A stage of the early *Miocene, about 20.43–15.97 Ma ago, underlain by the *Aquitanian and overlain by the *Langhian.

**buried soil** Soil covered by an *alluvial, colluvial (see COLLUVIUM), aeolian, glacial, or organic deposit, and being a product of a former period of pedogenesis. In US usage a buried soil is defined as lying beneath 30–50 cm if the covering layer is more than half the thickness of the buried soil, otherwise beneath more than 50 cm.

**Burma cedar** See TOONA.

**Burmanniaceae** A family of mostly small and saprotrophic (see SAPROTROPH), colourless herbs, in which leaves are often absent. The flowers are *trimerous, the fruit a *capsule. Principal genera are *Burmannia* and *Thismia*. The family is closely related to *Orchidaceae. There are 21 genera, with 165 species.

**Burma tulipwood** See DALBERGIA.

**Burseraceae** An important family of resiniferous, tropical timber trees, in which the leaves are *pinnate, with *stipules. The flowers are unisexual and small, their parts usually in threes or fives, with a *disc. The fruit is a *drupe or *capsule. There are 18 genera, with 540 species.

**bush** Wilderness or uncleared land, as contrasted with cultivated and settled land.

**bush veld** See VELD.

**butt** The base of a tree, or lower end of a log.

**butternut** See JUGLANS.

**butterwort** See PINGUICULA and LENTIBULARIACEAE.

**buttress root** A root that is similar to a *stilt root, but has a solid connection to the stem throughout its length. It is a means of providing stability to tall tropical trees in high winds. If the buttress root is detached from the trunk for part of its length it is known as a flying buttress root.

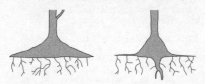

*Buttress root*

***Butyrospermum*** See VITELLARIA.

**butyrous** Like butter in consistency.

**Buxaceae** A family of evergreen shrubs or trees whose leaves are without *stipules. The flowers are small and regular, with 2 or 3 *sepals and *petals, 4–6 free *anthers, and a *superior *ovary of 3 fused *carpels. The fruit is a *drupe or *capsule. The seeds are black and shiny, sometimes with a *caruncle. The family is related to the *Euphorbiaceae. There are 5 genera, with about 60 species, occurring in temperate and tropical regions.

**Buxbaumiales (subclass *Bryidae)** An order of small, sometimes minute, *acrocarpous mosses. The leaves and shoots are often poorly developed, and the *capsule is disproportionately large; in some species there are no leaves and the branched, filamentous *protonema is persistent. They are found on acid soil on moorlands and in woods, and on soil-covered rocks in mountainous regions. (This characteristic group of mosses is sometimes elevated to subclass level: Buxbaumiideae.)

**Buxbaumiideae**   *See* BUXBAUMIALES.

*Buxus* **(family Buxaceae)** A widespread genus of trees that includes *B. sempervirens* (box) which has a valuable, very close-grained wood. There are 30 species.

**buzz pollination** The shedding of pollen from anthers that project from the flower, due to stimulation that occurs when the rapid vibration of a bee's wings resonates with the natural frequency of the anther structure. Some flowers, such as members of the Solanaceae (including nightshades and potato), have anthers projecting in this way.

**Byblidaceae** A small family of 2 genera and 4 species of *herbs and small shrubs, which are close to *Droseraceae.

*Byblis* **(family *Byblidaceae)** A genus of carnivorous *herbs in which the stems and leaves have glandular hairs and exude a sticky substance that attracts and attaches insects. The flowers are 5-petalled and usually purple. There are 2 species, in Australia and New Guinea.

*Byssochlamys* **(family *Eurotiaceae)** A genus of *fungi which are found chiefly in soil. *B. fulva* can cause spoilage of canned or bottled fruit. The *ascospores of this species can survive heating to a temperature of 84–87°C.

**C** *See* CARBON.

**¹⁴C dating** *See* RADIOCARBON DATING.

**C₃ pathway** The most common pathway of carbon fixation in plants. *See also* PHOTOSYNTHESIS and PHOTORESPIRATION.

**C₄ pathway** A pathway of carbon fixation, found most commonly in low-latitude plants, including many grasses, that are adapted to high temperatures and high light intensity. The first product formed as a result of the carboxylation by $CO_2$ of an acceptor molecule, phosphoenolpyruvate (PEP), is the four-carbon compound oxaloacetate (OAA); hence the name $C_4$. OAA is converted to another four-carbon compound that leaves the mesophyll cells just below the leaf cells containing chlorophyll, passes through spaces between the cells (called plasmodesmata), and enters bundle-sheath cells that are tightly packed around the leaf veins. There the compound gives up its $CO_2$, which enters the light-independent stage of photosynthesis (*see* DARK REACTIONS). PEP carboxylase, the enzyme catalyzing carboxylation, has no affinity for oxygen. Consequently $C_4$ plants suffer less than $C_3$ plants (*see* $C_3$ PATHWAY) from losses due to *photorespiration. Nevertheless, $C_4$ plants outcompete $C_3$ plants only under high light and high temperature conditions. Their efficiency is lower than $C_3$ plants in high latitudes. *Zea mays* (maize or corn) and *Saccharum officinarum* (sugar cane) are $C_4$ plants.

**Ca** *See* CALCIUM.

**caatingas** A thorn forest of semi-arid, tropical, north-eastern Brazil and similar or allied types of vegetation including thorn woods and thorn bush. They are differentiated from *savannah woodlands by the paucity or absence of grass, which reflects a more prolonged dry period and smaller total amounts of rainfall.

**cabbage** *See BRASSICA.*

**Cactaceae** The cactus family, comprising plants that have succulent stems, bearing tufts of spines, and leaves that are reduced or absent (except in *Pereskia*). Flowers are solitary, *sessile, showy, and *hermaphrodite, with many parts, and the *ovary is *inferior. The fruit is a juicy or leathery *berry with many seeds. The family has no close relatives. A few cacti are *epiphytes. There are 130 genera, with about 1650 species, occurring in the semi-deserts of America (with *Rhipsalis* in Madagascar and Sri Lanka).

**caducous** Soon dropping off.

**Caesalpinia (family *Fabaceae, subfamily *Caesalpinoideae)** A genus of 100 species of tropical plants, many of which are woody climbers. Wood of *C. sappan* yields a red dye (sappan or Brazilwood).

**Caesalpinoideae** One of the three subfamilies of *Fabaceae, comprising 162 genera and 2000 species, of mainly tropical and woody plants, many of which are climbers. Leaves are usually *pinnate with *stipules. Flowers are more or less *zygomorphic, with free *sepals and *petals, and are usually *imbricate, the uppermost petal innermost in the bud; usually they have 10 or fewer *stamens.

**Caesar's mushroom** The common name for the *mushroom-shaped, orange *fruit body of *Amanita caesarea*, which is edible and highly esteemed. It is found on the ground under broad-leaved trees, and occurs in parts of Europe (not Britain) and in the Americas.

**Caesia (family *Liliaceae)** A genus of small plants that have strap-like leaves and clusters of white or pale mauve, 6-petalled flowers. They are Gondwanic in distribution (*see* GONDWANA), with 10 species in Australia, New Guinea, Madagascar, and S. Africa.

**caespitose** Growing in dense tufts.

**Cainozoic** *See* CENOZOIC.

*Cajanus* (family *Fabaceae, subfamily *Papilionatae) A genus of 2 species of tropical African plants, one of which is *C. cajan* (pigeon pea), a *perennial shrub; the whole plant is useful as a cover crop and the seed is a source of dhal.

**cajeput oil** *See* MELALEUCA.

**Calabrian** An early *Pleistocene marine stage, about 1.806–0.781 Ma ago, which is split between two land-mammal stages (Upper *Villafranchian and Biharian).

*Caladenia* (family *Orchidaceae) A genus of ground orchids that have a small, narrow leaf, produced annually. The flowers are borne on a single stem and are very variable in size. There are about 70 species, found in Malesia, Australia, New Caledonia, and New Zealand.

**Calamitaceae (horse-tails)** A family of plants that were among the dominant vegetation of the *Carboniferous swampland floras. They had a *sporophyte that resembled a tree, with much *secondary thickening in the stem and branches. Some grew to a height of 20–30 m. They appeared in the Upper *Devonian and became extinct early in the *Triassic.

*Calamites cistiiformes* (horse-tails; family Calamitaceae) A species of jointed-stemmed plants that were an important component of *Carboniferous swampland floras. Unlike their small modern counterparts some Carboniferous species grew to 18 m tall. *Calamites cistiiformes* was the earliest representative of the Calamitaceae and was first described in 1877. Calamatids are known throughout the northern hemisphere.

*Calamus* (family *Arecaceae) A genus of palms in which the stem is clothed in a spiny tubular sheath, 10 m or more long, and *flexuous, rarely short and erect. Leaves are *pinnate, often with a long, protruding, spiny tip (cirrus), or with modified, sterile, spiny inflorescences (flagella), rarely both; the spiny whips grapple the surrounding forest. *Calamus* is *dioecious. The inflorescence has tubular, spiny sheaths. Fruits are scaly, often with a thin, sour pulp. Many species produce useful cane, a few entering international commerce. *Calamus* is the largest of all palm genera, and the most wide-ranging genus of climbing palms (rattans or rotans). There are about 370 species in Africa, India, and southern China, south through the Malay archipelago to northern Australia, and from the western Pacific islands to Fiji.

*Calandrinia* (parakeela; family *Portulacaceae) A genus of small, *annual or *biennial herbs that have pink flowers. They are represented by about 150 species in western N. America, S. America, and Australia.

**calcareous** Applied to substances containing calcium carbonate. *See also* CALCAREOUS SOIL.

**calcareous soil** Soil that contains enough free calcium carbonate to effervesce visibly, releasing carbon dioxide gas, when treated with cold 0.1 $N$ hydrochloric acid, and which could also be regarded as an *alkaline (basic) soil.

**calcic horizon** A *mineral soil *horizon with evidence of secondary calcium carbonate deposition which is more than 15 cm thick, with a calcium carbonate content of more than 15 per cent by weight, and with 5 per cent carbonate more than is in the parent material or horizons below it.

**calcicole** A plant species confined to, or most frequently found on, soils containing free calcium carbonate. *See also* ALKALINE SOIL. *Compare* CALCIFUGE.

**calcicolous** Applied to an organism that prefers to grow in, or can grow only in, *habitats rich in calcium ('lime').

**calcification** The process of redeposition of secondary calcium carbonate, from other parts of the soil *profile, which if sufficiently concentrated may develop into a *caliche, kunkar, or *calcic horizon (all similar and usually comprising more than 15 per cent weight of calcium carbonate in more than 15 cm soil thickness). Calcification involves limited upward and lateral movement of calcium salts in solution, and downward movement in less wet periods. When occasionally *leaching is deeper, some but not all of it may redissolve.

**calcifuge** A plant species not usually found on soils containing free calcium carbonate. *Compare* CALCICOLE.

**Calcisols** Soils that have a *calcic horizon within 125 cm of the surface. Calcisols are a reference soil group in the FAO *soil classification.

**Calcium (Ca)** An element that is necessary for plant growth. In *eukaryotic cells it is found mainly in the *apoplast where it preserves membrane integrity and strengthens cell walls. It protects roots against the effects of low *pH, ion imbalance, and toxic ions, so that reduced root growth is a symptom of deficiency, as is 'die back' of buds.

**calcrete** *See* CALICHE.

**caliche (calcrete)** A carbonate *horizon (the K horizon) formed in a soil in a semi-arid region, under conditions of sparse rainfall (20–60 mm/yr) and a mean annual temperature of about 18°C, normally by the precipitation of calcium carbonate carried in solution. It may become cemented and indurated on exposure, when it gives rise to a tabular landscape. *See also* CALCIC HORIZON; DURIPAN; and PETROCALCIC HORIZON.

**Caliciaceae (order *Caliciales)** A family of *fungi (mostly *lichen-forming) in which the *thallus is thin and *crustose or immersed in the substratum; the *asci are thin-walled and soon disintegrate, forming a *mazaedium. *Spores are brown or colourless. There are many genera. Usually they are found on bark or wood.

**Caliciales (subdivision *Ascomycotina)** A heterogeneous order of *fungi (mostly *lichen-forming) in which the *thallus is *crustose, *squamulose, *foliose, or *fruticose. There are 3 families. They are found chiefly on wood or bark, occasionally on rocks; there are a few *parasitic species.

**Callistemon (family *Myrtaceae)** A genus of plants in which the *inflorescences are conspicuous, stout spikes, beyond which a leafy shoot extends. Several species are ornamentals. There are 20 species, occurring in Australia and New Caledonia. They are close to *Metrosideros*.

**Callitrichaceae** A family of aquatic plants in which the stems are *filiform, and leaves are in *opposite pairs, entire, and *exstipulate. The flowers are tiny, *monoecious, borne in leaf *axils, and without a *perianth. There is 1 *stamen, the *ovary is 4-celled and 4-lobed, there are 2 long *styles, and the fruit consists of 4 tiny *drupes. There is 1 genus, *Callitriche*, with 17 species, of cosmopolitan distribution.

**callose** An insoluble *glucan that is found in higher plants, in which glycosidic (*see* GLYCOSIDE) linkages join the first and third carbon atoms of neighbouring beta-*glucose units. Structurally it is similar to the protozoal storage carbohydrates, paramylon and leucosin, and also to the algal product laminarin. In *angiosperms it is formed in response to injury, especially in *phloem and in germinating *pollen tubes.

**Calluna (family *Ericaceae)** A *monotypic genus (*C. vulgaris*) of low, evergreen shrubs that have tiny, *adpressed, *opposite leaves, and flowers borne in the *axils of upper leaves. The flowers are *tetramerous; the *calyx is coloured like the *corolla but is larger and more deeply lobed. There are 8 *stamens. The fruit is a *capsule. *Calluna* dominates large areas of heathland on acid soils in western Europe, forming the community known as a Callunetum. The genus is confined to Europe and Asia Minor.

**Callunetum** *See* CALLUNA and -ETUM.

**callus** *Tissue that forms over a wound, or that develops from actively dividing plant tissue in a tissue culture.

**Calocera (order *Dacrymycetales)** A genus of *fungi which form upright, cylindrical or tapering, branched or unbranched, orange or yellow, more or less gelatinous *fruit bodies. They are found on tree-stumps, logs, etc., in woodland; *C. viscosa* grows on softwood, while *C. cornea* is common on hardwood.

**Calophyllum (family *Clusiaceae)** A genus of tropical timber trees, in which the bark surface is usually fissured, releasing coloured, sticky latex. The fruit is a *drupe, the stone often opening by a plug. There are 187 species, occurring mainly in

Malesia, with a few in Africa, Madagascar, and the New World.

**Caloplaca** (order *Teloschistales) A genus of *lichens in which the *thallus is *crustose, sometimes with radiating lobes, and is frequently bright yellow or orange. The *apothecia are typically yellow, orange, or rust-coloured. *Spores are colourless, and generally have a thick wall across the middle which is perforated by a narrow canal. There are many species, found on a range of substrates, including rocks, bark, etc.

**calorie** The energy required to raise one gram of water through one degree Celsius. One calorie is equivalent to 4.128 Joules. The Calorie, used to measure the energy value of foods, is 1000 calories.

**calorific value** The gross calorific value of a substance is the number of heat units that are liberated when a unit weight of that substance is burned in oxygen, and the residual materials are oxygen, carbon dioxide, sulphur dioxide, nitrogen, water, and ash. The energy content of biological materials has been expressed traditionally in calories (c) or kilocalories (C) per gram dry weight. Sometimes results are expressed more significantly in terms of ash-free dry weight, i.e. in terms of organic constituents only. Contemporary studies of ecological energetics express results in terms of the SI energy unit, the joule (4.182 J = 1 calorie).

**Calospatha** (family *Arecaceae) A genus of climbing palms (rattans), in which the stem is solitary and robust, growing to a height of 10 m, with no grappling whips. The plant is *dioecious, the inflorescence being short with persistent, successively smaller, overlapping *bracts. The fruits are scaly, and contain 3 seeds embedded in a thin sheet of edible pulp (sarcotesta). The cane is not used. There is 1 species, endemic to Malaya, where it is very local and was overlooked for 50 years.

**Calothrix** A genus of filamentous *cyanobacteria (section IV) in which the tapered filaments are attached to a substrate at their bases only, typically occurring in felt-like tufts. They are found growing on rocks in freshwater and marine environments. *Compare* RIVULARIACEAE.

**Calvatia** (order *Lycoperdales) A genus of *fungi which includes the giant puffball *Calvatia gigantea* (also known as *Lycoperdon giganteum* or *Langermannia gigantea*). *Fruit bodies of this species are usually 30–50 cm across, but specimens measuring up to 1.5 m across have been recorded. The colour is initially white, becoming yellowish to brown. The flesh is white, becoming olive-brown and powdery as *spores develop. It is found in grassland, woods, gardens, etc., and is edible when young.

**Calvin cycle** A cyclic series of reactions occurring in the *stroma of *chloroplasts, in which carbon dioxide is fixed and reduced to glucose, using *ATP and NADPH formed in the light reaction of photosynthesis. The relatively stable product formed is a 3-carbon sugar.

**Calycanthaceae** A small family of trees and shrubs, most of them aromatic, with *simple, *opposite, *exstipulate leaves, and flowers that have numerous *tepals, *stamens, and free *carpels, the carpels and *achenes lying within the hollowed *receptacle. There are 3 genera, with 9 species, occurring in eastern Asia (*Chimonanthus* and *Sinocalycanthus*) and N. America (*Calycanthus*).

**calyciform** Shaped like a *calyx or cup.

**calyptra** A protective cap- or hood-like covering on the developing *capsule in a *moss or *liverwort. In mosses the calyptra usually remains until the capsule is almost ripe, when it usually falls. In liverworts the calyptra covers the capsule only very early in its development; later it may be visible as a collar at the base of the stalk.

**Calystegia** (bindweed; family *Convolvulaceae) A genus of herbaceous and woody plants which have long, trailing, and twining stems. They have alternate, *simple leaves without *stipules. The regular flowers are bisexual with an *involucre of large *bracteoles below the *calyx. There are 5 *sepals, and 5 *petals fused to form a funnel or bell shape. The *stamens are attached to the *corolla base. The flowers are generally large and *axillary. The *ovary is *superior, with 2 fused *carpels and 2 *locules, each with 2 *ovules. The fruit is a *dehiscent *capsule containing hairy seeds with large, folded *cotyledons. There are about 25 species, some of which are grown as ornamentals, found in temperate and subtropical regions of the world.

**Calytrix** (family *Myrtaceae) A genus of heath-like shrubs with *xeromorphic foliage that are found mainly on heaths and semi-arid sites of low fertility in *Eucalyptus* communities. The star-like, 5-petalled flowers have a persistent, reddish *calyx which remains showy after the *corolla has withered. There are about 40 species, endemic (*see* ENDEMISM) to Australia.

**calyx** A collective term for all the *sepals of a flower.

**CAM** *See* CRASSULACEAN ACID META-BOLISM.

**cambic horizon** A weakly developed *mineral soil *horizon of the middle part (B horizon) of soil *profiles, and one that has few distinguishing morphological characteristics except for evidence of weathering and sometimes of gleying. It is found in *brown earths and *gleys. It is a *USDA term. The name is from the Latin *cambiare*, to change.

**Cambisols** Soils that have a *cambic or *mollic horizon above a B *soil horizon with a *base saturation lower than 50% in the upper 100 cm, or an *andic, *vertic, or *vitric horizon with an upper boundary 25–100 cm below the surface. Cambisols are a reference soil group in the FAO *soil classification.

**cambium** In the stems and roots of *vascular plants, a layer of cells lying between the *xylem and *phloem. These retain the ability to divide, producing secondary xylem and phloem. This continues throughout the lifetime of woody plants, increasing the girth of the stem. *See also* BARK.

**Cambrian** The first of six periods of the *Palaeozoic Era, from about 542–488.3 Ma ago, during which sediments deposited include the first organisms with mineralized skeletons. The plant fossils are mainly of *algae and *fungi. Trilobites (arthropods) are important in the stratigraphic subdivision of the period.

**Camellia** (family *Theaceae) A genus of evergreen shrubs and small trees, many of which are grown for their showy flowers as ornamentals. *C. sinensis* is the tea plant; its variety *assamica* yields Indian tea. There are 82 species, occurring in eastern Asia.

**Camellia japonica** *See* THEACEAE.

**Camellia sinensis** (tea plant) *See* CAMELLIA and THEACEAE.

**c-AMP** (cyclic AMP) Cyclic form of adenosine monophosphate that is formed from *ATP (adenosine triphosphate) in a reaction mediated (catalysed) by adenyl cyclase. It has numerous functions in animal cells and micro-organisms, acting variously as a genetic regulator, as a mediator in the activity of some *hormones, as an *enzyme activator, as a secondary messenger, and as a chemical attractant. Its presence in higher plants has not been fully established.

**Campanula** (bellflowers) *See* CAMPANULACEAE.

**Campanulaceae** A family of (mostly) *herbs in which the leaves are alternate, *simple, and *exstipulate. The flowers are regular, usually bell-shaped, with the *corolla lobes fixed together below. The *inferior *ovary is 2–10-celled, with many *ovules. The fruit is usually a *capsule. *Campanula* (bellflowers) is a genus of more than 300 species, with erect stems, and blue, purple, or white flowers; many of these species are cultivated. There are 87 genera, with 1950 species; they are of almost cosmopolitan, but mostly temperate, distribution.

**campanulate** Shaped like a bell.

**camphor** *See* CINNAMOMUM.

**Campnosperma** (family *Anacardiaceae) A genus of tropical trees which are fast-growing and which yield pale, soft, useful timber. There are 10 species, with a Gondwanic range (*see* GONDWANA), occurring in tropical America, Madagascar, Seychelles, Malesia, and the western Pacific.

**campo cerrado** The *campos of Brazil comprises *savannah-type vegetation associated with very poor soils, although, in the absence of fire, woodland or scrub would no doubt prevail. There are several campos variants, including campo cerrado, which is a *tree-covered savannah, and campo sujo, in which the aspect is more open and trees and shrubs are sparse.

**campos** The general regional term applied to the tropical *savannah grasslands, with scattered, broad-leaved trees, of Brazil.

**campo sujo** *See* CAMPO CERRADO.

**Campsis** (family *Bignoniaceae) A genus of 2 species of showy, woody climbers; they are the trumpet vines of eastern Asia and N. America.

**Camptothecium sericeum** *See* HOMALO-THECIUM.

**Campylobacter** A genus of spirally shaped *bacteria in which the cells have a single *flagellum at one or each end. They are *micro-aerophilic. There are several species. They are found in the alimentary and reproductive systems in humans and other animals, and may be *pathogenic, causing abortion in cattle and sheep.

**Campylopus** *See* DICRANALES.

**campylotropous** Applied to the position of an *ovule that is horizontal, with the funicle attached midway between the *chalaza and *micropyle. *Compare* ANATROPOUS and ORTHOTROPOUS.

**Canada balsam** Resin distilled from the bark of *Abies balsamea* (balsam fir) and other N. American *Abies* species.

**canalization** The developmental process that is held within narrow bounds despite both genetic and environmental disturbing forces. Thus, cells will progress along particular developmental pathways until they become differentiated into their final, adult forms. Development is

such that all the different *genotypes have a standard *phenotype over the range of environments common to that species.

**canalizing selection** The elimination of *genotypes that render developing individuals sensitive to environmental fluctuations. Genetic differences may be revealed in organisms by placing them in a stressful environment, or if a severe *mutation stresses the developmental system.

**Cananga** (family *Annonaceae) A genus of trees that includes *C. odorata*, whose flowers yield the powerful perfume ylang ylang or macassar oil. There are 2 species, occurring from tropical Asia to northern Australia.

**Canarium** (family *Burseraceae) A genus of timber trees in which the leaves are *pinnate, with *stipules. The wood is resiniferous. The fruit is a *drupe, and is an important food for specialized, frugivorous birds. The seed is oily, those of *C. commune* and *C. indicum* (ngali) being important local foods. There are about 75 species, occurring in the Old World tropics.

**candelabra tree** *See* DISJUNCT DISTRIBUTION.

**Candelariella** (order *Lecanorales) A genus of *lichens in which the *thallus is *crustose and yellow. The *apothecia are *sessile and *lecanorine. Species are found mostly in nitrogen-rich habitats, e.g. on rocks or trees exposed to bird droppings, on farm buildings, etc.

**Candida** (family *Cryptococcaceae) A genus of *imperfect *fungi which may have a *yeast-like vegetative state, or may form a pseudomycelium or a true, septate *mycelium. There are numerous species. Some have perfect states classified either in the *Ascomycotina or in the *Basidiomycotina. Some, particularly *C. albicans*, are normal components of the flora inhabiting the mouth and gastrointestinal tract of humans and other animals and can sometimes cause disease (i.e. they can be opportunist *pathogens). Other species are *saprotrophic, occurring in soil, plant matter, etc.

**candlenut** *See ALEURITES.*

**candle-snuff fungus** The antler-shaped fruiting structures of the *ascomycete *Xylaria hypoxylon*.

**Candolle, Alphonse Louis Pierre Pyramus de** (1806–93) The son of Augustin Pyramus de *Candolle, who, in 1842, succeeded his father as professor of natural history at the University of Geneva and completed his father's work by publishing 10 volumes of his *Prodromus* (one of them in collaboration with his own son). In his own *Géographie botanique raisonnée* (1855), *La Phytographie* (1880), and *Origine des plantes cultivées* (1833, published in English as *Origin of Cultivated Plants* in 1886), Candolle considered why particular species occur in some places but not others and contributed greatly to establishing the basis of modern biogeography.

**Candolle, Augustin Pyramus de** (1778–1841) A Swiss botanist who studied in Geneva and settled in Paris in 1796. At the request of the French government, he conducted a botanical and agricultural survey of the whole of France, the results of which were published in 1813. In 1816 he returned to the University of Geneva as professor of natural history and devoted the remainder of his life to developing his own system of botanical classification, which he tried to make more 'natural' than previous classifications. He began his most famous work, *Prodromus systematis naturalis regni vegetabilis* in 1824. Seven volumes were published before his death and a further 10 by his son, Alphonse de *Candolle.

**cane blight** A disease that affects the raspberry cane (*Rubus idaeus*). The leaves shrivel and die, dark patches appear on stem bases, and the bark splits open. The disease is caused by a *fungus, *Leptosphaeria coniothyrium*, which enters the plant via wounds.

**canker** A general term for a localized disease of woody plants in which *bark formation is prevented. Cankers may be caused by *bacteria or *fungi, or occasionally even by viruses.

**Canna** (family Cannaceae) A genus of rhizomatous (*see* RHIZOME) herbs which are cultivated for their showy inflorescences. *C. edulis* of the Andes is cultivated for its edible rhizomes (Queensland arrowroot or achira). There are 25 species, occurring in tropical and subtropical America.

**Cannabidaceae** A small plant family, closely related to *Moraceae (but differing from it mainly in having its floral parts in fives), of *herbs without latex that have lobed, *stipulate leaves, and small, *dioecious flowers borne in *axillary clusters. The *perianth is *pentamerous, there are 5 *stamens, and the *ovary is *unilocular, with 2 *stigmas. The flowers are wind-pollinated. The fruit is a small *nutlet. *Cannabis* produces the fibre hemp and also a narcotic drug. (Its cultivation is now illegal in many countries.) *Humulus*, a climbing plant, is used to flavour beers. There are 2 genera, *Cannabis* (hemp) and *Humulus* (hop), of northern temperate distribution.

**Cannabis** (family *Cannabidaceae) A *monotypic genus comprising *C. sativa* (hashish, hemp, ganja, marijuana) which is native to central Asia. It is a large herb that provides a narcotic as well as hemp fibre.

**canopy (ecol.)** The part of a woodland or *forest community that is formed by the trees. In complex forests, e.g. in *tropical rain forest, the canopy is often arbitrarily subdivided into emergent, middle, and lower zones. The term may also be applied to the upper layer of shrub and scrub communities, or in any terrestrial plant community in which a distinctive habitat is formed in the upper, denser regions of the taller plants. *Compare* GROUND LAYER.

**cant** The section of a *coppiced woodland that is cut in a particular year in the rotation. *See also* COUPE and HAG.

**Cantharellales** (subclass *Holobasidiomycetidae) An order of *fungi in which the *fruit body may be more or less funnel-shaped or mushroom-shaped. The *spores may be borne on gill-like ridges or on a smooth or wrinkled surface. Genera include *Cantharellus* and *Craterellus*.

**Cantharellus** (order *Cantharellales)
A genus of *fungi in which the *fruit
body consists of a more or less funnel-
shaped *pileus with a central *stipe.
The *hymenium is borne on wrinkles,
or blunt *gill-like ridges on the lower,
or outermost, surface of the pileus. The
genus includes the edible *chanterelle.

**cap**  *See* PILEUS.

**Cape floral region**  A region synonymous with R. Good's (*The Geography of the
Flowering Plants*, 1974) S. African kingdom,
and in relation to its size one that has
perhaps the most remarkable and richest
flora in the world. There are about 2500
species, and an unusually high proportion
of the genera are endemic (*see* ENDEMISM).
Many well-known garden plants and
greenhouse succulents originated in this
floral region. *See also* FLORAL PROVINCE and
FLORISTIC REGION.

**Cape primrose**  *See* STREPTOCARPUS.

**caper**  A flower bud of *Capparis spinosa*.
*See also* CAPPARIS and CAPPARIDACEAE.

**capillarity**  *See* CAPILLARY ACTION.

**capillary action (capillarity)** The process
in which soil moisture moves in any
direction through the fine (capillary)
pores of the soil, under surface tension
forces with individual soil particles. Soil
moisture in this state is known as capillary
moisture. It exists as a film or skin of
moisture on the soil particles, and may
be drawn above the water-table by capillary action and into the plant roots by
the process of *osmosis. 'Capillary conductivity' is now obsolete in US terminology.

**capillary moisture**  Moisture that is
left in the soil, along with hygroscopic
moisture and water vapour, after the
gravitational water has drained off.
Capillary moisture is held by surface tension (known in the US as 'water potential')
as a film of moisture on the surface of soil
particles and peds, and may move (and
some of it may be used by plants) in this
form.

**capillitium**  The sterile threads that
occur among the *spores in the *fruiting
bodies of some *slime moulds and *fungi.

**capitulum**  An *inflorescence that
consists of closely packed flowers or
*florets which have no stalks and arise on
a flattened *axis, all at the same level. The
capitulum is surrounded or subtended
by an *involucre of *bracts giving it the
appearance of a single flower. Capitula are
typical of the *Brassicaceae.

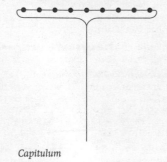

*Capitulum*

**Capparaceae**  *See* CAPPARIDACEAE.

**Capparidaceae (Capparaceae; order
Capparidales)** A family of dicotyledonous
(*Dicotyledoneae) plants most of which
are shrubs and small trees. The leaves are
alternate, *simple or *compound, with
*stipules which are sometimes spiny. The
flowers are usually borne in *racemes, and
have 4 *petals, 2 pairs of *sepals, 4–6 or
more *stamens, and a *superior *ovary of
2 *carpels. The fruit is a *berry, *nut or
*drupe. The flower buds of *Capparis spinosa*
(caper) are cooked and pickled for use as
flavouring. There are about 45 genera,
with 675 species, occurring in tropical
and warm temperate regions.

**Capparis** (family *Capparidaceae) A
genus of trees and shrubs with tough
leaves and prickly stems which are found
in dry regions. The flowers are 4-petalled,
with numerous *stamens, and are yellow
to white in colour. The fruits are fleshy.
They are found throughout the tropical
and warm regions of the world, with
about 250 species, only 2 of which occur in
Europe.

**Capparis spinosa** (caper) *See* CAP-
PARIDACEAE.

**Caprifoliaceae**  A family of shrubs
and small trees in which the leaves are

*opposite, sometimes with *stipules. The *inflorescence is *cymose. The flower has 4 or 5 *imbricate *sepals and *petals, and an *inferior *ovary. There are many ornamentals, including *Abelia, Diervilla, Leycesteria, Lonicera* (honeysuckles), *Symphoricarpos* (snowberry), and *Viburnum*. The family is related to *Rubiaceae. There are 16 genera, with 365 species, most of the northern temperate zone, but also of tropical mountains.

**Capsella bursa-pastoris** (shepherd's purse) *See* ALBUGO.

**Capsicum** (family *Solanaceae) A genus of plants that includes *C. annuum*, varieties of which provide paprikas, green or red peppers (capsicums), chillies, and cayenne pepper. There are 10 species, native to tropical America.

**capsule 1. (of *prokaryotes)** The gelatinous, outer surface layer to the cell (also known as the sheath in *cyanobacteria), which is composed primarily of *polysaccharides. In *pathogenic *bacteria, and possibly in others, it appears to serve a protective function against the defensive mechanism of the host, since such bacteria become noninfective in its absence. **2.** The *spore-bearing structure of a *moss or *liverwort. In most mosses the capsule may be spherical, cylindrical, oval, or pear-shaped, and is borne on a stalk or *seta; the seta in turn is anchored to the *gametophyte by an absorptive region called the foot. The capsule is protected, at least initially, by a *calyptra. In most cases the capsule has a mouth initially covered by a lid; when the spores inside the capsule are mature, the lid falls off and the spores are released, to be dispersed by wind. (*Compare* SPHAGNUM and ANDREAEIDAE.) In liverworts the capsule is usually ovoid or spherical and has no lid. On ripening, the capsule usually splits into four 'valves', which open out to release the spores. **3.** A dry *fruit, normally *dehiscent.

**carambola** *See* AVERRHOA.

**Carapa** (family *Meliaceae) A genus of 2 species of small trees that occur in tropical America and west and central Africa. Their fruits are big, spherical *capsules.

**carbohydrate** A generic term for molecules that have a basic empirical formula of $C_x(H_2O)_y$ and that are aldehyde or ketone derivatives of polyhydroxyl alcohols. They are normally classified as mono-, oligo-, or poly-saccharides, depending upon whether they are single sugars, short-chain molecules, or large polymers respectively.

**carbon (C)** A non-metallic element, which is unique in the number of compounds it is able to form that contain chains or rings of carbon atoms. This ability to form large, complex molecules in which other elements are bonded to carbon atoms is exploited by all living organisms, and carbon thus provides the basis for organic chemistry (the chemistry of organisms). Carbon is extracted from gaseous carbon dioxide by plants during *photosynthesis, and is incorporated in to living matter; when organic matter decomposes, its carbon is oxidized and so is returned to the atmosphere as carbon dioxide. Pure carbon occurs naturally as diamond, graphite, as the amorphous carbon black, and recently an additional polymorph has been recognized called buckminsterfullerene, in which 60 atoms are arranged to form an approximately spherical shape, rather like a football. Charcoal, produced by the destructive distillation of organic matter, is also a fairly pure form of carbon.

**carbon dioxide acceptor** A molecule (e.g. *ribulose-1,5 diphosphate) that assimilates carbon dioxide from the air into a plant.

**carbon dioxide method** Any non-destructive technique for measuring rates of *primary productivity that monitors changing carbon dioxide ($CO_2$) concentrations as a means of assessing the rate of carbon dioxide uptake in *photosynthesis and release in *respiration. A wide choice of carbon dioxide measuring techniques exists, some more applicable to laboratory than to field conditions. The most frequently used research methods include infrared gas analysis, conductivity measurements, and radiocarbon tracers. *See also* AERODYNAMIC METHOD.

**Carboniferous** The penultimate period of the *Palaeozoic Era, from 359.2–299 Ma ago, preceded by the *Devonian and followed by the *Permian. It is divided into two epochs, the Mississippian (359.2–318.1 Ma) and Pennsylvanian (318.1–299 Ma). During the Carboniferous very lush *swamp forests dominated the landscape in low-lying areas, where minor changes in sea-level alternately exposed land supporting forest then inundated and buried the vegetation. The forests were dominated by *Lycopsida and *Calamitaceae, some of which grew to the size of trees (see CALAMITES CISTIIFORMES and LEPIDODENDRON SELAGINOIDES), and the forest floor supported ferns and seed ferns. The buried vegetation was compressed and changed through time to form the rich coal measures of South Wales, England, Scotland, the USA, and many other areas world-wide, in which recognizable seed plants are common *fossils.

**carboxyl** The group $-CO_2H$ (also written as $-COOH$).

**carboxylase** An *enzyme that catalyses reactions in which a molecule of carbon dioxide is incorporated into an organic compound.

**carboxysome** One of the small, polyhedral bodies in the cells of *cyanobacteria that are involved in the *Calvin cycle.

*Cardamine* (family *Cruciferae) A cosmopolitan genus of *herbs found in temperate and cool temperate regions. There are about 130 species, some of which are cultivated as ornamentals.

**cardamom** See AMOMUM.

*Carex* (sedges; family *Cyperaceae) A genus of *perennial *herbs that have *rhizomes. The stems are characteristically triangular in cross-section, solid, and leafy. The leaves are linear, often inrolled or keeled, with a sheathing leaf base, and a *ligule. The *inflorescence varies from a branched *panicle to a single *spike, and is composed of unisexual flowers, usually clearly separated on the inflorescence, the plants being generally *monoecious. Each single-flowered *spikelet is subtended by a *glume. There is no *perianth, although the female flowers are surrounded by a triangular or flattened sac with a beak through which the 2 or 3 *stigmas project from the triangular or oval *ovary. The ovary is *superior, with 3 fused *carpels and a single *locule. The fruit, a small *nut, remains within the sac, and contains copious *endosperm. The male flowers have 2 or 3 *stamens and are generally held at the top of the inflorescence, often in the terminal or upper, later spikes. The female flowers are restricted to the lower branches or the bottom of the spike. Sedges are found throughout the world, often as components of marshy habitats. Each species has very precise ecological requirements. There are about 1000 species, of which a limited number are used locally as food or bedding for animals, and a few are cultivated as ornamentals.

**Caribbean floral region** Part of R. Good's (*The Geography of the Flowering Plants*, 1974) neotropical kingdom. It comprises isthmian America and the W. Indies. There are thought to be well in excess of 500 endemic genera (see ENDEMISM), and many thousands of endemic species. Some of the most diverse floras in the world are found here. For example, Cuba alone has roughly 8000 species, many of which are endemic. A very large number of valuable economic plants and some garden plants are from this region. *See also* FLORAL PROVINCE and FLORISTIC REGION.

*Carica* (family *Caricaceae) A genus of small trees which have pithy trunks supported by *phloem tissues. Several species have edible fruits, notably *C. papaya* (papaya), now very widely cultivated, whose leaves and unripe fruit contain a *proteolytic enzyme, used to tenderize meat. There are 22 species, found in the American tropics. See DISJUNCT DISTRIBUTION.

**Caricaceae** A family of small trees which have few branches or none. The sap is watery to milky. The leaves are *palmately lobed or divided. The flowers are regular, mostly unisexual. The fruit is a *berry. The family is related to *Passifloraceae. There are 4 genera, with 31 species, occurring in the tropics of America and Africa. See DISJUNCT DISTRIBUTION.

**carina** See KEEL.

**carinal canal** In *Equisetum* and some of its fossil relatives, one of a number of longitudinal channels inside the *metaxylem and formed by the disintegration of the *protoxylem. See also VALLECULAR CANAL.

**Carludovicia** (family *Cyclanthaceae) A genus of trees which have a short stem and tufted, *palmate leaves. The flowers are borne on a *spadix which at first is enveloped by *bracts, developing later into a head of *berries. Young leaves of *C. palmata* (Panama-hat palm) provide the fibre for Panama hats. There are 3 species, occurring in the northern part of tropical America.

**carnation** See DIANTHUS.

**carnauba wax** A major source of wax for polishes, obtained from the leaf surfaces of *Copernicia cerifera* (wax palm) of the dry north-eastern shoulder of Brazil, where it is exceedingly abundant in open savannah forest. A tribe of Indians, the Carnaubeira, was almost completely dependent on its products. See also COPERNICIA.

**carotene** One of a group of hydrocarbon *carotenoids.

**carotenoid** A generic term for water-insoluble, polyisoprenoid pigments, which often function as accessory photosynthetic pigments in higher plants and photosynthetic bacteria, with absorption peaks between 450 and 480 nm. The group includes the *carotenes, which are orange, and xanthophylls, which are yellow. During the *senescence of leaves, *chlorophyll breaks down faster than carotenoids and carotenoid colours are revealed. In the vertebrate liver, carotene is changed into (i.e. is a precursor of) vitamin A.

**carpel** One of the female reproductive organs of the flower, i.e. a unit of the *gynoecium, comprising an *ovary (containing 1 to many *ovules borne on a *placenta) and with a usually terminal *style tipped by the *stigma.

**Carpentaria** (family *Arecaceae) A *monotypic genus (*C. acuminata*) that is endemic (see ENDEMISM) to tropical Australia where it is found as a component of semi-evergreen rain forests.

**Carpinus** (hornbeams; family *Betulaceae) A genus of *monoecious, *catkin-bearing trees, in which the male flowers are without *bracteoles, and have about 10 forked *stamens. The female flowers are borne in terminal, drooping catkins. The fruit is a small *nut with a large, 3-lobed, *bract-like, or leafy *involucre on one side. Hornbeams are smooth-barked trees, with *simple leaves, at one time much *coppiced in Europe. The wood was also used formerly to make wooden gear wheels, on account of its toughness. There are about 35 species, occurring in the northern temperate zone, and in Central America.

**Carpobrotus** (Hottentot fig) See AIZOACEAE.

**carpogonium** A female sex organ (*gametangium) in *seaweeds belonging to the class *Rhodophyceae. Usually it consists of a swollen basal region, and a narrow, elongated, gelatinous tip which receives the male *gamete.

**carpospore** The *spore produced by the *carpogonium of *Rhodophyta.

**carposporophyte** The *sporophyte (see ALTERNATION OF GENERATIONS) of *Rhodophyta, which becomes part of a structure known as the *sporocarp (or cystocarp), comprising the *haploid *pericarp and *diploid carposporophyte; on maturity the sporocarp produces *carpospores.

**carr** A locally variable term of Scandinavian origin, used to describe forested, wet, *rheotrophic habitats, usually with some *peat development but with neutral (not extremely acid) waters and good nutrient status. In East Anglia, England, they are typically alder woods (*Alnus glutinosa*), but where nutrient levels are less favourable willow (*Salix* species), especially *S. cinerea* (grey willow), may be the principal tree species. See also FEN; compare BOG.

**carrageenan** See CARRAGHEEN.

**carragheen** (carrageenan) A complex *polysaccharide obtained from certain red seaweeds (*Rhodophyceae), especially from *Chondrus crispus* (Irish moss) and *Gigartina* species. Carragheen has a number of commercial uses as a stabilizer in paints, pharmaceutical products, and various types of food such as sauces and creams. (The term 'carragheen', but not 'carrageenan', may also be used to refer to the seaweed *Chondrus crispus* itself.)

**carrot**   See *DAUCUS*.

**carrying capacity**   The maximum population of a given organism that a particular environment can sustain; the *K (saturation) value for species populations showing *S-shaped population growth curves. It implies a continuing yield without environmental damage. It may be modified by human intervention to improve environmental potential, e.g. by applying fertilizers to range-land and reseeding it with nutritious grasses.

**caruncle**   A reduced *aril, in the form of a fleshy, often waxy or oily, outgrowth near the *hilum of some seeds. Usually it is brightly coloured. It acts as an aid to dispersal. *Viola seeds have an oily caruncle and are sought and dispersed by ants.

**Carya**   (hickory; family *Juglandaceae) A genus of trees in which the leaves are *pinnate and the wood is tough and springy. The fruit is a *drupe with an edible seed (pecan or hognut). There are 17 species, found in eastern Asia and N. America.

**caryogamy**   See KARYOGAMY.

**Caryophyllaceae**   A family, mostly of *herbs, in which the leaves are *simple, *opposite, normally entire, and *exstipulate. The flowers are usually borne in terminal, forked *cymes (dichasia) with a central flower in each fork; they are regular, with usually 4 or 5 free *petals, 4 or 5 *stamens, or twice as many stamens as petals, and a *superior *ovary which is *syncarpous, 1-celled with many *ovules on a free central *placenta, and has 2 to 5 *styles. The fruit is usually a *capsule. Some genera (*Dianthus, Saponaria*, etc.) are much cultivated for the flowers. There are 89 genera, with about 2070 species, most of them found in the northern temperate region.

**caryopsis**   A dry, nut-like *fruit typical of grasses, e.g. a cereal grain. It is an *achene with the *ovary wall united with the seed coat.

**Caryota**   (fishtail palms; family *Arecaceae) A genus of palms which are small to 30 m tall, and solitary or clump-forming; they are unique among palms in having leaves that are twice-*pinnate, with the leaflets fishtail-shaped, and pinnately nerved. The inflorescences develop *basipetally, the female flower flanked by 2 males. The fruits have an irritant, sappy *pericarp, and contain 1 or 2 seeds. There are 12 species, native in tropical Asia to Queensland, and in the Solomon Islands.

**Casearia**   (family *Flacourtiaceae) A genus of 180 species of small to medium-sized tropical trees in which the leaves have translucent points or lines.

**cashew**   See *ANACARDIUM*.

**Casparian strip**   A band of waterproof, corky tissue that is found on the side and walls of the *endodermis of roots. The strip prevents water from entering the *pericycle except through the *cytoplasm of endodermal cells; this may be important in producing *root pressure.

**cassava**   See *MANIHOT*.

**Cassia**   (family *Fabaceae, subfamily *Caesalpinoideae) A genus of small to medium-sized trees in which the leaves are *pinnate. The flowers are slightly *zygomorphic and showy. Several species are widely grown as ornamentals. The purgative senna is prepared from the leaves. There are 535 species, occurring in tropical to warm temperate regions.

**Cassytha**   An aberrant genus of *Lauraceae that comprises 16 species of slender, woody, parasitic twiners of the Old World tropics, which attach to host stems by haustoria (*see* HAUSTORIUM).

**Castanea**   (family *Fagaceae) A genus of trees that yield useful timber and fruit (*C. sativa* is the Spanish or sweet chestnut), and which are also suitable for tanning. The fruits are entirely enclosed in a spiny *cupule. There are 12 species, occurring in northern temperate regions.

**Castanopsis**   (family *Fagaceae) A genus of medium-sized trees in which the fruits are entirely enclosed in a warty or spiny *cupule. A few have edible fruits. There are 110 species occurring in tropical and subtropical Asia, and 2 species in south-western N. America.

***Castanospermum*** (family *Fabaceae, subfamily *Caesalpinoideae*) A *monotypic genus, *C. australe*, of trees that occurs in Australia and New Caledonia. It has edible seeds (Moreton Bay chestnut) and fine timber (black bean).

**castor oil** See RICINUS.

***Casuarina*** (family *Casuarinaceae) A genus of *xeromorphic trees in which the twigs are slender, cylindrical, green, and grooved with *whorls of minute scale leaves at the *nodes. The flowers are tiny, borne in *compound *spikes, and become woody in the fruit, like small cones. They are wind-pollinated. The roots have nitrogen-fixing nodules. The genus may perhaps be related to the *Hamamelidaceae. There are 4 genera and about 70 species, occurring from the Mascarenes to Polynesia, with most species in Australia.

**catabolism** The part of cellular metabolism that encompasses the reactions that yield energy through the degradation of food molecules.

***Catalpa*** (family *Bignoniaceae) A genus of small trees, several of which are cultivated for their showy flowers (e.g. *C. bignonioides*, Indian bean). There are 11 species, occurring in eastern Asia and southern N. America.

**catalysis** The acceleration of a chemical or biochemical reaction that is brought about by the action of a catalyst. The catalyst, which is not affected by the overall reaction, acts by either lowering the activation energy of the reaction through a reorientation of molecules in collision, or by facilitating an alternative mechanism that has a different activation energy. *Enzymes are naturally occurring catalysts which are universally present in all living cells. Enzymes are generally much more efficient and more specific than other catalysts.

**cataphyll (bud scale)** A small, scale-like leaf, often containing resin, that covers a dormant bud on a deciduous plant. Cataphylls protect buds, especially from desiccation.

**catastrophic evolution (catastrophic speciation)** A theory proposing that environmental stress might lead to the sudden rearrangement of *chromosomes, which in self-fertilizing organisms may then give rise *sympatrically to a new species. Recent research suggests that at best this explanation applies to some special cases.

**catastrophism** A theory that associates past geological change with sudden, catastrophic happenings. Early geologists, including Buckland, Cuvier, and Sedgwick, claimed that catastrophism was a sound scientific theory. Although it met with considerable scorn in more recent times, many modern geologists would describe themselves as 'neocatastrophists'. The collision of a bolide with the Earth at the Cretaceous/Tertiary boundary (65 Ma) is an accepted example of a singular catastrophic event that has had considerable influence on the Earth's history.

**catena** A topographic sequence of soils, of the same age and usually on the same parent material, that is repeated across larger landscape transects. Individual soil-*profile types are related to site conditions and to position on a slope. The term was introduced in E. Africa in the 1930s, and is mainly applicable in certain non-glaciated landscapes, particularly those with small, hilly relief (e.g. *loess areas).

***Catharanthus*** (family *Apocynaceae) A genus that comprises 8 species of tropical herbs, 7 of them endemic (*see* ENDEMISM) to Madagascar. *C. roseus*, a showy ornamental and often a weed, produces important medicinal alkaloids.

**Catherine's moss** The common name for the moss *Atrichum undulatum* (order *Polytrichales). The shoots are erect, up to 10 cm tall. The leaves are dark green, long and narrow, with characteristic transverse undulations, and toothed margins. The *capsule is cylindrical, and curved, with a long beak on the lid. The moss is found on the ground in woods, also on *heaths and waste ground.

**cation** An *ion that carries a positive electrical charge (e.g. the metallic element of salt compounds). A cation can combine with certain *anions (which have negative charges).

**cation exchange** A process in which cations in solution are exchanged with cations held on the exchange sites of mineral and organic matter, particularly on the surfaces of *colloids of clay and humus.

**cation exchange capacity (CEC)** The total amount of exchangeable cations that a particular material or soil can adsorb at a given pH. Exchangeable cations are held mainly on the surface of colloids of clay and humus, and are measured in moles per gram of material or soil.

**catkin** A pendulous *spike, usually of *simple, *unisexual flowers.

*Catkin*

**cat's-tail** **1.** *See* ACALYPHA. **2.** *See* TYPHACEAE.

**caulescent** In the process of becoming stalked.

**caulid** The main 'stem' of a bryophyte.

**cauliflorous** Borne on the trunk; applied especially to tropical trees in which flower shoots grow from the main trunk.

**cauliflower fungus** *See* SPARASSIS.

**cauline** Pertaining to the stem.

**Caulobacter** A genus of *Gram-negative *bacteria not assigned to any taxonomic family. The cells undergo a life cycle that is unusual in bacteria. Mature cells are straight or curved rods with a stalk-like extension at one end. The cell is attached to rocks, plants, other cells, etc. by means of an adhesive holdfast at the tip of the stalk. Asymmetric division of a mature cell leads to the formation of a *motile (flagellated) immature cell; this subsequently attaches to a substrate, loses its *flagellum, and develops a stalk, thus becoming a mature cell. These bacteria are *aerobic *chemo-organotrophs. They are found, for example, in soil and in *oligotrophic fresh water.

**cayenne pepper** *See* CAPSICUM.

**ceará rubber** *See* MANIHOT.

**CEC** *See* CATION EXCHANGE CAPACITY.

**cecidium (pl. cecidia)** A plant *gall.

**cecidization** The formation of a plant *gall (especially by gall midges of the family Cecidomyidae).

***Cecropia*** **(family *Urticaceae)** A genus of fast-growing, small trees in which the wood is very light, and the young stems hollow. Many have symbiotic ants. There are about 20 species, found in secondary forest in tropical America.

**cedar apple** A gall formed on juniper (*Juniperus*) trees infected with the *rust fungus *Gymnosporangium juniperi-virginianae*.

**cedars** **1.** *See* CEDRUS. **2.** **(Australia cedar, Burma cedar)** *See* TOONA. **3.** **(pencil cedar)** *See* JUNIPERUS. **4.** **(incense cedar)** *See* LIBOCEDRUS. **5.** **(Japanese cedar)** *See* CRYPTOMERIA.

***Cedrela*** **(family *Meliaceae)** A genus of deciduous trees in which the leaves are *pinnate. The flowers have a columnar *disc to which the *petals and *stamens are *adnate. The fruit is a small *capsule. Some yield valuable timber (especially *C. odorata*, used for cigar boxes), closely related to *Toona*. There are 8 species, occurring from Mexico and the Caribbean region to Argentina.

***Cedrus*** **(cedars; family *Pinaceae)** A genus of elegant, evergreen conifers, which are large, spreading trees with valuable timber, and often planted for ornament. The needles are clustered on short shoots. There are four species: *C. libani*, the cedar of Lebanon; *C. atlantica*, Atlantic (or Atlas) cedar; *C. brevifolia*, the Cyprus cedar; and *C. deodara*, the deodar of the Himalayas.

**Ceiba** (family *Bombacaceae*) A genus of trees whose seeds are enveloped in silky hairs. *C. pentandra* (kapok), naturalized in W. Africa and planted widely in Asia, is one of the tallest trees of the Amazon rain forests, attaining 70 m. There are 4 species, occurring in tropical America.

**Celastraceae** A family of trees and shrubs in which the leaves are *simple, and *opposite or alternate. The flowers are small, regular, and *tri- or *pentamerous, usually with a *disc, and the *ovary is *superior. The fruits take various forms. The seed commonly has a vivid *aril. *Euonymus europaeus* is the spindle tree or burning bush of Britain. There are 94 genera, with about 1300 species, with a cosmopolitan distribution, except for the Arctic and Antarctic.

**Celastrophyllum circinerve** One of the earliest flowering plants known to palaeobotanists, belonging to the family *Celastraceae which pre-dates most others. The family began with *C. circinerve* in the Early *Cretaceous and survives today in the form of *Euonymus europaeus*, the spindle tree, or burning bush.

**celeriac** See *APIUM*.

**celery** See *APIUM*.

**celery pine** See *PHYLLOCLADUS*.

**cell** The fundamental autonomous unit of plant and animal bodies, consisting of, at least, a *cell membrane containing *cytoplasm and nuclear material, but often having a more complex structure. Simple organisms are unicellular, but more complex organisms consist of many co-operating cells. Characteristically, in plants, but not in animals, cells are surrounded by a *polysaccharide wall.

**cellar fungus** The common name for *Coniophora puteana* (*C. cerebella*), which is a frequent cause of timber decay in damp buildings.

**cell culture** A mass of *cells that have been derived either from a single cell or from a small group of cells from the same tissue or organ, and that is maintained *in vitro* using solid or liquid nutrient media.

**cell cycle** The sequence of events that occurs between the formation of a *cell and its division into daughter cells.

**cell differentiation** The process by which descendants of a single *cell produce structural and functional specializations and maintain these during the course of the life of that individual. Specializations arise during early development of the individual by differential *DNA–*RNA transcription; within a species, particular cells differentiate into particular forms, each with their own functions.

**cell fusion** The experimental fusion of nuclei and *cytoplasm from different somatic *cells to form a single hybrid cell. Cells used for fusion often come from tissue cultures derived from different species; fusion is facilitated by the modification of the surface of cells by adsorption of certain viruses (e.g. Sendai virus).

**cell growth** An irreversible increase in the size of a *cell; it can be caused by a change in the *osmotic potential within the cell, or by a reduction of the pressure exerted by the *cell wall.

**cell line** A lineage (or pedigree) of *cells that are related through asexual division. A cell line may be derived in the laboratory from a primary *cell culture.

**cell membrane (plasmalemma, plasma membrane)** The sheet-like membrane, 7.5–10 nm thick, that encloses and delimits the contents of a *cell. It is a living structure consisting of *lipid molecules forming a fluid bi-layer and associated *protein. Water molecules can pass freely through this structure, but the passage of most other molecules can be controlled (i.e. it is selectively permeable). The cell membrane may also have other functions, as in *prokaryotic cells, where it is associated with oxidative metabolism and cell division.

**cell plate** The partition formed between daughter *cells during *cytokinesis in plant cells. It is a membrane pieced together from fragments which line up along the mid-plane of the *spindle and then fuse from the centre outwards. Subsequently this forms the framework for new *cell walls.

**cell sap** The contents of the large *vacuole that is often found in plant *cells.

**cell theory** The basic theory (proposed by M. K. *Schleiden and T. *Schwann in 1838) that all animals and plants are made up of *cells, and that growth and reproduction are due to division of cells.

**cellular slime moulds** Slime moulds in which the feeding stage consists of independent, uninucleate, amoeboid cells (*myxamoebae) that generally aggregate to form a *pseudoplasmodium prior to fruiting.

**cellulase** A highly specific, endo-glucosidase *enzyme that catalyses the *hydrolysis of *cellulose into *glucose by attacking beta-1,4 linkages.

**cellulolytic** Able to break down or digest *cellulose.

**cellulose** A straight-chain, insoluble *polysaccharide that is composed of *glucose molecules linked by beta-1,4 glycosidic (see GLYCOSIDE) bonds. It is the principal structural material of plants, and as such is the most abundant organic compound in the world. It has also been found in certain sea squirts.

**cellulytic** Able to break open (lyse) cells.

**cell wall  1. (of eukaryote)** A strong, often rigid, extraprotoplasmic layer in plant cells, whose growth is directed from within the cell. In addition to water, which may comprise by weight up to 70 per cent of the wall, it consists of a variety of *polysaccharides, notably *celluloses, *hemicelluloses, and *pectins, as well as variable (but smaller) amounts of *proteins, *lipids, *lignins, *tannins, and even mineral salts. Cell walls provide a skeletal support to the whole plant and also a barrier against injury and infection. **2. (of prokaryote)** A rigid wall structure that lies inside the capsular layer, but outside the plasma membrane of cells. In

*cyanobacteria it is composed primarily of *cellulose, but in bacteria it is composed of a mixture of materials not found elsewhere. Gram-positive (see GRAM-REACTION) bacteria have relatively thick walls (15–80 nm) composed principally of peptidogly-cans (40–90 per cent), techoic acids, and other complex polysaccharides, with little sign of differentiation into separate layers. Gram-negative bacteria have much thinner walls (10 nm) composed of several distinct layers which are chemically different from those of Gram-positive species. Peptidoglycans represent only 1–10 per cent of the wall material by weight; techoic acids are absent, but complex lipopolysaccharides are present.

***Celosia    cristata*** **(cockscomb)**  *See* AMARANTHACEAE.

***Celtis*** **(family *Ulmaceae)** A genus of trees in which the fruit is a *drupe. They yield useful timber, and in India their foliage is used for browse. There are 60 species, occurring mainly in the northern hemisphere, and extending to the Old World tropics.

**cemented** Applied to massive, infilled, and indurated *mineral soil: such soil has a hard and often brittle consistency because soil particles are joined together by cementing substances, e.g. calcium carbonate, silica, iron and aluminium oxide, or *humus. Cemented soil usually appears as a highly distinctive and resistant *horizon.

***Cenchrus*** **(family *Poaceae)** A genus of 22 species of grasses, medium in height, variable in habit, and with *culms that are not tuberous (see TUBER). They are widespread in the tropics and subtropics.

**Cenomanian** One of the stages of the Upper *Cretaceous Period in Europe, preceded by the *Albian, followed by the Turonian, and dated at about 99.6–93.5 Ma ago.

**Cenozoic (Cainozoic, Coenozoic, Kainozoic)** An era of geologic time that began about 65.5 Ma ago and continues to the present day. It includes the *Palaeogene, *Neogene, and *Pleistogene Periods; these are commonly grouped into the *Tertiary and *Quaternary Sub-eras, although

Tertiary and Quaternary are being abandoned as formal names. Molluscs and microfossils are used in the stratigraphic subdivision of the era. The Alpine orogeny (episode of mountain-building) reached its climax during the Cenozoic Era.

**Central Australian floral region** Part of R. Good's (*The Geography of the Flowering Plants*, 1974) Australian kingdom, which accounts for most of the Australian continent, including all the central parts. The flora is imperfectly known, although the great majority of it is likely to be endemic (*see* ENDEMISM). Ecologically it coincides with extensive *thorn forest, with much *Acacia aneura* (mulga), *A. harpophylla* (brigalow), and *Eucalyptus hemiphloia* (mallee). The flora is poor, partly because of the extensive deserts and semi-deserts that account for most of the region, and partly because central Australia was submerged for much of the *Tertiary Period. *See also* FLORAL PROVINCE and FLORISTIC REGION.

**central limit theorem** The theorem stating that the arithmetic mean values for a series of similar-sized, fairly large samples ($n > 30$) taken from a large population will be approximately normally distributed about the true population mean ($n > 30$), irrespective of the actual distribution pattern of the individual counts.

***Centranthus*** (spurred valerian) *See* VALERIANACEAE.

**centre of diversity** (gene centre) A geographical location or local region (often in the tropics) where a particular *taxon exhibits greater genetic diversity than it does anywhere else. Many authorities believe that centres of diversity are also the *centres of origin of the taxa concerned.

**centre of origin** A geographical location or local region (often in the tropics) where a particular group of organisms is believed to have originated. Many authorities believe centres of origin are also *centres of diversity. For example, there are 6 genera of Arecaceae (palms) in the lower part of the Amazon basin, twice the number found anywhere else. It is difficult, however, to reconcile this explanation of centres of origin with the concept of *allopatric *speciation.

**centric diatom** A diatom (*Bacillariophyta) that has radial symmetry. Most centric diatoms are marine. *Compare* PENNATE DIATOM.

**centric fusion** The whole-arm fusion of *chromosomes by the joining together of two *acentric chromosomes to form one *metacentric chromosome.

**centrifugal** Developing from the centre outwards, so the structures become younger with increasing distance from the centre. *Compare* CENTRIPETAL.

**centrifuge** An apparatus that is used for the separation of substances by the application of centrifugal force: this is generated by high-speed rotation of a vessel containing a fluid in which the substances are suspended but not dissolved. Separation occurs because different substances have different rates of sedimentation according to their molecular size and shape.

**centriole** In plants, a cylindrical *organelle occurring in flagellated or ciliated cells, where it acts as a precursor to the *basal body of each *flagellum or *cilium. Centrioles are absent from higher plants.

**centripetal** Developing from the outside towards the centre, so the structures become older with increasing distance from the centre. *Compare* CENTRIFUGAL.

**Centrolepidaceae** A family of tufted or cushion-like herbs (*see* CUSHION PLANT) that have bristle-like leaves, similar to grasses, rushes, or mosses. They may be *annual or *perennial. The inflorescence is terminal and in a *spike, with one or more *pseudanthia consisting of 1 or 2 male flowers and 2 or more female flowers. The female flowers have a *unilocular *ovary, 1 *style, and 1 *ovule. The fruit has a membranous *pericarp and dehisces by a longitudinal split. There is 1 seed in the locule; it has a large *endosperm. The most closely related family is the *Restionaceae. There are 3 genera, with 28 species, found in South-east Asia, the Pacific, and temperate S. America.

**centromere (spindle attachment)** The region of attachment on a *chromosome containing the *kinetochore that links its two halves, the *chromatids, to the *spindle at *mitosis or *meiosis. Its position determines its shape and *anaphase (a rod, or J, or V). In a few species the centromeric properties are distributed along the entire length of the chromosome; such species are said to be polycentromic (possessing a diffuse centromere).

**centrosome** In a *cell, the distinct part of the *cytoplasm that organizes the assembly and disassembly of *microtubules and which contains the *centriole. It occurs in only a few lower plants.

**century plant** *See* AGAVE.

**cep (cèpe, penny bun)** The common name for the *fruit body of *Boletus edulis*, an edible and highly esteemed *bolete often marketed in parts of Europe. It is used commercially as an ingredient of dried *mushroom soup.

**cèpe** *See* CEP.

**Cephaleuros (division *Chlorophyta)** A genus of epiphytic (*see* EPIPHYTE) and/or parasitic algae which grow beneath the leaf *cuticles of a host plant. A *Cephaleuros* species is responsible for the disease 'red rust' of tea and other plants.

**cephalin** A group of *phospholipid compounds, which includes phosphatidyl ethanolamine and phosphatidyl serine. Both are major phosphoglycerides in animals and higher plants, in which they are important constituents of membranes.

**cephalodium** A specialized region in a lichen *thallus in which occur *cyanobacteria (blue-green algae) that are capable of carrying out *nitrogen fixation. Lichens that have cephalodia have a green *phycobiont (i.e. a member of the *Chlorophyta) in the remainder of the thallus.

**Cephalosporium** *See* ACREMONIUM.

**Cephalotaxaceae (cow's tail pines)** A family of *gymnosperms, close to the *Taxaceae, that comprises small trees

with narrow, *pinnately arranged, yew-like foliage. They are *dioecious, like the yews. The male cones are small, and globular to oval. The female flowers usually occur as 2 pairs, or in bunches of 3–5, on curved stalks at the shoot bases, and produce olive-like 'fruits' which are erect and which have a fleshy *aril enclosing the seed (except at its tip). They are cultivated as ornamental trees. There is 1 genus (*Cephalotaxus*), with 4 species, which occur in South-east Asia.

**Cephalotus (family Cephalotaceae)** A genus consisting of 1 species, *C. follicularis*, which is native to the peat bogs of W. Australia. It is a *perennial *herb with a woody *rhizome and its leaves form a *rosette. Many of the leaves are modified as insect-catching, lidded pitchers (resembling *Nepenthes*). The flowers are small, their parts in sixes, apetalous, and borne on a long *scape. Perhaps they are related to *Saxifragaceae.

**Cephalozia (order *Jungermanniales)** A genus of leafy liverworts. *C. bicuspidata* is a very common species. It is very small, with thread-like stems bearing two ranks of pale green leaves; each leaf is deeply divided into two pointed lobes. *Flagella may be present. It is found in wet acidic habitats, including banks, woodland paths, *Sphagnum* bogs, etc.

**Ceramiales (division *Rhodophyta)** A large order of red seaweeds, most of which have a delicate membranous or filamentous *thallus. The order includes more than 250 genera, and is distributed world-wide.

**Ceratiomyxa (class *Ceratiomyxomycetes)** A genus of *slime moulds; in *C. fruticulosa* the *fruiting bodies are variable, but typically they are branched white columns 1–10 mm tall. *C. fruticulosa* is common and very widely distributed, being found on bark, rotting wood, etc., from Arctic to tropical regions. Two other species of *Ceratiomyxa* are found in tropical and subtropical regions.

**Ceratiomyxomyces (division *Myxomycota)** A class of *slime moulds in which *spores are borne externally on the surface of column-like fruiting structures.

**Ceratocystis** *See* OPHIOSTOMA.

**Ceratodon** (order *Dicranales) A genus of mosses in which the leaves are ovate to narrowly lanceolate with recurved margins. *C. purpureus* is very common in Britain. It grows on bare ground, forming extensive colonies which are particularly conspicuous in spring owing to the reddish-purple colour of the abundant *capsule stalks (*setae). It is found on *heaths, *moors, roadside banks, etc., and on wall tops. It is a cosmopolitan genus of about 22 species.

**Ceratolobus** (family *Arecaceae) A genus of climbing palms (rattans) in which the stems are clustered and the leaves have a spiny cirrus (whip-like tip), some of them with fishtail-shaped leaflets. They are *dioecious. The inflorescence is entirely enclosed in a single, flattened *bract with a slit-like opening. The fruits are scaly, and contain 1 seed with a thin fleshy *testa. The cane is not used. There are 6 species, occurring in Sumatra, Malaya, Java, and Borneo.

**Ceratophyllaceae** (hornworts) A family of submerged, aquatic *herbs, which have leaves in *whorls of narrow, forked segments. The flowers are small, unisexual, sessile, and solitary in the leaf *axils, with a tiny *perianth of many narrow lobes. The male flowers have 10–20 *stamens, the females a 1-celled *ovary forming a 1-seeded *nut. There is 1 genus (*Ceratophyllum*) and 2 species, cosmopolitan in distribution except for the Arctic.

**Cercidiphyllaceae** (order Hamamelidales) A *monotypic family, *Cercidiphyllum japonicum* (katsura tree), a large, deciduous tree, closely related to the *Magnoliaceae. The leaves are *simple, *opposite or alternate, with *stipules. The flowers are unisexual, with 4 *perianth parts, and the plants are *dioecious. The fruits are *follicles. Over 30 m tall, and an important source of fine-grained timber, it produces striking yellow and orange autumn foliage. The family is native to Japan and China.

**Cercidiphyllum japonicum** (katsura tree) See CERCIDIPHYLLACEAE.

**cerebriform** Resembling a brain; convoluted.

**Ceroxylon** (family *Arecaceae) A genus of palms in which the stem is solitary. They are the tallest of all palms, sometimes reaching 60 m, and some have waxy stems that shine white. The leaves are *pinnate with an erect *sheath; the inflorescences form among the leaves. The wax is of commercial value. They provide the official emblem of Colombia. There are 15 species, found in the Andes north from Lake Titicaca, to altitudes of more than 4000 m.

**cerradão** Areas of semi-evergreen woodland, in places with an almost closed *canopy, that occur in the *campos savannah grasslands of Brazil.

**Cetraria** (order *Lecanorales) A genus of *lichens in which the *thallus is *foliose to more or less *fruticose with erect, flattened branches. *Rhizines may be present. *Apothecia are *lecanorine and occur at or near the edges of the thallus in foliose species. The *spores are colourless and *aseptate. They are found on the ground, on trees, on rocks, etc., according to species. See also ICELAND MOSS.

**Chaenomeles** (japonica; family *Rosaceae) A genus of 3 species of small trees, close to quince (*Cydonia*) and to apple (*Malus*) but with many seeds per fruit cell, flowers (commonly scarlet) in clusters, thorns on twigs, and globose, yellow, downy fruits. There is 1 species in cultivation, *C. speciosa*, from China.

**chalaza** The base of an *ovule, bearing an embryo sac surrounded by *integuments.

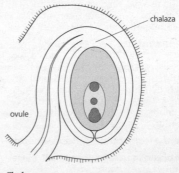

*Chalaza*

**chalybeate** Of natural waters, containing iron. (Chalybite is a synonym of siderite, an iron mineral, $FeCO_3$.)

**Chamaecyparis** (family *Cupressaceae) A genus of coniferous trees with *appressed, scale-like leaves in *opposite pairs, foliage in flattened, *dorsiventral sprays, and small cones. They are much cultivated and hybridized as ornamental trees and hedging shrubs. There are 7 species, which occur in N. America, Japan, and Taiwan.

**Chamaedorea** (family *Arecaceae) A genus of palms most of which are small, delicate, shade-loving palms of forest undergrowth. Many are clump-forming. The stems are slender, hoop- marked, and erect, or rarely climbing. The leaves are small, *pinnate, or *simple. The plants are *dioecious. Several are in cultivation as house plants. There are about 110 species, occurring in Central and S. America, and around the Caribbean.

**chamaephyte** One of *Raunkiaer's life-form categories, being a plant in which the perennating bud or shoot apices are borne very close to the ground. Four subcategories are recognized: (*a*) suffruticose chamaephyte, in which the aerial shoots die back partially at the onset of unfavourable conditions and buds arise on the lower, persistent stem portions; (*b*) passive chamaephyte, in which the aerial stems fall over as they die back, to give buds on horizontal axes near the ground; (*c*) active chamaephyte, which also produces buds on horizontal stems, but as the more normal growth form of the plant; and (*d*) cushion chamaephyte, which is essentially a compacted suffruticose chamaephyte. Compare CRYPTOPHYTE; HEMICRYPTOPHYTE; PHANEROPHYTE; and THEROPHYTE.

**Chamaerops** (family *Arecaceae) A *monotypic genus, *C. humilis*, one of Europe's two native palms, native to the western Mediterranean. It has a short, stout trunk, and may be solitary or growing in clumps. It has fan-shaped leaves with stiff leaflets. The inflorescences form among leaves. The flowers are hermaphrodite.

**Chamaesiphon** A genus of unicellular *cyanobacteria (section I) in which the ovoid cells reproduce by repeated budding.

**Chamaesiphonales** An outdated phycological order of 'blue-green algae' which corresponds roughly with section II of the *cyanobacteria (except for *Chamaesiphon*, which is included in section I).

**chañaral** A type of thorny, scrub vegetation, in which chañar (*Gourliaea decorticans*) is common, that occurs in southern-central parts of S. America.

**channelled wrack** The common name for the brown seaweed *Pelvetia canaliculata*. The *thallus is flattened and branched, the branches lacking a *midrib, and having inrolled margins which form moisture-retaining channels. This plant can grow higher on the shore than can any other seaweed: at and above high-water mark. It is used sometimes as fodder for sheep and cattle.

**chanterelle** The common name for the edible fruit body of *Cantharellus cibarius*. The cap is funnel-shaped, and 3–10 cm across; the lower surface bears shallow, branching, *gill-like ridges which extend down the short *stipe. The whole fruit body is a bright egg-yellow and has a distinctive odour resembling that of apricots. It is common on the ground in summer and autumn, especially under beech or oak trees, and occurs throughout Europe and N. America.

**chaparral** The *sclerophyllous scrub of west California and adjacent regions. Like other sclerophyllous scrub in regions with Mediterranean-type climates, much of the chaparral has been derived by a combination of burning and other factors from an earlier forest cover.

**Chara** An *alga, known from the *Triassic to the present day, in which the female *oogonia are protected by an envelope of calcium carbonate.

**character** Any detectable attribute or property of the *phenotype of an organism. Defined heritable differences in the character may exist between individuals within a species.

**characteristic species** See KENNARTEN SPECIES.

**charophytes** A group of plants the members of which are usually regarded as *algae, although they resemble *bryophytes in the structure of the male *gametes and in the presence of a sterile envelope enclosing the sex organs. Many species are heavily calcified. Fossil charophytes (from the *Triassic onwards) are known only by their calcified *oogonia, which often have a spiral ornament. Charophytes are aquatic, occurring primarily in fresh water, although some species are found in brackish water. Charophytes are sometimes included in a class within the division *Chlorophyta and sometimes as a distinct division (Charophyta). Genera include *Chara* and *Nitella*. See also STONEWORT.

**chase** 1. In Britain, a royal forest that has passed into private ownership. 2. A lane between two woods.

**Chattian** The final stage of the *Oligocene Epoch, preceded by the Rupelian, and dated, at about 28.4–23.03 Ma ago, by planktonic foraminifera and magnetic polarity in sea-floor rocks.

**chaulmoogra oil** See HYDNOCARPUS.

**cheese plant** See MONSTERA.

**chelate** 1. Pincer-like. 2. A ring structure formed as a result of the reaction of a metal ion with two or more groups on a *ligand.

**chelation** An equilibrium reaction between a metallic ion and an organic molecule in which more than one bond links the two components. The metallic ion is termed the complexing agent, the chelating organic molecule the *ligand. Chelation is a naturally occurring mechanism in soils, useful since it removes heavy-metal ions that are in solution in simple inorganic form where they may be directly toxic to plants or may interfere with the uptake of essential nutrients. Heavy-metal toxicity in waste land will tend to be reduced by the application of organic material.

**chemical oxygen demand (COD)** An indicator of water or effluent quality, which measures oxygen demand by chemical (as distinct from biological) means, using potassium dichromate as the oxidizing agent. Oxidation takes 2 hours and the method is thus much quicker than a 5-day *BOD assessment. Since the BOD:COD ratio is fairly constant for a given effluent, COD is used more frequently for routine monitoring of an effluent once this ratio has been determined.

**chemiosmotic theory** A theory concerning *oxidative phosphorylation in which it is proposed that the *electron-transport chain is arranged such that it generates an energy-rich proton gradient across the inner membrane of a *mitochondrion. This energy is then used to drive the phosphorylation of *ADP through a membrane-bound ATP-ase.

**chemo-autotroph** See CHEMOSYNTHETIC AUTOTROPH.

**chemo-heterotroph** A chemotrophic organism that obtains its carbon chiefly or solely from organic compounds.

**chemo-lithotroph** A chemotrophic organism that obtains its energy from the oxidation of inorganic compounds or elements.

**chemo-organotroph** A chemotrophic organism that obtains its energy from the metabolism of organic compounds.

**chemosynthetic autotroph (chemo-autotroph) 1.** An *autotroph that is capable of synthesizing complex organic materials from inorganic reactions (e.g. iron oxidation). 2. A *chemotroph that uses carbon dioxide as its main or sole source of carbon.

**chemotaxis** In a *motile organism or cell, a change in the direction of locomotion, made in response to a change in the concentrations of particular chemicals in its environment.

**chemotroph** An organism that obtains its energy by a reaction pathway that is independent of light.

**chengal** See NEOBALANOCARPUS.

**Chenopodiaceae (goosefoots)** A family, mostly of *herbs but including some shrubs, of plants that are frequently succulent, or covered with mealy hairs. The leaves are mostly *simple and alternate. The flowers are inconspicuous, borne in dense clusters, bisexual or unisexual, *tri- to *pentamerous, and have a *superior *ovary of 1 cell. The fruits are 1-seeded *nuts. *Beta* (beetroot, sugar beet) and *Spinachia* (spinach) are important vegetables or sources of sugar. There are about 120 genera, with 1300 species, cosmopolitan in distribution but occurring mainly in arid regions or saline habitats, and also as weeds in enriched cultivated or waste ground.

**cherimoya** *See ANNONA.*

**chernozem (black earth)** A freely draining soil *profile whose name is the Russian word for 'black earth'. Chernozems are associated with grassland vegetation in temperate climates, and identified by the deep and even distribution of *humus and of exchangeable *cations (calcium and magnesium) through the profile (now included in *Mollisols). Because of their richness in plant nutrients and their excellent *crumb structure, chernozems are among the most agriculturally productive soils in the world. Chernozems are a reference soil group in the FAO *soil classification.

**cherry** *See PRUNUS.*

**chervil** *See ANTHRISCUS.*

**chestnut 1. (Castanea)** *See FAGACEAE.* **2. (horse-chestnut)** *See HIPPOCASTANACEAE.*

**chestnut blight (chestnut canker)** A disease of the American chestnut (*Castanea dentata*) caused by the *fungus *Endothia parasitica*. Invasion of the vascular *cambium by the *pathogen results in wilting and the death of the tree. The introduction of the fungus to America from Asia led to the wholesale destruction of the hitherto commercially important American chestnut.

**chestnut canker** *See CHESTNUT BLIGHT.*

**chewing gum** *See DYERA.*

**chiasma (pl. chiasmata)** In genetics, a cross-shaped structure forming the points of contact between non-sister chromatids of *homologous *chromosomes, first seen

in the *tetrads of the *diplotene stage of meiotic *prophase I. Chiasmata are thus the visible expression of *crossing-over of *genes. There are usually one or more chiasmata per chromosome per *meiosis. *See also* BREAKAGE and REUNION.

**chiasma interference** The nonrandom frequency of more than one *chiasma in a *bivalent segment during *meiosis. If the frequency of occurrence is higher than that expected from purely chance events, then it is termed negative chiasma interference; if the frequency is lower than expected, it is referred to as positive chiasma interference.

**chiasmata** *See CHIASMA.*

**chickrassy** *See CHUKRASIA.*

**chicle** *See ACHRAS.*

**chicory** *See CICHORIUM.*

**chiku** *See ACHRAS.*

**Chilean firebush** *See EMBOTHRIUM.*

**Chilean wine palm** *See JUBAEA.*

**Chile pine** *See DISJUNCT DISTRIBUTION.*

**chillies** *See CAPSICUM.*

**chilling-sensitive plant** A plant that is badly injured or killed by temperatures above freezing point, up to about 20°C.

**chimaera** *See CHIMERA.*

**chimera (chimaera)** A tissue containing two or more genetically distinct cell types, or an individual composed of such tissues. It arises as a result either of: (*a*) *mutation or abnormal distribution of *chromosomes, affecting a particular cell during development and hence all its descendants; or (*b*) in the case of plants, by the artificial grafting together of two individuals with different *genotypes, to produce a mixing of characters. It is also referred to as a mosaic.

**China jute** *See ABUTILON.*

**Chinese lacquer** *See RHUS.*

**Chinese swamp cypress** *See GLYPTO-STROBUS.*

**Chinese water chestnut** *See ELEO-CHARIS.*

**chiropterophily** Pollination by bats (Chiroptera).

***Chisocheton*** (family *Meliaceae) A genus of rain-forest trees or treelets in which the leaves are remarkable, being *pinnate with an intermittently growing terminal bud. The fruit is a globose, usually velvety, thick-walled *capsule, and the seeds are large, with a fleshy jacket. There are about 51 species, occurring from Indo-Malesia to Vanuatu.

**chi-squared test** ($\chi^2$ test) A statistical test that is used to determine whether data obtained by sampling agree with those predicted hypothetically, and thus to test the validity of the hypothesis.

**chitin** A linear homopolysaccharide of N-actyl-D-glucosamine that is found as a major constituent of fungal *cell walls (and as the major component of the cuticle of an insect) in which the molecules are linked to form chains that are arranged in layers. Depending on their orientation, these chains can be cross-linked to yield a very strong, lightweight material.

***Chlamydia*** (order *Chlamydiales) A genus of *bacteria with the characteristics of the order. There are 2 species, both of which are *parasitic and capable of causing disease (e.g. trachoma and psittacosis).

**Chlamydiales** An order of *Gram-negative *bacteria which are obligately parasitic within the cells of vertebrates. The growth cycle involves the alternate formation of two distinct cell types: small, rigid-walled, infectious forms, and larger, flexible-walled, non-infectious forms which can undergo cell division. There is 1 family, with 1 genus.

***Chlamydomonas*** (division *Chlorophyta) A genus of unicellular green *algae in which each cell contains a single *nucleus and a *chloroplast the shape of which varies with species. The cell may be *motile or nonmotile according to conditions of growth: motile cells have a pair of similar *flagella at one end. There are many species. They are widely distributed and common, found in almost all types of aquatic environment and on moist soil. *C. nivalis* can grow on snow in mountains in N. America, the masses of cells giving rise to a reddish coloration known as 'red snow'. (*Chlamydomonas* is sometimes classified as a genus of protozoa in the class *Phytomastigophora.)

**chlamydospore** A thick-walled fungal resting *spore, resistant to adverse conditions, that is formed from a cell in a vegetative *hypha.

**Chloranthaceae** A family of trees, shrubs, or *herbs, which are often rather succulent. The leaves are *opposite. The flowers are tiny, reduced, and borne on spicate *racemes. The fruit is a tiny *drupe. The family is related to *Lauraceae. There are 4 genera, with 58 species, occurring in tropical America, and from China and India to Melanesia and New Zealand.

***Chlorella*** (division *Chlorophyta) A genus of non-*motile, unicellular green *algae. Each cell contains a single, usually cup-shaped, *chloroplast. Cells divide to form 2, 4, 8, or 16 daughter cells which resemble the parent cell and are initially enclosed within the parent cell wall. *Chlorella* is very common in a wide range of aquatic and terrestrial habitats, often colouring small bodies of standing water (e.g. drinking troughs) a pea-soup green.

**chlorenchyma** Tissue containing *chloroplasts.

**chlorine** (Cl) An element that is necessary for normal plant growth. Its main role seems to be that of controlling *turgor, but it may also be involved in the light reaction of *photosynthesis. If plants are deficient in chloride ions, wilting occurs and the young leaves become blue-green and shiny. Later they become bronze-coloured and chlorotic (*see* CHLOROSIS).

**chlorinity** A measure of the chloride content, by mass, of sea water. It is defined as the amount of chlorine, in grams, in 1 kg of sea water (bromine and iodine are assumed to have been replaced by chlorine). Chlorinity and salinity are both measures of the saltiness of sea water. The relationship can be expressed mathematically, as salinity is equivalent to 1.80655 times the chlorinity.

***Chloris* (family \*Poaceae)** One of the 9 genera of grasses that are represented in all arid regions of the world. They are typical components of arid grazed habitats, and several species are valuable pasture grasses. The genus is cosmopolitan, with about 10 species.

**chlorite** *See* CLAY MINERALS.

**Chlorobiaceae (green sulphur bacteria; order \*Rhodospirillales)** A family of \*anaerobic, anoxygenic, photosynthetic \*bacteria which use sulphide or sulphur as a source of reduction for photosynthesis. \*Sulphur is sometimes produced from sulphide, but never inside the cells. All species are non-\*motile. Cells of different species are spherical, curved or straight rods, or ovoid. Cells may be organized into chains, clumps, nets, spirals, etc. Photosynthetic pigments are attached to membrane vesicles linked to, but not continuous with, the cytoplasmic \*membrane. There are several genera. The organisms are found in sulphide-rich, anaerobic, aquatic environments, often forming greenish or brownish blooms.

***Chlorociboria* (order \*Helotiales)** A genus of \*fungi in which the \*hymenium lines cup-shaped \*apothecia. In *C. aeruginascens* (*Chlorosplenium aeruginascens*) the apothecia are stalked, shallowly cup-shaped, and 5–10 mm across. This species grows chiefly on oak wood, which consequently becomes stained a bright blue-green and is valued for decorative purposes.

***Chlorococcum* (division \*Chlorophyta)** A genus of single-celled green \*algae which produce biflagellate \*spores and autospores. Vegetative forms are non-\*motile. They are found, for example, on damp soil, bricks, etc.

**Chloroflexaceae (order \*Rhodospirillales)** A family of filamentous, thermophilic, photosynthetic \*bacteria which are capable of a gliding motility when in contact with a surface. Vesicles containing photosynthetic pigments are attached to, but not continuous with, the cytoplasmic \*membrane. There are several genera.

***Chloroflexus* (family \*Chloroflexaceae)** A genus of filamentous, gliding \*bacteria. They are predominantly \*photo-organotrophic under \*anaerobic conditions in the presence of light, but \*chemo-organotrophic under \*aerobic conditions. Their colour is green in the absence of air, but orange in the presence of air. *C. aurantiacus* is found in hot springs, where it forms dull-green or orange mats.

**Chloromonadophyceae** *See* RHAPHIDOPHYCEAE.

***Chlorophora* (family \*Moraceae)** A genus of trees that includes *C. excelsa* (iroko of Africa) and *C. tinctoria* (fustic, of the West Indies and S. America), both of which are important timber trees. There are about 5 species, occurring in tropical America and Africa.

**Chlorophyceae** *See* CHLOROPHYTA.

**chlorophyll** The green pigment in plants that functions in \*photosynthesis by absorbing radiant energy from the Sun, predominantly from blue (435–438 nm) and red (670–680 nm) regions of the spectrum. The light removes an electron from the chlorophyll molecule. This is used to produce either \*ATP or NADP (\*nicotinamide adenine dinucleotide phosphate) for carbon dioxide fixation. Chlorophylls are magnesium–porphyrin derivatives, the principal variants in land plants being designated chlorophylls *a* and *b*, and in marine algae *c* and *d*.

**chlorophyll method** A method that is used for estimating the \*primary productivity of \*ecosystems. Experimental evidence shows that, with appropriate calibration, the \*chlorophyll content of the community occupying a given area can form an index of the area's productivity. The method was pioneered for marine ecosystems, but has now been extended to other aquatic and to terrestrial communities.

**Chlorophyta (green \*algae)** A division of algae which are typically green in colour. They contain \*chlorophylls *a* and *b* and the storage product (starch) is formed in \*chloroplasts rather than in the \*cytoplasm. The organisms take many forms, ranging from unicellular to relatively complex multicellular plants. They are found mainly in freshwater habitats, and their distribution is cosmopolitan. They are known from the *Precambrian to the present, and the earliest *eukaryotes were probably of this class. The classification of the green algae is controversial. The division Chlorophyta is sometimes regarded as containing only 2 classes (Chlorophyceae and Charophyceae), but more recently it has been divided into 5 classes.

**chloroplast** A biconvex or planoconvex *plastid, typically 5–10 μm long and 2–3 μm wide, that has a complex internal structure comprising stacks of membranaceous discs (grana) that bear photosynthetic pigments embedded in a matrix (*stroma). It is a semi-autonomous *organelle, which contains some genetic material and has some ability to direct the synthesis of its own *proteins.

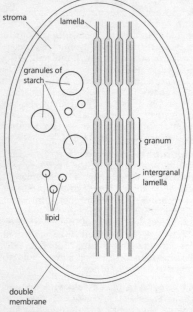

*Chloroplast*

**chloroplast-DNA (cp-DNA)** Circular *DNA, like that of *mitochondria but several times larger, which is found in *chloroplasts and other *plastids and contains *genes involved in the photosynthetic pathway. It is independent of nuclear DNA and is maternally inherited. Between 40 and 80 DNA molecules occur in each *organelle.

**chlorosis (adj. chlorotic)** A symptom of disease or disorder in plants, which involves a reduction in or loss of the normal green coloration; consequently, the plants are typically pale green or even yellow. Chlorosis is caused by conditions that prevent the formation of *chlorophyll (e.g. lack of light or a deficiency of iron or magnesium).

**chlorotic** *See* CHLOROSIS.

***Chloroxylon*** **(family *Rutaceae)** A *monotypic genus, comprising *C. swietenia*, which yields satinwood, an attractive pale timber used for veneer, and also a gum. It occurs in Sri Lanka and southern India.

**Choanephoraceae (order *Mucorales)** A family of *fungi which, typically, form both *sporangia and *sporangiola. The columellate sporangia contain dark brown spores. Species are found, for example, on fruit and flowers.

***Choisya*** **(family *Rutaceae)** A genus of 7 species of shrubs of south-western N. America; *C. ternata* is widely cultivated for its fragrant, white flowers.

**choke of grasses** A disease that can affect many types of grass, particularly cocksfoot (*Dactylis glomerata*) and timothy (*Phleum pratense*). The causal agent is the *fungus *Epichloe typhina*. A ring of white, felt-like *mycelium develops around the flower stalk and becomes yellow, then orange, as the fungal *perithecia develop. Flowering is partially or completely suppressed.

**chomophyte** A plant that grows on a rock ledge or within a rock fissure. *See also* ENDOLITHIC and PETROPHYLLOUS. *Compare* LITHOPHYTE.

***Chondrostereum*** **(family *Stereaceae)** A genus of lignicolous *fungi. *C. purpureum* (formerly *Stereum purpureum*) causes *silver leaf in trees of the *Rosaceae; it forms bracket-shaped *fruit bodies, the brackets often overlapping and becoming fused.

***Chondrus* (order \*Gigartinales)** A genus of red seaweeds which includes only one British species: *C. crispus* (Irish moss). The \*thallus is rather variable in both form and colour. The \*fronds are flattened and strap-like, extensive \*dichotomous branching giving rise to a bushy, tufted habit. They are usually purplish-red, but may be green in strong sunlight. They are found attached to rocks in pools on the lower part of the intertidal region, and below the low-water mark.

**-chore** A suffix adopted in a little-used classification scheme proposed for seed-dispersal mechanisms. Ten categories, or chores, are recognized, e.g. pterochore (winged seeds dispersed by the wind).

***Choretrum* (family \*Santalaceae)** A genus of shrubs that are root parasites. There are 6 species, endemic (*see* ENDEMISM) to eastern Australia.

**-chory** A suffix adopted in a little-used classification scheme of seed-dispersal mechanisms. *See* ANDROCHORY; ANEMOCHORY; AUTOCHORY; BARACHORY; BLASTOCHORY; BOLOCHORY; CRYSTALLO-CHORY; ENDOZOOCHORY; ENTOMOCHORY; EXOZOOCHORY; GYNOCHORY; HYDROCHORY; MYRMECOCHORY; PTEROCHORY; SAURO-CHORY; and ZOOCHORY.

**Chromatiaceae (purple sulphur bacteria; order \*Rhodospirillales)** A family of \*anaerobic, anoxygenic, photosynthetic \*bacteria that can use sulphur, sulphide, or other reduced inorganic sulphur compounds as sources of reduction for photosynthesis. Most species accumulate granules of sulphur inside their cells. Many species are \*motile with \*flagella. Cells are spherical, spiral, ovoid or rod-shaped, and range in colour from pinkish-red to orange-brown. Cells may occur singly or as aggregates embedded in slime. Photosynthetic pigments are attached to membranes which are continuous with the cytoplasmic membrane. There are several genera, with many species; they are found in sulphide-rich, \*anaerobic, aquatic environments, often forming pink to purple-red blooms.

**chromatid** One of the two daughter strands of a \*chromosome that has undergone division during \*interphase. Chromatids are joined together by a single \*centromere, usually positioned in the centre of the pair as they lie beside one another. When the centromere divides at \*anaphase of \*mitosis or anaphase II of \*meiosis, the sister chromatids become separate chromosomes.

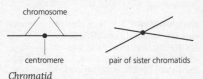

*Chromatid*

**chromatid interference** The non-random participation of non-sister \*chromatids of a \*tetrad in successive \*crossovers of \*meiosis: it results in a deviation from the expected $1:2:1$ ratio for the frequencies of 2-, 3-, and 4-strand double cross-overs.

**chromatin** The substance of \*chromosomes, which includes \*DNA, chromosomal proteins, and chromosomal \*RNA. Chromatin stains strongly with \*basic dyes. It is thought that the chromatin is most deeply stained when it is most condensed and inactive.

**chromatography** An analytical technique used for the separation of the components of complex mixtures, which is based on their repetitive distribution between a mobile phase (of gas or liquid) and a stationary phase (of solids or liquid-coated solids). The distribution of the different component molecules between the two phases is dependent on the method of chromatography used (e.g. gel-filtration, or ion-exchange), and on the movement of the mobile phase (which results in the differential migration and therefore separation of the components along the stationary phase).

**chromatophore 1.** A pigmented \*plastid of a plant cell. **2.** In \*prokaryotic organisms, a membrane-bounded \*vesicle that contains photosynthetic pigment.

**chromomere** A small beadlike structure visible in a *chromosome during *prophase of *meiosis and *mitosis, when it is relatively uncoiled (in particular at the leptotene and zygotene stages of meiosis). Chromomeres in corresponding positions on *homologous chromosomes pair during meiosis in many organisms.

**chromoplast** A *plastid that contains pigments other than *chlorophyll (e.g. *carotenoids).

**chromosome** A *DNA-histone protein thread, usually associated with *RNA, occurring in the *nucleus of a cell. Although chromosomes are found in all animals and plants, bacteria and viruses contain structures that lack protein and contain only DNA or RNA: these are not chromosomes, though they serve a similar function. Chromosomes occur in pairs, which associate in a particular way during meiosis. Each species tends to have a characteristic number of chromosomes (e.g. 20 in maize), found in most nucleated cells within most organisms. The presence of pairs of *homologous chromosomes is referred to as the diploid state and is normal for the sexual phase of an organism. *Gametes, and cells of the *gametophyte of plants, however, are *haploid with only one member of each pair in their nuclei. Usually chromosomes are visible only during *mitosis or *meiosis when they contract to form short thick rods coiled into a spiral. Each chromosome possesses *chromomeres and a *centromere, and some contain a nucleolar organizer. Chromosomes contain a line of different genes, a *spindle-attachment at some point along their length, and regions of *heterochromatin, which stains strongly with basic dyes.

**chromosome map** A map showing the locations (*loci) of *genes on a *chromosome, deduced from genetic recombination (see RECOMBINANT) experiments. For example, the frequency of *cross-overs between pairs of *genes indicate their relative positions or linear order, the distances being given in units of *cross-over frequency.

**chromosome polymorphism** The presence of one or more *chromosomes in two or more alternative structural forms within the same interbreeding population.

**chromosome substitution** The replacement of one or more *chromosomes by others (totally or partially *homologous) from another source (either a different strain of the same species or a related species that will permit hybridization; see HYBRID) by a suitable crossing programme.

**chromosome theory of heredity** The unifying theory put forward by W. S. Sutton in 1902 that *Mendel's laws of inheritance may be explained by assuming that *genes are located in specific sites on *chromosomes.

*Chromulina* See CHRYSOPHYTA.

**chronosequence** A sequence of related soils that differ in their degree of *profile development because of differences in their age. Chronosequences can be found in evolving landscapes such as those produced by deglacia-tion, volcanic activity, wind deposits, or sedimentation.

**chronospecies** (evolutionary species) According to one view of evolution (that of *phyletic gradualism), a new organism may be derived from its ancestor by a process of slow, steady evolutionary change. Conceivably, the descended organism might not be regarded as a member of the same species as its ancestor, in which case it would constitute a new species, in particular a chronospecies.

**chronostratigraphy** A branch of stratigraphy linked to the concept of time. In chronostratigraphy intervals of geological time are referred to as chronomeres. These may be of unequal duration. Intervals of geological time are given formal names and grouped within a Chronomeric Standard hierarchy. The formal terms are: eon, era, period, epoch, age, and chron. Of these, the last four are the equivalent of system, series, stage, and chronozone in the Stratomeric Standard hierarchy. The formal terms are written with initial capital letters when accompanied by the proper names of the intervals to which they refer. Some geologists hold that the term 'chronostratigraphy' is synonymous with *'biostratigraphy', but most agree that the two branches are separate.

**Chroococcales** An outdated phycological order of 'blue-green algae' which corresponds roughly with section I of the *cyanobacteria.

**Chroococcus** See GLOEOCAPSA.

**Chrysobalanaceae** A small family of forest trees and shrubs, related to *Rosaceae, in which the flowers are somewhat irregular and *perigynous. The fruit is a *drupe. Several yield *siliceous timber, others have edible fruit. The coco plum of the West Indies is *Chrysobalanusicaco*. There are 17 genera, comprising about 460 species, with a pantropical distribution.

**Chrysophyllum** (family *Sapotaceae) A genus of trees which includes *C. cuintto*, the star apple of the W. Indies, grown widely for its edible fruits. There are 80 species, occurring in the tropics and subtropics.

**Chrysophyta** (golden algae, golden-brown algae) A division of predominantly unicellular *algae (sometimes alternatively regarded as protozoa, class *Phytomastigophora) in which the *chloroplasts contain large amounts of the pigment fucoxanthin, giving the algae their brown colour. Most are flagellated, having one *flagellum of the tinsel type and a second flagellum of the whiplash type, which may be reduced to a short stub. Cysts or resting *spores are formed which are often characteristically ornamented with spines, etc.; the cyst walls contain silica. Chrysophytes are found mainly in freshwater habitats that are low in calcium. Genera include *Chromulina*, *Mallomonas*, *Ochromonas*, and *Synura*.

**Chukrasia** (family *Meliaceae) A *monotypic genus, *C. tabularis*, of deciduous trees in which the leaves are *pinnate. The flowers are large, the *stamens wholly fused. The fruit is a *capsule. They yield valuable timber (chickrassy and yom hin). They occur in India and from southern China to Malesia.

**chusan palm** See TRACHYCARPUS.

**Chytridiales** (class *Chytridiomycetes) An order of microscopic *fungi which

rarely form a true *mycelium. Most species are aquatic *saprotrophs or *parasites, but some (e.g. *Synchytrium) are important *pathogens of terrestrial plants.

**Chytridiomycetes** (subdivision *Mastigomycotina) A class of small to microscopic *fungi which form *sporangiospores that have single, posteriorly directed, whiplash *flagellum. They are *saprotrophs and/or parasites in aquatic or terrestrial habitats.

**Cibotum** See THYRSOPTERIDACEAE.

**Cicca** (family *Euphorbiaceae) A *monotypic genus, *C. acida*, which is a small tropical fruit tree that is widely cultivated and is probably native to north-eastern Brazil. Its leafy twigs resemble *pinnate leaves. The fruit is an acid, pulpy *berry.

**Cichorium** (family *Asteraceae) A genus of *perennial *herbs that have alternate leaves and flower heads of blue, strap-shaped *florets. The *achenes are angled, with blunt scales on top. The roots of *C. intybus* (chicory) are used for blending with coffee; *C. endivia* (endive) is used as a vegetable. There are 8 species occurring in the Mediterranean region, of which one reaches northern Europe, and one Ethiopia.

**cider sickness** The condition of cider that has become contaminated with the *bacterium *Zymomonas mobilis*. The bacterium produces acetaldehyde which spoils the flavour and appearance of the cider.

**cilia** See CILIUM.

**cilium** (pl. cilia) A short, hair-like appendage, normally 2–10 µm long and about 0.5 µm in diameter, usually found in large numbers on those cells that have any at all. Cilia have a microtubular skeletal structure enclosed by an extension of the *plasma membrane. The *microtubules are typically arranged in 9 sets of doublets around the circumference, with 2 single tubules in the centre, the so-called '9 + 2' construction. In certain protozoa, cilia function in locomotion and/or feeding. They generate currents in the fluid surrounding the cell by beating in a co-ordinated manner.

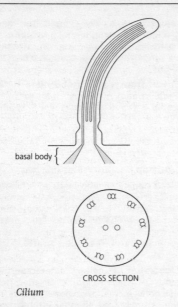

basal body

CROSS SECTION

*Cilium*

**Cinchona** (family *Rubiaceae) A genus of trees, several species of which were formerly widely cultivated for quinine and related drugs, contained in the bark. There are 40 species, native to the Andes.

**cincinnus** A *monochasium in which the branches occur on alternate sides of the stem and the *inflorescence is commonly bent to one side (e.g. in *Strelitzia reginae).

**Cinnamomum** (family *Lauraceae) A genus of trees in which the leaves are *opposite, and spicy when crushed. The fruit is a *berry contained in a cup-like *perianth. The spice cinnamon is the bark mainly of *C. zeylanicum*. Camphor is obtained by distillation of the wood of *C. camphora*. There are 250 species, occurring in eastern Asia and Indo-Malaysia.

**cinnamon** *See* CINNAMOMUM.

**circadian rhythm** The approximately 24-hourly pattern of various metabolic activities seen in most organisms. The rhythmic patterns may persist even when the organism is removed from exposure to 24-hour cycles of light and dark. In a natural habitat the rhythm is 24-hourly;

in constant conditions it becomes slightly longer or shorter than 24 hours. The rhythm is thought to be controlled by an endogenous biological clock, though little is known about the sites and modes of action of the mechanisms involved. The word is derived from the Latin *circa*, meaning 'about' and *dies*, meaning 'day'.

**circalittoral zone** The area of the continental shelf sea-bed that lies below the zone of periodic tidal exposure. It is approximately equivalent to the *sublittoral zone. *Compare* INFRALITTORAL; MEDIOLITTORAL; *see also* LITTORAL ZONE.

**circinate** *See* VERNATION.

**circle of vegetation** The highest classificatory unit used by the *Zurich–Montpellier school of phytosociology, this refers to a definite geographical region with a distinctive floristic element. This approximates to a major plant geographical region.

**circumaustral distribution** The distribution pattern of organisms found around the high latitudes of the southern hemisphere in the *austral region, i.e. next to the Antarctic zone. An example of such a pattern is provided by the southern species of *Danthonia* (poverty grasses and wild-oat grasses).

**circumboreal distribution** The distribution pattern displayed by organisms around the high latitudes of the northern hemisphere in the boreal zone. The boreal zone lies next to the Arctic zone, and many circumboreal organisms are in fact also circumarctic. An example of a plant with both types of distribution is *Saxifraga oppositifolia* (purple saxifrage).

**circumnutation** The spiral movement of a shoot apex as it grows.

**circumpolar distribution** Denotes the range of organisms distributed around the North or South Poles. Strictly, the term should be reserved for organisms that extend into the polar regions. *Eutrema edwardsii* is a good example of a flowering plant with such a distribution around the North Pole; an animal example is the reindeer or caribou (*Rangifer tarandus*).

**circumscissile** Opening all the way around along a transverse slit.

**cirri** In *Fungi, pustules that produce asexual *spores.

**cirrus** *See* CALAMUS; CERATOLOBUS; and *DAEMONOROPS*.

**Cistaceae** A family mostly of low shrubs but also of a few *herbs, in which the leaves are usually *opposite. The flowers are regular and bisexual, with either 5 unequal or 3 equal *sepals and 5 crinkled, showy *petals. There are many *stamens. Usually the *ovary is 1-celled, and contains many seeds. The *style is simple, the fruit a *capsule. *Cistus* species (which have a 5-valved capsule) are much cultivated for their large flowers, as are some *Helianthemum* species (which have a 3-valved capsule). There are 7 genera, with about 175 species, centred in the Mediterranean region, but extending through Europe, central Asia, and America.

**cisternae** Large, flattened, membranaceous sacs that occur in a number of cell *organelles (e.g. the *Golgi body) and in the *endoplasmic reticulum.

**cis-trans test** *See* COMPLEMENTATION TEST.

**cistron** A section of the *DNA molecule that specifies the formation of one *polypeptide chain. A *complementation test is performed to determine whether two mutant sites of a *gene are in the same *cistron or in different ones. In the *cis* configuration, both mutants are on one *homologous chromosome and both *wild-types are on the other ($a^1a^2/++$), producing a wild-type *phenotype. In the *trans* configuration each homologue has a mutant and a non-mutant ($a^1+/a^2+$), producing a mutant phenotype.

**Cistus** *See* CISTACEAE.

**citric acid cycle (Krebs' cycle, tricarboxylic acid cycle)** The cyclic series of reactions which represent the principal means by which most living cells provide hydrogen and electrons for the generation of *ATP via the electron-transport chain and *oxidative phosphorylation. The hydrogen is taken up by NAD (*nicotinamide adenine dinucleotide) or FAD (*flavin adenine dinucleotide) and passed into the *electron-transport chain. The reactions result in *oxidative decarboxylation. The sequence is initiated when acetyl groups, fed in by acetyl *coenzyme A, are dehydrogenated (i.e. oxidized), to release 4 pairs of hydrogen atoms, and decarboxylated to form 2 molecules of carbon dioxide. This yields oxaloacetic acid and then citric acid. The citric acid passes through a series of reactions wherein oxaloacetic acid is reformed, and two moles of $CO_2$ and water and one mole of ATP are synthesized. Since two molecules of acetyl coenzyme A are formed for each glucose molecule that is oxidized the cycle rotates twice, forming two moles of ATP.

**Citrus** **(family *Rutaceae)** A genus of plants, many of which are in cultivation for their fruit: sweet orange is *C. sinensis*, sour (or Seville) orange *C. aurantium*, grapefruit *C. paradisi*, lemon *C. limon*, lime *C. aurantiifolia*, and there are others. The leaves are *simple, the stalk often winged. The fruit is a *hesperidium. There are about 16 species, native to South-east Asia.

**Cl** *See* CHLORINE.

**clade** A term derived from the Greek *klados*, meaning a twig or branch. In *cladistics, or *phylogenetic systematics, it refers to a lineage branch that results from splitting in an earlier lineage. A split produces two distinct new taxa, each of which is represented as a clade or branch in a phylogenetic diagram.

**cladistics** A special taxonomic system applied to the study of evolutionary relationships. In the branching diagrams or 'cladograms' used to portray these relationships, cladogenesis, or splitting of an evolutionary lineage, always creates two equal sister taxa: the branching is *dichotomous. Thus, each pair of sister taxa constitutes a monophyletic group with a common stem taxon, unique to the group. *See* PHYLOGENETIC SYSTEMATICS; *see also* AREA CLADISTICS.

**Cladium** **(family *Cyperaceae, *tribe Rhynchosporoideae)** A genus of rhizomatous (*see* RHIZOME) *herbs which have smooth, leafy stems. The leaves are long and linear, a number of them arising at the base of the stem. The flowers are bisexual, or sometimes the upper or lower are male. There is no *perianth; there are 2 or 3 *stamens and the same number of *stigmas. The *ovary is *superior and *unilocular, with 2 or 3 fused *carpels. The fruit is a globular or *trigonous *nut with much *endosperm. There are 2 species, found in tropical and temperate marshy regions. *C. mariscus* (saw sedge) is used for thatch in Europe.

**cladode** *See* PHYLLOCLADE.

**cladogenesis** In *cladistics, the derivation of new taxa (*see* TAXON) that occurs through the branching of ancestral lineages, each such split forming 2 equal sister taxa that are taxonomically separate from the ancestral taxon.

**cladogram** A diagram that delineates the branching sequences in an evolutionary tree. *See also* DENDROGRAM.

***Cladonia*** (order *Lecanorales) A genus of *lichens in which the *thallus usually consists of a *crustose or *squamulose primary thallus (which may be *evanescent), and a *fruticose 'secondary thallus' called a *podetium. Podetia may be erect, and branched or unbranched, or may be shaped like miniature cups or wine glasses (cup lichens). *Apothecia, when present, are borne at the tips of the branches, or on the edges of the cups, and may be red or brown. *Cladonia* is a large genus consisting of over 300 species, found on the ground, on rocks, on wood, etc. *See also* REINDEER MOSS.

***Cladophora*** (division *Chlorophyta) A genus of filamentous, branched, green *algae in which the cells are multinucleate. Species of the genus *Cladophora* are common. They form long, tangled, rough-textured skeins (blanket weed), usually attached to rocks, in standing or running fresh water. They may be a troublesome weed in garden ponds. Some species are marine, growing (often abundantly) on rocks on the lower shore, and in rock pools.

***Cladosporium*** (class *Hyphomycetes) A *form-genus of *fungi that form *septate *mycelium and chains of dark-coloured *conidia. They are *saprotrophic, or *parasitic in plants or animals, sometimes causing disease. *See also* CREOSOTE FUNGUS.

**clamp connection** In certain types of *fungi, a bulge on one side of a *dikaryotic hypha at the site of a *septum. It is formed during cell division, as a result of a mechanism that ensures maintenance of the dikaryotic state.

**Clanwilliam cedar** *See* WIDDRINGTONIA.

**classification methods** Any scheme for structuring data that is used to group individuals, or sometimes attributes. In ecological and taxonomic studies especially, quite sophisticated numerical classification schemes have been devised, and an extensive theoretical literature exists relating to them. The methods developed in these disciplines are being applied increasingly in other fields, notably pedology and palaeontology. Various classificatory strategies may be used, e.g. hierarchical or non-hierarchical; and where hierarchical schemes are used, these may be divisive or agglomerative, and monothetic or polythetic, with the divisive polythetic approach being considered the optimum strategy. *Compare* ORDINATION METHODS.

***Clavaria*** (family *Clavariaceae) A genus of *fungi in which the *hymenium is borne on the surface of cylindrical or club-shaped fruit bodies. *C. vermicularis* forms clusters of white, pointed, cylindrical fruit bodies in grassy places. *C. fumosa* is similar, but yellowish-grey in colour. *C. argillacea* forms club-shaped, unbranched, whitish fruit bodies on mossy, peaty ground, and is common in summer and autumn.

**Clavariaceae** (order *Aphyllophorales) A family of *fungi in which the *fruit bodies are erect, unbranched and club-shaped, branched, or coralloid, and often brightly coloured. Spores are produced on the surface of the fruit body. There are many genera, found chiefly on the ground, or on rotting wood. *See also* CLUB FUNGI and CORAL FUNGI.

**clavate** Club-shaped; thicker towards one end.

***Claviceps*** **(order** *Clavicipitales)* A genus of *fungi that are typically *parasitic on grasses. *C. purpurea* is the causal agent of *ergot. This species forms elongated, curved, purple–black *sclerotia which replace the grain in an infected host plant. A wide range of grasses and cereals can be affected.

**Clavicipitales** **(subdivision** *Ascomycotina)* An order of *fungi that form perithecioid (*see* PERITHECIUM) *ascocarps which may be superficial or immersed in a fleshy *stroma; the *ascospores are long and narrow. There are many genera. Some species are *parasitic on grasses, sometimes causing important plant diseases; some are parasitic on insects or on other fungi.

***Clavulina*** **(family Clavulinaceae, order** *Aphyllophorales)* A genus of *fungi in which the *hymenium is borne on the surface of small, club-shaped, branched or coralloid *fruit bodies. The *basidia bear 2 spores rather than the usual 4. *C. cristata* (white coral fungus) forms brittle, white, irregularly branched fruit bodies 3–8 cm tall. It is common on the ground in broadleaved and coniferous woods. *C. rugosa* is similar but is longitudinally wrinkled, and less extensively branched. *See also* CLUB FUNGI and CORAL FUNGI.

**clay** **1.** A soil separate comprising mineral particles less than 2 μm in diameter. **2.** A class of soil texture, usually containing at least 20 per cent by weight of clay particles. *Compare* CLAY MINERAL.

**clay films** *See* CUTAN.

**clay mineral** A member of a group of chemically related hydrous aluminium silicates, which generally occur either as very small platy or fibrous crystals. They have a layered structure and the ability to take up and lose water readily. It is difficult to distinguish various clay minerals, and therefore geologists employ sophisticated techniques (e.g. X-ray diffraction analysis and electron microscopy). The most important clay minerals belong to the kaolinite, hydrous mica, smectite, chlorite, vermiculite, and talc groups.

**clay pan** *See* PAN.

**clayskins** *See* CUTAN.

**cleavage** *See* SEGMENTATION.

**cleavers** *See* GALIUM.

**cleistocarp** *See* CLEISTOTHECIUM.

**cleistogamy** Having flowers that do not open and, therefore, the pollination and fertilization of an unopened flower.

**cleistothecium** **(cleistocarp)** An *ascocarp in which the *asci are enclosed completely. *Ascospores are released only on rupture of the wall of the cleistothecium.

***Clematis*** **(family** *Ranunculaceae)* An anomalous genus in the family, comprising plants which have a woody, *liane habit and *opposite, *compound leaves whose stalks or tendrils twine to support the climbing plants. The flowers usually have 4 *petaloid *perianth segments, and long, plumed *styles that aid in the wind dispersal of *achenes. There are about 230 species, most in the northern temperate zone and a few in the African mountains; many are cultivated for their flowers or foliage.

**Clements, Frederic Edward** **(1874–1945)** An American botanist and ecologist whose *Research Methods in Ecology* (1905) was the first work to deal fully with field experiments. Clements maintained that a plant community develops by constant adjustments of the relationships among species (i.e. a *succession) to an optimal state, the *climax community, and that a plant community may in some sense be regarded as a superorganism. *See also* TANSLEY, ARTHUR GEORGE.

***Clerodendron*** *See* VERBENACEAE.

**Clethraceae** A family of shrubs or trees, closely related to the *Ericaceae, which have *simple, alternate leaves. The fragrant flowers have a bell-like, persistent *calyx, 5 free, *imbricate *petals, 10–12 free *stamens, *anthers with apical pores; and a *superior, 3-celled *ovary. The fruit is a *capsule. There is 1 genus, with 64 species, occurring in tropical to warm temperate America, eastern Asia, and Madeira.

***Clianthus*** **(family** *Fabaceae, subfamily* *Papilionatae)* A genus of legumes that bear large, red, pea-like flowers on erect *peduncles and have soft, grey stems and leaves. *C. formosus* (Sturt's desert pea) is the emblem of the state of S. Australia. It grows rapidly in dry regions following rain and forms a scrambling plant up to 2 m high. The genus comprises 2 species, one found in Australia and the other in New Zealand.

**Cliftonia monophylla** (buckwheat tree) *See* CYRILLACEAE.

**Climacium** *See* ISOBRYALES.

**climacteric** The phase of increased respiration found at the ripening of *fruit and at *senescence.

**climatic climax** A plant community that is in equilibrium with a zonal climate. *See also* CLIMAX THEORY.

**climatic optimum** The period of highest prevailing temperatures since the last ice age; in most parts of the world it occurred about 4000–8000 years ago.

**climax adaptation number** A value, in the range 1–10, assigned to species in the *Wisconsin ordination scheme and based on the species' importance values. Stands with the same leading dominants are grouped together, and importance values for all species in the group are recalculated. By comparing the changing importance of species in groups with different leading dominants, the leading dominants and then other species can be placed in *phytosociological order. *See also* IMPORTANCE VALUE; and DENSITY-FREQUENCY–DOMINANCE.

**climax community** (climax vegetation) The final stage of a plant *succession, in which vegetation reaches a state of equilibrium with the environment. The community is self-perpetuating, except that changes may occur very slowly and over a time-scale that is extensive compared with the rapid and dramatic changes during the early stages of succession. *See also* CLIMAX THEORY.

**climax theory** A theory that embraces the idea of *successional development to an optimum sustainable vegetation community that is in equilibrium with its environment. Two conceptual models

of climax vegetation exist: monoclimax (F. E. *Clements, in 1904 and 1916) and polyclimax (A. G. *Tansley, in 1916 and 1920). The essential difference between them lies in the time-scale envisaged for the development of the climax. Monoclimax envisages the development of vegetation and soil towards a definite end-point controlled by climate. Many ecologists consider this approach unworkable in practice, since equilibrium can never be reached owing to variations in climate over long periods of time (e.g. the known climatic changes in historical and postglacial times in mid-latitudes). Evidence from climatically relatively uniform and stable tropical areas shows different, equally persistent, woodland communities *edaphically controlled. Insistence on a single uniform climax community for a major climatic region necessitates the introduction of an extensive terminology for subclimax communities, so tending to obscure an otherwise useful and simple concept. The alternative (polyclimax) approach proposes successional development to equilibrium with the environment, which may be controlled by climate or by some other factor, e.g. soil or fire, as often occurs in practice. *See also* CLIMAX COMMUNITY and SUCCESSION.

**climax vegetation** If a bare surface becomes available for colonization by plants, the vegetation will pass through a *succession of so-called *seral stages as more complex communities replace the earlier, simpler ones. Ultimately a state of equilibrium or climax is reached, which may reflect predominantly the control of climate, soil, or humans. *See also* CLIMAX COMMUNITY and CLIMAX THEORY.

**climograph** A two-dimensional, graphical representation of one major climatic element against another, e.g. rainfall against temperature. Typically, the average values for each month are plotted, and the points joined together sequentially to give a dodecagon of a shape that is characteristic for a particular area. This forms an easily prepared basis for quick visual comparison of the climates of different areas. Uses include the assessment of the suitability of a new environment for a proposed species introduction.

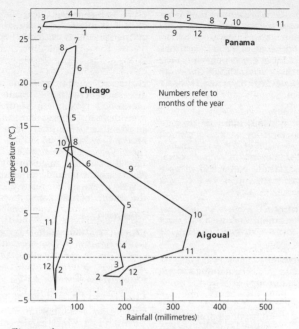

Temperature (°C)

Rainfall (millimetres)

Panama

Chicago

Numbers refer to
months of the year

Aigoual

*Climograph*

**climosequence** A sequence of soil *profiles, usually on the same parent material, that differ from each other in their profile development because of local or site differences in climatic conditions. Climosequences can be found along mountain slopes in certain highland areas.

**climotope** The climatic component of the *habitat or *ecotope of a *biogeocoenosis.

**clinal speciation** A type of *allopatric speciation in which a *geographic barrier falls across a *cline, cutting a species into two segments which are thus already somewhat different and which can continue to *diverge in their new isolation.

**cline** A gradual change in gene frequencies or character states within a species, across its geographic distribution.

**clinosequence** A sequence of soils in which soil-*profile development is related to the angle of slope of the soil surface. Clinosequences can be found on landforms such as escarpments and drumlins with varying surface angles of slope.

**clitochory** *See* BARACHORY.

***Clitocybe*** (family *Tricholomataceae) A genus of *fungi which form *mushroom-like fruit bodies in which the cap is more or less funnel-shaped and fleshy. The *gills are often *decurrent. There are many species, some of which are edible whereas others are extremely poisonous. They are usually found on the ground, especially under trees.

**clod** A compact, coherent block of soil, found *in situ* when soil is broken up by digging or ploughing. Clods are of varied sizes.

**clonal dispersal** A method of plant *dispersal in which the plant produces *stolons or *rhizomes from which new plants develop, the new plants being genetically identical to one another and to the parent (i.e. they comprise a *clone). Bracken (*Pteridium aquilinum*) and American quaking (or trembling) aspen

(*Populus tremuloides*) are among the plants that disperse in this way. Such clones often cover a large area (nearly 14 ha in the case of one clone of *P. tremuloides*) and sometimes achieve great longevity (1400 years in the case of a *P. aquilinum* clone). *See also* GUERILLA GROWTH FORM and PHALANX GROWTH FORM.

**clone** A group of genetically identical cells or individuals, derived from a common ancestor by asexual *mitotic division. If a section of *DNA is engulfed into the *chromosome of a *bacterium, *phage, or *plasmid vector and is replicated to form many copies, each copy is referred to as a DNA clone.

**closed canopy** The condition in which the crowns or canopies of individual trees overlap to form a virtually continuous layer. *See also* CANOPY.

**closteroviruses** A group of elongated and thread-like, *RNA-containing viruses which can, as a group, infect a wide range of plants, including both *monocotyledons and *dicotyledons. Most closteroviruses are transmitted from plant to plant by aphid vectors. Some can cause diseases of considerable economic importance.

**Clostridium** A genus of *endospore-forming, typically *Gram-positive *bacteria in which the cells are rod-shaped and usually *motile. They are *chemoorganotrophic, and most species can grow only in the absence of air. There are many species, found in soil, in aquatic habitats, in animal intestines, etc. The genus includes some important *pathogens, e.g. the causal agents of botulism, tetanus, and gas gangrene.

**cloud forest** Tropical *montane forests that are usually located more than 1000 m above sea-level, and are shrouded in cloud for much of the day. The moisture encourages profuse *epiphytic growth, especially of *ferns, filmy ferns, *lycopods, and *mosses.

**clove** A spice consisting of the young flower-bud of *Eugenia caryophyllus*, native to the Moluccas, and now produced mainly in Zanzibar.

**clover** *See* TRIFOLIUM.

**club fungi** (fairy clubs) Those *fungi of the families *Clavariaceae and Clavulinaceae that form club-shaped *fruit bodies.

**clubmosses** *See* LYCOPSIDA.

**clubroot** (finger and toe) A serious disease of cruciferous plants (*Brassicaceae), e.g. cabbages and broccoli. Symptoms include *wilting of plants in bright sunshine, and the leaves may take on a bluish or reddish coloration. The roots show characteristic *gall-like swellings. The disease is caused by the fungus *Plasmodiophora brassicae*. This organism can survive in the soil for many years, even in the absence of host plants.

**club rush** *See* SCIRPUS.

*Clusia* (family *Clusiaceae) A genus of woody trees or shrubs most of which are stranglers. There are about 145 species, occurring in tropical and subtropical forests in the New World.

**Clusiaceae** A family of trees and shrubs, formerly known as Guttiferae, whose leaves are *simple and *opposite and have oil glands. The flowers have cyclic or spiral *petals and *sepals, many *stamens, and a *superior, 3–5-locular *ovary. Many species are useful for timber, and some have edible fruits. There are 47 genera, with about 1350 species, occurring mainly in the tropics.

**cluster cup** The common name for the cup-shaped *aecium formed, for example, on the leaves of plants infected by certain *rust fungi (order *Uredinales).

**Cneoraceae** (spurge olives) A family of small shrubs which bear flowers in *axillary *cymes, the *peduncle fused below to the *petiole. The flowers are *tri- or *tetramerous, their parts free on an elongated *receptacle. The *ovary is 3- or 4-celled, forming 1–4 small *drupes in the fruit. There is 1 genus (*Cneorum*) of few species, occurring in the Mediterranean region.

**co-adaptation** The development and maintenance of advantageous genetic traits, so that mutual relationships can persist (i.e. both parties evolve adaptations that increase the effectiveness of the relationship). Predator–prey and flower–pollinator relationships often exhibit examples of co-adaptation, which is an aspect of *co-evolution. For example, the relationship between the ant *Pseudomyrex ferruginea* and the plant *Acacia hindsii* is obligatory and dependent on co-adaptations. The ant is active 24 hours a day (which is unusual for ants) and thereby provides continuous protection for the acacia. In a similar evolutionary gesture, the acacia bears leaves throughout the year (most related species lose their leaves), providing a continuous source of food for the ants. *See also* BELTIAN BODIES.

**coast redwood** *See* SEQUOIA.

**cobnut** *See* CORYLUS.

**cocaine** A drug obtained from *Erythroxylum coca*. *See* ERYTHROXYLUM.

**coccolithophorids (class *Prymnesiophyceae)** A family of unicellular, marine, planktonic *protists which are, at least at some stage in their life cycle, covered in calcareous plates (*coccoliths) embedded in a gelatinous sheath. They are spherical or oval, and less than 20 μm in diameter. They range in age from the Upper *Triassic to *Holocene, although dubious examples have been described from the Upper *Precambrian and *Palaeozoic.

**coccoliths** The microscopic calcareous plates or discs, often oval and commonly intricately patterned and ornamented, that occur as part of the protective covering of a group of the unicellular *algae called *coccolithophorids. Coccoliths are a major component of the modern deep-sea calcareous oozes, and were especially abundant in the *Mesozoic, particularly the *Cretaceous Period, in which they became a major component of the Chalk lithology.

**coccus** A spherical, *berry-like structure (e.g. the *fruit in *Coriariaceae), or a bacterial cell that is spherical, or nearly so. The word is derived from the Greek *kokkos*, meaning 'berry'.

**Cochlospermum (family *Cochlospermaceae)** A genus of trees in which the leaves are *palmately divided and the flowers showy and *pentamerous. The fruit is a *capsule with many hairy seeds. There are 12 species, occurring in the tropics and subtropics.

**cockscomb (*Celosia cristata*)** *See* AMARANTHACEAE.

**cock's-foot grasses** *See* DACTYLIS.

**cocoa** *See* THEOBROMA.

**coconut** *See* COCOS.

**coco plum** *See* CHRYSOBALANACEAE.

**Cocos (family *Arecaceae)** A *monotypic genus, *C. nucifera*, the coconut ('nature's greatest gift to man') which grows wild on sandy coasts from the east coast of Africa to Polynesia, and also in the Caribbean. It is very widely cultivated from 20°N to 20°S. It has a solitary stem, and leaves with many *leaflets. The fruit is a *nut with a fibrous mesocarp (the source of coir fibre) and stony endocarp (*see* PERICARP) (the source of a fine medicinal charcoal), lined by *endosperm (the source of copra).

**cocoyam** *See* COLOCASIA.

**COD** *See* CHEMICAL OXYGEN DEMAND.

**Codiaeum (family *Euphorbiaceae)** A genus of small trees and shrubs which includes *C. variegatum* (garden croton), a widely grown shrub with numerous varieties having differently coloured leaves. There are 6 species, occurring in the Indo-Malaysian tropics.

**Codium (division *Chlorophyta)** A genus of *siphonaceous green seaweeds; the *thallus is dichotomously branched, round in cross-section, thick, and spongy in texture. *C. tomentosum* occurs on sea shores in southern England; other species occur in the Mediterranean and Atlantic.

**codominant** A heterozygote that shows fully the *phenotypic effects of both *alleles at a *gene *locus. For example, humans of the AB blood group show the phenotypic effect of both $I^A$ and $I^B$ codominant genes.

**codon** The triplet sequence of *nucleotides in *messenger-RNA that acts as a coding unit for an *amino acid during protein synthesis (*see* RIBOSOME). It binds by base-pairing (*see* BASE PAIR) to a complementary sequence, the anticodon, in *transfer-RNA.

**coefficient of coincidence** The experimental value of the observed number of double *recombinants (crossovers) divided by the expected number.

**coefficient of inbreeding** The probability that two allelic *genes forming a *zygote are both descended from a gene found in an ancestor common to both parents. It is also used for the proportion of *loci at which an individual is *homozygous. The coefficient of consanguinity of an individual is the probability that two *homologous genes drawn at random, one from each of the two parents, will be identical, and therefore homozygous in an offspring. The term is sometimes known as Wright's inbreeding coefficient (F), after the geneticist Sewall Wright who formulated it. $F = 1/2$ for a selfed mating; $1/4$ for full-sib mating; $1/8$ for uncle × niece and aunt × nephew, or double first cousins (i.e. cousins by both parents); and $1/16$ for first cousins.

**Coelomycetes (subdivision *Deuteromycotina)** A class of *imperfect *fungi which produce *conidia within *acervuli or *pycnidia. The vegetative state is a septate *mycelium. They are *saprotrophic on plant debris, soil, etc; some can also cause disease in plants. Some have *perfect states classified in the *Ascomycotina.

**Coelostegia (family *Bombacaceae)** A genus of trees in which the young parts have silvery scales. The fruit is a big, spiny *capsule; the seeds are *arillate. There are 5 species, occurring in the rain forests of Malaya and Borneo.

**coenobium** An algal colony consisting of a definite number of cells in a specific arrangement. The colony as a whole behaves as an individual organism.

**coenocline** A gradient of communities (e.g. in a transect from the summit to the base of a hill), reflecting the changing importance, frequency, or other appropriate measure of different species populations. The term is applied most widely to vegetation records. *See also* ECOCLINE; *compare* COMPLEX GRADIENT.

**coenocyte** A cell or organism with many *nuclei which are not separated by *cell walls. This condition is caused by the repeated division of the nucleus, but not of the *cytoplasm, of the original cell.

**coenozone** *See* ASSEMBLAGE ZONE.

**coenozygote** A *zygote that contains many nuclei owing to the *fertilization of multinucleated *gametes.

**coenzyme** A non-*protein, organic substance that acts as a *cofactor for an *enzyme.

**co-evolution** A complementary *evolution of closely associated species. The interlocking *adaptations of many flowering plants and their pollinating insects provide some striking examples of co-evolution. In a broader sense, predator–prey relationships also involve co-evolution, with an evolutionary advance in the predator, for instance, triggering an evolutionary response in the prey.

**cofactor** A non-*protein component that is required by an *enzyme in order for it to function, and to which it may be either tightly or loosely bound. Cofactors that are tightly bound are known as prosthetic groups. Cofactors are required by many enzymes: they may be metal ions (activators) or organic molecules (coenzymes), and generally act as donors to or acceptors from the *substrate (or substrates) of functional groups of atoms. Examples include *nicotinamide adenine dinucleotide (NAD), *nicotinamide adenine dinucleotide phosphate (NADP), and *adenosine triphosphate (ATP).

**Coffea (family *Rubiaceae)** A genus of 40 species of shrubs and small trees which occur in the Old World tropics, especially Africa. The fruit is a *berry. *C. arabica* gives Arabian coffee, of the highest quality; *C. liberica*, grown mainly at lower elevations, gives Liberian coffee; and *C. canephora* gives robusta coffee. Beans of the latter two are blended with beans from *C. arabica*.

**coffee** *See* COFFEA.

**cohort** **1.** A group of individuals of the same age. **2.** In plant taxonomy, a little-used term meaning a group of related families.

**cohune palm**   See ORBIGNYA.

**coir**   Fibre produced from *Cocos nucifera*. See *Cocos*.

**Coix** (family *Poaceae) A genus of 6 grass species which occur in tropical Asia. *C. lachryma-jobi* is Job's tears, a cultivated cereal of subtropical Asia.

**Cola** (family *Sterculiaceae) A genus of small trees, several species of which provide kola nuts which contain much caffeine and are widely traded in W. Africa as a stimulant to combat fatigue. There are 125 species, occurring in Africa.

**colchicine**   An *alkaloid drug that is obtained from meadow saffron (*see* COLCHICUM). It has a disruptive effect on microtubular activity (*see* MICROTUBULE), though not on that of *microfilaments. Thus, it affects tissue metabolism generally and *mitosis in particular. It can be used to induce *polyploidy in plants, because it prevents the development of the *spindle and so blocks the separation of *chromosomes during mitosis.

**Colchicum** (family *Liliaceae) A genus of 65 species of herbs with underground *corms, occurring from the Mediterranean region to central Asia and India. *C. autumnale* (meadow saffron, autumn crocus) is the source of the drug colchicine.

**coleoptile**   In a grass seed, a sheath surrounding the apical *meristem that protects it as it grows toward the soil surface.

**coleorhiza**   In a grass seed, a protective sheath surrounding the *radicle.

**colinearity**   The correspondence between the order of *nucleotides in a section of *DNA (*cistron) and the order of *amino acids in the *polypeptide that the cistron specifies. The evidence for this is that the positions of mutation within a particular cistron correspond to, and are in the same order as, the positions at which amino acid substitutions are found in the polypeptide coded by the cistron.

**collateral bundle**   A *vascular bundle in which *phloem lies on only 1 side of the *xylem.

**collateral protostele**   See HYPOPHLOIC HAPLOSTELE.

**Collema** (order *Lecanorales) A genus of *lichens in which the *thallus is gelatinous, and typically black when wet. The *phycobiont is a *cyanobacterium of the genus *Nostoc*. The thallus range is *foliose to *fruticose. Species are found on various types of substrate in moist habitats.

**collenchyma**   Tissue that provides strengthening and support for primary structures (i.e. those lacking *secondary thickening), such as young shoots and leaves. It consists of elongated cells with unevenly thickened walls.

**collet**   In vascular plants (*Tracheophyta), the point where the root and stem meet.

**Colletotrichum** (order *Melanconiales) A *form-genus of *fungi which includes a number of important plant *pathogens. *Colletotrichum* species can cause disease (e.g. in legumes, cotton, grasses (including cereals), and flax).

**colloid**   **1.** A substance that is composed of two homogeneous phases, one of which is dispersed in the other. **2. (pedol.)** Soil colloids are substances of very small particle size, either mineral (e.g. clay) or organic (e.g. humus), which therefore have a large surface area per unit volume. Colloids usually provide surfaces with high *cation exchange capacity, and also exhibit an instability controlled by soil chemistry.

**colluvium**   Weathered rock debris that has moved down a hill slope either by creep or by surface wash.

**Collybia** (family *Tricholomataceae) A genus of *fungi in which the *fruit body is *mushroom-shaped with a convex cap, and (when young) an inrolled margin. The *stipe is tough and fibrous. The *spore print is white or pale-coloured. There are many species, some of which are edible, found on the ground, on rotting wood, etc.

**Colocasia** (family *Araceae) A genus of giant, rhizomatous (*see* RHIZOME) herbs, which includes *C. esculenta* (taro, dasheen, or cocoyam). These herbs are cultivated extensively, especially in Melanesia and Polynesia, as a staple carbohydrate. There are 6 species, occurring from India to Polynesia.

**colpate** Applied to a *pollen grain that has 1 or more colpi (*see* COLPUS).

**colpi** *See* COLPUS.

***Colpomenia*** *See* SCYTOSIPHONALES and OYSTER THIEF.

**colpus (pl. colpi)** A germinal groove or aperture on the surface of a *pollen grain, which is elliptical or approximately rectangular in shape and at least twice as long as it is wide. The shape and arrangement of colpi are diagnostic in the identification of pollen. *Compare* PORE.

**columbine** *See* AQUILEGIA.

**columella** The central column of sterile *tissue within a *spore-bearing structure, found in some *fungi and *gymnosperms (e.g. *Diselma*).

**Combretaceae** A family of trees, shrubs, and climbers, in which the flowers are *pentamerous, with an *inferior *ovary surmounted by a *disc. The fruits are often winged, and sometimes are fleshy and edible. *Lumnitzera* and *Laguncularia* grow in mangrove forests and are *viviparous. There are several ornamentals (e.g. *Quisqualis*), and many useful timber trees (e.g. *Combretum*, *Lumnitzera*, and *Terminalia*). There are 20 genera, with about 500 species, occurring mainly in the tropics.

***Combretum*** *See* COMBRETACEAE.

**Commelinaceae** A family of often rather succulent herbs of moist places, which have scented, and often creeping, stems. The flowers are *trimerous. There are many ornamentals. *Tradescantia* is spider plant or wandering Jew, and *Rhoeo* is also cultivated. There are 42 genera, with about 620 species, of mainly tropical distribution.

**commensalism** The interaction between species populations in which one species, the commensal, benefits from another, sometimes called the host, but this other is not affected. *Compare* MUTUALISM and PARASITISM.

**Commons, the** The concept that the major resources of the planet (land, air, and water) are commodities to which all people have equal right of access and use, and which no one has a right to spoil.

**community** A general term applied to any grouping of populations of different organisms found living together in a particular environment; essentially, the *biotic component of an ecosystem. The organisms interact (by *competition, *predation, *mutualism, etc.) and give the community a structure. Globally, the *climax communities characteristic of particular regional climates are called *biomes. Plant ecologists often use the term to cover merely the botanical components of a total biotic community. *See also* INDIVIDUALISTIC HYPOTHESIS.

**community ecology** An approach to ecological study which emphasizes the living components of an ecosystem (the *community). Typically it involves description and analysis of patterns within the community, employing methods of classification and ordination, and examines the interactions of community members, e.g. in the partitioning of resources and in succession. *See also* SYNECOLOGY.

**companion cells** Modified *parenchyma, linked to *sieve cells by *plasmodesmata. Unlike sieve cells, companion cells are nucleated and have many *mitochondria. Companion cells may regulate *translocation.

**companion species** A term formerly used in the *Braun–Blanquet *phytosociological scheme to mean 'indifferent species'. Companion species at the association level sometimes emerge as characteristic (*kennarten, or *faithful) species when communities are classified at higher levels, such as alliances or orders.

**compartment model (box model)** A modelling approach that emphasizes the quantities and materials in different compartments of a system, and may also express connections between compartments by some form of transfer coefficient. The approach is frequently used for studies of whole ecosystems.

**compensation level** The depth at which light penetration in aquatic ecosystems is so reduced that oxygen production by photosynthesis just balances oxygen consumption by respiration. Generally this implies a light intensity of about 1 per cent of full daylight.

**compensation light intensity** The light intensity at which the amount of oxygen released by plants through *photosynthesis is equal to the amount absorbed through *respiration.

**compensation point** **1.** The light intensity at which plants do not accumulate carbon through *photosynthesis, because the rate at which they fix carbon is equal to the rate at which they release carbon by *respiration. **2.** At a given light intensity, the atmospheric concentration of carbon dioxide at which the rate of carbon fixation by photosynthesis is equal to the rate at which carbon is released by respiration.

**competition** The interaction between individuals of the same species, or between different species populations at the same trophic level, in which the growth and survival of one or all species or individuals is affected adversely. The competitive mechanism may be direct (active), as in *allelopathy and *mutual inhibition, or indirect, as when a common resource is scarce. Competition leads either to the replacement of one species by another that has a competitive advantage, or to the modification of the interacting species by selective adaptation (whereby competition is minimized by small behavioural differences, e.g. in feeding patterns). Competition thus favours the separation of closely related or otherwise similar species. Separation may be achieved spatially, temporally, or ecologically (i.e. by adaptations in behaviour, morphology, etc.). The tendency of species to separate in this way is known as the *competitive exclusion, or Gause, principle. Some ecologists differentiate between *interference competition (where space is substituted for a resource) and *exploitation competition (where organisms compete for a resource by enhancing their efficiency in gaining access to it).

**competitive exclusion principle (Gause principle)** The principle that two or more resource-limited species, having identical patterns of resource use, cannot coexist in a stable environment: one species will be better adapted and will out-compete or otherwise eliminate the others. The concept was derived mathematically from the logistic equation by Lotka and Volterra, working independently, and was first demonstrated experimentally by G. F. *Gause (1934) using two closely related species of *Paramecium. When grown separately, both species populations showed normal *S-shaped growth curves; when grown together, one species was eliminated.

**competitive inhibition** The reversible inhibition of the activity of an *enzyme, brought about by the presence of an inhibitor molecule, structurally similar to the normal *substrate, which competes with the substrate for the *active site.

**competitor** In the classification of *plant strategies proposed by J. P. Grime, a plant species that exploits conditions of low stress and low disturbance. *Compare* RUDERAL and STRESS-TOLERATOR.

**complementary genes** Mutant *alleles at different *loci which complement one another to give a *wild-type *phenotype. Dominant complementarity occurs where the *dominant alleles of two or more *genes are required for the expression of a particular trait. Recessive complementarity is the case of suppression of a particular trait by the dominant allele of either gene, so that only the *homozygous double *recessive displays the trait.

**complementation map** A gene map in which each *mutation is represented by a line or 'bar' that overlaps the bars for other mutations which it will not complement. Non-complementing mutants are represented by overlapping, continuous lines. Each bar probably represents the region of the *polypeptide that is distorted by an *amino acid substitution. Complementation maps are usually linear, and the positions of mutants on the complementation and genetic maps usually accord.

**complementation test (*cis-trans* test)**
A test to determine whether two mutant sites of a *gene are on the same
*cistron. It is performed by introducing
two *mutant *chromosomes into the
same cell and observing whether the
*wild-type *phenotype will be expressed,
which it will if each chromosome *com-
plements the defect of the other. If complementation occurs, the two sites must
be in different cistrons. The *genotype of
the complementing *hybrid would be
symbolized as a+/+b (where + represents
the wild-type allele).

**complete penetrance** The case in
which a specified *genotype always mani-
fests itself at the *phenotypic level. For
this to happen, not only must a *dominant gene always produce a *phenotypic
effect; so must a *recessive gene in the
*homozygous state.

**complex gradient** A gradient of environmental factors linked in a complex
fashion, e.g. the interrelated changes in
rainfall, wind speed, and temperature,
found along a transect from high to
low elevation. *See also* ECOCLINE; *compare*
COENOCLINE.

**Compositae** *See* ASTERACEAE.

**compound** Applied to flowers or
leaves that have two or more parts.
*Compare* SIMPLE.

**compression wood** *See* REACTION WOOD.

**conceptacle 1.** An *urceolate cavity
in which *gametes are formed. It is found
inside the inflated tip of the *thallus of
certain brown *algae (e.g. *Fucales), and
has a small opening called the ostiole.
*See also* RECEPTACLE (3). **2.** In the *thallus
of some members of the *Phaeophyta,
a cavity in which gametangia (*see* gametangium) are formed.

**conchocelis stage** In the *life cycle
of *Porphyra*, a *haploid stage preceding
the stage that is recognized as the
*gametophyte.

**concretion** A localized concentration
of material (e.g. calcium carbonate or iron
oxide) in the form of a nodule of varying
size, shape, or colour.

**conduplicate** *See* VERNATION.

**cone** An inflorescence of Coniferophyta. *Ovules and seeds are borne on
sterile scales. *See also* STROBILUS.

**confined aquifer** *See* AQUIFER.

**congenital** Existing at birth; applied
to inherited traits that become apparent
only during growth.

**conidiophore** A specialized *hypha
upon which one or more *conidia are
borne.

**conidiospore** *See* CONIDIUM.

**conidium** (conidiospore) A thin-walled
*spore produced asexually by many types
of *fungi.

**Coniferales** In some classifications,
the order comprising the conifers. *See*
CONIFEROPHYTA.

**Coniferophyta** (Pinophyta) The biggest division of *gymnosperms, with a
long fossil history, comprising trees and
shrubs, nearly all of which are evergreen,
commonly with *monopodial crowns.
Most are resinous. The wood lacks vessels
(*see* TRACHEA and VESSEL ELEMENT). Conifers
are extremely important for timber and
paper production. The leaves are often
needle- or scale-like. Fertile parts occur in
unisexual cones, variously containing
sterile scales. *Stamens are borne on
commonly *peltate scales. The *ovule
and seed are naked and borne on a
scale. They first appear as fossils in
*Carboniferous rocks. There are about
550 extant species.

**Coniferopsida** In some classifications,
a subdivision of the *gymnosperms, comprising 4 orders, the oldest of which,
the *Cordaitales, appeared in the *Carboniferous. They became extinct at the
end of the *Permian, but their place
was taken by 2 other coniferopsid orders:
the Ginkgoales (ginkgos) and Coniferales
(conifers). These are now usually
ranked as divisions, the *Ginkgophyta
and *Coniferophyta.

**Coniophora** (family *Coniophoraceae) A
genus of *fungi in which the *fruit bodies
are *resupinate and more or less *convoluted. In *C. puteana* (*C. cerebella*) the fruit

bodies are thin, flattened, irregular in shape, and initially creamy-white, becoming greenish-yellow on maturation. This species is common on wet, decaying wood, and on timbers in damp buildings.

**Coniophoraceae (order \*Aphyllophorales)** A family of \*fungi in which the \*spores are coloured, smooth, and have double-layered walls. The family includes some important timber-rotting fungi.

**conjugated protein** A \*protein that contains a non-protein component; this may be a metal ion or an organic substance.

**conjugation** A process whereby organisms of identical species (but opposite mating types) pair and exchange genetic material (\*DNA); i.e., each individual may both give and receive material. This process of \*sexual reproduction is found only in unicellular organisms, although the term is sometimes also applied to the union of \*gametes, particularly in \*isogamy (found in \*fungi and some green \*algae). The gametes are not released into the external environment. (E.g., in *Spirogyra* a conjugation tube, formed by the fusion of protuberances from conjugating cells, acts as a passageway for the gamete of the cell to move into another cell and fuse with its nucleus.) Details of the process differ greatly between different organisms. For example, in bacteria only DNA is transferred from one cell to another, while in some \*eukaryotic microbes the process may involve the fusion of two entire \*protoplasts. The transfer of DNA may be unidirectional (as in the case of \*Escherichia coli), in which case one cell is called the donor and the other the recipient, or bi-directional (as in \*Paramecium aurelia).

**conjunctive symbiosis** \*Symbiosis that involves the union of 2 different species into a single body (e.g. the \*mycobiont and \*phycobiont in a \*lichen). *Compare* DISJUNCTIVE SYMBIOSIS.

**conk** A term used by foresters and wood-cutters to refer to the \*fruit bodies of certain wood-rotting \*fungi, particularly those of \*polypores.

**Connaraceae** A family of mostly woody climbing plants in which the leaves are usually \*pinnate, without \*stipules. The flowers are \*pentamerous. The fruit is usually a 1-seeded \*follicle, the seed \*arillate. There are 20 genera, with about 380 species, occurring throughout the tropics, mainly in the Old World.

**connate** Applied to similar organs (e.g. leaves or \*petals) that are joined together. *Compare* ADNATE.

**connecting band** During sexual reproduction in diatoms (\*Bacillariophyta), a band that connects the \*epithecium to the \*hypothecium.

**connective** A type of tissue that connects the \*pollen sacs of an \*anther.

**Conocephalum (order \*Marchantiales)** A genus of robust, \*thallose \*liverworts. *C. conicum* has a broad, ribbon-like \*thallus which is irregularly forked and up to 20 cm long and 2 cm wide. The upper surface is perforated with pores which are quite clearly visible to the naked eye; each pore occurs at the centre of a hexagonal marking beneath which is an air chamber (areola). These liverworts are common on rocks and on soil in habitats that are both moist and somewhat shaded, e.g. on banks of streams and ditches.

**Conocybe (family Bolbitiaceae, order \*Agaricales)** A genus of \*fungi that form \*mushroom-shaped \*fruit bodies with a regular cap and central \*stipe. The \*gills are yellowish to cinnamon-coloured. The \*spore print is bright \*rusty brown in colour. They are found mostly in grassy places.

**consequential dormancy** *See* DORMANCY.

**conservation** The maintenance of environmental quality and resources. The resources may be physical (e.g. fossil fuels), biological (e.g. tropical forest), or cultural (e.g. ancient monuments). In modern scientific usage conservation implies sound \*biosphere management within given social and economic constraints, producing goods and services for humans without depleting natural ecosystem diversity, and acknowledging the naturally dynamic character of biological systems. This contrasts with the preservationist approach, which, it is argued, protects species or landscapes without reference to natural change in living systems or to human requirements. *See also* BIOLOGICAL CONSERVATION.

**consistence (consistency)** The resistance of soil to physical impact such as ploughing, digging, or handling. It is controlled by the degree of adhesion between soil particles. It is described when dry as loose, soft, or hard, when moist as loose, friable, or firm, and when wet as sticky, or plastic.

**consistency** *See* CONSISTENCE.

**consociation** A phytosociological term of the British and American traditions, meaning a *community with a single dominant species, e.g. an oak or beech wood. The term is also used by the *Uppsala school. (A community with several dominants is a *plant association.)

**consocies** A little used term, derived from the Clementsian scheme for vegetation classification, that refers to a seral community in the succession to a *climax community with *consociation status. *See also* PHYTOSOCIOLOGY.

**conspecific** Applied to individuals that belong to the same species.

**constancy** *See* CONSTANT SPECIES.

**constant species (constancy)** In *phytosociology, a species common to a particular association or *community, but not necessarily confined to that community; a species of wide ecological amplitude compared with a *faithful species. A species of high constancy would be present in all, or almost all, of a series of relevés or field samples that describe an association or community.

**constellation diagram** A representation of species affinities based on $\chi^2$ as a measure of the association between species. The reciprocal of the $\chi^2$ value for each species pair is used to plot the diagram, so that highly positively associated species with large $\chi^2$ values are positioned closely together. Thus, clusters of similarly distributed species may emerge, while the transitional affinities of a species with those of another main focal area or cluster will be evident. The constellation diagram exemplifies a simply calculated ordination method.

**constitutive enzyme** An *enzyme that is always produced whether or not a suitable *substrate is present. Such enzymes are sometimes produced by particular regulatory mutants which, though not affecting the structure of the enzyme, instead affect the process by which its synthesis occurs. An example is the *lac*-operon, which controls the synthesis of three enzymes (beta-galactosidase, permease, and acetylase): enzymes that are involved in the lactose metabolism of the bacterium *Escherichia coli*.

**consumer organism** In the widest sense, a *heterotrophic organism that feeds on living or dead organic material. Two main categories are recognized: (*a*) macroconsumers, mainly animals (*herbivores, *carnivores, and *detritivores), which wholly or partly ingest other living organisms or organic particulate matter; and (*b*) micro-consumers, mainly bacteria and fungi, which feed by breaking down complex organic compounds in dead protoplasm, absorbing some of the decomposition products, and at the same time releasing inorganic and relatively simple organic substances to the environment. Sometimes the term 'consumer' is confined to macroconsumers, micro-consumers being known as *'decomposers'. Consumers may then be termed 'primary' (herbivores), 'secondary' (herbivore-eating carnivores), and so on, according to their position in the food-chain. Macroconsumers are also sometimes termed *phagotrophs or biophages, while microconsumers correspondingly are termed *saprotrophs or *saprophages. *Compare* PRODUCER.

**contact inhibition** The cessation *in vitro* of both movement and replication in a cell on making contact with other cells, such that a confluent monolayer is formed in the culture. Probably it occurs as a result of the formation of cytoplasmic (*see* CYTOPLASM) bridges between cells.

**contagious distribution** *See* OVER-DISPERSION.

**context** A type of tissue that constitutes a fungal *fruit body.

**contiguous grids** A system of adjacent quadrants. *See also* GRID ANALYSIS OF PATTERN.

**continental South-east Asia floral region** Part of R. Good's (*The Geography of the Flowering Plants*, 1974) *palaeotropical kingdom, which lies within the Indo-Malaysian subkingdom. Floristically it is a poorly documented region, transitional between the rich floras of China to the north and Malaysia to the south. There are relatively few endemic genera (*see* ENDEMISM), probably around 250, though the proportion of endemic species is thought to be high. This region is very likely the source of such important crop plants as rice, tea, and citrus fruit. *See also* FLORAL PROVINCE and FLORISTIC REGION.

**contingency table (two-way table)** A table of data for two methods of classification of the same individuals, e.g. leaf shape and flower structure. This type of data can then be analysed statistically for association between these properties using a $\chi^2$ test.

**continuous distribution** Data that yield a continuous spectrum of values e.g. the height of a plant.

**continuous variation** An assemblage of measurements of a phenotypic character which form a continuous spectrum of values. The continuity of phenotype is a result of two phenomena: (*a*) each *phenotype does not have a single phenotypic expression but a norm of reaction that covers a wide phenotypic range; (*b*) there may be many segregating loci whose alleles make a difference to the phenotype being observed.

**continuum** The idea that vegetation is continuously variable and cannot be classified into discrete entities, since it shows gradual change in response to environmental change. Such change may be analysed using *ordination methods, e.g. the 'continuum approach' of the *Wisconsin ordination scheme.

**continuum index** A measure of the total environment of a stand of trees, expressed in terms of species composition and their relative abundance. In the *Wisconsin ordination scheme the index

is calculated by multiplying the *importance value of each species in each sample by the adaptation number, and summing these results for each stand. *See also* CLIMAX ADAPTATION NUMBER and IMPORTANCE VALUE.

**contractile root** A specialized root, the upper part of which contracts as soon as the tip is firmly secured in the soil. This contraction pulls the bulb, corm, rhizome, or other structure to an appropriate depth.

**contractile vacuole** In many freshwater, unicellular organisms, a membranebound cavity that expands and contracts with a pulsating movement as it takes in water from the cell contents and expels it from the cell. Contractile vacuoles serve to regulate the water content of the cell.

**contraction limit** *See* ATTERBERG LIMITS.

**controlled pollination** A practice used in plant hybridization (*see* HYBRID). *Pistillate flowers are enclosed in bags (usually of muslin) to protect them from unwanted *pollen: when they are in a receptive condition, the flowers are dusted with pollen of the required type. This is commonly performed in horticulture and in genetic experiments.

**controlling gene** A gene that is involved in turning on or off the *transcription of structural *genes. Two types of genetic element exist in this process: a regulator and a receptor element. A receptor element is one that can be inserted into a gene, making it a *mutant, and can also exit from the gene. (The mutation is thus unstable.) Both of these functions are under the influence of the regulator element.

***Convallaria* (lily of the valley; family *Liliaceae)** A genus of plants that have long-stalked, oval leaves in pairs, and leafless, 1-sided *spikes of white, globose to bell-shaped, 6-lobed flowers. The fruit is a red, globular *berry. There are 3 species, occurring in northern temperate regions.

**convergent evolution** The development of similar characteristics in organisms that are unrelated (except through distant ancestors) as each adapts to a similar way of life. *Compare* DIVERGENT EVOLUTION and PARALLEL EVOLUTION.

**convolute**  *See* VERNATION.

**Convolvulaceae**  A family mostly of twining *herbs, but with some trees and shrubs, in which the flowers are regular, with 5 *sepals, and a 5-lobed *corolla which is often showy, bell- or funnel-shaped, and hardly lobed. There are 5 *stamens fixed to the corolla tube. The *ovary is *superior, and 1- to 4-celled. The fruit is a *capsule or (rarely) fleshy. There are 58 genera, with about 1650 species, occurring mainly in tropical but also in temperature regions.

**Conyza** (family *Asteraceae) A genus of *annual or *perennial *herbs which have *simple leaves arranged in a spiral along the stem. The flower heads are usually held in clusters or *corymbs. The central *florets are bisexual and the outer ones female, with tubular, simple *corollas. There are no broad *petals in the ray florets. The *achenes are usually softly hairy and compressed, with a simple row of *pappus hairs. Many species are N. American weeds. There are about 50 species, found in temperate and subtropical regions.

**Cooksonia hemispherica** (family **Rhyniaceae**) An extremely primitive plant from the Upper *Silurian and Lower *Devonian belonging to a family whose members did, however, possess an *epidermis and *stomata to control the passage of gases. They also had an underground rooting portion, the *rhizome, and they branched *dichotomously. *C. hemispherica* is the first representative of the Rhyniaceae, and is known from the Silurian of Europe.

**co-operativity**  Interaction between binding sites within a *protein molecule whereby the binding of a *ligand to one site influences the affinity of other sites on the same molecule for further ligands.

**Copernicia** (family *Arecaceae) A genus of big, solitary palms which have fan-shaped leaves. The inflorescences are *axillary and much-branched, with subtending

tubular *bracts. Several species are of spectacular beauty and are in cultivation. *C. cerifera* yields carnauba wax. They occur in the lowland seasonal tropics, mainly in the W. Indies (especially Cuba), with 29 species in total.

**copper** (Cu)  An element that is required in small amounts by plants, although high concentrations of it can be toxic. It is found bound to *proteins and is involved in oxidation–reduction reactions, especially those involving molecular oxygen. The signs of copper deficiency can be very varied: leaves may become chlorotic (*see* CHLOROSIS) or dark green; the bark of trees may blister; shrubs may become very bushy.

**copper moss**  A moss that grows on copper-rich rocks. Such mosses, including species of *Mielichhoferia*, *Dryptodon*, and *Merceya*, can survive levels of copper far in excess of those lethal to other mosses.

**coppice  1.**  A traditional European method of woodland management and wood production, in which shoots are allowed to grow up from the base of a felled tree. Trees are felled in a rotation, commonly of 12–15 years. A coppice may be large, in which case trees, usually ash (*Fraxinus*) or maple (*Acer*), are cut, leaving a massive stool from which up to 10 trunks arise; or small, in which case trees, usually hazel (*Corylus*), hawthorn (*Crataegus*), or willow (*Salix*), are cut to leave small, underground stools producing many short stems. The system provides a continuous supply of timber for fuel, fencing, etc.,

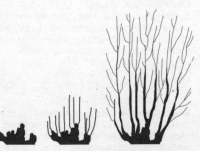

*Coppice*

but not structural timber. In Britain, coppicing has largely been abandoned now, except for conservation purposes, since high labour costs and alternative fuels and materials render the practice unprofitable. One consequence of coppicing is that the stool enlarges because each subsequent growth of shoots occurs on its outside. The diameter of a stool is thus directly related to its age. **2.** The smaller trees and bushes that regenerate from cut stumps and occasionally (e.g. in *Ulmus* species) from root suckering. **3.** An area of land in which underwood and timber is or was grown. **4. (copse)** Any type of wood in which the shrub layer predominates and is periodically coppiced. **5.** The action of cutting coppice.

**coppiced scrub** The regrowth of trees resulting from irregular coppicing (*see* COPPICE).

**coppice shoot** A shoot that arises from a *bud at the base of the *stool of a tree that has been cut near the ground.

**coppice stump (moot, stool)** The remnant of a tree that has been cut, usually to or near to ground level, and from which *coppice growth develops.

**coppice-with-standards** A *coppice system in which scattered trees, typically oak (*Quercus robur*), are allowed to grow to their full height (standards) for use as structural timber, while the understorey, commonly of hazel (*Corylus avellana*), is coppiced.

**copra** The *endosperm of *Cocos nucifera*. *See* Cocos.

**Coprinaceae (order *Agaricales)** A family of *fungi in which the *spores in masses appear dark brown or black. The *fruit bodies are often fragile. Species are found on dung, soil, or rotting wood.

***Coprinus* (family *Coprinaceae)** A genus of *fungi which form *mushroom-like *fruit bodies. The *gills are parallel-sided, and typically liquefy (*see* AUTOLYSIS) from the lower edge upwards as the *spores mature. There are many species. Most are found on dung, but some grow on rotting timber, tree-stumps, etc. Some species are edible. *See also* INK CAP.

**coprophilous** Growing on or in dung.

***Coprosma* (family *Rubiaceae)** A shrubby genus of 90 species with fleshy fruits of eastern Malesia, New Zealand, south-eastern Australia, and the Pacific islands; probably it is dispersed by birds.

**copse 1.** *See* COPPICE (4). **2.** In modern usage, any small wood, particularly a detached wood that is isolated from other woodland.

**copse-bank** A boundary between areas of land made from an earth bank, especially where the boundary separates an intermittently enclosed *coppice within a wood-pasture system.

**coral fungi** Those *fungi of the families *Clavariaceae and Clavulinaceae that form coralloid *fruit bodies.

**Corallinaceae (order Cryptonemiales, division *Rhodophyta)** A family of red seaweeds, most of which are heavily impregnated with lime, and hard and brittle. There are many genera. Some have a crustose *thallus which adheres closely to rocks or to other plants. *Corallina* has an erect, branching thallus that is characteristically jointed; *C. officinalis* is a common species fringing rock pools in the intertidal zone. Members of the family are found in all regions, from the Arctic to the tropics. Some important constituents of atolls and coral reefs, being able to survive pounding by surf better than true corals.

**coralloid** Coral-like; branching repeatedly.

**coral spot** The common name for the coral-pink *conidial pustules formed by the *fungus *Nectria cinnabarina* on twigs, branches, etc.

***Corchorus capsularis* *See* TILIACEAE.

**cord** A stack of wood, usually measuring 2.54 m (8 ft. 4 in.) long by 1.27 m (4 ft. 3 in.) high, but with local variations. It is usually composed of small-diameter material. A cord of this size measures about 3.6 m³ (128 cu. ft.) when stacked, and contains 2–2.8 m³ (75–100 cu. ft.) of wood. A 'short' cord is equal to half a cord.

**Cordaitales** An extinct *gymnosperm order, included in the Coniferopsida, which appeared in early *Carboniferous times and disappeared towards the end of the *Permian. The order produced trees up to 30 m high with strap-like leaves and primitive cones.

**cordate** Of a leaf, heart-shaped.

**cordwood** A small-diameter wood that has been cut into lengths suitable for use as domestic fuel or for making charcoal.

**Cordyline** (family *Agavaceae) A genus of plants which are unusual among monocotyledonous (see MONOCOTYLEDON) plants in possessing woody stems. The leaves are narrow and tufted, some yielding useful fibre, others decorative. There are 15 species, occurring in the tropics and warm temperate regions.

**coremium (mycol.)** A sheaf-like bunch, or fascicle, of *hyphae or *conidiophores.

**co-repressor** A metabolite that in conjugation with a *repressor molecule binds to the *operator gene present in an *operon and prevents the synthesis of a repressible enzyme.

**coriaceous** Having a leathery texture.

**Coriariaceae** A family of shrubs comprising 1 genus (*Coriaria*), with 5 species of plants that have *imbricate *sepals, persistent, internally keeled *petals to the regular flowers, and 5–10 free *carpels which form a fruit of several cocci (see COCCUS) enclosed in the enlarged petal bases. The range covers Eurasia, New Zealand, and Central and S. America.

**Coriolus** (family *Polyporaceae) A genus of *fungi in which the *hymenium lines tubes that open in small, irregular pores on the underside of a non-stipitate, typically *bracket-like *fruit body. Species are found on wood. *C. versicolor* (*Trametes versicolor, Polystictus versicolor*) is very common on dead wood of broad-leaved trees. The fruit bodies often grow in tiers, and are thin, semicircular brackets with a velvety upper surface usually marked with concentric bands of yellow, brown, grey, etc; the margin and lower surfaces are white.

**cork (phellem)** In woody plants, a layer of protective *tissue that forms below the *epidermis. It comprises dead cells, derived from the cork *cambium (*phellogen), and coated with a waxy substance (*suberin) that renders them waterproof. Cork develops abundantly in the bark layer of certain plants, e.g. *Quercus suber* (cork oak), and is removed for commercial use. *See also* QUERCUS.

**cork cambium** *See* PHELLOGEN.

**corkwood** *See* DUBOISIA.

**corm** In plants, an underground storage organ formed from a swollen stem base, bearing *adventitious roots and scale leaves. Often it is renewed annually, each new corm forming on top of the preceding one. It may function as an organ of *vegetative reproduction or in *perennation.

**corn** *See* ZEA.

**Cornaceae** A family of trees, shrubs, and a few *herbs, which are closely related to the *Apiaceae in having umbellate or *panicled inflorescences, an *inferior, usually 2-celled *ovary with 1 *ovule in each cell, and *tetra- or *pentamerous flowers; but, unlike the Apiaceae, they have leaves that are usually *opposite and entire, and fruits that are fleshy. *Aucuba* and *Cornus* species are cultivated for ornament. There are 12 genera, with about 90 species, mostly northern temperate in distribution, but also occurring in Africa, S. America, New Zealand, and South-east Asia.

**corn salad** (*Valerianella*) *See* VALERIANACEAE.

**corolla** A collective term for all the *petals of a flower. The corolla is a non-reproductive structure, often arranged in a *whorl. It encloses the reproductive organs and, with the *sepals when present, protects them. *Petals are often brightly coloured and attract pollinating animals.

**corona** A series of *petal-like structures in a flower, either outgrowths of the petals or modified from the *stamens.

**corpus** **1.** The body of a *pollen grain that has bladders, or air-filled sacs (e.g. *Pinus*). **2.** In *angiosperms, the cells below the *tunica of the apical *meristem.

**correspondence analysis**  *See* RECIP-
ROCAL AVERAGING.

**corridor farming**  *See* AGROFORESTRY.

**cortex**  An outer layer. In plants, *tissue
immediately below the *epidermis but
outside the *vascular bundles.

**cortical canal**  *See* VALLECULAR CANAL.

**corticated**  Containing *cortex tissue.

**corticolous**  Growing on or in tree
*bark.

**cortina**  A mesh of fine, web-like fibres
extending from the edge of the *pileus to
the *stipe in the *fruit bodies of certain
*agarics.

**Cortinariaceae (order *Agaricales)**  A
family of *fungi in which the *spores in
masses appear brownish or rust-coloured.
The *fruit body characteristically has a
fine, cobweb-like *cortina when young;
the *stipe (when present) and *pileus are
confluent. There are many genera, found
mainly on soil, and particularly in woods,
where many species form *mycorrhizal
associations with trees.

**Cortinarius (family *Cortinariaceae)**  A
very large genus of *fungi in which the
*mushroom-like *fruit bodies have a fine,
cobweb-like *cortina when young. The
*gills are rust brown at maturity, and the
*spore print is rust-brown. *Cortinarius*
species are found on the ground in woodland, where they often form *mycorrhizal
associations with trees. The numerous
species are not easy to distinguish from
each other in the field.

**Corylus (family *Betulaceae)**  A genus
of small trees or shrubs which often
reproduce by suckering. *C. avellana*, *C.
maxima*, and others have edible *nuts
(subtended by leafy *bracteoles), the
hazel, filbert, and cobnut. Formerly they
were coppiced to provide brushwood for
fuel and light constructions (*see* COPPICE).
There are about 10 species, which occur in
northern temperate forests.

**corymb**  A *racemose *inflorescence in
which the lower *pedicels are longer than
the upper so that the flowers lie as a dome,
or dish, and the outline is roundish or
flattish.

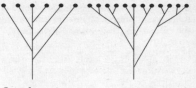

*Corymb*

**Corynebacterium (order Actino-
mycetales)**  A genus of *Gram-positive
*bacteria, in which the cells tend to be
irregular in shape, but are usually straight
or slightly curved rods. They are *chemo-
organotrophic, and are found in a wide
range of habitats. Some are *pathogenic
in animals (e.g. *C. diphtheriae* can cause
diphtheria).

**coryneform**  Applied to the club-
shaped and *pleomorphic cells of
*Corynebacterium* species and similar
*bacteria.

**Corypha (family *Arecaceae)**  A genus of
giant fan palms which are among the
most massive of all palms. The stem is
solitary, forming a gigantic, terminal,
tree-like inflorescence, and then dying.
The flowers are bisexual. The *petals are
free, and the *ovary *trilocular: only 1
*locule develops to fruit. The stem yields
a reddish sago. The leaf *sheaths yield
useful fibre and the blades are used for
writing material. The *endosperm is a
vegetable ivory. *C. umbraculifera* (talipot
palm), is the most massive species.
There are 8 species, found from India to
northern Australia in seasonally dry
climates.

**cosmopolitan distribution**  A distribution of an organism that is worldwide
or panendemic. Apart from *weeds, commensal animals, and some of the lower
groups of *cryptogams, there are relatively few organisms that occur on all 6
of the widely inhabited continents. *See* BI-
CENTRIC DISTRIBUTION; DISJUNCT DISTRIBU-
TION; and UNICENTRIC DISTRIBUTION.

**costa (pl. costae)**  A ridge, midrib, or
vein.

**Costus (family *Zingiberaceae)**  A genus
of plants which are distinctive among
the gingers for their spiralling stems with

spirally borne leaves. The inflorescence is a terminal head. There are about 90 species, with a pantropical distribution.

**cotidal line** A line joining points at which given tidal levels (such as mean high water or mean low water) occur simultaneously. The lines are shown on certain hydrographic charts. The same information is contained in tide tables, where the data are given as differences from the times of high or low water at a 'standard port'.

***Cotinus*** (family *Anacardiaceae*) A genus of shrubs whose resin is mildly irritant. *C. coggygria*, found from the Mediterranean region to China, is the smoke bush or wig tree, and is much cultivated. There are 3 species, occurring in warm temperate regions.

***Cotoneaster*** (family *Rosaceae*) A genus of trees and low shrubs which resemble *Crataegus* but are not thorny. Members of the genus have entire leaves and *carpels that are free from each other on the inner sides, with stony walls. The fruit has a mealy, fleshy case formed from the *receptacle cup. Several species are grown in gardens. There are about 50 species, with a palaearctic and temperate distribution.

**cotton** *See* MALVACEAE.

**cottonwood** *See* POPULUS.

**cotyl** In seed plants, the point where the *hypocotyl and *epicotyl meet and the *cotyledons arise. *Compare* COLLET.

**cotyledon** A seed leaf that is borne on a plant *embryo. Characteristically there is 1 leaf in the *monocotyledons, and 2 leaves in the *dicotyledons, but there are exceptions, especially in the latter group.

***Coula edulis*** (African walnut) *See* OLACACEAE.

**coupe** An area that is to be cut in a particular year of a *coppice rotation.

**covalent bond** A bond in which a pair (or pairs) of electrons is shared between two atoms. The bond is often represented by drawing a single line between the symbols of the two atoms that have bonded together. Sometimes the bonding is between atoms of different elements (e.g. hydrogen chloride, H—Cl), and sometimes between atoms of the same element (e.g. fluorine, F—F). The name 'molecule' is used to describe any uncharged particle containing covalently bonded atoms. *See also* HYDROGEN BOND, IONIC BOND, and METALLIC BOND.

**cover** In descriptions of plant communities, the proportion of ground, usually expressed as a percentage, that is occupied by the perpendicular projection down on to it of the aerial parts of individuals of the species under consideration. The most widely used visual scales are the *Domin scale and the *Braun-Blanquet 5-point scale. A more objective estimate may be obtained using a *pin-frame sample of a *point-quadrat.

**cover–abundance measure** A linked scheme for estimating cover visually. It is based on percentages at the top end, but uses abundance estimates for species with low-cover values. The most widely known is the *Domin scale.

**covered smut** A disease of cereals caused by *smut fungi (order *Ustilaginales) in which the fungal *spore masses are enclosed within the *glumes of the cereal and are not released until threshing. *Compare* LOOSE SMUT.

**cover–sociability scale** A method for recording vegetation, devised by *Braun-Blanquet, in which two scales are combined. The first describes the number and cover of a species, ranging from presence through 5 scales of cover; the second indicates the spatial arrangement of the individuals concerned, e.g. as 'isolated' or 'clumped'.

**cow's tail pine** *See* CEPHALOTAXACEAE.

**cow-tree** (milk tree) *See* BROSIMUM.

***Coxiella*** (family *Rickettsiaceae*) A genus of *bacteria which are *parasitic inside the cells of vertebrate or arthropod hosts. The cells are *pleomorphic, non-*motile rods. There is one species, *C. burnetii*. This is the causal agent of Q fever in humans. The organism is resistant to adverse environmental conditions, e.g.

high temperatures and desiccation, and can often survive pasteurization.

**cp-DNA**   See CHLOROPLAST-DNA.

**crab apple**   See *MALUS*.

**crab grass**   See *PANICUM*.

**cranberry**   See *VACCINIUM*.

**crassula (bar of Sanio)** In the *tracheids of *gymnosperms, a bar-like thickening in the cell wall and intercellular material. Crassulae occur in pairs, associated with the *bordered pits.

**Crassulaceae** A family of usually succulent *herbs or low shrubs, related to the *Rosaceae, in which the flowers are regular and usually *pentamerous, with the *stamens alternating with the *petals or twice their number. The *carpels are usually free, forming *follicles (or a *capsule) in the fruit. They show unusual (*crassulacean) respiratory metabolism. Many are cultivated, especially *Sedum* and *Sempervivum* species (stonecrops, house leeks). There are 33 genera, with about 1300 species. Most occur in dry, warm, temperate regions, but they are cosmopolitan.

**crassulacean acid metabolism (CAM)** A method of carbon dioxide fixation that conserves water in certain succulent, drought-resistant plants. At night, when the external temperature and therefore the evaporation rate are low, the leaf stomata open allowing carbon dioxide to enter leaf cells where it is incorporated in an organic acid. During the day the stomata remain closed, conserving moisture, while the acids are decarboxylated and the carbon dioxide is used for the *dark reactions of *photosynthesis. The initial fixation of carbon dioxide results in the formation of compounds with four carbon atoms; this process uses the enzyme phosphoenol pyruvate carboxylase, as in $C_4$ photosynthesis (see $C_4$ PLANTS). In CAM, however, there is no spatial movement of this product prior to processing by the Calvin cycle. This pathway of carbon fixation was first observed in members of the *Crassulaceae, hence the name.

***Craterellus* (order *Cantharellales)** A genus of *fungi in which the *hymenium is borne on the smooth or wrinkled lower surface of the *fruit body. In most species the fruit body is dark-coloured. They are found on the ground under trees. See also HORN OF PLENTY.

***Cratoxylum* (family *Clusiaceae)** A genus of medium-sized trees which produce a light, general-purpose timber. The fruit is a *capsule, the seeds are winged. There are 6 species, occurring in tropical South-east Asia.

**Crenarchaeota** The less derived (see APOMORPH) of the *Archaea, consisting principally of extreme *thermophiles and *psychrophiles. Members of the Crenarchaeota show a greater genetic similarity to *eukaryotes than to members of the *Eubacteria. In the widely used five-kingdom system of classification, the Crenarchaeota form a group within the Archaea subkingdom of the *Bacteria. In the three-domain system they comprise a kingdom within the *domain Archaea.

**crenate** With a round-toothed or scalloped edge or margin.

**crenulate** Finely notched.

**creosote fungus** *Hormoconis resinae* (formerly *Cladosporium resinae*), a *fungus which can grow in the presence of creosote.

***Crepidotus* (family Crepidotaceae, order *Agaricales)** A genus of *fungi in which the *fruit body has no *stipe, or a very short stipe on the underside of the cap. The cap is small (up to 2–3 cm) and typically kidney-shaped or shell-shaped; *gills are white or pale at first, becoming coloured cinnamon or pinkish. They are found on wood or on other plant debris.

**cress**   See *LEPIDIUM*.

**Cretaceous** The third of the three periods that comprise the *Mesozoic Era. The Cretaceous lasted from 145.5–65.5 Ma ago; its end is defined by the mass extinction of many invertebrate and vertebrate stocks associated with a bolide impact. The period is noted for the deposition of the chalk of the White Cliffs of Dover, England, much of the chalk being derived from the calcareous plates (*coccoliths) of marine *algae. *Angiosperms, which arose during the *Jurassic, came to dominance during the Cretaceous at the expense of such groups as cycads (*see* CYCADOPHYTA) and pteridosperms (*see* PTERIDOSPERMALES). Woody species evolved during the Cretaceous, and around 65 Ma stratified forests appeared and there was an increase in the variety of fruits.

**Crick, Francis Harry Compton (1916–2004)** The British geneticist who, with J. D. *Watson and M. Wilkins, won the 1962 Nobel Prize for Physiology or Medicine for their modelling of the DNA (*see* DEOXYRIBOSE NUCLEIC ACID) molecule. Crick and Watson worked at the Cavendish Laboratory, Cambridge; in 1977 Crick moved to the Salk Institute, California.

**Crinum (family *Amaryllidaceae)** A genus of large, bulbous plants with *umbels of showy flowers. The bulbs of some species are cultivated in non-tropical countries. There are about 130 species, occurring in the tropics, especially on coasts.

**criss-cross inheritance** The transmission of a *gene from mother to son or father to daughter.

**crista (pl. cristae)** An infolding of the inner membrane of a *mitochondrion that projects into the matrix of the *organelle. Cristae bear numerous mushroom-shaped bodies, variously called respiratory assemblies, stalk bodies, or elementary particles, which contain certain of the *enzymes involved in *ATP synthesis.

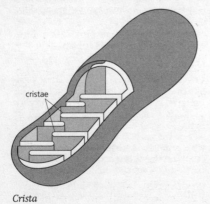

*Crista*

**Cristispira (family Spirochaetaceae, order *Spirochaetales)** A genus of spirally shaped *bacteria in which the cells each contain 100 or more periplasmic flagella. They are found in a variety of marine and freshwater molluscs and other aquatic organisms.

**Crocosmia (*Tritonia*, montbretia; family *Iridaceae)** A genus of plants that have *corms and 2-ranked, narrow leaves flattened in 1 plane, as in irises. There are 11 species, occurring in S. Africa.

**Crocus (family *Iridaceae)** A genus of herbs in which tufts of narrow leaves arise from *corms and have white midribs. The flowers are solitary, *subsessile on the corms, and regular. They have 6 similar *perianth segments and an extremely long perianth tube resembling a flowerstalk. There are three *stamens on the tube. The *ovary is 3-celled and *inferior, forming a *capsule lifted up on a stalk in the fruit. Many species are cultivated for their beauty. Saffron is produced from the style branches of *C. sativus*. There are about 70 species, occurring in the temperate Old World, especially in the Mediterranean region.

**cross-breeding** Usually, *outbreeding or the breeding of genetically unrelated individuals. In plants this may entail the transfer of *pollen from one individual to the *stigma of another of a different *genotype.

**cross-dating** The matching of *treering width patterns and other properties among the trees and fragments of wood from a particular area. This enables the year in which each ring was formed in living trees and recent stumps to be determined accurately, the presence of false rings or the absence of rings in individual specimens being made apparent. By matching ring series from living specimens with those from older (e.g. constructional) timbers, the chronology may be extended backwards (a dating procedure known as *dendrochronology).

**crossing-over** The exchange of genetic material between *homologous *chromosomes by *breakage and reunion. This occurs during pairing of *chromosomes at prophase I of *meiosis, and in some organisms (for example certain fungi, and insects such as *Drosophila*) may also occur at *mitosis. The temporary joins between chromosomes during crossing-over are called *chiasmata.

**cross-over region** The segment of a *chromosome lying between any two particular *genes used as markers, for example during gene mapping by recombination experiments (*see* RECOMBINANT).

**cross-pollination** The transfer of *pollen from the *stamen of a flower to the *stigma of a flower of a different *genotype but usually of the same species, with subsequent growth of the *pollen tube. *See also* CROSS-BREEDING.

***Croton*** (family *Euphorbiaceae) A genus of small trees and shrubs, commonly with *stellate hairs, producing a fruit which is a *capsule. Several are medicinal (croton oil from *C. tiglium* is a powerful purgative). There are 750 species, occurring throughout the tropics and subtropics.

**croton** *See* CODIAEUM.

**croton oil** *See* CROTON.

**crotovina** *See* KROTOVINA.

**crowberry** *See* EMPETRACEAE.

**crown gall** A common and widespread plant disease which can affect a very wide range of woody and herbaceous plants. The disease is caused by the bacterium *Agrobacterium tumefaciens*. *Galls are formed at the crown (stem–root junction) or, less commonly, on roots, stems, or branches of infected plants. On herbaceous plants the galls are usually soft, while on woody plants they tend to be hard. On trees, crown galls may reach sizes of 1 metre or more across.

**crozier** **1.** The hook-shaped tip, reminiscent of a bishop's crozier, of the ascogenous (*see* ASCOGONIUM) dikaryotic (*see* DIKARYON) hypha. Its 2 nuclei divide simultaneously, followed by *cytokinesis, and further divisions lead to the formation of an *ascus. **2. (fiddlehead)** In ferns, the tightly coiled leaf before it opens.

**cruciate (cruciform)** Cross-shaped.

**Cruciferae** *See* BRASSICACEAE.

**cruciform** *See* CRUCIATE.

**crumb structure** A type of soil structure in which the structural units or peds have a spheroidal or crumb shape. Crumb structure is often found in more porous than granular organo-mineral surface soil *horizons, and provides optimal *pore space for soil fertility.

**crustaceous** *See* CRUSTOSE.

**crustose (crustaceous)** Crust-like in appearance, usually used of certain types of *lichen *thallus.

**cryic** *See* PERGELIC.

**cryophilic** Applied to organisms preferring to grow at low temperatures.

**Cryosols** Soils that have a permafrost layer within 100 cm of the surface. Cryosols are a reference soil group in the FAO *soil classification.

**Crypteroniaceae** A family of trees in which the leaves are *simple, *opposite, and without *stipules. The flowers are *tetra- or *pentamerous. The fruit is a *capsule. The family is related to *Lythraceae. There are 5 genera, and 11 species, occurring from Assam to Malesia.

**cryptobiosis** Dormancy, used, for example, in relation to microbial *spores that may show no signs of life for extended periods of time.

**Cryptococcaceae (class *Blastomycetes)** A family of *imperfect fungi which do not form *ballistospores. The vegetative stage may be unicellular and *yeast-like, or some kind of *mycelium may be formed. There are many genera. Species are *saprotrophic on fruit, bark, etc. or *parasitic on skin and other tissue in humans and other animals.

***Cryptococcus*** (family *Cryptococcaceae) A genus of *imperfect yeasts in which no *mycelium is ever formed. There are many species, found in a wide range of habitats. *C. neoformans* can cause disease in humans.

**cryptogam** A plant that reproduces by *spores or *gametes rather than seeds, i.e. an *alga, *bryophyte, or *pteridophyte.

***Cryptogramma*** (family *Adiantaceae) A genus of ferns in which the fertile and sterile *fronds are different but both have 2 or more *pinnae. *Sori are borne on the tips of veins below, protected by the reflexed leaf margin, and merging when mature into a continuous band. There are 4 species native to the alpine and boreal parts of the northern hemisphere.

**Cryptomeria** (family *Taxodiaceae) A genus of conifers comprising one species, *C. japonica* (Japanese cedar), occurring in China and Japan, which is a tall tree with long, pointed, curved, spirally arranged leaves, and short, globose cones with several short, curved spines on each cone scale.

**cryptomonads** *See* CRYPTOPHYTES.

**Cryptophyceae** *See* CRYPTOPHYTES.

**cryptophyte** One of *Raunkiaer's life-form categories, being a plant in which the perennating bud lies below the ground or water surface. The groups geophytes, helophytes, and hydrophytes are distinguished by the environment in which the perennating bud is found, i.e. land, marsh, and water respectively. *Compare* HEMICRYPTOPHYTE; CHAMAEPHYTE; PHANEROPHYTE; and THEROPHYTE.

**cryptophytes** (cryptomonads) A small group of unicellular protists, sometimes regarded as *algae (class Cryptophyceae) and sometimes as protozoa (class *Phytomastigophora). The cells are typically dorsoventrally flattened, lack a *cell wall, and have 2 *flagella of equal length. Most species are photosynthetic, but a few are colourless and *heterotrophic. Sexual reproduction is unknown. They are found in freshwater and marine habitats.

**crystallochory** Dispersal of spores or seeds by glaciers.

**crystalwort** *See* RICCIA.

**Cu** *See* COPPER.

**cucumber** (*Cucumis sativa*) *See* CUCURBITACEAE.

**Cucumis melo** (melon) *See* CUCURBITACEAE.

**Cucumis sativa** (cucumber) *See* CUCURBITACEAE.

**Cucurbitaceae** A family of a few shrubby members, but mostly of tendril-climbing *herbs with weak, sappy stems and *palmate leaves. The tendrils are coiled spirally, and arise beside the leaf bases. The flowers are unisexual, often *dioecious, with five *petals often joined below. The fruits are *berries, often very large. There are many important food plants, e.g. *Cucumis melo* (melon), *C. sativus* (cucumber), *Cucurbita pepo* (gourd), marrows, pumpkins, and squashes. There are 121 genera, with 735 species, most of which are tropical or subtropical.

**Cucurbita pepo** (gourd) *See* CUCURBITACEAE.

**cudbear** A purple dye obtained from the *lichen *Ochrolechia tartarea*. This lichen forms thick, grey to brownish-grey, rough, warty crusts on a range of substrates, including trees, rocks, soil, other lichens, etc. The dye used to be extracted from the lichen *thalli by steeping them in urine.

**culm** In *Poaceae and *Cyperaceae, a jointed stem that may be hollow (in many grasses), filled with *pith (in some grasses), or solid (in most sedges (*see* CAREX and CLADIUM)).

**cultivar** Any variety or strain of plant which has been produced by horticultural techniques and is not normally found in wild populations.

**cultivation** Tilling the soil by ploughing, digging, draining, and/or smoothing, done in the course of seeding, transplanting, loosening soil, controlling weeds, or incorporating residues.

**cultural eutrophication** *See* EUTROPHICATION.

**culture** A population of microorganisms or of the dissociated cells of a tissue grown, for experiment, in a nutrient medium: they multiply by asexual division.

**cumquat** *See* FORTUNELLA.

**Cunninghamellaceae** (order *Mucorales) A family of *fungi in which 1-spored *sporangiola are formed; *sporangia are not formed. Species are found in soil, dung, vegetation, etc.

**Cunninghamia** (family *Taxodiaceae) A genus of tall, coniferous trees that have long, flat, fine-pointed leaves, spirally arranged on twigs, but twisted so as to be *pinnately arranged. Cones are ovoid and erect, with a small brown spine on the rounded tip of each white-fringed scale. There are 2 species, 1 of which occurs in south-western China, where it has long been cultivated, and the other in Taiwan.

**Cunoniaceae** A family of trees and shrubs, closely related to *Staphyleaceae, in which the leaves are *opposite and *simple or *compound with *stipules. Flowers are small, and *tetra- or *pentamerous. The fruit is usually a *capsule. There are 24 genera, with about 340 species, centred in the temperate southern hemisphere, and extending to the tropics.

**cup fungi** *Fungi that form cup-shaped *fruit bodies (apothecia), particularly members of the order *Pezizales.

**cup lichen** The common name for any species of *Cladonia in which the *podetia are shaped like miniature cups or wine glasses.

**cupola** See RAISED BOG.

**Cupressaceae** (cypress) A family of conifers in which the leaves are in opposite pairs, scale-like and *adpressed, or needle-like, and the cones are usually small, and globose to long. The scales of the cones have no spine tips. There are 17 genera, with 113 species, distributed in the northern temperate zone, with outlying species in tropical mountains and in temperate S. America.

**Cupressocyparis** (family *Cupressaceae) A hybrid between *Cupressus and *Chamaecyparis, which is a fast-growing, evergreen, columnar conifer, especially C. leylandii (Leyland cypress), which is very widely planted and reaches 20 m in 25 years.

**Cupressus** (cypress; family *Cupressaceae) A genus of trees in which the twigs are not flattened in one plane, the leaves are scale-like and in opposite pairs, and the scales of the cone are in 3–6 pairs with *peltate tips. The trees are much cultivated. There are 13 species, occurring in the northern temperate zone.

**cupulate** Cup-shaped.

**cupule** A cup-like sheath surrounding or enclosing certain *fruits, e.g. the acorn of *Quercus.

**curare** A plant extract containing *alkaloids that block the passage of nerve impulses at synaptic junctions by competing with acetylcholine for receptor sites on the post-synaptic membrane. It is used by some S. American Indians as an arrow poison. Curare is obtained from the bark of Strychnos toxifera and the root of Chondrodendron tomentosum. Curare is also used as a muscle-relaxant in surgery. See also STRYCHNOS.

**Curcuma** (family *Zingiberaceae) A genus of 40 species of aromatic *herbs which occur in subtropical and tropical Asia. The *rhizomes of C. longa yield turmeric.

**curry leaf** See MURRAYA.

**Cuscuta** (dodder; family *Convolvulaceae) A genus of climbing plants that are parasitic on many crop plants, to which they can cause serious damage. They have no green tissue, but thin, herbaceous stems and small flowers in *ebracteate clusters. There are about 145 species of tropical and temperate regions. They have become widespread along with the host plants.

**cushion chamaephyte** See CHAMAEPHYTE.

**cushion plant** A plant that has small, hairy, or thick leaves borne on short stems and forming a tight hummock. The habit is an adaptation to cold, dry, or windy conditions.

**cuspidate** Having a sharp tip or point.

**custard apple** See ANNONA.

**cutan** (clay films, clayskins, argillans, tonhäutchens) Deposited skin or coating of material on the surfaces of peds and stones, which is usually composed of fine, *clay-like soil particles which have been moved down through the soil.

**cuticle** A thin, waxy, protective layer covering the surface of the leaves and stems of plants.

**cutin** The complex mixture of fatty-acid derivatives with waterproofing qualities of which the *cuticle of plants is composed.

**cutinization** The deposition of *cutin in a *cell wall.

**cyanobacteria** A large and varied group of *bacteria which possess *chlorophyll *a* and which carry out *photosynthesis in the presence of light and air with concomitant production of oxygen. They were formerly regarded as *algae (division Cyanophyta) and were called 'blue-green algae'. Fossil cyanobacteria have been found in rocks almost 3000 Ma old and they are common as *stromatolite colonies in rocks 2300 Ma old. They are believed to have been the first oxygen-producing organisms and to have been responsible for generating the oxygen in the atmosphere, thus profoundly influencing the subsequent course of evolution. The organisms may be single-celled or filamentous, and may or may not be colonial. Some are capable of a gliding motility when in contact with a solid surface. There are no *chloroplasts: chlorophyll is carried on specialized membranes (*thylakoids) within the cells. Many species can carry out the fixation of atmospheric nitrogen. Cyanobacteria are widely distributed and are found in freshwater and marine environments, on soil, on rocks, and on plants as *epiphytes or as *symbionts. The taxonomy of the cyanobacteria is confused. Most of the phycological genera and species (defined when they were regarded as algae) are now known to be based on unreliable characteristics (e.g. characteristics that may depend on conditions of growth) and have been redefined or abandoned according to bacteriological criteria. Five 'sections' of cyanobacteria have been recognized. Section I includes unicellular cyanobacteria which reproduce by binary fission only. Section II includes unicellular species in which the cells are enclosed by an external, fibrous layer; reproduction occurs by multiple fission (*see* BAEOCYTE). Section III includes organisms that form unbranched, single (uniseriate) chains of vegetative cells only (no *akinetes or *heterocysts). Section IV includes

organisms forming unbranched, single chains of cells which may include akinetes and heterocysts. Section V includes the most complex cyanobacteria; they are filamentous but cell division may occur in more than one plane, resulting in filaments which are partly multiseriate.

**cyanogenic** Applied to plants that emit hydrogen cyanide when cut or bruised. Some *genotypes of bird's foot trefoil (*Lotus corniculatus*) are cyanogenic but not all. Bracken (*Pteridium aquilinum*) is cyanogenic in all except its very young stages.

**Cyanophyta** *See* CYANOBACTERIA.

**cyanophyte** A cyanobacterium (blue-green alga). *See* CYANOBACTERIA.

**Cyatheaceae** A family of ferns, most of which are massive tree-ferns, with *gradate sori arising at first marginally, and then deflecting to the under-side of the *fronds, where they are protected by a pair of *indusial flaps. There are two genera and 265 species, *Cyathea* and *Dicksonia* occurring mainly in the tropics and moist, warm, temperate southern hemisphere. *See also* THYRSOPTERIDACEAE.

**cycad** *See* CYCADOPHYTA.

**Cycadaceae** In some classifications, a family comprising the extant members of the order Cycadales, now included in the division *Cycadophyta.

**Cycadales** In some classifications, an order of *gymnosperms comprising the cycads, now included in the division *Cycadophyta.

**Cycadophyta** (cycads) A division of *gymnosperms comprising plants with leaves and habit similar to those of palm trees, although some species are quite small. Cycads are *dioecious, and most bear large, coloured, female or male cones. *Pollen grains have *motile *spermatozoa within them, which is a very primitive feature. Formerly they were much more important and, following their appearance in the *Permian, remained important members of the world's *Mesozoic floras. Their reduction was particularly marked in the Late *Cretaceous as they were progressively displaced by *angiosperm trees. The survivors are regarded as 'living fossils'. There are 9 or 10 genera, and about 100 extant species. All are tropical or subtropical. Four genera are American: *Zamia*, *Microcyas*, *Dioon*, and *Ceratozamia*; 5 are Old World: *Macrozamia* and *Bowenia* (Australian), *Stangeria* and *Encephalartos* (south-eastern Africa only), and *Cycas* (widespread).

**Cycas** (family *Cycadaceae) A genus of stout trees which differ from other Cycadaceae in that the female plants do not bear cones, but carry *ovules (and seeds) on massive, terminal, leaf-like *sporophylls, which is a primitive feature. There is a terminal crown of long, finely *pinnate leaves, in which the leaflets are flat. They are *dioecious. A few yield edible starch (sago) from the central pith, nowadays used only as a famine food. There are 16–20 species, occurring from Madagascar and tropical Asia to the Pacific islands and Australia.

**Cyclamen** (family *Primulaceae) A genus of *herbs with large corms, from which arise the *simple leaves and the solitary, nodding flowers on long, leafless stalks with large, *reflexed *corolla lobes. They are grown widely for their showy flowers. There are 17 species, occurring from Europe to Iran.

**Cyclanthaceae** A family of palm-like plants which may be rhizomatous (*see* RHIZOME) herbs, shrubs, or root-climbers. The leaves are always very deeply bilobed and *plicate. The flowers are unisexual, and borne on *axillary spadices (*see* SPADIX) enclosed in conspicuous *bracts (spathes). The fruit is a fleshy *syncarp, and brightly coloured. There are 11 genera, with 190 species, occurring in tropical America.

**cyclic AMP** *See* C-AMP.

**cyclic GMP** *See* GUANOSINE PHOSPHATE.

**cyclic photophosphorylation** The synthesis of *ATP during *photosynthesis, coupled to the cyclic passage of electrons to and from $*P_{700}$, the specialized form of *chlorophyll *a* which is involved in *photosystem I, using a series of carrier molecules.

**cycling pool** *See* ACTIVE POOL.

**cyclosis** The circulation of *protoplasm within a cell; the protoplasm moves at about 1–5 cm/h. *See also* CYTOPLASMIC STREAMING.

**cyme** An *inflorescence in which each *axis ends in a flower; the oldest flowers are in the centre; younger ones are produced successively from the *axils of *bracteoles. Growth is *sympodial.

**cymose** In the form of, or resembling, a *cyme.

**Cynometra** (family *Fabaceae, subfamily *Caesalpinoideae) A genus of trees with *pinnate leaves and small flowers which are borne in *racemes and are sometimes *cauliflorous. The pod is *indehiscent, with 1 seed. They yield useful timber. There are 70 species, with a pantropical distribution, occurring in lowland rain forests.

**Cyperaceae** A family of monocotyledonous (*see* MONOCOTYLEDON) *herbs of grass-like habit that have inconspicuous, wind-pollinated flowers but solid, often *trigonous stems and 3-ranked leaves. They occur mostly in wet places or on poor soils. The family contains some garden plants but they are of little economic importance. There are about 115 genera, with 3600 species.

**Cyperus** (family *Cyperaceae) A genus of usually *perennial *herbs that have *stolons, *tubers, or *rhizomes. They have erect stems with a leafy base. Flowers are borne in a terminal *umbel. The fruits are *trigonous *nuts. *C. papyrus* (paper-reed) is a riverside plant growing up to 4 m high whose stems used to be split into thin strips to make papyrus paper. There are about 600 species, found in tropical and warm temperate regions.

**cyphella** (pl. **cyphellae**) A minute, rimmed, cup-like depression or pore, found in *lichens of the genus *Sticta, which is visible in the lower surface of the *thallus as a small white pit.

**cypress 1.** *See* CUPRESSACEAE and *CUPRESSUS*. **2.** (swamp cypress) *See* ACTINOSTROBUS.

**Cypripedium** *See* ORCHIDACEAE.

C

**Cyrillaceae** (order Ericales) A small family of deciduous or evergreen shrubs and small trees in which the leaves are *simple and alternate. The flowers are borne in *racemes and have 5 *sepals and 5 *petals, which may be free or fused. The *superior *ovary develops into a *capsule or *drupe, sometimes winged, with a fleshy *endosperm. *Cyrilla racemosa* (leatherwood) and *Cliftonia monophylla* (buckwheat tree) are grown as ornamentals, and have attractive white flowers and reddish autumn foliage. There are 3 genera, with about 14 species, occurring in warm parts of America.

**Cyrilla racemosa (leatherwood)** *See* CYRILLACEAE.

**Cyrtosperma** (family *Araceae) A genus of rhizomatous (see RHIZOME) *herbs several of which are gathered wild for food. *C. merkusii* (giant swamp taro), cultivated in Micronesia and parts of Melanesia, thrives over coral. There are 11 species, occurring in the eastern tropics.

**Cyrtostachys** (family *Arecaceae) A genus of clump-forming palms which have *pinnate leaves and a prominent crown-shaft. The pendulous *inflorescence occurs below the crown-shaft, and has two big *spathes, the flowers being in triads of a female and twin males. *C. renda* (sealing-wax palm) of peat-swamp forest, is much cultivated for its scarlet crown-shaft. There are about 12 species, occurring in rain forests from Malaya to the Solomon Islands.

**cyst** A thick-walled resting cell formed by *bacteria of the genus *Azotobacter and by certain methane-utilizing bacteria.

**cysteine** An aliphatic, polar (see POLAR MOLECULE), *alpha amino acid that contains a *sulphydryl group.

**cystidium** A type of enlarged, sterile cell, which occurs among the *basidia in the *hymenium of certain *basidiomycetes.

**cystine** An *amino acid *dimer that is formed from two *cysteine residues linked by a *sulphydryl bond.

**cystocarp (sporocarp)** *See* CARPO-SPOROPHYTE.

**cystolith** A stalk-like crystal of calcium carbonate formed inside an epidermal cell by crystal growth on to an ingrowth from the cell wall. A cell containing a cystolith is known as a lithocyst.

**Cystopteris** (family *Aspleniaceae) A genus of ferns characterized by their orbicular (disc-shaped) *sori borne on *receptacles on their own vascular strands, with inflated, bladder-like *indusia. There are about 18 species, of cosmopolitan distribution.

**Cystoseiraceae** (order *Fucales) A family of brown seaweeds in which the *thallus is typically extensively branched and bushy in appearance. They are found chiefly in warmer regions in deep rock pools, or below low-water mark.

**cytidine** The *nucleoside that is formed when *cytosine is linked to *ribose sugar.

**cytochrome** One of a group of haemo-proteins, which are classified into four groups designated a, b, c, and d. They function as electron carriers in a variety of redox reactions in virtually all aerobic organisms.

**cytochrome oxidase** An *enzyme containing both copper and iron that, because of its ability (unique among the *cytochromes) to reduce molecular oxygen to water, catalyses the terminal reaction in *oxidative phosphorylation.

**cytogenetics** The scientific discipline that combines cytology (the study of the structure, function, and life history of the cell) with genetics (the study of heredity). This usually involves microscopic studies of *chromosomes.

**cytokinesis** During the division of a *cell, the division of the constituents of the *cytoplasm. In plants, it usually begins in early *telophase with the formation of a *cell plate which is assembled within the *phragmoplast across the equatorial plane. The cell plate grows, eventually to fuse with the parent *cell wall and effectively separates the two daughter cells. *See* MITOSIS.

**cytokinin** One of a group of *plant hormones, including *kinetin, that act *synergistically with *auxins to promote cell division but, unlike auxins, promote lateral growth.

**cytology** The scientific study of the cell, including its structure and function.

**Cytophagales** An order of *Gram-negative, rod-shaped or filamentous *bacteria in which at least some stages are capable of a gliding motility when in contact with a solid surface. They are found in soil and various aquatic habitats. Some bacteria often included in the Cytophagales are alternatively regarded as apochlorotic *cyanobacteria.

**cytoplasm** The part of a cell that is enclosed by the *plasma membrane, but excluding the *nucleus.

**cytoplasmic inheritance** A non-Mendelian (extra-chromosomal) inheritance via *genes in cytoplasmic *organelles. Examples of such organelles are viruses, *mitochondria, and *plastids.

**cytoplasmic streaming** The continuous, often rapid, movement of *cytoplasm within a cell. It is a process requiring expenditure of energy by the cell and is thought to involve *microfilament and microtubular (*see* MICROTUBULE) activity. *See also* CYTOKINESIS.

**cytosine** A pyrimidine base that occurs in both *DNA and *RNA.

**cytosis** The evagination, invagination, budding, or fusion of a *cell membrane. *See* ENDOCYTOSIS, EXOCYTOSIS, PHAGOCYTOSIS.

**2,4-D** (2,4-dichlorophenoxyacetic acid) *See* AUXIN.

**dabberlocks** Common name for the seaweed *Alaria esculenta*. The \*midribs of young plants are said to be edible.

***Dacrydium*** (family \*Podocarpaceae) A genus of coniferous trees in which the leaves are spiral or \*aplanate and scale- or needle-like. \*Pollen cones are small, and the female cones usually have only 1 \*ovule. The seed is solitary, and seated on a cup-like structure. There are 25 species, occurring from Hainan and Indo-Malaysia to Australia, New Zealand, and Melanesia, with 1 species in Chile.

***Dacrydium elatum*** (sempilor) A coniferous timber tree of the lowland tropical rain forests of the Malay archipelago.

***Dacrymyces*** (order \*Dacrymycetales) A genus of \*fungi which typically form flattened, cushion-like, or wrinkled gelatinous fruit bodies. *D. deliquescens* forms small, orange, jelly-like cushions on wet, rotting wood. It is very common throughout the year.

**Dacrymycetales** (subclass \*Holobasidiomycetidae) An order of \*fungi which form forked ('tuning fork') \*basidia. The \*fruit body is typically gelatinous. These \*fungi are \*saprotrophic and are found on decaying wood.

***Dacryodes*** (family \*Burseraceae) A genus of resinous trees in which the leaves are \*pinnate, with \*stipules. The fruit is a \*drupe. They produce heavy, often \*siliceous timber. There are 40 species, found in lowland tropical rain forests, with a pantropical distribution.

***Dactylis*** (cock's-foot grass; family \*Poaceae) A \*monotypic genus (*D. glomerata*) of important pasture grass in which the \*panicles are \*compound, with long, stiff lower branches, and in which the \*spikelets and leafy shoots are compressed. It is found in the northern temperate regions of the Old World.

***Dactylocladus*** (family \*Crypteroniaceae) A genus comprising 1 species (*D. stenostachys*) of trees of the Bornean peat-swamp forests, which yield the important timber jongkong.

***Daedalea*** (family \*Polyporaceae) A genus of \*fungi in which elongated, maze-like pores occur on the underside of a thick, brown, hoof-shaped, corky or woody fruit body. In *D. quercina* the upper surface is greyish-brown, often with concentric bands or furrows. This species is common throughout the year on stumps of oak and other trees.

***Daemonorops*** (family \*Arecaceae) A genus of climbing palms (rattans) which occur solitarily or as clumps, and which range from stemless \*rosettes to high-climbing plants. The leaf \*sheaths are spiny and the leaves \*pinnate with a terminal, spiny whip (cirrus). The plants are \*dioecious. The \*inflorescences have \*bracts that split along their length at maturity. The fruits are scaly, and the seed is usually single, with a \*sarcotesta. Some provide cane and a few the resin known as dragon's blood. In several, ants inhabit tunnels between spiny flanges on the leaf sheaths. *Daemonorops* is the second largest genus of climbing palms, with about 114 species, found in north-eastern India, southern China, and east to New Guinea, with most occurring in Sumatra, Malaya, and Borneo.

**daffodil** *See* NARCISSUS.

***Dahlia*** (family \*Asteraceae) A genus of herbs that have tuberous roots and stems up to 8 m long. Originally they were grown for food but now they are cultivated for their ornamental flowers. There are 28 species, occurring from the mountains of Mexico to Colombia.

**daisy bush** *See* OLEARIA.

***Dalbergia*** (family \*Fabaceae, subfamily \*Papilionatae) A genus of trees and woody climbers in which the leaves are usually *pinnate. The pod is *indehiscent, with 1 to a few seeds. Several species produce the fine cabinet timbers known as rosewood, Brazil tulipwood, and Burma tulipwood. There are about 100 species, occurring in the tropics and subtropics.

***Daldinia* (order *Sphaeriales)** A genus of *fungi; in *D. concentrica*, the dark-coloured, hemispherical *fruit bodies are about 5–10 cm in diameter and are formed annually. When cut open the fruit bodies show very characteristic concentric bands of black and white. Early in the season the surface is coated with pale-brown, powdery *conidia. Later, *perithecia are formed in the outer layer of the *fruit body. *D. concentrica* is found chiefly on ash (*Fraxinus excelsior*).

**Dalradian** The last, or youngest, stratigraphic unit of the *Precambrian of Scotland and Ireland.

**dammar** A resinous exudate of *Agathis*, some *Burseraceae, and many *Dipterocarpaceae in the rain forests of Malesia. It is still in demand for specialist varnishes.

**damping-off** A disease of young seedlings, in which the stems decay at ground level and the seedlings collapse. It may be caused by any of a number of *fungi, including e.g. species of *Pythium* and *Rhizoctonia*.

**dandelion** See TARAXACUM.

**Danian** The oldest of the *Cenozoic stages, dated at about 65.5–61.7 Ma ago, which in Denmark is characterized by chalky limestone rich in reef-dwelling organisms. It is a problematic stage once referred to the *Cretaceous but now placed at the base of the *Palaeocene.

***Danthonia* (family *Poaceae)** A genus of plants that bear *panicles of *spikelets with 2–10 flowers, and a *ligule that is a line of hairs. The genus comprises 10 species, found throughout the tropical and temperate world, but especially in S. Africa.

**Daphniphyllaceae** A small family of trees in which the leaves are *simple. The flowers are small, unisexual, and apetalous. The fruit is a *drupe. The family is related to *Magnoliaceae or *Hamamelidaceae, and is often erroneously placed near *Euphorbiaceae. There is 1 genus, comprising 10 species, occurring in eastern Asia and Malesia.

**dark bottle** A bottle covered with tape, or similarly adapted, to exclude light: it is used to monitor respiration rates in aquatic productivity experiments. See also OXYGEN METHOD.

**dark mildew** See BLACK MILDEW.

**dark reactions** Photosynthetic reactions (*see* PHOTOSYNTHESIS) that involve the reduction of carbon dioxide in the *Calvin cycle and which can take place in darkness provided there is sufficient *ATP and *NADPH.

**Darwin, Charles Robert** (1809–82) An English naturalist who is remembered mainly for his theory of *evolution, which he based largely on observations made in 1832–6 during a voyage around the world on *HMS Beagle*, which was engaged on a mapping survey. In 1858, in collaboration with Alfred Russel Wallace (who had reached similar conclusions), he published in the third volume of the *Journal of the Linnean Society* a short paper, 'On the tendency of species to form varieties: and on the perpetuation of varieties and species by natural means of selection', and in 1859 he published a longer account in his book, *On the Origin of Species by Means of Natural Selection*. In this he presented powerful evidence suggesting that change (evolution) has occurred among *species, and proposing natural selection as the mechanism by which it occurs. The theory may be summarized as follows: (*a*) The individuals of a species show variation. (*b*) On average, more offspring are produced than are needed to replace their parents. (*c*) Populations cannot expand indefinitely and, on average, population sizes remain stable. (*d*) Therefore there must be competition for survival. (*e*) Therefore the best adapted variants (the 'fittest') survive. Since environmental conditions change over long periods of time, a process of natural selection occurs which favours the emergence of different variants and ultimately of new species (the 'origin of species'). This theory is known as Darwinism. The subsequent discovery of *chromosomes and *genes, and the development of the science of genetics, have led to a better understanding of the ways in which variation may be caused. Modified by this modern knowledge, Darwin's theory is called 'neo-Darwinism'.

**darwin** A measure of evolutionary rate, given in units of change per unit time.

***Darwinia*** (family *Myrtaceae) A genus of heath-like shrubs found in semi-arid regions on heaths of low fertility. The shrubs are prostrate to upright in form, with small, crowded leaves, and small flowers with a conspicuous *pistil which in some species is surrounded by coloured *bracts. There are 23 species, endemic (*see* EN-DEMISM) to Australia.

**Darwinian fitness** *See* ADAPTIVE VALUE and SELECTION.

**dasheen** *See* COLOCASIA.

**dasycladalean algae** (division *Chlorophyta) A group of green *algae in which the vegetative *thallus contains a single *nucleus, becoming multinucleate immediately prior to reproduction; the thallus consists of an erect *axis with branches, the whole showing radial symmetry. Most members show some degree of calcification. Fossil dasycladalean algae are known from the Lower *Palaeozoic onwards. Modern genera include *Acetabularia*.

**date palm** *See* PHOENIX.

**Datiscaceae** A small, disparate family of herbs and trees in which the leaves are large, *palmately nerved, and without *stipules. The flowers are unisexual, the *ovary *inferior, and *placentation parietal. The fruit is a *capsule, with many seeds. The family is perhaps related to the *Begoniaceae and *Cucurbitaceae. There are 3 genera, with 4 species: *Datisca* species, herbs of America and Asia, and *Octomeles sumatrana* and *Tetrameles nudiflora*, trees of Indo-Malaysia.

***Datura*** (family *Solanaceae) A genus of robust *annual *herbs which grow up to 1.5 m tall, and often have purplish stems.

The leaves are alternate, but often *opposite towards the top of the stem; they have *petioles and are generally *glabrous. The solitary flowers are held in the *axils, with the flower stems elongating at the fruiting stage. The *calyx is angular and sharply toothed. The 5 *petals are fused into a funnel and are white or purple. The 5 *stamens are enclosed. The *ovary has 4 *locules and forms a 4-valved, spiny *capsule. Several species are used medicinally, some are cultivated as ornamentals, and others are poisonous weeds. *D. stramonium*, the common thorn-apple, has been spread from America to many parts of the world, including Britain and Australia. There are 8 species, found in tropical and temperate regions of America, and many have now become widespread.

***Daucus*** (carrot; family *Apiaceae) A genus of plants that have stout, spiny ridges on the fruits and tap roots which, in cultivated forms, become fleshy, nutritious vegetables. There are 22 species, found in northern temperate regions.

**daughter cells** The cells that result from the division of a single cell. The term is usually applied to the two cells formed by *mitosis. *See also* DAUGHTER NUCLEI.

**daughter nuclei** The nuclei that result from the division of a single nucleus. The term is usually applied to the two nuclei resulting from *mitosis. *See also* DAUGHTER CELLS.

**Davalliaceae** A family of *leptosporangiate ferns with mixed *sori which are marginal, flat, and enclosed in separate, tubular *indusia. There are 13 genera and 220 species, widespread in warm and tropical regions.

**dawn redwood** *See* METASEQUOIA and TAXODIACEAE.

***Dawsonites arcuatus*** A species of plant, petrified *axes of which are found in the Lower *Devonian of South Wales. It is a primitive fern (Trimerophytaceae) whose relatives died out before the beginning of the Upper Devonian.

**day degrees** The departure of the average daily temperature from a defined base (e.g. the minimum recognized temperature for the growth of a plant species). The number of day degrees may be totalled to assess the accumulated warmth of a particular year's growing season for crops. *See also* ACCUMULATED TEMPERATURE; AGROMETEOROLOGY; and MONTH DEGREES.

**day length**   *See* PHOTOPERIOD.

**day-neutral plant**   A plant in which flowering is independent of the *photoperiod (e.g. dandelion). *Compare* LONG-DAY PLANT and SHORT-DAY PLANT.

**dead man's fingers**   The common name for the finger-like *stromata of the *ascomycete *Xylaria polymorpha*.

**deal**   Softwood planking imported to Britain, especially from the Baltic region. Usually it is composed of Scots pine (red deal) or Norway spruce (white deal).

**deamination**   The removal of an *amino group from an organic compound.

**death cap**   The common name for the deadly poisonous, *mushroom-shaped *fruit body of *Amanita phalloides*. The cap is pale greenish or white; the *gills are white. It can be peeled like an edible *mushroom. It is common on the ground in broad-leaved woods in late summer and autumn.

**decarboxylase**   An *enzyme that facilitates the removal of a molecule of carbon dioxide from the *carboxyl group of an organic compound.

**deceiver**   The common name for the *fruit body of *Laccaria laccata*, which is notoriously variable in colour, size, and shape. Typically, it is reddish-brown with thick, well-spaced, flesh-coloured gills. The spore print is white. The base of the stipe bears short, white, woolly hairs. It is very common in woodland and heathland from summer to winter.

**deciduous**   Applied to parts of a plant or animal that are shed seasonally (e.g. deer antlers, leaves of certain plants), to trees that shed their leaves seasonally, and to the *perianth of a flower if this is shed after fertilization. In trees, this is not an indicator of taxonomic status; although deciduous trees are generally

*angiosperms, some (e.g. larch) are *gymnosperms.

**deciduous summer forest**   One of the two kinds of broad-leaved forest in the middle latitudes, the other being broad-leaved and *evergreen. The deciduous summer forest is by far the more important of the two in the northern hemisphere, and is absent from the southern hemisphere. The *deciduous nature of the forest is believed to be an adaptation to drought, when the soil is frozen in winter.

**decomposer**   A term that is generally synonymous with 'microconsumer'. In an ecosystem, decomposer organisms (mainly bacteria and fungi) enable nutrient recycling by breaking down the complex organic molecules of dead *protoplasm and *cell walls into simpler organic and (more importantly) inorganic molecules which may be used again by *primary producers. Recent work suggests that some macroconsumers may also play a role in decomposition (for example, detritivores, in breaking down litter, speed its bacterial breakdown). In this sense 'decomposer' has a wider meaning than that traditionally implied. *See also* CONSUMER ORGANISM.

**decorticate**   To remove the bark from a woody stem or branch; a stem or branch from which the bark has been stripped.

**decumbent**   Applied to a shoot that is prostrate as it grows.

**decurrent**   Applied to the *gills in fungal *fruit bodies in which the edges nearest the *stipe are attached to and extend down the stipe. It is also used of *analogous arrangements in other organisms: e.g. of a leaf-base that extends down the plant stem.

**decussate**   Applied to the arrangement of leaves in which pairs of leaves arise at each *node and each pair is at right angles to the pairs above and below it.

**defoliation**   The process of leaves being removed from a plant.

**deforestation (disafforestation) 1.** The permanent clear-felling of an area of forest or woodland. On steep slopes this can lead to severe soil erosion, especially where heavy seasonal rains or the melting of snow at higher levels cause sudden heavy flows of water. In the humid tropics it may also lead to a release of carbon dioxide from the soil (owing partly to the loss of gases as soil structure deteriorates and partly to the decomposition of organic material, including tree roots). **2.** A legal process whereby an area of forest land ceases to be regarded as forest under the terms of *forest law.

**degenerate code** A term applied to the *genetic code because a given *amino acid may be encoded by more than one *codon.

**degenerate phase** The stage in the cyclical pattern of changes typical of *grassland and *heathland *communities. In grassland, the term refers to the extended colonization by *lichens of the small, grassy hummocks of the mature phase. As the frequency of soil-binding grass plants declines, erosion of the hummocks starts and a new cycle begins. In heathland, the term refers to *Calluna vulgaris plants that are 20, 30, or more years old. A gap appears and widens gradually in the centre of the *C. vulgaris* bush, which becomes very straggly, with new shoots confined to the branch tips. In the gap and on the bare central stems lichens, especially *Cladonia species, and *mosses colonize. Eventually new *C. vulgaris* seedlings invade the central gap and a new cycle begins. *Compare* BUILDING PHASE; HOLLOW PHASE; MATURE PHASE; and PIONEER PHASE.

***Degeneria*** (family Degeneriaceae) A genus comprising 1 species of small trees, endemic (*see* ENDEMISM) to Fiji, in which the flowers have many free parts and are primitive, and in which the *ovary is open when young. They are related to the *Winteraceae.

**degree-day** *See* ACCUMULATED TEMPERATURE.

**dehiscent** Bursting or splitting open at maturity. Usually used of a *fruit that bursts open to release its *seeds (e.g. pea pod).

**dehydrogenase** An *enzyme that catalyses the removal of hydrogen from a *substrate.

**Delesseriaceae (order *Ceramiales)** A family of red seaweeds in which the *thallus is delicate, membranaceous, and brilliant red. In some genera there is a distinctive pattern of *'midribs' and 'veins', which resemble those of higher plants but are distinct in structure and function. These seaweeds are found attached to rocks or to other seaweeds, typically near or below low-water mark, or in shady pools.

**deletion** The loss of a chromosomal segment from a *chromosome set. The size of the deletion may vary from a single *nucleotide to sections containing several *genes. If the deletion is from the end of a chromosome, it is called 'terminal'; if it is from elsewhere, it is termed 'intercalary'.

***Delonix*** (family *Fabaceae, subfamily *Caesalpinoideae) A genus of trees which includes *D. regia* (flame of the forest or flamboyant), one of the loveliest of tropical flowering trees, known in the wild from just one specimen. There are 10 species, occurring in tropical Africa, Madagascar, and India.

***Delphinium*** (larkspur; family *Ranunculaceae) A genus of *annual or *perennial *herbs that have *palmately lobed leaves and irregular flowers in *racemes, comprising 5 *petaloid *sepals (the posterior one with a spur, giving the genus its common name), 4 *petal nectaries, many *stamens, and 1–3 *follicles which form the fruit. Many *Delphinium* species are cultivated for their attractive flowers. There are about 250 species, found in the northern hemisphere.

**deme** A spatially discrete interbreeding group of organisms with definable genetic or cytological characters, i.e. a subpopulation of a species. There is very restricted genetic exchange, if any, with other demes, although demes are usually contiguous with one another, unlike subspecies or races, which are generally isolated by some *geographical or habitat barrier. All possible male and female pairings within a deme have an equal chance of forming, for one breeding season at least. Populations that fulfil only one of the two key criteria, i.e. very occasional or no cross-breeding and free pairing, are also referred to as demes by some authors.

**denaturation** Reversible or irreversible alterations in the biological activity of *proteins or *nucleic acids, which are brought about by changes in structure other than the breaking of the primary bonds between *amino acids or *nucleotides in the chain. This may be accomplished by changes of solvent, *pH, or temperature, or through the physical abuse of the molecules.

*Dendrobium* (family *Orchidaceae) A genus of orchids, some of which are popular, cultivated, and sold throughout the world. There are more than 900 species, found in Asia, New Guinea and the south-western Pacific.

*Dendrocalamus* (family *Poaceae, subfamily Bambusoideae) A genus of giant bamboos in which the fruit is a *nut. *D. giganteus* is the biggest known bamboo. There are about 30 species, occurring from China to Malesia.

**dendrochronology** (tree-ring analysis) **1.** The science of dating by means of *tree-rings. **2.** All aspects of the study of annual growth layers in wood.

**dendroclimatology** A branch of *dendrochronology dealing with the relationships between annual growth increment and climate, and especially with the reconstruction of past climates from dated *tree-ring series. It is assumed that, by studying present tree-ring patterns in relation to climate, older tree-ring chronologies may be used to indicate climatic conditions experienced before detailed climate and weather records were kept.

**dendroecology** A branch of dendrochronology dealing with the relationships between patterns in dated *tree-ring series and all the ecological factors that may influence those patterns. As well as climate, it considers *competition, *predation, etc.

**dendrogeomorphology** The use of dated *tree-ring series to study land-forms and geomorphological processes.

**dendrogram** A diagram that represents relationships among groups of taxa, with the highest *taxon at the base of a vertical line from which lower taxa branch at appropriate levels. There are two principal types: (a) the *phenogram, which is based solely on similarities in *phenotypes; and (b) the *cladogram.

**dendrograph** An instrument that provides a continuous record of the circumference of a tree stem (often called a 'girthing'). It is especially useful for recording *diurnal changes in stem size related to differences in hydration. There are several designs. *See also* DENDROMETER.

**dendrohydrology** The use of dated tree-ring series to study hydrological questions, especially relating to the periodicity of river flow and flooding.

**dendroid** Tree-shaped (from the Greek *dendron*, meaning 'tree').

**dendrometer** An instrument for measuring the size of tree stems. Most commonly it consists of a thin metal band with a vernier scale: it allows precise measurement of the stem circumference.

*Dendrophthoe* (family *Loranthaceae) A genus of plants that are heterotrophic (*see* HETEROTROPH) *stem hemiparasites. There are 30 species, found in Africa, Asia, and tropical Australia.

**denitrification** The conversion of nitrate or nitrite to gaseous products, chiefly nitrogen ($N_2$) and/or nitrous oxide ($N_2O$), by certain types of *bacteria (called *denitrifying bacteria). Denitrification occurs mainly under *anaerobic or *micro-aerobic conditions. It is also usually associated with higher pH conditions.

**denitrifying bacteria** *Bacteria that can carry out *denitrification; they occur, for example, in soil and in freshwater and marine environments, and include, e.g., certain species of *Bacillus, *Hyphomicrobium, *Paracoccus, *Pseudomonas, and *Thiobacillus.

**Dennstaedtiaceae** A family of ferns in which the *sori are marginal, and either are continuous or form discrete units within tubular *indusia. They are intermediate in sorus type between the wholly *gradate sorus of Dicksoniaceae and the wholly mixed type of *Davalliaceae.

There are 24 genera and 410 species, with a world wide distribution, although they are mainly tropical.

**density dependence** The regulation of the size of a population by mechanisms themselves controlled by the size of that population, i.e. *environmental resistance factors, whose effectiveness increases as population size increases. Density dependence may act through increasing mortality within a population or by decreasing fecundity. See also S-SHAPED GROWTH CURVE.

**density–frequency–dominance** (DFD measure) A combined abundance estimate used in the early N. American ordination schemes but now rarely used. Usually expressed as relative values, relative density is the number of a given species expressed as a percentage of all species present, relative frequency is the frequency of a given species expressed as a percentage of the sum of frequency values for all species present, and relative dominance is the basal area of a given species expressed as a percentage of the total basal area of all species present. These three measures are summed to give the importance value, which may lie between 0 and 300.

**density independence** See J-SHAPED GROWTH CURVE.

**density measure** An estimate of the abundance of a particular plant species, as the number of individuals per unit area. In practice, problems arise with those plants, including many grasses, for which identification of single individuals is difficult.

**dentate** Toothed or serrated.

**denticulate** Having very small teeth or serrations.

**deoxyribonucleic acid** (DNA) A nucleic acid, characterized by the presence of the sugar deoxyribose, the *pyrimidine bases cytosine and thymine, and the *purine bases adenine and guanine. It is the genetic material of organisms, its sequence of paired bases constituting the *genetic code. See also WATSON–CRICK MODEL.

**deplasmolysis** The entrance of water into a plasmolysed (see PLASMOLYSIS) plant cell, causing the *cell membrane to return to the *cell wall.

**depside** A lipoid group of substances, often brightly coloured, found in *lichens.

**depsidone** A lipoid group of substances found principally in *lichens.

**derived** (evol.) See APOMORPH.

**dermal** Pertaining to *epidermis or *periderm.

**Dermatophilus** (order Actinomycetales) A genus of *aerobic or facultatively *anaerobic, *mycelium-forming *bacteria, in which the filaments divide transversely and longitudinally to form masses of cells that bear *flagella and are *motile. Species can cause skin diseases in humans and other animals.

**dermatophyte** A fungus that lives *parasitically on skin.

**Derris** (family *Fabaceae, subfamily *Papilionatae) A genus mainly of climbers, some of which are important as the source of fish poisons and are a cheap, safe insecticide. There are 40 species occurring throughout the tropics.

**desert** An area within which the rate of evaporation exceeds the rate of precipitation for most of the time. The rate of evaporation depends on temperature, but a desert is likely to form within any temperature range if the average precipitation is less than 250 mm/yr, and typically very erratic. Plants and animals are either absent or sparsely distributed, and are adapted to long droughts or to a lack of access to free water.

**desert biome** The characteristic *biotic *community of warm, arid regions, generally defined as areas with rainfall of less than 250 mm/yr. Characteristically, such areas have high evaporation rates and a large *diurnal temperature range. Organisms commonly show adaptations to drought and heat, e.g. water storage in succulents such as cacti, and the frequent use among desert mammals of the burrowing habit. A high percentage of annual species is characteristic of desert flora.

**desertification** The process of desert expansion or formation, which may occur as a direct consequence of climatic change (e.g. shifts in the location of the major planetary pressure and wind systems), poor land-use policy (e.g. overgrazing), or some complex interaction of these factors (e.g. overgrazing leading to albedo change, favouring climatic change in the form of increased dryness).

**Desfontainea** (family *Loganiaceae) A *monotypic genus (*D. spinosa*) of bushes that have small, holly-like leaves and scarlet, tubular flowers. It occurs from Costa Rica to Cape Horn, is the national flower of Bolivia, and is sometimes cultivated in Britain.

**desiccation** The drying-out of an organism that is exposed to air.

**desmids** A group of green *algae (division *Chlorophyta) which are basically unicellular but have two distinct halves, or 'semicells'. The cells are usually solitary but in a few species they may form irregular or filamentous colonies. The cells may vary in shape within a single species, making identification and classification difficult. Sexual reproduction occurs by *conjugation, similar to that in *Spirogyra. They are found mainly in freshwater habitats and are usually indicators of clean (unpolluted) water.

**Desmoncus** (family *Arecaceae) A genus of climbing palms in which all the parts are very spiny. The leaves are *pinnate, the *apical *leaflets often modified as tough *reflexed hooks. There are 65 species, occurring in the New World where they are the counterpart of the rattans of the Old World.

**destroying angel** The common name for the deadly poisonous *fruit body of *Amanita virosa. It is *mushroom-like, with white cap, white *gills, and a shaggy *stipe. It is found on the ground in broadleaved woods in late summer and autumn.

**Desulfovibrio** A genus of *bacteria which is not assigned to a taxonomic family. The cells are *Gram-negative, rod-shaped, curved, and *motile with *flagella at one end. They are *chemoorganotrophic or *mixotrophic and can grow only in the absence of air. Energy is obtained by the oxidation of inorganic compounds with the concomitant reduction of sulphate (and other inorganic sulphur compounds) to sulphide. *Desulfovibrio* species are found in aquatic environments, waterlogged soils, etc.; they produce hydrogen sulphide (which smells like rotten eggs), which is responsible for the blackening of mud (owing to the formation of black metal sulphides) in such habitats.

**detrital pathway** (detritus food-chain) Most simply, a food-chain in which the living primary producers (green plants) are not consumed by grazing herbivores, but eventually form litter (detritus) on which decomposers (microorganisms) and detritivores feed, with subsequent energy transfer to various levels of carnivore. Detritus from organisms at higher trophic levels than green plants may also form the basis for a detrital pathway, but the key distinction between this and a grazing pathway lies in the fate of the primary producers.

**detritivore** (detritus feeder) A *heterotrophic animal that feeds on dead material, e.g. leaf litter. The dead material most typically is of plant origin, but may include the dead remains of small animals. Since this material may also be digested by decomposer organisms (fungi and bacteria) and forms the *habitat for other organisms (e.g. nematode worms and small insects), these too will form part of the typical detritivore diet. Animals (e.g. hyena) that feed mainly on other dead animals, or that feed mainly on the products (exuviae, e.g. dung), of larger animals, are termed scavengers. *See also* FOOD-CHAIN.

**detritus** Literally, debris produced by rubbing; fragments of dead material (e.g. leaf litter, dung, moulted feathers, and corpses). In aquatic habitats, detritus provides habitats equivalent to those that occur in soil humus.

**detritus agriculture** The planned production of detritus as a source of food, e.g. silage production. Detritus agriculture is thought by some workers to have great potential as an alternative to harvesting the products of the \*grazing pathway (i.e. animal products). They argue that the economic and environmental costs of microbial conversion of the detritus to palatable human food would be less than the costs of pest and disease control, etc.

**detritus feeder** *See* DETRITIVORE.

**detritus food-chain** *See* DETRITAL PATHWAY.

**Deuteromycotina (Fungi Imperfecti; division \*Eumycota)** A subdivision that includes an assorted assemblage of imperfect fungi in which \*perfect states are either unknown or classified in other subdivisions. The subdivision and its constituent groups are convenient categories and not true phylogenetic taxa.

**Devensian (Weichselian, Würm)** The last glacial advance in northern Europe, approximately 75 000–10 000 years BP. It is approximately synchronous with the Wisconsinian glaciation in N. America.

**Devonian** The fourth of the six periods of the \*Palaeozoic Era, preceded by the \*Silurian and followed by the \*Carboniferous. It began about 416 Ma ago and ended about 359.2 Ma ago. In Europe there are both marine and continental facies present, the latter being commonly known as the Old Red Sandstone. Although originally described from the type area in Devon, England, the marine Devonian is subdivided stratigraphically into stages established in other parts of the world. These stages are: the Lochkovian (416–411.2 Ma, Czech Republic); Praghian (411.2–407 Ma, Czech Republic); and Emsian (407–397.5 Ma, Uzbekistan) of the Lower Devonian; the Eifelian (397.5–391.8 Ma, Germany) and Givetian (391.8–385.3 Ma, Morocco) of the Middle Devonian, and the Frasnian (385.3–374.5 Ma, France) and Famennian (374.5–359.2 Ma, France) of the Upper Devonian. The subdivision of the marine deposits is based on lithologies and the presence of an abundant invertebrate fauna including goniatites (Ammonoidea) and spiriferid brachiopods (Spiriferida). The continental Old Red Sandstone deposits contain a fauna of jawless fish and plants belonging to the primitive psilo-

phyte group. As a result of the Caledonian orogeny of late Silurian times, much of the British Isles was covered with continental red-bed facies. Fossils of vascular plants are abundant in Devonian beds; the Rhynie chert flora (Middle Devonian) consists of well-preserved psilophytes. Insects are also present within the Devonian and may have their origin in the preceding Silurian.

**dextran** A branched-chain \*polysaccharide, composed of D-\*glucose units, that acts as a storage compound in bacteria and yeasts.

**dextrorse** Growing or arising in a right-handed or clockwise spiral from the point of view of an observer. The term is applied to a climbing-plant stem, a chain of spores, etc.

**dextrose** *See* GLUCOSE.

**DFD measure** *See* DENSITY–FREQUENCY–DOMINANCE MEASURE.

**D-fructose** *See* FRUCTOSE.

**D-glucose** *See* GLUCOSE.

**diageotropism** A tropic response of a plant organ in which it takes up a position at right angles to the direction of the force of gravity. An example is the ivy-leaved toadflax (*Cymbalaria muralis*), which climbs vertical stone walls and has diageotropic fruits. The seeds are thus pushed laterally into crevices in the wall. *See* TROPISMS.

**diagnostic horizon** A soil layer that contains a combination of characteristics typical of that kind of soil.

**diakinesis** *See* MEIOSIS.

***Dialium* (family \*Fabaceae, subfamily \*Caesalpinoideae)** A genus of tropical trees in which the leaves are \*pinnate. The flowers are reduced, with 1, 2, or no \*petals. The fruit is \*indehiscent, brittle-skinned, and \*drupe-like, with an edible \*mesocarp. The trees yield useful timber. There are about 40 species occurring in western Malesia, Madagascar, Africa, and a single species in America.

**diallelic** Applied to a \*polyploid individual with more than two sets of \*chromosomes in which two different \*alleles exist at a particular \*gene \*locus.

**dialysis** The separation of dissolved crystalloids from colloidal *macromolecules by means of a partially permeable membrane that allows the passage of the former but not of the latter.

***Dianthus*** (family *Caryophyllaceae) A genus of mostly *perennial *herbs in which the *calyx is tubular and enclosed tightly at the base by an *epicalyx of 1–3 pairs of scales. There are 5 long-clawed *petals. The fruit is a 4-toothed *capsule. Many species (e.g. pinks and carnations) are grown for their beautiful, often sweet-scented flowers. There are about 300 species, found mainly in Eurasia but extending to S. Africa.

**Diapensiaceae** A family of plants, most of which are cushion-forming (*see* CUSHION PLANT), evergreen undershrubs, but a few of which are *herbs. The *calyx and *corolla are deeply 5-lobed, there are 5 *stamens alternating with the *corolla lobes, and the *ovary is 3-celled and lobed and ripens to a *capsule. There are 5 genera, with 13 species, found in Arctic and northern alpine regions.

**diaphototropism** A *tropic response of a plant organ in which it grows at right angles to the stimulus of light.

**Diaporthales** (subdivision *Ascomycotina) An order of *fungi in which the *ascocarps are perithecioid (*see* PERITHECIUM); the *asci are unitunicate and evanescent. Species are *saprotrophic or *parasitic on plants. Some can cause plant diseases.

**diarch** A root with 2 strands of *xylem.

**diaspore** A *spore, *seed, or other structure that functions in dispersal; a *propagule.

**diatom** *See* BACILLARIOPHYTA.

**diatomaceous earth** (**kieselguhr**) A deposit composed of fossil diatoms, which is mined for many industrial uses: as a mild abrasive in metal polishes, as a filtering medium (e.g. in sugar refineries), for insulation of boilers and blast furnaces, etc. Vast deposits of diatomaceous earth are mined at Lompoc, California.

**diatropism** A *tropic response of a plant organ in which it grows at right angles to a stimulus.

**diazotroph** An organism capable of utilizing ('fixing') atmospheric nitrogen. *See also* NITROGEN FIXATION.

**dicaryotic** (**dikaryotic**) The occurrence of 2 *haploid nuclei within each cell.

***Dicentra*** *See* FUMARIACEAE.

**dicentric** Applied to a *chromosome or *chromatid with two *centromeres.

**Dichapetalaceae** A family of shrubs, climbers, and small trees that have *simple, alternate, often *pubescent leaves without *stipules. The *inflorescence is a *cyme or cluster, with a stalk. The flowers are regular or irregular, bisexual or unisexual, and *monoecious. There are 5 overlapping *sepals and 4 or 5 bilobed or forked *petals which are often black. The 5 *stamens are united to the petals, or free. The *ovary is *superior, with 2 or 3 biovular *carpels, each with 2 *styles. The fruit is a *drupe, usually lobed and pubescent, with up to 3 *locules, each with a single seed. The seeds contain no *endosperm. There are 3 genera, with about 180 species, found throughout tropical regions. Some species are cultivated as ornamentals. Many are very poisonous, and are used locally to poison game or vermin.

**dichasium** (pl. **dichasia**) A *cymose inflorescence in which each branch gives rise to two more branches. *Compare* MONOCHASIUM.

**2,4-dichlorophenoxyacetic acid** (**2,4-D**) *See* AUXIN.

**dichogamy** The maturing at different times of the male and female organs of a flower.

**dichotomous branching** Repeated division into two parts (bifurcation).

**diclinous** In flowering plants, the occurrence of *stamens and *carpels in separate flowers on the same plant. *Compare* MONOCLINOUS.

**dicotyledon** An *angiosperm (flowering plant) in which the *embryo characteristically has two *cotyledons (though sometimes there are more). *Compare* MONOCOTYLEDON.

**Dicotyledoneae** A former division comprising the *dicotyledons. The name is no longer used.

**Dicranales** (subclass *Bryidae) An order of mosses, which may be large or small. The leaves are long, and narrow to ovoid; each has a single nerve. Genera include *Dicranella*, *Ceratodon*, *Dicranoweisia*, *Dicranum*, *Campylopus*, and *Leucobryum*. Species are numerous and widespread, found on a range of substrates.

**Dicranum** (order *Dicranales) A genus of medium-sized to large mosses in which the shoots are erect and unbranched. The leaves are long and slender, and in many species are curved to one side (falcatosecund). *D. majus* is a large, striking moss of mountain woodland, with strongly falcatosecund leaves up to 1.5 cm long. *D. scoparium* is similar but smaller; it is very common on the ground in woodland clearings, on *heathland, etc. The genus is cosmopolitan, with about 150 species.

**dictyosome** The *Golgi body in a plant cell.

**Dictyosperma** (family *Arecaceae) A *monotypic genus, *D. album* (the princess palm), which is very widely cultivated and of similar general appearance to *Archontophoenix alexandrae*. The trunk is solitary and prominently hoop-marked and the leaves are *pinnate with a conspicuous green crown-shaft below which appear the pendulous, simply branched inflorescences, at first enclosed in a large, broad, flat *spathe. It is native to Mauritius and the Rodrigues Islands, in the southwestern Indian Ocean.

**dictyostele** A *stele that is divided into several strands, called meristeles. A dictyostele may consist of *protosteles (*polystele and *plectostele) or *siphonosteles (*eustele and *atactostele). *Compare* MONOSTELE.

**Dictyosteliomycetes** (division *Myxomycota) A class of *cellular slime moulds in which the *myxamoebae have slender *pseudopodia. Prior to fruiting, the myxamoebae collect into streams, eventually aggregating to form a macroscopic *pseudoplasmodium.

**Dictyostelium** (class *Dictyosteliomycetes) A genus of *cellular slime moulds; *D. discoideum* has been studied particularly as a model differentiation system. In this

species, the slug-like *pseudoplasmodium varies in size (e.g. 0.5–2.0 mm in length), and the *fruiting body consists of a mass of *spores borne at the tip of an unbranched, tapering stalk. There are many species, found in soil, dung, decomposing vegetation, etc.

**Dictyotales** (division *Phaeophyta) An order of brown seaweeds in which growth occurs from a single apical cell or an apical margin of cells. The *thallus is flattened and there is an *isomorphic alternation of generations. Members are common in warmer regions. *Dictyota dichotoma* is yellowish-brown to olive-brown, and the flattened *fronds fork regularly; this species is found in intertidal regions in rock pools, and in deeper water at and below low-water mark.

**differential species** In *phytosociology, species that seem to be *mutually exclusive in a comparison of two *community types, but which examination of other communities reveals are not necessarily characteristic for either community. *See also* KENNARTEN SPECIES.

**differentially permeable** *See* SEMI-PERMEABLE.

**differentiation** The occurrence of changes in the structure and function of groups of cells owing to increased specialization in an organism.

**differentiation, theory of** On the basis of the different distribution patterns of the families, genera, and *species of flowering plants, it was suggested by some early plant geographers that *evolution in this group proceeded from the family level downward into genera and species, rather than the other way around.

**diffusion** The movement of molecules from a region of higher to one of lower solute concentration as a result of their random thermal movement.

**diffusion pressure deficit** *See* SUCTION PRESSURE.

**Digitalis** (foxgloves; family *Scrophulariaceae) A genus of tall *herbs that have alternate leaves and attractive, drooping, 2-lipped, bell-like flowers. They are cultivated for their flowers and for the *alkaloid digitalis, used for heart stimulation. There are about 20 species, found in Europe and Central Asia.

**digitate** *See* PALMATE.

**dikaryon** (dicaryon) A fungal *hypha or *mycelium in which each cell contains 2 nuclei which (usually) are genetically distinct.

**dikaryotic** *See* DICARYOTIC.

***Dillenia*** (family *Dilleniaceae) A genus of trees and shrubs in which the showy flowers have 5 *sepals and *petals, numerous *stamens developing centrifugally, and numerous *carpels. The seeds are often *arillate. Some species are cultivated as ornamentals, and some yield useful timber. There are about 60 species, occurring from the Mascarenes and Indo-Malaysia to the Pacific islands.

**Dilleniaceae** A family of trees, shrubs, and *herbs, in which the leaves are *simple, without *stipules. The flowers are showy and hermaphrodite, with free, *imbricate *sepals and *petals, usually many *stamens, and a *superior *ovary of 1 to many, usually free *carpels. The fruit may or may not be *dehiscent and is subtended by the persistent *calyx. The seeds are sometimes *arillate. There are 12 genera, with about 300 species occurring in the Old and New World tropics and subtropics, and well represented in Australia.

**dimension analysis** The detailed measurement of plant dimensions in productivity studies. Control studies enable estimation of allometric (*see* ALLOMETRY) relationships between external dimensions and dry-matter production, so that in future studies destructive sampling (*harvesting techniques) is unnecessary. Dimension analysis has particular relevance to forest productivity studies and has been much used.

**dimer** A *protein that is made up of two *polypeptide chains or subunits paired together. If the subunits are identical in *amino-acid sequence the protein is said to be homomeric; if they are different, it is heteromeric. Dimeric proteins may be detected by *electrophoresis. In monomeric *enzymes, the isozyme pattern of the heterozygote will represent a simple mixture of the two forms occurring by themselves in each of the corresponding homozygotes. In dimeric enzymes, there are homomeric forms representing the two homozygotes, but the heterozygote occurs in heteromeric form: when stained after electrophoresis, this results in 3 bands (instead of 2 bands as with a monomer). An example of a dimeric enzyme is *glucose-phosphate isomerase. Some enzymes are trimeric (comprising 3 components) or tetrameric (with 4 components).

**dimerous** Of a flower, with parts in twos.

**dimictic** Applied to a lake in which two seasonal periods of free circulation occur, as is typical of lakes in mid-latitude climates. In summer, thermal stratification occurs as surface waters are warmed and cease to mix with the denser, colder, deep waters. In winter, when they cool to below 4°C, surface waters expand, so becoming less dense than warmer waters beneath them, giving a reverse stratification. Free circulation through the depth of the lake is possible only in spring, when the surface temperature rises to above 4°C, and the water becomes heavier than that beneath and so sinks and mixes, and in autumn, when the surface (*epilimnion) waters cool to the temperature of the deep (*hypolimnion) waters.

**dimidiate** 1. Divided into two. 2. (Of a fungal *fruit body) Semicircular in outline.

**dimorphic fungi** *Fungi that can exist either as single, *yeast-like cells or as filaments (*mycelium), depending on conditions.

**dimorphism** The presence of one or more morphological differences that divide a species into two groups. Many examples come from sexual differences of particular traits, such as types of flowers in *dioecious plants. These result from sex-linkage of the *genes coding for the particular trait. However, some dimorphisms, such as the aerial or submerged leaves of some aquatic plants (e.g. water crowfoot), may not be sex-linked.

**dioecious** Possessing male and female flowers or other reproductive organs on separate, *unisexual, individual plants. *See also* ANDROECIUM; *compare* MONOECIOUS.

**d**

*Dioscorea* (family *Dioscoreaceae) A genus of plants in which the twining *annual stems arise from an often massive subterranean *tuber; the tuber contains much starch and is known as a yam. The leaves have *palmate main nerves and nervation as fine as a spider's web. Some species yield *steroids used for oral contraceptives. There are about 600 species, of the tropics and subtropics, 1 species occurring in Europe.

**Dioscoreaceae** A family of plants most of which are slender climbers. The leaves are commonly *cordate. The flowers are regular, small, inconspicuous, *tri- or *hexamerous and the *ovary is *inferior. The fruit is a *capsule or *berry. There are 6 genera, with about 630 species, most of which are tropical, but including *Tamus communis*, the black bryony of Europe.

*Diospyros* (ebony; family *Ebenaceae) A genus of small, dark trees which have *monopodial crowns and charcoal-like outer bark. The *heartwood is hard, dense, and dark brown streaked with black, and is a valuable timber. The plants are *dioecious. The fruit is a *berry, seated on the persistent *calyx. There are few seeds, which are bony. In several species the *pericarp is edible. The persimmons are *D. kaki*, *D. lotus*, and *D. virginiana*. There are about 500 species, occurring in the warm parts of the world.

**diphosphopyridine nucleotide (DPN)** *See* NICOTINAMIDE ADENINE DINUCLEOTIDE.

**diplanetism** The phenomenon, observed in some members of the *Oomycetes, of there being two distinct *motile phases, with morphologically different *zoospores formed in each.

**diplococcus** The arrangement of cells in certain types of *bacteria: *cocci joined together in pairs.

*Diploglottis* (family *Sapindaceae) A genus of trees of subtropical regions that includes *D. cunninghamii* (native tamarind), with edible fruit, of eastern Australia, which grows to 17 m high, has *pinnate, fern-like leaves, and small, yellow flowers borne in *panicles. The yellow fruit contains a red pulp used for making jam. The genus comprises 3 species, probably endemic (*see* ENDEMISM) to Australia, but perhaps reaching New Guinea.

**diplohaploplontic** Applied to a *life cycle having an *alternation of generations during which *spores are produced meiotically by *sporophytes and *gametes are produced mitotically by *gametophytes.

**diploid** A cell with 2 *chromosome sets, or an individual with 2 *chromosome sets in each cell (excluding the *sex chromosomes which may or may not be represented twice, according to the sex of the individual). A diploid state is written as $2n$ to distinguish it from the *haploid state of $n$. The *zygotes of many green *algae and *fungi are diploid, as are the *sporophytes of other algae, *mosses, *liverworts, and *vascular plants.

**diploidy** The *diploid condition.

*Diplophylum* (order *Jungermanniales) A genus of leafy *liverworts. *D. albicans* is a common and easily recognized species, with creeping primary stems from which arise numerous ascending or erect leafy shoots. The leaves are arranged in 2 ranks; each leaf is deeply divided into 2 lobes, the smaller of which is bent back and partly covers the larger, lower lobe, thus giving the appearance of 2 leaves rather than 1. A whitish 'vein' consisting of a band of long, narrow, clear cells runs down the middle of each leaf. They are found on a range of non-calcareous substrates: soil, *peat, rock, etc.

**diploplontic** Applied to a *life cycle that has no *alternation of generations, *gametes being produced meiotically by a *diploid organism.

**diplotene** *See* PROPHASE.

**dipole** A molecule with an uneven charge distribution, one pole having a net negative charge, the other a net positive charge.

**dipole moment**  See POLAR MOLECULE.

**Dipsacaceae**  A family of *herbs or low shrubs, resembling the *Asteraceae but with *opposite leaves. Each *floret in the head has its own *involucre of united *bracts, the *stamens are *exserted, their *anthers on long filaments. There are about 8 genera, with 250 species, occurring mainly in southern Europe and western Asia.

**Dipteridaceae**  A family of ferns with *dichotomous, fan-like divisions of the *fronds, and with superficial, non-*indusiate, radiate, mixed *sori. There is 1 genus and 8 species, occurring from tropical Asia to Polynesia.

**Dipterocarpaceae**  A family of resinous trees in which the leaves are *simple, alternate, and *stipulate. There are 5 twisted *sepals and *petals, and 5–10 *stamens, and the *ovary is *superior and *trilocular. The fruit is a *nut seated in the persistent *calyx, with several sepals enlarged as wings. There are 16 genera, with about 530 species found in the Old World tropics. They are abundant in the lowland rain forests of western Malesia, where there are many very lofty species.

**Dipterocarpus**  (family  *Dipterocarpaceae) A genus of lofty, resinous trees which yield heavy, valuable timber (keruing and apitong). They have large amplexicaul *stipules. The leaves are usually *plicate between the main nerves. The wood of some is tapped for gurjun or keruing oil. There are 69 species, occurring from Sri Lanka to western Malesia.

**Dipteronia**  See ACERACEAE.

**directed speciation**  A speciational trend recognized in plants: the *species do not conform to a continuum of *adaptive types, but rather to a stepwise succession of distinct species. The evolutionary significance of such a trend is unclear. It may relate to differential survival as opposed to differential speciation along an environmental gradient, i.e. to species selection rather than directed speciation.

**directional evolution**  See ARISTO-GENESIS.

**directional selection**  A selection that operates on the range of *phenotypes for a particular characteristic existing in a population, by moving the mean phenotype towards one phenotypic extreme. Directional selection usually occurs in response to a steady change in environmental conditions, with a consequent shift in selection pressure such that the frequency of particular *alleles will change in a constant direction. It is often used in agriculture and horticulture to produce a shift in the population mean of a trait derived by humans. For example, the breeder might select for plants that fruit only in a particular season. *Compare* DISRUPTIVE SELECTION and STABILIZING SELECTION.

**disafforestation**  See DEFORESTATION.

**disassortative mating**  Mating between individuals of unlike phenotype. *See also* ASSORTATIVE MATING.

**disc**  A fleshy outgrowth, often secreting nectar, that is developed from the *receptacle or *stamens of a flower.

**Dischidia**  (family *Asclepiadaceae) A genus of *epiphytes or bole climbers, many of which have concave leaves *adnate to the *bole, subtending *adventitious roots, and containing commensal ants plus their accumulated frass. There are about 80 species occurring in forests from Indo-Malaysia to Australia.

**disclimax**  In the *monoclimax model of vegetation development, a plant community replacing the climax community following an environmental disturbance, e.g. the introduction and maintenance of grazing pressure. Disclimax is analogous to the more widely used terms *plagioclimax and/or biotic climax.

**Discomycetes**  (subdivision Ascomycotina) A class of *fungi in which the *fruiting body is generally a typical *apothecium, i.e. cup-shaped or dish-shaped, with the *hymenium exposed. The class is no longer recognized in most taxonomic schemes.

**discontinuous distribution**  See DISJUNCT DISTRIBUTION.

**Diselma**  (family *Cupressaceae) A *monotypic genus comprising *D. archeri*, a gymnospermic (*gymnosperm) shrub which is usually about 2 m in height but can be up to 6 m. The leaves are small, thick, blunt, and keeled. They are held closely to the stem and incurved against it, the stem appearing to be 4-angled because of the positioning of the leaves. The male cones are solitary, terminal, and barely distinguishable from the vegetative branch tips. The female cones are solitary and terminal, with 2 pairs of scales and a central *columella. The inner pair of scales has the *ovules. The seeds have hardened and expanded wings. The genus is endemic to Tasmania, where it forms small stands in sub-alpine and wet conditions in the mountainous regions.

**disjunct distribution (discontinuous distribution)** The occurrence of closely related species in a limited number of locations separated by oceans. For example, the family Caricaceae comprises four genera and about 30 species of trees, one of which is *Carica papaya* (papaya or pawpaw). Members of the family are most abundant in S. America, with some in C. America, but one genus, *Cylicomorpha*, grows in tropical Africa. *Araucaria* species (*A. araucana* is the monkey puzzle tree, or Chile pine, and *A. angustifolia* is paraná pine or candelabra tree) are native to S. America, where they form forests. *A. bidwilli* (bunya-bunya) occurs naturally in New Guinea, north-eastern Australia, and on islands in the S. Pacific Ocean. Disjunct distribution is evidence supporting continental drift.

**disjunction** The separation of *homologous chromosomes at the *anaphase stage of *mitosis and *meiosis, and movement towards the poles of the nuclear *spindle.

**disjunctive symbiosis** A symbiotic relationship (*see* SYMBIOSIS) between 2 species in which there is no physical union. *Compare* CONJUNCTIVE SYMBIOSIS.

**disomy** The condition in which a particular *chromosome is represented by 2 members; i.e. the *diploid condition. *Compare* NULLISOMY, POLYSOMY, TETRASOMY, and TRISOMY.

**dispersal** The tendency of an organism to move away from either its birth site (natal dispersal) or its breeding site (breeding dispersal): the opposite of *philopatry. Rates of regional dispersal depend on the interaction of several factors, notably the size and shape of the source area, the dispersal ability of the organisms, and the influence of such other environmental factors as winds or ocean currents. Dispersal may be passive (e.g. of winged seeds or ballooning spiderlings), active (e.g. of many mammals), passive but involving an active agent (e.g. seeds carried on the coats of mammals), or *clonal; in practice these categories are difficult to define precisely. Mathematical modelling using these factors has practical applications in the design of nature reserves, and provides an insight into the present distribution of organisms.

**dispersal barrier (ecological barrier)** An area of unfavourable *habitat separating two areas of favourable *habitat, e.g. oceans in the case of terrestrial organisms, or a cereal *monoculture in the case of woodland organisms.

**dispersal biogeography** The term now applied to the traditional school of *biogeography, regarding organisms as arising in a centre of origin, and spreading out by stages.

**dispersal mechanism** The characteristic adaptation for dispersal which forms part of the reproductive strategy of many slow-moving or *sessile organisms. It is most characteristic of the dispersal of spores, seeds, and fruit from plants, but is found in other organisms, especially for the dispersal of larvae. Typical examples are the hooked seeds and fruits that attach themselves to the coats of animals.

**dispersion** 1. In statistics, the internal pattern of a population, i.e. its distribution about the mean value. In spatial statistics, the pattern relative to some specific location, or of individuals relative to one another, e.g. clumped or random. 2. In *pedology, the process of separating soil particles (as in *aggregates) from each other so that they may react as individual particles. Aggregates or *peds of soil particles are destroyed by dispersion (and formation is initiated by *flocculation).

**dispersion coefficient** The measure of the spread of data about the mean value, or with reference to some other theoretically important threshold or spatial location, e.g. the standard deviation. *See also* DISPERSION; UNDERDISPERSION; and OVERDISPERSION.

**disruptive selection** A selection that changes the frequency of *alleles in a *divergent manner, leading to the fixation of alternative alleles in members of the population. The result after several generations of selection should be two divergent *phenotypic extremes within the population. This may be achieved, for example, by selecting seeds from the longest and shortest ears of corn in a population over a number of generations. *Compare* DIRECTIONAL SELECTION and STABILIZING SELECTION.

**disseminule** Any part of a plant from which a new plant may arise.

**dissolved oxygen level** The concentration of oxygen held in solution in water. Usually it is measured in mg/l (sometimes in $\mu g/m^3$) or expressed as a percentage of the saturation value for a given water temperature. The dissolved oxygen level is an important first indicator of water quality. In general, oxygen levels decline as pollution increases.

**distal** Applied to the region of an organ that is furthest from the point by which it is attached to the plant.

**distichous** In two ranks.

**disulphide bridge** A *covalent bond formed between two sulphur atoms. It is a particular feature of *peptides and *proteins, where it is formed between the *sulphydryl groups of two *cysteine residues, helping to stabilize the tertiary structure of these compounds.

**dittany** *See* ORIGANUM.

***Diuris*** (family Orchidaceae) A genus of orchids that are found in habitats of low fertility. There are 38 species, all endemic to Australia and found mainly in the east or west, except for 1 that is found in Java.

**diurnal** **1.** During daytime (as opposed to nocturnal), as applied to events that occur only during daylight hours, or to species that are active only in daylight. **2.** At daily intervals, as applied to such daily rhythms as the normal pattern of leaf or flower opening and closing, or the characteristic rise and fall of temperature associated with the hours of light and darkness. *See also* CIRCADIAN RHYTHM.

**diurnal curve method** A technique for measuring oxygen production in aquatic ecosystems as a means of assessing gross *primary or community productivity. Dissolved oxygen measurements are taken throughout a 24-hour period so that oxygen production by day and use at night by the aquatic community can be assessed.

**diurnal temperature variation** Daily variations in temperature at a particular place, related to the local radiation budget. In mid-latitudes, for example, maximum temperatures usually occur after noon and minimum temperatures in the early morning. The range varies according to location, with high variation in continental areas, and low variation in maritime areas. The diurnal range in equatorial areas exceeds the annual variation in average temperature.

**divergent evolution** The situation in which descendants of an ancestral group of organisms split into two or more groups that become increasingly different as time passes. Genetic separation and differentiation occurs to such an extent that distinct derivative taxa may result. Divergence may be at the *species, genus, family, order, or higher level. For example, *gymnosperms and *angiosperms arose from a stem group and subsequently diverged.

**diversity** Most simply, the species richness of a *community or area, though it provides a more useful measure of community characteristics when it is combined with an assessment of the relative abundance of species present. Diversity in ecosystems has been equated classically with stability and *climax communities. Such generalizations may be criticized on many counts, however.

**diversity index** The mathematical expression of the species *diversity of a given *community or area, which includes due allowance for the relative abundance of different species present. Such indices (e.g. the *Shannon–Wiener index) are generally considered an important means for comparison of community structure and stability. A different and specialized case of a diversity index is the *'biotic index' used in water-pollution studies.

**divisive method** A system of hierarchical classification that proceeds by subdividing the whole into successively smaller and more homogeneous units.

**DNA** See DEOXYRIBOSE NUCLEIC ACID.

**docks** See POLYGONACEAE.

**dodder** See *Cuscuta*.

**Dodonaea** (hop bush; family *Sapindaceae) A genus of shrubs and small trees with *simple or *pinnate, alternate leaves which are often sticky. The small, solitary flowers are in *panicles, *racemes, or clusters, usually white to red in colour, and often unisexual. The fruit is a large, membranaceous *capsule, and 3- or 4-winged, similar to that of the hop, hence the common name. There are about 50 species, found in tropical and subtropical regions, and especially well represented in Australia.

**dog lichen** The common name for the *lichen *Peltigera canina* in which the *thallus is *foliose with broad, spreading lobes, and margins that usually turn downwards. The upper surface is dark grey to greyish-brown when wet, and pale ash-grey when dry; it is downy, especially near the edges. The underside is white or pale brown, with conspicuous, anastomosing 'veins' and long white *rhizines. It is common on the ground in woods, moors, and sand-dunes, and among mosses on rocks, walls, wood, etc.

**dog stinkhorn** See *Mutinus*.

**Dolichandrone** (family *Bignoniaceae) A genus of trees that have *pinnate leaves and large, tubular flowers. The fruit is a pod with many small winged seeds. There are 9 species, occurring in E. Africa, Madagascar, and from South-east Asia to Australia.

**doliform** Barrel-shaped or jar-shaped.

**Dollo's law** A law describing evolutionary irreversibility: once regarded as inevitable, but now considered to apply mainly in special cases. The potential for further useful *mutation may well be very limited in highly specialized organisms, since only those mutations that will allow the organism to continue in its narrow *niche will normally be functionally possible. In such cases there is therefore a self-perpetuating, almost irreversible, evolutionary trend, so much so that it is regarded virtually as a law, Dollo's law (after the palaeontologist Louis Dollo). The trend results from steady directional selective pressure, or *orthoselection reinforced by specialization.

**domain** The highest taxonomic category in a classification system based on comparisons of ribosomal RNA. There are three domains: *Archaea; *Eubacteria; and *Eukarya. Each domain comprises organisms that are not closely related genetically to members of either of the other domains. See also KINGDOM.

**domestication** The selective breeding by humans of plant and animal *species in order to accommodate human needs. Domestication also requires considerable modification of natural *ecosystems to ensure the survival of, and optimum production from, the domesticated species (e.g. the removal or competing weed species when growing cereal crops).

**dominance** See DOMINANT GENE.

**dominant** In ecology, the species having the most influence on *community composition and form. Sometimes the term is also used to refer to the largest and/or most abundant species in the community.

**dominant gene** In *diploid organisms, a *gene that produces the same *phenotypic character when its *alleles are present in a single dose (*heterozygous) per *nucleus as it does in a double dose (*homozygous). For example, if $A$ is dominant over $a$, then $AA$ (the homozygote) and $Aa$ (the heterozygote) have the same phenotype. A gene that is masked in the presence of its dominant allele in the heterozygote state is said to be *recessive to that dominant. Such a dominance-recessive relationship is common between two alleles, the gene most frequently present at a given *locus being usually dominant to its alleles.

**Domin scale** A system for describing the cover of a species in a vegetation *community. The scale ranges from simple presence through 10 grades of linked cover–abundance and *cover measures. The scheme is based on the original 5-point cover scale of *Braun-Blanquet, but the finer subdivisions allow more detailed interpretation. *See also* COVER–SOCIABILITY SCALE.

**Donatia** (family *Stylidiaceae) A genus of small, sub-alpine *cushion plants found in wet areas. The leaves are densely arranged, alternate, and linear. The flowers are terminal, solitary, and *sessile. The genus comprises 2 species, found in Tasmania, New Zealand, S. America, and the subantarctic.

**Donatiaceae** *See* STYLIDIACEAE.

**Donax** (family *Marantaceae) A small genus of *herbs which have spreading *rhizomes and reed-like stems. There are 6 species, occurring in the lowland tropics from Indo-Malaysia to the Pacific.

**donor** The *bacterium from which the *chromosome migrates during *conjugation.

**dormancy (hypobiosis)** A resting condition with reduced metabolic rate. This is found in non-germinating seeds and non-growing buds. Dormancy is predictive if it protects the organism against adverse conditions and occurs before their onset. Predictive dormancy most commonly occurs in environments that undergo regular seasonal change; in animals it is often called *diapause, in plants innate dormancy. Consequential (secondary) dormancy commences after the onset of adverse conditions.

**dormin** *See* ABSCISIC ACID.

**dorsal** In a plant, abaxial (i.e. facing away from the stem).

**dorsiventral** With upper and lower sides differing in structure.

**Dorstenia** (family *Moraceae) A genus of *herbs in which the flowers are borne on, or sunken in, a disc-like, flat, or concave *receptacle, often more than 2 cm across, and are unisexual and minute. When ripe, the fruit is ejected by the receptacle becoming turgid. There are 170 species, occurring in the tropics.

**Doryanthes** (family *Amaryllidaceae) A genus of large *herbs that grow in big clumps. They have sword-shaped, radical leaves up to 2 m long, a thickened underground stem, and contractile roots. The flowers are in a large head or *panicle on top of a thick stem up to 3.5 m long, and are lily-like and usually red, surrounded by brown *bracts. The fruit is a 3-celled, woody *capsule. The flat seeds germinate early and mature slowly, flowering only after several years. There are 2 species, endemic (*see* ENDEMISM) to the coastal areas of eastern Australia.

**Doryphora** (family *Monimiaceae) A genus of 2 species, endemic (*see* ENDEMISM) to Australia. *D. sassafras* (Australian or New South Wales sassafras) is a tall tree, up to 30 m, with elliptical leaves up to 7 cm long and toothed. The tree produces white flowers in winter. Both the bark and leaves are fragrant, and provide aromatic oils used in perfume. It occurs as a *dominant in the rain forests of Queensland and New South Wales or as an understorey to eucalypts in drier habitats.

**dosage compensation** A genetic process that compensates for *genes which exist in two doses in the *homozygous *dominants, so that the *heterozygotes produce the same amount of gene product as the homozygotes. In animals, dosage compensation occurs because of the location of the relevant genes on the X-chromosome. In plants, sex determination may be under direct genetic control or may result from the disposition of appropriate hormones (which are themselves genetically controlled), and in only a few plants (e.g. willow) does sex appear to be determined by two sex chromosomes, as it is in mammals.

**Dothideales** (subdivision *Ascomycotina) The largest order of ascomycetous *fungi, in which the *ascocarps are *ascolocular. Species may be *saprotrophic, plant-parasitic, or lichenized.

**double coconut** *See* LODOICEA and ARECACEAE.

**double fertilization** The production of 2 sperm nuclei, both of which contribute to fertilization. The male *gametophyte, comprising the *pollen grain and *pollen tube, contains 2 sperm nuclei and 1 vegetative nucleus. The vegetative nucleus degenerates once the pollen tube has penetrated the *embryo sac. The 2 sperm nuclei enter the embryo sac. One unites with the egg nucleus to form the *zygote, which develops into the *embryo. The other unites with both of the *polar nuclei or with the secondary (definitive) nucleus formed by their fusion, to form the primary *endosperm nucleus, from which the endosperm develops. Double fertilization occurs only in *angiosperms, with the single exception of the *gymnosperm genus *Ephedra*, in which 2 sperm nuclei are produced, 1 of which fuses with the egg nucleus. The other unites with an adjacent cell, but develops no further.

**Douglas fir** See PSEUDOTSUGA.

**doum palm (gingerbread palm)** See HYPHAENE.

**down** The name applied to *grassland in the lowland zone of Britain, which has been created and maintained by grazing. Typically such grassland occurs on chalk and limestone hills, but occasionally it is found on acidic rocks, such as the Old Red Sandstone of the Gower Peninsula in southern Wales.

**downy mildew** Either a *fungus of the order *Peronosporales or a plant disease caused by such a fungus. The leaves of infected plants typically show yellowish spots or patches, with whitish or purplish mould on the underside. A wide range of plants may be affected.

**DPN (diphosphopyridine nucleotide)** See NICOTINAMIDE ADENINE DINUCLEOTIDE.

**Dracaena (family *Agavaceae)** A genus mostly of trees whose stems branch and thicken by an extrafascicular cambium (see FASCICULAR CAMBIUM). *D. draco* is the dragon tree. Trunks of several species exude dragon's-blood resin. There are about 40 species, occurring in the Old World tropics and subtropics.

**Dracontomelon (family *Anacardiaceae)** A genus of big trees in which the leaves are *pinnate and the flowers

*pentamerous. The fruit is a *drupe with edible, fleshy pulp and a flattened, angular stone with 1 or 2 seeds. Several species yield attractively coloured, valuable timber. There are 8 species, occurring in rain forests from Indo-Malaysia to Fiji.

**Dracophyllum (family *Epacridaceae)** A genus of ornamental, winter-flowering shrubs that have stem-clasping, sharply pointed leaves. The tubular flowers are white to red in crowded, 1-sided, terminal *racemes. The fruit is a small *capsule and the seed is slow to germinate. The species of exposed habitats are small; some are *cushion plants. There are 48 species, found in Australasia and New Caledonia.

**dragon's blood** See DAEMONOROPS and DRACAENA.

**dragon tree** See DRACAENA.

**Drimys (family *Winteraceae)** A genus of small trees with spicy bark in which the flowers have a distinct *calyx and *corolla. It is the widest-ranging genus of the family. There are 9 species, occurring from Borneo to New Caledonia, to S. America.

**Drosera (family *Droseraceae)** A genus of insectivorous herbs, often with creeping or tuberous (see TUBER) *rhizomes, that have *rosette leaves, the leaf blade being circular or elongated with stalked glands on the upper surface which secrete a sticky fluid on which insects become trapped. The plants are able to live in very poor soils but most of the extra nutrient obtained from the insects is devoted to seed production and is not essential to the survival of the plant. *Drosera* are usually found in acid bogs. Several species are cultivated for interest and ornament. There are about 80 species, with cosmopolitan distribution, but they are especially well represented in Australia and New Zealand.

**Droseraceae (sundews)** A family of insectivorous *herbs in most of which the leaves are borne in *rosettes bearing glandular hairs which secrete protein-digesting *enzymes so that the soft parts of insects trapped on the sticky leaves are digested and absorbed. The plants occur mostly in boggy places or on poor, acid, stony soils, and they may benefit from extra mineral nutrients obtained from the insect bodies. The flowers are regular, in 1-sided *cymes. There are 4 genera, with about 85 species, with a cosmopolitan distribution.

**drought cycle** A temporary and repetitive phase of drier conditions in an otherwise favourable environment (e.g. the 22-year drought cycles of N. American grasslands).

**Drude, Carl Georg Oscar (1852–1933)** A German botanist and biogeographer, who described vegetation types in terms of *formations. He published many works, including *Handbuch der Pflanzengeographie* (1890) and *Oekologie der Pflanzen* (1913), and in 1910 began a collaboration with H. E. *Engler to produce *Die Vegetation der Erde*. Drude was professor of botany and director of the botanic garden at the University of Dresden.

**drupe** A fleshy *fruit, such as a plum, containing one or a few *seeds, each enclosed in a stony layer that is part of the fruit wall.

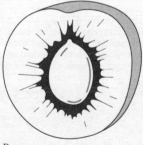

*Drupe*

**drupelet** The individual small *drupes of some *fruits, e.g. blackberry.

**druse (sphaeroraphide)** A spiky, globular mass of crystals, usually of calcium oxalate, formed around a centre of organic material and found free inside cells or attached to *cell walls in *cortex, *pith, and *phloem tissue.

**dryad's saddle** The common name for the *fruit body of *Polyporus squamosus.

This species forms handsome, semicircular, kidney-shaped or fan-shaped fruit bodies up to 60 cm across. The upper surface is yellowish-brown with dark brown, flattened, feathery scales arranged in more or less concentric rings. The lower surface is whitish with angular pores. The *stipe is lateral, sometimes very short. It is common on dead wood, but can also *parasitize living trees.

**Dryandra (family *Proteaceae)** A genus of small trees and shrubs that have narrow, toothed leaves. The flowers are yellow to red with a conspicuous *pistil, borne in dense heads with a basal *involucre of *bracts. The woody seed capsules are held vertically at the base, and the seed is sometimes retained on the plant following the winter flowering. The genus is similar to *Banksia. There are about 55 species, endemic to southwestern Australia.

**Dryas** Part of the characteristic threefold late-glacial sequence of climatic change and associated deposits following the last (*Devensian) ice advance and prior to the onset of the markedly warmer conditions of the current (*Flandrian) *interglacial. The type sequence was first described for Allerød in Denmark, and shows upper- and lower-clay deposits rich in remains of *Dryas octopetala* (mountain avens), and between them deposits of lake mud with remains of cool temperate flora, e.g. tree birches. The colder Dryas phases mark times of cold, *tundra-like conditions throughout what is now temperate Europe. The threefold Dryad–Allerød–Dryas sequence forms Pollen Zones I, II, and III of the widely accepted late and postglacial chronology of Europe. The basal, older, Dryas deposit forms Zone I; the Allerød Zone II; and the younger Dryas Zone III.

**dry-matter production** The expression of plant or animal productivity in terms of the dry weight of material produced per unit area during a specified time period. It is a more easily achieved, though technically less accurate, measure of organic *production than are calorific values; in the latter, the inorganic (ash) component can be separated. *See also* CALORIFIC VALUE.

**Dryobalanops** (family *Dipterocarpaceae*) A genus of lofty, resinous trees which yield the valuable timber kapur. *D. aromatica* also produces Borneo camphor, crystalline in the wood. There are 7 species, occurring in Sumatra, Malaya, and Borneo.

**Dryopteris** (buckler ferns, shield ferns; family *Aspleniaceae*) A genus of ferns in which the *rhizome is short, stout, and scaly, with leaves in a crown, much *pinnately divided. The *sori are superficial, rounded, and with kidney-shaped *indusia. There are about 150 species, found mainly in northern temperate regions but also widespread in the tropics.

**Dryptodon** *See* COPPER MOSSES.

**dry rot** **1.** Any of several plant diseases which are characterized by the formation of dry, shrivelled lesions; they are due usually to *fungal infection. **2.** A serious type of timber decay in buildings, caused by the fungus *Serpula lacrymans*. Typically, infected timber develops longitudinal and cross-grain cracking and bears a surface growth of whitish *mycelium; leathery *fruit bodies bearing rust-coloured spores may appear.

**dry season** A period each year during which there is little precipitation. In tropical climates (e.g. over much of India) the dry period is often in the winter season. In places in very low latitudes, two dry seasons may occur each year, between the northward and southward passage of the equatorial rains. In subtropical, Mediterranean, and in the climates found on the west coasts of continents, the dry season is in the summer.

**Duabanga** (family *Sonneratiaceae*) A genus of large trees which have a *monopodial crown and drooping branches with terminal clusters of white, nocturnal, bat-pollinated flowers. The fruit is a *capsule. *Duabanga* produce pale, light, useful timber. There are 2 species, occurring from the Himalayas to New Guinea.

**Duboisia** (corkwood; family *Solanaceae*) A genus of tall shrubs, up to 6 m, which have *obovate leaves and white, star-like flowers, and yield the *alkaloid

duboisine. *D. myoparoides* and *D. hopwoodii* were used by the aborigines as pituri, a chewed narcotic. There are 3 species in Australia and New Caledonia.

**duckweed** *See* LEMNACEAE.

**Duke of Argyll's tea-plant** *See* LYCIUM.

**dulse** The common name for the red seaweed *Palmaria palmata* (formerly *Rhodymenia palmata*). The *thallus is flattened and usually *dichotomously branched. It grows in the intertidal zone and at low-water mark, attached to rocks or to other seaweeds by a discoid *holdfast. It has been used widely as food, although it is reputed to be tough and somewhat flavourless. *See also* PEPPER DULSE.

**duplication** A chromosomal aberration in which more than one copy of a particular chromosomal segment is produced within a *chromosome set.

**duric horizon** A *soil horizon containing cemented silica. The name is from the Latin *durum* meaning 'hard'.

**duricrust** A weathered soil deposit, found especially in subtropical environments, which may ultimately develop into a hardened mass. A range of types occurs, each distinguished by a dominant mineral. Ferricrete and alcrete are dominated by sesquioxides of iron and aluminium respectively, silcrete by silica, and *caliche (calcrete) by calcium carbonate. *See also* DURIPAN.

**du Rietz, Gustaf Einar** (1895–1967) A Swedish ecologist who, in 1934, became professor of plant ecology at the University of Uppsala. He contributed to several branches of biogeography and was a leading figure in the *Uppsala school of phytosociology.

**Durio** (family *Bombacaceae*) A genus of rain-forest trees whose slender parts are covered with *peltate scales. The flowers have numerous *stamens, are often *cauliflorous or *ramiflorous, and are probably all bat-pollinated. The fruit is a big, woody, spiny *capsule. The seeds are prominently *arillate. *D. zibethinus*, cultivated for its creamy, pungent *arils, arouses strong passions, because of its repulsive smell but delicious flavour, and 'is worth a trip to the east' (A. R. *Wallace). There are 27 species occurring from Burma to the Philippines.

**duripan** A mineral *diagnostic soil *horizon which is cemented by silica and so will not slake or fall apart in water or hydrochloric acid. It may contain secondary cement (e.g. carbonates and iron oxide). Where duripans are exposed on the soil surface, they are called *duricrust. *Compare* CALICHE.

**Durisols** Soils that have a *duric horizon within 100 cm of the surface. Durisols are a reference soil group in the FAO *soil classification.

**durum wheat** *See* TRITICUM.

**Dutch elm disease** A devastating disease which can affect all species of elm (*Ulmus*). The causal agent is *Ceratocystis ulmi*, a *fungus which appears to have originated in Asia, not Holland. The fungus develops and spreads in the *xylem vessels; *tyloses are formed. Symptoms include wilting, with curling and yellowing of foliage, followed by rapid death of branches or the whole tree. The fungus is spread from tree to tree by elm-bark beetles (commonly *Scolytus* species). In the 1960s a new and more virulent strain of the *pathogen was introduced into Britain on logs imported from Canada; many millions of elms were killed, dramatically changing the landscape in many regions of Britain.

**Dutchman's pipe** *See* ARISTOLOCHIACEAE.

**dyad** In genetics, one of the products of the *disjunction of the *tetrads at the first *meiotic division, contained in the nuclei of secondary *gametocytes.

**Dyera** (jelutong; family *Apocynaceae) A genus of giant forest trees whose trunks are tapped for white latex, now used solely for chewing gum. They yield pale, light, close-grained, useful timber. There are 2 species, occurring in western Malesia.

**Dysoxylum** (family *Meliaceae) A genus of trees or treelets, most of which occur in lowland rain forest. A few attain timber size. The leaves are *pinnate, the flowers bisexual or functionally unisexual. The fruit is a leathery *capsule. There are few seeds; each has a fleshy jacket. There are about 75 species, occurring from Indo-Malaysia to Melanesia, with 1 species in New Zealand.

**dysphotic zone** The region of the *photic zone that lies below the *compensation level, and within which light penetration, is such that oxygen production by photosynthesis is exceeded by oxygen consumption by respiration. *Compare* EUPHOTIC ZONE.

**dystrophic** Applied to a lake that has become so shallow (through organic and inorganic sedimentation) and so depleted of oxygen (through the aerobic bacterial decomposition of the organic matter), that *bog begins to form and *peat to develop. A dystrophic lake may be regarded as the post-*eutrophic stage in the transitional sequence, over geological time, of lake sedimentation, productivity, and maturity. These sequential phases are termed oligotrophic, mesotrophic, eutrophic, and dystrophic.

**-eae** In plant *taxonomy, the suffix used to indicate a *tribe.

**eaglewood** (aloeswood) *See AQUILARIA.*

**early blight** *See* POTATO BLIGHT.

**earth ball** The common name for the spherical *fruit body formed by species of *Scleroderma. S. aurantium (S. citrinum) and S. verrucosum are common on the ground in woodland and heathland, and may form *mycorrhizal associations with trees such as birch and pine.

**earth fan** *See THELEPHORA.*

**earth star** The common name for a *fruit body formed by *fungi of the *Lycoperdales (e.g. *Geastrum species), in which the exoperidium (outer layer) splits and peels back from the endoperidium (inner layer) to give a star-shaped structure.

**earth tongue** The common name for the narrow, tongue-shaped *fruit bodies formed by certain *fungi (e.g. of the genera Geoglossum and Trichoglossum). They are found on the ground in grassland, etc.

**East African steppe floral region** Part of R. Good's (*The Geography of the Flowering Plants*, 1974) African subkingdom, which lies within his *palaeotropical kingdom. The flora includes about 150 endemic (*see* ENDEMISM) genera, nearly all of which are small. They include Saintpaulia, one species of which, S. ionantha, is now a popular house plant. *See also* FLORAL PROVINCE and FLORISTIC REGION.

**East Indian arrowroot** *See TACCA* and TACCACEAE.

**Ebenaceae** A family of mainly tropical trees in which the leaves are *simple, without *stipules. The flowers are usually unisexual. They have fused *sepals and *petals and *epipetalous *stamens. The *ovary is usually *superior. The fruit is a *berry. There are 2 genera, with 485 species, centred on the Indo-Malaysian rain forests, in which there are 200 species.

**ebony** *See DIOSPYROS.*

**ebracteate** Without *bracts.

**ecesis** The ability of some migrating plant species, having arrived at a new site, to germinate, grow, and reproduce successfully, while others fail to become established in the new environment. It represents the third in a series of 6 phases in plant *succession.

***Echinops*** (globe thistle; family *Asteraceae) A genus of plants that have a thistle-like habit and globular heads of flowers 4–6 cm wide, the individual flowers each having an *involucre of bristle-like *bracts. They are grown for their unusual blue flower heads. There are about 120 species, occurring in central and southern Europe, and western Asia.

**echinulate** Covered with small points or spines.

**ecocline** (ecological gradient) A gradation from one *ecosystem to another when there is no sharp boundary between the two. It is the joint expression of associated *community (*coenocline) and complex environmental gradients.

**ecological amplitude** The range of *tolerance of a species, diagrammatically forming a bell-shaped curve. Species with a narrow ecological amplitude often form good *indicator species.

**ecological and phytosociological distance** (D) *See* AFFINITY INDEX.

**ecological backlash** The unexpected and detrimental consequences of an environmental modification (e.g. dam construction) which may outweigh the gains anticipated from the modification scheme.

**ecological barrier** *See* DISPERSAL BARRIER.

**ecological efficiency** The ratio between energy flows measured at different points in a food-chain, usually expressed as a percentage. Many approaches have been devised to relate different aspects, e.g. intake, *assimilation, and *production. Two main categories of efficiencies are studied: (a) those of energy transfer between different *trophic levels; and (b) those of energy transfer within a single trophic level.

**ecological energetics** The study of energy transformations within ecosystems.

**ecological factor** See LIMITING FACTOR.

**ecological genetics** The study of genetics with particular reference to variation on a global and local geographic scale.

**ecological gradient** See ECOCLINE.

**ecological homoeostasis** The tendency of a plant population to adapt to environmental conditions.

**ecological indicator** Any organism or group of organisms indicative of a particular environment or set of environmental conditions. For example, *lichens may be used as indicators of air *pollution and fossil assemblages as indicators of past environments. See also INDICATOR SPECIES.

**ecological isolation** The separation of groups of organisms as a result of changes in their ecology or in the environment in which they live. This is one of the processes leading to *speciation, since there will be a restriction in the movement of *genes between groups thus separated, and changes in *gene frequencies may occur owing to local selection or drift until eventually the groups may be so *divergent that reproductive barriers exist.

**ecological niche** See NICHE.

**ecological pyramid (Eltonian pyramid)** A graphical representation of the *trophic structure and function of an *ecosystem. The first trophic level, comprising *producer organisms (usually green plants), forms the base of the pyramid, with succeeding levels of *consumer organisms arranged above it to the apex. Each level is represented by a horizontal bar; the bars are of equal thickness but of widths that vary to indicate the magnitude at that level. There are three types of pyramids: the *pyramid of numbers, the *pyramid of biomass, and the *pyramid of energy. Usually the bar at the base is the widest, with progressively narrower bars above it, creating the pyramid shape, but the concept is sometimes reversed in aquatic ecosystems. For example, the pyramid of biomass in a aquatic system stands on its head, having a higher biomass of predatory fish than of either primary producers or grazers. The alternative name for the concept is taken from the name of Sir Charles *Elton, FRS, the British ecologist who devised it.

**ecological system** See ECOSYSTEM.

**ecology** The scientific study of the interrelationships among organisms and between organisms and all aspects, living and non-living, of their environment. The German zoologist Ernst Heinrich *Haeckel (1834–1919) is usually credited with having coined the word 'ecology' in 1866, deriving it from the Greek oikos, meaning 'house' or 'dwelling-place'.

**eco-organ** A term used in some modern attempts to devise a system for describing vegetation types on the basis of life-form rather than species composition. An eco-organ is a characteristic morphological feature which reflects adaptation to external environmental conditions. Sunken stomata are one example; narrow, needle, or divided leaves are another. Compare LIFE-FORM and RAUNKIAER.

**ecosphere** See BIOSPHERE.

**ecosystem (ecological system)** A term first used by A. G. *Tansley (in 1935) to describe the interdependence of species in the living world (the *biome or *community) with one another and with their non-living (*abiotic) environment. Fundamental concepts include the flow of energy via food-chains and food-webs, and the cycling of nutrients *biogeochemically. Ecosystem principles can be applied at all scales; thus, principles that apply to an *ephemeral pond, for example, apply equally to a lake, an ocean, or the whole planet. In Russian and central European literature 'biogeocoenosis' describes the same concept.

**ecotone 1.** The narrow and fairly sharply defined transition zone between two or more different *communities. Ecotones arise naturally (e.g. at land–water interfaces) but elsewhere may often reflect human intervention, e.g. the agricultural clearance of formerly forested areas. *See* EDGE EFFECT. **2.** A physical gradient that causes a gradual change in biological composition in response to physical factors.

**ecotope** The *habitat component of a *biogeocoenosis.

**ecotron** *See* MICROCOSM.

**ecotype** A locally adapted population of a widespread species. Such populations show minor changes of morphology and/or physiology, which are related to *habitat and are genetically induced. Nevertheless they can still reproduce with other ecotypes of the same species. *Heavy-metal-tolerant ecotypes of common grasses such as *Agrostis tenuis* are an example.

**ectexine** *See* SEXINE.

**Ectocarpales (division *Phaeophyta)** An order of brown seaweeds in which the *thallus is usually composed of two parts: a prostrate disc which functions as a *holdfast, and an erect thallus which may be filamentous, foliose, or bulbous. There are many families. They are found mainly in the intertidal zone, attached to rocks or to other seaweeds.

**ectocrine (environmental hormone, exocrine)** A chemical substance, released by an organism into the environment during decomposition processes, which influences the activity of another organism.

**ectomycorrhiza** A type of *mycorrhiza in which the fungal *hyphae do not penetrate the cells of the root, but cover the root and grow between the root cells. This type of mycorrhiza is common in forest trees.

**ectophloic protostele** *See* HAPLOSTELE.

**ectophloic siphonostele** A *monostele type of *siphonostele in which a ring of *xylem occurs around the *pith, and a ring of *phloem outside the xylem. *Compare* AMPHIPHLOIC SIPHONOSTELE.

**ectoplasm** In plant cells (and some Protozoa), the outer, gel-like layer of the cell *cytoplasm, which lies immediately beneath the *cell membrane and contains packed layers of *microtubules.

**ectotrophic mycorrhiza** A *mycorrhiza in which the fungal component forms 2 sheath layers around the roots of a plant, the inner layer forming a sense mesh of *hyphae, called the hartig net. The close association of the fungal and plant components causes changes in the root *morphology giving rise to a palisade-like layer in the root *cortex and increased root branching. *Compare* ENDOTROPHIC MYCORRHIZA.

**edaphic** Of the soil, or influenced by the soil.

**edaphotope** The soil component of the ecotope (*habitat) of a *biogeocoenosis.

**edge effect** The change in the number of species occurring in the zone where two *habitats are in contact. Since this zone may contain biotic elements from both habitats and some unique to itself it may be rich in species, but because those species are ill-adapted to the immediately adjacent habitat, the rate of local extinction is usually high at edges. Predation, in particular, is greatest at a habitat edge. The effect occurs because the overlap region supports some species from both adjacent ecosystems and some peculiar to itself. Ecologists now regard the edge effect as a sign of ecological deterioration. The fragmentation of habitats causes an increase in edge areas, but a decrease in the internal areas of *ecosystems, leading eventually to a loss of species from all affected ecosystems and an increase in edge species, which are usually commonplace.

**Ediacaran** The final period of the *Proterozoic Eon, preceded by the Cryogenian, followed by the *Cambrian, dated at 600–542 Ma ago, and marked by a distinctive Ediacaran fauna. It is named after a site at Ediacara, South Australia, but *fossils are found in many parts of the world.

**effective population size** The average number of individuals in a population that actually contribute genes to succeeding generations by breeding. This number is generally rather lower than the observed, censused, population size, being reduced by the following factors: (*a*) a higher proportion of one sex may mate; (*b*) some individuals will pass on more genes by having more offspring in a lifetime than others; and (*c*) any severe past reduction in population size may result in the random loss of particular *genotypes.

**effector (inducer)** In the *operon theory of gene regulation, a chemical substance that is able to render a *repressor inactive by combining with it, thus permitting the *messenger-RNA for a particular *enzyme to be produced.

**effigurate** Having a margin with a definite form.

**effused** Loosely or irregularly spreading.

**effused-reflexed** Of a fungal *fruit body, for example: flat and spreading except at the edges, which grow outwards from the substratum.

**egg wrack (knotted wrack)** The common name for the brown seaweed *Ascophyllum nodosum*. The *thallus is flattened and strap-like, branching, with no *midrib; there are large, single air *bladders. It is found in the middle part of sheltered rocky shores, often in great abundance. It may also be found in salt-water lochs.

**EIA** *See* ENVIRONMENTAL-IMPACT ASSESSMENT.

**Eichhornia (family *Pontederiaceae)** A genus of herbs that includes *E. crassipes*, water hyacinth, a floating plant with bladder-like swollen *petioles that is a noxious weed in many parts of the tropics. There are seven species occurring in warm and tropical America.

**Eifelian** *See* DEVONIAN.

**eigen value (latent root, λ)** The components (latent roots) derived from the

data which represent that variation in the original data accounted for by each new component or axis.

**eigen vector (latent vector)** The loading of an attribute or variable on a component, as measured by the correlation between the original variable and the new component.

**einkorn (*Triticum monococcum*)** *See* TRITICUM.

**ektexine** *See* SEXINE.

**Elaeagnaceae** A family of shrubs in which the leaves are covered densely with scale-like hairs, giving them a silvery appearance. The flowers are tiny, solitary, or clustered, with 2 or 4 *sepals. The fruit is *drupe-like. They are often cultivated for their foliage. There are 3 genera, with about 45 species, found in northern temperate and tropical regions.

**Elaeis (oil palm; family *Arecaceae)** A genus of stout, solitary palms which have persistent leaf-bases, and a crown of numerous, large, *pinnate leaves. They are *monoecious. There are dense, *axillary, separate male and female inflorescences. The fruits have an oily, fleshy-fibrous *mesocarp and a stony *endocarp (stone) with 3 *apical pores, containing 1 seed, which has an oily *endosperm. There are 2 species, 1 found in Africa, and 1 in northern S. America. The former, *E. guineensis*, is now cultivated extensively as one of the world's major sources of vegetable oil.

**Elaeocarpaceae** A family of tropical and subtropical trees and shrubs in which the leaves are *opposite or alternate, with *stipules. The flowers have 4 or 5 free *sepals and *petals. The petals are often frilled, or absent. There is a *disc with many *stamens, and *anthers with pores. The *ovary is *superior. The fruit is a *capsule or *drupe. There are 11 genera, with 220 species.

***Elaeocarpus* (family *Elaeocarpaceae)** A genus of trees whose fruit is a *drupe, which in some species is edible. There are about 60 species, occurring from eastern Asia to the Pacific islands.

**elaioplast** A *leucoplast type of *plastid involved in the storage of oils.

**elaiosome** A structure on the surface of a seed that secretes and stores oil, usually as an attractant to ants, which assist in seed dispersal.

**elastic growth** The part of *cell-wall extension that is reversible.

**elater** In most liverworts (*Hepaticae), 1 of many long, thin cells mixed with the *spores in the *capsule of the *sporophyte that assist in spore dispersal. Helical thickening in the wall of an elater causes increasing tension as the capsule dries. When the capsule bursts the tension is released and the spores are forcibly ejected.

***Elateriospermum*** (family *Euphorbiaceae) A *monotypic genus, *E. tapos*, which is a rain-forest tree whose fruit is a big, bony *capsule. The seeds are edible after boiling or roasting and are collected but never cultivated. The hard, heavy timber is good for firewood and heavy construction. It occurs in the Malay peninsula, Sumatra, and Borneo.

**Elatinaceae** A family of small, aquatic or marsh *herbs in which the leaves are *opposite or in whorls, and *simple, with *stipules. The tiny flowers are solitary or clustered in the leaf *axils, with 3–5 *imbricate *sepals and *petals, and with twice as many *stamens as petals. There are 2 genera, with 32 species, with a cosmopolitan distribution.

**elder** See SAMBUCUS.

**electrode potential** See REDUCTION POTENTIAL.

**electron carrier** A compound that functions as an acceptor and donor of electrons and/or protons in an electron-transport system.

**electron-transport chain 1.** Respiratory chain. A system of redox compounds called electron carriers, present in *mitochondria, which sequentially transport electrons and/or protons previously removed from metabolites in *glycolysis, the *citric-acid cycle, and other metabolic reactions. **2.** A system

of redox compounds involved in electron transport in *metabolic pathways such as *photosynthesis.

**electrophoresis** The migration of charged particles under the influence of an electric field within a stationary liquid. The latter may be a normal solution or held upon a porous medium (e.g. *starch, acrylamide gel, or cellulose acetate). The rate at which migration occurs varies according to the charge on the particle and also its size and shape. The phenomenon is exploited in a variety of analytical and preparative techniques employed in studies of *macromolecules.

***Eleiodoxa*** (family *Aeecaceae) A *monotypic genus of spiny palms which form dense, extensive clumps in lowland swamp rain forest. They are similar to *Salacca*, but the inflorescences form a massive terminal head. The fruit pulp is sour. They occur in Malaya, Borneo, and Bangka and the Riau archipelago, Indonesia.

***Eleocharis*** (family *Cyperaceae) A genus of sedges, mainly *perennial, that have many flowered *spikelets and creeping underground *rhizomes. *E. dulcis* is the Chinese water chestnut, which is cultivated in China and Japan for its edible *tubers. There are about 150 species, distributed world-wide.

**elfin wood** A *facies of tropical upper *montane or subalpine forest, with dwarfed and gnarled trees, that grows on extreme or exposed sites. The term is sometimes applied outside the tropics, although in this context it is more usual to apply the name *kampfzone or *krummholz.

**elms** See ULMUS.

**El Niño** A weakening of the Equatorial Current, allowing warm water to accumulate off the S. American Pacific coast; it is associated with a change in the atmospheric circulation known as a southern oscillation, the two together comprising an El Niño—Southern Oscillation (ENSO) event. Its climatic effects are felt throughout the Pacific region. (A similar phenomenon may occur in the Atlantic.) About once every 7 years, during the Christmas season (midsummer in the southern hemisphere), prevailing trade winds weaken, the Equatorial counter-current strengthens, and warm surface waters that are normally driven west-wards by the wind to form a deep layer off Indonesia flow eastwards to overlie the cold waters of the northward-flowing Peru current. In exceptional years (e.g. 1953, 1972–3, 1982–3, and 1997–8) the severity with which the upwelling of nutrient-rich cold water is inhibited causes the death of a large proportion of the plankton population and a con-sequent decline in the numbers of fish.

**Elton, Charles Sutherland (1900–91)** A British zoologist who studied animal communities and made major contribu-tions to the development of ecological studies in general and of animal ecology in particular. One of his early books, *Animal Ecology* (first published in 1927) was very influential. He emphasized the importance of conserving species and habitats. He proposed the existence of *niches, occupied by species at particular points in a food-chain. From 1932 until 1967 he was director of the Bureau of Animal Population at the University of Oxford. Elton was elected a Fellow of the Royal Society in 1953 and was awarded many scientific prizes.

**Eltonian pyramid** See ECOLOGICAL PYRAMID.

**eluviation** The removal of soil materials in suspension or in solution from surface *horizons, and with partial deposition in the lower horizons of soil *profiles. Removal in solution is called *leaching, and hence the term 'eluviation' is often limited in use to removal in suspension.

**emarginate 1.** Applied to a leaf, petal, or sepal that has a notch at its apex. **2.** Of a fungal *gill, having a notch at the edge nearest the *stipe.

**Embden–Meyerhof pathway** See GLYCOLYSIS.

**Emblica (family *Euphorbiaceae)** A genus of small trees, including *E. officinalis* (*Phyllanthus emblica*) which has edible fruit. There are 4 species, occurring in Madagascar and Indo-Malaysia.

**Embothrium (family *Proteaceae)** A genus of 8 species, central to the southern Andes. *E. coccineum*, the Chilean firebush, is a widely cultivated, showy species with red flowers.

**embryo** A young plant developed from an *ovum sexually or asexually and, in *seed plants, contained within the seed.

**embryogenesis** See EMBRYOGENY.

**embryogeny (embryogenesis, embry-ony)** The sequence of events leading to the formation of an *embryo.

**embryony** See EMBRYOGENY.

**embryophyte** A plant that produces *embryos from multicellular repro-ductive organs. *Bryophyta and see plants are embryophytes. In some classifications, embryophytes are ranked as a subking-dom, Embryophyta. Compare THALLOPHYTE.

**embryo sac** The female *gametophyte (an oval structure in the nucellus of the ovule) of flowering plants, formed by the division of the *haploid *megaspore nucleus, and the site of *fertilization of the egg and development of the *embryo. It consists of 6 haploid cells without *cell walls (2 *synergidae, 3 *antipodal cells, and an egg cell) and 2 haploid nuclei (polar nuclei). Sometimes the 2 haploid, polar nuclei fuse to form a single, *endosperm mother cell. At fertilization, 1 male nu-cleus fuses with the egg nucleus to form a *zygote which develops into the embryo. The second male nucleus fuses with the primary endosperm nucleus to form the endosperm nucleus. This then divides to form the endosperm. See also DOUBLE FER-TILIZATION.

**emergence marsh** The upper zone of a *salt marsh, from the general mean-high-water level to the mean-high-water level of spring tides. Typically it has fewer than 360 submergences per annum, often with less than an hour of sub-mergence daily during sunlight. The minimum period of continuous exposure may exceed 10 days. Compare SUBMERGENCE MARSH.

**emergent aquatic** An aquatic herb that is rooted below the water surface, but with its shoot and/or leaves above the water. Examples include bulrushes and reeds.

**emergents** The individual trees, or clumps of trees, that stand prominently higher than the top of the continuous *canopy of many lowland tropical rain forests.

**Empetraceae** A family comprising 3 genera and 5 species of heath-like shrubs in which the leaves are evergreen with inrolled margins. The flowers are very small, borne in clusters, with 4–6 *perianth segments in 2 similar whorls and 2 or 3 *stamens, and there are black, juicy *drupes for fruits. *Empetrum* (crowberry) is an example. They occur in north temperate regions and in temperate S. America.

**Empetrum** **(crowberry)** *See* EMPETRACEAE.

**Emsian** *See* DEVONIAN.

**emu apple** The fruit of *Owenia acidula*. *See* OWENIA.

**emu bushes** *See* EREMOPHILA; MYOPORUM; and MYOPORACEAE.

**enation** **1.** The outgrowth from a leaf in a plant infected with a certain type of virus. **2.** A leaf-like structure, lacking a *leaf gap, which grows out laterally from a stem.

**Encalyptales** **(subclass *Bryidae*)** An order of *acrocarpous mosses, in which the leaves are nerved and have a short point at the tip. The *capsules are cylindrical and are held erect on long *setae; the entire *capsule is covered by a distinctive, large *calyptra which is often fringed at its lower edge. Mosses of the order are found chiefly on soils rich in lime, e.g. in crevices in limestone walls, chalk banks, limestone screes, etc. There is 1 family, 1 genus (*Encalypta*), and about 35 species, widely distributed.

**endarch** Applied to strands of primary *xylem in which the first elements to form lie closest to the centre of the axis. *Compare* EXARCH and MESARCH.

**endemic** *See* ENDEMISM.

**endemism** A situation in which a species or other taxonomic group is restricted to a particular geographic region, owing to factors such as isolation or in response to soil or climatic conditions. Such a *taxon is said to be endemic to that region. The size of the region in this context will usually depend on the status of the taxon; thus, a family will be endemic to a much larger area than a species, all other things being equal. Reference is frequently made to 'narrow endemics', i.e. taxa with markedly restricted ranges. Some of these are evolutionary relics, such as the maidenhair tree (*Ginkgo biloba*), a single species within a single genus which is confined to Chekiang Province, China, where it was discovered in 1758. Endemics are always native to the region in which they are found and to which they are confined. They may be neoendemics, in which case they have evolved recently and may be restricted in their distribution simply because they have not yet had time to spread further, or palaeoendemics, in which case they have a long evolutionary history and their confinement is caused by barriers to dispersal.

**endexine** *See* NEXINE.

**endive** *See* CICHORIUM.

**endobiotic** Growing within a living organism.

**endocarp** *See* PERICARP.

**endocytosis** The entry of particles or fluids into a cell by their enclosure in an invagination of the *cell membrane which is then detached.

**endodermis** The innermost layer of the *cortex surrounding the *stele in plants, which plays an important role in controlling the transport of substances within the plant.

**endogenetic** *See* EXOGENETIC.

**endogenous** **infection** Infection with a micro-organism that is normally resident in the body.

**Endogonales** **(class *Zygomycetes*)** An order of *fungi that forms *zygospores and (asexually) *chlamydospores or *sporangiospores. They are commonly found in *mycorrhizal association with higher plants.

**endolithic** Growing within stone, as do the thalli of certain lichens. Some desert *lichens and *protists avoid desiccation by living beneath the surface of a rock.

**endomitosis** A doubling of the *chromosomes within a *nucleus that does not divide, so producing a *polyploid. The doubling may be repeated a number of times in a single nucleus. It occurs in plant tissues.

**Endomycetaceae** (order *Endomycetales) A family of *fungi in which the vegetative stage is a septate *mycelium. Single, *budding, *yeast-like cells do not occur. They are mainly *saprotrophic.

**Endomycetales** (subdivision *Ascomycotina) An order of *fungi in which *asci develop directly from *zygotes or from single cells; *ascocarps are not formed. Most species are *saprotrophic, and are found, for example, on plants and animals.

**endomycorrhiza** A type of *mycorrhiza in which the *hyphae of the fungus actually penetrate the cells of the root. Endomycorrhizas are found in a wide range of plant types.

**endopeptidase** See PROTEASE.

**endoperidium** The inner layer of a double-layered *peridium in certain fungi of the *Gasteromycetes. Compare EXOPERIDIUM.

**endophloeodal** Growing within bark.

**endophyte** A plant that lives inside another plant and is not a parasite.

**endoplasm** In plant cells (and some Protozoa), the inner layer of *cytoplasm within which are embedded the principal cell *organelles.

**endoplasmic reticulum** A complex network of cytoplasmic (see CYTOPLASM) membranaceous sacs and tubules which appears to be continuous with both the nuclear and *plasma membranes. It occurs in two forms: that bearing *ribosomes is called rough ER; that without, smooth ER. Both are involved in the synthesis, transport, and storage of cell products.

**endopolyploidy** *Polyploidy that results from *endomitosis.

**endosperm 1.** A triploid structure found in many *seeds of angiosperms: it frequently stores food materials which are broken down during *germination. **2.** See EMBRYO SAC and DOUBLE FERTILIZATION.

***Endospermum*** (family *Euphorbiaceae) A genus of soft-wooded, fast-growing trees in which the leaves are large, often *peltate, with *stellate hairs. The fruits are globose, *indehiscent, and leathery. The trees produce a light, general utility timber. There are about 12 species, occurring from China and South-east Asia to Fiji.

**endospore** (bacterial) A type of resting cell that develops within a *vegetative cell under certain conditions. Endospores are extremely resistant to adverse environmental conditions.

**endosymbiosis** *Symbiosis in which 1 symbiont lives within the other.

**endosymbiotic hypothesis** The hypothesis that the *plastid and *mitochondrial *organelles evolved from *prokaryotic endosymbionts within *eukaryotic cells early in eukaryotic evolution. At least 2 endosymbiotic events are postulated, 1 each for plastid and mitochondrial origins. It is contentious whether 2 independent events could account for the mitochondria that occur in plants and those that occur in animals, the present-day mitochondria of the 2 kingdoms being quite distinct from each other genetically.

**endothecium 1. (fibrous layer)** In *angiosperms, the layer of cells lying beneath the *epidermis of the wall of the *anther. As the anther matures, thickenings often develop in the cell walls of the endothecium, probably aiding dehiscence. **2.** In *Bryophyta, the inner layer of cells in the young *sporophyte.

**endotoxin** A component (lipopolysaccharide) of the walls of Gram-negative (see GRAM-REACTION) bacteria: it is toxic to animals, including humans.

**endotrophic mycorrhiza** A type of *mycorrhiza in which the fungal component penetrates the plant root, either pathogenically or beneficially. The fungal component does not change the root *morphology. *Compare* ECTOTROPHIC MYCORRHIZA.

**endozoochory**  Dispersal of spores or seeds by animals after passage through the gut.

**energy flow**  The exchange and dissipation of energy along the food-chains and food-webs of an *ecosystem.

**energy of activation**  *See* ACTIVATION ENERGY.

***Engelhardia* (family *Juglandaceae)**  A genus of trees in which the leaves are *pinnate and the flowers tiny and borne in *panicles. The fruit is a small, hard *nut attached to an enlarged, wing-like *bract. The trees yield useful timber. There are 5 species, occurring from the western Himalayas to New Guinea.

**Engler, Heinrich Gustav Adolf (1844–1930)**  A German taxonomist and biogeographer who helped to develop a system for classifying plant families and genera. In 1910 he began a collaboration with C. G. O. *Drude to produce *Die Vegetation der Erde*. Engler was professor at the museum and botanical gardens in Berlin-Dahlem.

***Enneapogon* (nine-awn grasses; family *Poaceae)**  A genus of slender, tufted, often glandular-hairy grasses which are *perennial, *annual, or erratic, and are brought into flowering by rainfall. The leaves are narrow and small. The *panicle is compact, with the *spikelets grouped closely together. The *lemma of each *floret has a fringe of 9 short and fluffy *awns. The flower head is dark in colour. There are about 30 species, found in warm temperate regions of the Old World and Australia, in which there are 9 endemic species. Some species are highly valued pasture grasses (e.g. *E. cylindricus*, jointed nine-awn, found in the calcareous grasslands of Australia).

**enokitake**  *See* FLAMMULINA.

**enrichment**  In forestry, the planting of young trees within a forest, commonly after depletion by timber extraction.

***Ensete* (family *Musaceae)**  A genus of giant, banana-like *herbs, whose inflorescences are terminal, with big persistent *bracts. There are 7 species, occurring in Africa and Asia.

**ensiform**  Sword-shaped.

**ensilage**  *See* SILAGE.

**ENSO**  *See* EL NIÑO.

***Entada* (family *Fabaceae, subfamily *Mimosoideae)**  A genus of shrubs and big, woody climbers, several of which have gigantic flat pods, up to 2 m long. The seeds are big and flat, and are often carried by sea currents. Cut stems of many climbers yield potable water. There are about 30 species, with pantropical distribution.

***Entandrophragma* (family *Meliaceae)**  A genus of big trees in which the leaves are *pinnate, the fruit a *capsule, and the seeds winged. They are important for timber, notably sapele and utile. There are 11 species, occurring in tropical Africa.

**entelechy**  An outmoded theory holding that evolution proceeds by the realization of that which was always potential. The word is derived from the Greek *entelekheia*, meaning 'become perfect'. *See also* ARISTOGENESIS; NOMOGENESIS; and ORTHOGENESIS.

**Enterobacteriaceae**  A large family of *bacteria, which are *Gram-negative and rod-shaped. They are *chemo-organotrophic and can grow in the presence or absence of air. There are many genera, found as *parasites and *pathogens in a wide range of organisms from plants to humans. Many can also live *saprotrophically, e.g. in soil and water.

***Enteromorpha* (division *Chlorophyta)**  A genus of macroscopic green *algae in which the *thallus consists of elongated, hollow, tubular *fronds with walls only one cell thick. The fronds may be *simple or branched. The species are often difficult to distinguish from one another. *E. intestinalis* is a common species in Britain, occurring in rock pools on the upper shore and in estuaries; the thallus is typically irregularly inflated, giving the resemblance to an intestine, and may reach up to 1.5 m in length and 1.5–2.0 cm in diameter.

**enterotoxin**  A type of toxin, produced by certain bacteria, which affects the function of the intestinal mucosa, causing diarrhoea, gastroenteritis, etc. in animals.

**entire**   Applied to a leaf whose margin is undivided.

**Entisols**   Embryonic *mineral soils, including those that have no distinct pedogenic *horizons. Representing only the initiation of soil-*profile development, entisols are common on recent flood plains, steep eroding slopes, stabilized sand-dunes, and recent deep ash or wind deposits.

**Entoloma (family Entolomataceae, order *Agaricales)** A genus of *fungi in which the *fruit body is *mushroom-like with confluent *stipe and cap. The spore deposit is pink or brownish; under the microscope the spores appear angular. *Entoloma* is found on the ground, often under trees.

**entomochory**   Dispersal of spores or seeds by insects.

**entomophilous**   Applied to flowering plants with floral parts that are adapted to pollination by insects.

**Entomophthorales (class *Zygomycetes)** An order of *fungi in which the *sporangium functions as a single *conidium and is typically discharged forcibly at maturity. Most are *mycelial and *coenocytic. Many species are *parasitic in insects or other animals; some can *parasitize *desmids or fern *prothalli, while others are *saprotrophic.

**environment**   The complete range of external conditions, physical and biological, in which an organism lives. Environment includes social, cultural, and (for humans) economic and political considerations, as well as the more usually understood features such as soil, climate, and food supply.

**environmental hormone**   *See* ECTO-CRINE.

**environmental-impact assessment (EIA, environmental-impact statement)** An attempt to identify and to predict the impact on the biogeophysical environment and on human health and well-being of proposed industrial developments, projects, or legislation. EIA also aims to devise easily comprehended, universally applicable schemes for communicating the results of the assessment.

**environmental resistance**   The sum total of the environmental *limiting factors, both *biotic and *abiotic, that together act to prevent the *biotic potential of an organism from being realized. *See also* LOGISTIC EQUATION.

**environmental science**   The study of environments. This may be interpreted fairly strictly as the physical environment; or may include the biological environment of an organism; or, in its widest sense, it may consider social, cultural, and other aspects of the environment.

**environmental variance**   The portion of *phenotypic variance that is due to differences in the environments to which the individuals in a population have been exposed. The total amount of variance (*see* MEAN SQUARE) observed among individuals in a population will be made up by an environmental component, determined by environmental variation, and a genetic component, determined by the variation that is inherited.

**enzyme**   A molecule, wholly or largely *protein, produced by a living cell, that acts as a biological catalyst. Enzymes are present in all living organisms, and through their high degree of specificity exert close control over cellular metabolism.

**Eocene**   A *Tertiary epoch, from about 55.8–33.9 Ma ago, which began at the end of the *Palaeocene and ended at the beginning of the *Oligocene. It is noted for the expansion of mammalian stocks and the local abundance of nummulites (marine protozoans of the Foraminiferida). In southern Britain, a humid, subtropical climate allowed rain forests to flourish. The name is derived from the Greek *eos* meaning 'dawn', and *kainos* meaning 'new'.

**Epacridaceae**   A family of shrubs and small trees which are heath-like and the southern counterpart of the *Ericaceae. The flowers have free *sepals, the *stamens are *epipetalous, and the *anther opens by slits. The fruit is a *drupe or *capsule. There are 31 genera, with 400 species, occurring mainly in Australasia, but a few extending to Indo-China, and also occurring in southernmost America.

**Epacris** (family *Epacridaceae) A genus mostly of shrubs, similar to *Erica* and *Calluna* species in the northern hemisphere. They have narrow, *simple, *sessile, and alternate leaves. The flowers are small and borne in *spikes or *racemes. There are several *ovules to each *locule, and the fruit is a *capsule. There are 35 species, endemic to southeastern Australia, Tasmania, and New Zealand.

**Eperua** (family *Fabaceae, subfamily *Caesalpinoideae) A genus of trees in which the leaves are *pinnate. *E. falcata* (wallaba) yields a hard, valuable timber, and grows gregariously on white-sand soils. There are 14 species, occurring in tropical S. America.

**Ephedra** (family Ephedraceae) A genus of shrubs which have scale-like leaves and whip-like, slender green stems. They are *gymnosperms, related to *Gnetum, and their *ovule is naked within the *perianth, which becomes woody around the seed. There are 40 species, found in the warm temperate regions of each hemisphere.

**ephemeral** (ephemerophyte) A plant that completes its life cycle very rapidly. In favourable environments ephemerals may germinate (*see* GERMINATION), bloom, and set seed several times during a single year. Many weed species are ephemerals. Ephemerals are also particularly characteristic of *desert environments. One type of ephemeral is the winter annual, often found in summer-dry locations such as sand dunes. These germinate in the moist conditions of autumn, grow as a small *rosette through the winter, flower and fruit in the early spring, then enter a dormant state as a seed through the hot drought of summer.

**ephemerophyte** *See* EPHEMERAL.

**epibiontic** (noun epibionty) Applied to old endemic (*see* ENDEMISM) taxa which now have a restricted distribution although formerly they were more widespread.

**epibionty** *See* EPIBIONTIC.

**epibiotic** Growing on the surface of a living organism.

**epicalyx** A whorl of *sepal-like appendages which resembles the *calyx but is outside the true calyx.

**Epichloë** (order *Clavicipitales) A genus of *fungi that are *parasitic on grasses. *E. typhina* is responsible for the condition known as *choke. It forms a ring of *conidia-bearing *mycelium around the uppermost leaf sheath. The mycelium turns orange as *perithecia develop.

**epicotyl** The part of a seedling above the *cotyledons; it will develop into the shoot.

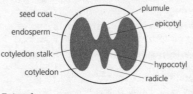

*Epicotyl*

**epidermis** The outermost layer or layers of cells in a plant or animal. It is one cell thick in plants.

**epigeal** (epigean) Growing or occurring above ground, commonly with reference to a mode of seed germination in which the *cotyledons are carried above the soil on an axis (the hypocotyl). *Compare* HYPOGEAL.

**epigean** *See* EPIGEAL.

**epigenetics** The study of the mechanisms by which *genes bring about their *phenotypic effects.

**epigynous** Applied to a flower in which the *calyx, *corolla, and *stamens are inserted near the tip of the *ovary.

**epilimnion** The upper, warm, circulating water in a thermally stratified lake in summer. Usually it forms a layer that is thin compared with the hypolimnion.

**epilithic** Growthing on or attached to the surfaces of rocks or stones.

**Epilobium** (family *Onagraceae) A genus of *herbs, usually of open habitats, in which the leaves are *simple, and *opposite or alternate. The flowers are generally solitary but some are in *racemes. Numerous seeds, without *endosperm, are produced from the *capsule with a plume of hairs at the *chalaza. There are about 200 species, of northern and southern temperate and Arctic regions.

**epinastic growth** (epinasty) Differential growth of the upper or *adaxial part of a plant organ. A well-known example is the growth of the upper side of the *petiole causing the leaf to be bent downwards. *Auxin and ethylene are involved in this response, which may also be induced by application of *herbicides. It may be a symptom of disease, in which there is a pronounced and atypical bending downwards of the leaves.

**epinasty** *See* EPINASTIC GROWTH.

**epipedon** A subsurface *soil horizon.

**epipelic** Growing on mud.

**epipetalous** Borne on the *petals, by concrescence. *See also* CONCRESCENT.

**epiphloeodal** Growing on or attached to the surface of tree bark.

**epiphyllous** (folicolous) Growing on leaves.

**epiphyte** A plant that uses another plant, typically a tree, for its physical support, but which does not drawn nourishment from it. Well-known examples include *Platycerium (staghorn fern) and many members of the *Bromeliaceae and *Orchidaceae. Epiphytes are a conspicuous feature of many kinds of tropical rain forest. Even the epiphytes of tropical rain forests need to be adapted to drought because of their total dependence on rainfall for a water supply, hence the water collecting systems of bromeliads and the occurrence of some cacti as epiphytes, e.g. *Zygocactus* (Christmas cactus).

**epiphytotic** An outbreak of disease (an epidemic) in a population of plants.

**episepalous** Borne on the *sepals, by concrescence (*see* CONCRESCENT).

**episodic evolution** The fossil record is characterized by extinction events and succeeding phases of rapid evolutionary innovation. The overall picture is thus one of episodic evolution. However, the term has recently acquired other connotations, and tends to be linked with *punctuated equilibrium.

**epistasis** A situation in which an *allele of one *gene (called the epistatic gene) prevents the expression of all allelic alternatives of another gene. The opposite situation is called hypostasis.

**epitheca** In diatoms (*Bacillariophyta), the older, outer half of the *frustule.

**epithecium** A layer of *tissue that covers the *hymenium in an *ascocarp. *See also* GIRDLE GROOVE and HYPOTHECIUM.

**epithelium** The lining of a resin canal in a *gymnosperm or a gum duct in a dicotyledon. Epithelial cells often secrete resin or gum.

**epixylous** Growing on wood.

**equatorial division** The division of each *chromosome during the *metaphase of *mitosis or *meiosis into two equal longitudinal halves which are then incorporated into separate *daughter nuclei.

**equatorial plate** An arrangement of the *chromosomes in which they lie approximately in one plane, at the equator of the *spindle. This is seen during *metaphase of *mitosis and *meiosis.

**equilibrium species** A species in which competitive ability (*see* COMPETITION), rather than *dispersal ability or reproductive rate, is the chief survival strategy: competition is the typical response to stable environmental resources. In unstable or extreme environments, e.g. deserts, equilibrium species survive unfavourable periods by living on stored food resources and reducing life processes to a minimum. *Compare* FUGITIVE SPECIES.

***Equisetum*** (class Sphenopsida, family Equisetaceae) The sole surviving genus of its class and family, comprising the horsetails, which typically are plants of damp places. They are vascular *cryptogams with hollow stems, tiny, scale-like leaves, branches in *whorls, and apical, *spore-producing cones. The spores germinate into *prothalli bearing the sex organs. The sperms are multiciliate. There are 29 species, almost cosmopolitan in distribution except for Australasia. In *Carboniferous times the Sphenopsida produced tree-like forms. One such is *Calamites*, a common coal-measure fossil, which grew to heights of 30 m and, apart from its size and woody stems, closely resembled *Equisetum*.

**Equisitites hemingwayi** The first known species of the family Equisetaceae, known from the *Carboniferous of Europe, and the direct ancestor of the only extant genus, *Equisetum*.

**equivalence point** See MINIMAL AREA.

**Eremophila** (emu bush; family *Myoporaceae) A genus of shrubs and trees that have alternate, entire leaves, often sticky or shiny. The flowers are large, colourful (red or white), tubular, 2-lipped *corollas, with the *calyx united in a bell-like tube which remains on the fruit. The fruit is a fleshy or dry *drupe. Eremophilas are good horticultural plants in hot, dry areas, and *E. mitchelli* provides useful timber. There are about 180 species, endemic (see ENDEMISM) to Australia.

**ergastic matter** Non-protoplasmic (see PROTOPLASM) products of cell metabolism (e.g. starch grains, oil droplets, crystals, and tannins).

**ergot** A disease that affects many grasses, including cereals. It is caused by the *fungus *Claviceps purpurea*; the conspicuous, hard, black *sclerotia of the fungus replace the grain in the *spikelets of an infected plant. These sclerotia contain *alkaloids which can cause severe poisoning or even death if ingested by animals or humans.

**Erica** (heaths; family *Ericaceae) A genus of shrubs and a few trees in which the leaves are leathery and small. The flowers are bell-like, with a *disc. *E. arborea* of southern Europe is *bruyère*, whose roots provide briar pipes. There are 665 species, occurring from Europe to Asia Minor and the African mountains, and species are very numerous in S. Africa, with 580 in the southern Cape, of which 520 are endemic (see ENDEMISM). Cape species are generally larger and more showy than European ones.

**Ericaceae** A family of shrubs and small trees in which the leaves are leathery and evergreen. The flowers have 4 or 5 free *sepals, a *corolla of 4 or 5 fused *petals, and 8–10 free *stamens. The *anthers are porous, and the *ovary may be *superior or *inferior. The fruit is a *capsule, *drupe, or *berry. There are 103 genera, with about 3350 species. They are cosmopolitan in distribution but centred in the northern hemisphere and almost absent from Australasia.

**ericeto** See MEDITERRANEAN SCRUB.

**Erigeron** (family *Asteraceae) A genus of mainly *perennial herbs with *simple leaves which are either in a *radical *rosette or alternate. The flower heads are stalked, solitary, or in *panicles. The herbaceous *involucre *bracts are numerous and either all of one length or in 2 or more rows, the ray-florets are female and numerous, in 2 or more rows. The disc-florets are bisexual, tubular, and 5-toothed. The *achenes are flattened with a *pappus of bristles. There are about 200 species, found throughout the temperate regions of the world, but especially in America. Some species are alpine. Many are now naturalized in regions to which they have been introduced.

**Erigonum** See POLYGONACEAE.

**Eriobotrya** (family *Rosaceae) A genus of small trees, including *E. japonica* which yields the edible fruit the loquat. There are 27 species, occurring in the warm parts of Asia.

**Eriocaulaceae** A family of monocotyledonous (see MONOCOTYLEDON) herbs, in which the leaves are linear, borne mostly in a basal *rosette. The flowers are tiny, in a bracteate head on a long stalk, unisexual but normally *monoecious. There are 14 genera, with about 1200 species. They have an almost cosmopolitan distribution but are absent from Europe except for western parts of the British Isles. Most are swamp or aquatic plants.

**Eriostemon (wax flowers; family
*Rutaceae)** A genus of shrubs that have
*simple, alternate leaves with oil glands.
The flowers are *axillary or terminal,
solitary, and white, pink, or blue. The
5-pointed *sepals form a large *calyx,
and the 5 *petals are spreading and waxy,
usually arranged in a star-like form. The
10 woolly *stamens are arranged closely
around the *pistil. The fruit is composed
of 5 separate *carpels with a single black,
shiny seed. There are 33 species, all of
them endemic (*see* ENDEMISM) to Australia,
where they are present in all temperate
regions.

**erumpent** Bursting through.

**Erwinia (family *Enterobacteriaceae)**
A genus of Gram-negative (*see* GRAM REAC-
TION), rod-shaped *bacteria which are typically *motile with many *flagella. They are
found mainly as *saprotrophs or *parasites in association with plants. The genus
includes some important plant pathogens,
e.g. the causal agents of *fire-blight and of
soft rots of carrots, potatoes, etc.

**Eryngium (sea holly; family *Apiaceae)** A
genus of shrubby, often spiny, *perennial
*herbs which are unlike the characteristic
umbellifers in appearance. The lobed or
deeply cut leaves are toothed and without
*stipules. The *sessile flowers are held in a
dense head surrounded by leafy, spiky, and
showy *bracts. The *calyx teeth are often
sharp, rigid, and longer than the *petals.
The 5 petals are narrow and notched, with
free *stamens and an *inferior *ovary
with 2 *unilocular *carpels. The fruit is a
2-celled *schizocarp which splits into 2
single-seeded fruits. There are about 230
species, found in temperate and tropical
regions, where they are often found in
arid conditions.

**Erysipelothrix** A genus of *bacteria
which are not assigned to a taxonomic family. The cells are rod-shaped to filamentous,
non-*motile, and typically Gram-positive
(*see* GRAM REACTION). There is 1 species,
*E. thusiopathiae*, which is parasitic on mammals, birds, and fish; it can be pathogenic.

**Erysiphales (subdivision *Ascomy-
cotina)** An order of *fungi in which
the septate *mycelium typically is colour-

less and the *fruit bodies are usually
dark-coloured, spherical or hemispherical
*cleistothecia. All members are *parasites
of higher plants and some can cause
diseases of economic importance.

**Erysiphe (order *Erysiphales)** A genus
of *fungi which includes some economically important plant *parasites.
*E. graminis* is responsible for *mildew
of grasses and cereal crops. The
species occurs in a number of forms,
each specifically attacking a particular
type of host plant.

**Erythrina (family *Fabaceae, subfamily
*Papilionatae)** A genus of small, quick-
growing, soft-wooded trees which have
spiny branches, and leaves with a few big
leaflets. The flowers are large and showy,
and are bird-pollinated. They are grown as
ornamentals and make good living fences.
There are 108 species, occurring in tropical and subtropical, mainly seasonal climates.

**Erythroxylum (family *Erythroxyl-
aceae)** A genus of trees and shrubs
in which the leaves are *simple with
persistent longitudinal folds. *E. coca* of
the Andes is the source of cocaine,
obtained from the leaves. There are about
250 species, occurring in tropical and
subtropical regions, mainly of America
and Madagascar.

**Escalloniaceae (order Rosales)** Often
classified as a subdivision of the *Saxifragaceae, the Escalloniaceae is a family of
shrubs with *simple, alternate leaves,
some species being evergreen. The flowers
are small, greenish, and borne in *catkins.
The fruit is a *capsule. Most are from S.
America and Australia.

**Escherichia (family *Enterobacteri-
aceae)** A genus of *Gram-negative,
rod-shaped *bacteria. *E. coli* is the
most thoroughly studied of all bacteria.
It is found mostly in the lower gut
of mammals, but also occurs in,
e.g., sewage-contaminated natural waters;
it is commonly used as an indicator
of sewage pollution. Some strains of *E. coli*
are *pathogenic, causing, for example,
dysentery in the old and in infants and
young animals.

**esculent** Suitable for human consumption.

**esparto grass** *See STIPA.*

***Espeletia*** **(family \*Asteraceae)** A genus of characteristically weird \*paramo plants which have a short, stout stem, and crowded, terminal, hairy leaves. There are about 80 species, occurring in the Andes.

**essential element** A chemical nutrient that is vital for the successful growth and development of an organism. Elements needed in relatively large amounts are termed macronutrients; those needed only in small or minute quantities are called micronutrients or trace elements. Macronutrient elements include carbon, hydrogen, oxygen, nitrogen, sulphur, phosphorus, potassium, magnesium, and calcium. Important micronutrient elements include iron, manganese, zinc, copper, boron, molybdenum, and cobalt.

**established cell line** Cells, derived from a primary culture, which may be subcultured indefinitely *in vitro.*

**ester** A compound that is formed as the condensation product (i.e. water is removed) of an acid and an alcohol; water is formed from the OH of the acid and the H of the alcohol.

**estivation** *See AESTIVATION.*

**estovers (common of) 1.** The right to gather lying and dead standing wood, usually up to a stated size, for firewood on common land. Dead standing wood and branches can be obtained 'by hook or by crook', i.e. by the use of a blunt tool. **2.** Firewood, timber, and other materials taken from common land as a privilege.

**-etalia** *See ZURICH–MONTPELLIER PHYTOSOCIOLOGICAL SCHOOL.*

**-etea** *See ZURICH–MONTPELLIER PHYTOSOCIOLOGICAL SCHOOL.*

**ethanol (ethyl alcohol)** A colourless liquid ($CH_3CH_2OH$) that is produced naturally as a by-product of the fermentation of sugary liquids during \*anaerobic respiration. Ethanol mixes completely with water and pure ethanol absorbs water vapour. Many organic compounds are soluble in ethanol and some gases are more soluble in it than they are in water.

**ethene (ethylene)** A gas ($C_2H_4$) that is produced naturally by plants, and which functions as a \*hormone is the control of such processes as germination, cell growth, fruit ripening, \*senescence, and \*abscission. It is also involved in the response of a plant to gravity and to stress.

**ethnobotany** The study of the use of plants by humans.

**ethyl alcohol** *See ETHANOL.*

**ethylene** *See ETHENE.*

**etiolation** The state of plants that have been grown in the dark: they are not green, having little or no \*chlorophyll, and have very long internodes and rudimentary leaf growth. These features associated with etiolation ensure, under natural conditions, that the shoot is carried towards the light as rapidly as possible.

**etioplast** An \*organelle that develops from \*chloroplast \*proplastids in the absence of light.

**-etosum** *See ZURICH–MONTPELLIER PHYTOSOCIOLOGICAL SCHOOL.*

**-etum** A suffix used in the \*Zurich–Montpellier phytosociological scheme to denote a \*community of association rank. It is applied to the generic name of the most characteristic species, i.e. that which differentiates the association from others in the same alliance. A \*heathland community typified by \*Calluna vulgaris thus becomes a Callunetum. *See also* DIFFERENTIAL SPECIES.

**eu-** A prefix meaning 'well', 'good', etc. It is used in ecology to denote, in particular, enrichment or abundance, e.g. \*'eutrophic', nutrient-rich; \*'euphotic', light-rich.

**Eubacteria** In the widely used five-kingdom system of \*taxonomy, one of the subkingdoms in the \*kingdom \*Bacteria, containing organisms that need to be distinguished from members of the other subkingdom, the \*Archaea. The Eubacteria comprises all of the 'true' bacteria, ranked as three divisions: Gracilicutes;

Tenericutes; and Firmicutes. Division Gracilicutes comprises the proteobacteria (the largest group, including such organisms as *Escherichia coli*), spirochaetes, *cyanobacteria, and other groups. In the three-domain classification system, one of the three *domains, containing the single kingdom Bacteria.

**Eucalyptus** (family *Myrtaceae) A huge genus of trees which impart a distinctive aura of Australian forests. Their leaves are held vertically, and are commonly *glaucous and resinous. The *calyx forms an *operculum which is thrown off when the flower opens. The *receptacle is hollow, and woody in the fruit. There are many *stamens. The fruit is a *capsule. There are many seeds. Many eucalypts are widely cultivated as fast-growing timber trees. Jarrah is *E. marginata*; karri is *E. diversicolor*; mountain ash is *E. regnans*, the tallest tree in the world. There are about 450 species, most of which are confined to Australia.

**Eucapsis** See SYNECHOCYSTIS.

**eucarpic** Applied to a *fungus in which only part of the *thallus differentiates to form a reproductive structure or structures.

**euchromatin** See HETEROCHROMATIN.

**Eucommiaceae** A *monotypic family comprising a tree that is native to China and is known only in cultivation. The flowers are unisexual, without a *perianth. The bark latex is a source of gutta percha.

**Eucryphia** (family Eucryphiaceae) A genus of trees or shrubs, related to *Cunoniaceae, which are evergreen, with *opposite, *simple or *pinnate, *stipulate leaves. The flowers are showy. Several are valuable ornamentals. There are 5 species, occurring in Chile and Australia.

**Eugeissona** (family *Arecaceae) A genus of clump-forming palms that have a stout, branched, underground *rhizome and mostly stemless rosettes of spiny, pinnate leaves. Most of the trees are massive. The inflorescence is tall and narrow. The flowers are paired, 1 male, 1 female, the petals woody, long, and pointed. The

*ovary has 3 *ovules, which are incompletely *trilocular. The fruits are finely scaly, and the outer fibrous and inner stony walls have a basal pore. The fruit contains a single seed. There are 7 species, occurring in rain forests of Malaya and Borneo.

**Eugenia** (family *Myrtaceae) A genus of trees, closely similar to *Syzygium*, in which the leaves are *opposite and have a distinct submarginal nerve and no *stipules. The *calyx cup has 4 or 5 *sepals, 4 or 5 *petals, and many *stamens, and the *ovary is *inferior. The fruit is a *berry. The timber is hard. Some produce an edible fruit. There are about 1000 species, with a pantropical distribution, but occurring mainly in America.

**Euglena** A genus of unicellular, green, photosynthetic protists, sometimes regarded as *algae (division *Euglenophyta), sometimes as protozoa (class *Phytomastigophora). The cell body is typically spindle-shaped but is capable of broadening and narrowing to some extent; it is *motile with a single *flagellum. *Euglena* is common in all freshwater habitats, especially when these are slightly polluted with organic matter.

**Euglenophyta** A division of typically unicellular protists, sometimes regarded as *algae, sometimes as protozoa (class *Phytomastigophora). They are characterized by the possession of a single *flagellum, the formation of *paramylum as a storage product, possession of *chlorophylls *a* and *b*, and the absence of sexual reproduction. Vegetative cells lack a *cell wall but possess a proteinaceous *pellicle. There are many genera, the best known of which is *Euglena*. They are found in a wide range of aquatic habitats: ditches, ponds, puddles, and rivers (especially those polluted with organic matter). Some species occur in brackish or marine waters.

**euhaline** See BRACKISH.

**Eukarya (Eukaryota)** In *taxonomy, the group that includes all *eukaryotes. In the widely used five-kingdom classification system, the Eukarya is ranked as a superkingdom containing the kingdoms

*Protoctista, Animalia, *Fungi, and *Plantae. In the three-domain classification system, Eukarya is the *domain containing the kingdoms Protoctista, Animalia, Fungi, and Plantae.

**Eukaryota** *See* EUKARYA.

**eukaryote** An organism whose cells have a distinct *nucleus (i.e. all protists, fungi, plants, and animals). The first eukaryotes were almost certainly green algae (*Chlorophyta) and what appear to be their microscopic remains occur in *Precambrian sediments dating from a little less than 1500 Ma ago.

**eulittoral zone** In marine ecosystems, the main area of the *littoral zone lying below the *littoral fringe. Barnacles are the most characteristic animals on rocky shores, with mussels and oysters also typical; the green *algae *Enteromorpha and *Ulva species are also common. On sandy shores burrowing animals are common, e.g. shrimps, crabs, and polychaete worms (such as the lug worms, *Arenicola* species).

**Eumycota** ('true fungi') A division containing those fungi that do not form a *plasmodium or *pseudoplasmodium, i.e. all fungi excluding the *slime moulds and their allies.

**Euodia** (family *Rutaceae) A genus of trees of shrubs in which the leaves are *opposite, usually with 3 leaflets, and are aromatic when crushed. The fruits are follicular (*see* FOLLICLE), with black, glossy seeds. There are about 100 species, occurring in the Mascarenes, and from Indo-Malaysia to China and the Pacific islands.

**Euonymus europaeus** (spindle tree) *See* CELASTRACEAE.

**Eupatorium** (family *Asteraceae) A genus of *herbs that includes *E. odoratum*, a noxious tropical weed (siamweed). There are 1200 species, mostly American, but a few occurring elsewhere; many are now split off into numerous other small genera.

**Euphorbia** (spurges; family *Euphorbiaceae) A genus of plants that have a very distinctive, *compound inflorescence of tiny flowers without *perianths grouped in *umbel-like *cymes; several male flowers, each with 1 *stamen, are grouped within an *involucre, and the single female flower consists of a stalked 3-celled *ovary. The fruit is an explosive *capsule. In temperate Europe most are herbs with leafy stems. In the tropics, especially in Africa, many are trees or shrubs, often stem succulents and cactus-like, but differing from cacti in their latex and twinned spines. All produce an acrid, milky juice. *Euphorbia* is a large genus, with about 1600 species, found in most temperate and tropical regions.

**Euphorbiaceae** A huge family, mostly of trees and shrubs but with a few *herbs, a minority of which have latex. The leaves are *simple, usually with *stipules. The flowers are usually tiny and unisexual, with *petals usually absent, and the *ovary having 3 chambers, each with 2 pendulous *ovules. The fruit is typically a 3-lobed woody *capsule. The seeds sometimes have a fleshy jacket. There are 326 genera, with about 7750 species, and they are cosmopolitan in distribution.

**Euphoria** (family *Sapindaceae) A genus of trees and shrubs that includes *E. longana* which is cultivated for its fruit, the longan. There are 15 species, occurring from India and southern China to Malesia.

**euphotic zone** The upper, illuminated zone of acquatic *ecosystems: it is above the *compensation level and therefore the zone of effective *photosynthesis. In marine ecosystems it is much thinner than the deeper aphotic zone, typically reaching 30 m in coastal waters but extending to 100–200 m in open ocean waters. In freshwater ecosystems it is subdivided into *littoral (shallow edge) and *limnetic zones. The term is occasionally used of the upper, well-lit strata of architecturally complex terrestrial ecosystems, such as rain forests. *Compare* DYSPHOTIC ZONE; PHOTIC ZONE; and PROFUNDAL ZONE.

**Euphrasia** (family *Plantaginaceae, subfamily Rhinanthoideae) A genus of annual *herbs, with small, *opposite or alternate leaves, which are *hemiparasitic on grasses. The bisexual flowers are usually white, blue, or pink, and are held in the upper leaf *axils. They are *sessile and often appear to form a terminal *spike with reduced leaves between each flower and the next. The bell-shaped *calyx is 4-lobed. The upper tip of the *corolla is incurved, with 2 *reflexed lobes; the lower lip is 3-lobed. There are 4 *stamens. The *ovary is *superior, with 2 fused *capsules, 2 *locules, and a single *style. The fruit is a dry capsule with numerous small seeds. The species hybridize readily. There are about 450 species, found in all temperate regions of the world.

**euploid** A cell having any number of complete *chromosome sets, or an individual composed of such sets. It is thus a *polyploid with a chromosome number that is an exact multiple of the basic number for the species from which it originated.

**Eupomatiaceae** A small family of shrubs and trees in which the leaves are alternate, *simple, and without *stipules. The flowers are showy, solitary, · and bisexual. The *perianth is fused; the upper part becomes a conical lid, and the lower a *receptacle with the *stamens attached to the rim. The inner stamens are petal-like and sterile, the others are pointed. The *carpels are sunk into the receptacle, each with several *ovules. The fruit is a *berry with oil glands and contains several angular seeds with copious *endosperm. There is only 1 genus (*Eupomatia*) and 2 species. *E. laurina* gives a decorative wood. Both species are restricted to the rain forests of eastern Australia and New Guinea.

***Eurhynchium*** See HYPNOBRYALES and *RHYNCHOSTEGIUM*.

**Eurosiberian floral region** Part of R. Good's (*The Geography of the Flowering Plants*, 1974) boreal kingdom: an extensive region, in which the flora of the western part is richer than that of the eastern part. Whereas Europe has roughly 150 endemic genera (*see* ENDEMISM), Siberia apparently has only about 12, almost all of which are *monotypic. The contrast between the two subregions reflects primarily the different climatic regimes. *See also* FLORAL PROVINCE and FLORISTIC REGION.

**Eurotiales (subdivision *Ascomycotina)** An order of *mycelium-forming *fungi which form cleistothecioid (*see* CLEISTOTHECIUM) *ascocarps and unitunicate, thin-walled *asci containing unicellular *ascospores. They are mainly *saprotrophic and are found in a very wide range of habitats, including soil, wood, dung, and animal debris (such as hair, horn, and feathers). Some members are important in that they cause spoilage of certain foods; others are used in certain industrial fermentations. A few members can cause disease in animals. The order includes the perfect (sexual) stages of important form-genera (e.g. *Aspergillus* and *Penicillium*).

**eury-** A prefix, derived from the Greek *eurus*, meaning 'wide', that is used in ecology to describe species with a wide range of tolerance for a given environmental factor, e.g. 'euryecious', having a wide range of habitats. *Compare* STENO-.

**Euryarchaeota 1.** In the widely-used five-kingdom system for classifying organisms, a phylum within the subkingdom *Archaea in the *kingdom *Bacteria. **2.** In the three-domain classification, the more derived (*see* APOMORPH) of the two kingdoms (sometimes called subdomains) within the *domain Archaea. The Euryarchaeota contains a broad range of *phenotypes, including *methanogens, *halophiles, *hyperthermophiles, *acidophiles, and *sulphur-reducing organisms. Genetically, members of the Euryarchaeota are more different from members of the domains *Eukarya and *Eubacteria than are members of the *Crenarchaeota.

**eurychoric** Having a widespread geographic distribution in various climates.

**euryecious** Having a wide range of habitats.

**euryhaline** Able to tolerate a wide range of degrees of salinity.

**euryhydric** Tolerant of a wide range of moisture conditions.

**euryoxic** Able to tolerate a wide range of concentrations of oxygen.

**euryphagic** Using a wide range of types of food.

**eurythermal** Able to tolerate a wide temperature range.

***Eusideroxylon*** (family *Lauraceae) A genus of trees that yield very hard, durable timber, valuable for house posts and shingles. There are 2 species, found in Borneo and Sumatra. *E. zwageri* is Borneo ironwood (belian).

**eusporangiate** In *ferns, applied to the supposedly primitive, massive, stout-stalked, thick-walled type of *sporangium. This arises from more than one parent cell, and lacks the specialized *annulus or ring of thickened cells that is present in the *leptosporangiate, supposedly more advanced ferns. (The latter have delicate, thin-walled, and thin-stalked sporangia.)

**eustele** A *dictyostele type of *siphonostele in which *ectophloic siphonosteles are arranged in a circle. *Compare* ATACTOSTELE, PLECTOSTELE, and POLYSTELE.

***Euterpe*** (family *Arecaceae) A genus of palms, many of which are large trees. The trunk is columnar, and sometimes slightly swollen. The leaves are *pinnate, the crown-shaft is prominent. The plants are *monoecious. *E. oleracea* of Brazil and Paraguay is the assai palm, whose fleshy fruit wall provides the popular plum-coloured beverage assai. There are 20 species, found in the W. Indies, and in Central and S. America.

**eutrophic** Originally applied to nutrient-rich waters with high *primary productivity but now also applied to soils. Typically, eutrophic lakes are shallow, with a dense plankton population and well-developed *littoral vegetation. The high organic content may mean that in summer, when there is stagnation resulting from thermal stratification, oxygen supplies in the *hypolimnion become limiting for some fish species, e.g. trout. *Compare* OLIGOTROPHIC.

**eutrophication** The process of nutrient enrichment in aquatic ecosystems. It occurs naturally over geological time, but may be accelerated by human activities, e.g. sewage disposal, or land drainage: such activities are sometimes termed 'cultural eutrophication'. The rapid increase in nutrient levels stimulates *algal blooms. On death, bacterial decomposition of the excess algae may deplete oxygen levels seriously. This is especially critical in thermally stratified lakes, since the decaying algal material typically sinks to the *hypolimnion where, in the short term, oxygen replenishment is impossible. The extremely low oxygen concentrations that result may lead to the death of fish, creating a further *oxygen demand, and so leading to further deaths.

**evagination** The release of the contents of membranaceous vesicles to the exterior.

**evanescent** Temporary; soon disappearing.

**evapotranspiration** A combined term for water lost as vapour from a soil or open water surface (evaporation) and water lost from the surface of a plant, mainly via the *stomata (*transpiration). The combined term is used since in practice it is very difficult to distinguish water vapour from these two sources in water-balance and atmospheric studies.

**evening primrose** *See* OENOTHERA.

**Everglades** The low, flat plains area of southern Florida, USA, which is subject to periodic freshwater flooding. In summer the area becomes swampy, but in winter it is extremely dry. *Cladium effusum* is widespread. Scattered trees include palms, and pines occur on higher ground. Within the same area small, isolated, *tropical rainforest communities, locally termed hammocks, are found. The coastal areas are renowned for their *mangrove swamps.

**evergreen** Applied to a tree or shrub that has persistent leaves, and whose crown is never wholly bare. Although the entire plant remains green throughout the year, each leaf has a limited life span, but is physically tougher and usually longer-lived than a deciduous leaf. Evergreen leaves have the advantage that where nutrients such as nitrogen are in short supply their longer life span allows a more efficient use of the limited resource.

**evergreen forest** A forest in which there is no complete, seasonal loss of leaves (i.e., trees shed old leaves and produce new ones throughout the year, rather than during particular periods). The trees may be *conifers or *hardwoods. The distribution of these forests ranges through *boreal, middle, and tropical latitudes. The northern coniferous forests and the equatorial rain forests (*see* TROPICAL RAIN FORESTS) are the most extensive evergreen forests. Evergreen coniferous forests predominate where the growing season is less than half the year; broad-leaved evergreen forest is found in regions that lack a prolonged dry season.

**evergreen mixed forest** A forest in which the dominant trees are both *evergreen broad-leaved hardwoods and *conifers. Such forests are particularly well developed in the southern hemisphere, in S. Africa, Tasmania, New Zealand, and Chile. In the northern hemisphere, eastern Asia and the Mediterranean basin once contained large tracts of such forest, but most of it has long been cleared.

**everlasting daisies** See *HELICHRYSUM* and *HELIPTERUM*.

**Evernia** (order *Lecanorales) A genus of *lichens in which the *thallus is more or less flattened, and sparsely to extensively branched, according to the species. Species are found on trees and, rarely, on rocks. *See also* OAK-MOSS.

**eversible** Capable of being everted (turned inside out).

**evolute** Unfolded or turned back. *Compare* INVOLUTE.

**evolution** The process by which new *species are formed from pre-existing species over a period of time. The phenomenon is amply demonstrated by the fossil record, for the changes over geological time are sufficient to recognize distinct eras, for the most part with very different plants and animals.

**evolutionary determinism** The change in *gene frequencies by directed or deterministic processes, in contrast with change due to random or stochastic processes. The relative importance of the two kinds of change in evolutionary development is uncertain.

**evolutionary lineage** A line of descent of a taxon from its ancestral taxon. A lineage ultimately extends back through the various taxonomic levels, from the species to the genus, from the genus to the family, from the family to the order, etc.

**evolutionary rate** The amount of evolutionary change that occurs in a given unit of time. This is often difficult to determine, for several reason: for example, should the unit of time be geological or biological (the number of generations)? How should morphological change in unrelated groups be compared? In practice it is necessary to adopt a pragmatic approach, such as the number of new genera per million years. *See also* DARWIN.

**evolutionary species** *See* CHRONOSPECIES.

**evolutionary trend** A steady change in a given adaptive direction, either in an *evolutionary lineage or in a particular attribute, e.g. height of shoot. Such trends are often apparent in unrelated taxa. Formerly they were attributed to *orthogenesis; now *orthoselection or the contending theory of species selection is invoked.

**exalbuminous** Applied to seeds that contain no *endosperm when mature. *Compare* ALBUMINOUS.

**exannulate** Lacking an *annulus.

**exarch** Applied to strands of primary *xylem in which the first elements to form lie furthest from the centre of the axis. *Compare* ENDARCH and MESARCH.

**exchangeable ions** Charged ions that are adsorbed on to sites, oppositely charged, on the surface of the *adsorption complex of the soil (mainly clay and humus colloids). Exchangeable ions can replace each other on this surface, and are also available to plants as nutrients. Although *cations (e.g. calcium and magnesium) are the most common, exchanging at negatively charged sites, some *anions (e.g. sulphate and phosphate) do exchange at positively charged sites. *See* ANION-EXCHANGE CAPACITY; and CATION-EXCHANGE CAPACITY.

**exchange capacity** The total ionic charge of the *adsorption complex in the soil that is capable of adsorbing *cations or *anions.

**exchange pairing** The type of pairing of *homologous chromosomes that allows genetic *crossing-over to take place.

**exchange pool** *See* ACTIVE POOL.

**excipulum** A cup-shaped layer of sterile *tissue that contains the *hymenium in an *apothecium.

**exclusive species** In *phytosociology, the optimum-*fidelity class 5, i.e. a species that is confined completely or almost completely to a particular *community. *Compare* ACCIDENTAL SPECIES; INDIFFERENT SPECIES; PREFERENTIAL SPECIES; and SELECTIVE SPECIES.

**excurrent** Of, for example, a leaf: having a *midrib that projects beyond the tip.

***Exidia*** (order *Tremellales) A genus of *fungi in which the *fruit body is gelatinous and variously shaped. There are many genera. *E. glandulosa* (*E. plana*) forms black, rubbery *fruit bodies ('witches' butter') on dead logs, stumps, etc.

**exiguous** Small and narrow; scanty.

**exine** The outer, decay-resistant coat of a *pollen grain. The exine is characteristic for different plant families and genera, and sometimes even for different species. Hence it forms the basis for the identification and quantitative analysis of the vegetation composition of *peats and other suitable sedimentary deposits dating back many thousands of years. *See also* PALYNOLOGY.

**Exobasidiales** (subclass *Holobasidiomycetidae) An order of *fungi that are obligately parasitic (see parasitism) in higher plants. *Basidia are arranged in a *hymenium, but there is no true *basidiocarp. Species are found on a wide range of plants; some cause disease in cultivated plants.

**exocarp** *See* PERICARP.

***Exocarpos*** (family *Santalaceae) A genus of shrubs and small trees in which the leaves are reduced to small scales. The roots usually cling to other plants, on which they are semi-parasitic. The scale-like leaves are *opposite or alternate. The flowers, held in short *spikes or clusters in the leaf *axils, are small and cream-coloured and have 4 or 5 *perianth segments. The fruit is a *nut with a fleshy stalk. There are 26 species, found in Southeast Asia, Australia, and New Zealand.

***Exocoecaria*** (family *Euphorbiaceae) A genus comprising 40 species of small trees, occurring in the Old World tropics. *E. agollocha*, of coastal regions of Asia, produces a blinding latex.

**exocrine** *See* ECTOCRINE.

**exocytosis** The process by which a membrane-bound *vacuole fuses with the *plasma membrane and thus discharges its contents outside the cell. As well as a mechanism for the removal of wastes, it is commonly employed for the secretion of cell products.

**exodermis** *See* HYPODERMIS.

**exoenzyme** An *enzyme that is discharged outside the cell.

**exogamy** The tendency for a *gamete to fuse with distantly related gameters rather than with closely related ones.

**exogenetic** Applied to the various processes of erosion, transport, and deposition that take place at the Earth's surface. Usually the term is used in contrast with the 'endogenetic' or internal mechanisms.

**exopeptidase** *See* PROTEASE.

**exoperidium** The outer layer of the *peridium in certain species of *Gasteromycetes. *Compare* ENDOPERIDIUM.

**exoskeleton** A hard outer layer, often of calcium carbonate, found in some species of *algae (e.g. stonewort, *Chara).

**exotic species** An introduced, non-native species.

**exotoxin** A toxin that is secreted by a living organism.

**exozoochory** Dispersal of spores or seeds by their being carried on the surface of an animal.

**experimental error** An error that arises because of variation between experimental samples. It may be attributed to differences in materials and/or techniques, rather than to real differences (e.g. in growth). Experimental error must be monitored in any statistical comparison of experimental data.

**exploitation competition (exploitative competition)** *Competition occurs where two species require the same resource and that resource is in short supply. Whichever of the two species is more efficient in accessing the resource is more likely to succeed. This is known as exploitation competition, the alternative being *interference competition.

**exploitative competition** See EXPLOITATION COMPETITION.

**explosive evolution** A sudden diversification or *adaptive radiation of a group. The term is an old one, rarely used nowadays. Phases of explosive evolution have occurred in all the higher taxonomic groups, i.e. in genera, families, orders, and classes.

**exponential growth** A form of growth in which the logarithms of a value increase linearly in any given period of time. This means that the value grows more rapidly than it would by linear growth. An example of exponential growth would be a population that grows by 10 per cent of its value in each unit of time. Thus, a population that starts with a value of 100 would grow as: 100, 110 (100 + 10% of 100), 121 (110 + 10% of 110), 132.1 (121 + 10% of 121), . . .

**expressivity** The degree to which a particular *genotype is expressed in the *phenotype.

**exserted** Protruding, e.g. of *stamens projecting beyond a *corolla, or of a *moss *capsule projecting beyond the leaves of the *gametophyte.

**exstipulate** Without *stipules.

**extant** Applied to a *taxon some of whose members are living at the present time. Compare EXTINCT.

**extinct** Applied to a *taxon no member of which is living at the present time. Compare EXTANT.

**extinction** The elimination of a *taxon. The term can be used of the local loss of a species or a population.

**extremophile** A *micro-organism (see ARCHAEA) that thrives under extreme environmental conditions of temperature, pH, or salinity. See also ACIDOPHILE, ALKALIPHILE, HALOPHILE, HYPERTHERMOPHILE, PSYCHROPHILE, and THERMOPHILE.

**extremozyme** One of a range of enzymes, present in *extremophiles, that continue to function at temperatures, salinities, acidities, or alkalinities at which other enzymes would fail.

**extrorse** Applied to *anthers that open away from the centre of the flower, so releasing their pollen outside the flower and promoting *cross-pollination. Compare INTRORSE.

**eyespot** **1.** In certain unicellular *algae, an *organelle consisting of closely packed lipid globules (usually or always containing orange or red carotenoids); it is generally believed to be involved in *phototaxis. **2.** A disease of cereals caused by the *fungus Pseudocercosporella herpotrichoides. Initially a dark smudge appears on the outer leaf-sheath, then an oval 'eye-shaped' lesion develops near the base of the plant, weakening the stem and leading to increasing susceptibility to lodging (i.e. being laid flat by weather). A similar disease, called sharp eyespot, is caused by Rhizoctonia solari.

**F** Notation, introduced by the geneticist Sewall Wright, for the inbreeding coefficient. *See also* COEFFICIENT OF INBREEDING.

**$F_1$** The first filial generation of animals or plants, produced by crossing two parental lines (which are referred to as *P*).

**$F_2$** The second filial generation of animals or plants, produced by selfing or intercrossing of $F_1$ individuals.

**Fabaceae** The third largest family of flowering plants, formerly known as Leguminosae, comprising trees, climbers, shrubs, and herbs, some of which are aquatic. Innumerable species are of great economic importance for timber, fodder, drugs, or food. The leaves have *stipules and are usually *pinnate. Many species have *root nodules containing nitrogen-fixing bacteria. The *inflorescence is basically *racemose. The flowers are regular or *zygomorphic, basically with 5 *sepals and *petals, the sepals often united. Typically there are 10 *stamens (but many in *Mimusoideae) which are sometimes fused. The *ovary is *superior, with a single *carpel. The fruit is a pod, usually several-seeded, and usually *dehiscent. There are 3 subfamilies, distinguished by their different flower construction. *Caesalpinoideae and Mimosoideae are mainly tropical; *Papilionatae has many temperate species and many *herbs. There are 657 genera with about 16,400 species, of cosmopolitan distribution.

**facies** Aspect or appearance.

**facilitated diffusion** A carrier-mediated membrane transport mechanism that is driven by a diffusion gradient.

**facilitation** The process whereby a plant so modifies a *habitat as to allow other species to invade (as in a *succession). For example, the grass *Elytrigia juncea* facilitates the invasion of *Ammophila arenaria* (marram grass) in the early stages of sand-dune formation. If the invader is more efficient in the new circumstances, this may lead to the decline and eventual exclusion of the facilitating species. *Compare* INHIBITION.

**factor** **1. (stat.)** One of a pair or series of numbers which when multiplied together yield a given product. **2. (limiting factor)** *See* LIMITING FACTOR. **3. (ecological factor)** *See* LIMITING FACTOR.

**facultative** Applied to organisms that are able to adopt an alternative mode of living. For example, a facultative anaerobe is an *aerobic organism that can also grow under *anaerobic conditions.

**FAD** *See* FLAVIN ADENINE DINUCLEOTIDE.

**Fagaceae** An important family of evergreen or deciduous trees in which the leaves are *simple, with *stipules. The flowers are unisexual, often borne in *catkins, and are apetalous, with an *inferior *ovary. The fruit is a *nut within a *cupule, the fruits held in clusters of 1–4. Many of these trees yield valuable timber. There are 7 genera, including *Castanea (sweet chestnut), *Castanopsis, *Fagus (beech), *Lithocarpus, *Nothofagus (southern beech), *Quercus (oak), and *Trigonobalanus*, with about 1050 species, concentrated in the northern hemisphere and absent from Africa.

**faggot** **(fagotts)** Small branch-wood tied into a bundle. Faggots are used mainly for fuel, but also for fencing or thatching.

***Fagopyrum esculentum*** **(buckwheat)** *See* POLYGONACEAE.

***Fagraea*** **(family *Loganiaceae)** A genus of trees or shrubs, with a few *stranglers or *epiphytes, in which the flowers are fragrant, and the *corolla white or cream, and up to 30 cm across in a few species. The fruit is a *berry seated on the persistent *calyx, and has many small seeds. There are 35 species, occurring from Indo-Malaysia to the Pacific islands.

***Fagus*** **(beeches; family *Fagaceae)** A genus of trees, each *cupule of whose fruits encloses two oily *nuts. The wood is hard, close-grained, and valuable. There are 10 species, occurring in northern temperate regions.

**fairy club**   See CLUB FUNGI.

**fairy ring**   A circle of dark-green grass (in a lawn or field) in which *toadstools may be found. The circle is formed as a result of the radial growth of the *fungus through the soil, away from the centre of the ring. Fairy rings are often formed by *Marasmius oreades.

**faithful species (fidelity)** In *phytosociology, a species confined, or nearly so, to a particular association. The species may or may not be constant to the association fidelity, and therefore faithful or characteristic species can be determined satisfactorily only if a full knowledge of other *community types is available. Five fidelity classes are commonly distinguished: (a) *accidental; (b) *indifferent; (c) *preferential; (d) *selective; (e) *exclusive. See also COMPANION SPECIES; KENNARTEN SPECIES; and PLANT ASSOCIATION.

**falcate**   Curved, as a sickle.

**falcato-secund**   Curved to one side.

**false branching** In filamentous *algae, a break in the filament with 1 or both ends projecting from the sheath, giving the appearance of branching.

**false chanterelle** The common name for the *fruit body of the *fungus *Hygrophoropsis aurantiaca, which is *mushroom-shaped, with a convex to funnel-shaped cap, and *decurrent *gills which are narrow and forked. The whole is bright orange-yellow, or peach-coloured. The *spore print is white. H. aurantiaca is common throughout the northern hemisphere under coniferous trees in late summer and autumn.

**false death cap** The common name for the *fruit body of the *fungus *Amanita citrina (A. mappa). A. citrina resembles the *death cap, A. phalloides, but is not poisonous. It is common throughout the northern hemisphere on the ground in woodland in late summer and autumn.

**false fruit**   See PSEUDOCARP.

**false morel** The common name for the *fruit body of *Gyromitra esculenta.

**false rings**   See TREE RING.

**Famennian**   See DEVONIAN.

**farinose** Floury or powdery in appearance or texture.

**fasciation** A malformation in plants in which shoots tend to be thick and flattened and may occur in masses. Fasciation is sometimes caused by infection with the *bacterium Rhodococcus fascians.

**fascicle** A tuft or bunch of branches or leaves all arising from the same place.

**fascicular cambium** The part of the *cambium that develops within the *vascular bundles.

**fat**   See LIPID.

**fatiscent** Cracked in appearance; gaping open.

**fatty acid** A long-chained, predominantly unbranched, carboxylic acid, in which a side-chain of carbon atoms is attached to the *carboxyl group, and hydrogen atoms to some or all of the carbon atoms in the side-chain. There is usually an odd number of carbon atoms in the chain, commonly 15 or 17. If the carbon atoms of the side-chain carry as many hydrogen atoms as they are capable of carrying the fatty acid is said to be saturated; if there are fewer than the maximum possible number of hydrogen atoms it is unsaturated; if more than two sites on the chain are unsaturated it is polyunsaturated.

**faunizone**   See ASSEMBLAGE ZONE.

**Fe**   See IRON.

**feather grass**   See STIPA.

**feedback loop (feedback mechanism)** A control device in a system. Homoeostatic systems have numerous negative-feedback mechanisms which tend to counterbalance positive changes and so maintain stability. For example, denitrifying bacteria counteract the effects of nitrogen-fixing bacteria. Positive feedback reinforces change and in natural systems may result in radical environmental alteration. For example, an exceptionally cool summer in high mid-latitudes of the northern hemisphere may impede the melting of snow, leading to unusually high albedo, which reduces absorption of solar energy, leading to further cooling, etc.

**feedback mechanism** See FEEDBACK LOOP.

**feedback regulation** The process by which the product of a *metabolic pathway influences its own production by controlling the amount and/or activity of one or more *enzymes involved in the pathway. Normally this influence is inhibitory.

**fell** An area of open mountainside with low-lying vegetation. The word is derived from the Old Norse *fiall*, meaning 'hill', and survives in a number of place names in northern England, probably as a result of Viking settlement.

**fell field** An area, within the *tundra belt, of frost-shattered stony debris with interstitial fine particles, that supports various plant species in a mixed *community. The vegetation is sparse, however, and typically occupies less than half the ground. Frequently fell fields display patterned-ground phenomena, resulting from freeze-thaw activity in the soil.

**fen** An area of wetland vegetation that receives its water by both rainfall and ground water flow (rheotrophic), and in which the summer *water table is at or below the surface of the sediment. Fens are peat-forming *ecosystems (mires) and can be divided into rich fens, which are usually neutral to acid in their reaction and are often fed by streams draining limestone rocks, or poor fens, in which case their water is slightly acid and the concentration of dissolved nutrients is low. Some fens are erroneously given the label *bog (e.g. string bogs and valley bogs), a term that should be reserved for *ombrotrophic mires.

**fenestrated** Perforated with small openings or transparent areas.

**fennel** See FOENICULUM.

**fen peat** See PEAT.

**fenugreek (foenugreek)** See TRIGO-NELLA.

**feral** Applied to a cultivated form living wild.

**fermentation** Anaerobic respiration. The term is usually applied to the formation of ethanol or lactate from carbohydrate, but more generally it includes any catabolic processes that produce *ATP in the absence of oxygen.

**fern** See FILICOPSIDA.

**ferrallization** Part of the *leaching process found in tropical soils, by which large amounts of iron and aluminium oxides accumulate in the B *horizon of such soils as *red podzolic soils.

**Ferralsols** Soils that have an iron-rich B *soil horizon with distinctive red mottling that is more than 15 cm deep. This horizon is highly weathered and contains high concentrations of iron and aluminium. Ferralsols are a reference soil group in the FAO *soil classification.

**ferricrete** See DURICRUST.

**ferrodoxin** A non-haem iron *protein with a low redox potential, which functions as an electron carrier in both *photosynthesis (in *prokaryotes and *eukaryotes) and *nitrogen fixation (in certain prokaryotes).

**ferruginous** Resembling rust, containing rust, or rust-coloured.

**fertility 1.** The condition of a soil relative to the amount and availability to plants of elements necessary for plant growth. Soil fertility is affected by physical elements, e.g. supply of moisture and oxygen, as well as by the supply of chemical plant nutrients. **2. (fecundity, fruitfulness)** The number of eggs that develop in a mated female over a specified period. It is usually calculated at the stage when this number is readily observable (i.e. in seed plants when seeds are borne), although strictly speaking it applies from the time that fertilization occurs. Sometimes the term 'fertility' is applied only to the production of fertilized eggs (ova), while 'fecundity' is used for the production of fruit (in angiosperms) so excluding those embryos which fail to develop.

**fertilization** The union of two *gametes to produce a *zygote, which occurs during *sexual reproduction. Fertilization involves the fusion of two *haploid nuclei (containing genetic material from two distinct individuals (cross-fertilization) or from one individual (self-fertilization). The resulting zygote then develops into a new individual. Some lower plants, e.g. mosses, release their male gametes externally, which then swim like spermatozoa to the female gamete. Most higher plants have the male gamete released internally from the *pollen grain directly to the female gamete and in many *double fertilization occurs.

**Festuca ovina** See HEAVY METAL TOLERANCE.

**Fibonacci series** A mathematical series that is obtained by adding the two previous elements of the series, as in 1, 2, 3, 5, 8, 13, 21, etc. The interlocking spirals found in many plant structures, such as the fruit arrangement in a sunflower head, or the sections on a pineapple, are described by this series. The series was discovered by the Italian mathematician Leonardo Fibonacci (c. 1180–c. 1250), who was also influential in introducing Indo–Arabic numerals to Europe. Fibonacci lived in Pisa.

**fibre** A type of plant cell in which the wall has been thickened to perform a structural role. Typically, fibres are elongated *sclerenchyma cells with tapered ends and have fewer cavities in the thickenings than do other types of thickened cells, though neither feature is diagnostic. Such cells can be found in the *cortex, *phloem, and *xylem. Commercially, the term refers to a strand of fibre cells, or to the epidermal cells (*epidermis) of cotton or kapok *seed *pods. Flax and hemp fibres come from the phloem. Sisal is extracted from highly *lignified leaf fibres of *Agave*. The name 'fibre' is also given to a fine or narrow root.

**fibril** A small *fibre or thread-like structure.

**Ficus** (figs; family *Moraceae) A big genus of trees, *stranglers, root-climbers, and *epiphytes which produce watery latex. The leaves usually have conspicuous *stipules enveloping the bud. The flowers are minute, unisexual, and inserted on a concave *receptacle forming a closed sphere. They are pollinated by specialized wasps. Many are *cauliflorous. The fruits are tiny *drupes enclosed in a *receptacle, which is often brightly coloured and is eaten by many birds and mammals, which disperse the seeds. The receptacle of *F. carica* is the fig eaten by humans. *F. religiosa* is the pipal-tree or bodh-tree. There are about 800 species, most of which are tropical.

**fiddlehead** See CROZIER (2).

**fidelity** See FAITHFUL SPECIES.

**field blewit** (*Lepista saeva*) See BLEWIT.

**field capacity** Water that remains in soil after excess moisture has drained freely from that soil. Usually it is measured as a percentage of the weight or volume of oven-dry soil.

**field layer** The *herb and small shrub layer of a plant *community. *Compare* CANOPY; GROUND LAYER; and SHRUB LAYER.

**field maple** See ACER.

**field mushroom** The common name for the *fruit body of the *fungus *Agaricus campestris*, which resembles the cultivated *mushroom. It is edible and highly esteemed, and is common in fields, meadows, lawns, etc., in summer and autumn.

**filament** **1.** One of the strands of *protein, variously grouped according to diameter (in the range 4–15 nm), found in many types of cell. Their functional significance is incompletely understood, but since they are largely composed of the contractile proteins actin and/or myosin it is presumed that the motility of the cell or its contents forms part of their role. **2.** The stalk of a *stamen, which bears the *anther. **3.** A line of algal cells forming a thread-like structure.

**filbert** See CORYLUS.

**Filicopsida** A subdivision of the *Pteridophyta, which comprises all living and extinct ferns. They arose in the *Devonian from the Trimerophytina group of psilophytes and made an important contribution to *Carboniferous floras. They are the most advanced, numerous, and varied of the pteridophytes. In most cases they have relatively large, much-divided leaves, and are still significant components of many different plant communities around the globe.

**filiform** Thread-like; long and slender.

**fimbriate** With the margin bearing a fringe, usually of hairs.

*Fimbristylis* (family *Cyperaceae) A genus of *perennial, rhizomatous (*see* RHIZOME) *herbs, which have solid stems. The basal tuft of leaves is sheathing, and the leaves are linear with a grass-like blade. The small, inconspicuous flowers are usually bisexual and held in *spikelets with a single *bract. The spikelets form *panicles or *umbels. The *perianth is reduced to scales or bristles. Usually there are 3 *stamens. The *ovary is *superior and the fruit a *nut. There are about 250 species, found in tropical and subtropical regions, usually in marshy habitats.

**fimicolous** Growing in dung or manure.

**finger and toe** *See* CLUBROOT.

**finite resource (non-renewable resource)** A resource that is concentrated or formed at a rate very much slower than its rate of consumption and so, for all practical purposes, is non-renewable. *Compare* RENEWABLE RESOURCE.

*Finschia* (family *Proteaceae) A genus of trees in which the fragrant flowers are borne in showy *racemes. *F. chloroxantha* has edible seeds. There are 4 species, occurring in Micronesia, Melanesia, and New Guinea.

**fir** **1.** (Douglas fir) *See* PSEUDOTSUGA. **2.** (hemlock fir) *See* TSUGA.

**fire-blight** An important disease which affects many trees of the *Rosaceae (e.g. apple, pear, and hawthorn). Infected trees have the appearance of having been scorched by fire; the tree may be killed within months. The causal agent is the *bacterium *Erwinia amylovora. The disease occurs in N. America and in many parts of Europe; it is notifiable in Britain.

**fire climax (pyroclimax)** A *climax community for which fire is the dominating control factor, as in the long-leaf pine forests of the USA. Fire is also thought to be an important determining factor, interacting in complex ways with grazing and trampling pressure, in the formation of the major *grassland areas, e.g. *prairies and *savannah. Fire is also a major influence, with grazing, on the development and maintenance of *heathland communities.

**fire scar** A scar that is often found in the annual rings of a tree that has been subjected to fire but has survived. The scar results from fire damage and is visible in a cross section through the trunk. Fire scars make it possible to assign dates to past fires and to calculate fire frequencies.

*Firmiana* (family *Sterculiaceae) A genus of trees in which the flowers are vivid. They have a long *calyx tube, no *petals, and 5 *carpels which are united but become free and then dehisce to become papery, leaflike *follicles with 1 or 2 marginal seeds. There are 9 species, occurring from E. Africa to Indo-Malaysia.

**Fissidentales (subclass *Bryidae)** An order of small to medium-sized mosses in which the leaves occur in 2 distinct rows or ranks. They are found on soil, rotting wood, etc. *Fissidens bryoides* is the commonest British species, often forming dense colonies on mildly acid soils.

*Fistulina* (family Fistulinaceae) A genus of *fungi in which the *fruit body is *bracket-shaped with a more or less pronounced lateral *stipe. The *hymenium lines densely crowded tubes on the underside of the fruit body. There are 3 species, found on wood. *See also* BEEFSTEAK FUNGUS.

**Fistulinaceae (order *Aphyllophorales)** A family of *fungi in which the *basidiocarps are formed annually, are pileate (*see* PILEUS) with a lateral *stipe, and are fleshy and moist. They grow on wood.

**fitness** *See* ADAPTIVE VALUE.

*Fitzroya* (family *Cupressaceae) A *monotypic genus, *F. cupressoides*, which is a large, rare, coniferous timber tree of southern Chile. It is hardy in Britain.

**fixation 1.** A soil process by which certain nutrient chemicals required by plants are changed from a soluble and available form into a much less soluble and almost unavailable form. **2.** The first step in making permanent preparations of tissues for microscopic study, by killing cells and preventing subsequent decay with as little distortion of structure as possible. Examples of fixatives are formaldehyde and osmium tetroxide, often used as mixtures. **3.** A term applied to gene frequencies when all members of a population are homozygous for a particular allele at a given locus. **4.** A biological process by which inoganic molecules can be incorporated into organic molecules within living organisms. In *photosynthesis the green plant fixes carbon dioxide from the atmosphere, and several microbes are capable of fixing atmospheric nitrogen into amino acids and hence proteins (*see* NITROGEN FIXATION).

**flabelliform** Shaped like a fan.

**flaccid** Wilted (*see* WILTING). The state of cells when they are short of water.

**Flacourtia** (family *Flacourtiaceae) A genus of small trees or shrubs in which the twigs are at first spiny. The flowers are unisexual. The fruits are fleshy *drupes with several stones; they are edible in some species. There are about 15 species, occurring in Africa, and from Asia to Polynesia.

**Flacourtiaceae** A family of trees, or rarely shrubs, in which the leaves are *simple, with small *stipules, and the stalk commonly kneed. The flowers are small, regular, unisexual or hermaphrodite, their parts free, the *ovary *unilocular and usually *superior, the *ovules parietal (*see* PLACENTATION). Fruits are various. There are 88 genera, with 875 species, occurring mainly in the tropics.

**flagella 1.** Plural of *flagellum **2.** *See* CALAMUS.

**Flagellaria** (family *Flagellariaceae) A genus of giant, slightly woody climbers,

with elongate, parallel-nerved, tendrillate leaves with sheathing bases. The flowers are hermaphrodite, and *petaloid. There are 4 species, found in the Old World tropics.

**Flagellariaceae** A small family of monocotyledonous (*see* MONOCOTYLEDON) herbs, in which the leaves are elongate, and parallel-nerved. The flowers are regular, with 3-plus-3 *tepals, and 6 *stamens; the *ovary is *superior and *trilocular. The fruit is a fleshy *drupe. There are 3 genera, with 7 species, found in the tropics.

**flagellated fungi** The Mastigomycopsida, in which some cells are propelled by whip-like structures.

**flagellum (pl. flagella) 1.** A threadlike *organelle that usually functions in locomotion. Flagella are found on some bacteria and on a range of *eukaryotic cells, including those of certain fungi, protozoa, and algae. Bacterial and eukaryotic flagella are very different both in structure and mode of operation. Bacterial flagella rotate; eukaryotic flagella undulate. The eukaryotic flagellum has the more complex structure, and there are two types: the whiplash type, which is smooth and whiplike, and the tinsel type, which bears numerous fine, hair-like projections along its length. **2.** In certain liverworts, a slender, filamentous, runner-like stem which carries much-reduced leaves.

**flame of the forest** *See* DELONIX.

**Flammulina** (family *Tricholomataceae) A genus of *fungi that form *mushroom-like *fruit bodies. In *F. velutipes* (velvet shank), the cap is slimy and reddish to orange or yellow; the *stipe is central and velvety. This species forms clusters of fruit bodies in winter on the trunks and dead wood of broad-leaved trees; it is cultivated for food (e.g. in Japan, where it is known as enokitake).

**Flandrian** A local geological term used in the British Isles to describe the current interglacial, starting about 10 000 years ago. Most geologists regard the Flandrian as a term equivalent to the *Holocene Epoch, representing the second epoch of the *Quaternary Period, following the *Pleistocene Epoch. It can reasonably be argued, however, that the current interglacial is not sufficiently distinct from previous interglacials within the Pleistocene to merit an elevation to the level of epoch. On the other hand the dominance and impact of the human species in this interglacial could be regarded as a reason for such a distinction. In Europe the warmest stage occurred during Atlantic times, about 6000 BP, when warmth-loving trees dominated the landscape (the Hypsithermal is the equivalent North American *climatic optimum). No consensus view exists as to when the ice will advance again, bringing extremely cold conditions to high mid-latitudes, nor as to how quickly these conditions may arise.

**flannel flower** See ACTINOTUS.

**flask fungi** *Fungi that form flask-shaped *ascocarps or *perithecia.

**flavescent** Yellow, or becoming yellow.

**flavin** A generic name for a group of light-sensitive *pigments that are believed to lengthen the wavelength of light from blue to red and thus influence the action of *phytochrome. Flavins are also involved in *phototropism.

**flavin adenine dinucleotide (FAD)** A *coenzyme derivative of the vitamin *riboflavin, which participates in dehydrogenation reactions mediated by *flavoproteins. It is an important intermediate in *oxidative phosphorylation.

**flavin mononucleotide (FMN)** A *coenzyme, derived from *flavin, which acts as a *prosthetic group to several *dehydrogenases.

**flavoprotein** A *conjugated protein in which the *prosthetic group is a *flavin *nucleotide *coenzyme such as *flavin adenine dinucleotide (FAD) or *riboflavin mononucleotide (FMN); some flavoproteins contain either haem or metal ions in addition. They are widely distributed in cells of all types and serve as electron transport agents.

**flax** See LINUM and FIBRE.

**Flexibacter** (order *Cytophagales) A genus of gliding *bacteria which occur as flexible rods or filaments. Usually they are *aerobic. Species are found in marine or freshwater habitats and some species are *pathogenic for fish.

**flexuous** Wavy or bending in a zigzag manner; the term is usually applied to a stem or other *axis.

**Flindersia** (family *Rutaceae) A genus of valuable timber trees in which the leaves are usually *pinnate and dotted with glands. The fruit is a *capsule, the seed winged. The genus is somewhat intermediate between Rutaceae and the *Meliaceae. Queensland maple is a fine dark timber from an Australian species (*F. brayleyana*). There are 17 species, occurring in the Moluccas, New Guinea, New Caledonia, and tropical Australia.

**floating chronology** A tree-ring chronology for a particular area that does not overlap with chronologies from living trees in that area, and therefore cannot be dated absolutely.

**floccose** Having a loose, woolly, or cottony surface.

**flocculation** A process in which *clay and other soil particles adhere to form larger groupings or *aggregates, thereby coarsening the soil texture and making heavier soils easier to cultivate. The reverse of this process is known as dispersion.

**floccus** From the Latin *floccus*, meaning 'tuft', filaments or hairs borne in tufts and having a woolly appearance.

**flora** (adj. **floral, floristic**) All the plant species that make up the vegetation of a given area. The term is also applied to assemblages of fossil plants from a particular geological time, or from a geographical region in a former geological time. Examples of all three types of usage, respectively, are: British flora, *Carboniferous flora, and *Gondwana flora.

**floral** See FLORA.

**floral formula** A conventional method for recording the structure of a *flower. It uses a series of capital letters to identify parts: K = *calyx; C = *corolla;

A = *androecium; G = *gynoecium. The number of components in each is indicated by a number; if the number exceeds 12, the symbol ∞ (infinity) is used. If the parts are fused the number is enclosed in brackets. If 1 whorl is fused to the next, they are enclosed by a single bracket drawn horizontally above them. The position of the *ovary is indicated by a line; this is below the number following G if the ovary is superior and above it if the ovary is inferior. The formula is preceded by the symbol ⊕ if the flower is *actinomorphic and ·|· or ↑ if it is *zygomorphic.

**floral kingdom**   See FLORAL PROVINCE.

**floral province (floral kingdom, floral realm, floral region)** A major geographical grouping of plants, especially flowering plants, identified on the basis of floristic distinctiveness, particularly with regard to the degree of *endemism at family and generic level. Some authorities distinguish between kingdoms or realms, which are accorded the highest status, and provinces, which have lesser status and go to make up the kingdoms. *Compare* FLORISTIC REGION.

**floral realm**   See FLORAL PROVINCE.

**floral region**   See FLORAL PROVINCE.

**floret**   One of the individual small flowers of a clustered *inflorescence, e.g. sunflower. *See* ASTERACEAE.

**floridean starch**   The storage product found in algae of the *Rhodophyta. The structure of the compound is similar to *amylopectin.

**florigen**   A hypothetical plant *hormone that may be transported from leaves that have been stimulated with light for a suitable length of time to the apex, where flower production is initiated.

**floristic**   See FLORA.

**floristic region**   The spatial categorization of plants recognized in terms of floristic composition. The term is often used synonymously with *floral province, kingdom, or realm, depending on the authority.

**floristics**   See PHYTOGEOGRAPHY.

**flower**   In *angiosperms, the structure concerned with *sexual reproduction, consisting of the *androecium (male organs) and *gynoecium (female organs), commonly surrounded by a *corolla (*petals) and *calyx (*sepals). The male and female parts may be in the same flower or in separate flowers. In many plants the term 'flower' is popularly applied to an *inflorescence that in fact comprises numerous small flowers (florets) grouped together.

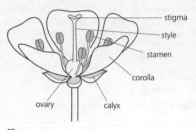

*Flower*

**flowers of tan**   The *myxomycete *Fuligo septica*.

**fluorescence**   A kind of luminescence, in which an atom or molecule emits radiation when electrons within it pass back from a higher to their former, lower-energy, state. The term is restricted to the phenomenon in cases where the interval between absorption and emission is very short (less than $10^{-3}$ s).

**fluorescent antibody technique**   A method for detecting the location of a specific *antigen in a cell by staining a section of the tissue with *antibody (specific to that antigen) that is combined with a fluorochrome (a substance that fluoresces in ultraviolet light). The antigen is located wherever fluorescence is observed.

**fluvic horizon**   A *soil horizon that is dark in colour and usually rich in volcanic rocks. The name is from the Latin *fulvus* meaning 'dark yellow'.

**Fluvisols**   Soils that have formed on recent *alluvial deposits and have a *fluvic horizon extending from 25 cm below the surface to a depth exceeding 50 cm. Fluvisols are a reference soil group in the FAO *soil classification.

**flux study** An examination of the productivity and *respiration of an entire *ecosystem by tracing changes in atmospheric gas (usually carbon dioxide) concentrations. Such studies are complicated by turbulence in the atmosphere and require the construction of detailed computer models of gas movements in order to calculate changes in gas levels. *See* AERODYNAMIC METHOD.

**fly agaric** The *mushroom-shaped *fruit body of *Amanita muscaria*. The cap is bright red, typically with remnants of the white *universal veil adhering to it; the *gills are white. It produces hallucinogenic poisons. It is common on the ground under birch (*Betula*) or pine (*Pinus*) in autumn.

**FMN** *See* FLAVIN MONONUCLEOTIDE.

**Foeniculum** (fennel; family *Apiaceae) A *monotypic genus of *herbs that have yellow flowers and finely divided leaves which contain a pungent, aromatic oil used in cooking and for flavouring. Fennel is native to southern Europe.

**foenugreek** (fenugreek) *See* TRIGONELLA.

**folicolous** *See* EPIPHYLLOUS.

**Folin–Ciocalteu reaction** The basis of a colorimetric method for the quantitative determination of *proteins. It is dependent upon the production of a blue colour as a result of the reaction of tyrosine in the sample with a phosphomolybdotungstic acid reagent.

**foliose** Leaf-like; made up of thin flat lobes, as are the *thalli of certain types of *lichen.

**follicle** A dry *fruit derived from a single *carpel which *dehisces along one side only.

**Fomes** (family *Polyporaceae) A genus of *fungi, in which the fruit bodies are often large, *perennial, and corky or woody in texture, and have light chestnut-brown 'flesh'. They are found on living or dead trees. Many species formerly belonging to this genus are now classified in other genera: for example, *F. annosus* is now *Heterobasidion annosum* and *F. igniarius* is now *Phellinus igniarius*. *See also* TINDER FUNGUS.

**Fomes annosus (Heterobasidion annosum)** *See* BRACKET FUNGUS.

**Fontinalis** (order *Isobryales) A genus of mosses, which are usually aquatic. The stems are long and irregularly branched. Leaves are ovate or lanceolate, concave or plane, and nerveless. There are about 20 species, found mainly in northern temperate regions. Two species are found in Britain. *F. antipyretica* (willow moss) grows on wood or rocks in slow-moving rivers and streams, and in lakes and ponds. The shoots are long (up to 70 cm) and little branched; leaves are borne in 3 ranks, and each leaf is folded and has a sharp keel. *F. squamosa* lacks the keel on the leaves and is found in fast-moving, non-calcareous waters.

**foolish seedling disease** *See* BAKANAE DISEASE.

**foot** That part of the *sporophyte *embryo, especially in *Bryophyta, which connects it to the *gametophyte and absorbs nutrients from it.

**foraging** Applied to plants that extend over new areas of soil by means of runners, and periodically take root when conditions are appropriate. It has been shown in some plants that they extend their runners faster when crossing areas of low nutrient content and slow down (and take root) when the runners encounter richer soils.

**forb** A non-grassy, herbaceous species, e.g. legumes and composites.

**Forbes, Edward** (1815–54) A British naturalist who studied medicine but abandoned it for natural history following the publication in the *Magazine of Natural History* (1835–6) of the botanical results of a tour he made of Norway in 1833. He was appointed curator of the museum of the Geological Society of London (1842), professor of botany at King's College, London (1843), palaeontologist to the

Geological Survey of Great Britain (1844), professor of natural history to the Royal School of Mines (1851), and professor of natural history at the University of Edinburgh (1854). It was in the *Memoirs* of the Geological Survey that, in 1846, he published his major contribution to the study of plant geography. In 'On the Connexion between the Distribution of the Existing Fauna and Flora of the British Isles, and the Geological Changes which have Affected their Area', he proposed that British plants may be considered as 5 distinguishable groups, most of which entered by migrating across land before, during, and after the Glacial (i.e. *Pleistocene) Epoch.

**forcipate**   Forked, like pincers.

**Fordia** (family *Fabaceae, subfamily *Papilionatae) A genus of small trees in which the leaves are *pinnate. The flowers are borne in *cauliflorous or *ramiflorous *racemes. There are 10 species, occurring from southern China to the Philippines.

**fore-dune**   In a coastal sand-dune system, the dune nearest to the sea; it is characterized by plants that can tolerate occasional flooding by sea water (e.g. the grass *Agropyron junceum*).

**forest**  **1.** A plant formation that is composed of trees the crowns of which touch, so forming a continuous canopy (*compare* WOODLAND). **2.** A collective noun describing the trees that comprise an area of forested land. **3.** In Britain, from Norman times, a district reserved for the hunting of deer, often belonging to the sovereign, to which special laws applied and which was administered by special officers. The word 'forest' is derived from the Latin *foris*, 'out of doors': the land lay beyond those enclosed for agriculture or parkland, and was unfenced. The land set aside as forest was not necessarily tree-covered, especially in the uplands, and might include open *heath, *grass, and *bog, as well as wooded areas. Most of the land formerly under forest law has been disafforested, although a district may still bear the 'forest' designation, e.g. 'Dartmoor Forest'. Some Crown forests, never disafforested, came to be used

for growing timber, especially for ship-building, and today are managed by the Forestry Commission, e.g. New Forest, Forest of Dean. **4.** To plant with trees.

**forestry**  **1.** The practice of growing and managing forest trees for commercial timber production. This includes the management of specially planted forests, of native or *exotic species, as well as the commercial use of existing forest, and the genetic improvement of timber trees for selected purposes. **2.** The scientific study of tree growth and timber production systems.

**forget-me-not**   *See* MYOSOTIS.

**formation**  **1.** (plant formation) In vegetation description and analysis, a classificatory unit which usually implies a distinctive *physiognomy rather than a distinctive species composition. In detail, the various *phytosociological traditions ascribe slightly different meaning and hierarchical status to the term. The early European schemes grouped associations into alliances and then formation groups (e.g. *sclerophyllous scrub), and then formation classes (e.g. *sclerophyllous woodland and scrub). The formation classes are roughly equivalent to the major world *biomes. **2.** *See* FORMATION TYPE.

**formation type**   A world vegetation type which has a relatively uniform appearance and life-forms. A formation is a geographically distinct component of a formation type. For example, the tropical rain forest is a formation type, which has a number of formations, each with a distinctive structure and physiognomy. The term *'biome' is nearly synonymous with 'formation type', and 'formations' are also known as 'regions'. A formation type is generally regarded as climatic *climax vegetation.

**form  genus**   A non-phylogenetic, artificial taxon of convenience used for the asexual states of certain *fungi. Such *imperfect fungi are grouped into form genera on the basis of similarities between asexual reproductive structures or other vegetative characteristics.

**formula of vegetation**   A shorthand system for the rapid, precise description of vegetation. For example, in Christian's and Perry's (1953) scheme, A, B, and C respectively denote trees, shrubs, and herbs; 3, 2, and 1 indicate size as tall, medium, and short; density is recorded as x, y, and z, indicating respectively dense, average, and sparse. Average layer heights are also appended. There are many other such schemes, of varying complexity.

**Fortunella** (family *Rutaceae) A genus of small trees in which the leaves have a winged *petiole. The fruit resembles an orange. Several are cultivated as kumquats. There are 4 or 5 species, occurring from eastern Asia to Malaya.

**fossil 1.** Generally, anything that is ancient, especially if it is discovered buried below ground (e.g. fossil fuel, fossil soil). In its original sense 'fossil' meant anything dug up from the earth, including mineral ores, precious stones, etc. **2.** In the modern sense, which dates from the late seventeenth century, a fossil is the remains or traces of a once-living organism. At one time the term was used only of material dating from before the end of the most recent glacial period, so fossils were more than 10 000 years old; remains and traces younger than 10 000 years were known as 'subfossils'. This restriction has been abandoned and now all ancient remains are called 'fossils' regardless of their age. Fossils include skeletons, tracks, impressions, trails, borings, and casts. Fossils are usually found in hard rocks, but not always; for example, small plants or parts of plants (e.g. leaves and flowers) have been found preserved in *amber of *Tertiary age.

**founder effect** A new population (e.g. on an oceanic island) may be derived from a single individual or a limited number of immigrants. This is the founder effect. The founder or founders will represent a very small sample of the *genetic pool to which it or they formerly belonged. *Natural selection operating on this more restricted genetic variety soon yields gene combinations quite different from those found in the ancestral population.

**four-o'clock plant** (*Mirabilis jalapa*) See PARTIAL DOMINANCE.

**four-winged sophora** See *SOPHORA*.

**fovea** A small depression or pit.

**foveate** Marked with *foveae.

**fox-fire** A light (*bioluminescence) emitted by moist decaying wood or by certain types of fungal *fruit body.

**foxglove** See *DIGITALIS*.

**Fragaria** (strawberry; family *Rosaceae) A genus of *perennial, stoloniferous (see STOLON) herbs in which the leaves are borne in basal rosettes, each leaf with 3 toothed leaflets. The flowers have a *pentamerous *epicalyx. The *receptacle becomes enlarged, juicy, and brightly coloured in the fruit, with the tiny *achenes on its surface. It is much cultivated for its delicious fruits. There are about 12 species, found in northern temperate and tropical regions, and in Chile.

**fragipan** A subsoil *horizon, found deep in a soil *profile and having a high bulk density. It is a dense, brittle, and compact layer, apparently with little or no cementation horizon, associated with acid soil conditions.

**fragmentation** *Vegetative reproduction in which a plant structure fragments, each fragment growing into a new plant.

**Francisella** A genus of *bacteria not assigned to a taxonomic family. The cells are ovoid to variable in shape, *Gram-negative, and non-*motile. They are *chemo-organotrophic, and can grow only in the presence of air. They are found, for example, as *parasites or *pathogens on animals, including humans.

**frangipani** (pagoda tree, temple tree) See *PLUMERIA*.

**Frangula alnus** (alder buckthorn) See RHAMNACEAE.

**Frankeniaceae** (sea heaths) A family of low shrubs or herbs that have narrow, often heath-like leaves in *opposite pairs without *stipules. The flowers are small, regular, and *tetra-, *penta-, or *hexamerous. There are 3 genera, with about 30 species, found mainly in saline areas of temperate and subtropical regions.

**Frankia** (order Actinomycetales) A genus of *mycelium-forming *bacteria. They are found in soil and in *root nodules in a variety of non-leguminous, dicotyledonous (*Dicotolydoneae) plants. The root nodules can bring about the fixation of atmospheric nitrogen. *F. alni* occurs in coralloid nodules on the roots of alder (*Alnus*). Nodules are generally present only near the soil surface.

**Frasnian** *See* DEVONIAN.

**Fraxinus** (ashes; family *Oleaceae) A genus of trees in which the leaves are *opposite and *pinnate. The flowers are borne in *racemes, are reduced, and are mostly wind-pollinated. The fruit is a *samara. The wood is elastic and useful. There are 65 species, found in the northern hemisphere, a few reaching the tropics.

**free central placentation** *Placentation in which the placentae develop around a central column or dome of tissue.

**french bean** *See* PHASEOLUS.

**fresh water** Water containing little or no chlorine. According to the Venice system, which classifies brackish waters by their percentage chlorine content, fresh water contains 0.03 per cent or less of chlorine. *Compare* BRACKISH.

**Freycinetia** (family *Pandanaceae) A genus of plants, nearly all of which are bole-climbers. The leaves are straplike, with spiny margins. The plants are *dioecious. The inflorescences often have showy *bracts, purportedly bird- or bat-pollinated. The fruits are fleshy *berries, in spherical or cylindrical heads. There are about 175 species, occurring in warm temperate and tropical forests, from Sri Lanka eastward to the Pacific islands.

**friable** Applied to the consistency or handling properties of soil, meaning that the soil crumbles easily.

**frigid** *See* PERGELIC.

**fringed waterlily** (*Nymphoides*) *See* MENYANTHACEAE.

**fringing forest** (gallery forest) A *forest which often extends along river banks from the rain-forest belt into adjacent *savannah. The ribbon-like tracts generally resemble the rain forest, but the luxuriance typical of the latter is normally less well developed.

**frizoles bean** *See* PHASEOLUS.

**frog-bit** (*Hydrocharis*) *See* HYDROCHARITACEAE.

**frond** A big compound leaf; a term applied to the leaves of *palms and *ferns.

**fructification** The process of forming a *fruit body, or the fruit body itself.

**fructose** (D-fructose, levulose, $C_6H_{12}O_6$) A sugar that occurs abundantly in nature as the free form; but also, with *glucose, in the form of the disaccharide sucrose.

**fruit** Strictly, the ripened *ovary of a plant and its contents. More loosely, the term is extended to the ripened ovary and *seeds together with any structure with which they are combined, e.g. the apple (a *pome) in which the true fruit (core) is surrounded by flesh derived from the floral *receptacle.

**fruit body** (fruiting body) A differentiated, *spore-bearing structure, particularly the *ascocarps and *basidiocarps of *ascomycetes and *basidiomycetes.

**Frullania** (order *Jungermanniales) A genus of leafy *liverworts, which have a characteristic type of leaf. Each leaf consists of 2 parts, the smaller of which is pitcher-like; these tiny pitchers hold water and often contain a characteristic microscopic fauna, including rotifers, which may in some way aid in the nutrition of the liverwort. The *thallus is often dark green to copper-brown or purplish in colour. *F. dilatata* and *F. tamarisci* are both common British species; *F. tamarisci* grows on rocks, on the ground, or less frequently on trees, while *F. dilatata* generally grows on trees.

**frustule** The silica wall of a *diatom. *See* BACILLARIOPHYTA.

**fruticose** Shrubby in habit, as in the *thalli of certain *lichens (e.g. *Usnea* species).

**Fucaceae** (order *Fucales) A family of brown seaweeds in which the *thallus is

**gaboon** See *Aucoumea*.

**Gaboon nuts** The edible seeds of *Coula edulis*. See Olacaceae.

***Gaeumannomyces*** (order *Diaporthales*) A genus of *fungi which form dark, thick-walled *perithecia. *G. graminis* (*Ophiobolus graminis*) is *parasitic on the roots of grasses and cereals, and causes 'take-all' disease of cereals.

***Gahnia*** (family *Cyperaceae) A genus of coarse, tufted, erect *perennials, which are rhizomatous (*see* RHIZOME). The plants grow in large clumps. The *culms are cylindrical, and up to 1.5 m in height, with dark-brown basal leaf *sheaths. The leaves and sheaths are slightly bristly, with inrolled margins. The *spikelets are 2-flowered with a terminal, bisexual, and fertile flower, and a single male flower. The *glumes are yellow-brown, turning darker on maturity. The *nut becomes blood-red and is 3-sided. There are about 30 species, found in swampy habitats in Asia, Australia, and the Pacific.

**galactolipid** A *lipid that contains *galactose. Galactolipids are characteristic and important constituents of plant membranes.

**galactose** An aldohexose monosaccharide that is not normally found naturally in the free form, but more usually is found as a unit in a larger molecule (e.g. a di- or *polysaccharide).

**galactosidase** See BETA GALACTOSIDASE.

**galactoside** See BETA-GALACTOSIDE.

**Galápagos Islands** A group of oceanic islands, about 970 km from the west coast of S. America, which *Darwin visited in 1835. He encountered a number of endemic (*see* ENDEMISM) species that were to prove influential in the development of his ideas on *evolution.

***Galbulimima*** See Himantandraceae.

***Galearia*** (family *Euphorbiaceae) A genus of trees or shrubs whose flowers are borne on long *spikes. The fruit is a *drupe with a leathery wall and sculptured stone. Sometimes the genus is segregated with the family *Pandaceae. There are 6 species, occurring from Indo-Malaysia to the Solomon Islands.

**galericulate** Covered or capped.

***Galerina*** (family *Cortinariaceae) A genus of *fungi, in which the *mushroom-shaped *fruit bodies are yellowish, rusty, or brown, and small and slender, with a thin, fragile, central *stipe. Species are common in mossy areas in woods, *tundra, etc. Their distribution is cosmopolitan.

***Galium*** (family *Rubiaceae) A genus of herbs which have *whorls of leaves and leaf-like *stipules. It is one of the few temperate and herbaceous genera of the family, and atypical of most. *G. aparine* is goosegrass or cleavers. There are 400 species, occurring in temperate regions.

**gall** (cecidium) An abnormal growth or swelling in a plant. The formation of a gall may be induced by infection of the plant with *bacteria or *fungi, or by attack from certain mites, nematodes, or insects. Galls may be formed on roots, stems, or leaves. Some galls (e.g. *clubroot and *crown gall) are symptoms of disease; others appear to do little harm to their hosts, while some may actually be beneficial to the plant (e.g. nitrogen-fixing *root nodules of legumes). Galls are variously structured, ranging from a simple outgrowth to a large and histologically complex structure with up to 5 distinct tissue layers with nutritive zones. Most gall-forming species are members of the insect family Cynipidae (Hymenoptera), and often have complex, heterogynous life cycles, utilizing different parts of the same host or different hosts, which are generally *Quercus* species. The mechanism of gall initiation and development is little understood, and the study of galls offers an unparalleled opportunity for physiological and ecological research. The gall community supports a large number of parasitoids, hyperparasitoids, and inquilines (species which use the gall but do not kill its occupant).

**gallery forest**   *See* FRINGING FOREST.

***Gallionella***   A genus of \*Gram-negative \*bacteria which is not assigned to any taxonomic family. The cells are ovoid or kidney-shaped and produce fibrous 'stalks' which commonly contain ferric hydroxide. They appear to derive energy from the oxidation of ferrous (to ferric) iron. They are \*micro-aerophilic, and are found in cool, iron-containing natural waters, and in soil.

**galochrous**   Milk-white in colour.

**galvanotaxis**   Change in direction of locomotion in a \*motile organism or cell, made in response to an electrical stimulus.

**gambier**   *See* UNCARIA.

**gamboge**   The solidified yellow resin of certain \*Garcinia species (especially *G. hanburyi*), used as a paint pigment. Cambodia, whence it comes, derives its name from this product.

**gametangium**   An organ in which \*gametes develop, found in \*fungi, \*mosses, and \*ferns. *See also* ANTHERIDIUM.

**gamete**   A specialized \*haploid cell (sometimes called a sex cell) whose nucleus and often cytoplasm fuses with that of another gamete (from the opposite sex or mating type) in the process of \*fertilization, thus forming a \*diploid \*zygote. In the lower plants the male gamete is a \*motile \*antherozoid, which needs water as a medium in which to move, similar to the spermatozoon in animals, and the female gamete is contained in the \*archegonium. In higher plants the male gamete is contained in the male \*gametophyte (\*microspore) and the female gamete is contained in the female gametophyte (\*megaspore); fertilization takes place only after the gametophytes come into contact (\*pollination) and the male gamete can be released without having to leave the plant. This adaptation, permitting fertilization

in the absence of water, is analagous to internal fertilization in animals, and, like internal fertilization of animals, is an adaptation to dry environments. Usually there are many small male gametes, but only a few or one female gamete.

**gametic equilibrium (linkage equilibrium)**   The condition in which the frequency of the \*gametes, formed by the association of \*alleles at different loci (*see* LOCUS), is equal to the product of the frequencies of the alleles that constitute them. *Compare* LINKAGE DISEQUILIBRIUM.

**gametocyte**   A cell that will undergo \*meiosis to form \*gametes. A cell giving rise to a male gamete is termed an antherocyte; a cell giving rise to a female gamete is termed an oocyte.

**gametogenesis**   The formation of \*gametes from \*gametocytes.

**gametophore**   In \*Bryophyta, the complex, \*thallose, or leafy structure that bears the sex organs.

**gametophyte**   A \*haploid phase of the \*life cycle of plants, during which \*gametes are produced by \*mitosis. It arises from a haploid \*spore produced by \*meiosis from a \*diploid \*sporophyte. In lower plants (such as mosses), the gametophyte is the dominant and conspicuous generation. *See also* ALTERNATION OF GENERATIONS.

**gamma-ray spectrometry**   An analytical method for the measurement of the intensities and energies of gamma radiation. Scintillation or semiconductor radiation detectors, coupled to various types of electronic circuitry, enable a spectrum to be accumulated. This may be used to identify the gamma-emitting radioisotopes, and their energy intensities can be used to determine the corresponding element concentrations.

**ganja**   *See* CANNABIS.

***Ganoderma***   (family   \*Ganodermataceae)   A genus of \*fungi which form stipitate or \*sessile, \*bracket-like \*fruit bodies. \*Spores are brown. They are found on dead wood, or as \*parasites of living trees (e.g. *G. applanatum* causes heart rot in various trees).

**Ganodermataceae (order *Aphyllophorales)** A family of *fungi that produce *spores with a complex wall structure; the spore wall is 2-layered, with ornamentations on the inner layer penetrating a colourless outer layer. Fruit bodies are *annual or *perennial, *bracket-like, and corky or woody in texture. They are found on wood, and the family includes important wood-rotting fungi.

**Ganua (family *Sapotaceae)** A genus of trees that yield a heavy timber. The terminal buds are scaly and the leaves often have *stipules. There are 17 species, found in rain forests of the Malay archipelago.

**Garcinia (family *Clusiaceae)** A genus of trees, and a very few shrubs, in which the crown is *monopodial. The leaf bases enclose the *apical bud. The flowers are unisexual on different trees, the male commonly with a distinctive *pistillode, the female with a big, conspicuous, varied *stigma. The fruit is a *berry seated on the usually persistent *calyx. Mangosteen is *G. mangostana*. There are about 200 species, occurring in the Old World tropics.

**Gardenia (family *Rubiaceae)** A genus of trees and shrubs in which the flowers are large and fragrant. Several are cultivated as ornamentals. There are about 200 species, occurring in the palaeotropics.

**gari** *See* MANIHOT.

**garigue** *See* GARRIGUE.

**garlic** *See* ALLIUM and LILIACEAE.

**garrigue (garigue)** A low-growing, secondary vegetation which is widespread in the Mediterranean basin and is derived from the original mixed forest. The dominant plants are aromatic *herbs and prickly dwarf shrubs, with drought-resistant foliage, many belonging to the mint family (*Lamiaceae) or *Fabaceae. Garrigue is a degraded, fire-prone form of vegetation produced by intensive grazing and other human-based activities. *See also* MEDITERRANEAN SCRUB.

**Garrya (family Garryaceae)** A genus of evergreen shrubs in which the leaves are *opposite. The flowers are unisexual in pendulous *catkins, the *ovary *inferior. The fruit is a *berry. Several species are cultivated as winter-flowering ornamentals. The genus is related to the *Cornaceae. There are 13 species, occurring in western temperate America to Panama and the W. Indies.

**Garuga (family *Burseraceae)** A genus of trees that have *pinnate leaves and whose fruit is a *drupe. There are 4 species, occurring in regions with seasonal climates from the Himalayas to Melanesia.

**gaseous exchange** The transfer of gases between an organism and the environment; it may occur in both *respiration and *photosynthesis.

**gas-exchange method** A method for measuring *primary productivity based on rates of carbon-dioxide uptake and oxygen release, these being the easily monitored gaseous raw material and by-product of photosynthesis. *See also* CARBON-DIOXIDE METHOD; and OXYGEN METHOD.

**Gasteromycetes (subdivision *Basidiomycotina)** A class of *fungi that form *epigean or *hypogean *basidiocarps which undergo angiocarpic (*see* ANGIO-CARPY) development; the *basidia are not *septate and *basidiospores are not forcibly discharged from the basidia. Most members are *saprotrophic; some form *mycorrhizal associations.

**gas vacuole** A small, gas-filled *vesicle, numbers of which are found in certain aquatic bacteria and *cyanobacteria. Their function appears to be that of giving buoyancy to the cells.

**Gause principle** *See* COMPETITIVE-EXCLUSION PRINCIPLE.

**GDP (guanosine diphosphate)** *See* GUANOSINE PHOSPHATE.

**Geastrum (order *Lycoperdales)** A genus of terrestrial *fungi in which *fruit-body development begins underground or at the soil surface. The *exoperidium splits open from the top and the sections open outwards in a star-like arrangement. The *endoperidium contains the spores, which are released via a pore in the tip. Species occur on the ground in woods, sand-dunes, etc. *See also* EARTH STAR.

**Geijera** (family *Rutaceae) A genus of trees and shrubs that have *simple, alternate, often pendulous, glandular leaves. The flowers, held in terminal *panicles, are small and creamy-yellow in colour. The 5 *sepals are united at the base, the 5 *petals star-like or cup-shaped, and the 5 *stamens are spreading. The fruit is a *carpel with 5 seeds. *G. parviflora* (wilga) is often cultivated. It is a small, bushy tree with long, narrow, linear, greenish-grey leaves. There are 7 species, found in New Guinea, Australia, and New Caledonia.

**geitonogamy** *Fertilization involving *pollen and *ovules from different plants on the same plant or *clone.

**gelatinous lichen** A *lichen in which the *phycobiont is a *cyanobacterium.

**gel filtration** A column-chromatographic technique normally employing as a stationary phase polymeric carbohydrate-gel beads of controlled size and porosity. Mixture components are separated on the basis of their sizes and rates of diffusion into the beads. Smaller molecules tend to diffuse more rapidly into the beads, thereby leaving the mainstream of solvent and so becoming retarded with respect to larger molecules. This method can also be used to determine the molecular weight of an unknown substance.

**Gelidiales** (division *Rhodophyta) An order of red seaweeds, in which the *thallus is filamentous, cylindrical or compressed, and typically is branched. Several species are of economic importance as sources of *agar. There are two British genera, *Gelidium* and *Pterocladia*. Three British species of *Gelidium* are currently recognized, but these are notoriously variable and difficult to identify. They are very common in intertidal regions, often forming a fringe around rock pools.

**gemma 1. (mycol.)** A thick-walled, *asexually derived *spore formed from a portion of a vegetative *hypha. **2. (bryol.)** A structure formed by certain *mosses and *liverworts, which functions in vegetative reproduction (*see* VEGETATIVE PROPAGATION). Gemmae may consist of 1, 2, or many cells. *See also* MARCHANTIA; LUNULARIA; and TETRAPHIDALES. **3.** A vegetative *propagule that is a modified organ of the parent plant (e.g. the small lateral *bulbils of *Agave americana*, the century plant).

**gemmation** *See* ASEXUAL REPRODUCTION; BUDDING; and GEMMA.

**gene** The fundamental physical unit of heredity. It occupies a fixed chromosomal *locus, and when transcribed has a specific effect upon the *phenotype. It may *mutate, and so yield various *allelic forms. A gene comprises a segment of *DNA (in some viruses it is *RNA) coding for one function or several related functions. The DNA is usually situated in thread-like *chromosomes, together with protein, within the nucleus; in bacteria and viruses, though, the chromosomes comprise simply a long thread of DNA. The part of a gene that functions as one unit is called a *cistron.

**gene bank** An establishment in which both *somatic and hereditary genetic material (*see* GERM PLASM) are conserved. It stores, in a viable form, material from plants that are in danger of extinction in the wild and *cultivars that are not currently in popular use. The stored genetic information can be called upon when required. For example, a crop may be needed that possesses a quality (e.g. tolerance to adverse climatic conditions) which cannot be found in currently exploited cultivars but was present in more antiquated varieties. The normal method of storage is to reduce the water content of seed material to around 4 per cent and keep it at 0°C. (Pollen material may also be used but its longevity is considerably less.) Stored this way, the material often remains viable for 10–20 years. When the desiccating process proves fatal, as is the case with tropical genera producing *recalcitrant seeds, where possible the material is maintained by growth. This may require considerable space, but in some cases the problem can be resolved using *tissue-culture methods. All stored stock is periodically checked by *germination.

**gene centre** *See* CENTRE OF DIVERSITY.

**gene conversion** A process whereby one member of a *gene family acts as a blueprint for the correction of the others. This can result in either the suppression of a new *mutation, or its lateral spread in the *genome.

**gene duplication** A process in *evolution where a *gene is copied twice; the two copies lie side by side along the same *chromosome.

**gene family** A group of similar or identical *genes, usually along the same *chromosome, that originate by *gene duplication of a single original gene. Some members of the family may work in concert, others may be silenced and become *pseudogenes.

**gene flow** The movement of *genes within an interbreeding group that results from mating and gene exchange with immigrant individuals. Such an exchange of genes may occur in one direction or both.

**gene frequency** The number of *loci at which a particular *allele is found divided by the total number of loci at which it could occur, for a given population, expressed as a proportion or percentage.

**gene library** *See* LIBRARY.

**gene pool** The total number of *genes, or the amount of genetic information, that is possessed by all the reproductive members of a population of sexually reproducing organisms.

**general adaptation** An adaptation that fits an organism for life in some broad environmental zone, as opposed to 'special adaptations' which are specializations for a particular way of life. Thus, leaves are a general adaptation, while the particular kind of leaf is a special adaptation. Major groups of organisms are differentiated very largely on the basis of general adaptations.

**generation time** The time that is required for a cell to complete one full growth cycle. If every cell in the population is capable of forming two *daughter cells, has the same average generation time, and is not lost through lysis (*see* LYTIC RESPONSE), the doubling time of the cell number in a population will equal the generation time.

**generic cycles, theory of** A theory that envisages a life cycle for a species or genus that resembles that of an individual. The first stage is characterized by vigorous spread; the second by maximum phyletic activity giving rise to new forms; the third marks a phase of decline in area owing to competition within the offspring species; and the fourth and final stage sees the extinction of the species. Although in many ways a useful analogy, there is no evidence that species become senile and die out spontaneously.

**genetic** Pertaining to the origin or common ancestor or ancestral type. The term is widely used outside the life sciences.

**genetic code** The set of correspondences between base (nucleotide pair) triplets in *DNA and *amino acids in protein. These base triplets carry the genetic information for protein synthesis (*see* RIBOSOME). For example, the triplet CAA (cytosine, adenine, adenine) codes for valine.

**genetic drift** The random fluctuations of *gene frequencies in a population such that the genes among offspring are not a perfectly representative sampling of the parental genes. Although drift occurs in all populations, its effects are most marked in very small isolated populations, in which it gives rise to the random *fixation of alternative *alleles, so that the variation originally present within single (ancestral) populations comes to appear as variation between reproductively isolated populations.

**genetic engineering** The manipulation of *DNA using restriction *enzymes which can split the DNA molecule and then rejoin it to form a hybrid molecule:

a new combination of non-homologous DNA (so-called recombinant DNA). The technique allows the bypassing of all the biological restraints to genetic exchange and mixing, and may even permit the combination of *genes from widely differing *species. Genetic engineering developed in the early 1970s, and is now one of the most fertile areas of genetics.

**genetic equilibrium (linkage equilibrium)** An equilibrium in which the frequencies of two *alleles at a given *locus are maintained at the same values generation after generation. A tendency for the population to equilibrate its genetic composition and resist sudden change is called genetic homoeostasis.

**genetic erosion** The loss of genetic information that occurs when highly adaptable *cultivars are developed and threaten the survival of their more locally adapted ancestors, which form the genetic base of the crop. For example, a new *hybrid strain of maize (Zea mays), developed largely by Donald F. Jones from a variety discovered in 1917, had yields 25 per cent greater than standard maize. By the 1960s it was economically very favourable to use this single hybrid type, and more traditional varieties rapidly receded in distribution. The widespread adoption of this hybrid led to the narrowing of the genetic base (i.e. genetic erosion). In 1962 workers in the Philippines noticed that the fungus *Helminthosporium maydis* (southern corn leaf blight) was highly virulent on this hybrid. By 1970, 80 per cent of the US crop was vulnerable to *H. maydis*, because of the heavy dependence on the one hybrid, and during the wet summer of 1970 around 20 per cent of the US crop was lost to the blight. Fortunately, in this case the *genetic resources needed to produce a resistant strain had not become completely obsolete and so a recovery could be made.

**genetic homoeostasis** See GENETIC EQUILIBRIUM.

**genetic load** The average number of *lethal mutations per individual in a population. Such *mutations result in the premature death of the organisms carrying them. Three main kinds of genetic load may be recognized: (a) input load, in which inferior *alleles are introduced into the *gene pool of a population either by mutation or immigration; (b) balanced load, which is created by selection favouring allelic or genetic combinations that, by segregation and recombination, form inferior *genotypes every generation; and (c) substitutional load, which is generated by selection favouring the replacement of an existing allele by a new allele. Originally called the 'cost of natural selection' by the geneticist J. B. S. Haldane, substitutional load is the genetic load associated with transient *polymorphism. The term 'genetic load' was originally coined by H. J. Muller in 1950 to convey the burden that deleterious mutations provide, but it is probably better recognized as a measure of the amount of *natural selection associated with a certain amount of genetic variability, which provides the raw material for continued *adaptation and *evolution.

**genetic map** The linear arrangement of mutable sites on a *chromosome, deduced from genetic recombination experiments. The percentage of *recombinants is used as a quantitative index of the distance between two *gene pairs, and this distance, together with those between other gene pairs, provides a map of their arrangement on the chromosome. One 'genetic map unit' is defined as the distance between gene pairs for which one product of *meiosis out of a hundred is recombinant (i.e. it equals a recombination frequency of 1 per cent).

**genetic polymorphism** An occurrence in a population of 2 or more *genotypes in frequencies that cannot be accounted for by recurrent *mutation. Such occurrences are generally long-term. Genetic polymorphism may be balanced (such that *allele frequencies are in equilibrium with one another at a given locus), or transient (such that a mutation is spreading through the population in a constant direction). In the former case, the different alleles may be maintained by different environmental conditions (in space or time), one being favoured under one set of circumstances and another under another set; or a *heterozygous *genotype may be in some way superior to the genotypes that are *homozygous at that locus; this is termed a 'heterozygous advantage'.

**genetic resources** The *gene pool in natural and cultivated stocks of organisms that are available for human exploitation. It is desirable to maintain as diverse a range of organisms as possible, particularly of domesticated *cultivars and their ancestors, in order to maintain a wide genetic base. The wider the genetic base, the greater the capacity for adaptation to particular environmental conditions (e.g. a pathogenic presence; *see* PATHOGEN). This has led to the establishment of *gene banks. *Compare* GENETIC EROSION.

**genetics** The scientific study of genes and heredity.

**genetic system** The organization of genetic material in a given *species, and its method of transmission from the *parental generation to its filial generations.

**genetic variance** A portion of *phenotypic variance that results from the varying *genotypes of the individuals in a population. Together with the environmental variance, it adds up to the total phenotypic variance observed among individuals in a population.

**geniculate** Bent abruptly, like a knee joint.

**genome** The total genetic information carried by a single set of *chromosomes (i.e. in a *haploid nucleus). A single representative of each of all the chromosome pairs in a nucleus will therefore bear the genome of an individual.

**genotype** The genetic constitution of an organism, as opposed to its physical appearance (*phenotype). Usually this refers to the specific *allelic composition of a particular *gene or set of genes in each cell of an organism, but it may also refer to the entire *genome.

**genotypic adaptation** *See* ADAPTATION.

**gentian** (*Gentiana*) *See* GENTIANACEAE.

**Gentianaceae** A family of hairless *herbs that have *simple, *opposite leaves, bell-shaped or wheel-shaped *corollas, and *superior *ovaries of 2 partly fused *carpels. There are about 74 genera, with 1200 species, centred on northern temperate regions, but extending far south down the Andes into S. America. *Gentiana* (gentian) has about 300 species, centred in the mountains of Europe. Many of them are prized for their splendid, mostly blue, flowers.

**Gentianella** (family *Gentianaceae) A genus of *annual and *perennial *herbs whose *opposite, entire leaves without *stipules are usually sessile. The plants are *glabrous. The flowers are regular and bisexual. The *calyx is lobed, the *corolla a cylindrical tube, usually white, blue, or purple, composed of 4 or 5 *petals. There are 4 or 5 *stamens. The *ovary is *superior and *unilocular with numerous *ovules. The *style is sometimes absent. The fruit is a *capsule containing many small seeds. There are about 125 species, found in all temperate regions except Africa. Most species occur on mountain grassland, calcareous pastures, and in coastal habitats.

**geobotanical exploration** (biogeochemical exploration) Traditionally, the use of indicator plant species or assemblages to detect the possible presence of metal-rich deposits. It is based on the principle of *limits of tolerance; i.e., it assumes that only specialized species can withstand metal-contaminated soils. In practice, plant response may be confusingly more complex (e.g., plants may respond to low availability of essential nutrients rather than to high presence of toxic minerals), which makes such indicators unreliable. In modern use the concept includes the collection and chemical analysis of plant materials or soil layers, especially humus, in which metal ions may accumulate. It is a supplementary rather than a primary prospecting method. Geobotanical exploration can also be used to detect reserves of elements in the soil, because some species of plant accumulate particular elements. For example, certain *Astragalus* (Fabaceae) species accumulate selenium. *See also* LIMITING FACTOR.

**geocarpic** Applied to plants that fruit below ground (e.g. *Arachis hypogaea*, peanut).

***Geoglossum*** **(order** *****Helotiales)** A genus of *fungi which form stalked, club-shaped, usually flattened *fruit bodies (commonly called 'earth-tongues'). They are *saprotrophic, and found mostly on the ground, but also on rotting wood or leaves.

**geographical floral element** A group of plant species that has marked geographical affinities. The flora of a given territory includes different geographical elements. For example, the Mediterranean element in the British flora comprises 38 species, concentrated in the extreme south and south-west of England. The majority are flowering *annuals or *biennials, categories of plants well represented in the sunny Mediterranean. When the full ranges of the 38 species are examined, they are all found to occur in the Mediterranean basin, with northerly extensions to Britain.

**geometric series** A numerical and graphical description of the relationship of species in a community according to their importance values. Curves approximating to a geometric series are common in pioneer or *ephemeral communities. In theory, the more successful *species restrict other species to the remaining *niche space (the pre-emption hypothesis) by occupying a larger proportion of the available niches. If the most successful species occupies 75 per cent of the available niche spaces, and the next species a similar proportion of the remaining spaces, and so on, then a geometric progression of importance values will result.

***Geonoma*** **(family** *****Arecaceae)** A genus of small, sometimes minute, abundant undergrowth palms, many of which form clumps. The leaves are pinnately nerved, *simple, or divided into broad or narrow leaflets. The palms are *monoecious. The inflorescence is branched or simple. Geonomas are little known in cultivation. There are 75 species, found in lowland and mountain forests of tropical America.

**geophyte** A land plant that survives an unfavourable period by means of underground food-storage organs (e.g. *rhizomes, *tubers, and *bulbs). *Buds arise from these to produce new aerial shoots when favourable growth conditions return.

**geotaxis** A change in direction of locomotion, in a *motile organism or cell, made in response to the stimulus of gravity.

**geotropism** A directional movement of a plant in response to the stimulus of gravity. Primary *tap roots show positive geotropism; vertical primary shoots show negative geotropism; horizontal stems and leaves are diageotropic (*see* DIAGEOTROPISM); and branches and secondary roots at oblique angles are plagiogeotropic.

**Geraniaceae** A family of herbs or low shrubs, rarely tree-like, that have alternate lobed or *compound *stipulate leaves. The flowers are regular, usually *pentamerous with free *petals and *sepals, and usually with twice as many *stamens as petals, a *superior 3–5-celled *ovary, and usually a long beak bearing the free *stigmas. The fruit is a lobed *capsule, the lobes usually 1-seeded, sometimes separating. There are 14 genera, with about 730 species, in temperate and tropical zones. Garden 'geraniums' are varieties of *Pelargonium* species from S. Africa, with slightly irregular, spurred flowers.

**germinal selection** The selection during *gametogenesis against induced *mutations that retard the spread of *mutant cells.

**germination** The beginning of growth of a *seed, *spore, or other structure (e.g. *pollen), usually following a period of *dormancy, and generally in response to the return of favourable external conditions, most notably warmth, moisture, and oxygen. The internal biochemical status of the seed or spore must also be appropriate. In seeds, germination may be *epigeal, with *cotyledons emerging above the ground, or *hypogeal, with the cotyledons staying below ground.

**germ line** Cells from which *gametes are derived, and which therefore bridge the gaps between generations, unlike *somatic cells, in the body of an organism.

**germ plasm** The hereditary material transmitted to offspring through the germ cells, and giving rise in each individual to the somatic (body) cells. The theory of the continuity of germ plasm (i.e. that germ plasm is unchanged from generation to generation) was proposed by August *Weismann; today *DNA is regarded as the molecular equivalent of germ plasm.

**germ plasm bank** An establishment concerned primarily with the conservation of hereditary genetic material which may be lost through the process of *genetic erosion. *Germ plasm loss is a major concern in Asia, parts of Africa, southern Europe, and countries bordering the Mediterranean, where antiquated *cultivars are rapidly replaced by new varieties. With the loss of older cultivars, qualities possessed by them may be lost permanently, and so cannot become incorporated in new varieties. *Gene banks are an important source of germ plasm.

**germ pore** On a *spore or *pollen grain, a thin-walled area through which the *germ tube or *pollen tube emerges on germination. The thin or absent *exine at the germ pore helps in the identification of some pollen grains.

**germ tube** The filament that emerges when a *spore germinates.

**Gesneriaceae** A large family of herbs and shrubs, some of which are epiphytic (see EPIPHYTE). The leaves are *simple, *opposite, alternate, or basal. The flowers are irregular, borne in inflorescences or singly, with 5 *sepals and 5 *petals forming a tube. Many are pollinated by humming-birds. The fruits are usually *capsules. The family includes many ornamentals (e.g. African violets and gloxinias). There are 146 genera, with about 2400 species, mostly tropical, mainly Old World.

**giane** See MEDITERRANEAN SCRUB.

**giant hogweed** See HERACLEUM.

**giant horsetails** See CALAMITACEAE and *CALAMITES CISTIIFORMES*.

**giant polypore** See MERIPILUS.

**giant puffball** See CALVATIA.

**giant swamp taro** See CYRTOSPERMA.

**giant taro** See ALOCASIA.

**gibberellin** The generic name of a group of *plant hormones that stimulate the growth of leaves and shoots. Unlike *auxins, they tend to affect the whole plant and do not induce localized bending movements. They are thought to act either at a transcriptional level or as inducers of *enzymes. Gibberellins were first isolated from a fungus, *Gibberella fujikuroi*, and were found to be the cause of *bakanae disease (a disease of rice, known in Japan as 'foolish seedling disease'), in which affected plants grow unusually tall but seldom survive to maturity.

***Gigantochloa*** (family *Poaceae, subfamily *Bambusoideae) A genus of 20 species of giant bamboos, occurring in Indo-Malaysia, many of which are used for building.

**Gigartinales** (division *Rhodophyta) An order of red seaweeds, in which the *thallus is *crustose (Cruoriaceae) or cylindrical, compressed or membranaceous, and branched or unbranched. There are many families and genera. *Gigartina stellata* is a common species which forms dense, dark-red tufts on rocks in the intertidal zone, often occurring in dense colonies. This species has been used commercially as a source of *agar.

**gill** A membranaceous, blade-like structure (lamella) which bears the *hymenium in the *fruit body of a *mushroom or other *agaric. Gills are usually arranged radially on the underside of the *pileus.

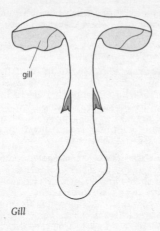

*Gill*

**ginger**  *See* ZINGIBERACEAE.

**gingerbread palm** (doum palm) *See* HYPHAENE.

***Ginkgo***  A genus whose sole living member, *G. biloba*, is the maidenhair tree, native in China. It is deciduous, has leaves with open *dichotomous venation, and is *dioecious. The naked seed is oily and edible but the *testa and *embryo are bitter. It is widely cultivated. Many fossil species are known. *G. digitata*, the earliest representative of the genus, is first recorded from the Middle *Jurassic and bears a close resemblance to its living relative. *See* GINKGOPHYTA.

**Ginkgoales**  Formerly an order of the *Coniferopsida, one of the *gymnosperm groups. Its members are now classified as the division *Ginkgophyta.

**Ginkgophyta**  The division of *gymnosperms that includes only the extant *Ginkgo biloba* (maidenhair tree) and its extinct relatives. The first undoubted maidenhairs occur in *Triassic rocks, and in the subsequent *Jurassic Period their distribution was practically worldwide. The surviving species is restricted (in the wild) to China, and its leaves are strikingly similar to fossil *Ginkgo leaves from the Triassic. The restricted geographical range, the unchanged appearance of the leaves, and the *motile male sperms (otherwise known only in living seed plants in the *Cycadales) have together led to the maidenhair being referred to as a 'living fossil'.

**ginseng**  *See* PANAX.

**girdle groove**  In diatoms (*Bacillariophyta), a groove around the cell corresponding to where the *epithecium and *hypothecium meet.

**girdling**  Cutting transversely across the *phloem in a stem so that downward transport of substances is unable to occur within the plant. For example, a tree is girdled by being cut right around its circumference to a depth that penetrates through the bark and into the wood; this kills the tree.

**Givetian**  *See* DEVONIAN.

**glabrous**  Smooth, lacking hairs.

**gladiate**  Sword-like in appearance.

***Gladiolus*** (family *Iridaceae) A genus of *herbs produced from *corms, which have showy, irregular flowers, each enclosed in its own *spathe on the spike-like inflorescence. There are some 180 species, in Europe, Asia, and Africa; African species are those most grown in gardens.

**glasswort**  *See* SALICORNIA.

**glaucous**  Sea-green or bluish-green in colour, or having a waxy, bluish-green bloom, as on a plum. Originally, the word meant bright, sparkling, greyish, like the eyes of Pallas Athena.

**Gleason, Henry Allan** (1882–1973) An American ecologist, who worked at the New York Botanical Garden and in 1917 challenged the *organismic *climax theory proposed by F. E. *Clements (and later by A. G. *Tansley in Britain), in favour of his own *individualistic hypothesis.

**gleba**  A spore-bearing *tissue within the *fruit bodies of members of the *Gasteromycetes and *Tuberales.

**Gleicheniaceae**  A family of thickset ferns forming long *rhizomes, and regularly forked, branching leaves up to 7 m long, with *gradate, naked, superficial, *rosette-like *sori. They are known from the Coal Measures of the *Carboniferous Period and are now of wide tropical distribution. There are 4 genera and about 140 species.

**gley** The product of waterlogged soil conditions, and hence an anaerobic environment; it encourages the reduction of iron compounds by micro-organisms and often causes mottling of soil into a patchwork of grey and rust colours. The process is known as gleying, or gleyzation (US usage).

**Gleysols** Soils with evidence of gleying (*see* GLEY) within 50 cm of the surface. Gleysols are a reference soil group in the FAO *soil classification.

**Gliricidia** (family *Poaceae, subfamily *Papilionatae) A genus of trees that have *pinnate leaves which smell of coumarin (the smell of freshly cut grass) when dry. *G. sepium* (Mexican lilac) is widely cultivated for shade, as firewood, and for living fences. Its powdered seeds or bark kill mice. There are 4 species, native to tropical America.

**globe thistle** *See* ECHINOPS.

**globose** Spherical.

**Globulariaceae** A small family of *herbs whose leaves are mostly in a basal rosette, and whose tiny flowers are aggregated into disc- or dome-shaped heads. The *ovary is *superior, the *corolla tube is 5-lobed and 2-lipped, and there are 4 *stamens and 2 *stigmas. There are 10 genera, with about 250 species, mainly from Europe, N. Africa, and the Near East. They are cultivated for their flowers.

**globular protein** A *protein in which at least one *polypeptide chain is folded in a 3-dimensional globular configuration. The stability of the structure is maintained by a number of intra-chain bonds. Globular proteins have a variety of functions (e.g. as *enzymes, facilitating transport, and providing storage).

**globulin** One of a group of *globular, simple *proteins, which are insoluble or only sparingly soluble in water, but soluble in dilute salt solutions. They occur in plant seeds, where they have a variety of functions.

**glochidia** *See* AZOLLACEAE.

**Gloeocapsa** A genus of unicellular *cyanobacteria (section I). The cells lie in small clusters embedded in mucilage. They are found in masses on wet rocks (e.g. on rocky coasts). (The genus includes organisms formerly comprising the phycological genus *Chroococcus*.)

**Gloriosa** (family *Liliaceae) A *monotypic genus, *G. superba*, of herbaceous, tendrillate climbers, which have swollen tubers and showy flowers. It is cultivated but contains colchicine and is highly poisonous. It was the national flower of Rhodesia. It occurs in tropical Africa and Asia.

**Glossopteris indica** An extinct plant, the last species referred to its genus and to the family Glossopteridales, which is known from the Trias of India. *Glossopteris* is characterized by a leaf with a fairly well defined midrib and a reticulate venation. The *Permian glacial deposits of S. Africa, Australia, S. America, and Antarctica are succeeded by beds containing a flora very different from that of N. America and Europe. The flora of the south grew in a cold, wet climate, while that of the north existed under warm conditions. Plants with elongate, tongue-shaped leaves dominated the southern flora, of which the genera *Glossopteris* and *Gangamopteris* are among the best known. Of these two, the genus *Glossopteris* gives its name to the flora.

**glucan** (glucosan) A *polysaccharide that is composed of glycosidically linked *glucose units. The types of linkage in a glucan chain may be mixed or may all be the same. Examples include *cellulose, *callose, and *starch.

**glucomannan** *See* HEMICELLULOSE.

**gluconeogenesis** The synthesis of *glucose from non-carbohydrate precursors. Any compound that can be converted into one of the intermediates of *glycolysis is potentially glucogenic. These include *amino acids, *fatty acids, *glycerol, and intermediates of the *citric acid cycle.

**Gluconobacter** (family *Acetobacteraceae) A genus of *bacteria in which the cells are ovoid or rod-shaped. They are *aerobic *chemo-organotrophs which can oxidize ethanol (ethyl alcohol) to acetic acid, and are found in soils and in plant materials, e.g. flowers, fruit, and products such as wine and cider. They can cause spoilage of alcoholic beverages.

**glucosan** See GLUCAN.

**glucose (dextrose, D-glucose)** $C_6H_{12}O_6$, an aldohexose monosaccharide that is a major intermediate compound in cellular metabolism. See also GLYCOGEN.

**glucoside A** *glycoside formed from *glucose.

**glume** In *Poaceae, one of the pair of *bracts that subtends each *spikelet. In reeds and sedges, a bract that subtends the inflorescence.

**Gluta** (family *Anacardiaceae) A genus of trees in which the leaves are *simple. The bark has a highly irritant black sap which in some species yields a lacquer. There are 30 species, 1 of which occurs in Madagascar, the rest in Indo-Malaysia.

**glutamic acid** A dicarboxylic, polar (see POLAR MOLECULE) *amino acid.

**glutamine** An *amino acid, the Υ-*amide of *glutamic acid.

**gluten** The principle *protein in wheat.

**glyceraldehyde** An aldotriose sugar, a phosphorylated derivative of which is an important intermediate in the *glycolate cycle.

**glyceride** An *ester formed from glycerol and between 1 and 3 *fatty-acid molecules, respectively designated mono-, di-, or tri-glyceride. Glycerides serve variously as sources of energy, and tri-glycerides (*lipids) also serve as thermal and mechanical insulators.

**glycerol** A 3-carbon, linear, trihydroxy alcohol. Its fatty *esters are a very important constituent of many *lipids, and some of its phosphorylated derivatives are intermediates in *glycolysis.

**glycine** The simplest *alpha-amino acid, and the only one not to exhibit optical activity.

**Glycine** (family *Fabaceae, subfamily *Papilionatae) A genus of 9 species of herbs that are native to Asia and Australia. G. max is the soya bean, which was probably first domesticated in north-eastern China and is now widely cultivated in warm temperate regions, mainly for its oil, the oilseed cake used for livestock feed, a flour made from the beans, and soy sauce, which is made from the fermented and processed seeds.

**glycogen** A highly branched homopolysaccharide that is composed of D-*glucose units. It is not found in plants, but does occur in some *bacteria, *cyanobacteria, and *fungi.

**glycolate cycle** A complex *metabolic pathway, parts of which occur in the *chloroplasts, *mitochondria, and *peroxisomes of plant cells. Its principal function is thought to be the formation of the *amino acids *serine and *glycine from non-phosphorylated intermediates of the carbon reduction cycle of *photosynthesis.

**glycolysis** (Embden–Meyerhof pathway) The stepwise anaerobic degradation of *glucose to produce as end-products ethanol and carbon dioxide in the cells of fungi and plants (or lactic acid in animal cells). One mole of glucose yields 1 mole each of ethanol and carbon dioxide in fungi and plants (or 2 moles of lactic acid in animals). In both cases the reaction sequence has a net yield of 2 moles of *ATP. However, in most cells, under aerobic conditions, the pathway serves primarily to provide pyruvate, which is oxidized via the *citric acid cycle, and intermediate compounds for biosynthetic processes.

**glycoprotein** A *conjugated protein that consists of a carbohydrate covalently linked to a *protein.

**glycoside** The product that is obtained when a sugar reacts with an alcohol or phenol.

**glyoxaline** See IMIDAZOLE.

**glyoxysome** A *microbody that occurs only in those micro-organisms and plant cells that contain the *enzymes of the glyoxylate cycle, whereby *lipids are converted into sugars. See also PEROXISOME.

**Glyptostrobus** (swamp cypress; family *Taxodiaceae) A *monotypic genus, *G. lineata* (Chinese swamp cypress), of deciduous conifers in which the soft leaves are arranged in 3 ranks and the cones are egg-shaped with shield-like scales. The trunk has a large basal buttress. It occurs in south-eastern China and is sometimes cultivated.

**Gmelina** (family *Verbenaceae) A genus of trees that yield good timber. *G. arborea yemane* is now extensively cultivated. There are 35 species, 2 occurring in Africa and the Mascarenes, the rest from Indo-Malaysia to Melanesia.

**GMP** (guanosine monophosphate) See GUANOSINE PHOSPHATE.

**Gnetales** See GNETOPHYTA.

**Gnetophyta** A remarkable and probably artificial division of *gymnosperms that comprises only the Gnetales. There are 3 constituent genera, embracing trees, shrubs, and *lianes (*Ephedra and *Gnetum), and even turnip-like plants (*Welwitschia*). Possible fossil material of the group is known from the early *Permian, and there is also some fossil pollen from the *Tertiary. With such a scant record it is not surprising that the evolutionary relationships of Gnetales remain unclear.

**Gnetum** The sole genus, with 28 species, of the pantropical *gymnosperm family Gnetaceae, of which 26 species are woody climbers with gouty nodes, and 2 are trees. They are *dioecious, and often *cauliflorous. The seeds are naked, borne on *racemes, and edible in many species, although some have irritant, hairy tissues.

**gnotobiotic** Applied to a culture in which the exact composition of organisms is known, down to the presence or absence of bacteria. Such cultures usually develop during the formation of experimental laboratory ecosystems (*microcosms) from *axenic cultures. The word is derived from the Greek *gnosis*, meaning 'knowledge', and *bio*, meaning '(human) life'.

**goat tang** The common name for the red seaweed *Polyides rotundus* (order *Gigartinales). The plants are *perennial; the *thallus is cylindrical and filamentous, tough and cartilaginous, dull to dark red, *dichtomously branched, and attached to a substrate by a discoid *holdfast. It is found in rock pools in the lower intertidal region and below low-water mark.

**golden algae** See CHRYSOPHYTA.

**golden-brown algae** See CHRYSOPHYTA.

**golden larch** See PSEUDOLARIX.

**golden rod** See SOLIDAGO.

**golden-yellow algae** See CHRYSOPHYTA.

**Golgi body** A system of flattened, smooth-surfaced, membranaceous *cisternae, arranged in parallel 20–30 nm apart and surrounded by numerous *vesicles. A feature of almost all *eukaryotic cells, this structure is involved in the packaging of many products of cell metabolism.

**Gondwana** A former supercontinent of the southern hemisphere from which S. America, Africa, India, Australia, and Antarctica are derived. Their earlier connection explains why related groups of plants and animals are found in more than one of the now widely separated southern land masses; examples include the monkey-puzzle tree (*Araucaria), common to S. America and Australia. Throughout Gondwana there existed floristic assemblages represented by a few species of plants that are thought to have grown in a cold climate. This view is supported by indications of glaciation in S. America, Africa, India, and Australia, which must then have been in a much

South Pole ●

*Gondwana*

higher latitude than they are today. *See* DISJUNCT DISTRIBUTION.

**gonidium** **1.** One of the algal cells in the *thallus of a *lichen. **2.** A *motile, single cell released by filamentous bacteria of the family Leucotrichaceae. **3.** A non-motile spore released by some *cyanobacteria. **4.** In some green algae (*Chlorophyta), a vegetative cell that enlarges and undergoes division. In *Volvox* it may be suspended in the hollow interior of the sphere.

**Gonyaulax** (division **Dinophyta**) A genus of dinoflagellates, species of which are responsible for the phenomenon known as *red tide. Species are capable of emitting light, and are a common cause of *bioluminescence in the sea.

**Gonystylus** (family *Thymelaeaceae) A genus of trees whose bark contains silky, irritant fibres. The leaves are dotted with glands. The flowers have 5 *sepals, no *petals, a *disc, many *stamens, and a *superior *ovary. The fruit is a *capsule, the seeds *arillate. The trees yield a pale, useful timber. *G. bancanus* provides ramin, much used for mouldings. There are about 20 species, found in rain forests of the permanently wet parts of Asia and the Pacific, from Malesia to the Solomons and Fiji.

**Goodenia** (family *Goodeniaceae) A genus of small herbs, or upright or scrambling shrubs, which have alternate or radical leaves. The flowers are held in terminal *spikes or *racemes and are usually yellow, white, or purple. The *calyx is lobed and the *sepals are united at the base but split open on one side only. The 5 *petals are fused at the base, often winged, and arranged in 2 lips with 3 lobes on the lower and 2 on the upper. The *stigma has a cup-shaped cover with stiff hairs around the lip. The 5 *stamens are free. The fruit is a small 2-celled *capsule containing the flat seeds. There are about 170 species, all endemic (*see* ENDEMISM) to Australia, except for 1 species found throughout south-western Asia.

**Goodeniaceae** A family of shrubs and herbs in which the flowers are *zygomorphic. The *corolla tube is split almost to its base, with a lobed limb. The *style has an outgrowth surrounding the *stigma. The *ovary is *inferior. There are 16 genera, with about 430 species, mainly confined to Australia, but a few occur on tropical coasts.

**gooseberry** *See* GROSSULARIACEAE and *RIBES*.

**goosefoot** *See* CHENOPODIACEAE.

**goosegrass (cleavers)** *See* GALIUM.

**goscojales** *See* MEDITERRANEAN SCRUB.

**Gossypium** (family *Malvaceae) A genus of shrubs or small trees whose fruit is a brittle *capsule. The seeds are hairy, providing cotton from several species (e.g. in America *G. barbadense* and *G. hirsutum*; and from Africa to India *G. arboreum* and *G. herbaceum*). There are 39 species, found in the tropics and subtropics.

**gourd** *See* CUCURBITACEAE.

**GPP** (gross primary productivity) *See* PRIMARY PRODUCTIVITY.

**Gracilariaceae** (order *Gigartinales) A family of red seaweeds in which the *thallus is cylindrical, strap-like, or broadly lobed. *Gracilaria* is the only British genus; the thallus is dark purplish-red,

g

cylindrical, and irregularly branched, with a discoid *holdfast. It is found attached to rocks at and below low-water mark.

**gradate sorus** A type of *sorus (or group of *sporangia) in *ferns (e.g. in *Hymenophyllaceae), in which there is a sequence of development of ripe sporangia from apex to base of the sorus. This contrasts with the simple sorus (e.g. in *Osmundaceae) in which there is simultaneous ripening of all the sporangia, which is regarded as primitive; or with the mixed sorus (e.g. in *Polypodiaceae) in which there is irregular ripening of later sporangia among earlier ones, which is regarded as advanced.

**gradient analysis** An ordination technique for the description of vegetation based on characteristics of the site rather than the composition of species. One or more environmental gradients are identified and *stands are arranged in order along them according to the characteristics of their sites. The vegetation is then examined to discover any related pattern.

**gradualism** *See* PHYLETIC GRADUALISM.

**graft** **1. (noun)** A small piece of *tissue implanted into an intact organism. **2. (verb)** To transfer a part of an organism from its normal position to another position on the same organism (autograft), or to a different organism of the same species (homograft), or an organism of a different species (heterograft). The source of the part that is grafted is called the scion and the organism to which it is united is called the stock. A graft *hybrid is an organism made up of two genetically distinct tissues resulting from the fusion of scion and stock after grafting. In cultivation, a stem from the plant (the scion) is fused with a rooted portion of another (the stock) to form a single plant. Most fruit trees are produced by grafting, the type of fruit being determined by the scion, but the size of the tree by the stock.

**Gramineae** *See* POACEAE.

**Grammatophyllum** (family *Orchidaceae) A genus of leafy *herbs, many

huge, including *G. speciosum*, the biggest orchid plant in the world, which has stout stems up to 2 m long and big, paniculate *inflorescences. There are 10 species, occurring from Malaysia to the Pacific islands.

**Gram-negative bacteria** *See* GRAM REACTION.

**Gram-positive bacteria** *See* GRAM REACTION.

**Gram reaction** A reaction obtained when *bacteria are subjected, in the laboratory, to a certain staining procedure called the Gram stain or *Gram's stain (after the Danish scientist Christian Gram (1853–1938) who first devised the technique in 1884). The bacteria are killed and stained, for example with crystal violet; the stained cells are then treated with an organic solvent such as acetone or ethanol. Bacteria fall into two categories: those that are readily decolorized under these conditions, and those that retain the stain. Cells of the former type are said to be Gram-negative; those of the latter type are said to be Gram-positive. The difference in reaction reflects a fundamental difference in the structure of the cell wall in the two types of bacteria.

**Gram's stain** The staining procedure used to determine the *Gram reaction of *bacteria.

**granadilla** *See* PASSIFLORACEAE.

**granite moss** The common name for a moss of the subclass *Andreaeidae, usually found on hard siliceous rocks in mountainous regions.

**granulose** Consisting of or covered with small granules or grains.

**granum** Part of the internal structure of a *chloroplast, consisting of 5–30 membranaceous discs (thylakoids), 0.25–0.8 μm in diameter. There may be 40–80 grana in a typical chloroplast.

**Granville wilt (southern bacterial wilt)** A disease of tobacco plants, caused by the *bacterium *Pseudomonas solanacearum*. Symptoms include wilting and yellowing of the leaves.

**grapefruit**  *See* CITRUS.

**Graphidales** (subdivision *Ascomy-cotina*) An order of *fungi; members include *lichens in which the *thallus is *crustose and the *ascocarps are apothecioid (*see* APOTHECIUM). There are many genera, found chiefly on bark in warmer regions.

***Graphis*** (order *Graphidales) A genus of *crustose *lichens in which the *apothecia are typically black, elongated, and slit-like. The *spores are colourless, elongated, and *multiseptate. There are many species, of which only 2 occur in Britain. They are found on trees.

**grass**  *See* POACEAE.

**grass heath**  *See* ACIDIC GRASSLAND.

**grassland** Ground covered by vegetation that is dominated by grasses. Grassland constitutes a major world vegetation type and occurs where there is sufficient moisture for grass growth, but where the environmental conditions, both climatic and anthropogenic, prevent tree growth. Its occurrence therefore correlates with a rainfall intensity between that of desert and that of forest, and the range of grassland is extended by grazing and/or fire to form a *plagioclimax in many areas that were previously forested. The extensive mid-latitude grassland is known as *steppe or *prairie, whereas the corresponding tropical vegetation is called *savannah.

**grass minimum temperature** The minimum temperature recorded in open ground at night by a thermometer whose bulb is exposed over the tips of short grass.

**grass tree**  *See* XANTHORRHOEA and XANTHORRHOEACEAE.

**graveolent** Strongly scented, often unpleasantly so.

**gravitational water** Water that moves through soil under the influence of gravity and must be removed from the soil before this can attain *field capacity.

**Gray, Asa (1810–88)** An American botanist and taxonomist who did much to popularize the study of botany and to expound, but also criticize, *Darwin's evolutionary theory. In 1842, he was appointed Fisher Professor of Natural History at Harvard University and founded the Gray Herbarium and a library. He was an original member of the National Academy of Sciences and in 1872 was elected president of the American Association for the Advancement of Science. In 1859 he published a memoir on the relationships between the floras of Japan and North America, one of the earliest studies of discontinuous plant distribution. His *Manual of the Botany of the Northern United States*, the first edition of which was published in 1848, was probably his most enduring work, although it was his *Statistics of the Flora of the Northern United States* (1856–7) that established his academic reputation. He collaborated with John *Torrey to produce the 2-volume *Flora of North America* (1838–43) and published his *Synoptical Flora* in 1878.

**grazing food-chain**  *See* GRAZING PATHWAY.

**grazing pathway** (grazing food-chain) A food-chain in which the *primary producers (green plants) are eaten by grazing herbivores, with subsequent energy transfer to various levels of carnivore. *Compare* DETRITAL PATHWAY.

**green algae** The common name for *algae of the division *Chlorophyta. Not all members are green; e.g., members of the genera *Trentepohlia* and *Cephaleuros* may be orange.

**greengage**  *See* PRUNUS.

**greenheart**  *See* OCOTEA.

**'greenhouse effect'** The effect of heat retention in the lower atmosphere as a result of absorption and reradiation of long-wave (more than 4 μm) terrestrial radiation by clouds and gases (e.g. water vapour and carbon dioxide). The insulating effect is not strictly analogous to that of greenhouse glass, since the higher temperature in a greenhouse is due partly to the reduction in air movement. The atmospheric effect is to alter the balance of incoming and outgoing radiation in the Earth's energy budget.

Marked increases in atmospheric carbon dioxide, generated for example by the combustion of fossil fuels, could result in a global increase of atmospheric temperatures if not offset by other (perhaps natural) changes.

**green oak** Oak wood that is stained bluish-green as a result of infection with the *fungus *Chlorociboria aeruginascens*. The wood is highly prized for inlay work in furniture-making, etc.

**green pepper** *See* CAPSICUM.

**green sulphur bacteria** **1.** *Bacteria of the Chlorobiaceae. **2.** Bacteria of either the Chlorobiaceae or the *Chloroflexaceae.

**Grenz horizon** *See* SUB-ATLANTIC.

***Grevillea** (family *Proteaceae) A genus of trees and shrubs, including *G. robusta* (silky oak), widely cultivated in the tropics as a shade tree and for its useful timber. There are 200 species, occurring in eastern Malesia, the western Pacific islands, and Australia.

**grex** A slug-like *pseudoplasmodium formed by *cellular slime moulds of the order *Dictyosteliales.

**grey-brown podzolic** A soil-*profile term describing eluviated (*see* ELUVIATION), freely draining soils that have a distinctive *clay-enriched B *horizon. The soil develops under temperate woodlands which have a moderate rainfall, and forms good pastureland after deforestation. *See also* ALFISOLS.

**grey mould** A common and widespread disease of plants caused by the *fungus *Botrytis cinerea*. A fluffy grey mould appears on flowers, leaves, or fruit. Infection is encouraged by poor ventilation, which produces conditions of still air and high humidity.

**grid analysis of pattern** The detection of pattern using a contiguous grid of *quadrats, rather than random sampling with a given quadrat size. By blocking adjacent quadrats in pairs, fours, eights, etc., the data may be analysed using increasingly larger quadrat sizes. This is

important since the detection of pattern relates to the quadrat size used, with the most marked demonstration of contagion occurring when the quadrat size is approximately equal to the clump area (i.e. the size of the clumps). When the presence of contagion is not immediately obvious, detection of the scale of non-randomness (i.e. the quadrat size at which it is evident) provides very useful information as it may suggest the likely cause of the clumping.

***Grifola** (order *Polyporaceae) A genus of *fungi in which the *fruit body consists of a branched stem, each branch ending in a flattened, fan-like *pileus. In *G. frondosa* the tuft of pilei may reach 40 cm across; each pileus measures 4–10 cm across, is leathery in texture, and typically has a greyish and wrinkled upper surface. The pores on the under-side are whitish and extend down the stem. This species has a distinctive odour of mice (acetamide). It is *parasitic on trees, especially oak and beech, and grows at the base of the trunk.

***Grimmia** (order *Grimmiales) A genus of mosses in which the stems are short and *dichotomously branched, usually forming dense cushions. The leaves typically end in a fine, silvery hair-point, giving the cushions a hoary appearance. It is a cosmopolitan genus of about 170 species; almost entirely *saxicolous. *G. pulvinata* forms hemispherical cushions on walls and roofs throughout lowland Britain. The *capsules are borne on *setae, which are initially strongly curved, but later straighten.

**Grimmiales** (subclass *Bryidae) An order of (mainly) *acrocarpous mosses of medium size; mostly tufted, or cushion-forming. The leaves often terminate in long, fine, hair-points. They are mostly *saxicolous. Genera include *Grimmia, *Racomitrium*, and *Schistidium. S. maritimum* and *S. apocarpum* were formerly known as *Grimmia maritima* and *Grimmia apocarpa*, respectively. *S. maritimum* is found only on rocks by the sea; *S. apocarpum* is common on rocks and walls, often in association with *G. pulvinata* and *Tortula muralis*.

**Grisebach, August Heinrich Rudolph** (1814–79) A German taxonomist whose identification of major vegetational units introduced the concept of the *floral province. He expounded this idea in *Vegetation der Erde* (1872).

**grisette** The common name for the edible *fruit body of *Amanita vaginata* (common grisette) or *A. fulva* (tawny grisette). Both species are common on the ground beneath broad-leaved trees in late summer and autumn.

**gross primary productivity** (GPP) *See* PRIMARY PRODUCTIVITY.

**Grossulariaceae** A family of shrubs that have alternate, *simple, often *palmately lobed leaves and regular, *tetra- or *pentamerous flowers in *racemes, with the same number of *stamens as *petals. The *ovary is *inferior, consisting of 2 fused *carpels, forming a *berry in the fruit. There are 3 genera, including *Ribes*, and 325 species, with a cosmopolitan distribution.

**ground cover** The area of ground that is covered by a plant when its *canopy edge is projected downwards perpendicularly.

**ground flora** A general term describing plants of the *field layer and *ground layer.

**ground layer** The lowest layer of a plant *community, comprising especially *mosses, *lichens, and *fungi, together with low-growing *herb species which often have trailing stems or *rosette forms.

**groundnut** *See* ARACHIS.

**groundsel (Senecio)** *See* ASTERACEAE.

**ground water** Water that occurs below the Earth's surface, contained in *pore spaces within *regolith and bedrock. It is either passing through or standing in the soil and underlying strata, and is free to move under the influence of gravity. *See also* VADOSE.

**group translocation** A mechanism that is widely utilized for the transportation of sugars across bacterial membranes and perhaps those of some higher cells. A donor compound is used to activate sugar molecules through the provision of a high-energy phosphate group. The activated sugar can thus more readily traverse the membrane. A number of membrane *proteins have been implicated in the transfer process.

**grove** **1.** A small wood, usually of less than 8 hectares. **2.** An area that is composed entirely of timber trees (without *underwood) within a larger woodland which contains underwood.

**growing season** In seasonal climates, the period of rapid growth. Various definitions are used. In Britain, the growing season is the period when the mean temperature exceeds 6°C, though there is no agreement whether daily, weekly, or monthly means should be used, or whether these should be of ground or air temperatures. In the USA, the growing season is defined as the period between the last *killing frost of spring and the first killing frost of autumn, a killing frost usually being regarded as one recorded on a thermometer exposed in a standard way in a Stevenson's screen.

**growth** The increase in size of a cell, organ, or organism. This may occur by cell enlargement or by cell division.

**growth form** **1.** The *morphology of a plant, especially as it reflects physiological *adaptation to the environment. *See also* LIFE-FORM. **2.** The shape of population growth, as expressed by a growth curve (e.g. *J-shaped or *S-shaped).

**growth regulator** A synthetic compound that, when applied to a plant, promotes, inhibits, or otherwise modifies the growth of that plant.

**growth retardant** One of a group of synthetic compounds that reduce stem elongation in plants by inhibiting the activity of the subapical *meristem. An important agricultural use is in the prevention of lodging in cereals. In horticulture retardants are used to produce more compact plants. Some of these compounds inhibit *gibberellin synthesis.

**growth ring** *See* TREE-RING.

**growth substance** A naturally occurring compound, other than a nutrient, that promotes, inhibits, or otherwise modifies the growth of a plant.

**GTP (guanosine triphosphate)** *See* GUANOSINE PHOSPHATE.

***Guaiacum* (family *Zygophyllaceae)** A genus of trees some of which yield an extremely hard, dense, oily timber. Lignum vitae comes from *G. officinale* (and others). This species also yields a medicinal resin, guaiacum. There are 6 species, occurring in the W. Indies and tropical America.

**guanido group** The basic-group configuration $C:(NH_2)_2$ found in arginine.

**guanine** A purine base that occurs in both *DNA and *RNA.

**guanosine** The *nucléoside that is formed when *guanine is linked to *ribose sugar.

**guanosine phosphate** A *nucleotide of the purine base *guanine. Guanosine phosphates are designated guanosine mono-, di-, and triphosphates (GMP, GDP, and GTP respectively). Cyclic GMP is believed to be antagonistic in cells to *c-AMP, while GTP is involved as a high-energy compound in *peptide-bond synthesis.

**guard cell** A specialized type of plant epidermal cell (*see* EPIDERMIS), 2 of which surround each *stoma. Changes in their *turgidity cause stomatal opening and closing. This procees is not fully understood but includes: (*a*) the effects of the overall water potential of the plant on the stomata; (*b*) a feedback mechanism, whereby when water is plentiful abscisic acid is formed and stomata open and when water is scarce abscisic acid is not released and the stomata close; (*c*) low carbon dioxide levels promote stomatal opening; (*d*) except in *CAM plants, stomata normally open in light and close in darkness; (*e*) generally, an increase in temperature tends to increase stomatal opening.

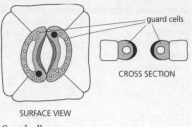

CROSS SECTION

SURFACE VIEW

*Guard cell*

***Guarea* (family *Meliaceae)** A genus of trees or treelets in which the leaves are *pinnate and the fruit is a *capsule. W. African species yield the timber guarea. There are 40 species, occurring in the American and African tropics.

**guava** *See* PSIDIUM.

**guerrilla growth form** The distribution that results when the *rhizomes or *stolons by which a plant spreads by *clonal dispersal are long and often short-lived. The clonal shoots are widely spaced and *fugitive, constantly appearing and disappearing in different territories, like a guerrilla army. *Compare* PHALANX GROWTH FORM.

**Guinea zone** The wettest and most wooded *savanna zone in the classic threefold model of savannah vegetation, which is based on examples taken from western Africa. These are the Sahel zone, consisting of dry scrub with occasional grasses and patches of bare earth; typical grassy savannah; and the wetter, semi-wooded Guinea zone.

**gullet** In unicellular organisms, the opening through which food enters.

**gum benjamin** *See* STYRAX.

***Gunnera* (family *Haloragidaceae)** A genus of plants, some of which are huge *herbs cultivated for their giant, palmately nerved leaves. The flowers are tiny, borne in big *panicles. The *petioles have hollows, inhabited by *cyanobacteria. There are about 40 species, occurring mainly in S. America, but also scattered in the southern temperate zone and the tropics generally. *G. manicata* is giant or prickly rhubarb, often grown in damp patches in gardens large enough to give it room.

**gurjun**   *See* DIPTEROCARPUS.

**gutta percha**   A latex, mainly of *Sapotaceae but also of *Eucommiaceae and some *Celastraceae. *Palaquium gutta*, of Sumatra, Java, Malaya, and Borneo, is the main source. The latex is obtained by tapping the bark or by crushing the leaves in hot water. It is a better thermal and electrical insulator than rubber, almost non-elastic, and plastic above 82°C. It is now used only for temporary dental stoppings, but formerly it was used for submarine telephone cables and golf balls.

**guttation**   The extrusion of water and sometimes salts from the aerial parts of plants, particularly at night when *transpiration rates are low. Guttation may occur in plants that are less than about 10 m tall, where hydrostatic pressure is insufficient to prevent the flow of water into the *xylem when the rate of transpiration is low. It also occurs in tropical plants, where high humidity inhibits transpiration. The process takes place through *hydathodes.

**Guttiferae**   *See* CLUSIACEAE.

**Gymnoascales (subdivision *Ascomycotina)**   An order of *fungi in which the *ascocarps are cleistothecioid (*see* CLEISTOTHECIUM), although sometimes without a definite wall or *peridium. The family includes the *perfect (sexual) stages of a number of important skin *pathogens (dermatophytes).

**Gymnocarpium (family *Aspleniaceae)**   A genus of ferns that have no *indusium to the rounded, superficial *sori; creeping *rhizomes; and fronds that often appear 3-lobed because the lowest pair of *pinnae are each nearly as large as the rest of the *frond. There are 6 species in the northern temperate zone (e.g. oak fern), which are sometimes cultivated.

**gymnocarpy**   A type of development of a fungal *fruit body in which the *spore-bearing tissue is naked and exposed during the whole of its development.

**gymnosperm**   A seed plant in which the *ovules are carried naked on the cone scales, in contrast to the *angiosperms, in which they are enclosed by an *ovary. Gymnosperms date from the *Carboniferous and subsequently they dominated the floras of the world until the *Cretaceous, since when they have been progressively displaced by the angiosperms (flowering plants).

**Gymnospermae**   Formerly, a subdivision of the seed plants (*Spermatophyta), but now regarded as a group of plants that are not closely related, and '*gymnosperm' is used only informally to describe what are now classified as 4 separate divisions: *Ginkgophyta, *Cycadophyta, *Coniferophyta (Pinophyta), and *Gnetophyta.

**gynaecium**   *See* GYNOECIUM.

**gynochory**   Dispersal of spores or seeds by *motile females.

**gynodioecious**   Applied to a *dioecious plant species in which female and hermaphrodite flowers occur on different plants. *Compare* ANDRODIOECIOUS, ANDROMONOECIOUS, and GYNOMONOECIOUS.

**gynoecium (gynaecium)**   The collective term for the female reproductive organs of a flower, comprising 1 or more *carpels. *Compare* ANDROECIUM.

**gynomonoecious**   Applied to a *monoecious plant species in which female and hermaphrodite flowers occur separately on the same plant. *Compare* ANDRODIOECIOUS, ANDROMONOECIOUS, and GYNODIOECIOUS.

**gynostemium**   The column formed when the *androecium and *gynoecium combine together (e.g. in *Stylidium).

**Gypsisols**   Soils that have a concentrated gypsum within 100 cm of the surface, or more than 15% gypsum (calcium sulphate) below 100 cm. Gypsisols are a reference soil group in the FAO *soil classification.

**gypsophilous**   Favouring growth in limestone soils.

**Gyromitra (order *Pezizales)**   A genus of *fungi. *G. esculenta* (false morel) resembles the edible *morel but is highly poisonous; it grows on the ground in association with conifers.

**gyrose** Sinuous, curved; marked with curved lines or grooves.

**gyttja (nekron mud)** A rapidly accumulating, organic, muddy deposit, characteristic of *eutrophic lakes. The precise nature of gyttja varies with the producer organisms involved, which include small algae or macrophytes.

**habitat** The living place of an organism or *community, characterized by its physical or *biotic properties.

**Hadean** The interval of geologic time that lasted from 4567.17 Ma until 3800 Ma and preceded the *Archaean eon. Divisions of geologic time are required to have firmly dated events at their bases. This is not possible with the Hadean and consequently its name is used only informally.

**Haeckel, Ernst Heinrich (1834–1919)** A German anatomist, zoologist, and field naturalist, who was appointed professor of zoology at the Zoological Institute, Jena, in 1865. He was an enthusiastic supporter of the Darwinian theory of evolution, and *Darwin credited him with the success that the theory enjoyed in Germany. In 1866 Haeckel coined the term 'ecology'.

***Haematoxylum*** (family *Fabaceae, subfamily *Caesalpinoideae) A genus of trees that have thorns in their leaf *axils. The *heartwood is coloured. *H. campechianum* (logwood) yields haemotoxylin, used for staining and dyeing. There are 2 species, occurring in Central and S. America, and 1 in south-western Africa.

**Haemodoraceae** A family of tropical and warm temperate herbs that have fibrous roots, *tubers, *rhizomes or *stolons. Leaves are linear with sheathing bases. Bisexual flowers are borne in *racemes, *cymes, or *panicles, and are often covered with woolly hair. The *perianth is persistent. The *ovary is *superior or *inferior, and *trilocular. The fruit is a *capsule; there is a small *embryo and a good-sized *endosperm in the seeds. Several species of *Angiosanthos* are cultivated. There are 16 genera, with 85 species, found in Australia, S. Africa, tropical and N. America.

***Haemophilus*** (family *Pasteurellaceae) A genus of *bacteria in which the cells are ovoid to rod-shaped or occasionally filamentous, *Gram-negative, and non-*motile. There are many species.

They are strictly *parasitic in animals including humans, and sometimes cause disease.

**hag 1.** A parcel of wood that has been marked off for cutting. *See also* CANT and COUPE. **2.** An exposed face of peat that has been cut or eroded.

***Hakea*** (family *Proteaceae) A genus of shrubs and trees that have alternate, rigid leaves, which are often *glabrous and aromatic, or silky. The flowers are held in crowded *axillary clusters or *racemes and are bisexual, paired, and on stalks. The bud scales are papery, folded, and conspicuous, but fall early. The *perianth segments are irregular and in the bud are fused into a tube, but later the lobes are free and spreading. The *stamens are found at the base of some of the perianth blades. The *ovary is *unilocular with 2 *ovules, and the *style protrudes from the perianth. The fruit is a woody *capsule, and the seeds are slightly convex with a broad terminal wing. There are about 125 species, endemic (*see* ENDEMISM) to Australia.

**half-inferior** Applied to the *ovary of an *inflorescence when the lower part is embedded in the *pedicel and the upper part is exposed (e.g. in *Tetragonia). *See also* INFERIOR and SUPERIOR.

**halo-** A prefix meaning 'pertaining to salt', derived from the Greek *hals, halos*, meaning 'salt'.

**Halobacteriaceae** A family of *archaebacteria which can grow only in the presence of high concentrations of salt (sodium chloride). Cells are spherical, rod-shaped, or discoid. They are predominantly *chemo-organotrophic. There are 2 genera, found in most saline environments: the sea, salt lakes, salted foods, etc.

***Halobacterium*** (family *Halobacteriaceae) A genus of *bacteria that require the presence of salt for survival and growth; in the absence of salt the cells may disintegrate. Some strains can carry out a unique type of energy-yielding process, obtaining energy directly from sunlight in a non-photosynthetic process which involves a specialized region of the cell membrane called the 'purple membrane'.

**halo blight** A bacterial disease of dwarf and runner beans (*Phaseolus* species) caused by *Pseudomonas phaseolicola*. Symptoms include the appearance of brown spots on the leaves, each spot being surrounded by a yellow ring or halo, and infected plants may be stunted.

**halocline** A zone in which there are rapid, vertical changes in the salinity. In low latitudes the halocline usually represents a decrease in salinity with increasing depth; in high latitudes it may represent the opposite.

**halophile** An *extremophile (*Archaea) that thrives in extremely saline environments.

**halophyte** A terrestrial plant that is adapted morphologically and/or physiologically to grow in salt-rich soils and salt-laden air (e.g. *Salicornia* species, glassworts). *See also* SALT MARSH.

**halopilic** Thriving in, or preferring to grow in, the presence of salt.

**Haloragidaceae** A family of aquatic or marsh herbs that have either *whorled, much-divided leaves (e.g. *Myriophyllum* and *Haloragis*), or enormous, radical, rhubarb-like leaves (e.g. *Gunnera*). The flowers are inconspicuous and solitary, in *axils of *spikes or *dichasia, with no or 2 or 4 tiny *sepals and *petals and short *stamens. The *ovary is *inferior and wind-pollinated. There are about 120 species, in 10 genera. Distribution is cosmopolitan, but mainly in the southern hemisphere.

**Haloragis (raspwort; family *Haloragidaceae)** A genus of rough, *perennial *herbs that have *opposite or alternate leaves which are entire, toothed, or lobed. The flowers are unisexual or bisexual, small, and held in terminal *panicles or *racemes. The 2–4 *sepals are united into a tube with erect but short lobes, and are usually persistent, remaining over the *ovary. The *calyx tube is usually ribbed. The 2–4 *petals are keeled and incurving,

often hairy on the outer side, and often absent in female flowers. There are twice as many *stamens as petals, and each has a short *filament and a long *anther. There are 2–4 *carpels, with the same number of *styles. The fruit is a small *nut with single-seeded *locules, which are sometimes abortive. There are 26 species, found mostly in Australia, New Caledonia, New Zealand, Easter Island, and Juan Fernandez.

**halosere** A characteristic sequence of *communities associated with the developmental stages in plant *succession on *salt marshes or salt desert. *See also* SUCCESSION.

**halo spot** A disease of barley and other cereals, caused by the *fungus *Septoria oxyspora* (*Selenophoma donacis*). Lesions on upper leaves have pale centres with dark margins. The disease is important in only a few regions of Britain, including the wetter western regions of England, Wales, and Ireland.

**Hamamelidaceae** A family of trees and shrubs that have *simple or *palmate, alternate leaves with *stipules. The flowers are either bisexual or unisexual, and the plants either *monoecious or *dioecious. The flowers are usually held in a *spike. The *calyx is 4 or 5 united *sepals and the same number of *petals. There are up to 14 *stamens. The position of the *ovary is variable from one species to another; it always has 2 *locules and 2 *styles. The fruit is a woody exocarp or a brittle endocarp (*see* PERICARP). There are 28 genera, with about 90 species, and representatives of the family are found throughout subtropical and temperate zones of the world, concentrated in eastern Asia. Several genera give useful timbers and aromatic gums (e.g. *Liquidambar*). *Hamamelis* is the genus of witch hazel of which *H. virginiana* gives the soothing lotion used in medicine. Other species are ornamentals.

**Hamamelis virginiana (witch hazel)** *See* HAMAMELIDACEAE.

**hamate** Hooked at the tip.

**Hanguana (family *Flagellariaceae)** A genus of 1 or a few species of giant herbs that occur in the swamps and forests of the tropical Far East. They have *petiolate leaves in a rosette or on a short stem, and their flowers are unisexual.

**haplobiont** A plant that lacks 1 of the 2 generations in its life cycle and exists only as the *gametophyte or *sporophyte generation.

**haplocheilic** In some *gymnosperms, applied to a type of *stoma in which the 2 *guard cells are derived from a single mother cell and the subsidiary cells are derived from a different initial. *Compare* SYNDETOCHEILIC, *see also* PERIGENOUS.

**haploid** Applied to a cell nucleus that contains one of each type of *chromosome, i.e. one set of chromosomes, designated *n* *Gametes are haploid, in contrast with most *somatic cells, which usually have some multiple of this number, usually 2*n* (diploid), but sometimes 3*n* (triploid), 4*n* (tetraploid), or many-*n* (polyploid). A haploid cell thus has only one chromosome set, and a haploid organism contains only haploid cells.

**haplontic** Applied to a *life cycle in which all the cells involved are *haploid, except for the *zygote.

**haplostele (ectophloic protostele)** A *monostele type of *protostele in which in cross-section the *xylem occurs as a central strand surrounded by the *phloem. *Compare* ACTINOSTELE, HYPOPHLOIC HAPLOSTELE, and SOLENOSTELE.

**hapteron** In *phycology, a term sometimes used as a synonym of *holdfast; it may also be used to refer only to root-like types of holdfast.

**haptonema** *See* PRYMNESIOPHYCEAE.

**Haptophyceae** *See* PRYMNESIOPHYCEAE.

**haptotropism** *See* THIGMOTROPISM.

**hardening** *See* ACCLIMATIZATION.

**hardpan** A hardened soil *horizon, usually found in the middle or lower parts of the *profile, that may be indurated (*see* INDURATION) or cemented by a variety of possible cementing materials. *See also* CALICHE; DURICRUST; and PAN.

**hardwood** The wood of *angiosperm trees. The designation is arbitrary and some hardwoods are softer than most *softwoods; e.g., balsa wood (from *Ochroma* species) is a hardwood.

**Hardy–Weinberg law** The law that states that in an infinitely large, interbreeding population in which mating is random and in which there is no selection, migration, or *mutation, gene and *genotype frequencies will remain constant from generation to generation. In practice these conditions are rarely strictly present, but unless any departure is a marked one, there is no statistically significant movement away from equilibrium. Consider a single pair of *alleles, A and a, present in a *diploid population with frequencies of $p$ and $q$ respectively. Three genotypes are possible, AA, Aa, and aa, and these will be present with frequencies of $p^2$, $2pq$, and $q^2$ respectively.

**haricot bean** *See* PHASEOLUS.

**harlequin chromosome** *See* SISTER CHROMATID EXCHANGE.

**hartig net** *See* ECTOTROPHIC MYCORRHIZA.

**harvest method** A productivity measuring technique, most commonly used for estimates of *primary productivity, especially in situations in which predation is low, e.g. among *annual crops, on certain *heathlands, in colonizing *grasslands, and sometimes in pond *ecosystems. Sample areas are harvested at intervals throughout the growing season, and the material is dried to estimate dry weight or *calorific value. The method may also be used for woodlands, although usually only one final felling and dry-weight estimation is feasible. In such situations it is generally more reliable and ecologically more desirable to use indirect, non-destructive estimates, e.g. by monitoring carbon dioxide profiles. The harvest method is usually used only for above ground *biomass and therefore neglects the large and important development of root biomass below ground level. *Compare* AERODYNAMIC METHOD.

**hashish** *See* CANNABIS.

**haulm** A stem, e.g. those of bamboos (*Bambusaceae) and grasses (*Poaceae).

**haustorium** **1.** In certain parasitic *fungi, an outgrowth from a hypha that penetrates a host cell in order to absorb nutrients from it. **2.** In some parasitic flowering plants (*angiosperms), outgrowths of the roots.

**Hawaiian floral realm** Part of R. Good's (*The Geography of the Flowering Plants*, 1974) *Polynesian subkingdom, which falls within his *palaeotropical kingdom. This is the most isolated floral region, which accounts for its great distinctiveness. About 20 per cent of the genera and over 90 per cent of the species are endemic (*see* ENDEMISM). No valuable economic or horticultural plants belong to this region. *See also* FLORAL PROVINCE and FLORISTIC REGION.

**hay** **1.** An enclosure or hedge. **2.** Part of a forest that has been fenced off for hunting. **3.** Grass that has been conserved by drying and that is to be used later as feed or bedding for livestock.

**hazel** *See* CORYLUS.

**heartwood** The dead, woody centre of the trunk of a tree. The cells become impregnated with various organic compounds which cause a change in colour, so that this tissue is distinguished easily from the remainder of the wood.

**heath forest** *See* KERANGA.

**heathland** Most typically, a lowland *community dominated by dwarf shrubs belonging to the family *Ericaceae. Sometimes, especially in Britain, 'heathland' is used more loosely to denote any community developing on acidic (*See* ACID SOILS), *podzolized soils, e.g. grass heaths and lichen heaths. The term was first used in a narrow sense to describe shrubby vegetation dominated by the genus *Calluna* (heath or ling). *Compare* MOOR.

**heaths** *See* ERICA.

**heavy metal tolerance** A biochemical and physiological adaptation to heavy metals (i.e. metals, e.g. copper and zinc, that have a density greater than $5 \, g/cm^3$) shown by plant species or *genotypes: such plants may therefore be found growing successfully on soils contaminated by metals, where other species or genotypes would fail. Many common grasses (e.g. *Agrostis tenuis* and *Festuca ovina*) have developed strains tolerant of heavy metals naturally on soils derived from geological strata rich in heavy metals. Seeds from these genotypes are used commercially in revegetating soil contaminated by metals.

**Hebe** (family *Plantaginaceae) A genus of shrubs, very closely related to *Veronica but, unlike that genus, truly shrubby and with the *capsule dorsally (not laterally) compressed. They are much cultivated for the flowers and foliage. There are about 75 species in the extreme south of S. America, New Zealand, and Australia.

**hebetate** Having a soft or blunt point.

**Hedera** (ivy; family *Araliaceae) A genus of woody root climbers whose leaves are often dimorphic (*see* DIMORPHISM). There are 4–5 species, found in the Canary Islands, Europe, and western Asia.

**hedgehog fungus** *See* HYDNUM.

**hekistotherm** A cold-tolerant plant of polar regions, according to A. L. P. de *Candolle's (1874) classic temperature-based scheme of world vegetation zones.

**Helianthemum** *See* CISTACEAE.

**Helianthus annuus** (sunflower) *See* ASTERACEAE.

**Helianthus tuberosus** (Jerusalem artichoke) *See* ASTERACEAE.

**Helichrysum** (everlasting daisies; family *Asteraceae) A genus of *herbs and shrubs that have entire, usually alternate, leaves. The flower heads may be large (up to 5 cm in diameter), and solitary, or in *panicles or clusters. The papery, dry, shiny *involucre *bracts extend beyond the flowers and have the appearance of *petals; they are yellow, white, brown, or pink. The shrubby species often have smaller flowers held in a terminal *corymb. The disc-florets are mostly bisexual, tubular, and yellow. The *anthers have short tails, and the *style has cylindrical lobes. The fruit is a *nut or *achene with a single seed. Many species are cultivated, as the flowers can be dried and keep their colour for a long period. There are about 500 species, found in south-eastern Europe, Africa, and Australia. The genus is centred in S. Africa, where more than 200 species are present.

**Helicia** (family *Proteaceae) A genus of 87 species of trees and shrubs that have *simple leaves. They are too small to yield useful timber. They occur from southern China and Indo-Malaysia to Melanesia.

**heliciform** Coiled, like a snail shell.

**Heliciopsis** (family *Proteaceae) A genus of trees in which the leaves are dimorphic (*see* DIMORPHISM), and either *simple or deeply lobed. There are 7 species, occurring in Burma, southern China, and western Malesia.

**Heliconia** (family Heliconiaceae) A genus of giant, *soboliferous herbs that have banana-like leaves, and flowers borne in *racemes, subtended by showy *bracts. The fruit is a *capsule. There are about 100 species, occurring in tropical America and the south-western Pacific.

**heliophyte** A plant that is characteristic of, and showing adaptation to, bright, sunlit *habitats, as opposed to shade-tolerant or shade-preferring species (i.e. *sciophytes).

**heliosis** *See* SOLARIZATION.

**heliotropic** (phototropic) Applied to a plant or part of a plant that shows a directional growth movement in response to light.

**Helipterum** (everlasting daisies; family *Asteraceae) A genus of mostly *annual herbs which have alternate leaves. The flower heads are small or large, and solitary or in *corymbs. The petal-like *involucre *bracts are soft and papery, and are white, yellow, brown, or pink; sometimes they are sensitive to moisture and close in damp weather. The disc-florets may be black. The *pappus of the *achene is branched and feathery. Several species are cultivated commercially for cut flowers. There are about 60 species, found in S. Africa and Australia.

**Helleborus** (family *Ranunculaceae) A genus of rhizomatous (*see* RHIZOME) herbs that contain poisonous *alkaloids. Many species are cultivated for winter and spring flowers. There are 20 species, occurring in Europe and the Mediterranean region.

**Helminthosporium maydis** *See* GENETIC EROSION.

**Helminthostachys** (family *Ophioglossaceae) A genus of ferns of 1 species, with a creeping *rhizome, the sterile leaf-blade ternately compound, and a dense, fertile spike. It is confined to the Indo-Malaysian region and Australasia, and is probably of very ancient origin.

**helophyte** A plant, typical of marshy or lake-edge environments, in which the *perennating organ lies in soil or mud below the water level, but the aerial shoots protrude above the water (e.g. *Phragmites communis*, the common reed). *Compare* HYDROPHYTE.

**Helotiales** (subdivision *Ascomycotina) An order of *fungi, in which the *ascocarps are apothecioid (*see* APOTHECIUM) and the *asci are more or less club-shaped; the *ascospores are varied in shape but are typically not elongated. Species are *saprotrophic or *parasitic on plants and found in a variety of habitats.

**Helvella** (order *Pezizales) A genus of *fungi, in which the *fruit body consists of a roughly saddle-shaped fertile head borne on a long, ribbed and furrowed stalk. *H. crispa* is whitish or cream-coloured, 4–10 cm tall, and found in autumn on the ground in broad-leaved woods.

**hemera** A period of geological time determined by the maximum development of a fossil plant or animal.

**Hemiascomycetes** (subdivision *Ascomycotina) A class of *fungi in which the *asci are formed singly and not in *ascocarps. The class is now not generally recognized; most of the 'hemiascomycetes' are now included in the *Endomycetales.

**hemicellulose** A heterogeneous group of compounds that in plant-*cell walls form part of the matrix within which *cellulose fibres are embedded. The two principal structural types are polymers of D-xylose and of *glucose and *mannose, known respectively as xylans and glucomannans.

**hemicryptophyte** One of *Raunkiaer's life-form categories, being a plant whose perennating buds are at ground level, the aerial shoots dying down at the onset of unfavourable conditions. Three subcategories are recognized: (a) protohemicryptophytes, in which the lowest leaves on the stem are smaller than others, or scale-like, giving added protection to the bud (e.g. *Rubus idaeus*); (b) partial rosette plants, in which the best-developed leaves form a basal rosette, but some leaves are also present on aerial stems (e.g. *Ajuga reptans*, bugle); and (c) rosette plants, in which the leaves are confined to a rosette at the base of the aerial shoots (e.g. *Bellis perennis*, daisy). *Compare* CHAMAEPHYTE; CRYPTOPHYTE; PHANEROPHYTE; and THEROPHYTE.

**hemiparasite** A plant parasite (*see* PARASITISM) that has *chlorophyll and photosynthesizes, but that augments its nutrient supply by feeding on its host (e.g. *Viscum album*, mistletoe).

**hemlock fir** *See* TSUGA.

**hemp** *See* CANNABIS and CANNABIDACEAE.

**Hepaticae** *See* HEPATOPHYTA.

**Hepatophyta** (liverworts) A division of plants, formerly ranked as the class Hepaticae, characterized by a combination of features. The *capsule is usually ovoid or spherical and does not have a lid; when ripe, it usually splits into 4 'valves' to release the spores. A tubular *perianth often surrounds the developing capsule. The *seta is colourless and semi-transparent; it lengthens after the capsule has reached its full size, and is structurally much weaker than a moss seta. Liverworts may be 'thallose', i.e. flattened and showing no differentiation into stem and leaves (*Anthoceratales, *Marchantiales, and Metzgeriales), or 'leafy' (*Jungermanniales), with leaves normally arranged in 2 or 3 distinct ranks. The leaves never have a thickened nerve or *midrib, and

are often lobed or segmented. Typically the *thallus is attached to a substrate by means of unicellular *rhizoids. Liverworts are found in a variety of habitats, particularly in moist conditions.

**heptamerous** Having seven parts.

**Heracleum** (family *Apiaceae) A genus of coarse, *biennial or *perennial herbs. *H. mantegazzianum* (giant hogweed) grows to 3 m tall with *umbels 1 m in diameter and produces a highly irritant sap; it is native to the Caucasus but naturalized in Europe, and is a noxious, invasive weed. There are about 60 species, occurring in northern temperate regions and tropical mountains.

**herb** A small, non-woody seed-bearing plant in which all the aerial parts die back at the end of each growing season. *Compare* SHRUB; SUBSHRUB; and TREE.

**herbage** **1.** The growing plants on which domestic animals feed. **2.** The ground vegetation (*herbs), especially grass, when these are considered as an agricultural crop. **3.** The payment made to the owner of an area of land in return for permission to graze livestock on that land.

**herbarium** A collection of dried plants together with collection data that might be used in taxonomic studies based on their anatomy or preserved biomolecules.

**herbicide** A chemical substance which suppresses, and is usually designed to eliminate, plant growth. It may be a non-selective weed-killer (e.g. paraquat); or selective, for example eliminating dicotyledonous (*see* DICOTYLEDON) plants from among monocotyledonous (*see* MONOCOTYLEDON) stands (e.g. phenoxyacetic acids) or vice versa (e.g. dalapan). A notable consequence of this kind of chemical control in cereal crops is the decline in typical weeds, e.g. poppy, and the increase of weed grasses, e.g. *Avena fatua* (wild oat) and *Alopecurus myosuroides* (black grass).

**Hericium** (family Hericiaceae, order *Aphyllophorales) A genus of *fungi in which the *hymenium is borne on tooth-like projections. The *fruit body is typically extensively branched, and white or cream-coloured, with teeth that typically hang downwards. The *spores are colourless, spherical or nearly so, and turn a dark violet colour when treated with iodine. Species are *parasitic on trees, or grow on dead wood, stumps, etc.

**heritability** A measure of the degree to which a *phenotype is genetically influenced and can be modified by selection. It is represented by the symbol $h^2$: this equals $V_a/V_p$ where $V_a$ is the variance due to *genes with additive effects (known as the additive genetic variance) and $V_p$ is the phenotypic variance. The variance $V$ may also be written as $s^2$. Parent–offspring correlations are estimates of familiality and not of heritability: they cannot account for environmental correlations between relatives. This definition of heritability is a narrow one: heritability in the broad sense (represented by $H^2$) is the fraction of total phenotypic variance that remains after exclusion of the variance resulting from environmental effects. Estimates of heritability are used widely by plant breeders to predict the likely effects of selection. If heritability estimates are low for a particular character, this indicates that the character is mainly influenced by the environment and suggests that the response to selection would not be rapid.

*Heritiera*   See MENGKULANG.

*Hernandia* (family *Hernandiaceae) A genus of plants in which the leaves are *simple and the flowers unisexual and borne in long *panicles. The fruit is a *nut enclosed in an inflated, papery envelope. There are 24 species, with a pantropical distribution.

**Hernandiaceae** A small family of shrubs, trees, and climbers, in which the leaves are palmately nerved, there are 6–10 free *perianth lobes and 3–5 *stamens, and the *ovary is *inferior and *unilocular. The fruit is *indehiscent. They are related to *Monimiaceae. There are 4 genera, containing 68 species, with a pantropical distribution.

**herpokinetic mobility** In some *algae, the snake-like movement of a filament.

**hesperidium** The *berry of a citrus fruit (*Rutaceae), i.e. a fruit whose fleshy parts are divided into segments, the whole being surrounded by a separable skin.

**hetero-** A prefix meaning 'different from', derived from the Greek *heteros*, meaning 'other'.

**heteroalleles**   See HOMOALLELIC.

*Heterobasidion* (family *Polyporaceae) A genus of *fungi in which clamp connections are absent and *fruit bodies are *resupinate to bracket-shaped. *H. annosum* (formerly *Fomes annosus*) is an important *pathogen of trees, particularly conifers, in which it causes decay of heartwood and roots. Woody *fruit bodies occur at the base of the infected tree and are orange-brown with a white margin. They are found throughout temperate regions.

*Heterobasidion annosum* (*Fomes annosus*)   See BRACKET FUNGUS.

**heteroblastic development** A succession of organs, e.g. leaves, in which the form changes during the development of the plant, usually becoming larger and more complex.

**heterocaryon**   See HETEROKARYON.

**heterochromatin** The *chromosome material that accepts stains in the *interphase nucleus (unlike *euchromatin). Such regions, particularly those containing the centromeric and nucleolus organizers, may adhere to form a chromocentre. Some chromosomes are composed primarily of heterochromatin: these are termed heterochromosomes. In many species, the *Y chromosome is a heterochromosome.

**heterochromosome**   See HETEROCHROMATIN.

**heterochrony** The dissociation, during the development of factors of shape, size, and maturity, so that organisms mature in these respects at earlier or later growth stages.

**heterocyclic** Applied to a compound whose molecules contain a ring system composed of atoms of different elements.

**heterocyst** A specialized cell found in certain types of *cyanobacteria; heterocysts are sites of *nitrogen fixation.

***Heterodendrum* (Australian rosewood; family *Sapindaceae)** A genus of small trees which have small, *lanceolate or linear leaves. The flowers are held in *panicles and have no *petals. The fruit is 5-lobed, each lobe being single-seeded. There are 4 species, all endemic (*see* ENDEMISM) to Australia.

**heterodimer** A *protein consisting of paired *polypeptides that differ in their *amino acid sequences.

**heteroecious (heteroxenous)** Applied to a *parasitic organism (e.g. the rust fungus *Puccinia graminis*) in which part of the life cycle occurs obligatorily in one host and the remaining part obligatorily in another. Heteroecism occurs in several groups of animal parasites but, apart from the rust fungi, in only a few plant species. *Compare* AUTOECIOUS.

**heterogamous** Applied to the condition in which two different kinds of flowers are borne on the same plant (e.g. in some *Asteraceae the discoid *florets are perfect and the radiate florets bear *pistils or are neutral).

**heterogamy** Reproduction involving two types of *gamete.

**heterograft** *See* GRAFT.

**heterokaryon (heterocaryon)** A fungal *mycelium or *hypha that contains nuclei of two or more genetic types.

**heterokont** Having *flagella of different lengths. *Compare* ISOKONT.

**heteromeric** *See* DIMER.

**heteromerous** **1.** Composed of units (e.g. cells) of different types. **2.** Having unequal numbers of parts. **3.** Layered, as in a *lichen *thallus in which the *algae are confined to a distinct layer.

**heteromorphic** Literally, differing in form, and applied to: (*a*) phases/stages of organisms in which there is *alternation of generations, particularly in *algae where generations are vegetatively dissimilar; and (*b*) a *chromosome pair that have some *homology but that differ in size or shape, e.g. the *X and *Y sex chromosomes.

**heterophylly** The state of having leaves of different shapes on the same plant. *Compare* ISOPHYLLY.

**heterosis (hybrid vigour)** The increased vigour of growth, survival, and fertility of *hybrids, as compared with the 2 *homozygotes. It usually results from crosses between 2 genetically different, highly inbred lines. It is always associated with increased *heterozygosity.

***Heterospathe* (family *Arecaceae)** A genus of palms in which the leaves are *pinnate with a crown-shaft. The plants are *monoecious. There are about 20 species, found in Malesia, Australia, and the western Pacific islands.

**heterospory** The production of *spores of 2 different types on the same plant. *Microspores are small and numerous; from these the male *gametophytes develop. *Megaspores, within which the female gametophytes develop, are larger and fewer. *Compare* HOMOSPORY; *see also* MICROSPORANGIUM.

**heterostyly** A *polymorphism among *angiosperm flowers that ensures *cross-fertilization through *pollination by visiting insects. Flowers have *anthers and *styles of different lengths; usually the anthers of one morph are at the same level as the *stigmas of another. A well-known example is the pin-eyed (long style) and thrum-eyed (short style) forms of *Primula vulgaris* (primrose).

**heterothallic** Applied (e.g.) to a *fungus or *alga that is self-sterile. Sexual reproduction requires the participation of 2 *thalli of compatible mating types.

**heterotrichous** Applied to a filament part of which is prostrate and part erect.

**heterotroph** An organism that is unable to manufacture its own food from simple chemical compounds and, therefore, consumes other organisms, living or dead, as its main or sole source of carbon. Often, single-celled *autotrophs (e.g. *Euglena*) become heterotrophic in the absence of light.

**heteroxenous** *See* HETEROECIOUS.

**heterozygosity** The presence of different *alleles (forms of a given *gene) at a particular gene *locus. *Heterozygosity provides a measure of the genetic variation, either in a population (the frequency of individuals heterozygous at a particular locus), or in an individual (the proportion of gene loci that are heterozygous).

**heterozygote** A *diploid or *polyploid individual that has different *alleles (forms of a given *gene) at at least one *locus. Its *phenotype is often identical to that of an individual that has one of these alleles in the *homozygous state. (For example, the phenotype of $A_1A_2$ may be identical to that of $A_1A_1$, but different from that of $A_2A_2$.) If so, that allele ($A_1$) is said to be *dominant over the other ($A_2$), which is said to be *recessive. When there exist only 2 alleles of a gene and when one of these is dominant over the other, one may use a lower-case letter for the recessive allele; e.g., one may write $Aa$ rather than $A_1A_2$. However, alleles are not always dominant one over another (*see* PARTIAL DOMINANCE); and often there are more than 2 alleles at a given locus (*see* MULTIPLE ALLELISM), especially in *polyploids. Because it has two or more different alleles at a given locus, a heterozygote does not *breed true.

**heterozygous advantage** *See* GENETIC POLYMORPHISM.

**Hevea** (family *Euphorbiaceae) A small genus of trees whose bark contains latex. The leaves are *trifoliate. The flowers are unisexual, male and female being in the same *panicle. The fruit is an explosively *dehiscent *capsule. Rubber (or para rubber) is *H. brasiliensis*, now planted widely in the Far East. There are 9 species, found in the rain forest of Amazonia.

**hexamerous** With parts in sixes.

**hexaphosphoinositol** *See* PHYTIC ACID.

**hexaploidy** *See* POLYPLOIDY.

**hexose** A monosaccharide sugar that contains 6 carbon atoms (e.g. *glucose, *fructose, and galactose).

**hexose monophosphate shunt** (pentose phosphate shunt) A *metabolic pathway, alternative to that of *glycolysis, of carbohydrate interconversion: hexose-6-phosphate is converted into pentose phosphate and carbon dioxide. The principal functions of the pathway are the production of deoxyribose and ribose sugars for *nucleic acid synthesis; the generation of reducing power in the form of NADPH for *fatty acid and/or *steroid synthesis; and the interconversion of carbohydrates. Parts of the pathway are involved in the *Calvin cycle in *photosynthesis.

**Hibbertia** (family *Dilleniaceae) A genus of *xeromorphic shrubs, sometimes climbing, some with *phylloclades, which have a curious range, believed to indicate former continental connections. There are 122 species, occurring mostly in Madagascar, but also in Malesia, Australia, New Caledonia, and Fiji.

**Hibiscus** (family *Malvaceae) A genus of trees and shrubs whose showy flowers have a prominent *stamen tube. They are bird-pollinated. Several are widely cultivated, especially *H. rosa-sinensis*. Okra is the mucilaginous young fruits of *H. (Abelmoschus) esculentus*. There are about 200 species, with a pantropical distribution.

**hickory** *See* CARYA.

**hierarchical and non-hierarchical classification methods** The grouping of individuals by a series of subdivisions or agglomerations to form a characteristic 'family tree' or dendrogram of groups. Alternatively, classification may be non-hierarchical, i.e. proceeding not by an organized series of progressive joinings or subdivisions, but instead achieving groupings by a series of simultaneous trial and error clusterings (successive approximation) until an optimum and stable pattern is found. A possible scheme, for example, is to select at random a number of starter individuals, equal to the number of groups required, and to which other individuals are added or from which they are removed, according to their characteristics and the classificatory philosophy used (that of seeking to minimize internal group heterogeneity, or of seeking to maximize differences between groups). The chief advantage of hierarchical over non-hierarchical methods is the clarity with which the routes to the final groupings may be followed, facilitating explanation of those groups. However, since most hierarchical classifications either dichotomize or pair at each division or join, natural clustering patterns may be distorted or poorly represented. Hierarchical classifications are also more likely to be unduly affected by irrelevant background information.

**high Arctic tundra** The most northerly sector of the *tundra, distinguished by a lack of complete vegetation cover, except in the most favoured, and usually very restricted, *habitats. Such tundra vegetation as exists tends to be marshy, with little but *lichen in the more exposed situations.

**higher categories** In taxonomy, categories higher than species, which are defined arbitrarily according to observed similarities among species, and which provide a useful hierarchical framework by which organisms may be described succinctly.

**higher fungi** *Fungi belonging to the subdivisions *Ascomycotina, *Basidiomycotina, and *Deuteromycotina. *Compare* LOWER FUNGI.

**high forest** A forest comprising trees of all ages from which it is planned to obtain a crop of timber. The term does not apply to *underwood or *coppice. It applies equally whether the trees were planted by humans, were self-sown, or were derived from coppice stools. The trunks of the trees remain unbranched up to considerable heights, and grow in strong competition with each other for light. As is implied by this, the density of trees is relatively high.

**Hiller peat-borer** *See* PEAT-BORER.

**hillock tundra** A poorly drained, marshy *tundra with numerous hummocks about 25 cm high, which give better drainage and may therefore support heathy plants and lichens. A range of microhabitats is provided by the hummocks and these are exploited by plants with different ecological requirements.

**Hill reaction** A reaction, discovered by Robert Hill in 1939, in which isolated *chloroplasts produce oxygen and hydrogen when illuminated in the presence of an oxidizing agent (e.g. a ferric salt).

**hilum** **1.** The point on a fungal *spore at which it was attached to its *sporophore. **2.** The scar on a *seed that marks the point at which it was attached to the plant.

**Himantandraceae** A family of aromatic trees with *fimbriate, *peltate scales, in which the leaves are *simple and without *stipules. The flowers have 2 fused *sepals, 4 *petals, and numerous *stamens and *carpels. The fruit is a globose *syncarp. Himantandraceae are close to *Magnoliaceae. There is 1 genus, *Galbulimima*, with 3 species, occurring in eastern Malesia and northern Australia.

**Himanthaliaceae (order *Fucales)** A family of brown seaweeds, which includes a single genus, *Himanthalia*. The plant lives for 2 years. During the first year a stalked, disc-like *thallus is formed: this measures several centimetres across. In the second year 1–4 long, strap-like branches grow from the disc, and these bear the *conceptacles within which *gametes are formed. There is 1 British species, *H. elongata* (sea thong), found at or near the low-water mark on exposed shores.

**Hippocastanaceae** A family of trees or shrubs that have opposite, palmately *compound leaves without *stipules and large terminal *panicles of showy, irregular flowers. The flowers have usually 5 *sepals and 4 *petals and a *superior *ovary that ripens to a large, thick-walled *capsule containing 1 or 2 large, leathery, glossy, nut-like, inedible seeds. Flowers are *zygomorphic, with a *disc. Winter buds have big scales. There are 2 genera. *Aesculus* (deciduous), with 13 species, occurs in the northern temperate zone, and 2 species of *Billia* (evergreen) in tropical Central and S. America. *A. hippocastanum* (horse-chestnut), is often grown for ornament. *Aesculus* also includes the buckeyes of N. America.

**Hippocrateaceae** A group of plants that were formerly regarded as a separate family, but are now included in the family *Celastraceae.

**Hippuridaceae** A *monotypic family consisting solely of *Hippuris vulgaris* (mare's-tail) which is a *perennial aquatic herb with linear leaves in *whorls. It is superficially like *Equisetum but is a flowering plant with inconspicuous flowers, each of 1 *stamen and an *inferior *ovary, in the *axils of those leaf-whorls that are emergent from the water. It is distributed in Europe, western Asia, and N. Africa.

**Hippuris vulgaris (mare's-tail)** See HIPPURIDACEAE.

**hispid** Having short, stiff hairs or bristles.

**histic epipedon** A surface soil *horizon, not less than 1 metre in depth, high in organic carbon, and saturated with water for some part of the year. See also HUMUS (2). The name is from the Greek histos meaning 'tissue'.

**histidine** A basic, polar (see POLAR MOLECULE) *amino acid that contains an imidazole group ($C_3H_4N_2$).

**histochemistry** The study of the chemistry of tissues and cells, using combined techniques from *histology and biochemistry.

**histocompatibility genes** The *genes for *antigens that are responsible for the acceptance or rejection of foreign bodies in the form of transplanted tissues (*grafts).

**histology** The scientific study of organic tissues.

**histone** One of a group of basic, *globular, simple *proteins that have a high content of the *amino acids arginine and lysine. Histones form part of the chromosomal material of *eukaryotic cells and appear to play an important, but as yet incompletely understood, role in gene regulation.

**Histosols** Organic soils, an order of the *USDA Soil Taxonomy, composed of organic materials. Histosols must have a thickness of more than 40 cm when overlying unconsolidated mineral soil, but may be of any thickness when overlying rock. Histosols are a reference soil group in the FAO *soil classification.

**hognut** See CARYA.

**hog plum** See SPONDIAS.

**Holarctica** A formerly unified, circumpolar biogeographic region, embracing what are now N. America, Europe, and Asia (i.e. Laurasia). The legacy of this region is attested to by the great floral and faunal similarities between the 3 northern continents.

**holdfast** A differentiated structure in a *seaweed or other *algae, the function of which is to attach the *thallus to a substrate (e.g. rocks, other plants, or shells). A holdfast may be superficially root-like or may be discoid and sucker-like.

**holistic (holological)** Relating to the whole. In *ecology, the term is applied to studies that aim to understand *ecosystems as a whole (i.e. as entire systems), rather than examining their component parts. Compare MEROLOGICAL APPROACH.

**hollow phase** A stage in the cyclical pattern of vegetation change in *grassland, at which a hollow forms from the erosion of an old grass hummock during the *degenerate phase. Colonization of the bare soil by grass seedlings triggers the development of a new hummock. Compare BUILDING PHASE; MATURE PHASE; and DEGENERATE PHASE.

**holly (Ilex aquifolium)** See AQUIFOLIACEAE.

**hollyhock** See ALTHAEA and MALVACEAE.

**Holobasidiomycetidae (class *Hymenomycetes)** A subclass of *fungi in which the *basidia are *aseptate. The subclass includes *saprotrophs and parasites (see PARASITISM) of higher plants.

**holocarpic** Applied to a *fungus in which the whole *thallus differentiates to form one or more reproductive structures.

**Holocene (Recent, Post-glacial)** The epoch that covers the last 10 000 years. See also ANTHROPOGENE and FLANDRIAN.

**holocentric** Applied to *chromosomes with diffuse *centromeres such that the properties of the centromere are distributed over the entire chromosome.

**holoenzyme** A entire conjugated *enzyme, comprised of a *protein component or apoenzyme, and its non-protein (*prosthetic) group.

**hologamete** In *Protista, a *gamete formed from the entire cell body.

**holological** See HOLISTIC.

**holophyletic** Of a *taxon: including all descendants of the common ancestor. The term is a special case of 'mono-phyletic'.

**holotype** See TYPE SPECIMEN.

***Homalothecium*** (order *Hypnobry-ales) A genus of medium-sized to large mosses, in which the stems are irregularly to sub-pinnately branched. The leaves are triangular, tapering from the base to a long point at the tip; a single nerve extends about three-quarters of the way up each leaf. It is a small genus, with about 11 species, found in Europe, Asia, Africa, and America. *H. sericeum* (formerly *Camptothecium sericeum*) is a prostrate, creeping plant which forms extensive mats on walls, rocks, trees, roofs, etc.; it is yellowish-green to brown, with a characteristic silky sheen especially noticeable when the plant is dry. *H. lutescens* (formerly *C. lutescens*) is not prostrate; it forms loose tufts, is yellowish-green in colour, and is found chiefly on calcareous soil.

**homeostasis** See HOMEOSTASIS.

**homo-** A prefix meaning 'the same', derived from the Greek *homoios*, meaning 'like'.

**homoallelic** Applied to *allelic mutants of a *gene that have different *mutations at the same site; as opposed to heteroallelic mutants, which have mutations at different sites within the one *gene. Recombination between hetero-alleles can yield a functional *cistron; recombination between homoalleles cannot.

**homodimer** A *protein that is made up of two identical *polypeptides paired together; as opposed to a heterodimer, in which the polypeptides are not identical.

**homoeology** The study of *chromo-somes that are related, but derived from different *genomes. Homoeologous

chromosomes may occur within the same organism, e.g. in an *allopolyploid, where the constitutive genomes are derived from different, but often closely related species.

**homoeostasis** (homeostasis) The tendency of a biological system to resist change and to maintain itself in a state of stable equilibrium.

**homograft** See ALLOGRAFT and GRAFT.

**homoiomerous** Uniform in structure; composed of units (e.g. cells) all of the same type. The term is applied to a *lichen *thallus in which the *algae are distributed uniformly throughout the thallus.

**homokaryon** (homocaryon) A fungal *mycelium or *hypha in which all the nuclei are genetically identical.

**homologous** A term applied to organs or *chromosomes. An organ of an animal is said to be homologous with that of another when both are thought to have the same evolutionary origin, although their functions may differ widely. Homology is generally deduced from similarity of structure and/or position of the organ relative to other organs, seen particularly during embryonic development; e.g., in the plant kingdom the fertilization of the lower plants (e.g. *Bryophyta) is often effected by *motile male *gametes; this is homologous to the male gamete of *Spermatophytes which is non-motile and transferred in the *pollen grain. Homologous chromosomes are those that contain identical linear sequences of *genes, and which pair during *meiosis. Each homologue is therefore a duplicate of one of the chromosomes contributed by one of the parents; and each pair of homologous chromosomes is normally identical in shape and size. *Compare* HETEROLOGOUS.

**homology** The basic similarity of a particular structure in different organisms: it usually results from their descent from a common ancestor.

**homomeric** See DIMER.

**homonym** In nomenclature, 1 of 2 or more separately published names for the same taxon or identical names for different taxa.

**homoplasy** The occurrence of similar features in distantly related taxa (*see* TAXON) as a result of convergent or parallel evolution.

**homospory** The production by a *cryptogamic plant of *spores that are all of uniform type and size and from which develop *gametophytes which usually contain both male and female cells. *Compare* HETEROSPORY; *see also* MICROSPORANGIUM.

**homothallic** Applied (e.g.) to a *fungus or *alga that is self-fertile (i.e. one in which sexual reproduction can occur in a single *thallus).

**homozygosity** The presence of identical *alleles at one or more *loci in *homologous chromosomal segments.

**homozygote** An individual having the same *alleles at one or more *loci. The *phenotype of the alleles for a particular locus will always be expressed since the 2 alleles at the *homologous loci are identical. An individual will *breed true at those homologous loci that are homozygous.

**honey fungus** (honey tuft) *See* ARMILLARIA.

**honey** *See* NECTAR.

**honeysuckle** (*Lonicera*) *See* CAPRIFOLIACEAE.

**honey tuft** (honey fungus) *See* ARMILLARIA.

**hoof fungus** *See* TINDER FUNGUS.

**Hooker, Sir Joseph Dalton (1817–1911)** A British botanist, who graduated in medicine at the University of Glasgow in 1839 and was then appointed assistant surgeon on board the *Erebus*, as a member of the Antarctic expedition led by Sir James Ross. On his return, in 1843, he published *Flora Antarctica* (1844–7), *Flora Novae Zelandiae* (1853–5), and *Flora Tasmanica* (1855–60). He explored the northern frontiers of India (1847–51), and published the *Flora of British India* (1855–97). The large number of rhododendrons he brought from India became popular ornamentals, transforming many British gardens. He prepared the fifth and sixth editions of *Bentham's *Handbook of the British Flora*, which then became known to generations of students as 'Bentham and Hooker'. He was appointed assistant director of the Royal Botanic Gardens at Kew in 1855 and succeeded his father as director in 1865.

**Hooker, Sir William Jackson (1785–1865)** A British botanist and authority on cryptogamic botany, who became the first director (1841–65) of the Royal Botanic Gardens at Kew. He studied the botany of Iceland (1809) and of France, Switzerland, and northern Italy (1814). He was appointed regius professor of botany at the University of Glasgow in 1820. He wrote prolifically, his works including *Tour of Iceland* (1811), two volumes of *Musci Exotica* (1818–20), *Flora Scotica* (1821), *Icones Filicum* (with R. K. Greville, 1829–31), *British Flora* (with G. A. W. Arnott *et al.*, 1830), and *British Ferns* (1861–2).

**Hookeriales** (subclass *Bryidae) An order of *pleurocarpous mosses, which are found mainly in tropical regions. There are several genera, including *Hookeria*. The only British species of *Hookeria* is *H. lucens*: a characteristic plant with large, translucent, bright green leaves, and flattened shoots. The leaf cells are very large and thin-walled: they are almost visible to the naked eye and easily visible under a hand lens. There is no nerve. It is found in deep shade in moist habitats, mainly in northern and western regions.

**hop** *See* HUMULUS and CANNABIDACEAE.

**hop bush** *See* DODONAEA.

*Hopea* (family *Diptocarpaceae) A genus of resinous trees, many of which have heavy timber. There are 102 species, occurring from southern China to New Guinea.

*Hordeum* (barley; family *Poaceae) A genus of grasses that were domesticated in very ancient times (they were grown in ancient Egypt) and that are used today mainly as livestock feed and for malting to make beer and whisky. *H. vulgare* is the most widely cultivated of about 20 species, native to the eastern Mediterranean region and cultivated in temperate regions throughout the world.

**horizon** In pedology, a relatively uniform soil layer that lies, at any depth in the soil *profile, parallel, or nearly so, with the soil surface, and which is differentiated from adjacent horizons above and below by contrasts in mineral or organic properties.

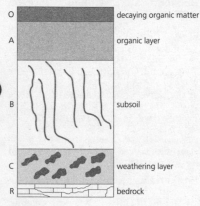

| | |
|---|---|
| O | decaying organic matter |
| A | organic layer |
| B | subsoil |
| C | weathering layer |
| R | bedrock |

*Horizon*

***Hormoconis resinae*** See CREOSOTE FUNGUS.

**hormogonium** A short length of a filament or *trichome in certain filamentous *cyanobacteria that becomes detached and functions as a *propagule.

**hormone** A regulatory substance, active at low concentrations, that is produced in specialized cells but that exerts its effect either on distant cells or on all cells in the organism to which it is conveyed via tissue fluids.

**hornbeam** See CARPINUS and BETULACEAE.

***Hornea*** See TMESIPTERIS.

**horn of plenty** The common name for the edible *fruit bodies of the *fungus *Craterellus cornucopioides*. The cap is funnel-shaped with a wavy edge; the inside of the funnel is black to brownish-black, while the outer surface is dark grey. The *spore print is white. It is found in groups on the ground under deciduous trees (especially beech) in late summer and autumn.

**hornwort** The common name for a *liverwort of the unique order *Anthocerotales; sometimes regarded not as a liverwort but as a distinct type of plant (class Anthocerotae).

**horotelic** See HOROTELY.

**horotely (adj. horotelic)** A normal or average rate of *evolution per million years, of genera within a given taxonomic group. Thus, slowly or rapidly evolving lines may be horotelic during certain episodes in their history.

**horse-chestnut** See *AESCULUS* and HIPPOCASTANACEAE.

**horse mushroom** The common name for the *fruit body of *Agaricus arvensis*, which resembles the cultivated *mushroom but is larger and smells of aniseed or almonds. It is edible and highly esteemed. It is found in pastures, on road verges, etc., during summer and autumn.

**horseradish tree** See *MORINGA*.

**horsetails** See CALAMITACEAE; *CALAMITES CISTIIFORMES*; and *EQUISETUM*.

***Horsfieldia*** (family *Myristicaceae) A genus of trees with *monopodial crowns, whose fruit is a leathery *capsule containing 1 big seed with an entire *aril. There are about 80 species, occurring from southern China to Queensland and the islands of the western Pacific.

**host** See GRAFT and PARASITISM.

**Hottentot fig (***Carpobrotus***)** See AIZOACEAE.

**houseleek (***Sempervivum***)** See CRASSULACEAE.

***Howeia*** (sentry palm; family *Arecaceae) A genus of palms that have a single, closely hoop-marked trunk. The leaves are *pinnate. The trees are *monoecious, with *inflorescences among the leaves. They are very commonly cultivated. There are 2 species, both endemic (see ENDEMISM) to Lord Howe Island (Australia).

***Hoya*** (family *Asclepiadaceae) A genus of twining and root-climbing plants that have fleshy leaves and flowers which are waxy, highly fragrant, showy, and borne in *umbels. They are delightful hothouse ornamentals. There are about 90 species, native from southern China to the Pacific.

**Humboldt, Friedrich Heinrich Alexander, Freiherr von (1769–1859)** A German naturalist, physical geographer, biogeographer, geologist, vulcanologist, and mining engineer, who began travelling extensively after the death of his mother in 1796. Accompanied by the French surgeon and botanist Aimé Bonpland (1773–1858), he set out to join Napoleon in Egypt, but their plans changed and they went instead to Madrid. Their experiences in Spain made them decide to explore Spanish America. They embarked in 1799 and spent 5 years in the South American tropics, exploring the drainage basins of the Amazon and Orinoco, investigating the properties of guano, measuring the temperature of the ocean current that bears his name, and studying the flora and fauna of the forests and *savannah, returning with more than 30 cases of botanical specimens. He wrote on many subjects, his most important botanical work being the 7-volume *Nova genera et species plantarum* (1815–25), written in collaboration with A. J. A. Bonpland and C. S. Kunth as part six of the 30-volume *Voyage de Humboldt et Bonpland* (1805–34). In *Ideen zu einer Physiognomik der Gewächse* (Ideas on a physiognomy of plants, 1806) he proposed the concept of biogeography.

**humic acid** A mixture of dark-brown organic substances, which can be extracted from soil with dilute alkali and precipitated by acidification to *pH 1–2 (in contrast with *fulvic acid, which remains soluble in acid solution). Its analysis in peat profiles provides a measure of the degree of *humification or decomposition, which in turn can provide an index of past wetness, possibly associated with climatic conditions.

**humidity** An expression of the moisture content of the atmosphere. Measures of humidity include statements of the total mass of water in 1 m³ of air (absolute humidity), the mass of water vapour in a given mass of air (specific humidity), relative humidity, vapour pressure, and the mixing ratio.

**humification** The development of humus from dead organic material, by the action of *saprotrophic organisms which use this dead material as their food source. Humification is essentially an oxidation process in which complex organic molecules are broken down into simpler organic acids which may subsequently be mineralized into simple, inorganic forms suitable for uptake by plants. Humification is therefore a vital stage in the cycling of nutrients. The degree of humification in a *peat *profile is often associated with the general colour of the peat, darker peats being better humified. The degree of humification relates to the wetness of the surface conditions at the time of peat formation, which can provide an index of climatic wetness at the time. Peat humification studies are now a major source of information about past climate change.

**Humulus** (family *Cannabidaceae) A genus of *perennial climbing herbs whose fruit is an *achene enclosed by a persistent *perianth. *H. lupulus* is the hop. There are 2 species, occurring mainly in the northern temperate zone.

**humus** **1.** Decomposed organic matter of soils that are aerobic for part of the year: it is dark brown and amorphous, having lost all trace of the structure and composition of the vegetable and animal matter from which it was derived. **2.** A surface organic soil *horizon that may be divided into types, e.g. *mor (acid and layered) or *mull (alkaline and decomposed). It is the *histic epipedon. *See also* HUMIFICATION.

**hyaline** Translucent or transparent; glass-like.

**hyaloplasm** The ground substance of cell *cytoplasm, within which the various sub-cellular *organelles and membranaceous components are embedded.

**Hybanthus** (family *Violaceae) A genus of *perennial, often creeping, *herbs and shrubs, whose leaves are alternate or in clusters, occasionally *opposite, *sessile, or with a short *petiole, and entire or slightly toothed, with or without *stipules. The flowers are held in *axillary *racemes or *cymes, or are solitary. They are irregular and bisexual. The *sepals are of similar length, the *petals irregular, with a broad, spurred, and clawed lower lobe. The *stamens are often attached to the *corolla. The fruit is a *dehiscent *capsule. There are about 150 species, found in tropical and subtropical regions of the world, often in mountainous areas. The roots of *H. calceolaria* yield a drug, white ipecacuanha, and are used as an emetic.

**hybrid** **1.** An individual plant or animal resulting from a cross between parents of differing *genotypes. Strictly, most individuals in an *outbreeding population are hybrids, but the term is more usually reserved for cases in which the parents are individuals whose *genomes are sufficiently distinct for them to be recognized as different species or subspecies. A good example is *Spartina townsendii*, produced by cross-breeding *Spartina maritima* (British cord grass) and the North American species *Spartina alterniflora* (each of which can breed true as a species). Hybrids may be fertile or sterile depending on qualitative and/or quantitative differences in the genomes of the two parents. Hybrids like *Spartina townsendii*, whose parents are of different species, are sterile, but generally reproduce vegetatively. **2.** By analogy with (1), any *heterozygote. Each heterozygote represents dissimilar *alleles at a given *locus, and this difference results from a cross between parental *gametes possessing differing alleles at that locus. **3. (graft hybrid)** *See* GRAFT.

**hybrid swarm** A continuous series of *hybrids that are morphologically distinct from one another, resulting from the hybridization of 2 species followed by the crossing and *backcrossing of subsequent generations. The hybrids are very variable owing to segregation of *alleles at each *locus.

**hybrid vigour** *See* HETEROSIS.

**hybrid zone** A geographical zone in which the *hybrids of two geographical races may be found.

**hydathode** A modified *stoma, usually on a leaf, which remains open and secretes water during *guttation.

**Hydnaceae (order** *Aphyllophorales)** A family of *fungi, in which the *fruit bodies are typically *mushroom-shaped and bear the *hymenium on pendulous, tooth-like spines. They are found on wood, soil, etc. They are mostly rare, often producing *fruit bodies only at intervals of several years. Many genera formerly included in this family are now considered to belong to other families, e.g. the *Auriscalpiaceae.

***Hydnocarpus*** **(family** *Flacourtiaceae)** A genus of trees in which the flowers are unisexual and small. The fruits are large, globose, woody, *indehiscent, and contain many seeds set in pulp. The seeds contain fatty acids, from which chaulmoogra oil (from *H. anthelmintica* and *H. kurzii*) is obtained. It is used to treat skin diseases and was formerly used to treat leprosy. There are about 40 species, found in rain forests from India to the Philippines and Celebes.

***Hydnophytum*** **(family** *Rubiaceae)** A genus of epiphytic (*see* EPIPHYTE) herbs, which have a swollen stem base with ant-inhabited chambers. There are about 60 species, occurring from the Andamans to the Pacific islands.

***Hydnum*** **(family** *Hydnaceae)** A genus of *fungi, in which the *fruit bodies are fleshy and *stipitate and the *hymenium is borne on spines which hang down from the underside of the cap. The *spore print is white. Species grow exclusively on the ground, typically in woodland. *H. repandum* (hedgehog fungus) is cream-coloured with a thick, eccentric *stipe; it is edible and is commonly sold in markets in Europe.

***Hydrangea*** **(family Hydrangeaceae)** A genus of erect or climbing, evergreen or deciduous shrubs. The leaves are opposite, the flowers in *cymes, often as flat heads and with showy, sterile, marginal flowers. Several species are cultivated as ornamentals. There are 23 species, occurring from the Himalayas to Japan and the Philippines.

**hydrarch succession** *See* HYDROSERE.

**hydric** *See* MESIC.

**hydrocarbon** A naturally occurring compound that contains carbon and hydrogen. Hydrocarbons may be gaseous, solid, or liquid, and include natural gas, bitumens, and petroleum.

***Hydrocharis* (frog-bit)** *See* Hydrocharitaceae.

**Hydrocharitaceae** A family of aquatic, monocotyledonous (*Monocotyledon) *herbs that have regular, usually *dioecious flowers within *spathes, usually with 3 *sepals and 3 *petals, 1–3 or more *stamens, and an *inferior 1-celled *ovary. The family includes *Hydrocharis* (frog-bit) and *Elodea*. In modern classifications there are 16 genera, with about 70 species. They are mostly tropical but also found in the temperate zone.

**hydrochory** Dispersal of spores or seeds by water.

**hydrogenase** An *enzyme that catalyses reactions involving the addition of hydrogen to a *substrate.

**hydrogen bond** The force of attraction (the hydrogen force) that exists between *polar molecules containing hydrogen atoms, or between one part of a molecular chain and another part that contains bonded hydrogen. It occurs because the single electron in the hydrogen atom is held only weakly, so hydrogen readily forms *ionic bonds, which allow a hydrogen atom to link other atoms. Some of the properties of water are due to the hydrogen bond that links water molecules. *DNA (*see* DEOXYRIBONUCLEIC ACID) molecules are linked by hydrogen bonds, and hydrogen bonding is also important in linking other organic molecules. *See also* COVALENT BOND; IONIC BOND; and METALLIC BOND.

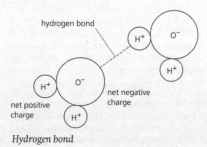

*Hydrogen bond*

**hydroid** In some *Bryophyta, an elongated, non-lignified cell that conducts water, analogous to a *tracheary element. *Compare* LEPTOID.

**hydrolase** An *enzyme that catalyses reactions involving the *hydrolysis of a *substrate.

**hydrolysis 1. (biochem.)** A reaction between a substance and water in which the substance is split into 2 or more products. At the points of cleavage the products react with the hydrogen or *hydroxyl ions derived from water. **2. (pedol.)** The process of enriching the soil *adsorption complex with hydrogen after exchangeable metallic ions have been replaced by hydrogen ions. *Compare* WEATHERING.

**hydronasty** A *nastic movement induced in plant organs by changes in atmospheric humidity.

**hydrophilic** Applied to a molecule or surface that can become wetted or solvated by water. This ability is characteristic of polar compounds.

**hydrophily** Pollination by the transport of pollen on or beneath the surface of water.

**hydrophobic** Applied to a molecule or surface that can resist wetting or solvation by water. The ability is characteristic of non-polar compounds.

**Hydrophyllaceae** A family of *annual or *perennial *herbs and small shrubs which have alternate, hairy or glandular, *simple or *compound leaves, without *stipules. The flowers are regular, bisexual, usually blue or purple, and borne in *cymes. They have 5 free *sepals, often with *auricles between the *calyx lobes. The *corolla is fused at the base into a bell or funnel shape, with 5 *stamens inserted into the base. The *ovary is *superior with 2 fused *carpels and 1 or 2 *locules, each with 2 or more *ovules. There are 1 or 2 *styles. The fruit is an elongated *capsule which splits along its length; it contains numerous seeds with fleshy *endosperm. There are about 22 genera, with 275 species, found throughout the world except Australasia. Several species are ornamentals.

**hydrophyte** A plant that is adapted morphologically and/or physiologically to grow in water or very wet environments. Adaptations include the development of finely divided submerged leaves, large floating leaves, the presence of *aerenchyma, and the reduction of root systems. The *perennating bud lies at the bottom of fairly open water. With the leaves submerged or floating, only the inflorescence protrudes above the water surface (e.g. *Nuphar lutea*, yellow water-lily). *Compare* HELOPHYTE.

**hydroponics** Plant growth in a liquid culture solution rather than in soil. This technique is used commonly in experimental studies of mineral nutrient deficiencies or excesses and their effects. Hydroponics also has commercial applications, although to date it has not been used extensively in this way.

**hydrosere (hydrarch succession)** A sequence of *communities that reflects the developmental stages in a plant *succession which commences on a soil submerged by fresh water.

**hydrotaxis** The locomotion of an organism in response to the stimulus of water.

**hydrotropism** A directional growth of a plant organ towards wetter regions in response to the stimulus of water.

**hydrous mica** (illite) *See* CLAY MINERALS.

**Hygrocybe (family *Hygrophoraceae)** A genus of *fungi in which the *mushroom-like *fruit bodies are small, brightly coloured (red, yellow, or green) or white, and have slimy or greasy caps and waxy gills; in some species the fruit body blackens with age. There are numerous species, found in grassland.

**hygrophanous** Translucent and watery in appearance.

**hygrophilic** *See* HYGROPHILOUS.

**hygrophilous (hygrophilic)** Growing in or preferring moist habitats.

**Hygrophoraceae (order *Agaricales)** A family of *fungi, which form waxy, often brightly coloured *fruit bodies (wax caps) bearing thick but sharp-edged gills. *Spores are colourless. Species are found on the ground in woods or grassland.

**Hygrophoropsis (family Paxillaceae, order *Boletales)** A genus of *fungi, which form *mushroom-shaped *fruit bodies. *H. aurantiaca* forms bright orange fruit bodies which are sometimes mistaken for *chanterelles, hence the common name *false chanterelle. This species is found on the ground in conifer woods and heathland.

**Hygrophorus (family *Hygrophoraceae)** A genus of *fungi, in which the *mushroom-like *fruit bodies are white, pale-coloured, or brown, and have slimy or greasy caps, and waxy gills. There are many species. They are found in woodland, where they form *mycorrhizal associations with trees.

**hygroscopic** Able to absorb water from the surroundings, including the absorption of atmospheric moisture.

**hygroscopic water** Water absorbed from the atmosphere and held very tightly by the soil particles, so that it is unavailable to plants in amounts sufficient for them to survive. *Compare* CAPILLARY MOISTURE.

**hygrotaxis** The movement of an organism in response to the stimulus of humidity or moisture.

**Hylocomium** *See* HYPNOBRYALES.

**hymenium** A layer of *spore-bearing structures (*asci or *basidia) on or within a fungal *fruit body.

**Hymenochaetaceae (order *Aphyllophorales)** A family of *fungi in which the *fruit bodies range from *resupinate to pileate (*see* PILEUS). The *hymenium may or may not line pores. Species are usually found on wood.

**Hymenomycetes (subdivision *Basidiomycotina)** A large class of *fungi in which the *basidia are typically organized into a *hymenium that is exposed only at maturity. *Basidiospores are discharged forcibly when mature. There are two subclasses. The class includes most of the fungi familiar as *mushrooms, *toadstools, *bracket fungi, *jelly fungi, etc.

**Hymenophyllaceae** A family of filmy ferns with *gradate sori on the *frond margins, and thin, membranaceous fronds with the *lamina only 1 cell thick. At present 33 genera and some 465 species are accepted. They are found mostly in the moist tropics, but with a few genera extending into mild oceanic temperate areas, where they occur mainly in very humid, shady habitats.

***Hymenophyllum*** (family *Hymeno-phyllaceae) A genus of filmy ferns with 2-lipped *indusia to the marginal *sori, widespread in the moist tropics but also occurring in humid temperate areas, including Europe. There are 25 species.

**hypanthium** A cup-like or tube-like enlargement of the floral *receptacle or base of the *perianth that surrounds the *gynoecium and *fruits.

**Hypericaceae** A former plant family whose members are now included in the family *Clusiaceae.

**hyperthermic** *See* PERGELIC.

**hyperthermophile** An *extremophile (*Archaea) that thrives in environments where the temperature is extremely high, in some cases preferring a temperature of about 105°C, tolerating 113°C, and failing to multiply below 90°C. *Compare* THER-MOPHILE.

**hypertonic** Applied to a cell in which the *osmotic pressure is higher than that in the surrounding medium. *Compare* HYPOTONIC and ISOTONIC.

**hypha** A thread-like filament that is the structural unit in many *fungi and *actinomycetes.

***Hyphaene*** (family *Arecaceae) A genus of fan palms whose trunks branch by a true dichotomy (*see* DICHOTOMOUS BRANCHING). They are *dioecious. The *inflorescences are stout and much-branched, and are borne among the leaves. The fruits have a juicy pulp, the *stigma is basal, and there is 1 seed with a stony wall. In Egypt *H. thebaica* (doum or gingerbread palm) has been cultivated since time immemorial for its gingerbread-flavoured fruits. There are 41 species, most of them African, extending to Madagascar, Arabia, and India.

**Hyphochytriomycetes** (subdivision *Mastigomycotina) A class of microscopic *fungi which form *zoospores having an anteriorly directed, tinsel *flagellum. They are found mainly in aquatic habitats as *parasites in *algae and aquatic fungi, and as *saprotrophs on dead plant and animal remains.

***Hypholoma*** (family *Strophariaceae) A genus of *fungi in which the *fruit bodies are *mushroom-shaped, and are typically some shade of yellow or brown. The spores are dark brown or purplish-brown. *Hypholoma* species are found on peaty ground among mosses, or on wood, especially on tree-stumps. *See also* SULPHUR TUFT.

***Hyphomicrobium*** A genus of *bacteria which are not assigned to any taxonomic family. The cells are variable in shape (spindle-shaped, egg-shaped, bean-shaped, etc.), and develop thread-like extensions which may branch. From the tips of these extensions, *motile (*flagellated) cells are produced by *budding; these subsequently become attached to a substrate, lose motility, and develop thread-like extensions, thus completing the cycle. These bacteria are *chemo-organotrophs, which grow *aerobically or *anaerobically in the presence of nitrate. They are found in soil and water.

**Hyphomycetes** (subdivision *Deutero-mycotina) A class of *imperfect fungi in which the vegetative stage is typically a well-developed *mycelium. In some species the sexual states are unknown; others are the conidial states (*see* CONID-IUM) of fungi classified in other (true) subdivisions. There are about 1000 genera. Sometimes the class is considered to contain a single order, sometimes it is not subdivided into orders and families. It includes *predatory, *parasitic, and *saprotrophic fungi, found in aquatic or terrestrial habitats.

**Hypnobryales** (subclass *Bryidae) An order of *pleurocarpous, slender to robust mosses. The leaves are lanceolate to rounded, and may be nerveless or have a single or double nerve. The *capsule is ovoid or cylindrical and inclined or horizontal; the *peristome is double.

There are many genera, including *Brachythecium*, *Eurhynchium*, *Homalothecium*, *Hylocomium*, *Hypnum*, *Rhynchostegium*, and *Rhytidiadelphus*. They are found in a wide range of habitats.

**Hypnum** (order **\*Hypnobryales**) A genus of mosses in which the stems are creeping to erect, and almost unbranched to closely \*pinnately branched; branch leaves are similar in shape to the stem leaves, but are smaller. The \*capsule is erect to horizontal, and ovoid to cylindrical; and is borne on a long, reddish \*seta. *Hypnum* is a cosmopolitan genus, with about 220 species. *H. cupressiforme* is one of the commonest British mosses. It is very variable in form, with a number of distinct varieties. The typical form is a prostrate moss with pinnate branching; the leaves are concave, curved, and overlap one another to give some resemblance to a cypress branchlet. The leaves are nerveless and the leaf cells are long and narrow. The \*capsule is cylindrical and curved and has a beaked lid. The leaves are characteristically all turned downwards, especially at branch tips, although this character is lacking in some of the varieties. *H. cupressiforme* is found in a very wide range of habitats.

**hypobiosis** A synonym for \*dormancy.

**hypocotyl** Part of the embryonic shoot (plumule) or seedling located below the \*cotyledon and above the radicle (young root). *See* EPIGEAL.

**Hypocreales** (subdivision **\*Ascomycotina**) An order of \*fungi that form perithecioid (*see* PERITHECIUM) \*ascocarps, frequently on or within a \*stroma, which are often brightly coloured. Species are \*saprotrophic or parasitic on plants or on other fungi.

**hypodermis** In the leaves of certain plants, the layer of cells immediately below the \*epidermis. It may be strengthened to provide protection or may store water.

**hypogeal** *See* GERMINATION.

**hypogean** Growing or occurring underground.

**Hypogymnia** (order **\*Lecanorales**) A genus of \*lichens, in which the \*thallus is \*foliose to sub-\*fruticose with radiating lobes; the upper surface is typically grey-green. Both surfaces are covered by a \*cortex. \*Rhizines are absent. Species are found on a range of substrates in cooler regions. *Hypogymnia physodes* is one of the commoner British lichens, and is fairly insensitive to atmospheric pollution.

**hypogyny** (adj. **hypogynous**) In flowers, the condition in which the \*calyx, \*corolla, and \*stamens are inserted on the \*receptacle or \*axis, below and free from the \*ovary.

**hypolimnion** The lower, cooler, non-circulating water in a thermally stratified lake in summer. If, as often occurs, the thermocline is below the compensation level, the dissolved oxygen supply of the hypolimnion depletes gradually: replenishment by photosynthesis and by contact with the atmosphere is prevented. Re-oxygenation is possible only when the thermal stratification breaks down in autumn.

**hypolith** A photosynthetic organism that lives on the underside of a rock in a desert, where it is protected from scouring by wind-blown dust and sand, and from ultraviolet radiation, and where trapped moisture provides them with water. A community of hypoliths is known as a hypolithon.

**hypolithon** *See* HYPOLITH.

**hyponasty** A differential, \*nastic growth of the lower part of a plant organ (e.g. the growth of the lower side of a leaf so that the leaf blade curves upwards).

**hypophloeodal** Growing or living beneath (or within) tree bark.

**hypophloic haplostele** (collateral protostele) A \*monostele type of \*protostele in which in cross-section the \*xylem occurs as a single strand with the \*phloem beneath it. *Compare* ACTINOSTELE, HAPLOSTELE, and SOLENOSTELE.

**hypostasis** *See* EPISTASIS.

**hypothallus** **1.** In \*myxomycetes that form \*sporangia, a layer of material upon which sits the sporangium or its stalk.

**2.** In certain *lichens, a layer of fungal *tissue beneath the *thallus and extending beyond the edges of the thallus.

**hypothecium** In diatoms (*Bacillariophyta), the inner *frustule or valve. Compare EPITHECIUM and GIRDLE GROOVE.

**hypothesis** An idea or concept that can be tested by experimentation. In inductive or inferential statistics the hypothesis is usually stated as the converse of the expected results, i.e. as a null hypothesis ($H_0$). For example, if a specified feature were being compared in 2 samples, the null hypothesis would be that no difference existed in the populations from which the samples were taken. This helps workers to avoid reaching a wrong conclusion, since the original hypothesis $H_1$ will be accepted only if the experimental data depart significantly from the values predicated by the null hypothesis. Working in this negative way carries the risk of rejecting a valid research hypothesis even though it is true (a problem with small data samples); but this is generally considered preferable to the acceptance of a false hypothesis, which would tend to be favoured by working in the positive way.

**hypothesis-generating method** A term often applied to data-structuring techniques, such as classification and ordination methods which, by grouping and ranking data, suggest possible relationships with other factors. Appropriate data may then be collected to test these hypotheses statistically. For example, a classification or ordination of plant data from a heathland might suggest a relationship with soil-water status. If suitable soil data were then recorded, this hypothesized relationship could be tested.

**hypothesis testing** An evaluation of an hypothesis using an appropriate statistical method and significance test.

See also STATISTICAL METHOD and NUMERICAL METHOD.

**hypotonic** Applied to the condition in a cell where the *osmotic pressure is lower than that in the surrounding medium. Compare HYPERTONIC and ISOTONIC.

**Hypoxidaceae** A former plant family whose members are now included in the family *Amaryllidaceae.

**Hypoxis** (family *Amaryllidaceae) A genus of herbaceous plants that have a bulbous root stock and long, linear leaves arising from the base of the plant. The *inflorescence is an *umbel on a long, leafless, flower stalk. There are about 80 species. The genus is represented in America, Africa, eastern Asia, and Australia.

**Hypoxylon** (order *Sphaeriales) A genus of *fungi, in which the *fruit body consists of a *stroma in which *perithecia are embedded. The fruit body is hard and dark-coloured and is often hemispherical but may be flattened and crust-like. The tips of the perithecia are just visible as tiny warts on the surface. The spores are dark brown. *H. fragiforme* is common from late summer to spring on dead beech wood; the fruit bodies are hemispherical, 0.1–1.0 cm across, crowded, salmon-pink in colour, becoming brick-red and finally black.

**Hypsithermal** See FLANDRIAN.

**hysginous** Red in colour.

**hyssop** See HYSSOPUS.

**Hyssopus** (hyssop; family *Lamiaceae) A genus of aromatic, *perennial herbs which are woody below, with narrow to linear leaves, and flowers in *whorls. The *calyx has 5 nearly equal teeth, and a 2-lipped *corolla. Formerly hyssop was much used as a flavouring, and it is still sometimes grown for ornament. There are 5 species, occurring from the Mediterranean region to central Asia.

**IAA** *See* INDOLE-ACETIC ACID.

**ianthinus** Blue to purple in colour.

**IBP** *See* INTERNATIONAL BIOLOGICAL PROGRAMME.

**Icacinaceae** A family of trees, shrubs, woody climbers, and a few herbs, in which the leaves are *simple, usually alternate, and without *stipules. The flowers are mostly unisexual and *tetra- or *pentamerous, and the *ovary is *trilocular. The fruits are mostly *drupes. There are 60 genera, with about 320 species, occurring mainly in the tropics, mostly in rain forests.

**Iceland moss** The common name for the *lichen *Cetraria islandica*. The *thallus is erect and tufted, with flattened, shiny, reddish-brown to brown branches bearing small spines on the margins. It is found on moors in northern Britain and in temperate alpine regions. The lichen is edible and can be used, for example, to make jellies.

**idioblast** A cell that differs markedly from those surrounding it.

***Ilex* (family *Aquifoliaceae)** A genus of trees and shrubs in which the leaves are leathery, and mostly evergreen. The flowers are usually *tetramerous. The fruit is a *berry. The genus includes ornamentals (holly is mainly *I. aquifolium*) and the source of yerba maté, a tea-like infusion from the leaves of *I. paraguariensis*. Several have close-grained, white timber. There are about 400 species. Their distribution is cosmopolitan, but they are absent from N. America.

**Illiciaceae** A family, related to the *Winteraceae, that consists of 1 genus, *Illicium*, of shrubs and small trees in which the leaves are *simple, sometimes occurring in pseudo-whorls. The flowers are regular, with numerous spiral parts. The fruit consists of spreading *follicles. There are 42 species, occurring in South-east Asia, N. America, and the W. Indies.

***Illicium*** *See* ILLICIACEAE.

**illuviation** A process of deposition (inwashing) of soil materials, either from suspension or solution, and usually into a lower *horizon, after removal from above or from a lateral source.

**imbibition** The adsorption of liquid, usually water, into the ultramicroscopic spaces or pores found in materials such as *cellulose, *pectin, and cytoplasmic (*see* CYTOPLASM) *proteins in seeds.

**imbricate** Overlapping, like tiles on a roof.

**imidazole (glyoxaline, iminazole)** A compound ($C_3H_4N_2$) whose molecule forms a pentagonal ring of C and H atoms, with an N and NH group attached.

**iminazole** *See* IMIDAZOLE.

**imine** A compound that contains the imino group NH.

**imino acid** An acid derived from an *imine in which the nitrogen of the imino group and the *carboxyl group are attached to the same carbon atom. Proline and hydroxyproline are imino acids, normally classified with *amino acids.

**imino group** *See* IMINE.

**immigration** In genetics, the movement or flow of *genes into a population, caused by immigrating individuals which interbreed with the residents. This is the usual source of new variation in a population, although the fundamental sources of all variation are *gene *mutation and *recombination. *See also* MIGRATION.

**immobilization** The conversion of a chemical compound from an inorganic to an organic form as a result of biological activity; the compound is thereby removed from the reservoir of compounds available to plant roots.

**immunity** A natural or acquired resistance of an organism to a *pathogenic micro-organism or its products.

**immuno-electrophoresis** A technique for the differentiation of *proteins in solution, based on both their electrophoretic (see ELECTROPHORESIS) and immunological properties. Initially the proteins are separated by gel electrophoresis: they are then reacted with specific antibodies by double diffusion through the gel. The pattern of precipitating arcs thus formed can be used to identify the proteins.

*Impatiens* (balsam, touch-me-not) See BALSAMINACEAE.

*Imperata* (family *Poaceae) A genus of plants that includes *I. cylindricum*, a noxious, weedy, fire-resistant grass of tropical Africa and Asia, which has tough *rhizomes. There are 8 species, occurring in the tropics and subtropics.

**imperfect cycle** See BIOGEOCHEMICAL CYCLE.

**imperfect flower** A *flower that lacks functional *stamens or *carpels.

**imperfect state** The asexual state of a *fungus, i.e. the state in which no sexual reproduction occurs.

**importance value** See DENSITY-FREQUENCY–DOMINANCE MEASURE.

**-inae** A standardized suffix used to indicate a ranking of plants at the subtribe level in the recognized code of classification.

**inbreeding depression** The decline in vigour in the offspring of organisms that are closely related genetically; it is often due to a homozygous recessive state.

**incense cedars** See LIBOCEDRUS.

**Inceptisols** An order of *mineral soils that have one or more *horizons in which mineral materials have been weathered or removed. Inceptisols are in the early stages of forming visible horizons, and are only beginning the development of a distinctive soil *profile. The term embraces *brown earths.

**incomplete dominance** In genetics, the situation of *partial dominance in which a *heterozygote shares a *phenotype that is quantitatively intermedi-

ate between those of the corresponding *homozygotes.

**incubous** (of liverwort leaves) See JUNGERMANNIALES.

**incumbent** Resting, or leaning upon a support.

**indehiscent** Applied to *fruits (e.g. *berries) that do not open to release their *seeds.

**independent assortment** (random assortment) The random distribution in the *gametes of separate *genes. If an individual has one pair of alleles A and a, and another pair B and b (this *genotype being represented as AaBb) then it should produce equal numbers of four types of gamete: AB, Ab, aB, and ab. The assortment of alleles of one gene occurs independently of that of the alleles of the other gene. This is found experimentally with many pairs of genes, and is asserted in *Mendel's second law, known as the law of independent assortment: in fact, though, it applies only to distantly linked or unlinked genes (see LINKAGE).

**Indian bean** See CATALPA.

**Indian floral region** Part of R. Good's (*The Geography of the Flowering Plants*, 1974) Indo-Malaysian subkingdom within his *palaeotropical kingdom, and more a region of floral mixing than a distinct entity, notwithstanding the extent of the subcontinent. There are only about 150 endemic genera (see ENDEMISM), and most of these are *monotypic. Many varied and useful plants are thought to be native to the region. See also FLORAL REGION and FLORISTIC PROVINCE.

**Indian spinach** See BASELLA.

**indicator species** **1.** A species that is of narrow *ecological amplitude with respect to one or more environmental factors and that is, when present, therefore indicative of a particular environmental condition or set of conditions. In geobotanical surveys, species or ecotypes with high *heavy-metal tolerance have been used as indicators of metallic ore. **2.** In plant community classification, the term is used more loosely to denote the most characteristic community members. In this case 'indicator species' may include species typical of and vigorous in a particular environment and is not necessarily restricted to species of narrow ecological amplitude. *See also* GEOBOTANICAL EXPLORATION. **3.** A *fossil species used in palaeontology to provide evidence of past conditions or as a chronological guide.

**indicator species analysis** In general, a classificatory scheme in which the final groups are characterized by *indicator species derived from the data in the course of group definition. More specifically, the term refers to a *polythetic divisive classificatory scheme proposed by M. O. Hill in 1975. Sites are ranked by a reciprocal averaging ordination and divided into 2 groups at the mid-point (the 'centre of gravity') of all the weighted data values of the ordination. Indicator species (usually 5) are then identified as those species exclusively, or most nearly so, associated with one or other side of this division (positive and negative indicators). The site-indicator scores, effectively a rough secondary ordination, determine their final classification; and the process may then be repeated within the groups identified. The indicator species form a key, enabling new sites to be added easily into the classificatory framework without excessive recalculation.

**indifferent species** A species with no real affinity for any particular *community, but that is not rare (as an accidental species would be). It is fidelity class 2 in the *Braun-Blanquet phytosociological scheme. *Compare* ACCIDENTAL SPECIES; EXCLUSIVE SPECIES; PREFERENTIAL SPECIES; and SELECTIVE SPECIES.

**indigenous** *See* NATIVE.

**indigo** *See* INDIGOFERA.

**Indigofera** (family *Fabaceae, subfamily *Papilionatae) A genus of small, shrubby or herbaceous legumes that are adapted to dry conditions and have small, *xeromorphic leaves. The seeds are dispersed in an explosive manner. The dye indigo can be extracted from the plants. There are about 700 species, and the genus is represented in all arid zones of the world.

**indirect inhibition** *See* MUTUAL INHIBITION COMPETITION TYPE.

**individualistic hypothesis** The view, first proposed by H. A. *Gleason in 1917, that vegetation is continuously variable in response to a continuously varying environment. Thus, no two vegetation *communities are identical. It implies also that vegetation cannot be classified, and that recognition of particular individual communities will be difficult (the problem arising because of the difficulty of defining boundaries). This viewpoint underlies one of the two polarized approaches to the description and analysis of vegetation communities that were much debated in the 1950s and 1960s. The individualistic hypothesis favours a continuum view of vegetation, for which ordination rather than classification methods are appropriate. *Compare* ORGANISMIC.

**indole-acetic acid (IAA)** A substance that acts as a *growth hormone or *auxin in plants, where it controls cell enlargement and, through interaction with other *plant hormones, also influences *cytokinesis.

**induced fit theory** A variation of the *lock-and-key theory of enzymatic function. It is proposed that the *substrate causes a conformational change in the *enzyme such that the active site achieves the exact configuration required for a reaction to occur. The overall effect would be a tighter binding for the substrate and enzyme.

**inducer** *See* EFFECTOR.

**inducible enzyme** (adaptive enzyme) An *enzyme, substantial amounts of which are formed only in the presence of its *substrate or of a substance resembling its substrate.

**induration** A process of forming indurated *horizons or *hardpans that have a high bulk density and are hard or brittle. Cementing materials may be present and responsible for the induration.

**indusium** The covering of the *sorus of a *fern.

**infarctate** Turgid or solid.

**infection** The invasion of the *tissues of a plant by a *pathogenic micro-organism. Sometimes the term is used as a synonym for a disease caused by a micro-organism.

**infectious** Applied to a disease that can be transmitted from one individual to another.

**inferior** Applied to an *ovary when the other organs of the flower are inserted above it. Compare SUPERIOR.

**infiltration** The downward entry of water into soil.

**inflorescence** A flowering structure that consists of more than a single flower.

**information analysis** An agglomerative classificatory scheme for describing vegetation *communities. It uses the *information statistic as the basis for the joining of groups.

**information index** See SHANNON–WIENER INDEX OF DIVERSITY.

**information statistic** A measure of the extent to which members of a group differ from one another (i.e. of disorder); it is zero when all individuals within the group are identical. In information analysis (an agglomerative, hierarchical classificatory technique, devised by W. L. T. Williams and others, and described in 1966) a hierarchy is constructed by repeatedly joining together those individuals or groups that exhibit the smallest increase in heterogeneity (disorder), and therefore the smallest change in information.

**informosome** See MATERNAL MESSAGE.

**infralittoral fringe** In marine ecology, a term sometimes used to distinguish the intertidal region exposed only at the lowest, i.e. equinoctial, tides. It lies between the *medio- and *circalittoral zones. In the more widely used shore zonation scheme, it corresponds to a transitional area between the *littoral and *sublittoral zones. In this zone *Zostera species are very characteristic plants on soft substrates. The *brown seaweeds (*Laminaria species) start to become abundant in this zone.

**infralittoral zone** See SUBLITTORAL ZONE.

**infructescence** A fruiting structure that consists of more than a single *fruit.

**infuscate** Brown in colour.

**initial** A *meristem cell that is actively dividing. Each division yields 1 differentiated cell which joins the main body of specialized plant tissue and 1 cell that retains its meristematic (see MERISTEM) qualities and is, therefore, available to be the initial for a further division.

**initiator** The *transfer ribonucleic acid (tRNA) that in *eukaryotes carries methionine and in *prokaryotes N-formylmethionine, that binds to the small unit of a *ribosome-bearing *messenger ribonucleic acid (mRNA), thus forming an initiation complex. This, in the presence of 3 protein initiation factors and GTP, which is hydrolysed, enables the larger ribosomal subunit to associate with the complex and peptide-chain synthesis to proceed.

**ink cap** The common name for any *Coprinus *fruit body in which the *gills of the *fungus undergo autodigestion (*autolysis), dissolving into a black, ink-like fluid.

***Inocarpus*** (family *Fabaceae, subfamily *Caesalpinoideae) A genus of trees in which the leaves are *simple. The flowers are regular with 5 narrow *petals. The pod is stout, leathery, and *indehiscent, with 1 big seed, which is edible in I. *fagifer* (Tahitian chestnut). There are 3 species, occurring from Malaya to the Pacific islands.

***Inocybe*** (family *Cortinariaceae) A genus of *fungi in which the *fruit bodies are small, dull-coloured, and *mushroom-shaped, with conical or umbonate caps. This is one of the commonest and largest genera of fungi in the world; however, identification of the species frequently requires microscopic examination of the spores. Most species grow on the ground. Some are very poisonous; none is edible.

***Inonotus*** (family *Hymenochaetaceae) A genus of *fungi in which the *fruit body is typically bracket-like with rusty-brown, fibrous 'flesh'. *I. dryadeus* (*Polyporus dryadeus*) is parasitic on oak trees (*\*Quercus* species).

**inositol** A carbocyclic or sugar alcohol ($C_6H_{12}O_6$) that is widely distributed in both plants and animals. It is often classed with the B vitamins, as it has been reported to be essential in the diets of some organisms.

**input load** *See* GENETIC LOAD.

**integument** The coats of the *\*ovule (usually 2 in flowering plants), which develop into the seed coat (\*testa) after *\*fertilization.

**intercalary meristem** *See* MERISTEM.

**interference competition** The *\*competition that occurs when two organisms demand the same resource and that resource is in short supply, and one of the organisms denies its competitor access to the resource. In essence, space is substituted for the resource as the prime object of competition and dominance of space provides an alternative to efficiency in resource exploitation. Territorial animals exhibit this type of competition, as do robust and aggressive plants (e.g. the mat grass, *Nardus stricta*). *Compare* EXPLOITATION COMPETITION.

**interflow** (**throughflow**) The lateral movement of water through the upper *\*soil horizons, most commonly during or following significant precipitation. If the interflow is shallow it may emerge at and flow for some time across the surface at the bottom of slopes (when it is known as 'return flow').

**interglacial** A period of warmer climate that separates two glacial periods. Mid-latitude interglacials show a characteristic sequence of vegetation change. *\*Pollen of healthy *\*tundra is replaced in the pollen record by abundant herbaceous pollen, which in turn is replaced by *\*Boreal and subsequently *\*deciduous forest, including pollen of *\*thermophilous species, e.g. *\*Tilia* (lime). From this peak the sequence reverses as the trend to colder conditions predominates.

**intergrade** A soil or soil *\*horizon that has the properties of two soils or horizons

that do not share a common origin. An intergrade can be regarded as transitional between two distinctive soils or horizons.

**intermediate filament** A filament or fibre, approximately 10 nm in diameter, that contributes to the cytoskeletal structure of *\*eukaryotic cells. There are several types, each composed of a different, non-*\*motile, structural *\*protein such as *\*keratin, vimentin, or desmin.

**International Biological Programme (IBP)** An international research programme conducted approximately from 1966 to 1975 with the aim of understanding the dynamics of whole *\*ecosystems from a range of world environments, e.g. *\*desert, and *\*tropical rain forest. Objectives included the development of predictive mathematical models for ecosystem structure and function to help assess the effects of as yet unexperienced environmental impacts, as well as aiding the understanding of changes associated with past disruptions. Although not all the projects were equally successful and a complete reference set of models has not been produced and accepted, the models that do exist provide a rational basis for evaluating the likely consequences of proposed human interventions, for remedying the effects of previous human use of an area, and for adapting management systems to improve productivity and environmental quality.

**International Code of Botanical Nomenclature** A set of rules for the formal naming of plants, accepted by botanists, in which the underlying principle is the allocation of a single, unambiguous name to each *\*taxon. The Code comprises a set of 6 Principles to guide those who are selecting a new name, 75 Articles, and a number of Recommendations. Observance of the Articles is mandatory, but not of the Recommendations. This section is followed by translations in French and German and appendixes listing conserved names. The starting-point for naming plants is taken as *Species Plantarum*, published by *\*Linnaeus in 1753; fungi, fossil plants, and bacteria start from different dates and authorities. The Code allows for the naming of *\*cultivars.

**International Union for Conservation of Nature and Natural Resources (IUCN)** An international, independent organization, founded in 1948, with headquarters at Morges, Switzerland, which promotes and initiates scientifically based conservation measures. Its members include more than 450 government agencies and conservation organizations in more than 100 countries, and it works closely with United Nations agencies. It publishes data books listing endangered species, and in 1980 it published the *World Conservation Strategy* in collaboration with the UN Environment Programme, the World Wide Fund for Nature, the Food and Agriculture Organization, and UNESCO.

**internode** Part of a stem that lies between 2 *nodes.

**interphase** A stage in the *cell cycle in which there is no visible evidence of nuclear division but in which there is intense activity, including replication of *chromosomes. Because of the lack of visible evidence of this activity, this phase has been inappropriately called a 'resting phase'.

**interstadial** A phase of warmer climate within a glacial period, but of shorter duration (and thought to be less warm) than an *interglacial. Species demanding warmth, e.g. *Tilia* (lime), are not represented in the pollen record, which shows *boreal affinities. The absence of *thermophilous species may, however, be as much a consequence of the shorter timespan of an interstadial as of the lack of warmth.

**intertidal zone** An area between the highest and lowest tidal levels in a coastal region. *See* LITTORAL ZONE.

**intine** The innermost layer of the wall of a *pollen grain or spore.

**intrinsic rate of natural increase** *See* BIOTIC POTENTIAL.

**introgression** The incorporation of *genes of one species into the gene pool of another. If the ranges of 2 species overlap and fertile *hybrids are produced, the hybrids tend to *backcross with the more abundant species. This results in a population in which most individuals resemble the more abundant parents but also possess some of the characters of the other parent species.

**introrse** Applied to *anthers that open towards the centre of the flower, so releasing their pollen into the flower and promoting self-pollination. *Compare* EXTRORSE.

***Intsia* (family *Fabaceae, subfamily *Caesalpinoideae)** A genus of trees that have *pinnate leaves consisting of a few big *leaflets. The flowers are showy, with 4 *sepals and 1 *petal. The pod is big, woody, and flat, with several seeds. The trees yield valuable, hard, heavy timber (merbau). There are 3 species, occurring in Madagascar, and from Malesia to Australia and the Pacific islands.

**inulin** A *polysaccharide in which about 32 beta-fructose units are joined in a chain by glycosidic (*see* GLYCOSIDE) linkages between the first and second carbon atoms on neighbouring sugar units. Each chain is initiated by a sucrose residue. Inulin is found as a storage compound, especially in the roots, rhizomes, and tubers of many members of the family *Asteraceae.

**inverse analysis** The grouping of attributes based on an analysis of the individuals that possess or lack those attributes; e.g. in plant ecology the grouping of species according to their presence, absence, or relative abundances at different sample sites. The term is used particularly in numerical vegetation classification, and is sometimes referred to as a species classification. Ordination methods may similarly be described as either plot (individual) or species (attribute) ordinations. In plant ecology especially, inverse classifications are often used to complement normal analysis. Thus, data will be analysed using both approaches and coincidence between the final groups examined. It is assumed that high coincidence implies the recognition of an important 'type' *community or *'nodum'. In its concept, nodal analysis is a method based on these principles. *See also* Q-TECHNIQUE; *compare* NORMAL ANALYSIS; and R-TECHNIQUE.

**inversion** A change in the arrangement of genetic material, involving the excision of a chromosomal segment that is then turned through 180° and reinserted at the same position in the *chromosome. The result is a reversal in the order of *genes in that segment of chromosome.

**invertase** See SUCRASE.

*in vitro* Literally, 'in glass', but applied more generally to studies on living material that are performed outside the living organism from which the material is derived. Examples include the use of tissue cultures, cell homogenates, and subcellular fractions.

*in vivo* Applied to studies of whole, living organisms, on intact organ systems therein, or on populations of micro-organisms.

**involucre** A *whorl of *bracts below an *inflorescence.

**involucrellum** In many fungi, a dark structure surrounding the *ostiole.

**involute** Rolled inward, e.g. of a leaf. Compare EVOLUTE.

**ion** An atom that has acquired an electric charge by the loss (*cation; positive charge) or gain (*anion; negative charge) of one or more electrons.

**-ion** See ZÜRICH–MONTPELLIER PHYTOSOCIOLOGICAL SCHOOL.

**ionic bond** The bond formed when an electron is transferred from one atom to another. The atom that loses the electron becomes a positively charged *ion and the atom that gains the electron becomes a negatively charged ion. A strong electrostatic force then bonds the 2 ions together. The bonding in a sodium chloride crystal (NaCl) is ionic, the crystal lattice containing $Na^+$ ions and $Cl^-$ ions. See also COVALENT BOND, HYDROGEN BOND, and METALLIC BOND.

**ionophorous antibiotic** A generic term for a group of non-polar, *lipid-soluble *antibiotics, which function by facilitating the passage of alkali-metal *cations (e.g. potassium) through biological membranes, thus disrupting transmembrane potentials in bacteria.

**ipoh (upas tree)** See ANTIARIS.

*Ipomoea* (family *Convolvulaceae) A genus, chiefly of climbing herbs, that have showy flowers which are sometimes noctural. The genus includes *I. batatas* (sweet potato) which has sweetish, edible *tubers. There are about 500 species, occurring in tropical and warm temperate regions.

**Iré rubber** See FUNTUMIA.

**Irian division** The lowland and hill component of the New Guinea biota, including the rain forests of the Cape York Peninsula, Australia, nearly all of whose animals are identical at the species level with those of the lower altitude forests of New Guinea. See also TUMBUNAN DIVISION.

*Iriartea* (family *Arecaceae) A genus of palms in which the trunk is tall, surmounting a mass of stout, spiny, *stilt roots. The leaves are *pinnate, the *leaflets *praemorse. The trees are *monoecious. There are 7 species, found in the rain forests of tropical America.

**Iridaceae** A family of monocotyledonous (see MONOCOTYLEDON), *perennial *herbs that grow from *bulbs, *corms, or *rhizomes. The flowers have *perianths with 2 similar or dissimilar *whorls, each of 3 segments, 3 *stamens, and an *inferior, 3-celled *ovary. They are cosmopolitan, with some 92 genera, comprising about 1850 species. Many genera are cultivated, especially *Iris*, *Crocus*, and *Gladiolus*.

*Iris* (family *Iridaceae) A genus of plants that have *rhizomes or *bulbs, leafy shoots usually flattened in one plane, and showy, characteristic flowers in which the 3 outer *perianth segments (the 'falls') usually arch down and are larger than the 3 inner perianth segments (the 'standards') which are often more or less erect and stalked. The 3 style-branches are broad and petal-like, and forked at the apex. There are some 300 species, throughout the northern temperate zone, many being cultivated (as are many hybrids) for their beautiful flowers.

**Irish moss** See CHONDRUS.

**iroko** See *Chlorophora*.

**iron (Fe)** An element required by plants. It is used in reactions in which rapid oxidation reductions occur by the transfer of electrons, as in *photophosphorylation and *oxidative phosphorylation. Other roles are not understood. Iron-deficient plants have chlorotic (*see* CHLOROSIS) young leaves; at first the veins remain green but later they too become chlorotic.

**iron bark** A group of *Eucalyptus* species, named from their distinctive bark.

**iron pan** An indurated (*see* INDURATION) soil *horizon, found usually at the top of the B horizon, in which iron oxide is the main cementing material.

**ironwood** See *Mesua*.

**irregular** See ZYGOMORPHIC.

**irritability** The responsiveness of an organism to outside stimuli.

**irruption** A sudden change or oscillation in the population density of an organism.

***Irvingia*** (family *Simaroubaceae) A genus of trees in which the leaves are *simple, with *stipules fused as a cap over the *apical bud. The fruit is a *drupe. There are 3 species, occurring in the African and Asian tropics.

**isabelline** Greyish in colour, drab.

**isidium** A small, branched or unbranched growth that develops from the *thallus in certain *lichens. Isidia may function as *propagules.

**island biogeography** The study of the distribution of plant and animal species on islands or in areas that are sufficiently isolated to resemble islands. Islands are numerous, and their biotas (*see* ISLAND BIOTAS) are often small enough to be quantified. Accordingly it has been possible to determine a relationship between area and species number, as an equilibrium between immigration and extinction (*see* EQUILIBRIUM THEORY; RESCUE EFFECT). This is the basis of island biogeography, and it also has great relevance to the continents, where plant and animal communities are effectively reduced to islands in a sea of cultivation or urbanization.

**island biogeography, theory of** A theory, advanced in 1967 by R. H. MacArthur and E. O. Wilson and now largely discredited, that the number of species on an island will reach a dynamic equilibrium between the continual immigration of species from a mainland source and the extinction of species already present. Once equilibrium is reached the species number will remain constant, but with a continually changing composition. The theory goes on to state that, if the immigration rates and the extinction rates are known, then the species number at equilibrium can be calculated. The theory fails, however, to take account of species interaction or of habitat diversity which is, *de facto*, usually greater on big islands; it also makes a major and possibly erroneous assumption that immigration is independent of island size. The theory also compares topographically identical near and far islands in relation to the mainland source, maintaining that near islands would have a higher rate of immigration than far ones because the biogeographical barrier presented by the sea would be smaller. It also states that the extinction rate would be independent of island location, and therefore the resulting equilibrium point for the number of species would be higher for the nearer island. The assumption that the extinction rate is independent of island location may be incorrect because of the *rescue effect. The concept has been extensively used as a basis for the selection of natural reserves, the assumption being that large areas with minimal boundaries located close to other similar areas will prove the richest and the most sustainable type of reserve. This use of the theory has been extensively tested with mixed results.

**island biotas** The plants and animals of oceanic islands. Because of their isolation, the biotas normally include numerous endemic (*see* ENDEMISM) taxa. Island biotas are generally fragile, unbalanced in that they lack plants and animals with poor transoceanic dispersal capacities, and vulnerable to disturbance by humans and introduced species.

i

**island hopping** The colonization of an island or islands by plants and animals from an adjacent island or islands. Birds are particularly likely to hop from one island to another. Over geological time, islands drift away from their areas of origin. The descendant biotas maintain themselves in the ancestral environment by island hopping on to successively younger islands as these emerge.

**Isobryales** (subclass *Bryidae*) An order of *pleurocarpous mosses in which the primary stems are prostrate and creeping but the secondary stems may be erect; branching is irregular to *pinnate or dendroid. The leaves are usually ovoid to lanceolate, may be nerved or not, and have pointed tips. *Capsules are round to cylindrical; the *peristome is double. Genera include *Climacium*, *Fontinalis*, and *Neckera*. They are found in a variety of habitats.

**isoelectric focusing** A technique for the electrophoretic (*see* ELECTROPHORESIS) separation of amphoteric (i.e. able to combine with either an acid or a base) molecules in a gradient of *pH, usually formed from a combination of *buffers held on a polyacrylamide gel support medium. The molecules will move in the gradient, under the influence of an electric field, until they reach their isoelectric pH (i.e. the pH at which they are electrically neutral), where they form a sharp band. Separation is achieved because the various molecular species will have different isoelectric pHs.

**isoenzyme** *See* ISOZYME.

**Isoetaceae** A family of *Pteridophyta, mostly aquatic *vascular cryptogams with 2 types of *spore (*see* HETEROSPORY). They have a short, stout, basal stock from which arise the roots below, and a dense *rosette of usually tubular, quill-like leaves, which bear the *sporangia sunken in their expanded bases. The first leaves of the season produce *megasporangia in their bases which contain large *megaspores that develop into female *gametophytes. *Microsporangia are produced in later leaves and contain tiny *microspores which develop into minute male gametophytes. There are 2 genera, *Isoetes* (quillworts), with some 70 species of cosmopolitan distribution, and *Stylites*, with 1 species in the Peruvian Andes.

***Isoetes* (quillworts)** *See* ISOETACEAE.

**isofrigid** *See* PERGELIC.

**isogamy** The fusion of *gametes that are morphologically alike. This is an uncommon condition, found in some green *algae, *fungi, and protozoa. *Compare* OOGAMY.

**isogeneic (syngeneic)** Applied to a *graft that involves a scion and stock that are genetically identical.

**isohyperthermic** *See* PERGELIC.

**isokont** Having *flagella of equal length. *Compare* HETEROKONT.

**isomer** Either or any of 2 or more compounds that have the same molecular composition but different molecular structure. Isomers differ from each other in their physical and chemical properties.

**isomerase** A *enzyme that catalyses a reaction involving the interconversion of *isomers.

**isomesic** *See* PERGELIC.

**isomorphic** Applied to an *alternation of generations in which the *sporophyte and *gametophyte generations are morphologically similar.

**isonome** *See* TRANSECT.

**isophylly** The state of having leaves all of the same *morphology on 1 plant. *Compare* HETEROPHYLLY.

**isoprene** (2-methyl butadiene) A 5-carbon compound that forms the structural basis of many biologically important compounds (e.g. the *terpenes).

**isoprenoid** A compound that consists of 2 or more *isoprenes or their derivatives.

**isothermic** *See* PERGELIC.

**isotonic** Applied to a cell in which the *osmotic pressure is equal to that in the surrounding medium. *Compare* HYPERTONIC and HYPOTONIC.

**isotope** One of 2 or more varieties of a chemical element whose atoms have the same numbers of protons and electrons but different numbers of neutrons. There are 300 naturally occurring isotopes, but only 92 naturally occurring elements. Isotopes may be produced by various nuclear reactions. Frequently the products are radioactive. *See also* ISOTOPIC DATING.

**isotopic dating** A means of determining the age of certain materials by reference to the relative abundances of the parent *isotope (which is radioactive) and the daughter isotope (which may or may not be radioactive). If the decay constant (the 'half-life' or disintegration rate of the parent isotope) and the concentration of the daughter isotope are known, it is possible to calculate an age. *See also* DATING METHODS; RADIOACTIVE DECAY; RADIOCARBON DATING; and RADIOMETRIC DATING.

**isozyme (isoenzyme)** One of 2 or more *enzymes that have identical or similar functions, but are encoded by different loci (*see* LOCUS).

**istle fibre** *See* AGAVE.

**iterative evolution** A repeated *evolution of similar or parallel structures in the development of the same main line. There are many examples of iterative evolution in the fossil record, spanning a wide range of groups. This evolutionary conservatism is probably due to the overriding morphogenetic control exerted by certain *regulatory genes.

**IUCN** *See* INTERNATIONAL UNION FOR CONSERVATION OF NATURE AND NATURAL RESOURCES.

**ivory nut** *See* PHYTELEPHAS.

**ivy** *See* HEDERA.

**iwatake** The lichen *Umbilicaria esculenta*, which is edible and considered a delicacy in Japan. The species does not occur in Britain.

**Ixora** (family *Rubiaceae) A genus of shrubs and small trees in which the flowers are showy with a long tube, and borne in heads. Several species are ornamentals. There are about 400 species, with a pantropical distribution.

**Jacaranda** (family *Bignoniaceae) A genus of trees in which the leaves are *bipinnate and *opposite. The flowers are bluish-lilac in colour and borne in *panicles or clusters. The fruit is a short pod. *J. filicifolia* is a widely grown ornamental. There are 30 species, occurring in tropical America.

**jackfruit (jak)** *See* ARTOCARPUS.

**Jacksonia** (family *Fabaceae, subfamily *Papilionatae) A genus of plants, most of which are shrubs with *simple or *palmate leaves. They are adapted to arid conditions but a few species occur in the rain forests of central Australia. There are about 50 species, all of them endemic (*see* ENDEMISM) to Australia, and mainly confined to the west.

**jak (jackfruit)** *See* ARTOCARPUS.

**Japanese cedar** *See* CRYPTOMERIA.

**Japanese umbrella pine** *See* SCIADOPITYS.

**japonica** *See* CHAENOMELES.

**japweed (strangleweed)** The common name for the brown seaweed *Sargassum muticum*; the seaweed originated in Japan but has become established on the coasts of Europe and N. America. It is regarded as a pest, fouling boat propellers and fishing nets and disturbing the natural balance of the indigenous marine flora and fauna.

**jarrah** *See* EUCALYPTUS.

**jarrah die-back** A disease of jarrah (*Eucalyptus marginata*, the mahogany gum-tree) caused by the *fungus *Phytophthora cinnamoni*. The disease is of considerable economic importance in Australia, where *E. marginata* is an important timber tree.

**Jatropha** (family *Euphorbiaceae) A genus of shrubs, a few of which are *xerophytes with swollen stems. Male and female flowers are borne in the same *panicle. The fruit is a 3-shouldered,

leathery, *indehiscent *capsule. There are about 170 species, occurring in the tropics and subtropics, and in N. America and S. Africa.

**jelly fungus** The common name for any *fungus of the *Phragmobasidiomycetidae (particularly members of the *Tremellales) in which gelatinous *fruit bodies are formed.

**jelutong** *See* DYERA.

**Jerusalem artichoke** (*Helianthus tuberosus*) *See* ASTERACEAE.

**jew's ear fungus** The common name for the ear-shaped *fruit bodies of *Auricularia auricula* (Hirneola auricula-judae).

**Job's tears** *See* COIX.

**Johannesteijsmannia** (family *Arecaceae) A genus of solitary palms, all but one of which are stemless, with huge, spectacular, diamond-shaped or *lanceolate, undivided leaves, up to 2.5 m long, with essentially *palmate nervation. The *inflorescences are among the leaves. The flowers are small, scented, and hermaphrodite, with 3 free *carpels. The fruits have a thick, knobbly wall. There are 4 species, occurring in lowland rain forests in Malaya (where 3 of the species are found), Sumatra, and Borneo.

**jointed nine-awn** *See* ENNEAPOGNON.

**Joinvillea** (family *Flagellariaceae) A genus of giant, grass-like herbs which have *plicate leaves with a sheathing base. The flowers are bisexual, *petaloid, and borne in *panicles. The fruit is a *drupe. There are 2 species, occurring from Malesia to the Pacific islands.

**jongkong** *See* DACTYLOCLADUS.

**jorales** *See* MEDITERRANEAN SCRUB.

**J-shaped growth curve** A curve on a graph that records the situation in which, in a new environment, the population density of an organism increases rapidly in an exponential or logarithmic form, but then stops abruptly as environmental resistance (e.g. seasonality) or some other factor (e.g. the end of the breeding phase) suddenly becomes effective. The actual rate of population change depends on the *biotic potential and the population size. It may be summarized mathematically as: $dN/dt = rN$ (with a definite limit on N) where N is the number of individuals in the population, $t$ is time, and $r$ is a constant representing the intrinsic rate of increase (biotic potential) of the organism concerned. Population numbers typically show great fluctuation, giving the characteristic 'boom and bust' cycles of some insects, or the ones seen in *algal blooms. This type of population growth is termed 'density-independent' as the regulation of growth rate is not tied to the population density until the final crash. *Compare* S-SHAPED GROWTH CURVE.

**Juan Fernandez floral region** Part of R. Good's (*The Geography of the Flowering Plants*, 1974) *neotropical kingdom, which corresponds geographically to the islands of Juan Fernandez and the Desventuradas Islands, 650–1000 km off the coast of Chile. There is a remarkable degree of *endemism, including 1 family, 18 genera, and nearly 160 species. *See also* FLORAL REGION and FLORISTIC PROVINCE.

**Jubaea** (family *Arecaceae) A *monotypic genus, comprising *J. chilensis* (*spectabilis*), the Chilean wine palm, which is the southernmost of all palms. Very hardy, it has probably the stoutest of palm trunks, 1.2–1.8 m diameter. The leaves are *pinnate, their bases persistent. It is *monoecious. The tree is felled for its sugary sap (it yields up to 300 l) which makes a honey or wine, and it is now rare.

**Juglandaceae** A family of trees in which the leaves are *pinnate, with no *stipules. The flowers are small, unisexual, and *bracteate, and borne in *spikes or *catkins. They are wind-pollinated, and have an *inferior *ovary. The fruit is a *drupe or *nut, sometimes attached to a wing-like *bract. The family yields several useful timbers (e.g. walnut

and hickory) and fruits (e.g. walnut and pecan). There are 7 genera, with 59 species, most of which occur in northern temperate regions, but a few of which are tropical.

**Juglans** (walnut; family *Juglandaceae) A genus of deciduous, *nut-bearing trees, up to 30 m tall, that have twisting, spreading branches. They have large, oily, deeply lobed *cotyledons. The leaves are *compound. Male and female *catkins occur on the same tree, the male catkins dangling in the wind, the female catkins upright. They produce oval, green fruits (a *drupe) with a fleshy exocarp and a bony endocarp (the shell, *see* PERICARP), which splits down the midrib. The wrinkled, woody nut is eaten fresh or pickled, and used in desserts, cakes, and confectionery. The most popular garden species are *J. regia* (Persian walnut), *J. nigra* (black walnut), and *J. cinerea* of N. America, the butter-nut. The wood is used for furniture. There are 21 species, occurring mainly in northern temperate regions, but extending to the tropics and native to Asia and America.

**jujube** *See* ZIZYPHUS.

**Juncaceae** A family of monocotyledonous (*see* MONOCOTYLEDON) *herbs that are close to *Liliaceae in the structure of the *trimerous flowers, but these are small and inconspicuous, with tiny, chaffy *perianth segments, and adapted to wind pollination. The leaves are long and narrow, and either grass-like or nearly cylindrical in form. They are of little economic value but are important ecologically, especially in poorly drained areas. There are 10 genera, with some 325 species, of cosmopolitan distribution, but centred in temperate zones.

**Juncaginaceae** A family of monocotyledonous (*see* MONOCOTYLEDON) *herbs most of which have linear, *radical, sheathing leaves, and erect *spikes or *racemes of tiny *di- or *trimerous flowers without *bracts. The *ovary is *superior, with 4 or 6 *carpels, usually joined together, and a short *style. *Triglochin* (arrow-grass) is widespread and occurs in wet places. There are 4 genera, with 18 species, distributed in temperate and cold, marshy habitats in the northern and southern hemispheres.

**Juncus** (rush; family *Juncaceae) A genus of tufted, usually *perennial *herbs that have entire leaves and open leaf *sheaths. The *inflorescence is small, with bisexual, usually brown flowers producing *capsules and many seeds. There are about 225 species. Distribution is cosmopolitan, but the plants are mainly restricted to cold and wet places and are therefore rare in the tropics.

**June berries** *See* AMELANCHIER.

**Jungermanniales** (class *Hepaticae) An order containing the leafy *liverworts. The shoots are usually prostrate or ascending and have a dorsiventral organization. The leaves commonly overlap one another in one of two ways; in the 'succubous' arrangement the upper edge of each leaf (i.e. that nearest the shoot apex) is hidden beneath the lower edge of the leaf above; in the 'incubous' arrangement the lower edge of each leaf is hidden by the upper edge of the next leaf down. These arrangements are constant characters in genera. Genera include *Cephalozia, *Diplophyllum, *Frullania, *Lophocolea, *Nardia, *Plagiochila, *Pleurozia, *Porella, and *Scapania.

**jungle** A *subclimax *tropical rain forest, consisting of a tangled growth of *lianes, bamboo scrub, palms, etc. It forms an impenetrable barrier to travel. It is characteristic of former clearings and of riversides, where light penetration is greater than in the forest interior. By contrast, true climax rain forest has little undergrowth, since light penetration is poor, and, contrary to the popular image, it is easily negotiated.

**juniper** *See* JUNIPERUS.

**Juniperus** (family *Cupressaceae) A genus of evergreen conifers containing trees and shrubs with needle-like (often juvenile) or scale-like leaves. The plants are *monoecious or *dioecious, female cones having 3–8 fleshy scales which coalesce to resemble berries, containing 1 to many hard seeds. There are about 50 species native to the northern hemisphere, mountains of Africa, or the West Indies. Several are cultivated for timber used for pencils or furniture, like *J. virginiana*, the pencil cedar or red cedar of eastern N. America; its fragrant wood repels insects and is used for chests or panelling. The aromatic fruits of *J. communis* are used to flavour gin, and many cultivars of this species are popular ornamental plants.

**Juniperus procera** (African pencil cedar) A valuable, small to medium-sized coniferous tree, which is widespread and common in the mountains of E. Africa.

**Jurassic** One of the three *Mesozoic periods, about 199.6–145.5 Ma ago, that followed the *Triassic and preceded the *Cretaceous. The Jurassic Period is subdivided into 11 stages, with clays, calcareous sandstones, and limestones being the most common rock types. The first birds, including *Archaeopteryx*, appeared in the Upper Jurassic and the flora included many forms that are still extant (e.g. conifers, cycads, ferns, and ginkgos).

**jute** The common name given to 2 species of *Corchorus* (family *Tiliaceae) and the fibre derived from their *bark. The fibre is used in the manufacture of fabrics for industrial and agricultural bags, canvas, and strong cordage. When compared with other similar fibrous raw materials, its consumption is second only to cotton and it is favoured because of its low cost.

**K** *See* POTASSIUM.

**kampfzone** An altitudinal belt of stunted and often prostrate trees, found between the upper limit of tall, erect trees growing in forest densities (waldgrenze; *timber-line) and the extreme upper limit of tree growth (the species limit or *baumgrenze). *See also* ELFIN WOOD.

**kangaroo grass** *See* THEMEDA.

**kangaroo paw** *See* ANIGOZANTHOS.

**kapok** *See* CEIBA.

**kapur** *See* DRYOBALANOPS.

**karri** *See* EUCALYPTUS.

**karyogamy (caryogamy)** The fusion of 2 (usually *haploid) nuclei during sexual reproduction, resulting in the formation of a *diploid *zygote. This occurs immediately after *plasmogamy.

**karyotype** The entire chromosomal complement of an individual or cell, which may be observed during *mitotic *metaphase.

**Kastanozems** Soils that have a *mollic horizon more than 20 cm deep and also concentrations of calcium compounds within 100 cm of the surface. Kastanozems are a reference soil group in the FAO *soil classification.

**katsura tree** *See* CERCIDIPHYLLACEAE.

**kauri pine** *See* AGATHIS.

**kava** See *PIPER* and PIPERACEAE.

**keel** 1. (carina) A cup-like petal (e.g. in *Pisum* flowers) formed by the fusion of two petals. 2. Any structure resembling a ridge, or the keel of a boat.

**kelp** 1. Brown seaweeds that grow below the low-tide level. Large brown *algae, e.g. *Laminaria* species, which anchor themselves firmly to the seabed, are typical. 2. The ash obtained by burning various large, brown seaweeds, used as an agricultural fertilizer and as a source of iodine, potash, and soda.

**kempas** *See* KOOMPASSIA.

**kennarten species (characteristic species)** Like 'faithful species', a former collective term for species of fidelity classes 3–5 (i.e. preferential, selective, and exclusive species). As *phytosociological work progressed, the need became evident for finer distinction within the kennarten species. *See also* FAITHFUL SPECIES.

**keranga (heath forest)** Forest-heath vegetation found in south-eastern Asia, the Amazon basin, and Guyana, and on a small area of the coast of Gabon, central Africa. It grows on siliceous, *podzolic soils. Much of the nutrient supply for this dense vegetation is furnished by rainfall rather than by rock weathering.

**keruing** *See* DIPTEROCARPUS.

***Keteleeria* (family *Pinaceae, *tribe Abietineae)** A genus of conifers that have spine-tipped leaves. It is close to *Abies*, but has erect, persistent cones, ovoid buds with scales forming an overlapping sheath, and male flowers in short clusters. There are 2 species, native to the mountains of southern China and Taiwan.

**keto-** A prefix attached to the name of a chemical compound to indicate that its molecule includes a carbon atom attached by a double bond to an oxygen atom and by single bonds to two other carbon atoms.

**ketone** An organic compound that contains a ketone group, $>C = O$.

**ketose** A monosaccharide or its derivative that contains a *ketone group.

**kidney bean** *See* PHASEOLUS.

**kieselguhr** *See* DIATOMACEOUS EARTH.

**Killarney bristle-fern** *See* TRICHOMANES.

**killing frost** A sharp fall in temperature that damages a plant so severely as to cause its death, or that prevents the reproduction of an *annual, *biennial, or *ephemeral plant. As the falling temperature approaches freezing-point, some water is lost from the *vacuole into the intercellular spaces, where a further drop in temperature causes it to form small crystals of ice. If the temperature then rises slowly, the water will be reabsorbed by the cell as the ice melts and the cell will recover, but if the thaw is rapid the water will be lost and the cell will die from dehydration. If the freezing temperature is prolonged, ice may be lost by sublimation (i.e. the direct change from the solid to the gaseous phase). This will also cause dehydration. Reproduction will be prevented if freezing causes such damage to flowers or to developing fruit that has not yet produced viable seed. Seeds themselves contain little water and are seldom damaged by frost. *See also* GROWING SEASON.

**Kimmeridgian** The penultimate stage of the *Jurassic, dated at about 156–150 Ma ago, underlain by the Oxfordian and succeeded by the Portlandian.

**Kimura, Motoo** *See* NEUTRALITY THEORY OF EVOLUTION.

**kinase** An *enzyme that catalyses reactions involving the transfer of phosphates from a *nucleoside triphosphate, particularly *ATP, to another *substrate.

**kinesis** A change by a *motile organism or cell of its rate of locomotion in response to a particular stimulus. The direction of locomotion remains random and is unrelated to the direction of the stimulus. *Compare* TAXIS.

**kinetin** (6-fururylaminopurine) A degradation product of animal DNA, which does not occur naturally and which has properties similar to those of *cytokinins. Applied to certain leaves, kinetin delays senescence in its vicinity and attracts nutrients.

**kinetochore** A dense, plaque-like area of the *centromere region of a *chromatid, to which the *microtubules of the *spindle attach during cell division.

**King Charles's apple** *See* OAK-APPLE GALL.

**kingdom** In *taxonomy, one of the major groups into which organisms are placed. In the widely-used five-kingdom system of classification the kingdoms are: *Bacteria; *Protoctista; Animalia; *Fungi; and *Plantae. The kingdoms are grouped into two super-kingdoms: *Prokarya, containing the kingdom Bacteria; and *Eukarya, containing the remaining four kingdoms. In the three-domain classification system, kingdoms are ranked below the *domains.

**king palm** *See* ARCHONTOPHOENIX.

**kino** *See* PTEROCARPUS.

***Klebsiella*** (family *Enterobacteriaceae) A genus of *Gram-negative, rod-shaped, non-*motile *bacteria, which occur as *saprotrophs, and as *parasites or *pathogens of humans and other animals.

**knotted wrack** *See* EGG WRACK.

***Koenigia islandica*** *See* BIPOLAR DISTRIBUTION.

**kola-nut** *See* COLA.

***Koompassia*** (family *Fabaceae, subfamily *Caesalpinioideae) A genus of giant trees in which the leaves are *pinnate. The flowers are tiny, regular, and *pentamerous. The pod is flat, tardily *dehiscent, and contains 1 seed surrounded by a papery wing. There are 3 species, occurring in lowland rain forests, but with a disjunct range, 2 species being found in western Malesia and 1 in New Guinea.

**Kornberg enzyme** The enzyme DNA polymerase, isolated from *Escherichia coli* in 1958 by Arthur Kornberg and his colleagues. It functions in repair synthesis of damaged *DNA.

***Korthalsia*** (family *Arecaceae) A genus of climbing palms (rattans) in which the leaves have an *ochrea and a terminal spiny whip, and the leaflets are diamond-shaped, their upper margins toothed. The *inflorescences are terminal (the stem dying after flowering), and burst through the leaf sheath. The flowers are hermaphrodite, and single. The fruits are scaly. The stems are hard and durable, and occasionally branched. They are widely utilized, and sometimes sold. There are 25 species, found in rain forests from Burma to New Guinea.

**kowhai**   *See* SOPHORA.

**Kranz anatomy**   A special structure in the leaves of plants that have a *$C_4$ pathway of carbon dioxide fixation. The leaves contain a ring of mesophyll cells, containing a few small *chloroplasts concerned with the initial fixing of carbon dioxide, surrounding a sheath of *parenchyma cells (the bundle sheath) which has large chloroplasts involved in the *Calvin cycle.

**krasnozem**   *See* RED PODZOLIC SOIL.

**krummholz**   Gnarled, stunted, and usually bush forms of trees, typically conifers, which grow in the *kampfzone between waldgrenze and *baumgrenze.

Krummholz trees can also be found at relatively low altitudes, whenever there is marked exposure to strong winds.

**K-selection**   The selection for maximizing competitive ability, the strategy of *equilibrium species. Most typically this is a response to stable environmental resources. This implies selection for low birth rates and high survival rates among the offspring, and prolonged development. $K$ represents the *carrying capacity of the environment for species populations showing an S-shaped population growth curve. *See* S-SHAPED GROWTH CURVE; *compare* R-SELECTION.

**kumquat**   *See* FORTUNELLA.

**kunkar**   *See* CALCIFICATION.

**labellum** In *Orchidaceae, the lowest of the 3 flower *petals, which differs from the other 2. In lipped flowers produced by species belonging to other families, the platform formed by the lowest petal or fused petals.

**Labiatae** See LAMIACEAE.

**Laboulbeniales (subdivision *Ascomycotina)** An order of *fungi, which are obligately *parasitic, mainly on insects but also on other arthropods. They appear not to cause their hosts significant harm. The *ascocarp is a *perithecium.

**Laburnum** (family *Fabaceae) A genus of trees that have *trifoliate leaves and leafless, pendulous *racemes of yellow, pea-like flowers. The fruits are rounded pods containing poisonous seeds. They are cultivated for ornament. There are 2 species, native to southern Europe.

**Laccaria** (family *Tricholomataceae) A genus of *fungi in which the *fruit bodies are *mushroom-like and very variable in colour. *L. laccata* (the deceiver) is tawny to brick red; *L. amethystea* (the amethyst deceiver) is deep purple-lilac, becoming paler on drying. The *spore print is white; *spores typically appear spiny under the microscope. There are many species, found on the ground in woods, heaths, etc.

**laciniate** Deeply cut, into irregular, narrow segments or lobes.

**Lac operon** The *operon that controls the synthesis of acetylase, permease, and beta galactosidase, 3 *enzymes involved in the metabolism of lactose in *Escherichia coli*. The Lac operon is a cluster of structural *genes specifying the above enzymes. These genes are controlled by the co-ordinated action of *cis*-dominant *promoter and *operator regions, the activity of which is in turn determined by a *repressor molecule that is specified by a separate *regulator gene. Most of the detailed discovery of this system was worked out by François Jacob and Jacques Monod during the late 1950s and early 1960s; the Lac operon *DNA was isolated in pure form in 1969.

**Lactarius** (family *Russulaceae) A genus of *fungi in which the *mushroom-shaped *fruit bodies exude a milky latex when damaged (hence the common name, milk caps). The *gills are decurrent. There are numerous species, some of which are edible, others poisonous. They are found on the ground under trees.

**lactic acid** A 3-carbon hydroxy-acid that is formed as the major metabolic product of certain bacteria (and also from pyruvic acid in animal cells when *glycolysis occurs under anaerobic conditions).

**Lactobacillus** A genus of *Gram-positive *bacteria in which the cells are ovoid or rod-shaped; most are non-*motile. They are *chemo-organotrophic, have complex nutritional requirements, and are found in various habitats where carbohydrates are available as nutrients. They rarely cause disease. Some strains are used in the manufacture of certain dairy products, e.g. yoghurt and cheese.

**Lactococcus** See STREPTOCOCCUS.

**Lactoridaceae** (order *Laurales) A *monotypic family, *Lactoris fernandeziana*, which is a shrub endemic to the Juan Fernandez Islands, off the S. American coast. The leaves are small and alternate, with *stipules. The plants are *monoecious but the flowers may be unisexual or bisexual. They comprise just 3 *sepals, 6 *stamens, and 3 *carpels, and develop into *follicles.

**Lactuca sativa** (lettuce) See ASTERACEAE.

**lacuna** A gap between cells, often filled with air. See LEAF GAP; compare LUMEN.

**lacunose** Having a surface pitted with cavities or indentations.

**LAD**   *See* LAST APPEARANCE DATUM.

**lady's mantle**   *See* ALCHEMILLA.

*Laetiporus* (family *Polyporaceae) A genus of *fungi in which the *fruit body is spongy or softly corky in texture, and bears sulphur-yellow or coral-red pores. The generative *hyphae lack clamp connections. Species are found on living or dead timber. *See also* SULPHUR FUNGUS.

**laevigate**   Smooth, as if polished.

*Lagerstroemia* (family *Lythraceae) A genus of trees in which the leaves are *opposite, with conspicuous, pointed, *axillary buds. The flowers are held in big terminal *panicles and are showy, purple to pink in colour, with 6 crinkled *petals and many *stamens. The fruits are woody *capsules seated on a persistent *calyx. Several species are widely planted as showy ornamentals. *L. speciosa* also produces the timber pyinma, which is ring-porous and prized for boat-building. There are 53 species, occurring from India to northern Australia. They are most common in seasonally dry South-east Asia.

**lagg**   The *rheotrophic *fens that surround a *raised bog.

**Lagos rubber**   *See* FUNTUMIA.

*Laguncularia*   *See* COMBRETACEAE.

**LAI**   *See* LEAF-AREA INDEX.

**lake forest**   A forest dominated by conifers, which occurs in the eastern half of N. America, in Minnesota, Michigan, northern Pennsylvania, southern Ontario, and northern New England. Although a distinct formation, it is in many respects transitional between *boreal coniferous forest and southern *deciduous forest. Very little lake forest survives, having been cut in the latter part of the nineteenth century.

**Lamarck, Jean Baptiste Pierre Antoine de Monet, chevalier de (1744–1829)** A French naturalist who, in 1809, advanced the theory that evolutionary change may occur by the inheritance of characteristics acquired during the lifetime of the individual. This theory was also the basis for *Lysenko's arguments on the inheritance of acquired plant characteristics. It is interesting to note that the theory of the inheritance of acquired characteristics did not hold a central position in Lamarck's own writings. His cardinal point was that evolution is a directional, creative process in which life climbs a ladder from simple to complex organisms. He believed the inheritance of acquired characteristics provided a mechanism for this evolution. Lamarck explained that this progress of life up the ladder of complexity is complicated by organisms being diverted by the requirements of local environments; thus, cacti have reduced leaves (and giraffes have long necks). *Compare* DARWIN, CHARLES ROBERT.

**Lamarckism**   The theory of *evolution propounded by Lamarck.

**lamella   1.** One of the membranes that comprise the *thykaloid discs in *chloroplasts. Those in adjacent *grana often appear interconnected. **2.** *See* GILL. **3.** A leaf, leaf-like structure, or leaf blade.

**Lamiaceae**   A family, formerly known as Labiatae, of *herbs or low shrubs, comprising a natural group whose members have quadrangular stems and *simple leaves without *stipules in *opposite pairs. Commonly there are apparently *whorled *spikes of flowers; actually branched *cymes are borne in the *axils of opposite pairs of *bracts but are usually very condensed. The flowers are strongly irregular, with a 5-toothed or 2-lipped *calyx and a tubular *corolla which is often clearly 2-lipped but may have only the lower lip developed. There are 2 or 4 *stamens and a 4-lobed *ovary with the *style inserted between the 4 lobes. The fruit is composed of 4, separating, 1-seeded *nutlets, a clear distinction from those genera of *Scrophulariaceae which are superficially similar. Many genera and species (e.g. mints and sages) are used as flavourings, as they contain aromatic, volatile, essential oils. Some of these oils (e.g. menthol and thymol) are used as mild antiseptics. Many species are cultivated for their flowers, others beautify northern temperate woodlands in springtime. There are 224 genera, with about 5600 species, and their main distribution is in tropical and warmer temperate regions.

**lamina** A flat, sheet-like structure (e.g. the blade of a leaf).

**Laminaria** (order *Laminariales) A genus of brown seaweeds in which the *thallus consists of a *holdfast, a *stipe, and a flattened blade which may or may not be divided. They are found mainly on lower shores and in shallow coastal waters, usually attached to rocks; they are usually exposed only at the very low spring tides. *See also* OARWEED, SEA-BELT, and TANGLE.

**Laminariales** (division *Phaeophyta) An order of brown seaweeds in which there is a *heteromorphic *alternation of generations, with a large *sporophyte and a *microscopic *gametophyte. The sporophyte is usually differentiated into *holdfast, *stipe, and blade. There are 4 families, and many genera, found mainly in cold and temperate seas.

**laminarin** A storage product found in brown algae (*Phaeophyta). It is a polysaccharide, comprising about 20 β-D-glucopyranose units linked through carbon atoms in the 1:3 position.

**laminate** **1.** Flat and broad, shaped like a leaf. **2.** Comprising layers of material.

**lanceolate** Broad, but tapering to a point at both ends, like the blade of a lance.

**land bridge** A hypothetical connection between two land masses, especially continents, that allowed migration of plants and animals from one land mass to the other. Before the widespread acceptance of continental drift, the existence of former land bridges was often invoked to explain faunal and floral similarities between continents now widely separated.

**landnam** A Danish word meaning a primary forest clearance, producing a characteristic horizon in the *pollen chronologies of Britain and western Europe, and also found elsewhere, though at different stratigraphic dates. The absolute quantity of tree pollen in relation to that of other plants is greatly reduced. Sometimes a charcoal layer is found at a similar stratigraphic level. The non-tree pollen shows an increase in arable weed species, e.g. plantain and nettles, and cereal pollens are first recorded in some abundance at this level. Together with archaeological evidence, the horizon is strongly suggestive of a first phase of deliberate forest clearance by Neolithic agriculturists. The landnam phase is dated to about 5000 BP throughout north-western Europe.

**landrace** A distinct crop variety or cultivar developed and maintained agriculturally.

**landscape architecture** A branch of architectural studies that is particularly concerned with design in relation to the scenic environment, e.g. the use of appropriate tree species to blend with buildings or landscape morphology, the harmonious design of way-marking, shelters, etc. The term is often used to include the design of gardens, i.e. synonymously with landscape gardening.

**landscape evaluation** The assessment of landscape as a scenic resource. It is controversial, since such concepts as 'scenic beauty' are neither generally agreed nor readily quantifiable. Nevertheless, evaluations are vital for the conservation of high-quality and/or traditional landscapes, especially in densely populated, developed regions.

**Langermannia gigantea** See CALVATIA.

**Langhian** That interval in the *Miocene Epoch which is preceded by the *Burdigalian and followed by the Serravallian Ages. Most authors subdivide the Langhian into the Early Langhian Age (Early Miocene) and Late Langhian Age (Middle Miocene) at about 14.4 Ma.

**lanose** Woolly.

**Lansium** (family *Meliaceae) A genus of small trees in which the leaves are *pinnate. The fruit is a *berry with 1–5 seeds surrounded by fleshy pulp. There are 3 species occurring in western Malesia, several of which are cultivated for their edible fruit.

**Lantana** (family *Verbenaceae) A genus of shrubs which produce showy flower heads. Several are cultivated as ornamentals. There are 150 species, occurring in tropical America and Africa. Several are now pantropical, and *L. camara* is a pernicious weed.

***Laportea*** (family *Urticaceae) A genus of trees or bushes whose leaves have highly irritant stinging hairs. There are 22 species, which are mainly tropical and subtropical in distribution, with some occurring in the tropics and warm temperate regions.

**larch 1.** *See* LARIX. **2. (golden larch)** *See* PSEUDOLARIX.

***Larix*** (larches; family *Pinaceae, *tribe Abietineae) A genus of deciduous and coniferous trees that have soft, deciduous leaves in tufts, mostly on short spur shoots, and oval, persistent cones. There are about 9 species, found in the cold temperate parts of the northern hemisphere.

**larkspur** *See* DELPHINIUM.

***Lasiopetalum*** (family *Sterculiaceae) A genus of soft-wooded trees and shrubs that have *simple or slightly lobed, alternate leaves. The flowers are borne in complex *cymes. There are 37 species, all of them endemic (*see* ENDEMISM) to Australia.

**last appearance datum (LAD)** The last recorded occurrence of a key taxon in biological history.

***Latania*** (family *Arecaceae) A genus of big fan palms in which the leaves are greyish and stiff. The plants are *dioecious. The *inflorescence consists of pendulous *spikes. The fruits contain 1–3 stony seeds. There are 3 species, occurring in the Mascarene Islands (Indian Ocean). They are widely cultivated.

**late blight of potato (potato blight)** A widespread and serious disease affecting the potato and related plants. Symptoms include the appearance of brown patches on the leaves, often with white mould on the undersides. Under damp conditions the entire foliage may collapse. Brown lesions also develop on tubers, spreading to involve the entire tuber in a dry *brown rot. The disease is favoured by wet weather. It is caused by the *oomycete *Phytophthora infestans* and was responsible for the Irish potato famine in the 1840s.

**Late Devensian Interstadial** *See* WINDERMERE INTERSTADIAL.

**late glacial** A term usually applied to the time between the first rise of the temperature curve after the last minimum of the *Devensian glaciation, and the very rapid rise of temperature that marks the beginning of the post-glacial, or *Flandrian period. The late Devensian extends from about 15 000–10 000 BP and shows a characteristic threefold climatic and hence depositional sequence, from cold, older *Dryas deposits, to warmer *Allerød, to colder, younger Dryas. In Europe there is some evidence for an additional warmer phase, the *Bølling interstadial, dating about 13 000 BP, i.e. during older Dryas times.

**lateral meristem** *See* MERISTEM.

**laterite** A weathering product of rock, composed mainly of hydrated iron and aluminium oxides, hydroxides, and *clay minerals, but also containing some silica. It is related to bauxites and is formed in humid, tropical settings by the weathering of such rocks as basalts.

**latex** A white, commonly sticky substance produced in specialized *tissues within a plant.

***Lathyrus*** (sweet pea, vetchling; family *Fabaceae) A genus usually of *herbs that climb by leaf tendrils and commonly have 2 pairs or 1 pair of *leaflets (sometimes none) on each leaf. The stems are often winged or angled. The *style is flattened and bearded above. Many are cultivated for their flowers, or more rarely for the edible seeds of some species. There are about 150 species, occurring in the northern temperate zone, and in mountains of African and S. American tropics.

**laticifer** A cell, or complex of cells, that contains *latex.

**latifoliate** Broad-leaved.

**latitudinal vegetation zone** A major vegetation belt that coincides with a broad, global, climatic regime. Thus, rain forest relates to the equatorial climate, and *savanna to the seasonal tropics roughly 10°–20° on either side of the equator. The simple pattern is complicated by the effects of altitude, so that in each latitudinal vegetation zone there are found vegetation types more characteristic of latitudes nearer the poles.

**lauan** *See* SHOREA.

**Lauraceae** A family of trees and shrubs (but *Cassytha* is a twining parasite) that have evergreen, alternate or rarely *opposite, *simple leaves without *stipules. The tissues contain aromatic oils. The flowers are regular, small, unisexual or hermaphrodite, and usually *trimerous. The *ovary is usually *superior and surrounded by a cup-like *receptacle. The fruit is a *berry or *drupe-like, and often enclosed by a *receptacle. Many species are valuable as ornamentals, or for timber, oil, or spices. There are 45 genera, with 2000–2500 species, occurring in the tropics and subtropics, and centred in Amazonia and South-east Asia.

**Laurasia** The northern continental mass produced in the early *Mesozoic by the initial rifting of *Pangaea along the line of the northern Atlantic Ocean and the *Tethys sea. Laurasia included what was to become N. America, Greenland, Europe, and Asia, while the large, southern continental mass (called *Gondwana) was later to divide into S. America, Africa, India, Australia, and Antarctica. Fossil evidence indicates that the Laurasian floral assemblage included many species of tropical plants that were incorporated into sediments to form the extensive coal measures that are mined throughout Europe and the eastern USA.

***Laurus*** **(family *Lauraceae)** A genus of small trees whose flowers are borne in *cymes enclosed in leafy *bracts. There are 2 species, 1 native to southern Europe (*L. nobilis*, sweet bay, whose dried leaves are used for flavouring and berries in veterinary medicine), the other to the Canary Islands and Madeira (*L. canariensis*).

***Lavandula*** **(family *Lamiaceae)** A genus of *xerophytic shrubs which have narrow, leathery leaves. The nectar from their flowers makes good honey and the plants produce an aromatic oil that is used in perfumery. There are 20 species, occurring on islands in the Atlantic, the Mediteranean region, Somalia, and India.

**laver (laver bread)** The common name for the edible seaweed *Porphyra umbilicalis* (Bangiales). The *thallus is thin and sheetlike, and attached to rocks and other objects in the intertidal zone. *Ulva lactuca* is sometimes known as green laver.

**leaching** Removal of soil materials in solution. Water may percolate downwards through a soil, removing humus and mineral bases in solution before depositing them in underlying layers by *illuviation. The upper layer of leached soil becomes increasingly acidic and deficient in plant nutrients.

**lead** *See* RADIOMETRIC DATING.

**leader** The growing tip of the main shoot, especially of a tree.

**leading dominant** According to the *Wisconsin ordination scheme, the species in any given stand or *quadrat that has the highest importance value.

**leaf** A thin, usually green, expanded organ borne at a *node on the stem of a plant, typically comprising a *petiole (stalk) and blade (lamina) and subtending a *bud in the *axil of the petiole. The leaves are the main site of *photosynthesis. Sometimes, in classification, the term is restricted to the leaves that are *diploid structures of the *sporophyte generation.

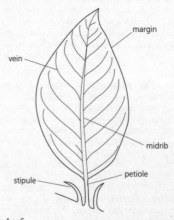

*Leaf*

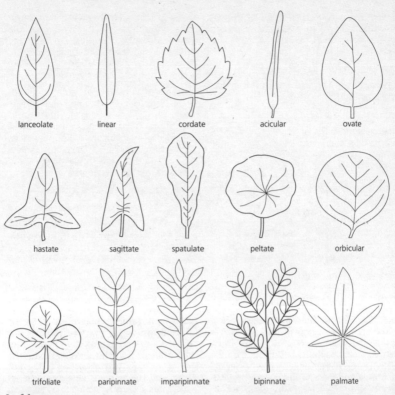

*Leaf shapes*

**leaf-area index (LAI)** Of a plant, the total leaf surface area exposed to incoming light energy, expressed in relation to the ground surface area beneath the plant (e.g. an LAI of 4 means that the leaf area exposed to light is 4 times the ground surface area).

**leaf-area ratio** The photosynthetic surface area per unit dry weight of a plant. It is a measure of the efficiency with which a plant deploys its photosynthetic resources. Typically, it is increased by low light intensities.

**leaf gap (lacuna)** In the *stele of many vascular plants, an area of *parenchyma cells associated with a *leaf trace.

**leaflet** Each leaf-like part of a *compound leaf that is not subtending a *bud in the *axil of its *petiole.

**leaf trace** A strand of vascular tissue connecting a leaf to a region within the *stele. *See* LEAF GAP.

**leatherwood** *See* CYRILLACEAE.

***Leathesia*** **(family Corynophlaeaceae, order *Ectocarpales)** A genus of brown seaweeds which form small (2.5–5 cm) bulbous growths on rocks or on other seaweeds; at first the bulbs are solid, but later they become hollow. The surface is shiny in living plants. *Leathesia* is common around British coasts, being found on the middle and lower shore from spring to early autumn.

***Lecanora*** **(order *Lecanorales)** A genus of *lichens in which the *thallus is *crustose, sometimes with radiating lobes. The *apothecia are characteristi-

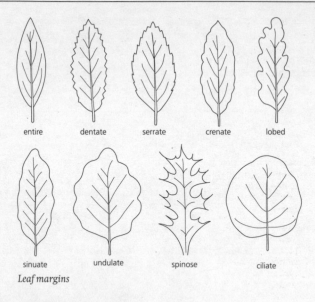

entire    dentate    serrate    crenate    lobed

sinuate    undulate    spinose    ciliate

*Leaf margins*

cally *lecanorine. There are many species, found in a wide range of habitats. Some species are more resistant than most lichens to atmospheric pollution. *L. coniza-eoides* is one of the lichens most tolerant of pollution; this species forms greyish-green, granular crusts on bark, wood, rocks, walls, and the like, even within large cities, but is uncommon in areas with clean air.

**Lecanorales** (subdivision *Ascomy-cotina) An order of *fungi most of which are lichenized but which may also be *saprotrophic or *lichenicolous. The *apothecia are typically solitary with open, round discs bearing the *hymenium; the *asci are thick-walled and normally turn blue when treated with iodine. There are many families, members of which are found in a wide range of habitats in most regions of the world.

**lecanorine** Applied to *apothecia of the type produced by *lichens of the genus *Lecanora*; this type of apothecium is surrounded by a rim of *tissue of the same colour as the *thallus, the rim being called the thalline margin.

*Leccinum* (order *Boletales) A genus of *fungi in which the *fruit bodies are *mushroom-shaped and the *hymenium lines tubes opening by pores on the under-side of the cap. The cap surface is dry, not slimy; and the *stipe is long and covered with lines or scales which darken with age. There are many species, found on the ground in association with trees.

**lecideine** Usually applied to *apothecia of the type produced by *lichens of the genus *Lecidea*; this type of apothecium lacks a margin of thalline *tissue, although the upper edge of the *excipulum may be visible as a rim (the proper margin) similar in colour to the disc of the apothecium.

*Lecidia* (order *Lecanorales) A genus of *lichens in which the *thallus is *crustose and may be more or less immersed in the substratum. Spores are colourless and *aseptate. *Apothecia are characteristically *lecideine. The genus originally included numerous species but has now been divided into many different genera. Individual species can usually be recognized only by microscopic examination. They are found on a wide range of substrates.

**lecithin (phosphatidyl choline)** One of a group of *phospholipid compounds that are found in higher plants and animals.

**lecotropal** Shaped like a horseshoe.

**lectin** A generic term for *proteins extracted from plants (especially legumes), and also from some molluscs and fish, that exhibit *antibody activity in animals.

**lectotype** One of a collection of *syntypes which, subsequent to publication of the original description, is chosen and designated through published papers to serve as the *type specimen.

**Lecythidaceae** A family of trees in which the leaves are *simple and crowded at the twig tips, and rarely have *stipules. The flowers are hermaphrodite and sometimes *cauliflorous. There are 4–6 *sepals and *petals and many *stamens, fused at the base and sometimes strongly one-sided in development. The *ovary is *inferior, and fused to the *receptacle. The flowers are frequently nocturnal and bat-pollinated. The fruit is a *berry or woody *capsule, opening by a lid. There are 20 genera, with about 280 species, occurring in the tropics and concentrated strongly in S. American rain forests.

***Lecythis** (family *Lecythidaceae)* A genus of trees whose flowers have a strongly asymmetric *androecium. Their oily, edible seeds are known as sapucaia nuts. There are 25 species, occurring in tropical America.

**Leguminosae** *See* FABACEAE.

**lemma** In *Poaceae, the lower of the two *bracts situated beneath each *floret.

**Lemnaceae (duckweeds)** A monocotyledonous (*see* MONOCOTYLEDON) family of small, freshwater, aquatic, often floating *herbs which are reduced relatives of *Araceae and have a disc-like, oval or elliptical, flattened or swollen plant body, not differentiated into stem and leaves, few or no roots, and minute flowers often enclosed in a tiny sheath. The *spathe subtends 1 female and 2 male, *apetalous flowers. The sexes are separate on any one plant and not always produced. There are 6 genera, with about 30 species, of cosmopolitan distribution.

**lemon** *See* CITRUS.

**Lentibulariaceae** A family of insectivorous *herbs, some of which are aquatic with insect-trapping bladders or finely divided leaves (e.g. *Utricularia*, bladderworts); some are bog or terrestrial plants with basal rosettes or sticky, insect-catching leaves (e.g. *Pinguicula*, butterworts); and some are terrestrial or epiphytic (*see* EPIPHYTE). The flowers are solitary or in *racemes on leafless stems, *pentamerous, 2-lipped, and spurred, the *petals joined into a tube below (resembling *Linaria*, etc., in the *Scrophulariaceae), but the *ovary is 1-celled. There are 4 genera, with some 250 species, or world-wide distribution.

**lentic** Applied to a freshwater habitat characterized by calm or standing water (e.g. lakes, ponds, swamps, and bogs).

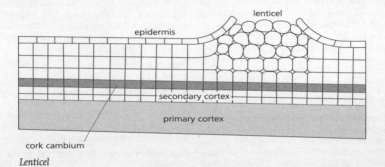

*Lenticel*

**lenticel** A pore in the stem of a woody plant, showing as a raised spot that may be filled with a powdery substance. The pore permits air to reach *tissues below the surface.

**lenticular** Shaped like a biconvex lens.

***Lentinus*** (family *Polyporaceae) A genus of *fungi in which the *fruit bodies are tough and have *gills with serrated margins. *L. lepideus* causes decay of timber; *L. edodes* is edible and is cultivated in some parts of the world. *See also* SHIITAKE and STAG'S-HORN FUNGUS.

***Lenzites*** (family *Polyporaceae) A genus of *fungi, which form *bracket-like *fruit bodies on the under-side of which are *gills bearing a *hymenium. The spores are colourless, smooth, and cylindrical. *L. betulina* forms semicircular or fan-shaped brackets, often in tiers, on *deciduous trees and dead wood. *See also* TRAMETES.

**Leonardian** The second of the four stages of the *Permian in N. America. In some areas it is zoned by the use of fusulinid foraminiferids. It is the N. American equivalent of the Rotliegende.

***Lepidium*** (family *Brassicaceae) A genus of *herbs that have alternate leaves without *stipules. The *inflorescence is usually a *raceme and flowers have 4 *sepals, 4 *petals, and 6 *stamens. The genus is cosmopolitan and found in all arid regions. There are about 150 species, which include *L. sativum*, the common garden cress, which is grown for food.

***Lepidodendron selaginoides*** An important species of *Palaeozoic plant, characterized by a *dichotomous branching, by diamond-shaped leaf scars, and by large cones. During the Upper *Carboniferous *Lepidodendron* species flourished on several continents and grew up to 30 m before branching.

**lepidoid** Scaly.

***Lepidophloios kilpatrickense*** The earliest known representative of the important *Palaeozoic family of plants, the Lepidodendraceae (Lycopsida). Closely related to *Lepidodendron, Lepidophloios kil-*patrickense* is distinguished by its internal anatomy. It is recorded from the Upper *Carboniferous of Scotland.

***Lepidosperma*** (family *Cyperaceae) A genus of sedges that have a creeping underground *rhizome and aerial stems which are often 3-angled in cross-section. They occur in eastern Asia and Australia. There are a total of 40 species, occurring in Malesia and Australia. *L. gladiatum* (sword sedge) is used to stabilize sand dunes in Australia.

**lepidote** Covered in small scales (e.g. the leaves of *Aetoxicaceae).

***Lepiota*** (family *Agaricaceae) A genus of *fungi in which the *fruit bodies are initially enclosed in a *universal veil; there is no *volva. *Spores are typically colourless. Many species are edible. They are found on the ground, on manure, on compost heaps, etc. *See also* PARASOL MUSHROOM.

**lepis** The tiny scales on the skin of the fruits of *Myrialepis*.

***Lepista*** (family *Tricholomataceae) A genus of *fungi, which form *mushroom-like *fruit bodies. The *spore print is pale pink; *spores appear rough or minutely spiny under the microscope. There are many species, some of which are edible and highly esteemed (*see* BLEWIT). They are found on the ground in woods, hedgerows, gardens, grassland, etc.

***Lepraria*** (form order Agonomycetales) A *form genus of *imperfect *lichens, in which the *thallus is powdery and indeterminate. *L. incana* forms pale greenish-grey powdery crusts in damp, shady places, e.g. on tree roots; it is common. *L. candelaris* forms bright, golden-yellow patches on tree trunks, especially in crevices in the bark of oak trees.

**leprose** Consisting of or bearing powdery or scurfy granules.

***Leptaspis*** (family *Poaceae) A genus of grasses that have broad, ovate leaves. There are 7 species, occurring in tropical rain forests in W. Africa, the Mascarenes, and from Indo-Malaysia to Fiji.

**leptodermous** Thin-skinned or thin-walled.

**leptoid** In some bryophytes (*Bryophyta), an elongated cell in the stem specialized for conducting nutrients, analogous to a *sieve cell. *Compare* HYDROID.

**leptokurtic** Applied to a distribution that is more peaked than a *Gaussian distribution (i.e. a few points occur far from the origin, but most are very close to it). This is typical of wind-dispersed *propagules.

**Leptosols** Soils in which there is hard rock within 25 cm of the surface, or that overlie material containing more than 40% calcium carbonate within 25 cm of the surface, or that have less than 10% fine-grained material to a depth of 75 cm. These are weakly developed soils. Leptosols are a reference soil group in the FAO *soil classification.

***Leptospermum*** (family *Myrtaceae) A genus of evergreen bushes or small trees whose leaves are small, spiral, and *sessile. The flowers are solitary. The fruit is a *capsule. There are about 30 species, occurring from Malaya to New Zealand and the Caroline Islands, but most are found in Australia, where they are known as tea-trees.

***Leptospira*** (order *Spirochaetales, family **Leptospiraceae**) A genus of flexible, spiral-shaped *bacteria in which one or both ends of the cells are typically bent or hooked. *Aerobic *chemo-organotrophs, they may be free-living in aquatic environments, or may be *parasitic or *pathogenic in vertebrate hosts. Strains of *L. interrogans* can cause Weil's disease (infectious jaundice).

**leptospire** A *spirochaete of the *genus *Leptospira*.

**leptosporangiate** In *vascular plants, applied to the condition in which *sporangia have walls comprising a single layer of cells and arise from a single parent cell. *Compare* EUSPORANGIATE.

**leptotene** *See* MEIOSIS.

***Leptothrix*** A genus of *Gram-negative *bacteria, not assigned to any taxonomic family. Cells are rod-shaped, occurring in chains enclosed by a sheath, or as single cells or pairs of cells. Sheaths often contain oxides of iron or manganese. Free cells are *motile with one or more *flagella. They are *aerobic *chemo-organotrophs. They are found in various aquatic habitats, especially in slow-moving, unpolluted streams containing iron, etc.

**lessivage** The *eluviation of insoluble particles to a deeper level within the soil. This process often produces *cutan.

**lethal mutation** A *gene *mutation whose expression results in the premature death of the organism carrying it. Dominant lethals kill *heterozygotes, whereas recessive lethals kill *homozygotes only. *See also* VISIBLE.

**lettuce** (*Lactuca sativa*) *See* ASTERACEAE.

***Leucaena*** (family *Brassicaceae, subfamily *Mimosoideae) A genus of shrubs or small trees which have finely divided, *bipinnate leaves. The flowers are *axillary, solitary or paired, with 10 *stamens. *L. leucocephala* is widely planted as shade for coffee and the like, browse (although if eaten in excess it causes hair to fall out), fuel wood, and timber. There are about 40 species, most of which occur in tropical America.

**leucine** An aliphatic, non-polar, neutral *amino acid that, unlike most amino acids, is sparingly soluble in water.

***Leucobryum*** (order *Dicranales) A genus of cushion-forming mosses, found mainly in tropical regions. *L. glaucum* is common in Britain; it is a distinctive moss, forming pale, glaucous to greyish-green cushions which become whitish on drying. The leaves can absorb large quantities of water. The cushions are frequently attached only loosely to the ground, and often become detached to form almost spherical moss balls. *Leucobryum* is found in woods and on wet moors; it is *calcifuge. The genus contains about 150 species.

***Leuconostoc*** (family *Streptococcaceae) A genus of *Gram-positive *bacteria in which the cells are spherical or lenticular and usually occur in pairs or chains. *Chemo-organotrophic, they can grow in the presence or absence of air. There are several species, found for example in fruit and vegetable matter, and dairy products; none is *pathogenic. Some strains are used in the manufacture of certain dairy products, e.g. butter.

**leucoplast** A colourless *plastid involved in storage functions. Leucoplasts can be subdivided into *aleuroplasts, *amyloplasts, and *elaioplasts.

***Leucopogon*** (family *Epacridaceae) A genus of *xeromorphic shrubs that have alternate, *simple, *sessile leaves with *stipules. They have small, regular, bisexual flowers in *spikes or *racemes, and the fruit is a *loculicidal *capsule. There are about 150 species, mainly represented in Australia but found also in Malesia, New Zealand, and New Caledonia.

**levulose** *See* FRUCTOSE.

***Leycesteria*** (family *Caprifoliaceae) A genus of deciduous shrubs that have flowers in *whorls in *axils of *bracts, forming drooping terminal *spikes. The flowers are regular, *pentamerous, and funnel-shaped. *L. formosais* is cultivated for ornament. There are 6 species occurring in Himalayan and Chinese mountains.

**Leyland cypress** *See* CUPRESSOCYPARIS.

**liana** *See* LIANE.

**liane (liana)** Any wiry or woody, free-hanging, climbing plant.

**Lias** A commonly used term for the lowest stage of the *Jurassic. The Liassic is synonymous with the Hettangian/ Sinemurian Stages of Europe. Blue-grey shales and muddy limestones are typical of the Lias, with extensive outcrops in England and France.

***Libocedrus*** (incense cedars; family *Cupressaceae, *tribe Cupressinae) A genus of trees that have leafy branches compressed on 2 sides and bearing scale-like leaves in *opposite pairs. The cones are oblong with elongate valvate scales, of which only 1 pair is fertile; the seeds are unringed. The leaves, rich in resin, smell of turpentine when crushed. There are 8 species occurring around the Pacific: in N. and S. America, eastern Asia, and New Zealand.

**library (gene library)** A random collection of cloned (*see* CLONE) *DNA fragments

in a number of *vectors that ideally includes all the genetic information of that species.

***Licania*** (family *Chrysobalanaceae) A genus of small, evergreen, rain-forest trees whose fruit is a *drupe and which yield useful but *siliceous timber. There are 171 species, all found in tropical America, except for 2–3 in western Malesia and one in Africa.

**lichen** A type of composite organism, which consists of a *fungus (the mycobiont) and an *alga or *cyanobacterium (the phycobiont) living in *symbiotic association. A lichen *thallus may be crust-like (crustose), scaly or leafy (foliose), or shrubby (fruticose), according to the species. Lichens are classified on the basis of the fungal partner; most belong to the *Ascomycotina. Specialized asexual reproductive structures may be produced (*see* e.g. SOREDIUM and ISIDIUM). Many lichens are extremely sensitive to atmospheric pollution and have been used as pollution indicators.

**lichenicolous** Growing on *lichens.

**lichenin** A *glucan that is found in lichens, containing 80–200 beta-*glucose units. Of these, about 30 per cent are glycosidically (*see* GLYCOSIDE) linked to the third carbon atom, and the remainder to the fourth carbon atom of the neighbouring sugar units.

**lichen woodland** The facies of the boreal conifer zone, situated south of the forest *tundra *ecotone and north of the closed forest proper. The characteristic is that of open woodland, with a sparse stand of conifers set in a ground layer dominated by lichens (e.g. *Cladonia*).

**lichen zone** An area with a characteristic lichen flora ranging from complete absence to a full complement of foliose and fruticose species, depending on the level and type of air pollution present in the area. The standardized lichen zones for England and Wales range from zone 0 (complete absence), typical of heavily polluted industrial areas, to zone 10, in which bearded lichens (e.g. *Usnea* and *Ramalina* species) are present in association with many other species. Zone

10 is typical of unpolluted areas in which virtually no sulphur dioxide, fluorine compounds, or other air pollutants are present.

**lichi** See LITCHI.

***Lichina*** **(order \*Lecanorales)** A genus of \*lichens in which the \*phycobiont is a \*cyanobacterium. The \*thallus is \*fruticose, branching, more or less gelatinous, and typically dark to black in colour. *L. confinis* and *L. pygmaea* are common on rocks by the sea at or around high-water mark.

***Licuala*** **(family \*Arecaceae)** A genus of small fan palms with windmill-like fronds, of forest undergrowth, which are sometimes stemless, and often clumped. Leaflets have multiple folds, and are blunt-tipped. The \*inflorescences are held as \*axillary \*panicles. The flowers are hermaphrodite with 3 free \*carpels. The fruits are brightly coloured, and thinly fleshy. There are about 100 species, found from southern China and Burma to Queensland and Vanuatu.

**Liebig, Baron Justus von (1803–73)** A German chemist who contributed much to the systematization of organic chemistry, to the early development of biochemistry, and to agricultural chemistry. In 1840 he published *Die organische Chemie in ihrer Anwendung auf Agrikulturchemie und Physiologie* ('Organic chemistry in its application to agricultural chemistry and physiology'), in which he showed that plants take up nutrients in simple chemical form and that nutrient deficiencies in soils may be remedied by the application of mineral fertilizers. He maintained that plant growth is limited by the availability of the scarcest essential nutrient (his law of the minimum). Others expanded on this work later to produce a broader appreciation of the 'limits of tolerance', recognizing that maximum as well as minimum thresholds exist for all commodities (not only chemicals) that are essential to plant and animal growth. *See also* SHELFORD'S LAW OF TOLERANCE.

**Liebig's law of the minimum** The concept first stated by J. von \*Liebig in 1840, that the rate of growth of a plant, the size to which it grows, and its overall health depend on the amount of the scarcest of its essential nutrients that is available to it. This concept is now broadened into a general model of \*limiting factors for all organisms, including the limiting effects of excesses of chemical nutrients and other environmental factors. *See also* SHELFORD'S LAW OF TOLERANCE.

**life cycle** A series of developmental changes undergone by the individuals comprising a population, including \*fertilization, reproduction, and the \*death of those individuals, and their replacement by a new generation. The life cycle in fact is linear with respect to individuals but cyclical with respect to populations. In many plants there is a succession of individuals in the entire cycle, with \*sexual or \*asexual reproduction linking them.

**life-form** The structure, form, habits, and life history of an organism. In plants, especially, characteristic life-forms, in particular morphological features, are associated with different environments. This observation has formed the basis of several attempts at life-form classifications of vegetation. \*Raunkier's scheme (1934) is the best-known and most widely applied life-form scheme.

**life-zone** An original concept of C. H. Merriam (1894) describing the way in which changing vegetation forms give a series of life-zones in relation to temperature gradients. In modern ecology these life-zones are defined by reference to a range of interacting environment gradients, and reflect animal as well as plant characteristics. On a world scale life-zones are thus synonymous with the major \*biomes. However, life-zone is more frequently used for more local changes (e.g. the altitudinal zonation of \*communities on mountains).

**ligand** **1.** An atom, \*ion, or molecule that acts as the electron-donor partner in one or more co-ordination bonds. A \*heterocyclic ring is formed if the ligand is an organic compound, and the product is termed a chelate. **2.** A molecule (e.g. an \*antibody) that can bind to specific sites on \*cell membranes.

**ligase** An *enzyme that catalyses a reaction which joins 2 *substrates, using energy derived from the simultaneous *hydrolysis of a *nucleotide triphosphate.

**light-and-dark bottle technique** *See* OXYGEN METHOD.

**light-dependent stage** *See* LIGHT RE-ACTIONS.

**light reactions** (light-dependent stage) During *photosynthesis, those reactions which require the presence of light (e.g. *photophosphorylation and *photolysis). During the light reactions, light strikes *chlorophyll molecules, which absorb its energy. This allows one electron to escape for each photon of light absorbed. The electron attaches to a neighbouring molecule, thereby ejecting another electron and causing electrons to move along an *electron-transport chain of molecules. Some of the transported energy is used to attach phosphate groups to molecules of *adenosine diphosphate (ADP) converting them to *adenosine triphosphate (ATP); this reaction is called photophosphorylation. ATP is used to carry energy to wherever it is needed in the organism. Energy not used in the ADP→ATP reaction is used to split a water molecule into $H^+$ and $OH^-$ ions; the reaction is called photolysis. The $H^+$ attaches to *nicotinamide adenine dinucleotide phosphate (NADP), converting it to NADPH. The $OH^-$ passes one electron to the chlorophyll molecule, restoring the neutrality of both chlorophyll and hydroxyl. Hydroxyls then combine to form water ($4OH \rightarrow 2\,H_2O + O_2 \uparrow$).

**lignicolous** Growing on *decorticate wood.

**lignified** Applied to cells that have a large amount of *lignin deposited in their *cell walls, giving them a rigid, woody structure.

**lignin** A complex, cross-linked polymer, comprising phenyl propene units, that is found in many plant-cell walls. Its function appears to be to cement together and anchor *cellulose fibres and to stiffen the *cell wall. Lignin reduces infection, rot, and decay. It is among the most chemically inert of plant substances and survives in fossils of woody stems.

**lignum vitae** (timber of *Guaiacum officinale*) *See* ZYGOPHYLLACEAE.

**ligulate** Strap-like or tongue-shaped; having *ligules.

**ligule** **1.** A scale-like membrane that covers the surface of a *leaf. **2.** In some *Compositae, a strap-shaped *corolla.

**lilac** *See* SYRINGA.

**Liliaceae** A large family of *monocotyledons, most of which are *herbs (lilies, onions, etc.) with elongated leaves springing from *rhizomes, *corms, or *bulbs, but some are shrubs or trees. The leaves may be all from the base, or alternate up the stem, or *whorled. The inflorescence is a *raceme or *umbel, the flowers mostly regular and *trimerous, with 2 usually similar whorls of *petaloid *perianth segments. There are usually 6 *stamens, the *ovary is *superior and normally 3-celled. The fruits are *capsules or *berries. Many (e.g. *Lilium* and *Tulipa*) are cultivated for their flowers, others as vegetables or for flavouring (e.g. *Allium*, onions and garlic). There are 294 genera, comprising about 4500 species, with a cosmopolitan distribution.

***Lilium*** (lily; family *Liliaceae) A genus much cultivated for the flowers, which are in *racemes on leafy stems and have 6 recurved *perianth segments. There are about 100 species and many garden varieties, occurring in the northern temperate zone, extending to the Philippines.

**lily** *See* LILIUM.

**lily-of-the-valley** *See* CONVALLARIA.

**lima bean** *See* PHASEOLUS.

**lime** **1.** *See* CITRUS **2.** (linden) *See* TILIA. **3.** Compounds mostly of calcium carbonates, but also other basic (alkaline) substances, used to correct soil acidity and occasionally as a fertilizer to supply magnesium.

**Limes convergens** A well-defined boundary zone between two fairly uniform major *habitat types. An example is a flood meadow zone, which is subject to periodic catastrophic flooding: this includes species characteristic of wet conditions, as well as those typical of trampled and grazed areas (which reflect the normal use when the area is not flooded).

**Limes divergens** A diffuse boundary zone in which one major *habitat type changes gradually into another, typically with both types showing internal variation. It is more stable than the *Limes convergens, since the change in environment is spatially and temporarily gradual and persistent, rather than catastrophic and/or intermittent.

**limestone forest** A distinctive forest type, found within tropical rain-forest regions of south-eastern Asia, and also in the Caribbean region, growing over limestone hills (karst). There are a few endemic (*see* ENDEMISM) genera (e.g. the palms *Liberbaileya* and *Maxburretia*, which are restricted to limestone forests in Malaysia).

**limicolous** Growing in mud.

**limiting factor** (ecological factor) Defined originally as whichever essential material is available in an amount most closely approaching the critical minimum needed, but now used more generally to describe any environmental condition or set of conditions that approaches most nearly the limits (maximum or minimum) of tolerance for a given organism.

**limits of tolerance** The upper and lower limits to the range of particular environmental factors (e.g. light, temperature, availability of water) within which an organism can survive. Organisms with a wide range of tolerance are usually distributed widely, while those with a narrow range have a more restricted distribution. *See also* SHELFORD'S LAW OF TOLERANCE.

**limnetic zone** The area in more extensive and deeper freshwater *ecosystems that lies above the *compensation level and beyond the *littoral (lake-edge) zone. This zone is mainly inhabited by *plankton and nekton with occasion *neuston species. The limnetic and littoral zones together comprise the *euphotic or well-illuminated zone. In very small and shallow lakes or ponds the limnetic zone may be absent.

**limnology** The study of freshwater *ecosystems, especially lakes.

***Limonium vulgare*** (common sea lavender) *See* POLYHALINE.

**Linaceae** A family of *herbs, shrubs, and trees that have regular *penta- or *tetramerous flowers and the same number of *sepals, *petals, and *stamens. The *ovary is 3–5-celled, the fruit a *capsule. They are cultivated for flowers or useful products (e.g. *Linum* species yield flax and linseed oil). Modern classifications recognize some 300 species, within 15 genera.

**Lindeman's efficiency** The ratio of energy assimilated at one *trophic level to that assimilated at the preceding trophic level; the ratio of energy intake at successive trophic levels. It is one of the earliest and most widely applied measures of *ecological efficiency.

**linden** *See* TILIA.

**lineage** *See* EVOLUTIONARY LINEAGE.

**linkage** The association of *genes that results from their being on the same *chromosome. Linkage is detected by the greater association in inheritance of two or more non-allelic genes than would be expected from *independent assortment. The nearer such genes are to each other on a chromosome, the more closely linked they are, and the less often they are likely to be separated in future generations by *crossing-over. All the genes in one chromosome form one linkage group.

**linkage disequilibrium** In genetics, the non-random association of *alleles at different *gene loci (*see* LOCUS) in a population (e.g. when two loci occur close together on the same *chromosome and selection operates to keep the allele combinations together).

**linkage equilibrium** *See* GAMETIC EQUILIBRIUM.

**linkage map** An abstract map of chromosomal *loci, based on experimentally determined *recombinant frequencies, which shows the relative positions of the known *genes on the *chromosomes of a particular species. The more frequently two given characters recombine, the further apart are the genes that determine them.

**Linnaeus, Carolus (Carl von Linné) (1707–78)** A Swedish naturalist remembered for his large contributions to plant classification and his introduction of the binomial system of nomenclature. Trained as a botanist and physician in Uppsala, he went to Holland to continue his studies. While there he published the *Systema Naturae* (1735), a classified list of plants, animals, and minerals. The list grew in later editions, and the 10th edition (1758) is the starting-point of zoological nomenclature. In the first edition of his *Genera Plantarum* (1737) he gave more details of his *artificial classification for plants based largely on the number of *stamens and *pistils in a flower and the manner in which they occurred. The system was very popular, because it allowed students to catalogue and recognize vegetation quickly; its popularity caused some difficulties when it had to be replaced by *natural classifications. In 1741 he was appointed to the chair of medicine at Uppsala, but exchanged it within a year for the chair of botany. He then began listing species, grouping them into genera, genera into classes, and classes into orders. In 1749 he introduced binomial classification, a system that used a Latin generic noun followed by a specific adjective. Until this time, polynomial plant names comprised concise Latin descriptions, which restricted the growth of classification. Ironically, it was largely this new system that allowed the development of ideas about the evolution of species, a concept to which Linnaeus was opposed. Linnaeus published the specific names in his most important botanical work, *Species Plantarum* (1753), still the official starting-point of current botanical nomenclature. Altogether he completed around 180 works and was considered to be an excellent tutor. After his death, his collections and library were sold in 1784 to Sir James Edward Smith, the first president of the Linnean Society of London (founded in 1788) and in 1828 the Society purchased the collection, which it still holds. *See also* INTERNATIONAL CODE OF BOTANICAL NOMENCLATURE.

**linseed oil** *See* LINUM.

*Linum* **(family *Linaceae)** A genus of herbs that have narrow, linear, *sessile leaves, and *cymose *inflorescences along the stems. The flowers have 5-clawed *petals that soon fall and globular 10-valved *capsules. Some species are grown for ornament, for the fibres (flax), or for the oil in the seeds (linseed oil). There are some 200 species, mostly in temperate and subtropical parts of the northern hemisphere.

**lipid** A member of a heterogeneous group of small organic molecules that are sparingly soluble in water, but soluble in organic solvents. Included in this classification are fats, oils, waxes, *terpenes, and *steroids. The functions of lipids are equally diverse and include roles as energy-storage compounds, as *hormones, as vitamins, and as structural components of cells, particularly membranes.

**lipoic acid** **(6–8-dithio-n-octanoic acid)** A compound that functions as a *coenzyme in the oxidative decarboxylation of pyruvic acid to alpha-ketoglutaric acid and of the latter to succinic acid. It is sometimes classed as a vitamin as it is an essential requirement for some micro-organisms.

**lipoprotein** A water-soluble, *conjugated protein in which the *prosthetic group is a *lipid.

*Liquidambar* **(family *Hamamelidaceae)** A small genus of 4 species of trees and shrubs which have maple-like leaves. Male flowers are held in *spikes, female flowers in heads. *L. styraciflua* (sweet gum) has superb autumn colours and provides the timbers American red gum (*heartwood) and American sap gum (*sapwood). They occur in temperate areas of western N. America, Asia Minor, and eastern Asia.

**liquid limit** *See* ATTERBERG LIMITS.

*Liriodendron* **(family *Magnoliaceae)** A genus of trees whose leaves have a curiously truncate or bilobed apex. There are 2 species: *L. tulipifera* (tulip tree or tulip poplar), which provides the timber American white wood, occurs in eastern

N. America; and *L. chinense* occurs in eastern Asia. They have now been hybridized. *See* BICENTRIC DISTRIBUTION.

**Litchi** (lichi, lychee; family *Sapindaceae) A *monotypic genus of small trees, closely related to *Nephelium* and producing a similar edible *arillate fruit, but in which the *pericarp is smooth. They are native to China and the mountains of western Malesia and cultivated in warm temperate climates.

**Lithocarpus** (family *Fagaceae) A genus of trees, related to the oaks (*Quercus*) and similar in their acorns, bark, and wood with broad rays, but which produce bisexual *inflorescences. The timber is used mainly for firewood because it is difficult to work. There are about 300 species, found in the tropics and subtropics from India to Japan and Malesia, plus one species in western N. America.

**lithocyst** *See* CYSTOLITH.

**lithophyte** A plant that grows on a rock surface. *See also* ENDOLITHIC and PETROPHYLLOUS. *Compare* CHOMOPHYTE.

**Lithops** (living stones; family *Aizoaceae) A genus of glabrous succulents which have a pair of massive, fleshy, sessile leaves at ground level, well camouflaged to their stony habitat. Many are cultivated. There are 37 species, occurring in southern Africa.

**lithosere** The sequence of plant *communities, proceeding through all the stages of a *succession to a *climax vegetation, that begins on a bare rock surface.

**lithotroph** An organism that obtains energy from the oxidation of inorganic compounds or elements (*compare* ORGANOTROPH). Sometimes the term is inaccurately used as a synonym of *autotroph.

**litmus** A dye obtained from *lichens of the genus *Roccella*. Litmus is used as a *pH indicator, being red under acidic conditions and blue under alkaline conditions.

**Litsea** (family *Lauraceae) A genus of evergreen trees in which the leaves

are alternate and the *inflorescence is subtended by an *involucre. The fruit is inserted in the *calyx tube. The genus provides many traditional medicines. There are about 400 species, found in warm temperate, subtropical, and tropical regions.

**litter** (L-layer) An accumulation of dead plant remains on the soil surface.

**littoral** Pertaining to the sea-shore.

**littoral fringe** The uppermost reaches of the *littoral zone of marine ecosystems. On rocky shores it is the main area for *Littorina* species (periwinkles), and the black lichen, *Verrucaria*, and is distinguished from the lower, *eulittoral zone by the absence of barnacles (e.g. *Balanus* species and *Chthamalus* species). On sandy shores crabs, beach amphipods, and some land insects are characteristic.

**littoral zone** 1. The area in shallow, fresh water and around lake shores where light penetration extends to the bottom sediments, giving a zone colonized by rooted plants. 2. In marine ecosystems the shore area or intertidal zone where periodic exposure and submersion by tides is normal. Since the precise physical limits of tidal range vary constantly, a biological definition of this zone, which essentially reflects typical physical conditions rather than rarely experienced events, is generally more useful. Thus, in Britain the littoral zone is defined as the region between the upper limit of species of the seaweed *Laminaria, and the upper limit of *Littorina* species (periwinkles), or of the lichen *Verrucaria*.

**live oak** An American term covering all evergreen species of oak, especially *Quercus virginiana* (despite its name, the state tree of Georgia), and *Q. geminata* (sand live oak).

**liverwort** (hepatica) The common name for a plant belonging to the class *Hepaticae.

**living stones** *See* LITHOPS.

**Livistona** (family *Arecaceae) A genus of tall, solitary fan palms which have fibrous leaf *sheaths. The *inflorescence is a hanging *panicle. The flowers are hermaphrodite, with 3 *carpels, united at the *styles; the *stigma is basal. There is 1 fleshy, brightly coloured fruit to each flower, containing a seed with a thin stone. There are 29 species, found from north-eastern Africa, Arabia, and Assam and southern China to the Solomon Islands (where there is 1 species), and Australia (12 species).

**Lixisols** All soils that have an *argic horizon within 100–200 cm of the surface, apart from *Acrisols, *Albeluvisols, *Alisols, and *Luvisols. Lixisols are a reference soil group in the FAO *soil classification.

**llanos** The *savannah grasslands of the Orinoco basin of Venezuela. Some authorities have suggested that the paucity of trees is due to the heavy subsoil of the region, which produces very wet conditions during the rainy season. It is more likely, however, that fire and grazing are responsible for preventing the development of woodland.

**L-layer** *See* LITTER.

**loam** A class of soil texture that is composed of *sand, *silt, and *clay, which produces a physical property intermediate between the extremes of the 3 components. It is an easily worked soil, much prized by farmers.

**Lobaria** (order *Peltigerales) A genus of *lichens in which the *thallus is *foliose, often large, with short, felt-like hairs on the lower surface. The *phycobiont may be a green *alga or a *cyanobacterium, according to the species. *Spores are colourless or pale brown, fusiform to elongated, with 1 to many *septa. *Lobaria* is found chiefly in moist districts (in Britain in the west) on trees or occasionally on rocks.

**lobed spinifex** *See* TRIODIA.

**Lobelia** (family *Lobeliaceae) A genus comprising mainly *herbs and some shrubs in which the flowers, in leaf *axils or in *racemes, have 2-lipped *corollas split down the back, and beards on some *anthers. The fruit is a *capsule. There are 365 species, throughout most tropical and temperate areas of the world. The genus shows an extraordinary degree of

*endemism in the E. African mountains, where many species with huge columnar *inflorescences occur, each mountain block tending to have 1 or more endemic species in the alpine zones.

**Lobeliaceae** A family of *herbs or small trees, mainly tropical but with some temperate species, that are close to *Campanulaceae but have irregular, 2-lipped *corollas. Many (e.g. *Lobelia*) are grown for ornament. Modern classifications recognize about 25 genera, with about 400 species.

**Lochkovian** *See* DEVONIAN.

**Loch Lomond Stadial** A relatively cold period that occurred towards the end of the last (*Devensian) glaciation in Scotland. The event took place about 11 000–10 000 BP. It is characterized by the development of small ice caps and cirque glaciers in the Highlands.

**loci** *See* LOCUS.

**lock-and-key theory** A theory to explain the mechanism of enzymatic reactions, in which it is proposed that the *enzyme and *substrate(s) bind temporarily to form an enzyme–substrate complex. The binding site on the enzyme is known as the 'active site' and is structurally complementary to the substrate(s). Thus, the enzyme and substrate(s) are said to fit together as do a lock and a key.

**locule** A cavity of the *ovary. *See* CARPEL.

**loculicidal** Of a *seed-bearing *pod, splitting longitudinally along the midrib.

**Loculoascomycetes** (subdivision *Ascomycotina) A class of *fungi, in which the *fruiting body is an *ascostroma and the *asci are *bitunicate. The class is not now generally recognized.

**locus (pl. loci)** A specific place on a *chromosome where a *gene is located. In *diploids, *loci pair during *meiosis and, unless there have been *translocations, *inversions, etc., the *homologous chromosomes contain identical sets of loci in the same linear order. At each locus is 1 gene; if that gene can take several forms (*alleles), only 1 of these will be present at a given locus.

**lodging** In plants, a state of permanent displacement of a stem-crop stem from its upright position. This can cause considerable reduction in yield. Normally it is caused by storm damage, but it may be due to rots, insects, or excess nitrogen.

***Lodoicea*** (double coconut; family *Arecaceae) A *monotypic palm genus (*L. maldivica*) that is endemic (*see* ENDEMISM) to the Seychelles, rare in the wild, and often cultivated. It is a huge fan palm, with a trunk up to 30 m tall and leaves 3 m or more long. It is *dioecious. The *inflorescences are unbranched, emerging through split leaf bases. Female flowers are 5 cm across. The fruit is like a huge bilobed coconut 45 cm long, weighing 14–22 kg, and containing 1 giant seed with a stony wall, the biggest seed in the world.

**loess** Unconsolidated, wind-deposited sediment composed largely of *silt-sized quartz particles (0.015–0.05 mm diameter) and showing little or no stratification. It occurs widely in the central USA, northern Europe, Russia, China, and Argentina. It can give rise to a rugged topography with steep slopes (up to 70°). The soils derived from loess are of a very high quality and support excellent crop yields.

***Logania*** (family *Loganiaceae, *tribe Loganieae) A genus of small trees and shrubs that have opposite, entire leaves, *phloem tubes scattered through the *xylem, and superficial *cork development. The flowers are regular in terminal *cymes, the fruit a *capsule or *berry. The seeds have a fleshy *endosperm. There are 15 species, restricted to Australia, New Zealand, and New Caledonia.

**Loganiaceae** A family of trees, shrubs, and climbers, in which the leaves are *simple, *opposite, and often with *stipules. The flowers are hermaphrodite and regular, with 4 or 5 fused *sepals and *petals with *imbricate lobes, 4 or 5 *epipetalous *stamens, and a *superior *ovary with 2 *locules. The fruit is a *capsule. Many species contain poisonous *alkaloids, some of which are medicinal. There are 29 genera, with about 600 species, occurring in temperate, subtropical, and tropical regions.

**logistic equation** (logistic model) A mathematical description of growth rates for a simple population in a confined space with limited resources. The equation summarizes the interaction of *biotic potential with environmental resources, as seen in populations showing the *S-shaped growth curve, as: $dN/dT = rN(K - N)/K$ where $N$ is the number of individuals in the population, $T$ is time, $r$ is the biotic potential of the organism concerned, and $K$ is the saturation value or *carrying capacity for that organism in that environment. The expression $(K - N)/K$ is the term that ensures the slowing down of the growth rate as $N$ approaches $K$. When $N = K$ the term is zero and population growth ceases. The resulting growth rate or logistic curve is a parabola, while the graph for organism numbers over time is sigmoidal. *Compare* J-SHAPED GROWTH CURVE.

**logistic model** See LOGISTIC EQUATION.

**logwood** See *HAEMATOXYLUM*.

***Lomandra*** (family *Xanthorrhoeaceae) A genus of woody rhizomatous (*see* RHIZOME) *perennials that have tall stems, *simple, linear sheathing leaves, and regular flowers, which are often rigid and coloured. The plants are *dioecious. There are about 35 species, found in Australia, New Zealand, New Guinea, and New Caledonia.

**longan** See *EUPHORIA*.

**long-day plant** A plant in which flowering is favoured by long days (i.e. days when there are more than 14 hours of daylight) and correspondingly short dark periods. There are 2 groups of such plants, species in which there is an absolute requirement for these conditions (such that flowering will not begin without them) and others in which flowering is merely hastened by them. Spinach, lettuce, and grasses are long-day plants; in Britain all of them flower in summer.

**longevity** The persistence of an individual for longer than most members of its species, or of a genus or species over a prolonged period of geological time.

***Lonicera*** (honeysuckles; family *Caprifoliaceae) A genus of 180 species of shrubs or woody climbers, most of which are deciduous. The flowers are usually very fragrant and *pentamerous, with a *corolla that is tubular below and usually strongly 2-lipped, but may be almost regular. The *ovary is 2-celled, ripening to a few-seeded *berry. Honeysuckles are much grown for ornament or, in the case of the evergreen *L. nitida*, for hedges. They occur in the northern hemisphere as far south as Mexico and the Philippines.

**lontar** See *BORASSUS*.

**loofah** See *LUFFA*.

**loose smut** A disease of plants, caused by a *fungus of the *Ustilaginales (e.g. species of *Ustilago*), in which the masses of *spores are exposed at maturity and can be dispersed freely by wind, etc.

**lop and top** The branches and top cut from a tree that has fallen or been felled or, more rarely, from a standing tree.

***Lophocolea*** (order *Jungermanniales) A genus of leafy *liverworts. The stems are slender and mainly prostrate. The leaves are obliquely set on the stem and overlap *succubously. In *L. bidentata* and *L. cuspidata* the leaves are divided into 2 sharp points; in *L. heterophylla* some leaves are bidentate but others are more or less entire and rounded. *L. bidentata* grows mainly on soil, in woods, on lawns, etc. *L. cuspidata* and *L. heterophylla* grow mainly on moist rotting wood.

**loquat** See *ERIOBOTRYA*.

**Loranthaceae** A family composed of small shrubs that are partially parasitic on trees, with *opposite or *whorled, entire leaves, and unisexual regular flowers, usually with 2-*perianth whorls. The *ovary is *inferior, the fruit a *berry. *Viscum* (mistletoe) has tiny *tetramerous flowers and white berries. In modern classifications there are some 70 genera, with 940 species. They are mostly tropical, but some are temperate.

**lorica** In some *algae, an open, siliceous case that surrounds a cell. See also NAKED CELL.

**lotic** Applied to a freshwater *habitat characterized by running water, e.g. springs, rivers, and streams.

**Lotka–Volterra equations** Mathematical models of competition between resource-limited species living in the same space with the same environmental requirements. They have been modified subsequently to simulate simple predator–prey interactions. The competition model predicts that coexistence of such species populations is impossible: one is always eliminated, as was verified experimentally by G. F. Gause. The predation model predicts cyclic fluctuations of predator and prey populations. Reduction of predator numbers allows prey to recuperate, which in turn stimulates the population growth of the predator. Increasing predator numbers depress the prey population, leading eventually to a reduction in the predator population. This was also tested experimentally by Gause (1934) and more recently and more convincingly by S. Utida (1950, 1957). *See also* COMPETITIVE EXCLUSION PRINCIPLE.

**lotus fruit** See *ZIZYPHUS*.

**low Arctic tundra** The southernmost latitudinal sector of the Arctic *tundra, distinguished from the more northerly belts by the presence of a continuous vegetation cover in most areas. The plant *communities form a vegetation mosaic, which reflects the importance of microhabitats in this harsh, exposed environment.

**lower fungi** *Fungi belonging to the *Mastigomycotina and the *Zygomycotina.

**Lowiaceae** A small family comprising 1 genus, with 7 species, of forest *herbs related to bananas (*Musaceae) that are rhizomatous (*see* RHIZOME) and stemless. Their flowers are bisexual and *trimerous, with 1 enlarged *sepal and *petal, and a long *hypanthium. They occur in Indomalesia.

**Loxsomaceae** A family of ferns in which the much-divided *fronds arise from creeping, hairy *rhizomes and have their *sori, which are of *gradate type, in cup-like *indusia terminally on the leaf veins on the leaf margins. The *sporangia have a complete oblique *annulus. There are 2 genera: *Loxsoma*, with only 1 species, in New Zealand, and *Loxsomopsis*, with 3 species, in Central and S. America.

**LPP group** A group of closely related, filamentous *cyanobacteria (section III), including the poorly defined genera *Lyngbya*, *Phormidium*, and *Plectonema*, as well as *Microcoleus* and *Schizothrix*. All form straight filaments which may be sheathed or non-sheathed, *motile or non-motile, and tend to aggregate into dense mats in freshwater and marine habitats.

**Luffa** (family *Cucurbitaceae) A genus of 6 species of tendrillate climbers of the tropics, including *L. aegyptiaca*, which furnishes the loofah, the vascular skeleton of the *pericarp.

**lumen** A space enclosed by *cell walls. *Compare* LACUNA.

**luminescence** Light emission produced by metabolic processes at ambient temperatures.

**Lumnitzera** (family *Combretaceae) A genus of big trees of mangrove swamps, whose fruits float and which yield useful timber. There are 2 species, occurring in E. Africa and from the Malay archipelago to the Pacific islands.

**lunate** Shaped like a half-moon.

**Lunularia** (order *Marchantiales) A genus of *thallose *liverworts, which are immediately recognizable by the crescent-shaped ridges of tissue on the surface of the *thallus; these protect small discoid gemmae. *L. cruciata* is very common in gardens, on flowerpots, moist brick, rockery stones, paths, etc.; it is also found on the ground in woods and by streams, etc.

**lupin** *See* LUPINUS.

**Lupinus** (lupin; family *Fabaceae, subfamily *Papilionatae) A genus mostly of *herbs (a few of which are shrubby) that have *palmately *compound leaves and erect, terminal, leafless *racemes of showy flowers. Lupins are much cultivated in gardens and some are used for cattle fodder. There are some 200 species, mostly in N. America, but a few are native in Europe.

**Lusitanian floral element** A geographical element of the British flora, including plants such as *Erica mackaiana* (Mackay's heath), that are found in western Ireland and that reappear again only far to the south, in the Iberian peninsula. These plants are generally thought to have migrated north to Ireland early in the post-glacial period, before sea-level was fully restored.

**Lutz phytograph** In the *phytosociological assessment of woodland *communities (especially tropical woodlands), a polygonal figure representing four structural characteristics of each major tree species present, usually (a) the percentage of the total number of trees that are larger than 25 cm diameter at breast height (dbh); (b) the percentage frequency of a particular species in the total number of trees larger than 25 cm dbh; (c) the occurrence of the species in each of 5 size classes reflecting maturity; and (d) the dominance of the species as reflected by its percentage of the total tree basal area.

**Luvisols** Soils that have an *argic horizon with a *cation exchange capacity greater than 24 cmol$_c$/kg, and with illuvial (*see* ILLUVIATION) accumulations of *clay. Luvisols are a reference soil group in the FAO *soil classification.

**lyase** An *enzyme that catalyses non-hydrolytic reactions in which groups are either removed or added to a *substrate, thereby creating or eliminating a double bond, especially between carbon atoms (C = C) or between carbon and oxygen (C – O).

**lychee** *See* LITCHI.

**Lycium** (Duke of Argyll's tea-plant; family *Solanaceae) A genus of often spiny shrubs that have entire, clustered or alternate leaves. The flowers have funnel-shaped, 5-lobed *corollas, and the *stamens inserted at the mouth of the corolla are long and protruding. The fruit is a *berry, retained in the *calyx. They are much cultivated, and naturalized in Europe and elsewhere. There are about 100 species, occurring in temperate and subtropical areas.

**Lycogala** (class *Myxomycetes) A genus of *cellular slime moulds whose *fruiting bodies (aethalia) resemble small *puff-balls. *L. epidendrum* is one of the commonest myxomycetes and is cosmopolitan in distribution. It is found on dead wood in woodland habitats.

**Lycoperdales** (class *Gasteromycetes) An order of (typically) terrestrial *fungi in which the mature *fruit body occurs above ground although its development may begin below ground. The fruit body is typically more or less spherical. The *spores are contained within a *peridium which is usually composed of at least 2 layers: the *exoperidium and *endoperidium. There are many genera. The order includes the *puff-balls and *earth stars.

**Lycoperdon** (order *Lycoperdales) A genus of *fungi in which the *fruit bodies typically grow on the ground (*L. pyriforme* grows on rotten wood). The *exoperidium is spiny or scaly; the *endoperidium opens by a regular pore at the apex through which the *spores escape. *See also* CALVATIA.

**Lycopersicon** (tomatoes; family *Solanaceae) A genus comprising 7 species of weedy herbs that are native to western parts of S. America and the Galápagos Islands. *L. esculentum* is the commercial tomato.

**Lycopodiaceae** A family of *Lycopsida within the *Pteridophyta, with small, *simple, usually narrow leaves, which clothe the forking stems or, as in *Phylloglossum*, arise from the base of the naked, unbranched stem. The stems bear *sporangia in the *axils of either normal or highly modified leaves, either along their length in alternating sterile and fertile zones, or (in most species) in clearly differentiated terminal cones. All sporangia are alike and contain numerous minute *spores, which germinate to produce sexual *gametophytes. Unlike *Selaginellaceae, the leaves have no *ligule on their upper side. The family is very ancient and similar plants have been found as fossils in *Devonian and *Carboniferous rocks. They are now usually divided into 5 genera, which contain some 450 species, some of which are cosmopolitan in distribution.

**Lycopsida** A subdivision of the *Pteridophyta, represented today by the club mosses (families *Lycopodiaceae, *Selaginellaceae, and *Isoetaceae), although in the *Carboniferous they included large trees. The oldest lycopods date from the *Devonian and derived from the most primitive of vascular plants, the *Psilophytopsida.

**Lyngbya** A genus of filamentous *cyanobacteria in which each filament is surrounded by a sheath. The filaments are capable of gliding motility. They may be free-floating or mat-forming. They are found in all types of aquatic environment and on wet rocks, damp soil, etc. *Heterocysts and *akinetes are not formed, and the organisms are difficult to distinguish from *Phormidium* or *Plectonema* (the *LPP group). *Lyngbya majuscula* produces several toxins, and blooms of this species can cause dermatitis in swimmers, a condition known as 'swimmers' itch'.

**Lysenko, Trofim Denisovich** (1898–1976) A fanatic anti-geneticist, who attained extraordinary power in the USSR under Stalin, and retained it under Krushchev. He sought successfully the suppression of research in genetics and was responsible for the imprisonment and execution of a number of noted Soviet geneticists. The failure of his own attempts, based on a kind of *Lamarckism, to increase grain production led to his own downfall and that of Krushchev.

**lysimeter** A device for the direct estimation of *evapotranspiration. Typically it comprises a vegetated block of soil 0.5–1 m$^3$, to which the amount of water added is known, and from which the amount lost as run-off or *percolation may be measured. Recording the changing weight of the soil vegetation system (keeping vegetation change resulting from growth static or monitored) reveals the amount of water retained by the system, and thus by difference the amount lost as evapotranspiration. For geographic comparisons, easily standardized, short, grass vegetation cover is used. For water-budget experiments, vegetation cover may be varied to simulate different crop types or *semi-natural communities.

**lysine** An aliphatic, basic, polar (*see* POLAR MOLECULE) *amino acid that is of limited occurrence in plant *proteins (but generally abundant in animal proteins).

**lysis** *See* LYTIC RESPONSE.

**lysogeny** A stable, non-destructive relationship between a *bacteriophage and its host bacterium; the genome of the bacteriophage may become integrated with that of its host. Under certain conditions the stable relationship breaks down and the phage may then destroy the bacterium.

**lysosome** A membrane-bound *vesicle in a cell that contains numerous acid *hydrolases capable of digesting a wide variety of extra- and intracellular materials. Lysosomes apparently originate from the *Golgi body, and function by fusing with and discharging their contents into a *vacuole containing the material to be digested. Lysosomes have been difficult to isolate from plant cells, but a number of biochemical studies have demonstrated clearly that plant cells contain a wide range of particles, varying in size and internal contents, which may be termed lysosomes. They are all surrounded by a single membrane and contain hydrolytic enzymes. In many plants, the main cell vacuole may assume the functions of lysosomes.

**Lythraceae** A family of *herbs, with a few shrubs and trees, in which the leaves are *opposite or *whorled and the flowers regular, with 4–6 (or no) free *sepals and *petals, 4–12 *stamens, and a 2–6-celled, *superior *ovary. The fruit is a *capsule. There are 26 genera, with about 580 species, with a pantropical distribution, plus a few temperate species.

***Lythrum*** (family *Lythraceae, *tribe **Lythreae**) A genus of *herbs and shrubs in which the leaves are either in *whorls or *opposite, usually *simple, and with small or no *stipules. The flowers are regular and bisexual, usually solitary, with 12 *stamens in 2 whorls of different lengths. The fruit is a dry *capsule, with numerous seeds without *endosperm. There are 38 species, found throughout the world, except S. America.

**lytic response** The rupture and death (lysis) of a bacterial cell following its infection by a *bacteriophage which then reproduces inside the cell, as opposed to a lysogenic response in which the infecting bacteriophage does not multiply but instead behaves as a *prophage.

**Ma** Millions of years.

**Maastrichtian (Maastrichtian)** The final stage of the *Cretaceous in Europe, dated at about 70.6–65.5 Ma ago. The stratotype is described from the Maastricht area of Holland. Throughout western Europe, the Maastrichtian is characterized by chalk limestones.

**Macadamia (family *Proteaceae)** A genus of shrubs and trees that yield oily, edible seeds. There are 9 species. One is found in Madagascar, 1 in Celebes, 3 in New Caledonia, and 5 in Australia. *M. integrifolia* and *M. tetraphylla* are cultivated, providing the delicious Queensland nut, or macadamia.

**Macaranga (family *Euphorbiaceae)** A genus of trees in which the leaves are often *palmately lobed or nerved with conspicuous *stipules. Some have hollow, ant-inhabited twigs. Many are fast-growing pioneer trees with soft wood. There are about 240 species, occurring in tropical and subtropical forests from Africa to Polynesia, southern China, and Queensland.

**Macaronesian floral region** A region comprising the Canaries, Madeira, and the Azores, in which few genera are endemic, perhaps less than 30 (*see* ENDEMISM); surprisingly, the most remote island group, the Azores, has no endemics. The flora of the Canaries displays a notably high frequency of succulents belonging to the *Crassulaceae, in particular the genus *Sempervivum. See also* FLORAL PROVINCE and FLORISTIC REGION.

**macassar oil** *See CANANGA.*

**macchia** *See MAQUIS.*

**mace** A spice obtained from the *aril of *Myristica fragrans. See MYRISTICA.*

**machair** An area of low, undulating tracts, supporting stable, herb-rich grassland growing on shell sand, that has developed over a long period by the accumulation of blown sand behind coastal sand-dunes,

occurring most typically in the Hebrides and along the north-west coast of Scotland.

**macrobiota (macrofauna, macroflora)** A general term for the larger soil organisms which may be hand-sorted from a soil sample. 'Macrofauna' in particular refers to burrowing vertebrate animals (e.g. rabbits and moles), while 'macrobiota' generally also includes larger plant material (e.g. tree roots). Some workers also include larger insects and earthworms in this category, but others consider them part of the mesobiota. *Compare* MESOBIOTA and MICROBIOTA.

**macroclimate** The climate character of a large region.

**macroconsumer** *See* CONSUMER ORGANISM.

**macrocyst** In *Dictyosteliomycetes, an aggregation of *myxamoebae within a membrane. In the macrocyst, cells divide and myxamoebae are then released.

**Macrocystis (order *Laminariales)** A genus of large brown seaweeds in which the *thallus may reach lengths of 50 m or more. The upper portions of the plant float on the sea surface, buoyed up by air *bladders. *Macrocystis* can grow in much deeper waters than most other seaweeds, e.g. to depths of 30 m or more in clear seas. It occurs in Antarctic regions, on the Pacific coast of N. America, and off the shores of S. Africa. In the USA, *Macrocystis* is harvested as a valuable source of *algin.

**macroecology** A term coined by James H. Brown of the University of New Mexico that is used in relation to biogeographic studies (*see* BIOGEOGRAPHY) of population and species interactions on a large rather than a local scale. It brings to bear both geographic and historical considerations in understanding the local abundance, distribution, and diversity of species.

**macroevolution** Evolution above the species level, i.e. the development of new species, genera, families, orders, etc. There is no agreement as to whether macroevolution results from the accumulation of small changes resulting from *microevolution, or whether macroevolution is uncoupled from microevolution.

**macrofauna**   See MACROBIOTA.

**macroflora**   See MACROBIOTA.

**macrofossil**   See MEGAFOSSIL.

**macromolecule**   A molecule that has a high molecular weight, often a polymer.

**macronutrient**   An organic or inorganic element or compound which is needed in relatively large amounts by living organisms. Organic macronutrient groups include *amino acids, *carbohydrates, and fats. _Compare_ MICRONUTRIENT; _see also_ ESSENTIAL ELEMENT.

**macrophyll**   See MEGAPHYLL.

**macrotidal**   Applied to coastal areas where the tidal range is in excess of 4 m. Tidal currents dominate the processes active in macrotidal areas (e.g. the coast of the British Isles).

**_Macrozamia_**   (family *Zamiaceae) A genus of cycads that occur as an understorey in *_Eucalyptus_ forests and woodlands. There are 14 species, found in Australia. Their distribution may have been more extensive in the past.

**Madagascan floral region**   Part of R. Good's (_The Geography of the Flowering Plants_, 1974) *palaeotropical kingdom, which geographically includes Madagascar, the Seychelles, and the Mascarenes. A high proportion of the genera, perhaps exceeding 200, is endemic (_see_ ENDEMISM); and, depending on the taxonomic system adopted, there are 2 or 3 endemic families. _See also_ FLORAL PROVINCE and FLORISTIC REGION.

**made ground**   (made land) An area of land that has been man-made, generally through the reclamation of marshes, lakes, or shorelines. An artificial fill is used, consisting of natural materials, refuse, etc.

**made land**   See MADE GROUND.

**_Madhuca_**   (family *Sapotaceae) A genus of small to big evergreen trees in which the leaves are spiral, and sometimes clustered. The flowers have 4 *sepals, and 6–16-lobed *petals. There are 85 species, occurring in South-east Asia, and centred on Malesia.

**Madras thorn**   See PITHECELLOBIUM.

**madroña laurel**   See ARBUTUS.

**Maestrichtian**   See MAASTRICHTIAN.

**magnesium** (Mg) An element that is found in high concentrations in plants. It plays an important role in the chemical structure of *chlorophyll and of membranes and is involved in many *enzyme reactions, especially those catalysing the transfer of phosphate compounds. Deficiency can produce various symptoms, including *chlorosis and the development of other pigments in leaves. Magnesium is a relatively abundant component of sea water and is thus quite plentiful in the rainfall of oceanic regions.

**_Magnolia_**   (family *Magnoliaceae) A genus of trees and shrubs with showy flowers, many of which are cultivated as ornamentals. The *tepals are *petaloid. The *stamens and *carpels are in an elongate *receptacle, and fruiting *carpels are follicular. _Magnolia_ is one of the most ancient of flowering plants, noted for the relatively primitive structure of its flower. _Magnolia_ leaves are first recorded from Kansas, USA, in the lower *Cretaceous. There are about 125 species occurring from the Himalayas to Japan and western Malaysia, and in southern N. America to Venezuela and the W. Indies.

**Magnoliaceae**   A family of trees and shrubs that have aromatic bark. The leaves are *simple, with large *stipules that enclose the *apical bud and leave a scar. The flower parts are numerous, free, and spirally arranged. The fruits are usually *dehiscent, and the seeds mostly large and *arillate. They yield useful timber and ornamental flowers. There are 7 genera, with 220 species, centred on temperate and tropical Asia, but also occurring in N. and S. America.

**Mahlanobis' _D_²**   A measure of generalized distance between samples based on the means, *variances, and covariances of various properties of replicate samples in multivariate analysis. The larger is $D^2$, the greater is the difference between the samples (or the properties measured).

**mahogany** *See* SWIETENIA.

***Mahonia*** (family *Berberidaceae*) A genus of evergreen shrubs that have *pinnate leaves and an *inflorescence in the form of a *raceme or *panicle of *trimerous flowers. The flowers usually have 5 *whorls of similar *perianth segments, 2 whorls of *stamens, and an *ovary of 1 *carpel which develops into a *berry. They are cultivated for their ornamental foliage, flowers, and fruits, which are like tiny, blue-black grapes (Oregon grapes), winter-flowering varieties being especially popular. There are some 70 species, occurring in eastern Asia and N. America.

**maiden** A tree arising from a seed, or more rarely a sucker, that has not been *coppiced or *pollarded.

**maidenhair fern** *See* ADIANTUM.

**maidenhair tree** *See* GINKGO and GINKGOALES.

**maintenance evolution** (**stabilizing selection, normalizing selection**) The stabilizing influence of *natural selection in an environment that changes little in space and time. It occurs in populations that have a *normal distribution for variations of a particular phenotypic characteristic; the extremes tend to be eliminated. It tends to inhibit evolutionary innovation, and accounts for the fact that many fossil groups changed very little over long periods of time. *See also* DIRECTIONAL SELECTION and DISRUPTIVE SELECTION.

***Maireana*** A former plant genus that is now included in the genus *Kochia*.

**maize** *See* ZEA.

**major gene** A *gene with pronounced *phenotypic effects, in contrast to a *modifier gene, which modifies the phenotypic expression of another gene.

**malacca cane** A walking stick made from a single internode of the rattan *Calamus scipionum*, the value increasing markedly with length.

***Malassezia*** (***Pityrosporum***; class *Hyphomycetes*) A *form genus of *yeast-like *fungi found on the skin of humans and other animals. It can cause skin disease.

**Malaysian floral region** Part of R. Good's (*The Geography of the Flowering Plants*, 1974) Indo-Malaysian subkingdom within his *palaeotropical kingdom, and including the Malay peninsula and archipelago of south-eastern Asia. The flora is one of the richest, if not the richest, in the world. It has yielded many plants of value to humans, of which the banana and various spice plants are perhaps the best known. *See also* FLORAL PROVINCE and FLORISTIC REGION.

**Malesian flora** A term used by some authorities for the tropical flora of Australasia. Strictly, however, the name 'Malesia' was introduced to replace 'Malaysia', which was used formerly, and so to distinguish a *floral province from the state of Malaysia, and it should be restricted to the flora of the islands on and between the Sunda and Sahul shelves, an area comprising Malaysia, Papua–New Guinea, and the islands of Indonesia and the Philippines. It is in this sense that the term is used in this dictionary. *See also* FLORAL PROVINCE and FLORISTIC REGION.

**male sterility** The condition in which functional *pollen is not produced; usually the *pollen tube fails to grow down the *style. This is often used in plant breeding to ensure cross-fertilization. *See also* S ALLELES.

**malic acid** A dicarboxylic acid that is formed during the *citric acid cycle by the reversible hydration of *fumaric acid.

**mallee** A southern Australian *sclerophyllous scrub community, about 2–3 m high, in which most of the species belong to the genus *Eucalyptus*. The scrub is very similar to that found in other continents where *Mediterranean type climates occur, e.g. the *chaparral of California and the *maquis of Europe.

***Mallomonas*** *See* CHRYSOPHYTA.

**mallow** *See* MALVACEAE.

**Malpighiaceae** A dicotyledonous (*see* DICOTYLEDON) family of shrubs, trees, and climbers, whose *simple, usually *opposite leaves are glandular below or on the leaf stalk. The *inflorescence is a *raceme with regular or irregular flowers, usually bisexual. The 5 *sepals and 5 *petals are both overlapping. There are 10 *stamens, the *ovary is *superior, 3 fused *carpels, and the *styles are distinct. The fruit is a *schizocarp with winged segments, or a fleshy or woody *drupe. The seeds contain no *endosperm. The family is divided into 2 tribes, Pyramidotorae and Planitorae, and is thought to be an evolutionary ancestor of the *Trigoniaceae and *Tremandraceae. It is regarded as a primitive family within the Polygalale group. There are 68 genera, with about 1100 species, found mainly in S. America and other tropical regions throughout the world. A number of species are ornamentals.

**maltose** A disaccharide that consists of two alpha–*glucose units linked by an alpha–1,4-glycosidic (See GLYCOSIDE) bond.

***Malus*** **(apples, crabs; family *Rosaceae)** A genus of usually thorny (in wild forms) deciduous trees in which the leaves are *simple and the flowers, in *umbel-like clusters, are regular with 5 *sepals, 5 *petals, numerous *stamens, and 3–5 *carpels which are fused completely with each other and to the *receptacle cup. In the fruit (a *pome) the receptacle swells to form the flesh of the apple, enclosing the cartilaginous *carpel walls. Each carpel contains 2 seeds (pips). In various cultivars they are much grown for their fruit, or for ornament. There are some 25 species, native to the northern temperate zone.

***Malva*** **(mallow)** *See* MALVACEAE.

**Malvaceae** A family of *herbs, with some shrubs and trees, in which the leaves are *palmately veined, often lobed, and rich in mucilage. The flowers are regular, usually *pentamerous, with free *petals, numerous *stamens joined into a tube below, and numerous *carpels joined into a ring to form the *superior *ovary. The fruit is a *capsule, or else breaks into many one-seeded *nutlets. Cotton is made from the seed-coat fibres of *Gossypium* species; species of *Hibiscus*, *Malva* (mallow), and *Althaea* (especially *A. rosea*, hollyhock), are cultivated for their flowers, while *A. officinalis* (marsh mallow) has been used in confectionery and medicine. In modern classifications there are 121 genera, with about 1500 species, occurring in most temperate and tropical regions.

***Mammea*** **(family *Clusiaceae)** A genus of small, evergreen trees, whose flowers have *sepals that are fused, later splitting with 2 valves, usually 4 *petals, and many stamens. The fruit is a *drupe (sometimes edible). There are about 50 species, with a pantropical distribution centred on South-east Asia.

**mandioca (cassava)** *See* MANIHOT.

**manganese** **(Mn)** An element that is required in small amounts by plants. It is involved in the light reactions of *photosynthesis and also binds to *proteins. The leaves of plants deficient in manganese show interveinal *chlorosis and may become malformed.

***Mangifera*** **(family *Anacardiaceae)** A genus of small to large evergreen trees which have *simple leaves borne spirally in tight terminal clusters; these leaves are red when young. The fruit is a *drupe, variously fibrous, and that of several species is edible (mangoes) *M. indica* is widely cultivated throughout the tropics. The timber is an attractive dark brown with blackish markings. There are about 35 species, native to tropical Asia.

***Manglietia*** **(family *Magnoliaceae)** A genus of evergreen trees which are like *Magnolia* but have 4 *ovules rather than 2 per *carpel. There are 25 species occurring in South-east Asia.

**mango** *See* MANGIFERA.

**mangosteen** *See* GARCINIA.

**mangrove forest** A swamp forest, of brackish or saline water, that develops on tropical and subtropical tidal mud-flats (*see* TIDAL FLAT), particularly in quiet creeks and estuaries. Characteristically, mangrove forest is low and dense with a tangle of aerating roots projecting above the mud. Occasionally, substantial, lofty trees occur in the mangrove interior.

**mangrove swamp** A characteristic vegetation of tropical, muddy coasts, and typically associated with river mouths where the water is shallow and the load of suspended sediment is high. The aerial roots of the mangrove trees trap the sediment, favouring the gradual seaward extension of the land area.

**Manihot** (family \*Euphorbiaceae) A genus of shrubs and herbs which produce white latex. The leaves are deeply palmately lobed. *M. esculenta* (*M. utilissima*) provides cassava (manioc, mandioca, tapioca, gari) from swollen \*tubers which, in many varieties, have a high cyanide content. *M. glaziovii* is a small tree whose latex provides ceará rubber. There are 98 species, occurring from the south-west of N. America to tropical America.

**Manila copal** *See* AGATHIS.

**Manila hemp** *See* ABACÁ.

**Manilkara** (family \*Sapotaceae) A genus of evergreen trees that includes *M. zapota* (or *Achras zapota*) which is native to S. America and is now widely cultivated. It yields the delicious, edible fruit known as zapote or, in the East, chiku. There are about 70 species, occurring throughout the tropics.

**man-induced turnover** The additional flow of an element through the active part of a \*biogeochemical cycle that results from human activity. For example, by burning fossil fuels humans add an extra 6.7 billion tonnes per year of carbon to the turnover of the carbon cycle, which is naturally about 75 billion tonnes per year. (One billion is equal to one thousand million, $10^9$.)

**manioc** *See* MANIHOT.

**mannan** A \*polysaccharide that is composed mostly of glycosidically linked \*mannose units. Mannans are storage polymers in some species and perform a structural role in others. A great range of linkage has been reported and some mannans also contain other sugars.

**mannitol** A polyhydroxy alcohol, which can be synthesized chemically by the reduction of \*mannose. It is found in many higher plants and in some algae.

**mannose** An aldohexose monosaccharide, which occurs in a wide variety of organisms including vegetable gums, fungi, bacterial capsular layers, and some immunoglobulins in animals.

**maquis** (Italian *macchia*; Spanish *matorral*) A French term for drought-resistant \*Mediterranean scrub, taller than garrigue, and composed of evergreen shrubs and small trees with thick, leathery leaves (sclerophylls), or spiny foliage, e.g. *Olea europaea* (wild olive), \*Cistus species (cistus), \*Erica species (heather), and *Genista* species (broom). For the most part this \*sclerophyllous formation has been derived by a combination of burning and grazing from the original mixed evergreen \*Mediterranean forest.

**Maranta** (family \*Marantaceae) A genus of \*herbs that includes *M. arundinacea*, whose \*rhizomes provide W. Indian arrowroot. There are 20 species, occurring in tropical America.

**Marantaceae** A family of \*herbs most of which have \*rhizomes or \*tubers. The \*petiole base is sheathing, and kneed at the top, the leaves \*pinnately nerved. The inflorescence is subtended by a large \*bract. The \*epipetalous flowers are hermaphrodite, \*zygomorphic, and \*trimerous, with 1 \*stamen. The \*ovary is \*inferior, the fruit a \*capsule or \*berry. Marantaceae are related to \*Musaceae. There are 31 genera, with about 550 species, occurring in the tropics, mainly in America.

**Marasmius** (family \*Tricholomataceae) A genus of \*fungi in which the \*fruit bodies are \*mushroom-like and are usually small; they shrivel on drying but revive when moistened. The cap is typically tough and pliable, and the \*gills are usually well spaced. Species grow in clusters or rings on the ground in grassland, woodland, etc. *See also* FAIRY RING.

**Marattiaceae** A family of large, primitive, \*eusporangiate ferns. Fossils that clearly belong in this family have been found in the \*Carboniferous rocks. At present there are about 7 genera and 100 species, widely distributed in the humid tropics and warm regions.

**Marchantia** (order \*Marchantiales) A genus of \*thallose \*liverworts. *M. polymorpha* is a common species in Britain. It is immediately recognizable by the presence on the upper surface of symmetrical, cup-shaped structures containing small discoid \*gemmae. The plant is particularly conspicuous when fertile; the male reproductive structures resemble miniature *mushrooms, while the female structures are stalked with many radiating, spoke-like lobes at the top of the stalk. It is common in gardens, on paths, flowerpots, etc., and is also found on *moors, *heaths, river banks, etc.

**Marchantiales** (class *Hepaticae) An order of *thallose *liverworts in which the *thallus is generally flat and ribbon-like, and branched by simple, regular forking. The upper surface is green. Within the thallus are air chambers (areolae) which open to the surface via pores that may or may not be visible to the naked eye. *Rhizoids on the lower surface attach the plant to a substrate and absorb water and mineral salts. Members of the Marchantiales are found mainly in damp habitats on a variety of substrates.

**mare's-tail**   See HIPPURIDACEAE.

**marijuana**   See *CANNABIS*.

**marl**   Lime-rich *clay, usually found as an *alluvial deposit and containing a high proportion of soft calcium carbonate.

**marsh**   A more or less permanently wet area of *mineral soil, as opposed to a peaty area (see PEAT). Marsh often occurs around the edges of a lake or on the flood-plain of a river. In North American usage, a marsh is a herbaceous wetland in which the *water-table is permanently above the soil surface, equivalent to the British 'swamp'.

**marsh fern**   See THELYPTERIDACEAE.

**marsh gas**   See METHANOGEN.

**marshmallow**   See *ALTHAEA* and MALVACEAE.

**marshy tundra**   Marshy tracts (see MARSH) that are found as important components in the low-, middle-, and high-Arctic *tundra belts, because drainage is frequently poor as a result of the widespread presence of permanently frozen ground at no great depth. Grasses, *sedges, and dwarf willows are the most prolific vascular plants of marshy tundra, often accompanied by luxuriant growths of *moss.

**Marsileaceae**   A family of highly specialized aquatic or semi-aquatic ferns. The *rhizomes bear alternate leaves which are either cylindrical and taper to a point, e.g. *Pilularia*, or bear 2 or 4 broad leaflets at their tips, e.g. *Marsilea* and *Regnelidium*. The *sporangia are of two kinds, *megasporangia and *microsporangia, and are borne in globose, hairy *sporocarps at the base of the leaf stalks; 2 or more *sori occur in each sporocarp, and each sorus contains both types of sporangium. There are 3 genera, two of which are almost cosmopolitan, and some 70 species.

**Martyniaceae** (unicorn plants) A dicotyledonous (see DICOTYLEDON) family of herbs with tuberous roots that are known as unicorn plants because of their horned fruits. Characteristically, the plants are covered in sticky hairs, and have *simple, *opposite or alternate leaves and no *stipules. The terminal *inflorescence is composed usually of showy flowers which are 2-lipped. The *corolla is cylindrical, or funnel- or bell-shaped, and the *stamens are attached. The *ovary is *superior with a basal nectary disc, 2 fused *carpels, and 1 *locule. There is a single *style and a forked *stigma which is sensitive to touch. The fruit is a *capsule, with the persistent style forming a horn or a hooked projection which is a useful aid to dispersal by animals. The seeds are compressed, with a straight *embryo and no *endosperm. Some species are cultivated for their unusually shaped fruits, which can be pickled and eaten. The family is related to the *Bignoniaceae. There are 3 genera, with 13 species, found mainly in tropical and subtropical America.

**marvel of Peru**   See PARTIAL DOMINANCE.

**mass spectrometry**   A technique that allows the measurement of atomic and molecular masses. Material is vaporized in a vacuum; ionized; and then passed first through a strongly accelerating electric potential, and then through a powerful magnetic field. This serves to separate the *ions in order of their charge:mass ratio; detection is commonly made using an electrometer, which measures the force between charges and hence the electrical potential.

**massulae** In aquatic *ferns of the family *Azollaceae, mucilaginous extensions of the *tapetum which enclose the *microspores and *megaspores. In the *microsporangium, a diverse number of massulae develop, each containing a number of microspores. The megaspore is enclosed by four massulae. These structures are believed to contribute to buoyancy.

**mast** A *fruit, especially of beech but also of oak and other forest trees, often used as a food for pigs.

**master chronology** An average tree-ring chronology for a particular region, or one derived locally from a number of closely matching individual tree-ring chronologies. The master chronology forms the reference against which new ring series may be compared and dated.

**Mastigomycotina (division *Eumycota)** A subdivision of *fungi which form *motile (flagellated) cells called *zoospores. The subdivision contains 3 classes.

**mastigoneme** A hair-like projection occurring along the length of a *flagellum.

**mast year** A year in which there is a particularly high production of *mast. In oak, for example, there are often several years of poor acorn production between mast years.

**maté** *See ILEX.*

**maternal message** *Messenger-RNA synthesized during *oogenesis and deposited in the *egg cytoplasm, with *ribosomes, as an inactive complex often termed an *informosome'. It is activated after fertilization and helps to control the earliest stages of development.

**mating type** The equivalent in lower organisms (especially micro-organisms) of sexes in higher organisms. Micro-organisms may be subdivided into mating types on the basis of their physiology and mating behaviour. Different mating types are usually identical in physical form, although individuals of one mating type possess on their surfaces *proteins that will bind to complementary proteins or *polysaccharides found only on the coats of individuals of the opposite mating type. In this way, only individuals of different mating types will undergo conjugation.

**Matoniaceae** A family of ferns, whose *fronds, which are borne on *rhizomes, have a characteristic *dichotomous branching which produces a fan-like form. The *sori are *gradate, superficial, rounded, and on the underside of the *pinnae; unlike the related *Gleicheniaceae, they have rounded *indusia. The anatomy of the rhizome is a complex *solenostele, with several concentric cylinders of vascular tissue. There are 2 genera, with 4 species, found in Malesia.

**matorral** *See MAQUIS.*

**mature phase** The stage in the cyclical pattern of *community change in *grasslands and *heathlands. In grasslands, the mature phase is the stage in which the grass hummocks attain their maximum height (about 4 cm) and begin to be colonized by lichens. In *heathlands, it is the stage at which *Calluna vulgaris bushes, about 12–18 years old, start to become less dense, allowing other species, especially *bryophytes, to grow. *Compare* BUILDING PHASE; PIONEER PHASE; HOLLOW PHASE; and DEGENERATE PHASE.

**Mauritia (family *Arecaceae)** A genus of big fan palms which are *dioecious (female with some hermaphrodites). The paniculate *inflorescences are borne among the leaves. The fruits are scaly. *M. flexuosa* of swamp forest provides useful fibre. There are 16 species, found in tropical S. America.

**Maxburretia (family *Arecaceae)** A genus of fan palms which grow in small clumps. Male and hermaphrodite flowers occur in *panicles on different plants. The flowers are fragrant, and have 3 free *carpels. The fruits are thinly fleshy. There are 3 isolated species, each restricted to a few limestone hills in the Malay peninsula.

**maximum sustained yield (MSY)** *See* OPTIMUM YIELD.

**maximum thermometer** Thermometer that records the highest temperature to which it has been exposed (e.g. by allowing a rise of mercury past a restriction in the tube, but preventing the mercury's return on contraction). It is most commonly used to record maximum daily temperatures.

**maximum velocity of enzyme** The fastest rate at which a given quantity of *enzyme can catalyse a reaction under defined conditions. It is achieved only when the enzyme is saturated with *substrates and cosubstrates.

**May apple** See *PODOPHYLLUM*.

**mazaedium** A type of *ascocarp in which the ascocarp contents (*asci, *paraphyses, etc.) disintegrate to form a powdery mass.

**meadow grass** See *POA*.

**meadow saffron** See *COLCHICUM*.

**meadow steppe** A variant of Eurasian *steppe adjacent to the forest to the north, and comprised primarily of various sod-forming grasses and subordinate tussock grasses. The grasses grow to over a metre, and associated with them are broad-leaved flowering herbs, the whole giving a meadow-like aspect in summer. Meadow steppe has long been greatly reduced in extent by agricultural practices.

**mealy hairs** Hairs that, collectively, have the consistency of meal and give the appearance of a mealy covering.

**mean square** (variance) The square of the mean variation of a set of observations around the sample mean.

**medieval woodland** A woodland that is known to have existed prior to the seventeenth century. Some authorities use 1600, others 1650, as the base date. See also ANCIENT WOODLAND.

**mediolittoral** Applied to that area of a shore that is approximately equivalent to the intertidal or *littoral zone, but excluding the lowest reaches that are uncovered for only very limited periods. The term is used in an alternative system for subdividing the near-shore area of marine ecosystems. Compare INFRALITTORAL and CIRCALITTORAL; see also LITTORAL ZONE.

**Mediterranean floral region** Part of R. Good's (The Geography of the Flowering Plants, 1974) boreal kingdom, and geographically a relatively small region, being restricted to Iberia and the Mediterranean coast, but with an important extension northward along the western coast of Europe. The vegetation is specialized, reflecting the distinctive climatic setting, and consequently contains a high proportion of endemics (see ENDEMISM). The Mediterranean has been called 'the cradle of civilization', and its long association with mankind has yielded many garden, horticultural, and agricultural plants. See also FLORAL PROVINCE and FLORISTIC REGION.

**Mediterranean forest** The *climatic climax of most of the Mediterranean basin, now virtually destroyed and over vast areas replaced by *garrigue and *maquis. The trees were *evergreen, both conifers and broad-leaved hardwoods. They were probably mixed, but there may well have been segregations of the two types.

**Mediterranean scrub** Two principal varieties are recognized on the basis of the stature of the constituent shrubs. The tall variety is called *maquis, the short variety *garrigue; but the distinction is somewhat arbitrary and there are many intergrade types. Both occupy vast areas and were derived by burning and grazing from an earlier forest cover. Indeed, constant use of fire has created distinctive plant associations within the scrub, such as the *giane* (rosemary and *Genista* species) and *ericeto* (tree heath) in southern Italy. In Iberia such associations are even more extensive and include the *brezales* (heathers), *jorales* (*Cistus* species), *goscojales* (*Quercus coccifera*, kermes oak), and *bujedales* (*Buxus sempervirens*, box).

**medlar** See *MESPILUS*.

**medulla** 1. The central part of a structure, e.g. the *pith of a stem. 2. A tangle of loose fungal *hyphae. 3. A layer of loose hyphae in the *thallus of a *lichen.

**medullary rays** A plate of *parenchyma that extends between the *medulla and the *cortex across the vascular region.

**megafossil** (macrofossil) A fossil that is large enough to examine without the aid of a microscope.

**megaphyll (macrophyll)** A large leaf, usually with *leaf gaps associated with the *leaf traces. *Compare* MICROPHYLL.

**megasporangium** A *spore sac that contains *megaspores. In flowering plants, this is known as the *ovule. *See also* HETEROSPORY and MICROSPORANGIUM.

**megaspore** The larger of two kinds of *spore produced by *heterosporous *ferns, or the first cell of the female *gametophyte generation of such ferns and *seed plants (angiosperms and gymnosperms). In angiosperms, it is one of 4 *haploid cells formed from a *megasporocyte during *meiosis, the other 3 degenerating. The megaspore of angiosperms divides to produce the female gametophyte (*embryo sac).

**megasporocyte** A *diploid *megaspore mother cell in an *ovule that forms *haploid megaspores by meiotic division (*see* MEIOSIS).

**megasporophyll (macrosporophyll) 1.** A leaf-like structure that bears megasporangia (*see* MEGASPORANGIUM). Despite the name (-phyll), it is not necessarily derived from a modified leaf. **2.** In *angiosperms, the carpel. *Compare* MICROSPOROPHYLL.

**megatherm** In Alphonse de *Candolle's (1874) classic temperature-based scheme of world vegetation zones, a plant of the most warm (i.e. tropical) environments, where each month has a temperature mean of no less than 18°C and moisture supply is not limited. *Compare* XEROPHILE.

**meiosis (reduction division)** A type of nuclear division that occurs at some stage in the life cycle of sexually reproducing organisms. It is a mechanism whereby the number of *chromosomes is halved to prevent doubling in each generation. Genetic material can be exchanged between *homologous chromosomes. Two successive divisions of the *nucleus occur, with corresponding cell divisions, following a single chromosomal duplication. This produces *gametes or sexual *spores that have one half of the genetic material or chromosome number of the original cell. This halving of the chromosome number ($2n$ to $n$) compensates for its doubling when the gametes ($n + n$) unite to form a zygote ($2n$) during *sexual reproduction. The process occurs during gamete or spore formation (some plants produce spores asexually); in many *fungi, and in green *algae, it occurs immediately after fertilization or on germination of the *zygote. Chromosomes first appear in the first stage of prophase I (leptotene) of meiosis, as single threads. The 2 homologous members of each chromosome pair associate side by side with corresponding *loci adhering together: this is called pairing. It occurs during the zygotene stage, each resulting pair being called a *bivalent. Thus the *apparent* number of chromosome threads is half what it was before, being the number of bivalents rather than the number of single chromosomes. During the pachytene stage, each bivalent separates into 2 sister *chromatids (except at the region of the *centromere), with some localized breakage and *crossing-over of genetic material of both maternal and paternal origin. There are now $n$ groups of 4 chromatids lying parallel to each other and forming a *tetrad. During the diplotene stage, 1 pair of sister chromatids in each of the tetrads begins to separate from the other pair except at the sites where exchanges have taken place. In these regions the overlapping chromatids form a cross-shaped structure called a *chiasma (pl. chiasmata) and these chiasmata slip towards the ends of the chromatids so that their position no longer coincides with that of the original cross-overs. This process continues until, during diakinesis, all the chiasmata reach the ends of the tetrads and the homologues can separate during *anaphase. At diakinesis the chromosomes coil tightly, so shortening and thickening to form a group of compact tetrads which are well spaced out in the nucleus, and the nucleolus disappears. This ends the *prophase stage of meiosis. During the first division (metaphase through to telophase), the *nuclear envelope disappears, with the tetrads arranged at the equator of the *spindle. The chromatids of a tetrad separate in such a way that maternal chromosomal material is kept distinct from paternal material except at regions distal to the points of crossing-over. This first division produces

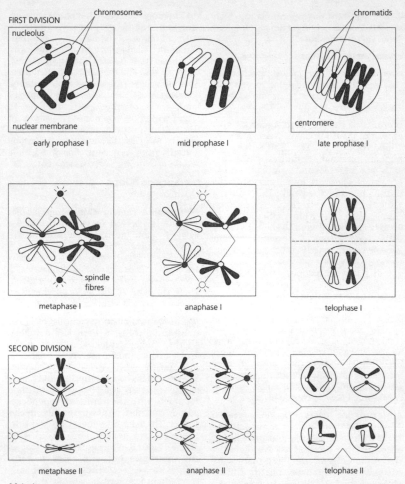

FIRST DIVISION

chromosomes

nucleolus

nuclear membrane

early prophase I

mid prophase I

chromatids

centromere

late prophase I

spindle fibres

metaphase I

anaphase I

telophase I

SECOND DIVISION

metaphase II

anaphase II

telophase II

*Meiosis*

2 *daughter nuclei containing dyads (a dyad is half a tetrad), each of which becomes surrounded by a nuclear envelope. In many plants there is no *telophase, *cell-wall formation, or *interphase, and the cell passes directly from anaphase I into prophase of the second meiotic division. In the second division the nuclear membrane disappears once more and the dyads arrange themselves upon the *metaphase plate, the chromatids of each dyad being equivalent to one another (except for those regions distal to points of crossing-over). The centromere divides and so allows each chromosome to pass to a separate cell and the process is complete. *Compare* MITOSIS.

**meiosporangium** A *sporangium in which *meiosis occurs. For example, in some species of *Allomyces* (order *Blastocladiales) the *sporophyte forms thick-walled, resistant meiosporangia in which meiosis occurs, giving rise to *haploid *zoospores (meiospores).

**meiospore** *See* MEIOSPORANGIUM.

**meiotic drive** Any meiotic mechanism that results in the unequal recovery of the 2 types of *gamete produced by a *heterozygote.

***Melaleuca*** **(family \*Myrtaceae)** A genus of woody shrubs or small trees that have *opposite, entire leaves. The flowers are regular, bisexual, *cymose or solitary, with 4 or 5 *sepals and a similar number of *petals. There are numerous *stamens in tufts opposite the petals. The *ovary is *inferior, with several *locules and a long, simple *style. The fruit is a fleshy *berry or dry *capsule, and the seeds have little *endosperm. The plants exhibit xerophytic (*see* XEROPHYTE) characters (e.g. thickened leaves and sub-epidermal oil glands). Many are halophytic (salt tolerant) but they are typically present in tropical areas that experience heavy summer rainfall, and occur in many *Eucalyptus forests in areas of low fertility. Some species are important for aromatic oils (e.g. *M. cajuputi* gives cajeput oil). There are about 150 species, of which most occur only in Australia, although they are cultivated elsewhere for their flowers. *Melaleuca* and the related *Callistemon* are commonly known as bottle-brushes.

***Melampsora*** **(class \*Urediniomycetes)** A genus of *rust fungi, in which the *teliospores are *sessile and unicellular, often forming a crust-like layer beneath the *epidermis of the host plant. The *aecia are not cup-shaped but are diffuse. *M. lini* causes rust disease in flax (*Linum usitatissimum*).

**Melanconiales** **(class \*Coelomycetes)** An order of *imperfect microscopic fungi which produce *conidia in *acervuli or on *sporoduchia. The vegetative *mycelium occurs within the substratum or host. The order includes *saprotrophic and plant-*parasitic fungi; some are important plant *pathogens.

**Melanesia and Micronesia floral region** Part of R. Good's (*The Geography of the Flowering Plants*, 1974) *Polynesian subkingdom, within his *palaeotropical kingdom, and geographically coinciding with Micronesia (Marianas, Carolines, Marshall, Kiribati, and Tuvalu) and Melanesia (New Guinea, Solomon Islands, Vanuatu, and Fiji, but not New Caledonia). Although there is an appreciable endemic (*see* ENDEMISM) component at the species level, most of the species belong to large, widespread genera. The status of the endemism is not high, therefore, and reflects the fact that the flora is largely derived from adjacent floras. *See also* FLORAL PROVINCE and FLORISTIC REGION.

**melangeophilous** Growing in dark *loam or *alluvial soil.

***Melastoma*** **(family \*Melastomataceae)** A genus of shrubs in which the leaves have 3 main nerves. The flowers are showy, mauve, *pentamerous, with 5 big and 5 small *stamens. The fruits are fleshy *capsules. *M. malabathricum*, a weedy shrub, is the straits rhododendron. There are about 70 species, occurring from southern China to the Pacific islands.

**Melastomataceae** A family of small trees and shrubs, with a few climbers, *herbs, and *epiphytes, in which the leaves are *opposite, without *stipules, and mostly with several main nerves. The flowers are hermaphrodite, and have 4 or 5 free *sepals and *petals, *geniculate *stamens with a prominent appendage, porous *anthers, and usually an *inferior *ovary. The fruit is a *berry or *capsule. There are 215 genera, with about 4750 species, found mainly in the tropics and subtropics, but with a few in warm temperate regions; they are especially common in S. America.

***Melia*** **(family \*Meliaceae)** A genus of trees or treelets which have *bipinnate or tripinnate leaves. The flowers are hermaphrodite, the fruit a *drupe. Some species have useful timber (e.g. *M. azedarach* (pride of India), native to the Himalaya but now naturalized throughout the tropics and subtropics). There are 3 species, found in the Old World tropics.

**Meliaceae** A family of trees and treelets, most of which have spirally *pinnate leaves. The flowers are bisexual or unisexual, and regular, with the *sepals and *petals united at the base, the *stamens usually forming a tube, and the *ovary *superior, with 2–6 *locules. Meliaceae produce a variety of fruits. Many are valuable timber trees. There are 51 genera, with about 575 species, occurring in the tropics and subtropics of both hemispheres.

**melon** (*Cucumis melo*) See Cucur-BITACEAE.

**membrane** A sheet-like structure, 7–10 nm wide, that forms the boundary between a cell and its environment and also between various compartments within the cell. It is composed mainly of *lipids, *proteins, and some *carbohydrates, the structural arrangement of which is still subject to speculation. Membranes function as selective barriers and also as a structural base for *enzymes, which may form an integral part of the membrane itself.

**Memecylon** (family *Melastomataceae) A genus of small trees that have exceedingly hard wood and very thin bark. The leaves have a single main nerve and innumerable close, faint, secondary nerves. The fruit is a *berry. There are about 150 species, occurring in the Old World tropics.

**memnonius** Brownish-black in colour.

**Mendel, Gregor Johann** (1822–84) An Austrian Augustinian monk in the monastery at Brünn (now Brno), whose interest in natural science led him to conduct the experiments that resulted in his discovery of the underlying principles of heredity. These experiments began in 1856 and were conducted in the monastery garden. They involved crossing varieties of garden peas, each of which had certain distinctive features, and recording the appearance of those features in the progeny. Mendel took part in the meetings of the Naturforscher Verein (natural science society) in Brünn and it was there that he reported his results on 8 March 1865. In 1900, while searching the literature in connection with their own research, K. E. Correns, E. Tschermak von Seysenegg, and H. de Vries came across Mendel's paper describing results similar to their own. Mendel was elected abbot of his monastery in 1854.

**Mendelian character** A character that follows the laws of inheritance formulated by Gregor *Mendel.

**Mendelian population** An interbreeding group of organisms that share a common *gene pool.

**Mendel's laws** Two general laws of inheritance formulated by the Austrian monk Gregor *Mendel. Re-expressed in modern terms, the first law, of segregation, is that the two members of a gene pair segregate from each other during *meiosis, each *gamete having an equal probability of obtaining either member of the *gene pair. The second law, of independent assortment, is that different segregating gene pairs behave independently. This second law is not universal, as was originally thought, but applies only to unlinked or distantly linked pairs (see LINKAGE). At the time of Mendel, genes had not been identified as the units of inheritance: he considered factors as a pair of characters segregating and members of different pairs of factors assorting independently.

**mengkulang** The timber of *Heritiera* from western Malesia. It makes an attractive veneer for plywood.

**Menispermaceae** A family of *lianes or shrubs which have alternate or *peltate leaves with no *stipules and unisexual, small, white or green flowers. The plants are *dioecious. The flowers have 2 or 3 *petals and often 2 rows of 3 *sepals. The *stamens and *carpels can be very numerous. The *stigma is entire or lobed. The fruit is a *drupe, often in a horseshoe shape. There are 8 *tribes, separated by seed structure, and 78 genera, with 525 species, which are mainly confined to tropical rain forests throughout the world. Several species are useful in medicine and the preparation of drugs.

**Mentha** (mint; family *Lamiaceae) A genus of herbs that have creeping *rhizomes and flowers with a 10–13-nerved *calyx with 5 nearly equal teeth, a *corolla with 4 nearly equal lobes, and

4 equal *stamens. The plants contain pleasantly fragrant volatile oils, much used in flavouring and for sauces. The oil is extracted and also used as a mild antiseptic. There are 25 species and many hybrids, occurring in temperate Eurasia and in S. Africa.

**menthol** *See* LAMIACEAE.

**Menyanthaceae (bogbean)** A family of aquatic or wetland (bog) *herbs that have creeping *rhizomes or tufted root-stocks. They are mainly *perennial. They are related to *Gentianaceae but are found in different habitats and have leaves that are mostly alternate. The regular flowers are borne in *cymes or dense clusters. The *corolla is 5-lobed, with valvate lobes when in bud. The *sepals (which remain in the fruit) and the united *petals are pink, yellow, or white, with a fringe of hairs in the flower. The *stamens are attached to the corolla tube. The *ovary is *superior, with 1 *locule and several *ovules, the *style is simple and forked. The fruit is a fleshy *nut or dry *capsule which is often *dehiscent. The seeds contain much *endosperm and are often winged. *Menyanthes* (buckbean) has *trifoliate leaves; *Nymphoides* (fringed water lily) has *simple, round leaves. There are 5 genera, with some 40 species, widespread in the world, although 2 of the genera are restricted to the northern hemisphere and 2 to the southern hemisphere. Some species are garden ornamentals, and species of *Menyanthes* are used in medicine.

**Menyanthes (buckbean)** *See* MENYAN-THACEAE.

**meranti** *See* SHOREA.

**merawan** The hard, heavy timber of some *Hopea* species from western Malesia, which is good for construction.

**merbau** The timber of *Intsia* from the Malay archipelago, which is dark, heavy, and good for furniture and flooring.

**Merceya** *See* COPPER MOSSES.

**meremium** Timber obtained from standing trees.

**Meripilus (family *Polyporaceae)** A genus of *fungi that form large,

fan-shaped *fruit bodies in *rosette-like clusters. The upper surface is marked with concentric zones of light and dark brown; the pores on the lower surface are small. Species include *M. giganteus* (giant polypore). They are found at the bases of living or dead, broad-leaved trees.

**Merismopedia** *See* SYNECHOCYSTIS.

**meristele** *See* DICTYOSTELE.

**meristem** A group of plant cells that are capable of dividing indefinitely and whose main function is the production of new growth. They are found at the growing tip of a root or a stem (apical meristem); in the *cambium (lateral meristem); and, in grasses, also within the stem and leaf sheaths (intercalary meristem).

**meristoderm** In some brown algae (*Phaeophyta), the outer layer of the *stipe, which resembles *meristem in that its cells divide continually to replace tissue damaged by abrasion against rocks.

**mermaid's cup (mermaid's wineglass)** The common name for the stalked, wineglass-shaped *thallus of the *green alga *Acetabularia*.

**mermaid's wineglass** *See* MERMAID'S CUP.

**merocenosis** *See* MEROTOPE.

**merological approach** An approach to *ecosystem studies in which the component parts of an ecosystem are studied in detail in an attempt to compose a picture of the whole system. *Compare* HOLISTIC.

**meromictic** Applied to lakes whose waters are stratified permanently, usually because of some chemical difference (e.g. contrasting salinities, and hence densities) between *epilimnion and *hypolimnion waters.

**meromixis** A unidirectional transfer of genetic material; the term is often used to describe genetic exchange in the Schizophyta.

**merotope** A microhabitat that forms part of a larger unit, e.g. a fruit or a pebble.

Organisms colonizing the merotope form a merocenosis.

**Merremia** (family *Convolvulaceae) A genus of large, tropical creepers, several of which have stout stems. Several are noxious weeds of disturbed forest. There are about 70 species.

**mersawa** See *ANISOPTERA*.

**mesarch** Applied to strands of *xylem in which the first elements form at the centre and subsequent elements form on both the inside and outside of them. *Compare* ENDARCH and EXARCH.

**mescal** See *AGAVE*.

**mesic** Applied to an environment that is neither extremely wet (hydric) nor extremely dry (xeric). *See also* PERGELIC.

**meso-** A prefix, derived from the Greek *mesos*, meaning 'middle', e.g. *mesotrophic, neither nutrient-poor nor nutrient-rich; and *mesobiota, the medium-sized soil organisms.

**mesobiota** (mesofauna, mesoflora) A general term for soil organisms of intermediate size. 'Mesofauna' in particular refers to small, invertebrate animals found in the soil, characteristically annelids, arthropods, nematodes, and molluscs. These organisms are readily removed from a soil sample using a tullgren funnel or similar device. *Compare* MACROBIOTA and MICROBIOTA.

**mesocarp** See PERICARP.

**mesoclimate** A general term applied to the characteristics of a relatively small region (e.g. a valley or an urban area).

**mesocotyl** In *Poaceae, the structure forming the axial part of the *embryo, below the *coleoptile, resulting from the fusion of parts of the *cotyledon with the *hypocotyl.

**mesofauna** See MESOBIOTA.

**mesoflora** See MESOBIOTA.

**mesogenous** In some *angiosperms, applied to a *stoma in which the 2 *guard cells and the subsidiary cells are all derived from a single mother cell. *Compare* PERIGENOUS; *see also* SYNDETOCHEILIC.

**mesohaline** See BRACKISH.

**mesophile** An organism that grows best at moderate temperatures, often quoted as 20–45°C.

**mesophyll** **1.** Internal *parenchyma tissue of a plant *leaf that lies between epidermal layers. Its main functions are *photosynthesis and the storage of starch. **2.** Plant tissue composed of unspecialized cells that lies immediately beneath the onter layer of leaf cells. Photosynthesis takes place in mesophyll cells and starch is stored in them.

**mesophyte** A plant adapted to environments that are neither extremely wet nor extremely dry. *Compare* HYDROPHYTE and XEROPHYTE.

**mesosome** A complex infolding of the *plasma membrane in *prokaryotic cells. It contains respiratory *enzymes and also appears to play a role in cell division, since the *chromosome is usually attached to it and it is often the site of initiation of *cytokinesis.

**mesotherm** A plant of warm temperate areas, where the hottest month has a mean temperature of more than 22°C and the coldest month a mean of not less than 6°C, according to Alphonse de *Candolle's (1874) classic temperature-based scheme of world vegetation zones.

**mesotidal** Applied to coastal areas where the tidal range is 2–4 m. Tidal action and wave activity both tend to be important in such areas.

**Mesozoic** The middle of three eras that constitute the *Phanerozoic period of time, about 248–65.5 Ma ago. The Mesozoic (literally 'middle life') was preceded by the *Palaeozoic Era and followed by the *Cenozoic Era. The Mesozoic comprises the *Triassic, *Jurassic, and *Cretaceous Periods.

**Mespilus** (medlar; family *Rosaceae) A *monotypic genus of trees, close to *Malus but with the *carpels exposed in a disc within the wide ring of the persistent *calyx at the top of the fruit, and with solitary flowers and large, leaf-like *sepals. Fruits can be eaten when they become soft and brown. They are native to southeastern Europe and central Asia.

**messenger-RNA (m-RNA)** A single-stranded *RNA molecule that is responsible for the transmission to the ribosomes of the genetic information contained in the nuclear *DNA. It is synthesized during *transcription and its base sequencing exactly matches that of 1 of the strands of the double-stranded DNA molecule.

**Messinian** An important stage of the Late *Miocene, marked by the presence of thick evaporite deposits in the Mediterranean. These may indicate that the Straits of Gibraltar were closed approximately 7.25–5.33 Ma ago and that the Mediterranean was reduced to a series of evaporite basins.

**Mesua** (family *Clusiaceae) A genus of trees or shrubs in which the fruit is either *dehiscent or not, and is seated on, or enveloped by, the enlarged woody *sepals. The timber (notably ironwood from *M. ferrea*) is hard, heavy, and durable. There are about 40 species, occurring in Indo-Malaysia.

**metabolic pathway** A sequential series of enzymatic reactions involving the synthesis, degradation, or transformation of a metabolite. Such a pathway may be linear, branched, or cyclic, and directly or indirectly reversible.

**metaboly** Ability to change shape.

**metacentric** A term applied to a *chromosome that has its *centromere in the middle.

**metallic bond** The chemical bond that links the atoms in a solid metal. The atoms are ionized and electrons move fairly freely among them as an 'electron gas', the bond being between the electropositive atoms and the electrons. It is the free electrons that give metals their high electrical and thermal conductivities. *See also* COVALENT BOND, HYDROGEN BOND, and IONIC BOND.

**Metallogenium** A genus of *bacteria not assigned to any taxonomic family. Cells lack rigid walls and develop flexible, tapering threads from which new cells probably arise by *budding. The threads are often heavily encrusted with oxides of iron and/or manganese. They are found in a range of freshwater habitats and in soil.

**metaphase** A stage of *mitosis or *meiosis at which the *chromosomes move about within the *spindle until they eventually arrange themselves in its equatorial region.

**metaphloem** *Primary phloem that develops after the *protophloem and before the *secondary phloem. It completes its elongation and persists, but may be obscured by secondary phloem.

**Metaphyta** *See* PLANTAE.

**Metasequoia** (family *Taxodiaceae) A *monotypic genus, *M. glyptostroboides* (dawn redwood), which is a deciduous conifer known (since 1941) as *Pliocene fossils from America and in 1948 discovered living in central China. It is planted as an ornamental, is fast-growing in wet sites, and strikes easily from cuttings.

**metaxenia** An effect exerted by *pollen on the tissues of female organs.

**metaxylem** *Primary xylem that develops after the *protoxylem and before the *secondary xylem.

**Methanobacteriaceae** A family of *archaebacteria which have a specialized type of energy-yielding metabolism, in which methane is an end-product. The organisms can grow only in the absence of air. They are found in *anaerobic habitats, e.g. sewage sludge, the intestines of animals, sediments in natural waters, and waterlogged soils.

**methanogen** A single-celled organism that derives metabolic energy by using hydrogen to reduce carbon dioxide to methane, emitting the methane as a by-product. Methanogens are obligatory *anaerobes that inhabit *swamps and *marshes where other organisms have consumed all the oxygen. The methane bubbling to the surface is known as marsh gas. Methanogens belong to the *Archaea, ranked as a subkingdom in the five-kingdom taxonomic system and as a *domain in the three-domain system.

**methanogenic** Methane-producing; applied, for example, to *Archaea that produce methane gas as a product of their metabolism.

**methanotroph** A bacterium that can use methane as a nutrient.

**methylation** The introduction of a methyl group (-CH₃) into an organic compound.

**methylotroph** An organism that can use (as its sole source of carbon and energy) organic compounds that contain only one carbon atom (i.e. compounds such as methane and methanol).

***Metrosideros*** (family *Myrtaceae) A genus of trees or shrubs, some of which are *stranglers. Some of them have showy *inflorescences from the numerous *exserted *stamens and are grown for ornament. There are about 50 species, occurring from southern Africa to the Pacific.

***Metroxylon*** (sago palms; family *Arecaceae) A genus of big palms, most of which form a huge, tree-like, terminal *inflorescence and die after fruiting. Some are clump-forming. The leaves are *pinnate, and weakly spiny. They are *monoecious. The flowers are held in *spikes subtended by *bracts, comprising a male, plus a female or hermaphrodite. The fruits are scaly. The *endosperm is hard, a vegetable ivory. There are 8 species, occurring from the Moluccas east to the Carolines and Fiji.

***Metroxylon sagu*** (sago palm) A clump-forming palm, native to the Moluccas and New Guinea, that is now extensively cultivated throughout the Malay archipelago for its leaves (which are used to make a superior thatch) and for the starch in its trunk (which is used as food for animals or for humans during famine, and which is still the staple in parts of its natural range in New Guinea). Formerly it was cultivated for its fruits, which yield a vegetable ivory.

**Mexican lilac** *See GLIRICIDIA.*

**Mg** *See MAGNESIUM.*

**Michaelis constant** The kinetic constant, $K_m$, characteristic of each *enzyme, which is numerically equal to the concentration, in moles per litre, of the given *substrate that gives half the *maximum velocity.

**Michaelmas daisy** *See ASTER.*

***Michelia*** (family *Magnoliaceae) A genus of trees and shrubs in which the fruiting *carpels are free and laxly spaced. *M. figo* is cultivated for its perfume, used in China for hair oil and scented tea. There are 45 species, occurring in subtropical and tropical Asia.

**micro-aerobic** Applied to an environment in which the concentration of oxygen is less than that in air.

**micro-aerophilic** Applied to an organism that grows best under *micro-aerobic conditions.

**microbe** An organism that can be seen only with the help of a microscope. The term covers *Bacteria, *Fungi, and *viruses, but not protozoa and *algae. *See also* MICRO-ORGANISM.

**microbial** (microbic) Of or pertaining to *micro-organisms.

**microbial genetics** The study of the genetics of *micro-organisms. This is an important discipline of genetics, since micro-organisms may be bred in captivity and since many generations can be obtained in short periods of time; their heredity can therefore be studied more readily than that of higher organisms.

**microbic** *See* MICROBIAL.

**microbiocoenosis** *See* BIOCOENOSIS.

**microbiology** The study of *micro-organisms and allied subjects.

**microbiota** (microfauna, microflora) The smallest soil organisms, comprising *bacteria, *fungi, *algae, and *protozoa.

**microbody** The name that is sometimes given to *peroxisomes and *glyoxysomes.

**microclimate** The atmospheric characteristics prevailing within a small space, usually in the layer near the ground that is affected by the ground surface. Special influences include the impact of vegetation cover on humidity (by evapotranspiration) and on temperature and winds.

**Micrococcaceae** A family of *bacteria in which the cells are *Gram-positive and spherical, and occur in regular or irregular groups. They may be *motile or non-motile. Resting stages are unknown.

They are *chemo-organotrophic. They are found in a wide range of habitats such as soil, aquatic environments, or the bodies of animals (including humans), etc.

**Microcoleus** *See* LPP GROUP.

**microconsumer** *See* CONSUMER ORGANISM.

**microcosm** A late nineteenth-century American term encompassing essentially the same ideas as the *ecosystem concept. Now it is applied especially to small-scale, simplified, experimental ecosystems, laboratory- or field-based, which may be either derived directly from nature (e.g. when samples of pond water are maintained subsequently by the input of artificial light and gas exchange) or built up from axenic (organism-free) cultures until the required conditions of organisms and environment are achieved. Such small-scale experimental ecosystems may also be called micro-ecosystems or ecotrons.

**Microcystis** A phycological genus of 'blue-green algae' (*cyanobacteria) in which the small cells are densely packed in mucilage to form irregularly shaped, often perforated colonies. They are planktonič in freshwater environments (e.g. lakes, bogs, etc.). Some strains of *M. aeruginosa* produce one or more toxins, and blooms of these strains in reservoirs have been responsible for the poisoning of water supplies. The genus will probably be incorporated into the cyanobacterial genus *Synechocystis*.

**micro-ecosystem** *See* MICROCOSM.

**microevolution** Evolutionary change within species, which results from the differential survival of the constituent individuals in response to *natural selection. The genetic variability on which selection operates arises from *mutation and sexual reshuffling of *gene combinations in each generation.

**microfibril** A more or less crystalline aggregation of *cellulose molecules, at least 60 nm long and about 200 nm² in cross-section, that is found in the *cell walls of higher plants.

**microfilament** A filament. 0.4–0.7 nm in diameter, that is composed of the *protein actin. Microfilaments often occur in abundance immediately beneath the *plasma membrane, and play a role in cell motility, *cytokinesis, and *cytoplasmic streaming.

**microflora** *See* MICROBIOTA.

**Micromonospora** (order Actinomycetales) A genus of *bacteria which form well-developed *mycelia with branched, *septate *hyphae. Dark-coloured *spores are formed singly. They are *saprotrophs, found in soil, compost, etc.

**micronutrient** An organic or inorganic element or compound that is needed in only relatively small amounts by living organisms. Vitamins are the main group of organic micronutrients. *Compare* MACRONUTRIENT; *see also* ESSENTIAL ELEMENT.

**micro-organism** Literally, a microscopic organism. The term is usually taken to include only those organisms studied in microbiology (i.e. bacteria, fungi, microscopic algae, protozoa, and viruses), thus excluding other microscopic organisms such as eelworms and rotifers.

**microphyll** A small leaf with a single *vein and no *leaf gap associated with the *leaf trace. *Compare* MEGAPHYLL.

**micropylar** *See* MICROPYLE.

**micropyle** (adj. micropylar) A canal in the coverings of the *nucellus through which the *pollen tube usually passes during *fertilization. Later, when the

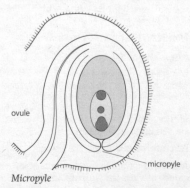

*Micropyle*

*seed matures and starts to germinate, the micropyle serves as a minute pore through which water enters.

**microsome** A small, membrane-bound *vesicle that is formed from the *endoplasmic reticulum in large numbers following the homogenization of cells. Microsomes may or may not bear *ribosomes, depending upon the type of endoplasmic reticulum from which they are derived.

**microspecies** A population of uniparental plants that is genotypically uniform and has recognizable phenotypic expression; in both respects it is distinctive within its own uniparental group.

**microsporangium** A *sporangium (e.g. in heterosporous (*see* HETEROSPORY) *ferns and the clubmoss *Selaginella*) that produces numerous minute *spores which germinate to produce only male *gametophytes. Those *pteridophytes that produce them also produce *megasporangia containing few (often only 1–4) large spores; these spores germinate to produce the female *gametophytes.

**microspore 1.** The first cell of the male *gametophyte generation of angiosperms and *gymnosperms (seed plants), later to form the *pollen grain. **2.** The smaller of the 2 kinds of *spore produced by heterosporous (*see* HETEROSPORY) forms.

**microsporocyte** A cell in a *microsporangium that undergoes *meiosis to yield 4 *microspores.

**microsporophyll 1.** A *leaf or modified leaf of plants that bear *microsporangia. **2.** In *angiosperms, the *stamen. *Compare* MEGASPOROPHYLL.

**microtherm** A plant of cool temperate environments, where temperatures in the warmest month should be between 10°C and 22°C and the coldest monthly mean temperature does not fall below 6°C, in Alphonse de *Candolle's (1874) classic temperature-based scheme of world vegetation zones.

**microtidal** Applied to coastal areas in which the tidal range is less than 2 m. Wave action dominates the processes active in microtidal areas (e.g. the Mediterranean Sea and the Gulf of Mexico).

**microtubule** A tubular structure, 15–25 nm in diameter, of indefinite length, and composed of subunits of the *protein tubulin. It occurs in large numbers in all *eukaryotic cells, either freely in the *cytoplasm, or as a structural component of *organelles (e.g. *cilia and *flagella). Microtubules appear to function in the motility of cells, the maintenance of cell shape, and the transport of materials within cells. In addition, they form part of the structure of the *mitotic spindle (responsible for the movement of *chromosomes during cell division), and have been implicated in sensory transduction in some receptor cells.

**middle Arctic tundra** The middle Arctic belt of the *tundra. The tundra vegetation at this latitude is intermediate in floristic composition and luxuriance of development between that of the low Arctic and high Arctic.

**middle lamella** A thin, gluey membrane separating two adjacent *cell walls that serves to cement them together. It consists of *pectins and other non-cellulose *polysaccharides.

**mid-latitude mixed forest** A forest comprising coniferous and broad-leaved trees, and belonging to 1 of 2 broad categories: (*a*) *ecotone mixed forest, which has characteristics transitional between those of the 2 great belts of *boreal coniferous forest and mid-latitude broad-leaved *deciduous forest; and (*b*) a second type that seems to have the status of true *climatic climax, in which both the conifers and broad-leaved trees are *evergreens. These evergreen mixed forests were once extensive in the Mediterranean basin, and in the southern hemisphere are found in Chile, southern Brazil, Tasmania, northern New Zealand, and the Cape Province (S. Africa).

**midrib** The central, thick, linear structure that runs along the length of a plant *thallus or *lamina. It occurs in true leaves as a vein running from the leaf base to the apex and in the leaf-like structures of mosses and seaweeds. It provides support and is a translocative vessel.

**Mielichhoferia**   *See* COPPER MOSSES.

**mignonette**   *See* RESEDACEAE.

**migration 1.** The movement of individuals or their propagules (seeds, spores, larvae, etc.) from one area to another. Three cases may be distinguished: (*a*) emigration, which is outward only; (*b*) immigration, which is inward only; and (*c*) migration, which in this stricter sense implies periodic two-way movements to and from a given area and usually along well-defined routes. Such migratory movement is triggered by seasonal or other periodic factors (e.g. changing day-length), and occurs in many animal groups. **2.** In plant *succession, specifically the arrival of migrating propagules (migrules) at a newly denuded area.

**migration route** The link between two *biogeographical regions that permits the interchange of plants and/or animals.

**migrule**   *See* MIGRATION.

**mildew** Any fungal disease of a plant in which the *mycelium of the causal agent is visible as white or pale-coloured, cottony or powdery patches on leaves etc. *See also* DOWNY MILDEW and POWDERY MILDEW.

**milfoil** *See* MYRIOPHYLLUM and HALO-RAGIDACEAE.

**milk cap** The common name for the *fruit body of a *fungus of the genus *Lactarius*. When damaged, such fruit bodies exude a milky latex.

**milk-tree (cow-tree)** *See* BROSIMUM.

**milkweed**   *See* ASCLEPIAS.

**millet**   *See* PANICUM.

**Millettia** (family *Fabaceae, subfamily *Papilionatae) A genus of trees and big woody climbers in which the leaves are usually *pinnate. The pods are sometimes tardily *dehiscent, the seeds large. The timber, with interlocking grain, is not much used. There are about 90 species.

**milli-equivalent** One-thousandth of an equivalent weight.

**Mimosa** (family *Fabaceae, subfamily *Mimosoideae) A genus mainly of *herbs

and shrubs, in which the *stipules are sometimes thorn-like. The leaves are multipinnate. *M. pudica* (the sensitive plant) has leaflets that droop on shaking. *M. invisa* is a noxious weed. There are about 400 species, occurring in the tropics, mainly in America.

**Mimosaceae**   *See* MIMOSOIDEAE.

**Mimosoideae** One of the 3 subfamilies of *Fabaceae (ranked sometimes as a family, Mimosaceae) of mainly woody plants, in which the leaves are often *bipinnate. The flowers are regular; the *sepals and *petals are free and *valvate; the *stamens are commonly numerous. There are 56 genera, found mainly in the tropics.

**Mimulus** (monkey flowers; family *Plantaginaceae) A genus comprising about 150 species of herbs that are much cultivated for their showy flowers. They occur in southern Africa, Asia, and especially in America.

**Mimusops** (family *Sapotaceae) A genus of trees in which the leaves are *stipulate. There are 8 *sepals, 24 *petals, and 8 *stamens. *M. elengi* is a handsome ornamental, and also has medicinal properties. There are 57 species, occurring in the Old World tropics, mainly in Africa.

**mineral cycle** A *biogeochemical cycle, in which elements move through the soil, living organisms, air, and water, or through some of these.

**mineralization** The conversion of organic tissues to an inorganic state as a result of decomposition by soil *microorganisms.

**mineral soil** Soil composed principally of mineral matter, in which the characteristics of the soil are determined more by the mineral than by the organic content.

**minimal area** The smallest area that can contain the species representative of a particular plant *community. When the number of species recorded in increasingly larger sample units is plotted graphically (to give a species–area curve), the minimal area is the point at which the curve becomes horizontal. In practice, the curve rarely becomes truly horizontal because of natural heterogeneity, and some subjective assessment of the minimal area is made, based on the species–area curve. Alternatively, the 'minimum quadrat number' may be defined as the 'equivalence point', i.e. the point at which the number of species and the number of *quadrats are equal.

**minimum quadrat number** See MINIMAL AREA.

**minimum temperature** The lowest temperature recorded—diurnally, monthly, seasonally, or annually, or the lowest temperature of the entire record. Daily air temperature minima are recorded by the screen minimum thermometer. See also GRASS MINIMUM TEMPERATURE.

**minimum thermometer**. A thermometer that records the lowest temperature to which it has been exposed (e.g. by allowing a fall of mercury past a restriction in the tube, but preventing the mercury's return on expansion). It is most commonly used to record minimum daily temperatures.

**mint** See MENTHA.

**mint-bush** See PROSTANTHERA.

**Miocene** The fourth of the 5 epochs of the *Tertiary Period, about 23.03–5.332 Ma ago, extending from the end of the *Oligocene to the beginning of the *Pliocene.

**Mirabilis jalapa** (four-o'clock plant) See PARTIAL DOMINANCE.

**mire** A *peat-producing *ecosystem.

**mirror yeast** A *ballistospore-forming *yeast of the family *Sporobolomycetaceae.

**mis-sense mutant** In genetics, a *mutant in which a *codon has been altered by *mutation so that it encodes a different *amino acid. The result is almost always the production of an inactive or possibly unstable *enzyme.

**missing-plot technique** A standard formula for the estimation of a missing datum observation (or, with suitable modification, the estimation of several missing values) in the analysis of the variance of data collected according to

a recognized experimental design. The missing observation $(x'_{ij})$ may be estimated as: $x'_{ij} = tT'_j + bB'_j - G'/(t-1)(b-1)$, where $T'_i$, $bB'_j$, and $G'$ are the treatment, block, and grand totals for the available observations, $i$ is the $i$th treatment, $j$ is the $j$th block, $t$ is the number of treatments, and $b$ is the number of blocks.

**Mississippian** See CARBONIFEROUS.

**mistletoe** (*Viscum*) See LORANTHACEAE.

**mitochondrial-DNA** (mt-DNA) Circular *DNA that is found in mitochondria. It is entirely independent of nuclear DNA and, with very few exceptions, is transmitted from females to their offspring with no contribution from the male parent. Mitochondrial-DNA codes for specific *RNA components of *ribosomes that are unique to those *organelles. It also codes for some of the respiratory enzymes found in mitochondria. Unlike animal mt-DNA, plant mt-DNA evolves very slowly and is capable of breaking into tripartate structures which can recombine.

**mitochondrion** An oval, or occasionally round or thread-shaped, *organelle, whose length averages 2 μm, and which occurs in large numbers in the *cytoplasm of *eukaryotic cells. It is a double-membrane-bound structure in which the inner membrane is thrown into folds (the cristae) that penetrate the inner matrix to varying depths. It is a semi-autonomous organelle containing its own *DNA and *ribosomes, and reproducing by binary fission. It is the major site of *ATP production and thus of oxygen consumption in cells, and houses the *enzymes involved in the *citric acid cycle and in *oxidative phosphorylation.

**mitogenic** Able to induce or to stimulate *mitosis.

**mitosis** The normal process of nuclear division (occurring at cell division) by which 2 *daughter nuclei are produced, each identical to the parent nucleus. Before mitosis begins each *chromosome replicates to form 2 *sister chromatids; these then separate during mitosis so that one duplicate goes into each daughter nucleus. The result is 2 daughter

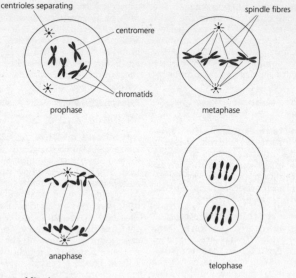

*Mitosis*

**m**

nuclei, each with an identical complement of chromosomes and hence of *genes. Mitosis is conventionally divided into 4 phases: *prophase, *metaphase, *anaphase, and *telophase. During prophase, chromosomes become visible within the *nucleus because they shorten, thicken, and coil up as a spiral. Each chromosome is longitudinally double except in the region of the *centromere, and each single strand of the chromosome is called a *chromatid. The *nucleolus and *nuclear membrane disappear. In lower plant cells, in metaphase the chromosomes move within the *spindle connecting the 2 *centrosomes (formed by division of the *centriole into 2 at the start of prophase, and the subsequent moving apart of these 2 daughter centrioles). The chromosomes finally arrange themselves along the equator of the spindle. During anaphase, the 2 chromatids making up each chromosome separate as the centromere becomes functionally double, and are thus converted to independent chromosomes moving to opposite poles. During telophase, the spindle disappears, a nuclear envelope reappears around each of the 2 groups of daughter chromosomes,

the chromosomes return to their extended state, and the nucleoli reappear. This may be followed by *cytokinesis. The final result is 2 daughter cells with identical nuclear contents. *Compare* MEIOSIS.

**mitotic spindle** The spindle-shaped system of *microtubules that, during cell division, traverses the nuclear region of *eukaryotic cells. The *chromosomes become attached to it and it separates them into 2 sets, each of which can be enclosed in the envelope of a separate *daughter nucleus.

***Mitrasacme*** **(family *Loganiaceae)** A genus of shrubs and herbs that have *opposite, entire leaves and regular flowers. The fruit is a *capsule or *berry. There are about 40 species, found in Australia, New Zealand, and eastern Asia.

**mixed sorus** *See* GRADATE SORUS and SORUS.

**mixed woodland** A woodland in which the minority of trees, but not less than 20 per cent of all the trees, are either coniferous or broad-leaved.

**Mlanji cedar** *See WIDDRINGTONIA.*

**Mn** *See* MANGANESE.

**Mnium** (order *Bryales) A genus of mosses, in which the ovate or lanceolate leaves have a border of long, narrow cells, with double teeth along the leaf margins (visible under the microscope). There is a nerve which ends in the leaf tip with a tooth at the back. There are about 12 species, found mostly in the northern hemisphere. *M. hornum* is one of the most abundant of British mosses, forming extensive carpets in woodland on *peaty acidic soil, rotting wood, etc. Normally it is a dull, dark green, although the young shoots that grow in spring are a contrasting bright green. (A number of species formerly included in this genus are now allocated to other genera: e.g. *M. undulatum* is now called *Plagiomnium undulatum* and *M. punctatum* is now *Rhizomnium punctatum*.)

**Mo** *See* MOLYBDENUM.

**mock orange** *See PHILADELPHUS*.

**model** A representation of reality in which the main features of some aspect of the real world are presented in simplified terms in order to make that aspect easier to comprehend, and often to facilitate the making of predictions.

**modern synthesis** (neo-Darwinism) Darwin showed that *evolution involves selection interacting with variation within populations, but he did not know that the bases of this variation are discrete units of heredity, or *genes. This was discovered by *Mendel. The fusion of Mendelian genetics and Darwin's *natural selection is referred to as the modern synthesis. A further synthesis has been achieved in recent years with the incorporation of knowledge of evolution at the molecular level.

**modifier** A *gene that modifies the *phenotypic expression of another gene.

**moisture balance** *See* MOISTURE BUDGET.

**moisture budget** (moisture balance) The balance of water, as represented broadly by the equation: balance = precipitation − (runoff + evapotranspiration + the change in soil-moisture). Over the year, for example in mid-latitudes, the budget is balanced by a high level of potential evapotranspiration and utilization of soil moisture in summer, compensated by a water surplus and recharge of soil moisture in winter, when evaporation is less and precipitation is sometimes greater.

**moisture index** The term used instead of 'moisture budget', e.g. by C. W. Thornthwaite (1955), and calculated from the aridity and humidity indices, as $I_m = 100 \times (S − D)/PE$, where $I_m$ is the moisture index, $S$ is the water surplus, $D$ is the water deficit, and $PE$ is the potential evapotranspiration.

**mold** *See* MOULD.

**molecular clock** The idea that molecular *evolution occurs at a constant rate, so that the degree of molecular difference between 2 species can be used as a measure of the time elapsed since they *diverged. Its accuracy depends on the validity of the *neutral mutation theory.

**molecular drive** The concept that changes within the *genome itself affect evolution, quite irrespective of natural selection, a mutation creeping through a population until a threshold is reached when a large number of mutated individuals appear to arise all at once.

**molecular evolution** The substitution of one *amino acid for another in protein synthesis (*see* RIBOSOME) as a result of *mutation of the *genetic code. According to the *neutral mutation theory of molecular evolution, the variability at the molecular level which results from mutation is caused by random drift of the *mutant *genes rather than by selection.

**molecule** *See* COVALENT BOND.

**mole drain** A drain made in soils by pulling a bullet-shaped device through the soil so that the compacted sides of the tunnel maintain that form for several years.

**mollic horizon** A surface *horizon of *mineral soil that is dark in colour, and relatively deep, and contains at least 1% organic matter, or 0.6% organic carbon, the determination of either being acceptable. It is the *diagnostic horizon of *Mollisols and is associated with base-rich materials and grassland vegetation. The name is from the Latin *mollis* meaning 'soft'.

**Mollicutes** A class of *bacteria in which the cells have no *cell walls. Species may be *saprotrophic but are usually *parasitic (sometimes *pathogenic) in animals or plants. There is 1 order, Mycoplasmatales, and 3 families, *Acholeplasmataceae, *Mycoplasmataceae, and Spiroplasmataceae, and at least 2 genera not included in any of the 3 families.

**Mollisols** *Mineral soils, an order identified by a deep *mollic surface *horizon (well decomposed and finely distributed organic matter) and base-rich mineral soil below. Mollisols form mainly grasslands in areas where moisture may be seasonally deficient (e.g. the Great Plains of N. America and the pampas of S. America). They are among the most fertile soils in the world and now produce most of the world's cereals.

**molybdenum** (Mo) An element that is required in small amounts by plants and is found largely in the *enzyme nitrate reductase. A symptom of deficiency is interveinal *chlorosis.

**molybdeus** Lead-coloured; drab grey.

**Monera** In some taxonomic schemes, one of the 5 kingdoms of life, comprising the *prokaryotic *Cyanophyta' (*cyanobacteria) and 'Schizomycophyta' (other *bacteria). Prokaryotes are currently placed into more fundamental categories (see ARCHAEA and EUBACTERIA). Bacteria are believed to have been the earliest forms of life on Earth, dating from at least 3300 Ma ago, while the first cyanobacteria appeared about 2600 Ma ago.

**moniliform** Necklace-shaped, like a string of beads.

**Monimiaceae** (order *Laurales) A family of evergreen trees and shrubs, with *opposite leaves, that includes many aromatic species. The flowers may be solitary or in *inflorescences, often unisexual, with variable numbers of overlapping *perianth segments. The fruit is a dry *achene. The family includes species used to produce aromatic oils, perfumes, medicines, and dyes. Some species are used locally for timber or edible fruits. There are 35 genera and about 450 species, occurring in warm and tropical climates.

**monkey flowers** See MIMULUS.

**monkeynut** See ARACHIS.

**monkey-puzzle** See ARAUCARIA and DISJUNCT DISTRIBUTION.

**monoallelic** In genetics, applied to a *polyploid in which all *alleles at a particular *locus are the same.

**Monoblepharidales** (class *Chytridiomycetes) An order of microscopic *fungi in which the *thallus is filamentous and *eucarpic; sexual reproduction is oogamous (see OOGAMY). Asexual reproduction involves the formation of *zoospores in cylindrical or flask-shaped *sporangia. Members are typically aquatic *saprotrophs.

**monocaryon** See MONOKARYON.

**monocentric** Applied to a *thallus that has a single reproductive centre.

**monochasia** See MONOCHASIUM.

**monochasium** (pl. **monochasia**) A *cymose *inflorescence that consists of a single branch bearing flowers and ending in a single terminal flower. Compare DICHASIUM.

**monoclimax theory** See CLIMAX THEORY.

**monoclinous** The condition in which active *stamens and *carpels occur in the same flower. Compare DICLINOUS.

**monocotyledon** An *angiosperm (flowering plant) in which the *embryo characteristically has one *cotyledon which is usually *amplexicaul and, like the later leaves, has parallel nervation. Compare DICOTYLEDON.

**Monocotyledoneae** A former division comprising the *monocotyledons. The name is no longer used.

**monoculture** The growing over a large area of a single crop species (e.g. *Triticum aestivum*, bread wheat), or of a single variety of a particular species. Monocultures are especially vulnerable to pest and disease infestation, but uniformity of height, development, etc., in a crop facilitates management, especially harvesting. The economic and ecological wisdom of monoculture is widely debated.

**monoecious** Applied to an organism in which separate male and female organs occur on the same individual (e.g. to a plant which bears male and female reproductive structures in the same flower or separate male and female flowers on the same plant. Some authors restrict the term to plants with separate male and female flowers; plants that bear male and female reproductive organs in the same flower are then called hermaphrodite. *See also* ANDROECIUM and UNISEXUAL FLOWER. *Compare* DIOECIOUS.

**monohybrid** A cross between two individuals that are identically *heterozygous for the *alleles of one particular *gene, e.g. $Aa \times Aa$.

**monohybrid heterosis** *See* HETEROSIS and OVERDOMINANCE.

**monoicious** Applied to the *gametophyte stage of *Bryophyta in which the *gametophore bears both *antheridia and *archegonia.

**monokaryon** (monocaryon) A fungal *mycelium or *hypha in which each cell contains a single *nucleus.

**monomictic** Applied to lakes in which only one seasonal period of free circulation occurs. In cold monomictic lakes, typical of polar latitudes, the seasonal overturn occurs briefly in summer and the water temperature never rises above 4°C, so inducing density stratification. In warm monomictic lakes, typical of warm temperate or subtropical regions, the seasonal overturn occurs in winter. At other times thermal stratification, with the formation of a distinct *epilimnion, prevents free circulation through the depth of the lake.

**monophyletic** Applied to a group of species that share a common ancestry, being derived from a single interbreeding

(or *Mendelian) population, as opposed to a polyphyletic group (*see* POLYPHYLETISM), which is derived from many such populations. If the members of a given *taxon are descended from a common ancestor, they are said to be monophyletic; e.g., the families within a class would be monophyletic if they were all descended from the same family or a lower taxonomic unit. Under the strictest definition, they would all have to be descended from a single species.

**monoplanetism** In some *Oomycetes, the occurrence of only one *motile phase and only one type of *zoospore.

**monopodial 1.** A type of branching in which lateral branches arise from a definite main, central stem. **2.** With a single *axis, an extension growth from the apex.

**monosomic genome** A genome that is basically *diploid but that has only one copy of one particular *chromosome type so that its chromosome number is $2n-1$.

**monospore** In the *Rhodophyta, a single *spore produced by the metamorphosis of the monosporangium, which is itself a single, vegetative cell.

**monostele** A *stele that is a single vascular vessel. *Compare* DICTYOSTELE.

**monothetic** In numerical classification schemes, the use of a single criterion or attribute as the basis for each subdivision of a sample population, as for example in association analysis. *Compare* POLYTHETIC.

**monotypic** Applied to any *taxon that has only 1 immediately subordinate taxon. For example, a genus that contains only 1 species would be described as monotypic, as would a family containing only 1 genus. *Compare* POLYTYPIC.

**monsoon forest** A tropical forest with a marked seasonal rhythm induced by alternating wet and dry seasons typical of the true monsoon lands. Most of the trees lose their leaves in the dry season, but, since many of them also flower at this time, the 'wintry' aspect is far less marked than in higher latitudes. Monsoon forests

m

contain many valuable hardwood species, e.g. the teak forests of Burma.

**Monstera** (family *Araceae) A genus of 22 species of tropical American climbers, some of which have holes in the leaves and were known formerly as Swiss cheese plants. The fruit of *M. deliciosa* (cheese plant), a common house plant, is edible.

**montane** Pertaining to a mountain or mountains.

**montane forest** A forest in the *montane zone of tropical and middle latitudes. It differs in floristic composition and ecological character from that found at lower elevations in the same latitude, and in both respects often has strong affinities with forest found in the lowlands of adjacent higher latitudes.

**montbretia** *See CROCOSMIA.*

**month degrees** The excess of mean monthly temperatures above 6°C (43°F), added together and used as accumulated temperature (indicative of conditions for vegetation growth) in some climate classifications (e.g. that of A. Miller, 1951).

**Montpellier school of phytosociology** *See ZURICH–MONTPELLIER SCHOOL OF PHYTOSOCIOLOGY; PHYTOSOCIOLOGY; compare UPPSALA SCHOOL OF PHYTOSOCIOLOGY.*

**moor** An acidic area, usually highlying and with *peat development, and most typically dominated by low-growing ericaceous shrubs (especially *Vaccinium myrtillus*, bilberry), though including grass and sedge-dominated areas. A. G. *Tansley (1939) maintained a traditional distinction between upland and lowland heaths, as opposed to heather moors, which have deeper peat development rather than any distinctive floral characteristics. *Compare* HEATHLAND.

**moot** *See COPPICE STUMP.*

**mor** A type of surface *humus *horizon that is acid in reaction, lacking in microbial activity except that of fungi, and composed of several layers of organic matter in different degrees of decomposition. It forms beneath conifer forest and on open heath and moorland in cool, moist climates and is very acidic.

**Moraceae** A family of trees and shrubs (except the herb *Dorstenia*) that produce milky latex. The leaves are *simple and spiral, with *stipules. The flowers are unisexual, regular, small, and grouped, often on the enlarged *receptacle. The fruits are often in infructescences with a fleshy receptacle or fleshy flower parts, and many are edible. There are 48 genera, with about 1200 species, occurring mainly in the tropics and subtropics.

**Morchella** (order *Pezizales) A genus of *fungi in which the *fruit body is several centimetres tall and consists of an ovoid head, extensively pitted with broad, shallow depressions (lined with the *hymenium), borne on a whitish hollow stalk. *M. esculenta* (morel) is a highly esteemed edible species, found growing on the ground in sandy soils in woods, on banks, etc.; it is seldom abundant in Britain. *M. vulgaris* (common morel) is also edible. *Compare GYROMITRA.*

**morel** The common name for the edible *fruit body of certain *fungi of the genus *Morchella.*

**Moreton Bay chestnut** *See CASTANOSPERMUM.*

**Morinda** (family *Rubiaceae) A genus of small tropical trees in which the flowers occur in heads, the fruits in a fleshy *syncarp. Several are useful for dyeing. There are about 50 species, occurring in the tropics, mainly in the Old World.

**Moringa** (family *Moringaceae) A genus of plants which includes *M. oleifera* (horseradish tree), widely cultivated in tropical Asia for its edible, pungent, young leaves and pods (rich in vitamin C), and for ben oil, extracted from the ripe seeds. There are 10 species.

**Moringaceae** A small family of small trees whose leaves are 2–4 times *pinnate. The flowers are bilaterally symmetrical, and *pentamerous, the *sepals fused, the *petals free. The fruit is a long pod splitting into 3 valves, and the seeds are large and 3-winged. There is 1 genus, occurring in tropical Africa, Madagascar, and from Arabia to India.

**morphology** The form and structure of individual organisms, as distinct from their *anatomy (which involves dissection). *Compare* PHYSIOGNOMY.

**morphospecies** A group of biological organisms whose members differ from all other groups in some aspect of their form and structure (*see* MORPHOLOGY), but which are so similar among themselves that they are lumped together for the purposes of analysis.

**Mortierellaceae (order *Mucorales)** A family of *fungi in which the *sporangia lack a *columella. There are many species, found in soil, dung, plant or animal matter, etc.

*Morus* **(family *Moraceae)** A genus of deciduous trees whose flowers are wind-pollinated. *Their infructescences (mulberries), fleshy from the swollen, juicy *perianths and borne in clusters like blackberries, are edible. They are widely cultivated for ornament, fruit, and as food for silkworms. There are 7 species, occurring widely in warm temperate and subtropical regions.

**mosaic** A general term for a virus disease of plants in which the symptoms include the appearance of angular areas of yellow colour on the leaves, forming a mosaic pattern.

**mosaic evolution** The differential rates of development of various adaptive attributes within the same evolutionary lineage. For example, a particular *taxon might show greatly different rates of change with respect to the leaves, shoots, and roots. This is a common phenomenon and makes the reconstruction of transitional fossil types very difficult.

**moschatel** *See* ADOXACEAE.

**moss** The common name for a plant belonging to the class *Musci.

**mossy forest** A tropical *montane forest, typically with contorted trees no more than 10–15 m high, their trunks, boughs, and twigs being festooned with *mosses, *lichens, and *liverworts. The growths of 'moss' are particularly luxuriant where mists prevail.

**mother-in-law's tongue** *See* SAN-SEVIERIA.

**motile** Capable of independent locomotion.

**mottle** A general term for a virus disease of plants in which the symptoms include the appearance of rounded or diffuse areas of yellow colour on the leaves.

**mottling** Patchwork of different colours in *mineral soil (usually orange or rust against a background of grey or blue) which indicates periods of anaerobic conditions.

**mould (mold) 1.** Any *fungus. **2.** Any fungus of 'mouldy' appearance, i.e. one with abundant, visible, woolly *mycelium upon which dusty or powdery *conidia can be seen (e.g. *Penicillium* species).

**mountain ash 1.** *See* SORBUS. **2.** *See* EUCALYPTUS.

**mountain fynbos** *See* FYNBOS.

**m-RNA** *See* MESSENGER-RNA.

**MSY** *See* OPTIMUM YIELD.

**mt-DNA** *See* MITOCHONDRIAL-DNA.

**mucilage** Any of a variety of complex, gum-like carbohydrates that are hard when dry, and slimy and jelly-like when wet. Mucilages are produced by many *bacteria and plants. In most cases they contribute to water retention. Some seeds have a mucilaginous coating that absorbs water and thus assists germination.

**muck 1.** Highly decomposed organic matter in which original plant material cannot be recognized. **2.** Farmyard manure (FYM) composed of animal faeces and urine mixed with straw and highly decomposed.

**mucoprotein** A class of mucopolysaccharides in which large numbers of disaccharide units are bound to a *protein chain.

*Mucor* **(family *Mucoraceae)** A genus of *fungi in which colourless, branched or unbranched *sporangiophores arise from the substrate *mycelium and bear at their tips spherical *sporangia containing many spores. *Zygospores are brown to black with warty surfaces. Species are common on soil, dung, fruit, vegetables, etc. and distribution is cosmopolitan.

They may cause spoilage of bread, fruit, and other foods.

**Mucoraceae (order *Mucorales)** A family of *fungi in which the *sporangia typically are all alike: spherical or pear-shaped, columellate, thin-walled, and containing numerous spores. There are many species, most of which are *saprotrophic in soil, dung, organic debris, etc. Some can *parasitize plants and some can cause disease in animals and humans.

**Mucorales (class *Zygomycetes)** An order of *fungi in which the vegetative stage is typically an extensive, usually *coenocytic, *eucarpic *mycelium. *Aplanospores are dispersed by wind, rain, insects, etc. Most species are terrestrial *saprotrophs common in soil, dung, plant debris, etc. Some are *parasites or *pathogens on other fungi, plants, or animals.

**mucro** A sharp point.

**mucronate** Applied to an organ (e.g. a leaf) that ends in a sharp point.

***Mucuna* (family *Fabaceae, subfamily *Papilionatae)** A genus of climbers, most of which have showy flowers in long *racemes. Several have pods with irritant hairs. *M. bennettii* (and others with red flowers) are the New Guinea creepers, extensively planted as ornamentals. There are about 100 species, occurring in the tropics and subtropics.

**mudflat** An area of a coastline where fine-grained *silt or sediment and *clay is accumulating. Its development is favoured by ample sediment, by sheltered conditions, and by the trapping effect of vegetation. It is an early stage in the development of a salt marsh.

***Muehlenbeckia* (family *Polygonaceae)** A genus of climbing or woody plants that have *simple alternate leaves and a sheathing membrane that unites the *stipules. The flowers are small, and white, pinkish, or green, and are held in a *raceme or solitary. There are no *petals, but the *sepals are often enlarged. The fruit is a triangular *nut and the seeds have a large *endosperm. The plants are either *dioecious or *polygamous.

There are 15 species. The genus occurs in Australia, New Zealand, and America. Several species are cultivated as ornamentals.

**mugwort** *See* ARTEMISIA.

**mulberry** *See* MORUS.

**mulch** A loose surface soil *horizon, either natural or man-made, composed of organic or mineral materials. It protects soil and plant roots from the impact of rain, temperature change, or evaporation.

**mull** A type of surface *humus *horizon that is chemically neutral or alkaline in reaction, that is well aerated, and that provides generally favourable conditions for the decomposition of organic matter. Mull humus is well decomposed and intimately mixed with mineral matter. It forms a ground-surface layer in deciduous forest and is typical of *brown earths.

**multifactorial (polygenic)** In genetics, a hypothesis to explain quantitative variation by assuming the interaction of a great number of *genes (polygenes), each with a small additive effect on the character.

**multiple allelism** The existence of several known allelic forms of a *gene. *See* ALLELE.

**multiple land-use strategy** The designed use of an area so that a range of compatible uses, or activities that can be rendered compatible by careful management, may be practised in a single locality (for example, forestry with rambling, camping, nature conservation, flood control, etc.).

**multiseriate** Arranged in many rows.

**murein** The main component of bacterial *cell walls; it consists of *polysaccharide molecules linked by short chains of *amino acids to form a rigid framework.

**muriform** Patterned like a brick wall.

***Murraya* (family *Rutaceae)** A genus of trees and shrubs, containing essential oils, in which the leaves are *pinnate and the fruit a *berry. Leaves of *M. koenigii* (curry leaf) are used in India to flavour curries.

There are 4 species, occurring from eastern Asia to the Pacific islands.

***Musa*** (family *Musaceae) A genus of giant, rhizomatous (*see* RHIZOME) *herbs which have an erect *pseudostem, formed from overlapping leaf bases, and terminal *inflorescences. The flowers are unisexual, the males terminal and subtended by coloured *bracts. The fruit is an elongate *berry with many stony seeds (absent in the edible bananas). Musas are bat-pollinated. Cultivated bananas are hybrids, and *triploid or tetraploid (*see* POLYPLOIDY). There are 35 species, occurring in the palaeotropics.

**Musaceae** A family of giant *herbs whose leaves have a midrib and parallel secondary nerves at right angles, often becoming tattered. There are 2 genera, and 42 species, occurring in the Old World tropics.

***Musanga*** (family *Urticaceae) A genus of 2 species. *M. cecropioides* is a soft-stemmed, fast-growing, pioneer tree occurring in humid tropical Africa.

***Musa textilis***   *See* ABACÁ.

**Musci** (mosses; division *Bryophyta) A class of plants all of which have a *gametophyte that is differentiated into stem and leaves. The leaves are generally only 1 cell thick (except at the *midrib); they are never stalked. The leaves may be finely toothed or entire; they come in a variety of shapes but are never deeply lobed, differing in this respect from some leafy *liverworts. Most mosses are either *acrocarpous or *pleurocarpous. Mosses can be found in a wide range of habitats, growing on rocks, soil, tree bark, etc. Their distribution is cosmopolitan.

**muscicolous** Growing on or among mosses (*Musci).

**mushroom** *Agaricus bisporus* (the cultivated mushroom) or any edible *fungus similar to it in appearance. *Compare* TOADSTOOL.

**muskeg** An ill-drained, often extensive, boggy tract, with at best sparse, stunted trees, which occurs within the *boreal forest. The vegetation is dominated by *Sphagnum* species (bog moss) and *Eriophorum* species (cotton sedge).

**Mussaenda** (family *Rubiaceae) A genus of small trees and shrubs in which the flowers have 1 *sepal, enlarged, leafy, and brightly coloured, rendering the *inflorescence showy and conspicuous. Several species are ornamentals. There are about 200 species, occurring in the palaeotropics.

**mutagen** An agent that increases the *mutation rate within an organism. Examples of mutagens are X-rays, gamma rays, neutrons, and certain chemicals, such as carcinogens.

**mutagenic** Causing a *mutation.

**mutant** 1. A cell or organism that carries a *gene *mutation. 2. A gene that has undergone mutation.

**mutation** 1. A change in the structure or amount of the genetic material of an organism. 2. A structural change in a *gene or *chromosome set. The majority of mutations are changes within individual genes (e.g. the substitution of a different *nucleotide at some point in the *DNA so that the *amino acid sequence is altered), but some are gross structural changes of chromosomes (e.g. *inversion or *translocation) or changes in the number of chromosomes per nucleus (e.g. *polyploidy). Mutations are the raw material for *evolution: they provide the source of all variation. For mutations to affect subsequent generations, though, they must occur in *gametes or in cells destined to be gametes, since only then will they be inherited. A mutation that occurs in a body cell is called a somatic mutation: it is transmitted to all cells derived, by *mitosis, from that cell. Most mutations are deleterious; evolution progresses through the few that are favourable. Some mutations are recurrent: they occur repeatedly within a population, or over long periods of time (as does haemophilia, for example).

**mutation rate** The number of *mutation events per *gene per unit of time (e.g. per cell generation). The term is also applied to the frequency with which mutation events occur in a given species or to the frequency with which a specified mutation event or mutational class occurs in a given population. Normally mutations occur at a constant, very low, rate: this rate can be greatly increased by irradiation with X-rays, gamma rays, neutrons, etc., or by treatment with carcinogens. *See also* MUTAGEN.

**Mutinus** (order *Phallales) A genus of *fungi that form *fruit bodies resembling those of *Phallus* species. In *Mutinus caninus* (dog stinkhorn) the fruit body has a pink or orange tinge, and the cap is initially covered with a dark green slime containing the *spores. The odour is faintly foetid, but less so than in *Phallus impudicus*. *Mutinus* species are found on the ground in woods, and are widespread.

**mutual inhibition competition type** A form of direct competition between the populations of two species in which both species actively inhibit one another. This is quite distinct from indirect competition, which arises from competition for a common resource that is in short supply.

**mutualism** The interaction of species populations that benefits both populations. Strictly, the term may be confined to obligatory mutualism, in which neither species can survive under natural conditions without the other. Sometimes the term is used more generally to include *facultative mutualism (*protocooperation).

**Mycelia Sterilia** *See* AGONOMYCETALES.

**mycelium** A mass of *hyphae, making up the vegetative stage of many *fungi and *actinomycetes.

**Mycena** (family *Tricholomataceae) A genus of *fungi in which the *fruit bodies are small and delicate, with a conical or bell-shaped cap and a long, slender, central *stipe. *Spores are smooth and colourless. There are many species, found on the ground, on wood, on plant debris, etc.

**mycetocyte** In certain insects, a cell that contains symbiotic *fungi or *bacteria.

**mycetozoans** The zoological term for *slime moulds and related organisms.

**Mycobacterium** (order Actinomycetales) A genus of *bacteria that typically form straight or curved, rod-shaped cells but also branched rods or fragile filaments. The genus includes many species, among which are several important *pathogens. *M. tuberculosis* (the tubercle bacillus) causes tuberculosis in humans, other primates, dogs, and certain other animals. *M. bovis* (bovine tubercle bacillus) also causes tuberculosis, and is generally more pathogenic in cattle and various other animals than is *M. tuberculosis*. *M. avium* (avian tubercle bacillus) causes tuberculosis in birds, including domestic fowl. *M. leprae* causes leprosy in man. The genus also includes some *saprotrophic species. *M. phlei* (timothy grass bacillus) is non-pathogenic and is found on grass and hay.

**mycobiont** The fungal *symbiont in a *lichen.

**mycocecidium** A *gall formed on a plant as a result of infection by a *fungus.

**mycology** The study of *fungi.

**Mycoplasma** *See* MYCOPLASMATACEAE.

**mycoplasmas** **1.** *Bacteria of the class *Mollicutes. **2.** Bacteria of the genus *Mycoplasma*.

**Mycoplasmataceae** (order Mycoplasmatales, class *Mollicutes) A family of *Gram-negative, *chemo-organotrophic *bacteria which require the presence of sterols (*see* STEROIDS) for growth. Cells are *pleomorphic, ranging from spherical to filamentous. Genera include *Mycoplasma*, of which there are many species; they are found as *parasites and *pathogens in a variety of mammals and birds. Some species can cause a type of pneumonia in humans and other animals.

**Mycoplasmatales** *See* MOLLICUTES.

**mycorrhiza** A close physical association between a *fungus and the roots of a plant, from which both fungus and plant appear to benefit; a mycorrhizal root takes up nutrients more efficiently than does an uninfected root. A very wide range of plants can form mycorrhizas of one form or another, and some plants (e.g. some orchids and some species of *Pinus*) appear incapable of normal development in the absence of their mycorrhizal fungi. *See also* ECTOTROPHIC MYCORRHIZA and ENDOTROPHIC MYCORRHIZA.

**mycosis** Any disease of humans or other animals in which the causal agent is a *fungus.

**mycotoxin** A toxin produced by a *fungus.

**mycotrophic** Applied to a plant that is associated with a *fungus in a *mycorrhiza.

**mycovirus** A *virus that infects and replicates in *fungi.

**Myoporaceae (emu bushes)** A family of ornamental trees and shrubs, with some timber trees, in which the leaves are alternate or *opposite, entire or toothed, often glandular or covered with woolly, glandular, or scaly hairs. The flowers are irregular, *axillary, and either solitary or in clusters. The *calyx and *corolla are both of 5 fused *sepals or *petals forming a 5-lobed tube. The 4 *stamens are fused to the corolla. The *ovary is *superior, the fruit a *drupe, and the seeds have almost no *endosperm. Several species are of horticultural interest. The family contains 5 genera, with about 220 species, most of them confined to Australia, but also found in the south Pacific and Indian Ocean islands and S. Africa.

*Myoporum* **(emu bushes; family *Myoporaceae)** A genus of small trees and shrubs that have alternate leaves and small creamy-white or lavender flowers with a short *corolla tube and cream, yellow, or purple *drupes. *M. laetum* is a timber tree. There are 32 species found mainly in Australia, but also in eastern Mauritius, eastern Asia, eastern Malesia, New Zealand, and Hawaii.

*Myosotis* **(forget-me-not; family *Boraginaceae)** A genus of hairy herbs that have *simple, alternate leaves, and curved (scorpioid) *cymes of small flowers each with a 5-lobed *calyx. The 5-lobed, wheel-shaped *corolla may be pink at first, then blue. Forget-me-nots are much cultivated for their flowers. There are some 50 species, occurring in temperate regions.

*Myrialepis* **(family *Arecaceae)** A *monotypic genus of climbing palms (rattans) which are clump-forming. They are *dioecious. The *inflorescences are produced from the uppermost leaf *axils, the stem then dying. The fruits have innumerable (myriad) tiny scales (lepis), like shark's skin. They are found in Malaya and Sumatra.

**Myricaceae** A family of shrubs whose alternate leaves are dotted with aromatic, resinous glands. The flowers are *monoecious or *dioecious and borne in *catkins. Male flowers have no *bracteoles and 2–16 *stamens, females have bracteoles and tiny, one-celled *ovaries. The fruit is a *nut or a *drupe. There are 3 genera, with about 50 species, found through most of the world.

*Myriophyllum* **(water milfoils; family *Haloragidaceae)** A genus of submerged, fresh-water aquatic herbs, with a few terrestrial species, in which the leaves are in *whorls, finely *pinnate when submerged and with no *stipules. The flowers are in *panicles or *spikes held above the water and are generally very small. The fruit is usually a small dry *nut, separating into 2 or 4 1-seeded *nutlets. The seed has a copious *endosoperm. The genus forms winter buds, which survive in the bottom of a lake or river, and from which the plant can develop in the spring. There are about 40 species, with cosmopolitan distribution.

*Myristica* **(family *Myristicaceae)** A genus of *monopodial trees in which the fruit has an *aril that is much divided, almost to the base. The seed of *M. fragrans* is nutmeg; its aril provides mace. There are about 80 species, occurring from India to the Pacific islands.

**Myristicaceae** A family of plants whose bark produces a watery, blood-like exudate. The leaves are *simple, alternate, and in flat sprays, without *stipules. The flowers are unisexual, usually *dioecious, tiny, and usually *trimerous; the *stamens are united as a column, the *ovary is *superior. The fruit is fleshy and *dehiscent, with 1 large seed enveloped in an often divided red *aril; the *endosperm is *ruminate. There are 19 genera, with about 400 species, occurring throughout the tropics but centred in Malesia, mostly in lowland rain forest.

**myrmecochory** Dispersal of spores or seeds by ants.

***Myrmecodia* (family \*Rubiaceae)** A genus of epiphytic (*see* EPIPHYTE) \*herbs in which the stem base forms a swollen, chambered, coarse \*tuber inhabited by ants. There are 45 species, occurring from Malesia to the Pacific islands.

**myrmecophily** A specialization by which a plant houses or provides food for ants.

**myrobalan** The common name for several different shrubs and trees of economic importance in warm climates or, in some cases, of their \*fruit. **1.** Two species of \*Terminalia (family \*Combretaceae) whose unripe, astringent fruits (also called myrobalans) are a source of tannins. The tropical or Indian almond is the edible kernel of the fruit of *T. catappa*. **2.** *Phyllanthus emblic* (emblic or myrobalan, family \*Euphorbiaceae) yields a fleshy, globular fruit about 2.5 cm in diameter which is used to make preserves in tropical Asia. **3.** *Prunus cerasifera* (myrobalan plum or cherry plum) produces edible fruit but is more valued as a grafting stock (*see* GRAFT) and as an ornamental.

**Myrsinaceae** A family of trees and shrubs that have alternate, \*simple, leathery leaves with no \*stipules and often with resin ducts or glands. The flowers are regular, small, and bisexual, or if unisexual then \*dioecious. They are held in clusters on short shoots in the leaf \*axils, or in terminal \*cymes or \*panicles. There are 4–6 \*sepals and \*petals which may be united at the base, forming a lobed \*corolla. The \*stamens are opposite the corolla lobes and of the same number. The \*ovary is \*superior with a few \*locules. The fruit is a fleshy \*drupe and the seed has a thick \*endosperm. There are two subfamilies, about 39 genera, with 1250 species. The family is widely distributed in tropical and subtropical regions of Africa, Asia, Australia, and America.

**Myrtaceae** A family of mainly tropical and subtropical plants in which the leaves are \*simple, commonly \*opposite, finely dotted with oil glands, and without \*stipules. The flowers are regular, with 4 or 5 \*petals, numerous \*exserted \*stamens, and the \*ovary more or less \*inferior. The fruit is a woody \*capsule or a \*berry. There are 1 to many seeds. Some are ornamentals with showy flowers. Many produce valuable timbers. There are 121 genera, comprising about 3850 species, occurring in tropical and warm regions, and abundant in Australia.

**myxamoeba** An amoeboid cell formed by a \*slime mould.

**Myxobacterales** An order of \*Gram-negative \*bacteria, which are capable of a gliding motility when in contact with a surface. Vegetative cells are rod-shaped, and often embedded in slime. Under suitable conditions cells aggregate to form a fruiting body containing resting cells; fruiting bodies are often brightly coloured and may be visible to the naked eye. These bacteria are \*aerobic \*chemo-organotrophs, obtaining nutrients mainly by producing enzymes that disrupt other bacteria or fungi. There are many species, found, for example, in soil, dung, and decaying plant matter.

**Myxomycetes (acellular slime moulds; division \*Myxomycota)** In some mycological classifications, a class of \*slime moulds in which the feeding stage of the organism is a multinucleate, \*diploid, \*motile \*plasmodium. Under appropriate conditions the plasmodium typically forms \*fruiting bodies in which \*spores are formed. The fruiting bodies are usually visible to the naked eye and in some cases may be quite large. The organisms are widely distributed and are found on decaying wood, humus, soil, bark, dung, etc.

**Myxomycota** In some mycological classifications, a division in which are placed the \*slime moulds and related organisms. In these organisms the feeding stage is either a \*plasmodium or many separate amoeboid cells (\*myxamoebae). The division includes several classes.

**Myxophyta** An old \*taxon formerly including the \*cyanobacteria.

**N** See NITROGEN.

**Na** See SODIUM.

**nacreous** Having a pearly lustre.

**NAD** See NICOTINAMIDE ADENINE DINUC-LEOTIDE.

**NADP** See NICOTINAMIDE ADENINE DINUCLEOTIDE PHOSPHATE.

**Najadaceae** A family of aquatic *monocotyledons comprising slender plants that have submerged, linear, *opposite to *whorled leaves with sheathing bases. The flowers are minute and unisexual. The male flower consists of 1 *stamen in a 2-lipped *perianth, the female of 1 *carpel which is usually without perianth. The fruit is a tiny *nut. There is 1 genus and 35 species, occurring in temperate and humid tropical regions.

**naked cell** A cell that is not surrounded by a *cell wall. Instead, *lorica or *periplast structures may lend the cell rigidity.

**nanoplankton** See PLANKTON.

**naphthalene acetic acid** A synthetic *auxin.

**Narcissus (daffodils; family *Amaryllidaceae)** A genus of bulbous *herbs whose regular flowers are borne singly or in groups on the tip of a leafless stem, and which have a papery *spathe around the flower or flower group. The flowers have 6 similar *perianth segments and also a cup or trumpet-shaped *corona surrounding the *stamens. *Narcissus* species are much cultivated (as wild species and as hybrids or cultivars) for the fine flowers. There are 27 species, occuring in Europe, western Asia, and N. Africa.

**Nardia (order *Jungermanniales)** A genus of leafy *liverworts, which includes *N. scalaris*, a species common in Britain. The stems are usually more or less prostrate, almost unbranched, often forming extensive patches. The leaves are more crowded towards the tips of the stems. Each leaf is broad, rounded, entire, and somewhat concave. *N. scalans* is common on gravelly or sandy soils on *heaths, *moors, mountains, etc.; it is *calcifuge.

**nascent** In the process of forming.

**Nasturtium (watercress; family *Brassicaceae)** A genus of plants that have *pinnate leaves and *spikes of white flowers. The elongated fruits have two convex valves. Watercress is cultivated for its edible leaves, used as a salad vegetable. There are 6 species, occurring from Europe to central Asia.

**nasturtium** See TROPAEOLUM.

**Nastus (family *Poaceae, subfamily *Bambusoideae)** A genus of bamboos, most of which are scramblers in the forest canopy. There are 9 species, occurring in the tropics, with a remarkably disjunct range in the Mascarenes, in western Malesia, and from New Guinea to the Solomons.

**nasty** The response of a plant organ to a non-directional stimulus (e.g. the opening or closing of flowers in response to changes in light intensity or temperature). The plant may respond by changes in cell growth or changes in *turgor.

**natant** Floating entirely under water.

**national park** As defined by the IUCN (1975), a large area of land containing *ecosystems that have not been materially altered by human activities, and including plant and animal species, landscape features, and *habitats of great scientific interest, or of beauty, or recreational or educational interest: it is under the direct control of the state, and the public is allowed to visit it for inspirational, cultural, and recreational purposes. Implicitly, a national park is an area set aside in perpetuity for *conservation, within which such public recreational activity is permitted as is compatible with the primary conservation objectives. British national parks do not comply strictly with the IUCN definition, since they comprise areas that have been altered significantly during a long period of human occupation and since they are still actively farmed and occupied, much of the land being owned privately rather than publicly. Land is so designated in order to maintain long-established, cultural landscapes. Similar schemes have been developed elsewhere, but they are often given distinctive and less confusing titles, e.g. *parcs naturels et régionaux* in France, and 'greenline parks' in the USA.

**native (indigenous)** Applied to a species that occurs naturally in an area, and therefore one that has not been introduced by humans either accidentally or intentionally. Of plants found in a particular place, the term is applied to those species that occur naturally in the region and at the site.

**native tamarind** *See DIPLOGLOTTIS.*

**natric horizon** A *mineral soil *horizon that is developed in a subsurface position in the *profile, that satisfies the definition of an *argic horizon, and that also has a columnar structure and more than 15 per cent saturation of the exchangeable *cation sites by sodium. The name is from the Latin *natrium* meaning 'sodium' (chemical symbol Na).

**natural** A term that is applied to a *community of *native plants and animals. 'Future-natural' describes the community that would develop were human influences to be removed completely and permanently, but allows for possible changes in climate or site. 'Original-natural' describes a community as it existed in the past, with no modification by humans. 'Past-natural' describes the condition in which the present features are derived directly from those existing originally, with relatively little modification by humans. 'Potential-natural' describes the community that would develop were human influence removed, but the consequent *succession completed instantly (future changes in climate or site are not taken into account). 'Present-natural' describes the commu-

nity that would exist now had there been no human modification. Because of the dynamic nature of any *ecosystem, this condition may not be identical to the last original-natural state before human intervention began.

**natural classification** The ordering of organisms into groups on the basis of their evolutionary relationships. *Compare* ARTIFICIAL CLASSIFICATION.

**naturalized** Applied to a species that originally was imported from another country but that now behaves like a *native in that it maintains itself without further human intervention and has invaded native populations.

**natural selection ('survival of the fittest')** A complex process in which the total environment determines which members of a species survive to reproduce and so pass on their *genes to the next generation. This need not necessarily involve a struggle between organisms.

**natural turnover rate** The normal rate of transfer of an element through the active part of a biogeochemical cycle. *Compare* MAN-INDUCED TURNOVER.

**natural woodland** A woodland comprising trees that have not been planted by humans, and where no human interference has occurred.

**nature and nurture** Synonyms for heredity and environment as they affect a character. Both may affect observed variation among individuals; only variations resulting from 'nature' are inherited, and it is these that form the subject of quantitative genetics.

**nature conservation** *See* BIOLOGICAL CONSERVATION and CONSERVATION.

**nature reserve** An area of land set aside for nature *conservation and associated scientific research, usually with strong legal protection against other uses. Public access may be restricted partially or completely. The precise status and definition of reserves varies from one country to another. In Britain, a key objective is to maintain a fully representative range of *habitats and their flora and fauna; this is done by means of national nature reserves (NNR), controlled by statutory bodies, complemented by local nature reserves (LNR), of local rather than national or international interest and importance, managed by various agencies, especially by country naturalists' trusts, on behalf of local government.

**navy bean**   See PHASEOLUS.

**nearest-neighbour analysis (nearest-neighbour measure)** A method for testing the pattern of distribution of individuals. The mean nearest-neighbour distance for all, or for a *random sample of, individuals in a given area is compared with the expected mean distance if the same individuals were randomly distributed throughout the area. The ratio of the two mean values ($R$) is a measure of the departure from randomness, and may be tested for statistical significance. The value of $R = 1$ indicates randomness; $R = 0$ indicates maximum aggregation; and $R = 2.149$ indicates maximum possible spacing, i.e. a hexagonal pattern.

**nearest-neighbour measure** See NEAREST-NEIGHBOUR ANALYSIS.

**nearest-neighbour sampling method** A method of plotless sampling in which the distance is measured from the first individual (the nearest to the *random sampling point) to its nearest neighbour. This permits the calculation of the density of individuals, or of its reciprocal, the mean area per individual.

**near-natural community** A vegetation *community consisting wholly or predominantly of native species which, though modified by human use, has not been deflected from its natural course of *succession. Examples include traditional wood-pasture and *coppiced woodland. Compare SEMI-NATURAL COMMUNITY.

**Neckera (order *Isobryales)** A genus of mosses in which the shoots are typically flattened, and *pinnately or *bipinnately branched. The leaves are usually glossy and in some species are wrinkled transversely; they either lack a *nerve or have a very short nerve visible under the microscope. There are about 120 species, found chiefly in tropical and subtropical regions. They grow on trees, rocks, and on the ground, usually preferring neutral or basic (alkaline) substrates.

**necrology** The scientific study of all the processes affecting dead animal and plant material, including decomposition and *fossilization.

**necromass** The mass of dead plant material lying as *litter on the ground surface.

**necrosis** The *death of a circumscribed piece of *tissue. Necrotic wounds are produced in tobacco plants by *infection with the *tobacco necrosis virus.

**necrotrophic** Applied to a parasitic organism (see parasitism) that obtains its nutrients from dead cells and tissues of its host organism.

**nectar** A liquid, secreted by a *nectary, that is up to 60 per cent sugar. It is from nectar that bees make honey.

**nectary** A gland that secretes *nectar. Nectaries are usually located at the bases of insect-pollinated flowers, where they serve as an insect attractant, but in some plants they may occur elsewhere (e.g. in the spines of some cacti) and may also attract insects that disperse seeds.

**necton**   See NEKTON.

**Nectria (order *Hypocreales)** A genus of *fungi whose brightly coloured fruiting structures are common on twigs and bark. *N. cinnabarina* forms pale pink *conidial pustules measuring 1–2 mm across, which are replaced later in the year by red *perithecial pustules. Some species can cause plant diseases (e.g. *N. galligena* causes apple canker). See also BEECH BARK DISEASE and CORAL SPOT.

**needle** A linear, commonly pungent *leaf (e.g. in many conifers).

**needle grass**   See STIPA.

**Neesia (family *Bombacaceae)** A genus of trees in which the fruit is a big, 5-angled, tessellated *capsule with the inner surface of the valves clothed in irritant hairs. There are 8 species, occurring in rain forests from Thailand to Borneo.

**negative feedback** In a system, the mechanism by which a process is limited internally, i.e. without reference to factors outside the system. Ecologically such mechanisms favour the maintenance of equilibrium in organisms, populations, and ecosystems. *Compare* POSITIVE FEEDBACK.

**Neisseria** (family *Neisseriaceae) A genus of *Gram-negative *bacteria in which the cells are typically spherical and often occur in pairs with adjacent sides flattened. They are *chemo-organotrophic. There are several species. They are *parasitic on the mucous membranes of mammals. Some species are *pathogenic: *N. gonorrhoeae* causes gonorrhoea, *N. meningitidis* causes meningitis. The organisms tend to be delicate and are easily killed by desiccation and by exposure to sunlight.

**Neisseriaceae** A family of *Gram-negative *bacteria, in which the cells are spherical, ovoid, or rod-shaped; *flagella are absent. They are *chemo-organotrophic. There are several genera, including *parasitic and *saprotrophic species.

**nekron mud** *See* GYTTJA.

**nekton (necton)** Aquatic organisms that swim actively (e.g. ciliated *plankton), rather than drifting passively.

**nematophagous fungi** *Fungi that are *parasitic or predatory on nematode worms; most can also grow *saprotrophically. Nematophagous fungi belong to various taxonomic groups and are found, for example, in soil and in decaying plant matter.

**nemorose** Inhabiting woodland habitats.

**Nenga** (family *Arecaceae) A genus of solitary palms in which the leaves are *pinnate, with a pronounced crown-shaft. The *inflorescence is sparsely branched. The flowers are spirally arranged, a female flanked by 2 males. The fruit wall is fleshy, and showy. There are 2 species, found in Sumatra and Malaya.

**Neobalanocarpus** (formerly *Balanocarpus*; family *Dipterocarpaceae) A genus of 1 species, *N. heimii*, a tree of lowland rain forests of Malaya and Sumatra, which provides the highly durable, heavy, hardwood timber chengal. It is close to *Hopea*.

**neocatastrophist** *See* CATASTROPHISM.

**neo-Darwinism** *See* DARWIN and MODERN SYNTHESIS.

**neoendemic** *See* ENDEMISM.

**Neogene** The middle of the three periods comprising the *Cenozoic Era, preceded by the *Palaeogene, followed by the *Pleistogene, and dated at 23.03–1.806 Ma ago. The Neogene is divided into the *Miocene and *Pliocene Epochs.

**neo-Lamarckism** Any modern variant of the theory of evolution by the inheritance of acquired characteristics that was proposed by *Lamarck.

**neomycin** An aminoglycoside *antibiotic, produced by *Streptomyces fradiae*, which functions by interfering with ribosomal activity and so causing errors in the reading of the *m-RNA.

**Neotethys** *See* TETHYS SEA.

**Neotropical floral kingdom (Neotropical floral realm)** One of the major floral subdivisions of the world, corresponding with central and S. America, with the exception of the southern tip which is included in the Antarctic kingdom. The floristic distinctiveness results from the isolation of S. America from other land masses for much of *Cenozoic time. *See also* FLORAL PROVINCE and FLORISTIC REGION.

**Neotropical floral realm** *See* NEOTROPICAL FLORAL KINGDOM.

**NEP** *See* PRIMARY PRODUCTIVITY.

**Nepenthaceae (pitcher plants)** A family of carnivorous and tropical herbs and shrubs which are often woody climbers or *epiphytes. The alternate leaves are without *stipules. Tendrils, formed by projections from the leaf midrib, assist in climbing. The pitcher forms on the end of the tendril, with a lid covering the mouth. The pitcher is inwardly curved, with nectar glands at the entrance and a slippery internal surface. It is often brightly coloured. The colour and nectar attract insects, which slip into the pitcher; they are unable to climb out and so drown in the water that has accumulated at the base of the pitcher. The plant is able to absorb the nutrients from the dead insects. The flowers are *dioecious and held in a *spike. They are small and brightly coloured, red, yellow, or green, with 3 or 4 *sepals and no *petals. The *stamens of the male flower are united into a column. The female flower has a short *style, and a *superior *ovary with 4 *locules, each with numerous *ovules. The fruit is a thickened *capsule and holds the small, light seeds with a fleshy *endosperm. The family is related to other insectivorous families, such as the *Droseraceae. There is 1 genus, *Nepenthes*, with about 70 species, occurring in Madagascar, the Seychelles, and from the Asian tropics to New Caledonia and Australia.

**Nepenthes** See NEPENTHACEAE.

**Nephelium** (family *Sapindaceae) A genus of small trees in which the leaves are *pinnate. The fruit is *indehiscent, and the *pericarp leathery, warty, or spiny, with 1 large seed enveloped in a juicy *aril. Several species are cultivated for their edible fruit, especially *N. lappaceum* (rambutan). There are 35 species, occurring from Burma to western Malesia.

**Nephroma** (order *Peltigerales) A genus of *lichens in which the *thallus is *foliose. The *apothecia are formed on the lower surface of the thallus on lobes that usually turn upwards. The *spores are brown and *multiseptate. Species are found in humid regions on a variety of substrates.

**neritic** Occurring between the coast and the edge of the continental shelf; the term is often applied to *plankton.

**neritic province** See NERITIC ZONE.

**neritic zone** (neritic province) The shallow-water, or near-shore, marine zone extending from the low-tide level to a depth of 200 m. This zone covers about 8 per cent of the total ocean floor and is the area most populated by benthic organisms (see BENTHOS), owing to the penetration of sunlight to these shallow depths.

**Nerium** (family *Apocynaceae) A genus of plants that includes *N. oleander* (oleander), with very poisonous latex (although oleanders are popular ornamentals, often grown indoors in pots). There are 2 species, occurring from the Mediterranean region to Japan.

**nerve** In mosses (*Musci), a bundle of cells occurring in the centre of the leaf.

**net ecosystem productivity** See PRIMARY PRODUCTIVITY.

**net plankton** See PLANKTON.

**net plasmodium** A *plasmodium-like structure consisting of non-living matter laid down by a labyrinthulomycete *cellular slime mould.

**net primary productivity** (NPP or $P_n$) See PRIMARY PRODUCTIVITY.

**net slime moulds** The name sometimes applied to organisms that form a net *plasmodium. Such organisms are variously regarded as *fungi (class Labyrinthulomycetes), as protozoa, or as 'protists' unrelated to other organisms.

**nettle** See URTICA.

**Neurospora** (order *Sordariales) A genus of *fungi, species of which are commonly used in genetic and biochemical studies on fungi. They form *perithecia and longitudinally ribbed *ascospores. They are *saprotrophic, and found chiefly on rotting vegetation or on burnt ground. *N. sitophila* (red bread mould) can cause spoilage of bread in bakeries.

**neuston** A collective term for organisms that are resting or swimming on the surface of an aquatic *ecosystem.

**neutralism** A situation in which two species populations coexist, with neither population being affected by association with the other.

**neutrality theory of evolution (neutral mutation theory)** A theory, proposed by the Japanese geneticist Motoo Kimura (1924–1994), which asserts that many genetic *mutations are adaptively equivalent (effectively neutral), and do not affect significantly the fitness of the carrier. Thus they can become fixed in the *genome at a random rate. Changes in their frequencies are due more to chance than to *natural selection. The theory applies only to protein evolution and does not deny the role of natural selection in shaping morphological and behavioural attributes, but Kimura made important contributions to the understanding of evolutionary change by constructing mathematical models based on it. *See also* MOLECULAR EVOLUTION.

**neutral mutation theory** *See* NEUTRALITY THEORY OF EVOLUTION.

**neutral soil** Soil with a *pH value of 6.6–7.3.

**neutral theory of biogeography** An approach to *biogeography, developed by Stephen Hubbell, a staff scientist at the Smithsonian Tropical Research Institute, that regards *communities as random collections of species cast together by the vagaries of history, dispersal, and chance. The theory of *island biogeography is an example of this approach, since it is based upon the assumption that all species within the system are equivalent to one another.

**neutrophilous** Preferring a habitat that is neither acid nor alkaline.

***Nevskia*** A genus of poorly characterized, *Gram-negative, aquatic *bacteria not assigned to any taxonomic family.

**New Caledonia floral region** Part of R. Good's (*The Geography of the Flowering Plants*, 1974) *Polynesian subkingdom, within his *palaeotropical kingdom, which also includes the Lord Howe and Norfolk Islands. It has a remarkably rich flora, with a qualitatively high degree of *endemism. Some of the endemic genera, of which there are well over 100, constitute separate families (or nearly so). *See also* FLORAL PROVINCE and FLORISTIC REGION.

**New Guinea creeper** *See* MUCUNA.

**New Guinea walnut** Timber of *Dracontomelum* (family *Anacardiaceae), one of the most valuable timbers of that region.

**New South Wales sassafras** *See* DORYPHORA.

**New Zealand flax** *See* PHORMIUM.

**New Zealand floral region** Part of R. Good's (*The Geography of the Flowering Plants*, 1974) Antarctic kingdom. Approximately 30 genera are known to be endemic (*see* ENDEMISM) to it, but the surprisingly large number of endemic species (about 75 per cent of the total) belong very largely to non-endemic genera. *See also* FLORAL PROVINCE and FLORISTIC REGION.

**nexine (endexine)** In a *pollen grain, the inner layer of the *exine.

**ngali** *See* CANARIUM.

**niacin** *See* NICOTINIC ACID.

**niche** **1. (ecological niche)** The functional position of an organism in its environment, comprising the *habitat in which the organism lives, the periods of time during which it occurs and is active there, and the resources it obtains there. **2. (evolutionary niche)** A way of life. In other words, its niche is the role that a species plays in a community.

***Nicotiana*** **(tobacco; family *Solanaceae)** A genus of *annual and *perennial *herbs whose leaves are of varying shape, without *stipules. The flowers are regular and composed of 5 *sepals and 5 *petals fused to form a tube. The *stamens are attached to the *corolla, the *ovary is *superior with 2 fused *carpels and 1 *style. The fruit is a *berry with seeds containing a copious *endosperm. *N. tabacum* is the cultivated tobacco. Many species contain the highly toxic *alkaloid nicotine, which is a powerful insecticide. Some species are cultivated for their bright flowers. There are 67 species, and the genus is represented in Australia, the south Pacific, subtropical regions of N. and S. America, and in south-western Africa.

**nicotinamide adenine dinucleotide (NAD; diphosphopyridine nucleotide, DPN)** A derivative of the vitamin nicotinic acid, which functions as a *coenzyme for various *dehydrogenase *enzymes involved in cellular respiration. It acts as an electron carrier in respiratory chain *phosphorylation.

**nicotinamide adenine dinucleotide phosphate (NADP)** A phosphorylated derivative of *nicotinamide adenine dinucleotide (NAD), which functions as a hydrogen acceptor in the light stage of *photosynthesis and is subsequently involved in the reduction of carbon dioxide in the dark reactions. It is a *coenzyme in the reaction involving the oxidation of ferrodoxin by NADP reductase.

**nictonasty (sleep movement)** A diurnal, nastic (see NASTY) movement (e.g. the opening of certain flowers in the day and their closing at night).

**Nidulariales (class *Gasteromycetes)** An order of *fungi, in which the *spore-bearing tissue is organized into 1 or more spherical or flattened structures, called peridioles (spore cases), contained within a fruit body which may be spherical, cup-shaped, or funnel-shaped. There are several genera, found on the ground, on dung, on wood, etc. See also BIRD'S NEST FUNGI.

**nine-awn**  See ENNEAPOGON.

**ninhydrin  (triketohydrindene hydrate)** A compound that reacts with the free *alpha-amino groups of *amino acids, peptides, and *proteins to yield coloured compounds, usually blue or purple. Ninhydrin is therefore used in the chromatographic detection and quantification of amino acids and peptides.

***Nitella***  See STONEWORT.

**Nitisols**  Soils that have a clay-rich horizon more than 39 cm deep, with a *cation-exchange capacity of less than 36 $cmol_c$/kg and no evidence of clay *lessivage within 100 cm of the surface. Nitisols are a reference soil group in the FAO *soil classification.

***Nitraria* (family *Zygophyllaceae)** A genus of halophytic (see HALOPHYTE) shrubs and *herbs in which the leaves are thickened or leathery and usually *opposite, with *stipules. The flowers are bisexual, regular, and solitary, paired or in *cymes. There are 4 or 5 overlapping *sepals and a similar number of *petals. The *stamens are in *whorls of 5. The *superior *ovary usually has 5 fused and winged *carpels. The fruit is a *capsule, *drupe, or *berry, and the seeds contain the *embryo surrounded by *endosperm. Some fruits are edible. There are 7 species, occurring in salt deserts of south-eastern Australia, Siberia, southern Russia, Afghanistan, and the Sahara.

**nitrification**  The oxidation of ammonia to nitrite, and/or of nitrite to nitrate, by *chemolithotrophic *bacteria of the family *Nitrobacteraceae.

**Nitrobacteraceae**  A family of *Gram-negative *bacteria in which the cells are variously shaped, and may or may not bear *flagella. They are characterized by their ability to obtain energy from the oxidation of either ammonia or nitrite (according to species), and typically they use carbon dioxide as their sole source of carbon. Growth can occur only in the presence of air. There are several genera, found in soil, and in freshwater and marine environments.

**nitrogen (N)**  An element that is essential to all plant and animal life. It is found reduced and covalently bound in many organic compounds, and its chemical properties are especially important in the structures of *proteins and *nucleic acids. Nitrogen-deficient plants are chlorotic (see CHLOROSIS) and etiolated (see ETIOLATION), with the older parts becoming affected first.

**nitrogen cycle**  A description of the balance, changes, and nature of the nitrogen-containing compounds circulating between the atmosphere, the soil, and living matter. For plants, *nitrogen fixation by soil bacteria, which renders nitrogen readily available for assimilation by the plants, is an essential and crucial part of the nitrogen cycle.

**nitrogen fixation**  The reduction of gaseous molecular nitrogen and its incorporation into nitrogenous compounds. In nature this occurs during thunderstorms by means of the electrical energy released as lightning, by photochemical fixation in the atmosphere, and by the action of nitrogen-fixing *micro-organisms. Free-living nitrogen-fixing soil and aquatic *bacteria include *Azotobacter species, *Bacillus species, *Clostridium species, and *cyanobacteria (e.g. *Nostoc). Symbiotic (see SYMBIOSIS) nitrogen-fixing bacteria include *Rhizobium and Bradyrhizobium

**n**

species, which form the characteristic *root nodules of leguminous plants. The bacteria supply the legume with ammonia and receive carbohydrate from the legume. Certain non-leguminous plants, e.g. *Alnus* species (alder), *Myrica* species (bog myrtle), and *Casuarina*, typically plants of poorly drained and nutrient-depleted habitats, form symbiotic associations with nitrogen-fixing *Actinobacteria. The other main type of symbiotic nitrogen-fixing association occurs in certain *lichens, in which a nitrogen-fixing cyanobacterium may be the main phycobiont or may occur in specialized cephalodia (*see* CEPHALODIUM).

**nitrophilous** Growing in potassium-rich *alkaline soils.

**noble rot** The rot of white grapes that is caused by the *fungus *Botrytis cinerea*; infected fruit is used to make high-quality sweet wines.

***Nocardia* (order Actinomycetales)** A genus of *aerobic *bacteria which form a branching *mycelium that sooner or later fragments into rod-shaped or ovoid cells. There are several species. They are *facultative *parasites or *saprotrophs, found in soils, marine sediments, etc.

***Noctiluca* (division Dinophyta)** A genus of marine, bioluminescent (*see* BIOLUMINESCENCE), *heterotrophic dinoflagellates. Some strains contain symbiotic, unicellular, green flagellates in *vacuoles within their cells.

**nodal analysis** *See* INVERSE ANALYSIS.

**node** On the stem of a plant, the point of attachment of a *leaf or leaves.

**nodule** *See* ROOT NODULE.

**nodum** In plant ecology, a characteristic vegetation unit. The term was first applied by M. E. D. Poore (1956) to similar sites grouped together using modified, traditional, *phytosociological techniques. It was applied by J. M. Lambert and W. L. T. Williams in 1962 to vegetation units doubly defined in *nodal analysis, i.e. cases in which examination of groups identified by normal and inverse classification of a data set showed complete coincidence of *indicator sites

and species. A sub-nodum occurs when coincidence of indicator parameters occurs in one direction only. *See also* INVERSE ANALYSIS.

**Nolanaceae** A family of *herbs and low shrubs that have fleshy, alternate, *simple leaves, often with glandular hairs, and without *stipules. The regular flowers are bisexual, solitary, and *axillary. There are 5 fused and persistent *sepals. The *corolla is 5-lobed, blue, pink, or white, and unfolds from a pleated bud to form a bell or funnel shape. The *stamens are unequal. The *style is single with a lobed *stigma. The *superior *ovary has 5 *carpels, divided into many segments. The fruits are *nutlets and the seeds contain a curved or spiral *embryo, and *endosperm. Some species are ornamentals. The family is related to the *Solanaceae. There is 1 genus, comprising 18 species, represented on the west coast of S. America and the Galápagos, often as seashore plants.

**nomen abortivum** In taxonomy, a name that contravened the Code in operation at the time.

**nomen ambiguum** In taxonomy, a name that is ambiguous, because different authors apply it to different taxa.

**nomen conservandum** In taxonomy, a name, otherwise unacceptable under the rules of nomenclature, which is made valid using specified procedures, with either the original or altered spelling.

**nomen correctum** In taxonomy, a name whose spelling is required or allowed to be intentionally altered under the rules of nomenclature but which does not have to be transferred from one *taxon to another.

**nomen dubium** In taxonomy, a name that cannot be attached certainly to any particular taxon, and is therefore dubious.

**nomen illegitimum** In taxonomy, a name that must be rejected under the rules of the Code and is therefore illegitimate.

**nomen imperfectum** In taxonomy, a name that, as originally published, meets all the mandatory requirements of the rules of nomenclature, but which contains a defect needing correction (e.g. incorrect original spelling).

**nomen invalidum** In taxonomy, a name that has not been published properly, or is unavailable, and is therefore invalid.

**nomen inviolatum** In taxonomy, a name that, as originally published, meets all the mandatory requirements of the rules of nomenclature and is not subject to any sort of alteration.

**nomen neglectum** In taxonomy, a name that was published at some time in the past, but has subsequently been overlooked.

**nomen novem** In taxonomy, a new name that is proposed as a replacement or substitute for an existing name.

**nomen nudum** In taxonomy, a name that, as originally published, fails to meet all of the mandatory requirements of the rules of nomenclature and thus has no status in the nomenclature, even if corrected.

**nomen oblitum** In taxonomy, a forgotten name; i.e. the name of a senior *synonym that has not been used in the botanical literature for at least 50 years. Such names are not to be used unless permission is sought first from the *ICBN.

**nomen perfectum** In taxonomy, a name that, as originally published, meets all of the requirements of the rules of nomenclature, needing no correction of any kind, but which nevertheless is validly alterable by a change of ending.

**nomen substitutum** In taxonomy, a new, replacement name, published as a substitute for an invalid one (e.g. a junior *synonym).

**nomen translatum** In taxonomy, a name that is derived by the valid change of a previously published name as a result of a transfer from one taxonomic level to another within the group to which it belongs.

**nomen triviale** In taxonomy, a species (or trivial) name.

**nomogenesis** An evolutionary model holding that the direction of evolution operates to some degree by sets of rules, or laws, independently of natural selection. For a long time it was regarded as an outmoded hypothesis but it has recently been maintained that it corresponds rather well with observations of evolution in the fossil record and that such mechanisms as *heterochrony and *molecular drive would produce nomogenetic effects.

**noncompetitive inhibition** The irreversible inhibition of the activity of an *enzyme, brought about by the presence of an inhibitor that is generally structurally unrelated to the normal *substrate.

**non-cyclic photophosphorylation** The light-requiring part of *photosynthesis in higher plants, in which an electron donor is required, and oxygen is produced as a waste product. It consists of two photoreactions, resulting in the synthesis of *ATP and NADPH$_2$. The hydrogen needed for the reduction of NADP (*nicotinamide adenine dinucleotide phosphate) is made available from the breakdown of water.

**non-hierarchical classification method** *See* HIERARCHICAL CLASSIFICATION METHOD.

**non-parametric test** *See* STATISTICAL METHODS.

**non-polar molecule** A molecule in which the electrons are shared equally between the nuclei. As a result, the distribution of charge is even and the force of attraction between different molecules is small. Non-polar molecules show little reactivity.

**nonsense mutation** A mutation that alters a *gene so that a nonsense *codon is inserted. Such a codon is one for which no normal *tRNA molecule exists: the codon therefore does not code for an *amino acid. Usually a nonsense codon causes the termination of *translation (i.e. the end of the *polypeptide chain). Three nonsense codons are recognized, and are called the amber, ochre, and opal codons.

**Norfolk Island pine** *See ARAUCARIA.*

**nori** The Japanese name for the edible seaweed *Porphyra* which, in Japan, is cultivated on bamboo stakes submerged in shallow seas.

**normal analysis** Groupings based on analysis of the attributes describing individuals (e.g. in plant ecology, the grouping of sample sites by analysis of their species composition). The term is particularly applied to numerical vegetation classification, and sometimes implies a plot, as distinct from species, classification. *See also* R-TECHNIQUE; *compare* INVERSE ANALYSIS and Q-TECHNIQUE.

**normalizing selection** *See* MAINTENANCE EVOLUTION.

**North American lowland coniferous forest** N. America once possessed great tracts of lowland coniferous forest to the south of the *boreal conifer zone. They comprised several formations, namely the lake forest, Pacific Coast forest, and the pine barrens of the north-eastern and south-eastern USA. The lake forest has been all but destroyed.

**north-east African highland-and-steppe floral region** Part of R. Good's (*The Geography of the Flowering Plants*, 1974) African subkingdom, within his *palaeotropical kingdom. It contains about 50 endemic (*see* ENDEMISM) genera, although the island of Socotra accounts for nearly half of these despite the fact that it is not notably isolated. *Coffea arabica* is native to this region. *See also* FLORAL PROVINCE and FLORISTIC REGION.

**Nostoc** A genus of filamentous *cyanobacteria (section IV), in which the filaments have a distinctly beaded appearance under the microscope. The filaments typically aggregate to form gelatinous colonies. They are capable of *nitrogen fixation. They are found in aquatic and terrestrial habitats. *Nostoc punctiliforme* occurs in *symbiotic association with the plant *Gunnera*, living in the leaf bases where it fixes atmospheric nitrogen. *Nostoc* is also found in coralloid root nodules in cycads, as the *phycobiont in certain *lichens (e.g. the *dog lichen, *Peltigera canina*), and in lichen *cephalodia. Some species of *Nostoc* are used as food in China and Japan.

**Nothofagus (southern beeches; family *Fagaceae)** A genus of evergreen or deciduous, big or small trees, in which the male *inflorescences are *sessile, or few-flowered, and the *styles short. They are wind-pollinated. The fruits are subtended by a valved spiny cupule. Antarctic beech is *N. antarctica*, which forms extensive low, impenetrable forests in southern Chile and Tierra del Fuego. There are 35 species of the southern hemisphere, occurring in the mountains of New Guinea, New Caledonia, Australasia, and temperate S. America. *See* BICENTRIC DISTRIBUTION.

**NPP ($P_n$)** Net primary production. *See* PRIMARY PRODUCTIVITY.

**nucellus** The mass of *tissue in the *ovule of a plant that contains the *embryo sac. Following *fertilization, it may be absorbed by the developing *embryo or persist to form a *perisperm. The size and shape of the nucellus is a diagnostic feature of some species.

**nuciferous** Bearing nuts.

**nuclear envelope (nuclear membrane)** The structure that separates the *nucleus of *eukaryotic cells from the *cytoplasm. It comprises 2 unit membranes each 10 nm thick, separated by a perinuclear space of 10–40 nm. At intervals, the 2 membranes are fused around the edges of circular pores (nuclear pores) which allow for the selective passage of materials into and out of the nucleus.

**nuclear membrane** *See* NUCLEAR ENVELOPE.

**nuclear pore** *See* NUCLEAR ENVELOPE.

**nuclear pore complex** A complex that is formed from the nuclear pore (*see* NUCLEAR ENVELOPE) of a *eukaryotic cell and the granular or fibrous annular material with which it is filled. The structure of the filling material, in particular, is still ill-defined, though it is thought to be a hollow cylinder. Its function is equally obscure, though probably it is concerned with regulating the transport of materials through the pores.

**nucleic acid** *Nucleotide polymers, with high relative molecular mass, produced by living cells and found in both the nucleus and cytoplasm of cells. They occur in 2 forms, designated *DNA and *RNA, and may be double- or single-stranded. DNA embodies the genetic code of a cell or *organelle, while various forms of RNA function in the transcriptional and translational aspects of protein synthesis.

**nucleoid** One of a number of names given to the area of a *prokaryotic cell that contains the chromosomal *DNA.

**nucleolus** A clearly defined, often spherical area of the *eukaryotic nucleus, composed of densely packed fibrils and granules. Its composition is similar to that of *chromatin, except that it is very rich in *RNA and *protein. It is the site of the synthesis of ribosomal RNA, which forms a major part of *ribosomes. The assembly of ribosomes starts in the nucleolus, but is completed in the *cytoplasm.

**nucleoprotein** A *conjugated protein, composed of a histone or protamine bound to a *nucleic acid as the non-protein portion.

**nucleoside** A *glycoside that is composed of ribose or deoxyribose sugar bound to a purine or pyrimidine base.

**nucleosome** A particle, approximately 10 nm in diameter, that is found in large numbers in isolated *chromatin. Nucleosomes are thought to consist of a piece of *DNA wound around the outside of a core of 8 *histone molecules.

**nucleotide** A *nucleoside that is bound to a phosphate group through one of the *hydroxyl groups of the sugar. It is the unit structure of *nucleic acids.

**nucleus** The double-membrane-bound *organelle containing the *chromosomes that is found in most non-dividing *eukaryotic cells; it is essential to their long-term survival. It is variously shaped, although it is normally spherical or ovoid. It disappears temporarily during cell division. The chromosomes, though probably intact, are not visible when the cell is in a resting stage (i.e. not dividing).

**nudation** The initiation of new plant *succession by a major environmental disturbance (e.g. a volcanic eruption).

**null allele** See SILENT ALLELE.

**null hypothesis** See HYPOTHESIS.

**nullisomy** The condition in which a particular *chromosome is not represented by any members. Compare DISOMY, POLYSOMY, TETRASOMY, and TRISOMY.

**numerical method** (numerical taxonomy) Any numerical description of an individual or community, or comparisons of these (such as similarity indices), or data-structuring methods (such as principal components analysis and classification). The essential point is that numerical methods make no assumptions about the data, e.g. the sampling approach used, the distribution of the data, or the probability of particular patterns or frequencies. In contrast, statistical methods do make certain assumptions about the form of the data, on which the validity of significance testing depends. Compare STATISTICAL METHOD.

**numerical taxonomy** See NUMERICAL METHOD.

**nut** A dry, *indehiscent, woody *fruit.

**nutlet** A little *nut (e.g. in *Lamiaceae).

**nutmeg** The seed of Myristica fragrans. See MYRISTICA.

**nutmeg tree** See TORREYA.

**nutrient cycle** See BIOGEOCHEMICAL CYCLE.

**Nuytsia** (family *Loranthaceae) A genus of small, bushy-crowned trees which are root parasites, at least when young. The flowers are vivid orange. There is 1 species, confined to W. Australia.

**nyatoh** See PALAQUIUM; PAYENA; and PLANCHONELLA.

**Nyctaginaceae** (bougainvilleas) A family of *herbs, shrubs, and trees that have *opposite or alternate, *simple leaves with no *stipules. The *inflorescence is *cymose, with bisexual or unisexual flowers sometimes surrounded by coloured *bracts resembling a *calyx. The *sepals are petal-like and tubular. There are no *petals. Usually 5 *stamens alternate with the lobes of the calyx. The *ovary is *superior with a single *carpel, a single *ovule, and a long *style. The fruit is an *achene, sometimes enclosed by the persistent calyx. The seeds contain *endosperm, *perisperm, and *embryo. Several *Bougainvillea* species are used as hedging plants in warm temperate regions, and many of the species of the family are cultivated as ornamentals. Several species are used in medicine, e.g. *Mirabilis jalapa*, which gives a purgative drug. There are 34 genera, comprising about 350 species. They are found throughout the tropical regions of the world, but more than 200 of the species are centred in America.

**nyctigamous** Applied to flowers that are closed by day and open at night.

***Nymphaea* (water lilies; family \*Nymphaeaceae)** A genus of *perennial, aquatic plants, found in shallow water, that have large *rhizomes and *peltate or *cordate, usually floating leaves. The flowers are solitary, with green or coloured *sepals, and 3 to many *petals. The *ovary is *half-inferior, and the seeds have an *aril. The flowers are often pollinated by beetles. Many species are cultivated, and the seeds and rhizomes can be eaten. There are 35 species, found in temperate and tropical regions of the world.

**Nymphaeaceae (water lilies)** A family of aquatic *herbs that have large, rounded, floating leaves on long stalks, attached to *rhizomes rooted on the bottom of the river or lake. The flowers are large and solitary, with 3–6 green *sepals, 3 to numerous *petals, numerous *stamens, and many *carpels which are usually joined into a many-celled *superior *ovary of flask-shaped or globose form. There are some 6 genera, with 60 species, found throughout tropical and temperate areas of the world.

***Nymphoides* (fringed water lily)** *See* MENYANTHACEAE.

***Nypa* (family \*Arecaceae)** A *monotypic genus of palms, and one of the oldest fossil flowering plants. It has an underground, *dichotomously branched, horizontal *rhizome, erect tufts, and big, *pinnate leaves, useful for thatch. The *inflorescences are erect, with separate, clustered, male and female flowers. There are 3 *carpels. The fruits are angular *nuts held in big, globose heads. Sugar or toddy is obtainable from the sap. *M. fruticans* inhabits mangrove swamps from Bangladesh to the Solomon and Mariana Islands, with fossils in Africa, America, and England (in the London Clay).

**oak** 1. *See QUERCUS.* 2. (African oak) *See* OCHNACEAE.

**oak-apple gall (King Charles's apple)** A *gall formed in *Quercus robur* by the unisexual generation of the wasp *Biorrhiza pallida.* Wingless females arise from root galls and climb the trunk of the tree in spring to lay many eggs in the bases of leaf *buds. The eggs cause the formation of *multilocular, pale pink, spongy galls that resemble small apples. The galls mature by midsummer and give rise to the sexual insect generation whose members emerge and mate in June and July. Mated females penetrate the soil around the tree and lay eggs on rootlets, giving rise to small, brown, rounded galls whose occupants emerge at the end of their second winter. May 29 is Oak Apple Day and commemorates the restoration of the English monarchy in 1660 and the birthday of Charles II; according to legend, the king once hid from his pursuers by climbing an oak tree.

**oak-marble gall** A dark brown, hard, spherical *gall on *Quercus robur* formed by the sexual generation of the gall wasp *Andricus kollari.* The galls contain members of the unisexual generation (all females) which emerge in spring to form 'ant-pupae' galls on the axillary buds of *Quercus cerris* (Turkey oak). The insect was introduced into Britain in 1830 for use in the dyeing and ink industry, the galls containing up to 17 per cent dry weight of tannic acid.

**oak-moss** The common name for the *lichen *Evernia prunastri.* The *thallus consists of flattened, strap-like branches which are greenish-grey on the upper surface and white beneath. Whitish, powdery soredia (*see* SOREDIUM) are often present. Oak-moss is common in areas of relatively low air pollution, on a range of substrates: trees, fences, rocks, walls, etc. It is extensively used in the perfume industry as a fixative and as a scent component. *Compare* PSEUDEVERNIA.

**oak-nut** A general common name given to any *gall that forms on an oak tree.

**oak-spangle gall** A generic name for the unisexual stage *galls found on the leaves of *Quercus robur* and produced by 4 species of wasps of the genus *Neuroterus.* The eggs laid by mated females give rise to generally small (1 mm diameter), flat, red galls. Gall densities may be high but competition seems to be avoided by niche separation within a single leaf and host tree, some species preferring leaf margins or apices, others the peripheral foliage. The galls overwinter in the leaf litter, often having become detached from the leaves.

**oak wilt** A disease of oak trees (*Quercus* species) that is caused by the *fungus *Ceratocystis fagacearum.* The fungus is carried by insects, particularly sap-sucking beetles.

**oarweed** Either any seaweed of the genus *Laminaria* or, more specifically, *L. hyperborea.* This is a large seaweed, with *fronds that can measure up to 3.5 m long. The *stipe is stiff, tapers towards the top, and has a rough-textured surface which provides a suitable substrate for many *epiphytic seaweeds. The blade is divided into a number of strap-like ribbons and is oar-shaped in outline. The *holdfast consists of numerous thick, root-like branches. It is found just below low-water mark, attached to rocks.

**oasis** 1. In an arid region, an isolated area that supports water-loving plants throughout the year. Oases most commonly occur in depressions, where the water table lies close enough to the surface to be within the reach of plant roots. The groundwater supplying the vegetation is often enriched in salts and the vegetation can be zoned according to the concentration and type of salts present. 2. By analogy, a small area supporting vegetation of one type (e.g. woodland) that is surrounded on all sides by a much larger area supporting vegetation of a quite different type (e.g. an arable field).

**oat** See AVENA.

**obeche** See TRIPLOCHITON SCLEROXYLON.

**obligate** Applied to an organism that can survive only if a particular environmental condition is satisfied. For example, an obligate aerobe can survive only in the presence of air, an obligate parasite only in association with its host.

**obovate** Applied to a leaf that has the stalk at the narrow end, the leaf widening towards the tip (e.g. in *Duboisia*).

**Ochnaceae** A family of trees, shrubs, and *herbs, that have *simple, alternate leaves with *stipules. The regular, bisexual flowers are held in *panicles, *racemes, or *umbels. The *calyx is either united at the base or free. The *corolla has 5 twisted or convoluted *petals. There are 5 or more *stamens. The *ovary is *superior with several *carpels which, although not usually united, have a common *style. The fruit is a cluster of *drupes, a *berry, or a *capsule. The seed often contains *endosperm. The family is losely related to the *Dipterocarpaceae and includes a few species of commercial timber trees known as the African oaks (*Lophira* species). There are 37 genera, comprising about 460 species, found throughout the tropics but concentrated in Brazil.

**ochrea** (ocrea) A structure around the base of a stem that is formed from the joining of *stipules or leaf bases.

**ochric horizon** A light-coloured, mineral *soil horizon, usually at the soil surface, and characteristic of arid-environment soils. The name is from the Greek *ochros* meaning 'pale'.

**Ochrolechia** (order *Pertusariales) A genus of *lichens in which the *thallus is *crustose. The *apothecia are *lecanorine. The *spores are large, colourless, and *aseptate. There are many species, found mainly on rocks and walls, sometimes also on bark. Some species were formerly used for dyeing cloth (*see* CUDBEAR).

**Ochromonas** (division *Chrysophyta) A genus of single-celled *algae in which

the cells are ovoid, somewhat pointed at the anterior end and blunt at the posterior end; there are typically 2 brown *chloroplasts, and 2 *flagella: a long, forward-pointing, tinsel flagellum, and a shorter, backward-pointing, whiplash flagellum. Although normally photosynthetic, the cells can also take in food *phagotrophically and require organic nutrients to survive. They are found in puddles, ponds, lakes, etc., especially in spring.

**Ocotea** (family *Lauraceae) A genus of trees, which includes *O. rodiaei* (greenheart). It and others are valuable for their heavy, hard, durable timber. There are about 200 species, occurring in the tropics, mainly in America, but a few in Africa.

**ocrea** See OCHREA.

**Octomeles** (family *Datiscaceae) A *monotypic genus (*O. sumatrana*) which is a huge, very fast-growing tree with leaves that are *palmately nerved and that yields useful timber (binuang). It occurs in most parts of the Malay archipelago to the Solomon Islands.

**octoploidy** See POLYPLOIDY.

**Oedogonium** (division *Chlorophyta) A genus of freshwater, filamentous (unbranched) *green algae in which each cell has a single nucleus. Asexual reproduction occurs by the formation of *motile cells which have a whorl of *flagella at one end. Cell division occurs by a unique mechanism.

**Oenothera** (evening primrose; family *Onagraceae) A genus of flowering plants that have spirally arranged leaves and 4-petalled, large, usually yellow to orange flowers in a leafy *spike. The seeds, unlike those of some other genera in the family, are not plumed. The unusual chromosome morphology and genetic mechanism of *Oenothera* have been the subject of much study by cytogeneticists. There are about 80 species, native to N. America, but some are much cultivated in gardens for the flowers, both as species and as hybrids, and are now well naturalized in Europe and elsewhere.

**offal** Refuse; waste. In forestry, the term is applied to the waste from squared timber, and to lop, top, and branches.

**offshore zone** The zone extending seaward from the point of low tide to the depth of wave-base level or to the outer edge of the continental shelf.

**oidium** A thin-walled cell formed by the fragmentation of a *hypha that is capable of functioning as a *gamete, or of *germinating to form a new hypha.

**oil palm** *See* ELAEIS.

**okoume** *See* AUCOUMEA.

**okra** *See* HIBISCUS.

**Olacaceae** A family of shrubs, trees, and climbers, whose members have alternate, *simple leaves without *stipules and with a rough texture. The flowers are green or white, and regular, with the *calyx reduced to narrow lobes. The *petals lie closely together, but do not overlap, and are of the same number as the calyx lobes, usually 4–6. There may be up to twice as many *stamens. The *ovary is half *inferior, with 1–3 uniovular *locules. There is a single *style with a lobed *stigma. The fruit is a single-seeded *nut or *drupe. The seed contains a copious *endosperm. The family includes several timber trees, and some with edible fruits. (*Heisteria* species yield a hard timber, used in building, and *Coula edulis* yields edible seeds (Gaboon nuts) as well as timber (African walnut) used in building.) The family consists of 29 genera, with about 200 species, found throughout the tropics, but with the largest number of species in Africa and Asia.

**old-field ecosystem** An ecosystem that develops on abandoned farmland. The term is most commonly used in the USA.

**old-growth forest** In N. America, a forest at a late seral stage and, by implication,[a] *primary forest that existed prior to European settlement. *Compare* ANCIENT WOODLAND.

***Olea*** (olive; family *Oleaceae) A genus of about 20 species of trees and shrubs in which the leaves are usually grey-green and elliptical or linear, and the fruits are *berries rich in oil. They are widely distributed in the Old World. *O. europaea* (common olive) is much cultivated in Mediterranean climates for its fruits, for eating, and for olive-oil production, but its native home is probably Africa.

**Oleaceae** A family of trees or shrubs that have *opposite, *exstipulate leaves and usually *tetramerous flowers with 2 *stamens attached to the *corolla tube. The *ovary is *superior and 2-celled. Besides *Olea*, *Fraxinus* (ash) is an important timber tree, with reduced flowers that normally lack a *perianth. There are some 24 genera, with 900 species, widely distributed but centred on Asia.

***Olea europaea*** (common olive) *See* OLEA.

**oleander** *See* NERIUM.

***Olearia*** (daisy-bushes; family *Asteraceae) A genus of shrubs or small trees with alternate leaves, which differ from many members of the Asteraceae in having a series of *involucre *bracts, a naked *receptacle, and *anthers that are blunt-ended instead of having tails. There are about 130 species found in Australia, New Zealand, and New Guinea.

**oligo-** A prefix meaning few or small, derived from the Greek *oligos*, meaning 'small' or (*oligoi*) 'few'; in ecology it is often used to denote a lack, e.g. *oligotrophic, nutrient-poor; *oligomictic, subject to little mixing.

**Oligocene** An epoch of the *Tertiary Period, about 33.9–23.03 Ma ago, that follows the *Eocene and precedes the *Miocene Epochs.

**oligohaline** *See* BRACKISH.

**oligomictic** Applied to lakes that are thermally almost stable, mixing only rarely. This condition is characteristic of tropical lakes with very high (20–30°C) surface temperatures.

**oligopeptide** A linear *peptide of 2–10 *amino acids. *Compare* POLYPEPTIDE.

**oligosaccharide** A linear or branched carbohydrate of 2–10 monosaccharides.

**oligotrophic** Applied to waters poor in nutrient and with low *primary productivity. Typically, \*oligotrophic lakes are deep, with the \*hypolimnion much more extensive than the \*epilimnion. The low nutrient content means that plankton blooms are rare and littoral plants are scarce. The low organic content means that dissolved oxygen levels are high. By comparison with \*eutrophic lakes, oligotrophic lakes are considered geologically young, or little modified by weathering and erosion products.

**oligotrophication** The process of nutrient depletion, or reduction in rates of nutrient cycling, in aquatic ecosystems. It arises as a consequence of acidification, typically the result of \*pollution and most notably associated with air pollution and acid rain.

**olive** *See* OLEA.

**olive knot** A type of \*gall that develops on olive trees (\*Olea) infected with the \*bacterium *Pseudomonas savastanoi*.

**Olpidium (order \*Chytridiales)** A genus of fungi that live \*parasitically on many types of plant; some species are involved in the transmission of certain plant diseases (e.g. tobacco necrosis virus).

**ombrogenous bog** A peat-forming vegetation community lying above \*ground water level: it is separated from the ground flora and the \*mineral soil, and is thus dependent on rain water for mineral nutrients. The resulting lack of dissolved bases gives strongly acidic conditions, and only specialized vegetation, predominantly *Sphagnum* species (bog mosses), will grow. Two types of ombrogenous bogs are commonly distinguished: raised bogs and blanket bogs. The elevated bog forest found in the coastal regions of some of the islands of southeast Asia comprises a third type of ombrogenous bog. These bog forests form extensive domes, often more than 10 km in diameter and their surfaces are elevated 5–10 m above the ground water table. *Compare* VALLEY BOG.

**ombrotrophic** Applied to a \*mire that is fed by rain water. *Compare* RHEOTROPHIC.

**Omphalina (family \*Tricholomataceae)** A genus of \*fungi in which the \*fruit bodies are small; the cap is thin and con-

vex (often with a central depression), the \*gills are \*decurrent, and the \*stipe is thin and central. The \*spores are smooth and colourless, and do not stain dark violet when treated with iodine. They are found chiefly on peaty ground in \*heathland.

**Onagraceae** A family of shrubs and \*herbs that have \*tetramerous (or, rarely, \*dimerous) flowers, with twice as many \*stamens as \*petals, and an \*inferior \*ovary of 4 fused \*carpels and 4 cells. Many members are now grown for ornament, e.g. *Oenothera* and *Fuchsia*. There are some 24 genera, comprising about 650 species, with a cosmopolitan distribution, especially in America.

**one-gene-one-polypeptide hypothesis** The hypothesis that a large class of structural \*genes exists in which each gene encodes a single \*polypeptide, which may function either independently or as a subunit of a more complex protein. Originally it was thought that each gene encoded the whole of a single enzyme, but it has since been found that some enzymes and other proteins derive from more than one \*polypeptide and hence from more than one gene.

**onion (*Allium*)** *See* LILIACEAE.

**onomatologia** The conventions used in the construction of plant names. *See also* INTERNATIONAL CODE OF BOTANICAL NOMENCLATURE.

**ontogeny** The development of an individual from \*fertilization of the egg to adulthood.

**Oodinium (division Dinophyta)** A genus of dinoflagellates which can cause disease in fish (velvet disease); *O. ocellatum* is marine, *O. limneticum* freshwater.

**oogamy** \*Fertilization by the fusion of a large, non-motile, female \*gamete with a small, usually \*motile, male gamete. It is an extreme form of \*anisogamy. *Compare* ISOGAMY.

**oogenesis** The formation of eggs or \*ova including, in \*angiosperms, the formation of the \*embryo sac.

**oogonium** The female sex organ of certain *algae and *fungi, containing one or more *oospheres.

**Oomycota (water mould, downy mildews)** Fungus-like organisms in which the *zoospores have two *flagella, one of the whiplash type and one of the tinsel type. The cell walls of *oomycetes are atypical of fungi in that in most species they contain cellulose but not chitin. The majority of species are aquatic, but some are important *pathogens of terrestrial plants. The class contains 4 orders.

**oosphere** The female *gamete contained within an *oogonium. *Fertilization of an oosphere is followed by the development of an *oospore.

**oospore** A thick-walled resting *spore formed following the *fertilization of an *oosphere within an *oogonium.

**Opegraphales (subdivision *Ascomycotina)** An order of lichenized and *lichenicolous *fungi. Lichenized members are *crustose or *fruticose. The *ascocarps are apothecioid (see APOTHECIUM). Species are found on rocks, bark, etc.

**open canopy** Forests or woodlands in which the individual tree crowns do not overlap to form a continuous canopy layer but are more widely spaced, leaving open sunlit areas within the woodland.

**open pollinated population** An outbred population in which pollination is uncontrolled.

**open population** A population that is freely exposed to gene flow, as opposed to a closed one in which there is a barrier to gene flow.

**open-space area** Any land area to which the public has unrestricted access (although access may extend only to permission to move freely over the area on foot).

**operator** A region of *DNA at one end of an *operon that acts as the binding site for a specific *repressor protein, and so controls the functioning of adjacent *cistrons.

**operculum** A little lid. In plants, the term usually refers to the *dehiscent cap present on certain *moss *capsules or other types of *sporangia.

**operon** A set of adjacent structural *genes whose *messenger-RNA is synthesized in one piece, together with the adjacent regulatory genes that affect the *transcription of the structural genes. The operon is under the control of an *operator gene, lying at one end of it. Operons are concerned in the control of gene transcription leading to the formation of particular *enzymes used in *metabolic pathways. To date they have been found only in *prokaryotes (e.g. *Escherichia coli*), but a number of *homologous systems exist in lower *eukaryotes such as *fungi.

**Ophioglossaceae** A family of primitive, *eusporangiate ferns, in which the *frond is forked into a sterile vegetative part and a fertile part which bears the massive, more or less *sessile *sporangia. There are 4 genera. *Botrychium* and *Ophioglossum* are cosmopolitan in distribution; *Helminthostachys* is confined to the Indo-Malaysian region; and *Rhizoglossum* is S. African.

***Ophioglossum* (adder's tongue, snake's tongue; family *Ophioglossaceae)** A genus of ferns with the sterile blade(s) usually *simple (rarely lobed) with reticulate veins, and the *sporangia sunken along the opposite sides of a simple spike, opening by transverse slits. There are about 40 species, found in most parts of the world.

***Ophiostoma* (order *Ophiostomatales)** A genus of *fungi (formerly *Ceratocystis*) which includes a number of important plant *pathogens. *O. ulmi* is the causal agent of *Dutch elm disease. Other species cause disease in oaks (*oak wilt), sweet potatoes, coffee, rubber, and sugar-cane, and some can cause staining of timber.

**Ophiostomatales (subdivision *Ascomycotina)** An order of *fungi in which the *ascocarp is a black, long-necked *perithecium, the *asci are spherical and *evanescent, the ascus wall disintegrating within the ascocarp. Species are *saprotrophic or *parasitic in plants.

**Opiliaceae** A family of evergreen trees and shrubs, plus a few climbers, very similar to *Olacaceae and often included in that family, most members of which are root parasites. They have alternate, *simple, entire leaves, without *stipules. The flowers are regular, usually bisexual but sometimes *dioecious, and held in simple or *compound *spikes or *racemes. If present, the *calyx is of 4 *sepals and the *corolla of 4 or 5 free *petals. The fruit is a *drupe, and the seed contains an oily *endosperm. There are 9 genera and 28 species, found throughout the tropics, but centred in Asia.

**opportunist species** *See* FUGITIVE SPECIES.

**opposite** Applied to the leaf arrangement in which leaves arise in pairs, one pair at each *node.

**optimum yield (maximum sustained yield, MSY)** The theoretical point at which the size of a population is such as to produce a maximum rate of increase. If the population has a symmetrical, *S-shaped growth curve this is equal to half the *carrying capacity. The concept is of practical use in farming and has been applied widely to commercial fisheries. It forms the basis for models that predict the stocking density required to maintain optimum fish production, and the harvesting methods and food supply needed to maintain production at that level.

**orange-peel fungus** The common name for the *fruit bodies resembling orange peel of the *ascomycete *Aleuria aurantia*. *See* ALEURIA.

**orbicular** Disc-shaped, circular, or globular.

***Orbignya* (family *Arecaceae)** A genus of palms, most of which are tall, spineless, and *pinnate-leaved. The *inflorescences are unisexual. The fruit has a woody endocarp (*see* PERICARP) with basal *germ pores. *O. speciosa* (babaçu) and *O. cohune* (cohune) produce oil from both mesocarp and *endosperm. There are 17 species, found in tropical S. America. *Compare* ELAEIS.

**Orchidaceae (orchids)** One of the most advanced and specialized families of monocotyledonous (*see* *MONOCOTYLEDON) plants whose members have a partial or total dependence on a *symbiosis with a fungus (a mycorrhizal relationship). Orchids are *perennial *herbs with *rhizomes, vertical stocks, or root *tubers, and sometimes totally saprophytic (*see* SAPROPHYTE). Many tropical species are *epiphytes with pendent roots which absorb moisture from the air. The leaves are *simple, entire, and parallel-veined. The flowers are borne in *racemes, *spikes, *panicles, or are solitary. They are irregular, with 3 *petaloid *sepals, and 3 *petals of which 1 is usually enlarged into a *labellum of complex and often very different form from the other 2. Usually there is only 1 *anther (2 in *Cypripedium* and related genera) borne with the *stigma on a column in the centre of the flower. The *ovary is *inferior and 1-celled, with numerous minute seeds in the ripe capsule. The most complex adaptations of structure, seen in the flowers of various genera and species, are concerned with intimate dependence on various insect pollinator species. Orchids form one of the largest *angiosperm families, with about 800 genera, with 17 500 species, centred in the tropics. Many tropical species and hybrids are cultivated for their remarkable flowers; even in temperate areas, the smaller, terrestrial species native there are universal favourites.

***Orchidantha* (formerly *Lowia*; family *Lowiaceae)** A genus comprising 7 species of plants, which occur from southern China to Borneo.

**orchids** *See* ORCHIDACEAE.

**orchil (orseille)** A purple dye obtained from *lichens of the genus *Roccella*.

**ordination method** A method for arranging individuals (or sometimes attributes) in order along one or more lines. The method is used, with many techniques, in the biological and earth sciences, and especially in biology. There is an extensive ecological literature that discusses ordination methods, their applicability to different situations, and the relative merits of ordination as opposed to classification schemes.

**Ordovician** The second of 6 periods that constitute the *Palaeozoic Era, named after an ancient Celtic tribe, the Ordovices. It lasted from about 488.3 to 443.7 Ma ago. The Ordovician follows the *Cambrian and precedes the *Silurian. *Algae were the predominant plants. Some may have been terrestrial, forming thick, moss-like mats on wet ground.

**oregano** *See* ORIGANUM.

**Oregon grape** *See* MAHONIA.

**organelle** Within a cell, a persistent structure that has a specialized function. In most cases, the organelle is separated from the rest of the cell by selectively permeable membranes.

**organic matter** In particular, the organic material present in soils; more generally, the organic component of any ecosystem.

**organic soil** Soil with a high content of organic matter and water. The term usually refers to peat. The *USDA defines an organic soil as one with a minimum of 20–30% organic matter, depending on the *clay content.

**organismic** Applied to groups of organisms, or *communities, that are thought to have properties (e.g. homoeostasis or reproduction) similar to those of a single living organism, making them 'supra-organisms'. The term is most used to describe plant communities by those who consider that discrete *climax vegetation entities, e.g. beech–oak woodlands, may be identified. These units will necessarily show a high degree of internal interdependence of species, and on the maturity and death of a community another identical plant association will replace it. This organismic concept forms the theoretical basis for a classificatory approach to the description of vegetation communities and their analysis. *Compare* INDIVIDUALISTIC HYPOTHESIS.

**organotroph** An organism that obtains energy from the metabolism of organic compounds, sometimes inaccurately used as a synonym of *heterotroph. *Compare* LITHOTROPH.

**Origanum** (family *Lamiaceae) A genus

of dwarf shrubs of *herbs that produce essential oils. Some are cultivated as pot-herbs (e.g. *O. onites*, pot marjoram; *O. majorana*, sweet marjoram; and *O vulgare*, oregano). There are 36 species, occurring in Eurasia.

**original-natural** *See* NATURAL.

**Ormosia** (family *Fabaceae, subfamily *Papilionatae) A genus of trees in which the leaves are *pinnate. The flowers have 5 subequal *petals. The pods contain a few large, shiny, scarlet seeds. The timber is hard and heavy. There are about 100 species, occurring in eastern S. America, Malesia, and north-eastern Australia.

**ornithocoprophilous** Preferring habitats rich in bird manure.

**ornithophily** Pollination by birds.

**Orobanchaceae** A family of totally parasitic herbs, devoid of *chlorophyll, that are often specific root parasites of particular *angiosperms. The flowers are borne in *spikes or *racemes, or rarely are solitary, on stems that bear only scale-like leaves. The flowers are irregular and 2-lipped, and very close in structure to those of the *Plantaginaceae, but the *ovary is 1-celled, though with a 2-lobed *stigma, and there are 4 *stamens. The fruit is a *capsule. There are some 17 genera, with 230 species, occurring chiefly in the warmer temperate parts of the Old World.

**orogeny** A mountain-building episode, especially one that is caused by the compression of a part of the Earth's crust to form a chain of mountains.

**orophilous** Applied to plants that grow in subalpine habitats.

**orseille** *See* ORCHIL.

**orthogenesis** Evolutionary trends that remain fairly constant over long periods of time and so appear to lead directly from ancestor organisms to their descendants. This was once explained as the result of some internal directing force or 'need' within the organisms themselves. Such metaphysical interpretations have been displaced by the concepts of *orthoselection and species selection.

**orthoselection** A primary selective pressure of a directional kind, which results in a self-perpetuating evolutionary trend. Species selection has been advanced as an alternative explanation for such trends. *See also* DOLLO'S LAW.

**Orthotrichiales** (subclass *Bryidae) An order of medium-sized *acrocarpous mosses in which the leaves are ovate to lanceolate, with or without a nerve. The *capsules are held erect and are usually ovoid with a double *peristome; each capsule may be covered with a bell-shaped, often hairy, *calyptra. They are found typically on trees; some species are found on rocks and walls. Genera include *Orthotrichum* and *Ulota*.

**orthotropic** A tropic response (*see* TROPISM) directly towards or away from a stimulus (e.g., a tree is negatively orthogravitropic).

**orthotropous** (atropous) Applied to the position of an *ovule that is upright, with the *micropyle directly above the short funicle. *Compare* ANATROPOUS and CAMPYLOTROPOUS.

**ortstein** An indurated *horizon in the B horizon of *podzols (*Spodosols), in which the cementing materials are mainly iron oxide and organic matter.

*Oryza* (rice; family *Poaceae) A genus of grasses that are cultivated for their grains. Rice is the most important cereal crop in the world in terms of the number of people depending upon it. There are 2 principal types: lowland (grown in fields inundated with water); and upland (grown on dry land). The unhusked grain is known as paddy. *O. sativa* is cultivated rice. There are 19 species, native to tropical Asia.

*Oscillatoria* A genus of filamentous *cyanobacteria (section III), in which the filaments are composed of disc-shaped cells and may reach lengths of more than 1 cm. They are capable of gliding motility. Sheaths are often not apparent, and *heterocysts and *akinetes are not formed. *Oscillatoria* species are common in all types of aquatic environment, often forming dark green mats on wet rocks, etc. *O. limnetica* can grow in the absence

of oxygen, carrying out anoxygenic photosynthesis, using hydrogen sulphide instead of water as the electron donor.

**osmophilic** Applied to organisms that grow best in *habitats containing relatively high concentrations of salts or sugars, i.e. those having a relatively high *osmotic pressure.

**osmoregulation** The process whereby an organism maintains control over its internal *osmotic pressure irrespective of variations in the environment.

**osmosis** The net movement of water or of another solvent from a region of low solute concentration to one of higher concentration through a *semi-permeable membrane.

**osmotic potential** (solute potential) The part of the water potential of a tissue that results from the presence of solute particles. It is equivalent to *osmotic pressure in concept but opposite in sign.

**osmotic pressure** The pressure that is needed to prevent the passage of water or another pure solvent through a *semipermeable membrane separating the solvent from the solution. Osmotic pressure rises with an increase in concentration of the solution. Where two solutions of different substances or concentrations are separated by a semipermeable membrane, the solvent will move to equalize osmotic pressure within the system.

**osmotrophic** Applied to an organism that absorbs nutrients from solution, as opposed to ingesting particulate matter.

*Osmunda* (family *Osmundaceae) A genus of large ferns with massive, erect *rhizomes, and bi-*pinnate leaves. The *sporangia are borne mostly on the upper margins of specialized *pinnules without *lamina or *chlorophyll, and are massive, short-stalked structures, cultivated as ornamentals (royal ferns). There are some 10 species, cosmopolitan in distribution but absent from Australasia.

**Osmundaceae** A family of *eusporangiate ferns of primitive type and ancient ancestry. The *sporangia, all developing together, are massive and short-stalked and have a group of thick-walled cells at their apices, instead of a ring (*annulus) of thickened cells as in more modern ferns. There are 3 genera, with about 18 species, found throughout most of the world.

**ostiole** An opening in a *fruit body (e.g. a *perithecium) through which *spores are discharged to the exterior.

**Ostropales (subdivision *Ascomycotina)** An order of *fungi in which the *asci are cylindrical and contain 8 smooth, colourless, commonly thread-like, multiseptate (see SEPTATE) *spores. The *ascocarps are apothecioid (see APOTHECIUM) or perithecioid (see PERITHECIUM). Species include *saprotrophic, plant-*parasitic, and lichenized fungi.

**Otidia (order *Pezizales)** A genus of *fungi in which the *fruit bodies are typically brown or yellowish, and are cupshaped or ear-shaped; the *hymenium lines the inner surface, and at maturity the tips of the *paraphyses are bent or hooked. *Otidia* species are found on the ground in a range of habitats.

**oubain** A cardiac glycoside, obtained from wood, which inhibits the transport of sodium and potassium *ions across *cell membranes.

**Oudemansiella (family *Tricholomataceae)** A genus of *fungi that form *mushroom-shaped *fruit bodies; the cap is white, smokey, or brownish, with broad, well-spaced gills, and a central, often long-rooting *stipe. They are found on wood. *O. mucida* (porcelain fungus) is common on beech trunks and is white, semi-translucent, and slimy. *O. radicata* (rooting shank) has a yellow-brown to olive-brown cap with white *gills, and a long-rooting stem. It is very common on roots or buried wood.

**outbreeding** The crossing of plants or animals that are not closely related genetically, in contrast to inbreeding, in which the individuals are closely related. See also CROSS-BREEDING.

**outlier** An organism that occurs naturally some distance away from the principal area in which its species is found.

**outwelling** The enrichment of coastal seas by nutrient-rich estuarine waters.

**ovary** Of a plant, the *gynoecium.

**overdispersion (contagious distribution)** In plant ecology, a situation in which the pattern formed by the distribution of individuals of a given plant species within a *community is not random but shows clumping, so that large numbers of both empty and heavily populated *quadrats are recorded. See also PATTERN ANALYSIS.

**overdominance (superdominance)** In genetics, the phenomenon in which the character of the *heterozygotes is expressed more markedly in the *phenotype than in that of either *homozygote. Usually the heterozygote is fitter than the two homozygotes: this can give rise to *monohybrid *heterosis, the *hybrid vigour obtained by crossing parents differing in a single specified pair of *allelic genes. See also BALANCED POLYMORPHISM.

**overgrazing** Pressure by grazing animals, either domestic or wild, which results in the degradation of pasture, leading to exposure of the bare soil surface and ultimately *erosion and even *desertification of the area.

**over-representation** In *palynology, the presence in considerable abundance of immediately local *pollen which tends to obscure the general pattern of changes in the regional *pollen-rain record, e.g. the predominance of pollen from bog plants at a peat bog site.

**overspecialization** An old theory which held that straight-line *evolution or *orthogenetic trends might proceed to the point at which the lineage was at an adaptive disadvantage. Overspecialization was therefore considered as one of the causes of extinction. There is no reason to believe, however, that *natural selection would permit evolution to proceed beyond maximum *adaptation. More recently the term has been applied to highly specialized organisms which have proved incapable of responding to environmental change and so have become *extinct.

**ovule** A structure in *angiosperms and *gymnosperms that, after *fertilization, develops into a *seed. In gymnosperms the ovules are unprotected, whereas in angiosperms they are protected by the *megasporophyll (which forms the *carpel). One or several ovules may be contained in the carpel and each is attached to the carpel wall by a stalk (funicle). When mature, the angiosperm ovule consists of a central mass of tissue, the *nucellus, surrounded by 1 or 2 protective layers, the *integuments, which eventually give rise to the seed coat. Within the nucellus is a large oval structure, the *embryo sac, which has developed from the *megaspore and contains the naked egg cell.

**ovum (pl. ova)** An unfertilized egg cell.

**Owenia (family *Meliaceae)** A genus of small, rain-forest trees in which the leaves are alternate, usually *pinnate, and without *stipules. The regular flowers are held in *cymes and have 3–5 *petals and the same number of *sepals. The fruit is usually a *capsule or *drupe. Some species have edible fruits. Several species appear to reproduce only vegetatively (e.g. *O. acidula*, emu apple, which has an edible pulp) and their seeds may be infertile. There are 6 species, endemic (*see* ENDEMISM) to tropical eastern Australia.

**Oxalidaceae** A family of *herbs (rarely shrubs or trees) that have alternate, *pinnate, *palmate, or *trifoliate leaves with entire *leaflets, and *pentamerous, regular flowers with 10 *stamens and a 3–5-celled *superior *ovary which lacks a beak (unlike those of the related *Geraniaceae). The fruit is a *capsule or rarely a *berry. Modern classifications recognize 8 genera, with 575 species, which occur mainly in tropical areas.

**Oxalis (family *Oxalidaceae)** A genus of *annual or *perennial *herbs that have alternate, *palmate leaves without *stipules. There are 1 to many *leaflets. The flowers are regular, bisexual, and usually solitary, with 5 free and persistent *sepals and 5 *petals, often fused at the base. There are 10 *stamens. The *superior *ovary is composed of 5 fused *carpels, each with a separate *style. The fruit is a *capsule. *Oxalis* species are known for their characteristically clover-like leaves

(e.g. *O. acetosella*, wood sorrel), and in many species the leaflets fold down at night. Several species are cultivated. Some can be eaten but others are troublesome weeds (e.g. *O. pes-caprae*, Bermuda buttercup, which is a weed of the bulb fields of the Scilly Isles, Great Britain). There are about 500 species, found throughout the temperate and tropical regions of the world, but centred on S. Africa, and Central and S. America.

**oxaloacetic acid** A dicarboxylic acid; a key intermediate in the *citric acid cycle, where it condenses with acetyl *coenzyme A to form citric acid and coenzyme A, thus initiating the cycle.

**oxic horizon** A *mineral subsoil *horizon that is at least 30 cm thick and is identified by the almost complete absence of weatherable primary minerals, and by the presence of kaolinite *clay, insoluble minerals such as quartz, hydrated oxides of iron and aluminium, small amounts of exchangeable bases, and low *cation-exchange capacity. It is the distinguishing subsoil horizon (B horizon) of an *Oxisol. The name is from the Latin *oxide* meaning 'oxide'.

**oxidase** An *enzyme that catalyses reactions involving the oxidation of a *substrate using molecular oxygen as an electron acceptor.

**oxidation** A reaction in which atoms or molecules gain oxygen or lose hydrogen or electrons.

**oxidative decarboxylation** *See* CITRIC ACID CYCLE.

**oxidative phosphorylation** The synthesis of *ATP from *ADP and inorganic phosphate, linked to the *redox reactions in the *electron-transport chains of *mitochondria. Oxidative phosphorylation occurs during aerobic respiration and produces a high yield of ATP moles.

**oxidative potential** The potential to cause a *redox reaction to occur through loss of electrons.

**oxidoreductase** An *enzyme that catalyses oxidation-reduction reactions. Included within this group are the *dehydrogenases and *oxidases.

**Oxisols** An order of *mineral soils that are distinguished by the presence of an *oxic *horizon within 2 m of the soil surface, or with plinthite close to the soil surface, and without a *spodic or *argillic horizon above the oxic horizon. Oxisols occur over a large part of the humid tropics, where they are usually heavily leached and exhausted by hand tillage, resulting in low fertility.

**oxygen demand** A measure of the amount of oxidizable material present in an effluent, stream, etc. (i.e. a measure of organic pollution). *See* BIOLOGICAL OXYGEN DEMAND.

**oxygen method** An estimation of *gross primary productivity by monitoring rates of production of oxygen, the by-product of *photosynthesis. In field practice this principle is most often used for productivity measurement in aquatic ecosystems. Paired 'light and dark' bottles are filled with water from the required sample depth and suspended at this level for several hours. The dark bottle records oxygen uptake by respiration of phytoplankton and other micro-organisms during the measurement period. Changes in oxygen levels in the light bottle reflect production in photosynthesis as well as loss in respiration. Assuming that respiration rates in the paired bottles are similar, gross primary productivity may be estimated by adding oxygen loss in the dark bottle to the change recorded in the light bottle. The light bottle alone records net (*ecosystem) productivity. *Net primary productivity estimates are possible only in dense phytoplanktonic communities, e.g. *algal blooms in which the oxygen loss resulting from respiration of other micro-organisms is proportionately very small. An important limitation and source of error in the application of the oxygen method to macrophytes arises from the internal storage and utilization of oxygen produced by photosynthesis.

**oxygen quotient (QO$_2$)** The volume, in microlitres, of oxygen taken up at normal temperature and pressure in 1 hour by 1 milligram of plant *tissue.

**oxygen sag** The characteristic fall and recovery in the curve of dissolved oxygen percentage saturation levels in streams, rivers, etc. downstream from a major *pollution source such as a sewage works, a brewery, or a paper-mill.

*Oxylobium* **(family *Fabaceae, subfamily *Papilionatae)** A genus of small, xerophytic (*see* XEROPHYTE) shrubs in which the leaves are *pinnate or *simple, usually alternate, and *stipulate. The flowers are irregular, coloured red or yellow, and held in terminal or *axillary *racemes. The *calyx lobes overlap; often the upper 2 are broader. The *petals are often clawed at the base, with a heel and wings all of the same length. There are 10 free *stamens. There are 24 species, all endemic (*see* ENDEMISM) to Australia and found throughout the continent in dry scrub and heathland.

*Oxyrrhis* *See* DINOPHYTA.

**oyster fungus** The edible *fruit body of *Pleurotus ostreatus*. The *pileus is more or less shell-shaped, and variable in colour from bluish-grey to yellowish-brown. The whitish *gills are strongly *decurrent, anastomosing near the base. The *fungus is typically found in clusters on trunks, tree-stumps, etc., and is common throughout the year.

**oyster thief** The common name for the brown seaweed *Colpomenia peregrina* (order *Scytosiphonales). The *thallus is a thin-walled, hollow sphere up to 20 cm in diameter; it is yellowish-green to olive-brown, and smooth but not shiny. It grows attached to seaweeds or shells on the middle to lower shore. The common name derives from the fact that at low tides the spheres tend to fill with air and act as floats, raising from the sea-bed any oysters that are attached to them.

**P**  *See* PHOSPHORUS.

**P**  *See* PARENTAL GENERATION.

**Pₙ (NPP)** Net primary productivity. *See* PRIMARY PRODUCTIVITY.

**P₆₈₀** The pigment that acts as the energy trap for *photosystem II in *chloroplasts. It is thought to be a form of *chlorophyll *a* which absorbs light most strongly at a wavelength of 680 nm.

**P₇₀₀** The pigment that acts as the energy trap of *photosystem I in *chloroplasts. It is thought to be *chlorophyll *a* in a special structural environment. It represents about 0.3 per cent of the total chlorophyll and absorbs light, most strongly at a wavelength of 700 nm.

**pachycaul** Having a thick stem.

**pachytene** *See* MEIOSIS and PROPHASE.

**Pacific Coast forest** The densest and most majestic coniferous forest in the world, extending from the north of California to the south of British Columbia. It is famed for its giant trees, the best known being *Sequoia sempervirens* (coastal redwood) and *Pseudotsuga taxifolia* (Douglas fir). All may exceed 100 m in height and even 110 m in the case of the big tree, and the girths of certain redwoods are often greater than 20 m.

**Pacific North American floral region** Part of R. Good's (*The Geography of the Flowering Plants*, 1974) boreal kingdom, distinguished by about 300 endemic genera (*see* ENDEMISM). Although the flora as a whole has contributed few economic plants, it has furnished numerous garden species. Like all floras of *Mediterranean-type climates, the Californian part of this region shows a marked degree of endemism, with 30–50 per cent of its species restricted to the state. *See also* FLORAL PROVINCE and FLORISTIC REGION.

**padang** A tropical heath-forest of south-eastern Asia, growing on *podzolic soils severely depleted of nutrient. It is a type of *keranga.

**padauk** *See* PTEROCARPUS.

**paddy** *See* ORYZA.

**Paeonia** (peony; family *Ranunculaceae) A genus of tall *herbs or shrubs that have large, terminal, usually solitary flowers with 5 *sepals, 5–13 large free *petals, many *stamens, and 2–5 free *carpels with fleshy walls, mounted on a fleshy *disc. The fruit is a group of *follicles containing several seeds. The *ovules have peculiar outer integuments and the basic *chromosome number is 5, features not seen in other genera of the family. There are 33 species, occurring in Eurasia and in western N. America.

**pagoda tree** (frangipani) *See* PLUMERIA.

**palaeo- (paleo-)** A prefix meaning 'of ancient times', derived from the Greek *palaios*, meaning 'ancient'.

**palaeobotany** The study of fossil plants.

**Palaeocene (Paleocene)** The lowest epoch of the *Tertiary Period, about 65.5–55.8 Ma ago. The name is derived from the Greek *palaios* 'ancient', *eos* 'dawn', and *kainos* 'new', and means 'the old part of the *Eocene' (the subsequent epoch).

**palaeoecology** The application or ecological concepts to fossil and sedimentary evidence to study the interactions of Earth surface, atmosphere, and *biosphere in former (prehistoric) times.

**palaeoendemic** *See* ENDEMISM.

**Palaeogene** The earliest of the three periods comprising the *Cenozoic Era, preceded by the *Cretaceous, followed by the *Neogene, and dated at 65.5–23.03 Ma ago. The Palaeogene is divided into the *Palaeocene, *Eocene, and *Oligocene Epochs.

**palaeolimnology** The study of the history and development of freshwater ecosystems, especially lakes.

**palaeontology** The study of fossil flora and fauna. Information thus gained may be used to establish the characteristics of ancient environments.

**palaeosol (paleosol)** A soil formed during an earlier period of *pedogenesis, and which may have been buried, buried and exhumed, or continuously present on the landscape until the current period of pedogenesis.

**palaeospecies** A group of biological organisms, known only from fossils, which differs in some respect from all other groups.

***Palaeotaxus rediviva*** The first recorded species of the *Taxaceae (yew) family, found in the Lower *Jurassic of Europe, and distinguished by the form of the leaf and the structure of the *cuticle.

**Palaeotropical floral kingdom (realm)** A region whose geographical extent varies with the authority consulted. R. Good's (*The Geography of the Flowering Plants*, 1974) definition embraces the African (excluding the Cape), Indo-Malaysian, and *Polynesian subkingdoms. Other authorities adopt a more restricted definition, embracing most of Africa, Arabia, and the north-west part of the Indian subcontinent. *See also* FLORAL PROVINCE and FLORISTIC REGION.

**Palaeozoic (Paleozoic)** The first of the 3 eras of the *Phanerozoic, about 542–251 Ma ago. The *Cambrian, *Ordovician, and *Silurian Periods together form the Lower Palaeozoic Sub-Era; the *Devonian, *Carboniferous, and *Permian the Upper Palaeozoic Sub-Era. It was an era of great evolutionary change among plants, which began to invade the land at its beginning. By the end of the era, amphibians and reptiles were major components of various communities and giant tree-ferns, horsetails (*Calamitaceae), and cycads (*Cycadaceae) gave rise to extensive forests. The name is derived from *palaeo- and the Greek *zoe*, meaning 'life'.

***Palaquium*** (family *Sapotaceae) A genus of medium to large rain-forest trees which yield moderate to very hard, sometimes *siliceous, timber (bitis or nyatoh).

*P. gutta* provides the rubber-like *gutta percha. There are over 115 species, occurring from Taiwan to Malesia and the Solomon Islands.

**pale** A boundary. Originally, a deer-proof fence erected around the perimeter of a park.

**palea** In the *inflorescence of *Gramineae, the upper of the two *bracts beneath each *floret. *Compare* LEMMA.

**paleo-** *See* PALAEO-.

**Paleocene** *See* PALAEOCENE.

**palisade mesophyll** *Chlorenchyma tissue, comprising tightly packed, columnar cells, each containing many *chloroplasts, in a leaf. In *mesophytes it is found together with *spongy mesophyll and is usually on the upper (adaxial) side of the leaf. In many *xerophytes it is found on both sides of the leaf and often forms the bulk of the *mesophyll.

**Palmae** *See* ARECACEAE.

**palmate (digitate)** Of leaves, *compound with 2 or more *leaflets arising from the top of a stalk or *rachis and spreading, like the fingers of a hand. The main nerves of a leaf may be palmate, a situation in which several more or less equally large ones diverge along the blade from an origin at the top of the *petiole. Such leaves commonly have a divided margin (i.e. are palmately lobed).

**palmately lobed** Applied to a leaf that is divided like the fingers of a hand. *See* PALMATE.

**palmelloid** Applied to an *algal colony composed of an indefinite number of non-*motile cells embedded in a gelatinous or mucilaginous matrix. It may occur as a non-motile phase in a normally *motile species.

**palmetto palm** *See* SABAL.

**palm wine** *See* RAPHIA.

**Palmyra palm** *See* BORASSUS.

**Palouse prairie (bunch grass prairie)** A distinctive tract of *prairie in the Palouse region of south-eastern Washington (USA), but similar prairie also extends into Oregon, southern Idaho, and northern Utah. A rich growth of bunch grasses, especially *Agropyron spicatum* (blue-bunch wheatgrass), occurs on the deep, highly fertile soils formed in aeolian parent materials. Much of the Palouse prairie is now cultivated for wheat.

**palsa** A mound or ridge, largely made from peat, that contains a perennial ice lens and is found in the damper sites of mires in periglacial areas. Widths are in the range 10–30 m, lengths 15–150 m, and heights 1–7 m. Probably palsas are a result of heaving associated with the growth of segregated ice.

**palsa mire** Peaty *tundra, in which ground ice is partly responsible for the formation of peat hummocks, called palsas or palsa mounds, up to 35 m long, 15 m wide, and 7 m high. Palsas have a core of frozen peat and develop cyclically, but the remainder of the mire remains clear of ice in summer. The palsa surface develops a cover of *lichen. This gives it a high albedo, preventing the interior from warming in summer and eventually allowing a block of permafrost to form. As the surface becomes unstable it erodes, removing the lichen and revealing the underlying peat, which is dark and reduces the albedo, causing the peat to absorb more heat. Pools form as the palsa warms and its core of ice melts, and the sequence begins again. Palsa mires are circumpolar in distribution. They are forming and decaying constantly, but there is some evidence that they have formed more rapidly at certain periods in the past, possibly due to climatic differences.

**palustrine** Inhabiting boggy ground.

**palynology (pollen analysis)** The study of *pollen grains and *spores, and especially their use in reconstructing the vegetational history of an area. The outer coat, or exine, of a pollen grain or spore is very characteristic for a given family, genus, or sometimes even species. It is also very resistant to decay, particularly under anaerobic conditions. Thus, virtually all spores and pollen falling on rapidly accumulating sediment, anaerobic water, or *peat are preserved. Since both pollen and spores are generally widely and easily dispersed, they give a better picture of the sur-

rounding regional vegetation at the time of deposition than do macroscopic plant remains, e.g. fruits and seeds, which tend to reflect only the vegetation of the immediate locality. With careful interpretation, palynology enables examination of climatic change, human influence on vegetation, and sediment dating, as well as direct study of vegetation character. The technique has also been applied, more controversially, to the pollen and spore contents of modern and fossil soil profiles. Studies of contemporary pollen and spores are useful in medicine (e.g. in allergy studies and patterns of disease spread), in commerce (e.g. for the examination and quality control of honey), in agriculture (e.g. for plant and animal disease control), and even in forensic science.

**pamirs** *See* PUNAS.

**pampas** Temperate S. American grasslands, the largest continuous area being in the eastern Argentinian province of Buenos Aires, together with parts of the adjacent provinces. Much of the moister sector of the pampas is now cultivated, but formerly it was covered by bunch grasses, in between the individual tussocks of which the soil lay bare. To the west and south-west drier conditions prevail, and here the pampas comprises a mixture of short grasses and *xerophytic shrubs.

**Pampas floral region** Part of R. Good's (*The Geography of the Flowering Plants*, 1974) *neotropical kingdom: it lies east of the Andes, and between the tropical flora of Brazil and the temperate flora of Patagonia. The flora is relatively poor, with only about 50 endemic genera (*see* ENDEMISM), all of them small. In terms of vegetation the region is dominated by the pampas. Few garden plants and no very important economic plants come from this region. *See also* FLORAL PROVINCE and FLORISTIC REGION.

**pan** A soil *horizon, usually in the subsoil, that is strongly compacted, indurated (*see* INDURATION), cemented, or very high in *clay content.

***Panaeolus* (family *Strophariaceae)** A genus of *fungi in which the *fruit bodies are *mushroom-like. The *gills are greyish, becoming black or mottled with black at maturity. The *spores are large, and appear smooth and dark brown under the microscope. Species are found on the ground, on dung, on rotting wood, etc.

**Panama-hat palm** *See* CARLUDOVICIA.

**Panax** (family *Araliaceae) A genus of 6 species of tropical, east Asian, and N. American herbs, the roots of several species of which provide the oriental tonic ginseng.

**panbiogeography** A term coined by L. Croizat to describe a new synthesis of the sciences of plant and animal distribution. The main features are that consistencies recur in distribution patterns and that analysis of these produces 'tracks' (joining areas of common floras and faunas) and 'nodes' (where different tracks meet). These ideas form the basis of the new school of *vicariance biogeography.

**Pandaceae** A small family of trees in which the leaves are *simple. The flowers are unisexual, small, regular, and *pentamerous. The fruit is usually a flattened *drupe. The family is related to *Euphorbiaceae. There are 4 genera, with 18 species, occurring in tropical America and Asia.

**Pandanaceae** A family of monocotyledonous (*see* MONOCOTYLEDON) trees and climbers in which the leaves are strap-like, leathery, with a spiny margin, in 3 or 4 conspicuous spirals. The flowers are small, unisexual, and held in heads subtended by a large, often coloured *bract (not *Sararanga*). The fruit is a *berry or *drupe, sometimes woody, united with the *perianth into a globose or cylindrical head. There are 3 genera, with 675 species, confined to the Old World tropics.

**Pandanus** (screw pines; family *Pandanaceae) A genus of trees which usually have *stilt roots and open, falsely *dichotomously branched crowns. The leaves are in 3 spiral ranks. The fruit heads are often reddish. Some fruits are edible (especially those occurring in New Guinea and Micronesia). Leaves are used for matting. There are about 600 species.

**Pandorea** (Australian bower plant; family *Bignoniaceae) A genus of *lianes, climbing shrubs, and woody plants, which

have *decussate, *compound leaves without *stipules. The flowers are brightly coloured and held in *cymes with a tufted, 5-lobed *calyx and a bell-shaped, lobed *corolla. The flat, winged seeds develop in a *capsule. Several species are grown as ornamentals. There are 6 species, occurring mainly in rain forests in central and eastern Australia, eastern Malesia, and New Caledonia.

**Pandorina** (division *Chlorophyta) A genus of green *algae in which the cells fit closely together to form *motile, spherical colonies (*coenobia) of 4–32 cells. They are common in puddles, ponds, lakes, rivers, etc.

**pandurate** Fiddle-shaped.

**Panellus** (family *Tricholomataceae) A genus of *fungi in which the *fruit body has a lateral or rudimentary *stipe, or no stipe at all. The spore print is white, and the spores stain dark violet when treated with iodine. They are found on wood.

**pan-endemic distribution** *See* COSMOPOLITAN DISTRIBUTION.

**Pangaea** A single supercontinent which came into being in late *Permian times and persisted for about 40 million years before it began to break up at the end of the *Triassic Period. It was surrounded by the universal ocean of *Panthalassa.

**Pangium** (family *Flacourtiaceae) A genus of trees in which the leaves have *palmate nervation. The flowers are unisexual. The fruits are globose, large, woody, and contain numerous large seeds, embedded in pulp. The plants are poisonous in all their parts, but the seeds are edible after leaching, and contain an edible oil. The dried fruits are used as rattles in New Guinea. There is 1 species, *P. edule*, which occurs in rain forests from Malesia to the Pacific Islands.

**panic grass** *See* PANICUM.

**panicle** Strictly, an *inflorescence that is a compound *raceme, i.e. one that comprises several *racemose parts. More loosely, the term is applied to any complex, branched inflorescence.

**Panicum** (panic grass, crab grass, millet; family *Poaceae) An important genus of tropical grasses, some of which reach

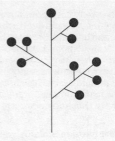

*Panicle*

large size. The genus includes fodder, grain, and ornamental species. There are about 600 species, occurring in tropical and warm temperate climates.

**pannage** The common of *mast: (a) the right to allow pigs to feed on common land where the land comprises oak or beech woodland; or (b) money paid for permission to allow pigs so to feed.

**Pannaria** (order *Lecanorales) A genus of *lichens in which the *thallus is mainly *squamulose. The *phycobiont is a *cyanobacterium (*Nostoc). Species are found in moist regions among mosses on bark, rocks, etc.

**pansy** See *VIOLA* and VIOLACEAE.

**Panthalassa** The name given to the vast oceanic area that surrounded *Pangaea when that supercontinent was in existence. *Tethys was a minor arm of this ocean. Once Pangaea began to split in the *Triassic, the names of all the modern oceans are normally applied to the developing ocean basins even though they were, at that time, still very small.

**panther cap** The common name for the poisonous *fruit body of *Amanita pantherina*, which is *mushroom-shaped. The cap is brown or olive-brown, with white patches adhering to the surface; the *gills are white. It is found on the ground under broad-leaved trees.

**pantothenic acid** Vitamin $B_3$, synthesized by green plants and micro-organisms, but not by animals, for which it is an essential dietary requirement. It forms part of the structure of the key metabolic compound *coenzyme A.

**pantropical distribution** The distribution pattern of organisms that occur more or less throughout the tropics. The plants in this category are mainly weedy, herbaceous species, but there are a few woody types, e.g. *Arecaceae (palms).

**Papaveraceae (poppies)** A family of herbs (or, rarely, shrubs) that have alternate, usually deeply lobed leaves; milky latex occurs in canals in all parts of the plants. The flowers are regular, commonly with 2 *sepals that soon fall, and usually with 4 overlapping, crinkly *petals. The *stamens are free and numerous, the *ovary *superior, of 2 to many fused *carpels and producing a *capsule that opens by valves or pores when ripe. Many species and genera are cultivated for their brilliant flowers; *Papaver somniferum* (opium poppy) is the source of opium and its derivative, heroin. There are 23 genera, with 210 species, occurring mostly in northern temperate regions.

**papaya** See *CARICA* and DISJUNCT DISTRIBUTION.

**paper mulberry** See *BROUSSONETIA*.

**paper-reed** See *CYPERUS*.

**Papilionatae (family *Fabaceae)** One of the 3 subfamilies of Fabaceae, comprising plants in which the leaves are *pinnate, *trifoliate, or *simple. The flowers are usually *zygomorphic, the *sepals and *petals *imbricate, the lowest petal innermost in the bud, and there are usually 10 *stamens. There are 437 genera and about 11 300 species, including many temperate, herbaceous plants.

**papilla** A small, blunt projection.

**pappus** A tuft of hairs or bristles, derived from the *calyx, that terminates the *fruit of many *Compositae and *Valerianaceae.

**paprika** See *CAPSICUM*.

**papyrus** See *CYPERUS*.

**paradigm** Essentially, a large-scale and generalized *model that provides a viewpoint from which the real world may be investigated. It differs from most other models, which are abstractions based on data derived from the real world.

**parakeela** *See* CALANDRINIA.

**parallel evolution** A similar evolutionary development that occurs in lineages of common ancestry. Thus, the descendants are as alike as were their ancestors. The nature of the ancestry imposes or directly influences the development of the parallelism.

**parametric test** *See* STATISTICAL METHOD.

**paramo** A humid Arctic-alpine meadow and *scrub in which mosses and lichens are common. It occurs in the Andes, situated between the tree-line and the snow-line, and deriving much moisture from cloud or mist.

**paramylum** A starch-like polysaccharide found as a storage product in *Euglenophyta.

**paraña pine** *See* ARAUCARIA.

**parapatric** Applied to species whose *habitats are separate but adjoining. *Compare* ALLOPATRIC and SYMPATRIC.

**parapatric speciation** A speciation that occurs regardless of minor *gene flow between *demes. In many species *selective pressures are sufficiently strong on the whole to prevent homogenization of the immigrant genes by interbreeding.

**paraphyletic** Of a *taxon, including some but not all descendants of the common ancestor, i.e. not *holophyletic.

**paraphysis** A sterile hair found among the reproductive structures of many *Fungi, *Bryophyta, and *algae. Paraphyses provide protection and in some species may be involved in dehiscence (*see* DEHISCENT).

**para rubber** *See* HEVEA.

**Paraserianthos** (family *Fabaceae, subfamily *Mimosoideae) A small genus, comprising 4 species of fast-growing, rainforest trees. *P.* (formerly *Albizia*) *fulcataria* is widely planted for ornamental shade and timber and is the world's fastest-growing tree. The genus occurs in the Moluccas, New Guinea, and the Solomon Islands.

**parasexual cycle** A type of *recombination, found in certain heterokaryotic fungi, that is based on *mitosis rather than *meiosis. Genetically distinct *haploid nuclei fuse in the *heterokaryon. The resulting *diploid nuclei multiply by mitotic division, with some *crossing over, and a diploid *homokaryon develops from a diploid *conidium. *Chromosomes may be shed by the diploid nucleus to produce a *haploid nucleus.

**Parashorea** (white seraya, white lauan; family *Dipterocarpaceae) A genus of 14 giant timber trees which occur from southern China to the Philippines.

**parasite** *See* PARASITISM.

**parasitism** The interaction of species populations in which one, typically small, organism (the parasite) lives in or on another (the host), from which it obtains food, shelter, or other requirements. Unlike a predator, a parasite does not have to kill its host in order to obtain food. Parasitism usually implies that some harm is done to the host, but this interpretation must be qualified. Effects on the host range from almost none to severe illness and eventual death, but even where such obvious immediate harm accrues to the individual host it does not follow that the relationship is harmful to the host species in the long term or in an evolutionary context (for example, it might favour beneficial adaptation in the host species population). Obligate parasites can live only parasitically. Facultative parasites may live as parasites or as independent saprophytes. Ectoparasites live externally on the host. Endoparasites live inside the body of the host. A partial parasite is a plant that carries out photosynthesis, but relies on its host for other nutrients and needs, e.g., support. *Compare* COMMENSALISM; MUTUALISM; and NEUTRALISM.

**parasol mushroom** The common name for the edible *fruit body of *Lepiota procera*. *L. rhacodes* is known as the shaggy parasol. Both species have a scaly cap, and an annulus (ring) that is loose and can be moved up and down the *stipe.

**parasol pine** *See* SCIADOPITYS.

**paratype** A specimen, other than a *type specimen, that is used by an author at the time of the original description, and designated as such by the author.

**parchment bark** *See* PITTOSPORUM.

**parenchyma** In plants, *tissue composed of the least specialized of plant *cells with a system of air spaces running between them. Parenchyma cells are regarded as the basic cells from which all other cell types have evolved; they form a large part of the bulk of many organs.

**parental generation** (*P*) The generation comprising the immediate parents of the *$F_1$ generation. The symbols $P_2$ *and* $P_3$ may be used to designate grandparental and great-grandparental generations respectively.

**parent material** The original material from which the soil profile has developed through *pedogenesis, usually to be found at the base of the profile as weathered but otherwise unaltered mineral or organic material.

**parietal** *See* PLACENTATION.

**park** **1.** In ancient farming systems, the enclosed fields lying between the inner fields next to the farm buildings, and the larger, outer fields used only seasonally for pasture. **2.** Enclosed land on which deer are or were kept. **3.** Land, usually wood-pasture, enclosed by a pale, and intended for the keeping of deer. **4.** In modern use: (*a*) an area of land set aside for public enjoyment and designed to resemble semi-natural land; (*b*) an enclosure for semi-wild animals.

**Parkia** (family *Fabaceae, subfamily *Mimosoideae) A genus of trees that usually have *bipinnate leaves. The flowers are in globular heads on long stalks held free from leaves, and are mostly nocturnal and bat-pollinated. Pods or seeds of some are eaten as a relish (petai). There are about 40 species, with a pantropical distribution.

**park woodland** A woodland in which there is an open canopy of mature trees over pasture.

**Parmelia** (order *Lecanorales) A large genus of *lichens in which the *thallus is *foliose and is frequently attached to its substrate by means of *rhizines on the lower surface. The *phycobiont is a *green alga. The *apothecia are *lecanorine. *Spores are colourless and *aseptate. *Parmelia* species are found on trees, rocks, etc.; some are very common.

**Parnassiaceae** *See* SAXIFRAGACEAE.

**Parrotia** (family *Hamamelidaceae) A genus of handsome, slow-growing, small, ornamental trees or shrubs which have fine autumn colouring. The flowers are held in clusters subtended by white *bracts. There is 1 species, *P. persica*, which is confined to the warm temperate forests of Iran.

**parsnip** *See* PASTINACA.

**Parsonsia** (family *Apocynaceae) A genus of climbing, woody plants that have *simple leaves, held in *whorls or *opposite. The sap is a milky latex. The regular, often showy and fragrant flowers are held in a *cyme. The *calyx is fused into a tube. The *corolla is also a tube but has 5 lobes. The 5 *stamens are also fused. The fruits are held in pairs. There are about 80 species, found in rain forests in South-east Asia, Australia, and New Zealand.

**parthenocarpy** The development of an *ovule to a fruit without fertilization.

**Parthenocissus** (Virginia creeper; family *Vitaceae) A genus of deciduous climbing shrubs that cling by means of branched tendrils with sucker-like tips. They have palmately *compound leaves. The tiny flowers have 5 *petals and grow in clusters. There are 10 species, of which *P. tricuspidata* from the Far East and *P. quinquefolia* from N. America are grown as ornamentals for their attractive red autumn foliage.

**parthenogenesis** The development of an individual from an egg without fertilization; it occurs in some plants (e.g. dandelion). *See* APOMIXIS.

**parthenospore** In *algae, a female *gamete that develops into a *spore without *fertilization.

**partial dominance** In genetics, incomplete or semi-dominance; the production of an intermediate *phenotype in individuals *heterozygous for the *gene concerned. Partial dominance is generally considered to be a type of incomplete dominance, with the heterozygote resembling one *homozygote more than the other. For example, in *Mirabilis jalapa* (the marvel of Peru or four-o'clock plant), individuals with red petals bred in a pure line when crossed with a pure line with white petals will yield individuals with pink petals in the first filial generation. The red phenotype (and its determining *allele) is said to be incompletely dominant over the white phenotype (and its allele). The impact of a gene at the level of the phenotype depends on its dominance relations, but also on the conditions of the rest of the *genome and on the conditions of the environment. *See also* PENETRANCE.

**partially permeable** *See* SEMI-PERMEABLE.

**partial parasite** *See* PARASITISM.

**partial random sample** *See* STRATIFIED RANDOM SAMPLE.

**partial rosette plant** *See* HEMICRYPTOPHYTE.

**partial veil** A membranous structure connecting the edge of the *pileus to the *stipe in the immature *fruit bodies of certain *agarics.

**particle density** The mass per unit volume of soil particles, usually expressed in grams per cubic centimetre. *Compare* BULK DENSITY.

**partridge cane** *See* RHAPIS.

**partridge wood** Oak wood that has been subjected to a *pocket rot caused by *Stereum frustulatum.

**Passifloraceae** (passion flowers, grandillas) A family of dicotyledonous (*see* DICOTYLEDON) shrubs, *herbs, or (mostly) climbers with tendrils which arise from the leaf *axils and correspond to sterile flower stalks. They have entire or lobed leaves, with small *stipules, and glandular *petioles. The flowers are regular and usually bisexual. There are 5 overlapping, persistent *sepals, sometimes basally fused. The 5 *petals are

rarely absent, and are overlapping and often basally fused. There may be a *corona of petal-like or stamen-like growths inside the *corolla. There are 5 or more *stamens held opposite the petals, sometimes in bundles. The *ovary is *superior and *unilocular, with many *ovules. The *styles are united, with 3–5 *stigmas. The fruit is a *berry or *capsule, and is *indehiscent, containing many seeds, each with a large *embryo, fleshy *endosperm, and surrounded by a pulpy *aril. Several species are cultivated either for their showy flowers (e.g. *Passiflora caerulea*, which is an indoor (or delicate outdoor) climber in Britain) or for the edible fruits (e.g. *P. quadrangularis*, granadilla, which is cultivated in the tropics). *P. edulis* is the passion fruit or purple granadilla of Brazil, which is widely cultivated. There are 18 genera, with about 530 species, found in all tropical regions but especially in America and Africa.

*Passiflora edulis* (passion fruit, purple granadilla) *See* PASSIFLORACEAE.

**passion flowers** *See* PASSIFLORACEAE.

**passion fruit** *See* PASSIFLORACEAE.

**passive chamaephyte** *See* CHAMAEPHYTE.

**passive dispersal** *See* DISPERSAL.

*Pasteurella* (family *Pasteurellaceae) A genus of *Gram-negative *bacteria in which the cells are ovoid or rod-shaped. They are *chemo-organotrophic, and found as *parasites in animals, including humans, and birds; they can cause disease. (The causal agent of plague, formerly called *Pasteurella pestis*, is now regarded as a species of *Yersinia.)

**Pasteurellaceae** A family of *Gram-negative *bacteria in which the cells are rod-shaped or ovoid, sometimes *pleomorphic and filamentous, and non-*motile. They are facultatively *anaerobic. Metabolism is *chemo-organotrophic and the organisms are nutritionally fastidious. All members are *parasitic or *pathogenic in humans, other animals, and/or birds and some can cause disease.

**Pastinaca** (parsnip; family *Apiaceae) A genus of plants that have *pinnate leaves, *compound *umbels, and either no *bracts or *bracteoles, or 1 or 2 that soon fall. The fruits are ovoid and dorsally compressed, with winged margins. There are 14 species, in temperate Eurasia and N. Africa. Cultivated forms have been developed with succulent, edible tap-roots.

**past-natural** *See* NATURAL.

**Patagonian floral region** Part of R. Good's (*The Geography of the Flowering Plants*, 1974) Antarctic kingdom: it has a small flora with a modest number of endemic genera (*see* ENDEMISM). It comprises the lowland regions of the extremities of S. America. The Falkland Islands (Malvinas) are also included, but despite their isolation they have no well-known endemic genera. There are clear relationships with the flora of New Zealand. *See also* FLORAL PROVINCE and FLORISTIC REGION.

**Patersonia** (family *Iridaceae) A genus of lily-like *herbs found in arid conditions, usually as a component of heath or grass communities. There are 13 species, 12 of which are endemic (*see* ENDEMISM) to Australia and Tasmania; the other also occurs in Borneo and New Guinea.

**pattern analysis** Any analysis for the detection of a non-random distribution of organisms, e.g. *nearest-neighbour analysis, or classification techniques. However, the term is applied more particularly to the detection of patterns in the distribution of individuals of a given plant species in a community, by comparing the observed number of individuals per quadrat with the expected number derived from the Poisson series (based on a population that is dispersed at random). The correspondence is tested for statistical significance either by comparing observed numbers with those predicted using the $\chi^2$ test, or by comparing the variance : mean ratio for the data with the value for the Poisson series of unity. Where this ratio is significantly greater or less than unity, the population is respectively overdispersed (contagious) or underdispersed (regular). Though first applied to studies of plant ecology, the technique is now widely used in other fields.

**Paulownia** (family *Plantaginaceae) A genus of small trees which have fox-glove-like flowers, and are widely grown as ornamentals or fast-growing timber trees. Sucker shoots bear enormous leaves. There are 6 species, occurring in eastern Asia.

**Paurotis** *See* ACOELORRAPHE.

**PAW** *See* PLANT AVAILABLE WATER.

**pawpaw** *Asimina triloba*. *See* ANNONACEAE and DISJUNCT DISTRIBUTION.

**Paxillus** (order *Boletales) A genus of *fungi in which the *spores are borne on *gills and in masses appear yellowish-brown to olive-brown. *P. involutus* is *mushroom-shaped, with a brown *pileus and *stipe; the rim of the cap is characteristically inrolled. This species, which is poisonous, is very common in broad-leaved woods and on acid *heathland.

**Payena** (family *Sapotaceae) A genus of small to large forest trees in which the leaves are alternate. They yield hard, heavy, sometimes *siliceous timber (nyatoh). There are 16 species, occurring from Burma to the Philippines.

**PE** *See* POTENTIAL EVAPOTRANSPIRATION.

**pea** **1.** (garden pea) *See* PISUM. **2.** (wild pea, sweet pea) *See* LATHYRUS. **3.** (Sturt's desert pea) *See* CLIANTHUS.

**peach** *See* AMYGDALUS.

**peach leaf curl** A disease that affects *Prunus* species, including peach, nectarine, almond, and (less often) apricot trees. The characteristic symptoms are the curling, thickening, and distortion of the leaves which typically become bright red. The disease is caused by a *fungus, *Taphrina deformans*.

**peanut** *See* ARACHIS.

**pear** *See* PYRUS.

**pear moss** The common name for the moss *Physocomitrum pyriforme* (order *Funariales). The name derives from the characteristic pear-shaped *capsules formed in spring; the shoots are small (to 5 mm tall) and inconspicuous. Pear moss is found on damp ground and on mud, on roadsides, by ditches, on river banks, etc.

**peat** An organic soil (*Histosol) in which the O *horizon is at least 40 cm thick (and often much deeper) and the dry weight contains a minimum of 65% organic matter, the remainder being of mineral material. *Ombrotrophic peat, which has developed in situations where the sole water supply is from rainfall, generally has a much higher proportion of organic matter, often as high as 99% dry weight. Peat formation occurs when decomposition of plant material is slow owing to the *anaerobic conditions associated with waterlogging. Decomposition of *cellulose and *hemicellulose is particularly slow for *Sphagnum plants (bog mosses), which are characteristic of such sites and hence among the principal peat-forming plants. In addition to Sphagnum the plant material consists mainly of moor grasses and heather. The extent of decomposition increases with depth, so there is a progressive transition from fibrous and often identifiable plant residues near the surface to highly humified (see HUMIFICATION) material lower down. *Fen and *bog peats differ considerably. In fen peats the presence of calcium in the *ground water neutralizes the acidity, often leading to the disappearance of plant structure, giving a black, structureless peat. Bog peats, formed in much more acidic waters, vary according to the main plants involved. It often remains possible to identify animal and plant species for a considerable time. Recent Sphagnum peat is light in colour, with the structure of the mosses perfectly preserved.

**peat-borer** An implement designed to extract peat cores with the minimum of disturbance. The most familiar is the Hiller peat-borer, which consists of a short screw auger head to ease penetration of the peat, backed by a chamber which can be opened and closed at the required sample depth, the sharp cutting edge of the chamber assisting detachment of the sample in more consolidated peats. The principal alternatives are the piston sampler, which is particularly good for loose peats, and the Russian borer, which allows easier removal of the complete peat core than is possible with the Hiller borer, but is more difficult to use in compacted peats since it has no screw auger head.

**peat moss** The common name for a moss of the genus *Sphagnum, common on wet *peat.

**peat podzol** A *podzol soil *profile distinguished by having a surface *mor (peaty) *humus up to a maximum thickness of 30 cm, and usually with an iron pan at the top of the B *horizons. The term occurs in most of the classification systems derived originally from the work of V. V. Dokuchaev, published in 1886. It has been superseded by the *USDA Soil Taxonomy, where podzols fall within the order *Spodosols. See SOIL CLASSIFICATION.

**pecan** See CARYA.

**pectin** One of a group of homopolysaccharides that contain a variety of monosaccharides, but are especially rich in galacturonic acid. They form a kind of cement, so contributing to the structure of plant *cell walls, being particularly abundant in young primary walls and fruits.

**pectinate** Resembling a comb in arrangement or shape.

**pectinolytic** Capable of digesting *pectin.

**ped** A unit of soil structure (e.g. an aggregate, crumb, granule, or prism) that is formed naturally. Compare CLOD.

**Pedaliaceae** (sesames) A family of *annual or *perennial *herbs and some shrubs, which have *opposite, *simple or lobed leaves, without *stipules but often with glandular hairs. The flowers are either solitary, or in small, *axillary *cymes, usually 3-flowered. They are irregular and bisexual, with a united *calyx and a tubular *corolla, each composed of 5 parts. The 4 *stamens are paired, and attached to the *petals. The *ovary is *superior, with 2 fused *carpels and a long *style with 2 *stigmas. There are 2–4 *locules, with several *ovules. The fruit is a *nut or *capsule, often hooked on the surface. The seeds contain little *endosperm. Some species of the genus Sesamum provide edible seeds and oil (sesame). S. indicum is widely cultivated in India for its seeds. There are 18 genera, with 95 species, found in arid and saline habitats, especially coasts, in Africa. Madagascar, southern India, the Malay archipelago, and Australia.

**pedicel**   The stalk of one *flower in an *inflorescence.

**pedogenesis**   The natural process of soil formation, including a variety of subsidiary processes such as humification, weathering, *leaching, and *calcification.

**pedology**   The scientific discipline that is devoted to the study of the composition, distribution, and formation of soils, as they occur naturally.

**pedon**   A 3-dimensional sampling unit of soil, with depth to the parent material and lateral dimensions great enough to allow the study of all *horizon shapes and intergrades below the surface.

**peduncle**   The *inflorescence stalk of a plant.

**pedunculate**   Possessing or pertaining to a *peduncle.

***Pelargonium***   See GERANIACEAE.

***Pellia* (order Metzgeriales, class *Hepaticae)**   A genus of *thallose *liverworts, which includes *P. epiphylla*, one of the commonest and most conspicuous of all liverworts, and often studied as a representative of this group of plants in elementary botany courses. The *thallus is robust, several centimetres long and up to 1 cm across. It has an ill-defined *midrib, and is irregularly branched, often spreading to form extensive carpets. Older parts are dark green, while the growing tips are pale green. *Pellia* species are found on moist banks by ditches and streams, also on damp *peat in *heaths, *moors, and on mountains.

**pellicle**   See PERIPLAST.

**peltate**   Shield-shaped; having a central rather than a lateral stalk.

***Peltigera* (order *Peltigerales)**   A genus of *lichens, in which the *thallus is *foliose and often large. The lower surface is loosely downy, and marked with more or less prominent, vein-like ridges. The *phycobiont may be a green *alga or a *cyanobacterium, according to species. *Apothecia are formed on the upper surface at the edges of the thallus. The

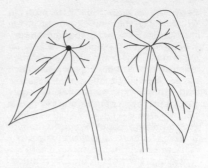

*Peltate*

*spores are *multiseptate, and colourless or pale brown. They are found on the ground or on mosses, trees, rocks, etc. *See also* DOG LICHEN.

**Peltigerales**   **(subdivision *Ascomycotina)**   An order of *lichens, in which the *thallus is *foliose with a *cortex on the upper surface and often also on parts of the lower surface, which is often marked with vein-like ridges, and may or may not bear *rhizines. The *asci turn blue when treated with iodine. *Spores have 2 or more transverse *septa, and are colourless to brown. Species are found on soil, rotting wood, etc.

***Pelvetia* (family *Fucaceae)**   A genus of brown seaweeds, in which the *thallus differs from that of *Fucus* species in having no *midrib, and from that of *Ascophyllum* in having no air *bladders. The only British species is *P. canaliculata* (*see* CHANNELLED WRACK).

***Pemphis* (family *Lythraceae)**   A genus of small trees or shrubs which comprises 2 species. *P. acidula* occurs on abrading coasts from Africa to the Pacific, and *P. madagascariensis* in the wet mountains of Madagascar.

**pencil cedar**   See *JUNIPERUS*.

**penetrance**   The proportion of individuals of a specified *genotype who manifest that genotype in the *phenotype under a defined set of environmental conditions. If all individuals carrying a *lethal *mutation die prematurely, then the *mutant *gene is said to show complete penetrance. An organism may not express the phenotype normally associated with its genotype because of the presence of modifiers, epistatic genes (see EPISTASIS), or suppressors in the rest of the *genome; or because of a modifying effect of the environment. Expressivity describes the degree or extent to which a given genotype is expressed phenotypically in an individual.

**penicillin** A type of antibiotic produced, for example, by fungi of the genus *Penicillium. Penicillins are active against certain types of bacteria (mainly Gram-positive species; see GRAM-REACTION) and are widely used in the treatment of diseases in animals caused by those bacteria. (Pencillin G was one of the first antibiotics to be used for the treatment of disease.) There is now a wide range of chemically modified penicillins, each with slightly different properties. They function by inhibiting the synthesis of bacterial *cell walls.

**Penicillium** (class *Hyphomycetes) A *form genus of *fungi which form a well developed, *septate *mycelium and branched or unbranched *conidiophores bearing *phialids. *Conidia appear as greyish-green or bluish-green powdery masses on white *mycelium. Most species are *saprotrophic and cosmopolitan in distribution. Many strains produce important *antibiotics, including *penicillins; many species form toxins. Some species are used in the manufacture of certain types of cheese.

**pennate diatom** A diatom (*Bacillariophyta) that has bilateral symmetry. Some pennate diatoms are marine, others inhabit fresh water, some are planktonic (see PLANKTON), others live attached to a *substrate. *Compare* CENTRIC DIATOM.

**Pennsylvanian** *See* CARBONIFEROUS.

**penny bun** *See* CEP.

**pentamerous** Having parts in fives. This is the most typical arrangement of the *flowers of *dicotyledons.

**Pentaphylax** (family *Pentaphylacaceae) A genus of small, bushy trees which have small, regular, hermaphrodite, *pentamerous flowers, porous *anthers, and

whose fruit is *a capsule. They are related to *Theaceae. There is 1 species, occurring from southern China to Malaya and Sumatra.

**pentose** A monosaccharide that consists of 5 carbon atoms.

**pentose-phosphate shunt** *See* hexose-monophosphate shunt.

**Pentoxylales** Extinct *gymnosperms of *Jurassic age, known from India and New Zealand, that probably grew as shrubs or very small trees.

**peony** *See PAEONIA.*

**PEP** *See* PHOSPHOENOLPYRUVATE.

**Peperomia** (family *Piperaceae) A genus of fleshy herbs, some of which are climbing or epiphytic (see EPIPHYTE). The leaves are without *stipules. Flowers are borne in *spikes and are hermaphrodite, with no *sepals or *petals and 2 *stamens. The fruit is a *berry. Some species are cultivated for their attractive foliage. There are 1000 species, occurring in the tropics, mainly in America.

**pepper** **1.** The fruit of *Piper nigrum. See* PIPER. **2.** *See CAPSICUM.*

**pepper dulse** The common name for the red seaweed *Laurencia pinnatifida* (family Rhodomelaceae, order *Ceramiales). The seaweed is variable in form and colour. When growing in bright sunlight on the upper shore, it usually grows in short, dense tufts and is yellowish-green in colour; but when growing further towards the low-water mark it is larger and dark reddish-purple. The plant is generally flattened and extensively branched in one plane. It has a strong, distinctive, pungent smell, and a peppery taste. It has been used as a condiment in some areas.

**peptide** A linear molecule that consists of 2 or more *amino acids linked by *peptide bonds.

**peptide bond** A chemical bond that links 2 or more *amino acids by a reaction between *carboxyl and *amino groups. According to the number of amino acids linked by such bonds, the resulting *peptide is designated dipeptide (2 amino acids), tripeptide (3 amino acids), oligopeptide (3–10 amino acids), or polypeptide (10 or more amino acids).

**peptidoglycan** An *amino-acid-containing *polysaccharide which forms the rigid component in the *cell walls of most bacteria and *cyanobacteria. It is never found in *eukaryotic cells.

*Peptococcus* A genus of *Gram-positive *bacteria which are not assigned to any taxonomic family. The cells are spherical and occur singly or in groups of 2 or more. They are non-*motile, *chemoorganotrophic, and can grow only in the absence of air. They are found in the bodies of animals (including humans).

**peptone** A partially hydrolysed *protein that is used as a nutrient in microbiological culture media.

**perambulation 1.** The established boundaries of a forest or parish. **2.** A legal document that defines an area of land by describing its boundaries. **3.** A walk around established boundaries.

**percentage cover** *See* COVER.

**perched aquifer** *See* AQUIFER.

**perched water table** The upper boundary of water that is held above the general *water table by a feature such as a layer of impermeable material beneath the material in which the water is held. For example, water in the upper layer of a *raised bog is held in an elevated position by the impermeable peat beneath. *See* AQUIFER.

**percolation** The downward movement of water through soil, especially through soil that is saturated or close to saturation.

**perennating bud (perennating organ)** The vegetative means whereby *biennial and *perennial plants survive periods of unfavourable conditions. The aerial parts die back to a minimum at the onset of unfavourable conditions, and food for the new shoots of the next growing season is stored in underground organs (e.g. *tubers, *bulbs, *rhizomes), or in *buds on the stems of woody plants. *Seeds may also be considered perennating organs.

**perennating organ** *See* PERENNATING BUD.

**perennial** A plant that normally lives for more than 2 seasons and, after an initial period, produces flowers annually.

*Pereskia* (family *Cactaceae) A genus of *lianes, small trees, or shrubs, which are unusual in the family in possessing true leaves. There are 20 species, occurring in tropical America.

**perfect cycle** *See* BIOGEOCHEMICAL CYCLE.

**perfect state** The sexual state of a *fungus (i.e. the state in which sexually produced *spores are formed).

**perforation plate** The end wall of a *vessel element, with 1 or more openings (perforations) to allow the passage of water and dissolved substances.

**pergelic** The lowest of the soil-temperature classes for family groupings of soils in the *USDA Soil Taxonomy system, applied to soils in temperate regions. The assessment of soil temperature is based on mean annual temperatures, and on differences between mean summer and mean winter temperatures, measured at a depth of 50 cm or at the surface of the underlying rock, whichever is shallower. Higher temperature classes in temperate region soils are called cryic, frigid, mesic, thermic, and hyperthermic, and in tropical regions the scale from cold to hot is isofrigid, isomesic, isothermic, and isohyperthermic.

**perianth 1.** Of a flower, the outer covering, composed of the floral leaves, usually an outer greenish *calyx, and an inner, brightly coloured *corolla. **2.** Of a *liverwort, a tube-like or purse-like sheath which surrounds the *archegonium.

**pericarp** The *fruit wall, often with 3 distinct layers, endocarp, mesocarp, and outer exocarp.

**pericentric inversion** The inversion of a *chromosome piece (containing a block of *genes) that involves the *centromere.

**perichaetium** In *Bryophyta, 1 of the enlarged leaves or bracts that surround the *archegonia and *antheridia, or the structure the perichaetia form. *See also* PERIGONIUM and PERIGYNIUM.

**periclinal** Parallel to a surface. *Compare* ANTICLINAL.

***Pericopsis*** (family *Fabaceae, subfamily *Papilionatae) A small genus of timber trees that yield extremely attractive and valuable cabinet timbers, notably afrormosia (*P. elata*) from tropical Africa. There is 1 species in Asia, and 3 in Africa.

**pericycle** The layer of plant *tissue, 1 to several cells thick, that lies between the *endodermis and the *phloem. It consists mainly of *parenchyma, from which the lateral roots originate.

**periderm** *See* BARK.

**peridium** The outer wall or membrane which encloses the *spores in the *fruit bodies of certain *fungi and *slime moulds.

**perigenous** In some *angiosperms, applied to a *stoma in which the 2 *guard cells are derived from a single mother cell and the subsidiary cells are derived from a different initial. *Compare* MESOGENOUS. *See also* HAPLOCHEILIC.

**perigonium** In *Bryophyta, 1 of the leaves or bracts that surround the *antheridium. *See also* PERICHAETIUM.

**perigynium** In *Bryophyta, 1 of the leaves or bracts that surround the *archegonium. *See also* PERICHAETIUM.

**perigynous** Applied *to flowers in which the *calyx, *corolla, and *stamens are inserted around the *ovary, on a disc-like structure.

**periphyses** Short, thread-like filaments that line the opening, or *ostiole, of a *perithecium in certain types of *fungi.

**periplasmodium** A fluid surrounding developing *microspores formed from the breakdown of the *tapetum.

**periplast** (pellicle) A proteinaceous wall inside the *cell membrane that gives the cell rigidity, often instead of a *cell wall.

**perisperm** In plants where the *endosperm does not replace the *nucellus completely, a nutritive tissue derived from the nucellus.

**perispore** In some species of vascular plants, an additional outer layer to the spore.

**peristome** A single or double ring of *setae (teeth) which fringes the mouth in many types of *moss *capsule; the peristome can be seen when the lid of a ripe capsule falls or is removed. The setae are hygroscopic (absorb moisture from the air). The fibres from which they are made are coiled helically; changes in humidity cause changes in the lengths of the fibres, twisting the capsule and producing jerky movements that aid in the dispersal of the *spores.

**perithecium** A rounded or flask-shaped *ascocarp from which ascospores, formed inside the perithecium, are discharged to the exterior via a small pore.

**permanent quadrat** A *quadrat repeatedly recorded in detail over a number of years in order that plant relationships and changes in them over time may be studied in detail.

**permanent wilting percentage** *See* PERMANENT WILTING POINT.

**permanent wilting point (PWP, permanent wilting percentage, wilting coefficient, wilting point)** As moisture is lost from the soil, the point at which the force with which the remaining moisture adheres to soil particles exceeds that exerted by plant roots. Plants are therefore unable to absorb moisture and *wilting results. Since this condition arises from the amount of water present in the soil, plants will not recover unless water is added to the soil, i.e. the wilting is permanent. The permanent wilting point usually occurs when soil moisture is held with a force of about 15 MPa (15 bar). It is also measured as the percentage of moisture remaining in the soil after a specified test plant has wilted under defined conditions and will not recover unless water is added to the soil. *Compare* TEMPORARY WILTING.

**permeability 1.** A property of a membrane or other barrier, being the ease with which a substance will diffuse or pass across it. **2.** The ease with which gases, liquids, or plant roots penetrate into or pass through a layer of soil.

**permeability coefficient**   A quantitative estimate of the rate of passage of a solute across a membrane. In a concentration of 10 moles/litre it represents the net number of solute molecules crossing 1 $cm^2$ of membrane per unit time.

**permease**   One of a class of *proteins that act as carrier molecules in cells, facilitating the active or passive transport of various substances across membranes. They have many of the characteristics of *enzymes with the exception that they may greatly alter the point of equilibrium of a reaction.

**Permian**   The final period of the *Palaeozoic Era, about 299–251 Ma ago. It is named after the central Russian province of Perm. The period is often noted for the widespread continental conditions that prevailed in the northern hemisphere and for the extensive nature of the southern hemisphere glaciation. Many groups of animals and plants vanished at the end of the Permian in one of the most extensive of all mass extinctions. It was during this period that the *Pteridophyta were superseded as the dominant vegetation by the *gymnosperms.

**p**

**Peronosporales** (class   *Oomycetes) An order of *fungi in which a well developed, *coenocytic, branching *mycelium is formed. Asexual reproduction occurs usually by the formation of *zoospores, sometimes by the formation of *conidia. Species are chiefly *parasites of vascular plants, but some can live *saprotrophically in soil or water. The order includes a number of important plant *pathogens, including the *downy mildews.

**peroxidase**   An *enzyme that catalyses the oxidation of certain organic compounds using hydrogen peroxide as an electron acceptor.

**peroxide**   An inorganic or organic compound that contains linked pairs of oxygen atoms (–O–O–). Peroxides may be regarded as derivatives of hydrogen peroxide ($H_2O_2$) in which the hydrogen atoms are replaced by other atoms or groups.

**peroxisome**   A round or oval, membrane-bound *organelle, 0.3–1.5 μm in diameter, that is found in numbers in most *eukaryotic cells. Typically, it has a finely granular, relatively dense internal structure, and contains numerous *enzymes associated with the production and degradation of peroxides, in particular the enzyme catalase. *See also* GLYOXYSOME.

***Persea*** (family *Lauraceae) A genus of tropical trees which includes *P. americana* (avocado pear), a widely cultivated fruit tree. There are 150 species.

**Persian walnut**   *See* JUGLANS.

**persimmon**   *See* DIOSPYROS.

***Persoonia*** (family (Proteaceae) A genus of small, xerophytic (*see* XEROPHYTE) trees and shrubs which have entire, alternate leaves. The yellow or white flowers are bisexual, solitary, and *axillary or in short *racemes. The *calyx is often cylindrical. The fruit is a *drupe with 1 seed, a juicy exocarp (*see* PERICARP), and a hard, thick endocarp. There are about 60 species, confined to Australia and New Zealand.

**perthophyte**   A plant-*parasitic *fungus that derives its nutrients from dead tissues within a living host plant.

**Pertusariales** (subdivision *Ascomycotina) An order of *fungi which includes *crustose *lichens and *saprotrophic and *lichenicolous fungi. *Apothecia are *sessile or immersed; they may have open discs or may be closed and *perithecium-like. *Asci are thick-walled; the *spores are large. Species are found on various types of substrate, but rarely on limestone.

**petai**   *See* PARKIA.

**petal**   In a *flower, one of the inner floral leaves, usually brightly coloured, and borne in a tight spiral, or *whorled. *See also* COROLLA.

**petaloid**   *Petal-like.

***Petalonia***   *See* SCYTOSIPHONALES.

**petiole**   The stalk by which a *leaf is attached.

**petiolule**   The stalk of a *leaflet in a *compound leaf.

***Petraeovitex*** **(family \*Verbenaceae)** A genus of slender, woody climbers, some of which are cultivated as ornamentals for their coloured flower \*bracts. There are 7 species, occurring from Malesia to New Zealand.

**petrified forest** **(submerged forest)** An area of \*peat containing eroded tree stumps that is exposed along a coastline at low tide. Its presence indicates a rise in sea level or a lowering in the level of the land. Post-glacial petrified forests are common in the estuaries of south-west England.

**petrocalcic horizon** An indurated (*see* INDURATION) \*calcic horizon that is cemented by a high concentration of calcium carbonate, often comprising 40 per cent by weight of the mineral material, and which is impenetrable to plant roots or to spades used for digging.

**petrogypsic horizon** A surface or subsurface soil \*horizon cemented by gypsum so strongly that dry fragments will not slake in water. The cementation restricts penetration by plant roots. This is a \*diagnostic horizon.

**petrophilous** Applied to plants that grow on rocks. Chomophytes grow on ledges or within fissures, lithophytes grow on rock surfaces. *See also* ENDOLITHIC.

**Pezizales** **(subdivision \*Ascomycotina)** An order of \*fungi in which the \*ascocarps are typically apothecioid (*see* APOTHECIUM); the \*asci have thin walls and typically open by means of an apical slit or operculum (lid). They are \*saprotrophic, and found on soil, wood, dung, plant litter, etc.

**PGA** *See* PHOSPHOGLYCERIC ACID.

**pH** A value on a scale 0–14 that gives a measure of the acidity or alkalinity of a medium. A neutral medium has a pH of 7. Acidic media have pH values of less than 7, and alkaline media of more than 7. The lower the pH, the more acidic is the medium; the higher the pH, the more alkaline. The pH value is the reciprocal of the hydrogen ion concentration expressed in moles per litre.

**Phacidiales** *See* RHYTISMATALES.

***Phaeoceros*** *See* ANTHOCEROTALES.

**Phaeophyta** **(brown algae)** A division of \*algae which includes no single-celled species; almost all are marine, growing mostly in the intertidal regions (but species of *Bodanella* and *Heribaudiella* occur in fresh water). They are the dominant seaweeds in the colder waters of the northern hemisphere. They are typically olive-brown or greenish in colour (at least when wet) owing to the presence of the pigment \*fucoxanthin in the \*chloroplasts. There are at least 13 orders, and many genera.

**Phaeozems** Any soil with a \*mollic horizon, apart from \*Chernozems and \*Kastanozems. Phaeozems are a reference soil group in the FAO \*soil classification.

**phage** *See* BACTERIOPHAGE.

**phagocytosis** The process by which a \*cell membrane can invaginate and enclose externally derived, solid material within a \*vacuole, without disrupting the continuity of the cell surface. Subsequently this vacuole will fuse with a \*lysosome and its contents will be wholly or partly digested. Although phagocytosis is a feature of animal cells, bacteria may be attacked by phagocytotic cells. The \*capsule may afford them some protection against such an attack.

**phagotroph** *See* CONSUMER ORGANISM.

**phagotrophy** A mode of nutrition in which particulate food is ingested.

**phalanx growth form** The distribution that results when the \*rhizomes or \*stolons by which a plant spreads by \*clonal dispersal are short and often long-lived. The clonal shoots are closely spaced and the clone advances along a densely packed front, like a Roman phalanx. *Compare* GUERILLA GROWTH FORM.

**Phallales** **(class \*Gasteromycetes)** An order of \*fungi in which the \*fruit body begins its development underground. It begins as a more or less spherical 'egg' stage, 0.5–5.0 cm in diameter, with a smooth, membranaceous \*peridium enclosing a gelatinous layer. In some genera the \*fruit body is mature at this stage. In others (the \*stinkhorns) a stem-like structure grows from the 'egg' and carries at its tip a cap of *spore-bearing tissue; the cap secretes a putrid-smelling slime which attracts flies, and these are responsible for spore dispersal. There are many genera, found on the ground, on rotting wood, etc.

**phallotoxin** A member of a group of poisonous substances present in the fruit bodies of *Amanita phalloides* and certain other species of *Amanita*. Ingestion of phallotoxins by humans causes vomiting and diarrhoea, and liver damage may occur.

**Phallus** (order *Phallales) A genus of *fungi in which the *spore-bearing tissue is raised above ground on the tip of a stalk-like structure growing from the under ground 'egg' stage. There are many species. *See also* STINKHORN.

**phanerophyte** One of *Raunkiaer's life-form categories, being a plant whose *perennating buds or shoot apices are borne on aerial shoots. Such plants are the least protected of those in Raunkiaer's scheme and therefore are most typical of environments where drought, cold, and exposure to strong winds are relatively infrequent. Several sub-categories are recognized, the most universal being: (*a*) evergreen phanerophytes without bud scales (tropical trees) or with bud scales; and (*b*) deciduous phanerophytes. These groups may be subdivided further according to height, as nano- (less than 2 m), micro- (2–8 m), meso- (8–30 m), and mega- (more than 30 m) phanerophytes. Other subgroups are epiphytic, stem succulent, and herbaceous phanerophytes, the last being confined to humid tropical environments. *Compare* CHAMAEPHYTE, CRYPTOPHYTE, HEMICRYPTOPHYTE, and THEROPHYTE.

**Phanerozoic** The period of geological time that comprises the *Palaeozoic, *Mesozoic, and *Cenozoic Eras. It began 542 Ma ago at the end of the *Proterozoic and is marked by the accumulation of sediments containing the actual remains of plant structures. The name is derived from the Greek *phaneros*, meaning 'visible' and *zoe*, meaning 'life'.

**phase diagram** A graphical method for examining stability in biological systems (e.g. host–parasite relationships). The system variables (the numbers of host and parasite organisms) at different times are plotted against one another, and change over time is shown by a line connecting the coordinate points for successive time values. The form of the resulting curve indicates the stability of the system. A line spiralling inward indicates damped oscillations favouring ultimate stability; an outward spiral suggests ultimate instability. By testing different values for the number of organisms, it is possible to establish limits for stability. The technique has considerable practical relevance in agriculture.

**Phaseolus** (family *Fabaceae, subfamily *Papilionatae) A genus of herbs or slender climbers, many of which produce edible seeds (beans) or pods. The genus includes: *P. coccineus* (runner bean); *P. lunatus* (lima bean); and *P. vulgaris* (kidney, french, navy, haricot, or frizoles bean). There are about 50 species, native to warm and tropical America.

**Phebalium** (family *Rutaceae) A genus of small, xerophytic (*see* XEROPHYTE) trees and shrubs which have *simple, usually alternate leaves, without *stipules. The flowers are bisexual with a united *corolla, and with twice as many *stamens as corolla lobes. The seeds contain copious *endosperm. There are 40 species, only 1 of which is found outside Australia, and then only in New Zealand.

**phellem** *See* CORK.

**Phellinus** (family *Hymenochaetaceae) A genus of fungi in which the *fruit bodies are typically *bracket-like or hoof-shaped, hard and leathery or woody in texture, with rust-brown 'flesh' and a layer of tubes on the underside. The spores are smooth. They are found on wood; some species are *parasitic on trees (e.g. *P. ignarius* is parasitic on deciduous trees, especially willow, causing a *white rot).

**phelloderm** *See* BARK.

**phellogen** (cork cambium) *Cambium located outside the vascular cambium in plants undergoing *secondary thickening. It comprises lateral *meristem and gives rise to *cork (phellem) and *bark (phelloderm).

***Phenakospermum*** (family *Strelitziaceae) A genus of unbranched trees whose banana-like leaves are held in a single plane, like a huge fan. The inflorescence is terminal. The fruit is a *capsule with brightly coloured seeds. There is 1 species found in the north of tropical S. America.

**phenetic classification** The grouping of biological organisms on the basis of observed physical similarities. *See* PHYLOGENETIC SYSTEMATICS.

**phenogram** A type of *dendrogram which is based on phenetic data (*see* PHENETIC CLASSIFICATION). Lines called phenon lines, drawn at right angles to the dichotomously branching dendrogram, represent lines of percentage similarity of phenetic features between organisms. Because the phenetic data are based on features that appear in different organisms regardless of whether they are *analagous or *homologous, organisms that are evolutionarily distant may be grouped closely together. For example, in evolutionary history the tree has evolved on several independent occasions. *Gymnosperm trees evolved in the *Carboniferous Period, some 355–290 Ma ago. The first *angiosperms evolved 120 Ma ago, but the first angiosperm trees did not evolve until 90 Ma ago. The 2 evolutions of the tree structure, then, occurred separately, and to group them together on the basis of the common structure would constitute an *artificial classification.

**phenol** An aromatic compound that bears 1 or more *hydroxyl groups.

**phenology** The study of the impact of climate on the seasonal occurrence of flora and fauna (dates of flowering, migration, etc.), and the periodically changing form of an organism, especially as this affects its relationship with its environment (e.g. the development of a tree seedling into a sapling and later a mature tree).

**phenon line** *See* PHENOGRAM.

**phenotype** The observable manifestation of a specific *genotype, i.e. those properties of an organism, produced by the genotype in conjunction with the environment, that are observable. Organisms with the same overall genotype may have different phenotypes because of the effects of the environment and of gene interaction. Conversely, organisms may have the same phenotype but different genotypes, as a result of *incomplete dominance, *penetrance, or*expressivity.

**phenotypic adaptation** *See* ADAPTATION.

**phenotypic variance** The total variance observed in a character.

**phenylalanine** An aromatic, *non-polar *amino acid.

**phialid (phialide)** A bottle-shaped structure within which or from which *conidia develop. It is formed by *imperfect stages of some *fungi.

**phialide** *See* PHIALID.

***Philadelphus*** (mock orange, syringa; family *Saxifragaceae) A genus of shrubs, formerly placed in the family *Hydrangeaceae, that have *opposite leaves and attractive *tetra-, *penta-, or *hexamerous flowers with white petals, many *stamens, and a 4-celled *ovary. The name 'syringa' is often incorrectly used for this genus; *Syringa* is actually the scientific

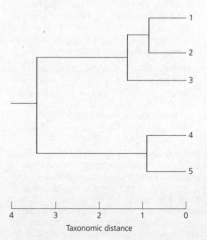

*Phenogram*

name for lilac. There are 65 species in the northern temperate zone, especially in eastern Asia and N. America. The genus is much cultivated for the fragrant flowers, which smell like and superficially resemble orange blossom.

**Philesiaceae** A family of monocotyledonous (see MONOCOTYLEDON) shrubs, climbers, and epiphytes, closely allied to the *Amaryllidaceae, which have alternate, *simple leaves without *stipules. The flowers are regular, bisexual, large, and pendulous; they are *axillary or held terminally, and are usually solitary. There are 3 *sepals and 3 *petals, the sepals and petals often resembling one another. There are 3 *stamens, with large *anthers. The *ovary is *superior, with 3 fused *carpels, and is either *uni-, or *trilocular, with a simple, 3-lobed *style. The fruit is *a berry. There are 7 genera and 9 species, all of them confined to the southern hemisphere.

**Philippia** (family *Ericaceae) A genus of tree heaths, comprising 70 species which occur in tropical and southern Africa and the Mascarenes, and are abundant in Madagascar.

**Philodendron** (family *Araceae) A genus of *herbs, mainly climbing, in which the leaf nervation is *pinnate and *reticulate, the blade sometimes being divided. Many are ornamentals. There are 500 species, occurring in tropical America.

**Philydraceae** A family of monocotyledonous (see MONOCOTYLEDON), erect, rhizomatous (see RHIZOME) herbs, which have 2-ranked, sheathing, linear leaves, clustered or in a basal rosette. The flowers are solitary, irregular, and bisexual, with a single *stamen. The *ovary is *superior with 3 fused *carpels forming a single *locule with numerous *ovules. The *style is simple. The fruit is a *capsule, with many seeds which contain *endosperm. There are 4 genera, and 5 species, restricted in their distribution to Japan, South-east Asia, and Australia.

**phloem** A *tissue comprising various types of cell, which transports dissolved organic and inorganic materials over long distances within *vascular plants, by

mechanisms that are still not understood fully. It can be distinguished from *xylem by the general absence of thickened cells and by the presence of cells containing areas resembling a sieve.

**phloroglucin** A dye that, with hydrochloric acid, stains *lignin bright red.

**Phlox** See POLEMONIACEAE.

**phobotaxis** The random change in direction of locomotion in a *motile micro-organism or cell, made in response to a given stimulus.

**Phoenix** (family *Arecaceae) A genus of palms in which the leaves are *pinnate (with a terminal leaf present, a feature unique among the palms), basal *leaflets often occurring as spines. They are dioecious. The *inflorescences form among the leaves as 1 big, *caducous *spathe. The flowers are single, with 3 *carpels. The fruits are fleshy, the seed hard and grooved. Several are widely cultivated. *P. dactylifera* is the date palm. There are 17 species, found from the Canary Islands and Africa to Sumatra and Malaya, mostly in seasonal or dry climates. *P. theophrasti*, of Crete and Turkey, is 1 of only 2 native European palms.

**Pholiota** (family *Strophariaceae) A genus of fungi which form *mushroom-shaped *fruit bodies. The *gills are initially pale but later become brown; the *stipe is central and may or may not bear a ring. The *spore print is brown to rusty-brown. Species are found on wood or on the ground, often in clusters.

**Phoma** (order *Sphaeropsidales) A *form genus of *fungi in which colourless *conidia are produced in dark-coloured, flask-shaped *pycnidia. The genus includes many important seed-borne plant *pathogens, which cause, for example, various types of stem and root rot.

**Phormidium** A genus of filamentous, sheathed *cyanobacteria (section III) in which the sheaths tend to *coalesce. They are capable of gliding motility (see MOTILE). They are found in all types of aquatic environment, and on wet rocks, damp soil, etc., often forming dense mats. They are frequently difficult to distinguish from *Lyngbya* and *Plectonema* (the *LPP group).

***Phormium*** (family *Agavaceae) A genus of plants that are *rosettes with a woody rootstock. There are two species, confined to Norfolk Island and New Zealand. *P. tenax*, New Zealand flax, is commercially important and several variegated or oddly coloured varieties are grown as ornamentals.

**phosphatase** An *enzyme that catalyses reactions involving the *hydrolysis of *esters of phosphoric acid.

**phosphatide** One of a large group of naturally occurring *phospholipids that are derivatives of glycerol phosphate and which normally contain a nitrogenous base.

**phosphatidyl choline** See LECITHIN.

**phosphodiester bond** A bond between two molecules that is formed by phosphoric acid, which is esterified (see ESTER) once to each molecule.

**phosphoenolpyruvic acid** A 4-carbon organic acid that is used as a substrate for fixing carbon dioxide in plants that have a *C4 pathway of *photosynthesis. It is also an important component of *glycolysis.

**phosphoglyceric acid (PGA)** A 3-carbon organic acid formed during *glycolysis that is the first relatively stable compound formed in the *Calvin cycle after the fixation of carbon dioxide in C3 plants. The first product, an unstable 6-carbon compound, is formed when carbon dioxide combines with ribulose diphosphate.

**phospholipid** A polar (see POLAR MOLECULE) *lipid that consists of a lipid derivative of *glycerol containing 1 or more phosphate groups. Phospholipids are particularly important in the structure and functioning of biological membranes.

**phosphorus (P)** An element that is required by plants in the oxidized form, as orthophosphate ($PO_4^{2+}$). The special chemical properties of orthophosphate are utilized in reactions in which energy is transferred, often involving *ATP. It is also a good *buffer, helping to maintain a neutral *pH. The leaves of plants deficient in phosphate become dark green or blue-green and a reddish pigment may develop. Growth is generally reduced.

**phosphorylation** The addition of a phosphate group to a compound, involving the formation of an *ester bond between the reactants. See LIGHT REACTIONS.

**phosphotransferase** An *enzyme that catalyses reactions involving the transfer of a phosphate group from one *substrate to another.

**photic zone** The zone of water within which organisms are exposed to sunlight. It is divided at the compensation level, with the *euphotic zone above and the *dysphotic zone below.

**photo-autotroph** A *phototroph that uses carbon dioxide as its main or sole source of carbon.

***Photobacterium*** (family *Vibrionaceae) A genus of *Gram-negative *bacteria in which the cells are rod-shaped and typically *motile. They are *chemo-organotrophic, and can grow in the presence or absence of air. Under appropriate conditions at least some species can emit light. They are found in marine environments and in the luminous organs of certain fish and cephalopods.

**photoblastic** Applied to seeds that use light to signal when conditions are right for germination.

**photochemical reaction** A chemical change induced by light. Such a reaction can be distinguished from other chemical reactions by being largely independent of temperature. An example is *photosynthesis, in which light causes electron shifts in the photoreceptors.

**photodissociation** The splitting of a molecule into atoms or other molecules as a result of its absorption of radiation.

**photo-heterotroph** A *phototroph that uses organic compounds as its main or sole source of carbon.

**photo-inhibition** The slowing or stopping of a plant process by light, e.g in the *germination of some *seeds.

**photokinesis** A change in the speed of locomotion in a *motile organism or cell made in response to a change in light intensity.

**photolithotroph** A *phototroph in which *photosynthesis is associated with the oxidation of an inorganic compound (or element, in the case of some photosynthetic bacteria). For example, green plants obtain energy from sunlight by photosynthesis, during which water is oxidized to yield oxygen.

**photolysis** The breaking down of water using light; it occurs in non-cyclic *photophosphorylation. See LIGHT REACTIONS.

**photomorphogenesis** The effect of light in regulating plant form and growth.

**photonasty** A nastic (see NASTY) response of a plant organ to the stimulus of light.

**photo-organotroph** A *phototroph in which *photosynthesis is associated with the oxidation of organic compounds. Certain photosynthetic *bacteria are photo-organotrophic.

**photoperiod** The relative lengths of the periods of light and darkness associated with day and night.

**photoperiodism** The response of an organism to periodic, often rhythmic, changes in either the intensity of light or, more usually, the relative length of day. Many plant activities, such as flowering and leaf-fall, are also photoperiodic in nature.

**photophosphorylation** The formation of *ATP, as in *photosynthesis, by processes that require light.

**photoreceptor** A pigment that absorbs the light used in various plant processes. Examples include *phytochrome and the photosynthetic pigments.

**photorespiration** A process in plants that reduces the efficiency of *photosynthesis by the *$C_3$ pathway. The active site of rubisco, the enzyme that catalyzes the fixation of $CO_2$ at the start of the *dark reactions, accepts either $O_2$ or $CO_2$. The two gases therefore compete and if the $CO_2$ concentration in the leaf is low, as it is during photosynthesis when $CO_2$ is being consumed, rubisco adds $O_2$ rather than $CO_2$ to ribulose biphosphate (RuBP). This alters the sequence of reactions, resulting in the release of some of the $CO_2$ that was fixed at the start of the cycle. The process is called respiration because it absorbs $O_2$ and releases $CO_2$, but it does so without yielding any energy. *Glycine max* (soybean) loses up to half of the $CO_2$ fixed during the dark reactions due to photorespiration.

**photosynthesis** The series of metabolic reactions that occur in certain *autotrophs, whereby organic compounds are synthesized by the reduction of carbon dioxide using energy absorbed by *chlorophyll from sunlight. In green plants, where water acts as both a hydrogen donor and a source of released oxygen, photosynthesis may be summarized by the empirical equation:

$$CO_2 + 2H_2O \xrightarrow[\text{light}]{\text{chlorophyll}} \left[CH_2O\right] + H_2O + O_2\uparrow$$

(oxygen being released as a gas). Photosynthetic bacteria are unable to utilize water and therefore do not produce oxygen. Instead they may use hydrogen sulphide (purple and green sulphur bacteria) or organic compounds (purple non-sulphur bacteria) as a source of hydrogen.

**photosynthetic quotient** The volume of oxygen released in *photosynthesis as a proportion of the volume of carbon dioxide used in that process.

**photosynthetic unit** A hypothetical light-harvesting unit of *photosynthesis that, in green plants, comprises about 300 light-absorbing molecules with a molecule of *chlorophyll acting as the reaction centre. There are thought to be 2 different types of photosynthetic unit, one for *photosystem I, the other for *photosystem II.

**photosystem I** (PSI) The system of photosynthetic reactions, involving the pigment *$P_{700}$, *chlorophyll $b$, and accessory pigments, that requires light of longer wavelength than *photosystem II: it results in the reduction of *NADP+ and the production of *ATP through *photophosphorylation.

**photosystem II (PSII)** The system of photosynthetic reactions, involving $*P_{680}$, chlorophyll *b*, and accessory pigments, that requires light of shorter wavelength than *photosystem I: it results in the dissociation of water and the evolution of molecular oxygen.

**phototaxis** A change in direction of locomotion in a *motile organism or cell made in response to a change in light intensity.

**phototroph** An organism that obtains its energy from sunlight. In the great majority of phototrophs, the process by which light energy is harnessed is *photosynthesis.

**phototropism (heliotropism)** A tropic response of a plant or plant organ to the stimulus of light.

***Phragmites*** **(reed-grasses; family *Poaceae)** A genus of tall, stout grasses which have robust stems and large plumes of slender *spikelets forming a nodding *panicle. They are *perennial with *rhizomes. The *ligule is a ring of hairs. All the *florets have a tuft of long, silky hairs arising from the base. The *lemma is 3-nerved and twice as long as the upper *glume and the *palea is less than one-third the length of the lemma. There are usually 3 *stamens, although often there are fewer in the lower florets. The *ovary is *glabrous, with a short *style. There are 3 species, found throughout the world, usually in swampy or marshy conditions.

***Phragmites cliffwoodensis*** An early representative of the reed-grasses (*Phragmites), recorded from the mid-*Cretaceous of New Jersey, USA, that has rather uncertain affinities but if accepted would be the first of the family *Poaceae.

**Phragmobasidiomycetidae** **(class *Hymenomycetes)** A subclass of *fungi in which the *basidia are divided by *septa. The *fruit bodies are typically gelatinous or waxy. There are 3 orders. Species include *saprotrophs, *parasites, and *symbionts. *See also* JELLY FUNGUS.

**phragmoplast** The region between the daughter nuclei of a dividing plant cell. At first it contains only *micro-tubules, but it develops into a barrel-shaped organ with microtubules, *Golgi apparatus, *ribosomes, and *endoplasmic reticulum. It assembles the *cell plate which functions in *cytokinesis.

**phreatic zone** *See* WATER TABLE.

**phycobilin** A generic term for a group of substituted open-chain *tetrapyrroles that occur in conjugation with specific *proteins. The resulting complex functions as a blue or red accessory pigment in the *chloroplasts of many algal species, especially those of the *Rhodophyceae.

**phycobilisome** An *organelle, occurring on the *laminae of *red algae and *cyanobacteria, that contains the red pigment phycoerythrin and/or the blue pigment phycocyanin.

**phycobiont** The principal *algal or *cyanobacterial *symbiont in a *lichen.

**phycocyanin** A blue accessory pigment, consisting of a *protein conjugated with a *phycobilin. It is found in the *chloroplasts of many algal species.

**phycoerythrin** A red accessory pigment, consisting of a *protein conjugated to a *phycobilin. It is found in the *chloroplasts of many algal species.

**phycology** **(algology)** The study of *algae.

**phycomycetes** The *lower fungi.

**phycovirus** A virus that infects and replicates in algae.

**phyletic evolution** An evolutionary change within a lineage, as a result of gradual adjustment to environmental stimuli.

**phyletic gradualism** **(gradualism)** A theory holding that *macro-evolution is merely the operation of *micro-evolution over relatively long periods of time. Thus, gradual changes eventually will accumulate to the point at which descendants of an ancestral population diverge into separate species, genera, or higher-level taxa.

***Phylica*** *See* FYNBOS.

***Phyllanthus*** **(family *Euphorbiaceae)** A genus of *herbs, shrubs, or mostly small trees, in which the leaves are often in 2 ranks, and the shoots resemble *pinnate leaves. The flowers are minute, and held in dense clusters. The fruit is dry and *dehiscent or fleshy and *indehiscent. Some species have green *phylloclades bearing flowers on their margins. There are about 600 species, with a pantropical distribution.

**phyllid** A flattened, leaf-like structure in *Bryophyta.

**phylloclade** (cladode) A specialized stem that resembles and functions like a leaf.

**phyllocladium** One of many small, scale-like projections that occur on the *pseudopodetium in *lichens of the family *Stereocaulaceae.

***Phyllocladus*** (family *Podocarpaceae) A genus of southern hemisphere conifers, whose foliage of *phylloclades is borne in the *axils of scale leaves. The seed is *arillate. The genus yields useful timber. There are 4 species, occurring from Borneo to New Zealand (where they are known as celery pines).

**phyllode** A *petiole that is leaf-like.

**phyllody** The condition in which parts of a flower are replaced by leaf-like structures; it is a symptom of certain plant diseases.

***Phylloglossum*** (family *Lycopodiaceae) A *monotypic genus of plants that have underground *tubers, a basal tuft of cylindrical pointed leaves, and a leafless stem bearing a single terminal cone. It is confined to Australasia.

**phyllosphere** The micro-environment on and below the surface of a leaf.

**phyllotaxis** In a plant, the arrangement of leaves on the stem (as *opposite pairs, *whorls, alternate, spirals, etc).

**phylogenetic systematics** The study of biological organisms, and their grouping for purposes of classification, based on their evolutionary descent. See PHENETIC CLASSIFICATION.

**phylogeny** Evolutionary relationships within and between taxonomic levels, particularly the patterns of lines of descent, often branching, from one organism to another (the relationships of groups of organisms as reflected by their evolutionary history).

**phylogerontism** (racial senescence) An outmoded view of *evolution which asserted that lineages proceeded through a life cycle, from youth to senility. Lineages on the verge of extinction were thus thought of as examples of racial senescence or phylogerontism.

**phymatodeus** Warted or verrucose.

***Physarum*** (class *Myxomycetes) A large genus of *slime moulds, including *P. polycephalum*, which can be cultivated and induced to fruit under laboratory conditions. This species forms a conspicuous yellow *plasmodium which gives rise to clusters of stalked yellowish or grey *sporangia. *P. polycephalum* is found on dead wood and on fleshy *fungi. It is distributed widely in the USA and has been reported from other countries, including France and Japan.

***Physcia*** (order *Lecanorales) A genus of *lichens in which the *thallus is *foliose and may be grey or greyish-brown. *Spores are brown and have 1 *septum. Species are found typically on substrates (e.g. bark or rocks) rich in nitrogenous matter (e.g. bird droppings).

**physical factor** A factor in the *abiotic environment that influences the growth and development of organisms or biological *communities.

**physiognomy** The form and structure of natural communities. *Compare* MORPHOLOGY.

**physiological ecology** The study of the functioning of organisms in relation to their environments.

***Physocomitrium pyriforme*** *See* PEAR MOSS.

***Phytelephas*** (family *Arecaceae) A genus of palms in which the stem is short, stout, and often *procumbent. The leaves are large and *pinnate. The flowers are *dioecious. The *inflorescences are stout, and held in unbranched spadices (*see* SPADIX). The fruits form in big, knobbly heads and the *endosperm is hard.

*P. macrocarpa* (ivory nut) is the principal source of vegetable ivory used for carving. *Phytelephas* species are very different from other palms. There are 12 species, occurring in northern tropical America.

**phytic acid (hexaphosphoinositol)** A substance that is present in some seeds, where it acts as a store for phosphate. It is the hexaphosphoric *ester of *inositol.

**phyto-alexin** An antifungal substance that is produced by a plant in response to damage or infection; phyto-alexins may help to protect the plant from disease.

**phytochrome** A photoreversible pigment, which occurs in every major taxonomic group of plants. It exists in 2 interchangeable forms with respect to absorption, a red and a far-red form. This phenomenon is utilized in plants as a switch mechanism, linked to environmental light signals, to control many activities. These include flowering, germination, vegetative growth, circadian rhythms, and many more physiological activities.

**phytocoenosis** *See* BIOCOENOSIS.

**phytogeography (floristics)** The study of the geography of plants, particularly their distribution at different taxonomic levels, i.e. family, genus, and species. Patterns of distribution are interpreted in terms of climatic and anthropogenic influence, but above all in terms of earlier continental configurations and migration routes.

**phytohormone** *See* PLANT HORMONE.

**phytol** A long-chain alcohol that comprises the major side-chain of *chlorophyll.

**Phytolacca (pokeweed; family *Phytolaccaceae)** A genus of trees, shrubs, and *herbs whose flowers are borne in *racemes opposite leaves, and which produce purple *berries. There are 25 species, occurring in tropical and warm climates, and sometimes cultivated for ornament or vegetables.

**Phytolaccaceae** A family of trees, shrubs, *herbs, or woody climbers in which the leaves are entire and alternate and the flowers regular, with a single *perianth of 4 or 5 usually separate segments, 4 to many *stamens, and 1–16 *carpels, free or united below. There are 18 genera, and about 65 species, confined to warm climates.

**Phytomastigophora (subphylum Mastigophora)** In protozoan classification, a class of flagellated, plant-like microscopic organisms which typically possess *chloroplasts and are photosynthetic. These organisms are often regarded as *algae (e.g. *Chlamydomonas, *Chrysophyta, *cryptophytes, Dinophyta, *Euglenophyta, *Prymnesiophyceae, and *Volvox).

**phytoncide** A substance that is produced normally by a plant and which has some antimicrobial action; phytoncides may play a role in preventing disease in the plant.

**phytopathology** The study of plant diseases.

**phytophagous** Feeding on plants.

**Phytophthora (order *Peronosporales)** A genus of *Oomycota, many of which are plant *pathogens of great economic importance (e.g., *P. infestans* causes late blight of potato and can also infect other plants of the *Solanaceae). There are many species.

**phytoplankton** The photosynthetic *plankton and *primary producers of aquatic ecosystems, comprising mainly *diatoms in cool waters with dinoflagellates being more important in warmer waters.

**phytosociology** The description of plant *communities, more especially their classification based primarily on floristic rather than life-form or other considerations. The most renowned and extensively used phytosociological scheme is that developed by J. *Braun-Blanquet (1927 and later) and his associates at Zurich and Montpellier. A similar, fairly widely applied, scheme was developed by G. E. *du Rietz (1921 and later) and others at Uppsala.

**phytosterol** A *steroid that occurs in plants. Although they are widespread, the functions of phytosterols are not clearly understood.

**phytotoxin** A substance toxic to plants, which is produced by a plant pathogen.

**piassava (African bass)** A fibre obtained from *petioles of the plam *Raphia hookeri* and *R. palma-pinus* by retting. Bahia piassava is obtained similarly from *Attaleafunifera*. *See* RAPHIA.

*Picea* **(spruce; family *Pinaceae)** A genus of large, evergreen conifers in which the leaves are stiff and are borne singly on peg-like projections which remain attached to the twig when the leaves fall. The cones are borne at the tips of the branches and hang downward when ripe. The genus includes many important timber trees (e.g. *P. sitchensis*, Sitka spruce, much planted in northern Britain). There are 34 species, found throughout the northern hemisphere except in Africa.

*Pigafetta* **(family *Arecaceae)** A *monotypic palm genus. *P. filaris* is a lofty, pinnate-leaved palm of disturbed forest, with stout, solitary, green, glistening trunks up to 15 m tall. It is *dioecious. The *inflorescences are *axillary, the fruits scaly. It grows very rapidly. It is a splendid, little-known ornamental, which occurs in central Malesia.

**pigeon pea** *See* CAJANUS.

**pigment** A compound (biochrome) that produces colour in the tissues of living organisms. The biological function of the compound may or may not be directly associated with this property.

**pileus** From the Latin *pileus*, meaning 'cap'. The fleshy or leathery structure upon which *hymenium-bearing *tissue occurs in a fungal *fruit body, e.g. the cap of an *agaric.

**piliferous** Bearing a hair or *trichome.

**piliferous layer (root-hair zone)** The region of the *epidermis of a *root, a short distance from the tip, that produces abundant *root hairs and is involved in the uptake of water and nutrients.

**Pilobolaceae (order *Mucorales)** A family of *fungi in which the *sporangia are discharged forcibly. Species are found on the dung of herbivores.

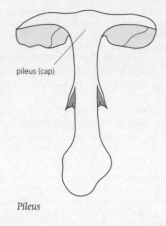

pileus (cap)

*Pileus*

**pilose** Covered with fine hairs or down.

*Pimelea* **(family *Thymelaeaceae)** A genus of shrubs and some *herbs in which the small leaves are *opposite or widely spaced along the stem. The bisexual flowers are protandrous (*see* PROTANDRY), held in terminal or *axillary heads, and are rarely solitary. The *calyx is fused and cylindrical with 4 lobes, the 2 *stamens being attached to the tube. The *ovary is *unilocular and the small fruit is enclosed in the base of the *calyx. The seeds contain no *endosperm but have a tough *testa. There are about 80 species, found only in New Zealand and Australia.

*Pimenta* **(family *Myrtaceae)** A genus of small, aromatic, evergreen trees which are functionally *dioecious. *P. dioica* yields pimento or allspice (reminiscent of cinnamon, nutmeg, and cloves), made from the dried, unripe fruits. It is exported solely from Jamaica. *P. racemosa* yields the oil, bay rum. There are 5 species, occurring in tropical America.

**Pinaceae (order *Coniferales)** A family of conifers that bear both male and female cones and in which the seeds are borne between the cone scales, with no fleshy *arils. The leaves are spirally arranged and linear. The cone scales are arranged spirally, each scale in female cones being a double structure with the *ovule-bearing scale borne in the *axil of *a bract that is fused to it only at the base. Female cone scales each bear 2 inverted ovules on their upper surfaces. Male cone scales each bear 2 pollen sacs on their lower surfaces, containing pollen grains each with (usually) 2 bladder-like sacs. There are 9 genera, comprising 194 species, occurring in the northern hemisphere, mainly in temperate climates.

**Pinanga** (family *Arecaceae) A genus of palms which grow in forest undergrowth and are locally abundant. They are solitary or form clumps. The leaves are entire or *pinnate and the crown-shaft is prominent. The *inflorescence is simple or sparsely branched, two male flowers flanking a female. The fruiting spikes and fleshy fruits are of bright, contrasting colours. There are about 120 species, occurring from India to Taiwan and New Guinea.

**pine 1.** See *PINUS*. **2. (Japanese umbrella pine)** See *SCIADOPITYS*.

**pineapple** See *ANANAS*.

**pine barren** An area of pine forest in which the various species of pine usually produce small or medium-sized trees. The barrens coincide with poor, sandy, and to a lesser extent marshy soil, and owe their ecological character in part to centuries of burning. They occur in the eastern USA, on the coastal plain of the Atlantic and Gulf of Mexico, from New Jersey to Florida, excluding its southern tip, and into Texas.

**pin-frame** A device for obtaining a quantitative estimate of vegetation *cover. Pin-frames are typically made from lightweight wood, aluminium, or plastic and comprise a cross-bar with pin-holes supported on legs of adjustable height. The pin-frame is set up so that the cross-bar sits above the vegetation to be sampled. Pins are lowered through the pin-holes and the plants hit by the pin-tips are recorded. Where the vegetation is of variable height, records of top cover (the first plant encountered by the pin) and bottom cover may be taken. The diameter of the pin used will affect the results and must be standardized for comparative work.

**Pinguicula** (butterwort) See LENTI-BULARIACEAE.

**pink** See *DIANTHUS*.

**pin mould** The common name for *Mucor* species (and other *Zygomycetes) in which the *sporangia, borne on straight *sporangiophores, resemble pins.

**pinna** One of the leaflets of a *pinnate leaf.

**pinnate** Of leaves, *compound, with leaflets displayed on either side of a central stalk or *rachis. The main nerves of most leaves are pinnate, arising along the *midrib.

**pinnatifid** Applied to leaves that are *pinnately divided, but not all the way down to the *rachis. *Compare* PINNATISECT.

**pinnatisect** Applied to leaves that are *pinnately divided all the way down to the *rachis. *Compare* PINNATIFID.

**pinnule** The ultimate divisions of a *fern *frond.

**Pinophyta** See CONIFEROPHYTA.

**Pinus** (true pines; family *Pinaceae) A genus of resinous, evergreen conifers in which the leaves (needles) are borne in groups of twos, threes, or fives on short shoots borne along the twigs. There are separate male and female cones, the latter with woody scales; the seeds are winged. Many pines are important for timber, also yielding resin and turpentine. There are 93 species, all in the northern hemisphere, occurring mainly in northern temperate regions, extending in America and eastern Asia to the seasonal tropics. The common name 'pine' is sometimes used to describe other pine-like trees of different genera (e.g. *Araucaria araucana*. Chile pine or monkey-puzzle tree), or, loosely, to include all conifers.

**Pinus aristata** See BRISTLECONE PINE.

**Pinus longaeva** See BRISTLECONE PINE.

**pioneer phase** A stage in the cyclical pattern of *community change in *heathlands, when individual *Calluna vulgaris plants are young (up to 10 years old) and provide little cover, so that there is maximum diversity of associated species. Although *C. vulgaris* *biomass is low, productivity in the new shoots is high.

*Compare* BUILDING PHASE; DEGENERATE PHASE; HOLLOW PHASE; and MATURE PHASE. The term may also be used more generally for the 'new plant' stage of any cyclical pattern of vegetation change.

**pioneer plants** *See* SUCCESSION.

**pioneer stage** A general term describing the early stages of a plant *succession.

**pipal-tree** (bodh-tree) *See* FICUS.

**Piper** (family *Piperaceae) A genus of *herbs, slender climbers, shrubs, and small trees in which the twigs have swollen *nodes, the leaves have *stipules, commonly tri-nerved, and the flowers are held in *spikes, are hermaphrodite, and have no *sepals or *petals. The fruit is a *berry. Pepper is the fruit of *P. nigrum*; the fruit is gathered early, and dried to give black peppercorns, or, if the outside is removed, white pepper. The betle pepper is *P. betle*. *P. methysticum* is kava, a narcotic which, when pulped and fermented in water, gives the national Polynesian sedative drink. There are more than 1000 species, with a pantropical distribution.

**Piperaceae** (peppers) A family of small trees, shrubs, and climbers, in which the leaves are alternate, *simple, and entire, with *stipules, and glands containing an aromatic oil. The *petioles are winged, and sheath the nodular, jointed stem. The stem form is a result of *sympodial growth. The flowers are small, bisexual or unisexual, and held in *spikes or *racemes opposite the leaves. The *ovary is *unilocular and *superior, with up to 4 fused *carpels surrounded by scale-like *bracts, but there is no *calyx or *corolla. The fruit is a small, fleshy *drupe with a single seed, often sunk into the stem. The *vascular bundles are peculiarly arranged for a *dicotyledon, often being irregularly scattered through the stem. *Piper nigrum* yields the condiment pepper, and is widely cultivated. There are 4 genera and more than 2000 species, found throughout the tropics and represented in most rain forests.

**Piper betle** (betle pepper) *See* PIPER.

**Piper methysticum** (kava) *See* PIPER.

**Piper nigrum** *See* PIPER.

**Piptocephalis** (order *Zoopagales) A genus of *fungi in which the *sporangia are cylindrical in shape and contain a single row of *spores. They are *parasitic on other fungi, often on other *Zygomycetes.

**Piptoporus** (family *Polyporaceae) A genus of *fungi which form thick, bracket-like, non-stipitate *fruit bodies which have a corky texture. *P. betulinum* is common on dead and dying birch trees (*Betula* species). The fruit body is semicircular to hoof-shaped, with a thick, rounded margin; the upper surface bears a thin greyish or brownish skin which may separate from the white flesh. *See also* RAZOR-STROP FUNGUS.

**Pisonia** (family *Nyctaginaceae) A genus of small trees and shrubs in which the leaves are usually *opposite and the flowers usually unisexual, and small. The seeds are sticky, clinging to bird feathers and thus dispersed; they are used as birdlime. There are 35 species, occurring in the tropics and subtropics, especially in America.

**pistachio** *See* PISTACIA.

**Pistacia** (family *Anacardiaceae) A genus of plants that includes *P. terebinthus* which yields chian turpentine, and *P. vera*, the source of edible pistachio nuts. There are 9 species, occurring in the Mediterranean region and subtropics.

**Pistia** (family *Araceae) A *monotypic genus, *P. stratiotes* (water lettuce), a floating water plant which has a *sympodial series of leaf rosettes. The *inflorescences are tiny. It is a noxious weed which occurs throughout the tropics and subtropics.

**pistil** The *gynoecium of a *syncarpous flower; each *carpel in an *apocarpous one.

**pistillate** Applied to a flower that has a *pistil or pistils, but no *stamens (i.e. it is a female flower). *Compare* STAMINATE.

**pistillode** A sterile, often reduced *pistil.

**piston sampler** *See* PEAT-BORER.

**Pisum** (garden pea; family *Fabaceae, subfamily *Papilionatae) A genus of *annual *herbs, many species of which climb by means of leaf tendrils. The leaves are *compound, with paired oval leaflets. The flowers are *zygomorphic with a distinctive petal arrangement: a large upright back petal and two smaller wing petals enclose a cup-like *keel petal. The round seeds are borne in a long pod, and expelled explosively. *P. sativum* (garden pea) and other species are grown for their edible seeds and sometimes pods, used as a protein source for humans and livestock. They can be dried for storage. There are 5 species, occurring in the Mediterranean region and western Asia.

**pit** A region in a *cell wall where the primary wall is not overlaid by *secondary thickening, through which substances can be exchanged between adjacent cells. The pit consists of a cavity, which is the area of thinning in the secondary wall, and a pit membrane, which is the primary cell wall covering the cavity. Pits usually occur in pairs. *See* BORDERED PIT, PRIMARY PIT, and SECONDARY PIT; *see also* PLASMODESMATA.

**pita fibre** *See* AECHMEA.

**pitcher plants** *See* NEPENTHACEAE.

**pith** A *tissue composed of *parenchyma cells that occupies the central part of a stem. *See* MEDULLA.

**Pithecellobium** (family *Fabaceae, subfamily *Mimosoideae) A genus of trees in which the leaves are *pinnate, the *stipules sometimes thorny. The pod is commonly coiled and *dehiscent. Some species have edible seeds. There are 20 species, confined to tropical America. *P. dulce* (Madras thorn) is widely cultivated as a shade tree and a thorny hedge.

**Pittosporaceae** A family of evergreen shrubs, trees, and a few climbers whose leaves are alternate, leathery, and without *stipules. The bark is resinous. The flowers are bisexual, regular, and *pentamerous. The fruit is a *capsule or *berry. There are 9 genera, comprising about 240 species, occurring in the Old World tropics but strongly concentrated in Australasia.

**Pittosporum** (parchment bark, Australian laurel; family *Pittosporaceae) A genus of evergreen shrubs and small trees that have leathery, entire leaves without *stipules. The flowers are

bisexual, regular, terminal, *axillary or lateral, and in 5 parts. Both the *sepals and *petals are united at the base. The *stamens are attached to the *calyx. The *ovary is *superior, with 2 or more fused *carpels, single or multilocular, and has a simple *style. The fruit is a leathery *capsule with numerous seeds which are sometimes sticky. Several species give valuable timber, and many are cultivated as ornamentals, with coloured, fragrant flowers, and attractive foliage. There are about 200 species, found in tropical and subtropical Africa, Asia, and Australasia and the Pacific.

**pituri** *See* DUBOISIA.

***Pityrosporum*** *See* MALASSEZIA.

***Pitys antiqua*** An early member of the *gymnosperm order *Cordaitales, recorded from the Upper *Carboniferous of Scotland. The Cordaitales dominated Upper *Palaeozoic floras with trees attaining a height of 30 m.

**placenta** In flowers, the part of the *ovary wall formed from the fused margins of the *carpel or carpels, on which are carried the *ovules.

**placentation** The position of the *placenta within the *carpel; it may be parietal (on the walls), axile (on the axis), basal, or *free-central.

**placic horizon** A subsurface soil *horizon, formed most readily in humid tropical or cold conditions, that is cemented by iron and organic matter, by iron and manganese, or by iron alone.

**placodioid** Applied to a *lichen *thallus which is roughly circular and *crustose, with a determinate margin and radiating peripheral lobes.

**plaggen** A man-made soil *horizon more than 50 cm deep, resulting from long continued manuring, often enriched by phosphate. The name is from the German *Plagge* meaning 'sod'.

***Plagiochila*** (order *Jungermanniales) A genus of leafy *liverworts including *P. porelloides*, which is common in moist, shady habitats. The shoots are robust (2–10 cm long), erect or ascending, and little branched. The leaves are large, and overlap in the *succubous arrangement. Each leaf is broad and rounded and often minutely toothed. *P. porelloides* is dull green, and found on banks in woods, on rocks and walls, etc., especially on neutral or calcareous substrates.

**plagioclimax (biotic climax)** Plagioclimax and *biotic climax are almost synonymous terms, which are sometimes given different meanings. Generally, they both refer to a stable vegetation *community arising from a *succession that has been deflected or arrested, either directly or indirectly, as a result of human activities. An important distinction exists between these two cases. In a deflected succession the resulting stable community, even when composed entirely of native species, is one that would not have occurred had the disturbing human intervention not taken place. For example, the lowland heath communities of Britain and western Europe arose as a consequence of forest clearance and subsequent grazing pressure and controlled burning. Associated changes in the physical environment may mean that even when these pressures are removed succession to the original *climax community is no longer possible. In an arrested succession, the stable community is a naturally occurring successional phase: following removal of the disturbing factor, continuation of the natural succession should be possible. For example, the cutting of reed beds arrests the natural succession to alderwood in a *hydrosere. Some authors restrict the term 'plagioclimax' to deflected successions; some use 'plagioclimax' where human intervention is more direct, reserving 'biotic climax' for more indirect effects, such as grazing by non-domesticated but introduced animal species, e.g. rabbits in Britain. 'Biotic climax' may also be applied to a natural, undisturbed succession in which the final community form is determined by a naturally occurring biological agent, e.g. grasslands on guano-enriched coastal areas.

**plagiogeotropic** *See* GEOTROPISM.

***Plagiomnium undulatum*** *See* MNIUM.

***Planchonella*** **(family *Sapotaceae)** A genus of rain-forest trees whose timber is hard, heavy, and sometimes *siliceous (nyatoh). There are about 60 species, one in the Seychelles, 2 in America, the rest in the eastern tropics.

**plane tree** *See* PLATANACEAE.

**plankton** Aquatic organisms that drift with water movements, generally having no locomotive organs. The phytoplankton comprise mainly diatoms, which carry out *photosynthesis and form the basis of aquatic food-chains. The zooplankton (animals), which feed on the diatoms, may sometimes show weak locomotory powers. They include protozoans, small crustaceans, and in early summer the larval stages of many larger organisms. Plankton are sometimes divided into net plankton (more than 25 μm diameter) and nanoplankton, which are too small to be caught in a plankton net. The word is derived from the Greek *plagktos*, meaning 'wandering'.

**planktonic geochronology** The use of planktonic organisms (e.g. microscopic algae) to provide a relative dating of sediments deposited in marine waters. Radioactive-decay methods applied to planktonic organisms may also yield an absolute date or information on palaeoclimates.

**planogamete** A *motile *gamete.

**Planosols** Soils that have within 100 cm of the surface a soil *horizon that has been exposed to stagnant water for a prolonged period. Planosols are a reference soil group in the FAO *soil classification.

**planospore** A *motile *spore.

**planozygote** A *motile *zygote.

**plan position indicator (PPI)** A radar display in which the echo from a target appears as a bright mark against a dark field, its position on the screen indicating the direction and distance of the target from the scanner.

**Plantae** (Metaphyta) A kingdom that includes all the plants. The earliest true plants were probably unicellular green *algae, which first appeared in the *Precambrian. The *Bryophyta (mosses and liverworts) are known from the *Devonian, and the first recorded vascular plants (*Tracheophyta) date from the *Silurian.

**Plantaginaceae** A family of dicotyledonous (see DICOTYLEDON) *herbs, with a few shrubs and trees. It shows great variation, but all its members have more or less irregular flowers, usually distinctly 2-lipped, or sometimes with flat *corollas and 4 unequal lobes (e.g. *Veronica*). There are usually 4 *stamens, rarely 5 (e.g. *Verbascum* and *Penstemon*), or sometimes only 2 (e.g. *Veronica*). The *ovary is distinctive. It is *superior, of 2 cells joined together, with *axile *placentation and a terminal *style that is simple or bilobed. The fruit is normally a *capsule, rarely a *berry, and the seeds are numerous in each capsule cell. Genera such as *Antirrhinum*, *Calceolaria*, *Digitalis*, *Mimulus*, *Verbascum*, and *Veronica* have flowers attractive enough to be garden favourites, and *Digitalis* (foxglove) is the source of an important heart stimulant, but otherwise the family is of little economic importance. There are 222 genera, with about 4500 species, of cosmopolitan distribution.

**plant association** (association) A term loosely applied in the classification of vegetation *communities and given varying interpretations in different phytosociological traditions, with resulting confusion. The current most widely accepted usage follows the *Zurich–Montpellier school. In this, the association is the basic vegetation unit, an abstract entity, floristically defined from field data or relevés. Each association has a distinctive *faithful species and a group of high-presence or *constant species which give the community or association a cohesive structure. Often these companion species of the association form the faithful species of the next and succeeding hierarchical levels, alliances, and orders into which similar associations are grouped. In British and American traditions, 'association' has tended to imply a community with *physiognomic as well as floristic unity, usually a *climax community in which the species *dominants are those of the upper vegetation layer. In this tradition an association usually has several co-dominant species. The term 'consociation' is reserved for single-species dominance. *See also* FORMATION and PHYTOSOCIOLOGY.

**plantation** A closely set stand of trees (other than an orchard), usually comprising 1 or 2 species, that has been planted by humans. Plantations do not maintain themselves. The ground is cleared and the trees planted, grown, and harvested like an arable crop.

**plant available water** (PAW) The amount of water present in the soil that is available to plants. The *field capacity marks the upper bound of PAW and the *permanent wilting point the lower bound.

**plant hormone** (phytohormone) A compound that is synthesized by a plant, but is not a nutrient, *coenzyme, or detoxification product, and which regulates growth, differentiation, or other specific physiological processes. It is classically defined as being 'made at one site and functioning at another'; but in fact it is made by various tissues, transported actively or passively and affects most tissues within the plant.

**plant sociability** In the description and analysis of plant *communities, a measure of the distribution pattern and organization of a species. The *Zurich–Montpellier phytosociological scheme uses a 5-point sociability scale: sociability (1) growing once in a place, singly; (2) grouped or tufted; (3) in troops, small patches, or cushions; (4) in small colonies, extensive patches, or forming carpets; and (5) in great crowds, or pure populations.

**plant strategies** A term introduced by J. P. Grime, who defines it as groupings of similar or analogous genetic characteristics which recur widely among species or populations and cause them to exhibit similarities in ecology. Grime first proposed a threefold broad division of plant strategies into *competitors, *ruderals, and *stress-tolerators. *See also* FUNCTIONAL TYPE.

**plaque** In bacteriology, a round, clear area forming on an otherwise opaque culture plate of *bacteria as a result of their lysis (*see* LYTIC RESPONSE) by a *bacteriophage (virulent virus) or other agent. *Temperate phages produce turbid plaques because of the growth of *lysogenized bacteria on the floor of the plaque.

**plasmagene** A *gene present in any cell structure other than the nucleus.

**plasmalemma** See CELL MEMBRANE.

**plasma membrane** See CELL MEMBRANE.

**plasmid** A particle found in the cytoplasm of a bacterial cell that carries 1 or more genes and can replicate itself autonomously. Genetic information on the plasmid is passed from each cell to its daughter cell, and sometimes to neighbouring cells. Plasmids normally remain separate from the chromosome, but some may become integrated into it temporarily and replicated with it incidentally. Plasmids are important in the genes they carry (e.g. they may confer resistance to particular antibiotics to their bacterial hosts).

**plasmodesmata** Cytoplasmic (*see* CYTOPLASM) bridges, lined with a *plasma membrane, that connect adjacent cells. At one time they were thought to be confined to plants, but they have now been observed in many animal cells. Almost certainly they provide major pathways of communication and transport between cells.

**plasmodium** In *acellular slime moulds, a vegetative (feeding) structure consisting of a non-cellular, mobile mass of naked *protoplasm containing many nuclei.

**plasmogamy** The fusion of *cytoplasm between 2 cells, usually *gametes or *protoplasts. Plasmogamy occurs shortly before *karyogamy and sometimes results in a *heterokaryon.

**plasmolysis** The result of placing plant cells in a hypertonic solution (e.g. about 10 per cent sugar solution) so that water is drawn out of the cell. The *cytoplasm shrinks and the cell membrane is pulled away from the *cell wall.

**plastic growth** The part of *cell-wall extension that is irreversible.

**plastic limit** See ATTERBERG LIMITS.

**plastid** An *organelle that is believed to have evolved from an autotrophic (*see* AUTOTROPH) *endosymbiont early in plant evolution. Plastids occur in a variety of *morphologies, including *chloroplasts, *chromoplasts, and *leucoplasts, which are capable of being interchangeable. (Chloroplasts can develop into chromoplasts, and leucoplasts into chloroplasts.)

**plastocyanin** An acidic *protein, containing 2 copper atoms per molecule, that acts as an electron carrier, linking *photosystems I and II during *photosynthesis.

**plastoquinone** A derivative of *quinone that acts as an electron carrier, linking *photosystems I and II during *photosynthesis.

**Platanaceae** A family of deciduous trees which comprises only 1 genus, *Platanus*. Plane trees have flaking bark, buds enclosed in *petiole bases, and alternate, *stipulate, *palmately lobed leaves. The flowers are tiny, in several globular heads arranged along a stalk and forming a *catkin, each catkin either of male or female flowers. Male flowers each have 3–5 *stamens within a tiny cup, female flowers have 5–9 *carpels in a *calyx cup. Numerous hairs surround

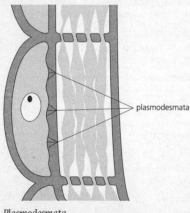

plasmodesmata

*Plasmodesmata*

the *nutlets in the globular fruiting heads. There are some 6 species in south-eastern Europe, south-western Asia, and N. America. They are much planted, especially in towns.

**Platanus** (plane tree)  See PLATANACEAE.

**Platycerium** (stag's-horn ferns; family *Polypodiaceae) A genus of ferns which have naked superficial *sori on erect, forked, antler-like, fertile *fronds. Sterile, broadly oval or rounded fronds bend down over the *rhizomes and help to trap *humus beneath them. They are *epiphytes of tropical rain forests, with 17 species, found mainly in the Old World.

**platyspermic** In seed plants, having seeds that are flattened and bilaterally symmetrical.

**pleated sheet** A structural configuration, normally found in fibrous *proteins, in which *polypeptide chains are partially extended and held together by inter-chain bonds between <NH and <CO groups on all *peptide bonds.

**Plectonema** A genus of filamentous *cyanobacteria (section III) which can carry out *nitrogen fixation. They are often difficult to distinguish from *Lyngbya and *Phormidium (the *LPP group). They are found in aquatic environments and on soil. Species growing on soil help to bind together soil particles and prevent erosion.

**plectostele** A *dictyostele type of *protostele in which in cross-section the *xylem and *phloem may form parallel, interwoven bands. The term is also applied to modified *actinosteles where the phloem strands separate the arms of the star-shaped xylem. Compare ATACTOSTELE, EUSTELE, and POLYSTELE.

**pleiotropy** The phenomenon of a single *gene being responsible for a number of different *phenotypic effects that are apparently unrelated.

**Pleistocene** The first of two epochs of the *Quaternary, preceded by the *Pliocene. It is held conventionally to have lasted from approximately 1.806 Ma ago until the beginning of the Holocene, about 10 000 years ago, but recent evidence from deep-sea cores may necessitate a revision of the earlier date. The epoch is marked by several glacial and *interglacial episodes in the northern hemisphere, during which the climate varied from very cold, with *tundra vegetation to middle latitudes in N. America and Eurasia, to warm temperate, with temperate forests.

**Pleistocene refugium** A favourable area where species survived periods of glaciation during the *Pleistocene Era. Such species are termed relic(t)s.

**Pleistogene** The most recent of the three periods comprising the *Cenozoic Era, preceded by the *Neogene. It began 1.806 Ma ago and continues to the present day. The Pleistogene is divided into the *Pleistocene and *Holocene Epochs.

**pleomorphism** The ability of an organism to exist in different forms or shapes.

**plesiomorph** Primitive (of a character state); the opposite of apomorph.

**plesiomorphic** Applied to features that are shared by different groups of biological organisms and are inherited from a common ancestor. The term means 'old-featured' and the features to which it is applied were formerly called 'primitive'.

**Pleurocapsales** (division *Cyanobacteria) An outdated phycological order of 'blue-green algae' which are now included in the *cyanobacteria, section II.

**pleurocarpous moss** A type of moss in which *archegonia, and hence *capsules, are borne on short, lateral branches, and not at the tips of stems or branches. Pleurocarpous mosses are usually monopodially branched, often pinnately so, and tend to form spreading carpets rather than erect tufts. Compare ACROCARPOUS MOSS.

**pleuropneumonia-like organisms** (PPLO) An early name for Mycoplasma species. See MYCOPLASMATACEAE.

**Pleurotus** (family *Polyporaceae) A genus of *fungi which form fleshy, *gilled *fruit bodies. The *stipe is short and usually eccentric, or absent. Pleurotus species are found mostly on wood, and some are edible. *See also* OYSTER FUNGUS.

**Pleurozia** (order *Jungermanniales) A genus of leafy *liverworts, of which *P. purpurea* is typical. It is a robust, distinctive plant with prostrate primary stems from which arise leafy shoots that are more or less erect, and deep purplish-red. Each leaf consists of 2 parts. One part is large, concave, with a shallowly cleft and toothed tip. The other part is much smaller, deeply concave and hooded; a flap at the base acts as a valve, allowing water to enter but preventing its escape. Microscopic animals may become trapped in these tiny 'pitchers' and may act as an additional source of nutrients for the *liverwort. The species is common on wet, *peaty *heathland and *moorland in western Scotland and Ireland.

**plexus** In data analysis, a diagram in which the lengths or widths of interconnecting lines reflect the relative similarity of the samples. It provides a useful visual presentation of similarity/dissimilarity indices.

**plicate** **1.** Folded or wrinkled. **2.** *See* VERNATION.

**plinthic horizon** A soil *horizon that is rich in *plinthite.

**plinthite** A constituent of some well-weathered tropical soils that results from prolonged *leaching and gleying (*see* GLEY). On drying plinthite changes irreversibly to an ironstone *hardpan.

**Plinthosols** Soils that have an iron-rich soil *horizon containing more than 25% *plinthite within 50 cm of the surface. The plinthic material hardens when exposed to the air. Plinthosols are a reference soil group in the FAO *soil classification.

**Pliocene** The last of the *Tertiary epochs, which began 5.332 Ma ago and ended 1.806 Ma ago.

**Ploiarium** (family *Bonnetiaceae) A genus of small, bushy, evergreen trees in which the flowers are borne in terminal clusters, with *stamens in 5 bundles, and opposite *petals. The fruit is a *capsule. There are 3 species, occurring from South-east Asia to New Guinea.

**plotless sampling** Sampling without the use of *quadrats. Individual *pin-frame records may be regarded as plotless samples. Plotless sampling is most often used for surveys of forest vegetation, especially where a rapid inventory is needed. Various approaches have been developed (e.g. the *nearest-neighbour sampling method).

**Plumbaginaceae** A family of *annual and *perennial *herbs, shrubs, and climbers, many of which are *halophytes. The leaves are *simple, either in a basal rosette or alternately placed along a branched stem. They have water or chalk glands in their surface, and no *stipules. The *inflorescence is a *cyme, *raceme, *spike, or densely clustered head, with the *bracts sometimes forming an *involucre. The flowers are bisexual and regular, with the parts in fives. The *calyx is fused, 5-lobed, and often coloured. The *petals are free or basally fused. The *stamens are either free, or fused in the base of the *corolla, and placed opposite the petals. The *ovary is *superior, with 5 fused *carpels, and *unilocular. The *stigmas and *styles are in fives. The fruit is a *nut, usually enclosed in the persistent *calyx, with the seeds containing a floury *endosperm. Several species are of medicinal value and many are grown as ornamentals, particularly in rock or scree gardens. There are 22 genera, with about 440 species, found throughout the temperate and tropical regions, mainly coastal.

**Plumeria** (family *Apocynaceae) A genus of small trees which includes *P. rubra* (frangipani), which has flowers with a heavy fragrance and is widely cultivated in the East. The genus comprises 7 species, found in the tropical regions of America.

**plum pox (sharka disease)** A disease of plum and related trees whose symptoms are variable according to the plant species, but which usually include the appearance of pale or dark rings or spots on leaves and *fruit. Plum pox is a disease of considerable economic importance, and is a notifiable disease in Britain. It is caused by a virus and is transmitted by aphids.

**plums** *See* PRUNUS.

**plumule** The shoot apex, or terminal bud, of an *embryo.

*Pluteus* (family **Pluteaceae, order** *Agaricales*) A genus of *fungi in which the *spore print is pink. The *fruit bodies are *mushroom-shaped. The *gills are not attached to the central *stipe. There are many species, found mainly on rotting wood.

**pneumatocyst** The hollow part of the *stipe, which acts as a float in some *Phaeophyta.

**pneumatophore** Specialized 'breathing' root developed in some plant species that grow in waterlogged or strongly compacted soils, e.g. mangroves. The aerial part of the root contains many pores, enabling gas exchange with the atmosphere. Internally, a well developed system of intercellular spaces allows gases to diffuse throughout the submerged portion of the roots.

**pneumococcus** The common name for the *bacterium *Streptococcus pneumoniae*, one of several causal agents of pneumonia.

*Poa* (meadow grasses; family *Poaceae) A genus of grasses that have *compound *panicles and *glabrous leaves. The *spikelets are keeled and usually of 2–5 *florets. The *lemma is keeled and nerved, often with a tuft of long, soft hairs arising from the base, and usually with a *hyaline tip, and no *awn. The lemma is longer than the awnless *glume. There are 3 *stamens with long *anthers, and a short, terminal *style. *P. annua* (annual meadow grass) is thought to be the most widely distributed and the commonest grass species in the world. It is found in all regions, but in the tropics is restricted to mountainous areas. There are more than 250 species, found in all temperate and cool regions, and at higher altitudes in the tropics. Many species are valuable pasture grasses.

**Poaceae** The grass family, known formerly as Gramineae; a very large and important family of *monocotyledons, most of which are *annual or *perennial herbs, but a few genera of which (e.g. the bamboos) are woody. From an ecological viewpoint, it is the most successful family of flowering plants. Where forests have been destroyed, grasses have tended to replace the trees as the dominant vegetation. The Poaceae includes the cereal grasses, including wheat, barley, oats, maize, rice, and millet, making it the most economically important of all plant families. Many grasses are important sources of fibres. Most grasses have hollow stems with solid *nodes containing intercalary *meristem and leaves in two opposite and alternating rows. Each leaf consists of a sheath around the *culm, a blade, and usually a flap or *ligule at the junction of sheath and blade. *Inflorescences are very varied, but are usually composed of *spikelets, each with a pair of sterile *glumes at the base. Each spikelet consists of 1 to many florets, each floret normally having 2 subtending scales (the lower chaffy *lemma and the upper membranaceous *palea), 3 *stamens with long *filaments and flexible *anthers (adapted to wind pollination), and an *ovary with 1 *ovule and 2 long, feathery *stigmas. There are 660 genera with about 10,000 species, distributed throughout the world.

**poached soil** *See* PUDDLED SOIL.

**pocket rot** A type of timber decay in which the decay occurs in small, isolated areas or pockets. *Heterobasidion annosum* is an important cause of white pocket rot in both conifers and hardwoods.

**poculiform** Shaped like a goblet.

**pod** A *fruit that *dehisces down both sides into 2 separate valves, which are most typically dry and somewhat woody. Pods are the characteristic fruit of the *Fabaceae.

**podetium** In certain *lichens (e.g. *Cladonia*), a *fruticose portion of the *thallus upon which *ascocarps (when present) are borne. A podetium may or may not be branched, and may be cup-shaped, awl-shaped, etc.

**Podocarpaceae** A family of conifers that is largely confined to the southern hemisphere, with a few species extending north across the equator. The seeds are borne inverted, singly or in pairs, not in cones, and each seed has a fleshy *aril or a fleshy *receptacle and stalk. The *pollen grains are winged. There are 12 genera, comprising about 155 species.

**Podocarpus** (family *Podocarpaceae) A genus of evergreen trees and shrubs which are *dioecious and whose mature seeds are fleshy, solitary, and often enclosed in the fleshy, enlarged *receptacle. They yield valuable timber. *P. totara* (totara) is a New Zealand timber tree with great importance in Maori mythology. There are 94 species, occurring in southern temperate regions and extending north into the tropics.

**Podophyllum** (family Boberidaceae) A genus of *herbs that are closely related to the *Berberidaceae. The fruit is a *berry, with the seeds enveloped in the enlarged *placenta. The *rhizome resin provides a drug used to treat warts. There are 2 or 3 species occurring in eastern Asia, and 1 species (*P. peltatum*, May apple) in N. America.

**Podostemaceae** A family of aquatic herbs whose structure is often modified into a *thallus with anchoring roots. The flowers are tiny, bisexual, regular, and reduced. The fruit is a *capsule. They are locally common and conspicuous, especially in tropical America, and rare and disjunct in the eastern tropics. They may be related to the *Saxifragaceae. There are 50 genera, comprising 275 species, occurring mainly in the tropics where they are confined to swiftly flowing rivers.

**podsol** *See* PODZOL.

**podsolization** *See* PODZOLIZATION.

**podzol (podsol)** A soil *profile formed at an advanced stage of *leaching by the process of *podzolization, and identified by its acid *mor *humus, eluviated (*see* ELUVIATION) and bleached E *horizon, and an iron-coloured B horizon, enriched with a variety of translocated materials. True podzols usually support heathland or coniferous forest. Podzols are a reference soil group in the FAO *soil classification. *See also* SPODOSOLS.

**podzolization** (podsolization) An advanced stage of *leaching, podzolization is the process of the removal of iron and aluminium compounds, *humus, and *clay minerals from the surface soil *horizons by an organic leachate solution, and the deposition of some of these translocated materials in lower B horizons.

**Pohlia** (order *Bryales) A genus of mosses in which the plants are usually tufted. The leaves are ovate to lanceolate, and are more crowded towards the stem apex. *Capsules are ovoid to ellipsoid and erect to pendulous. *Pohlia* is a cosmopolitan genus containing about 155 species. *P. nutans* is a very common British species. The shoots are green or yellowish-green, up to 7.5 cm tall, with stems *tomentose below. In spring and summer the unripe capsules are conspicuous: they are bright green and pendulous, and borne on long red *setae. They are found in *peaty or sandy soil and on decaying wood, and are *calcifuge.

**Poikilospermum** (family *Urticaceae) A genus of shrubs and climbers that are somewhat intermediate between Urticaceae and *Moraceae. There are 20 species, occurring from the Himalayas to Malesia.

**point mutation** A mutation that can be mapped to one specific *locus: it is caused by the substitution of one *nucleotide for another. A point mutation may also be caused by *deletion and *inversion.

**point quadrat** A *quadrat sampled by the *pin-frame method.

**poison ivy** *See* RHUS.

**Poisson distribution** The basis of a method whereby the distribution of a particular attribute in a population can be calculated from its mean occurrence in a *random sample of the population, provided that the population is large and the probability that the attribute will occur is less than 0.1. For a given mean, the distribution can be calculated, giving the probability that a sample will contain 0, 1, 2, 3, ... examples of the particular attribute. The distribution is named after the French mathematician S. D. Poisson (1781–1840).

**pokeweed** *See* PHYTOLACCACEAE.

**polar desert soil** Mineral soil without identifiable *horizons, and with almost no surface *humus. It is associated with arid, polar desert environments where precipitation is less than 130 mm annually, plant cover less than 25 per cent, and thawed, active soil is 20–70 mm deep.

**polarilocular** Applied to the *spores of certain *lichens, which have 2 cells separated by a partition that is perforated by a thin pore or channel.

**polar molecule** A molecule in which, though it does not carry a net electric charge, the electrons are unequally shared between the nuclei. In the water molecule, for example, the pull of the oxygen nucleus on the shared electrons is greater than the pull of the hydrogen nuclei. As a result, the oxygen end of the molecule is slightly negatively charged, and the hydrogen ends of the molecule are each slightly positively charged. The molecule is said to have a dipole moment and can attract other molecules with a dipole moment.

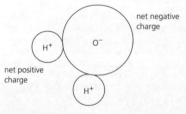

*Polar molecule*

**polar nuclei** Two *haploid nuclei inside the mid-region of the *embryo sac of a flowering plant which fuse with another nucleus (1 of 2 sperm nuclei) to form the *triploid *endosperm.

**polar transport** The movement in one direction within a plant, basipetally in shoots, of plant growth substances and plant growth regulators by a mechanism that requires metabolic energy and that does not involve intact *xylem or *phloem streams.

**pole** A single-trunked tree that is smaller and usually lower than a *standard. A *coppice pole is a stem arising from a coppice stool.

**Polemoniaceae (phloxes)** A dicotyledonous (*see* DICOTYLEDON) family, mainly of *perennial or *annual *herbs but including some shrubs, climbers, and trees. The leaves are *simple or *compound, *opposite or alternate, without *stipules. They are *glabrous or with short hairs. The *inflorescence is a *cyme or a solitary flower, and either *axillary or terminal. The flowers are regular and bisexual, with 5 *sepals and 5 *petals. The *calyx is a fused tube, and the *corolla is fused into a bell-shape or a plate-shape. The 5 *stamens are fused to the corolla tube. The *ovary is *superior, usually of 3 fused *carpels and 3 *locules, each with one or more *ovules. The *style is simple, with a 3-lobed *stigma. The fruit is a *capsule which splits to distribute the numerous, *endospermic seeds. The *testa is often sticky when wet. The pollination of the plants is usually carried out by bees, but several species rely on specific pollinators, such as humming-birds, bats, hawk moths, and other Lepidoptera. Many species, including *Phlox*, are cultivated for their showy flowers. There are 20 genera, comprising 275 species, found mainly in N. America, but with tropical species in Central America, and other species in Europe, northern Asia, and western S. America.

**pollard** To behead a tree at a convenient height, usually about 2 m above ground level, in order to produce a crown of small poles, suitable for firewood, fencing etc. This allows the production of small material out of the reach of deer and farm livestock.

bolling

*Pollard*

**pollarding** A system of wood management in which the main stem of a tree, usually a young one, is cut off about 2 m above ground level, thus favouring the development of lateral branches. With repeated pollarding a slightly swollen pollard boll develops in the main stem immediately beneath the lateral branches. Frequent pollarding, commonly seen in willows (*Salix* species), favours many relatively thin-stemmed lateral branches. Pollarding rather less frequently was the traditional management practice associated with wood-pastures, where grazing pressure made *coppicing, the more usual method for generating new growth, impracticable. Wood-pastures were particularly characteristic of slower-growing upland woodlands, and woodlands on infertile soils. Pollarding was also used to encourage particular sizes and shapes of timber needed for structural purposes. In Britain, the Ancient and Ornamental Woodlands of the New Forest include classic *relict wood-pastures with pollarded beeches and oaks.

**pollen** Collectively, the mass of *microspores or *pollen grains produced within the *anthers of a flowering plant (*angiosperm) or the male *cones of a *gymnosperm.

**pollen analysis** *See* PALYNOLOGY.

**pollen assemblage zone** *See* POLLEN ZONE.

**pollen diagram** A standardized pictorial summary of the *pollen record for a particular location. The vertical axis represents depth, and the proportions or absolute amounts of the various pollen types occurring at different levels are shown by bar histograms or by points on a continuous curve. Conventionally, similar patterns are grouped together on the diagram, with arboreal types shown first, followed in turn by shrubs, *herbs, and *spores.

**pollen grain** A *microspore in flowering plants, which germinates to form the male *gametophyte, a structure made up of the pollen grain plus a *pollen tube. The grain contains 3 *haploid nuclei (a *tube nucleus and 2 sperm nuclei), which pass down the tube to the *ovum. One of the

sperm nuclei fertilizes the ovum, and the second fuses with the 2 *polar nuclei forming the *endosperm. The tube nucleus (which is considered to be vestigial, having been completely functional earlier in the evolution of flowering plants) degenerates after *double fertilization (so called because of the two unions of nuclei).

**pollen mother cell** The *microsporocyte which undergoes 2 *meiotic divisions to produce 4 *microspores (the first cells of the male *gametophyte generation of *seed plants). Each microspore becomes a *pollen grain.

**pollen rain** The *pollen grains and *spores that fall on a particular site.

**pollen sac** The structure within which *pollen grains are formed in *angiosperms and *gymnosperms.

**pollen tube** The tube formed from a *germinating *pollen grain, and down which the 2 male *gametes (sperm nuclei) pass to the *ovum. In flowering plants each pollen tube penetrates the tissues of *stigma and *style into the *ovary where it grows towards an *ovule, passing down the *micropyle to the *nucellus and finally penetrating into the *embryo sac. Here the tip of the tube ruptures, releasing the 2 male gametes: one of these fuses with the nucleus of the ovum; the other fuses with the 2 *polar nuclei to give rise eventually to the *endosperm.

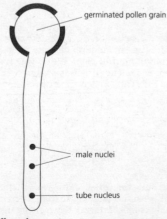

germinated pollen grain

male nuclei

tube nucleus

*Pollen tube*

**pollen zone (pollen assemblage zone)** One of the parts into which *pollen diagrams are divided on the basis of the total pollen assemblage found within them. Zone boundaries are placed at points where there is greatest evidence for change, hence the assemblage of pollen within a zone is relatively homogeneous. Pollen zones are defined by their content of pollen types and can be dated by reference to *radiocarbon dating or other methods, thus enabling them to be established as chronozones. Pollen assemblage zones are thus intended to delimit periods during which the surrounding vegetation has been relatively stable. This may be due to climatic stability within that period, or to a lack of change in human land management. Various schemes have been introduced in the past, particularly by K. Jessen in Denmark and Sir Harry Godwin in Britain, which have attempted to construct regional pollen zones. In Britain, for example, a series of eight pollen zones was established by Godwin to cover the late *Devensian (last glacial) and the *Holocene (*Flandrian). Lack of synchroneity even within the British Isles, however, has largely led to the abandonment of the scheme. Pollen assemblage zones are now constructed for each individual site and regional comparisons and correlations can then be made on the basis of chronology.

**pollination** The transfer of *pollen grains from the *anther (part of the *androecium) to the *stigma (part of the *gynoecium) of a flowering plant. This process facilitates contact between male *gametes and the female *ovum, leading to *fertilization, development of *seed, and thence a new plant. In *gymnosperms, the *pollen tube containing a pollen grain grows down and penetrates the neck of the *archegonium, facilitating contact between the sperm cells (male gametes) and the ova. Most archegonia contain many ova, so that multiple fertilization can occur, although only 1 sperm can fertilize an egg. Unlike *angiosperms, which do not possess archegonia, pollen cones and seed cones mature at different times within a season, so that there is usually a long interval between pollination and fertilization.

**pollinium** A coherent mass of *pollen grains, the product of a single *anther lobe, transported as a single unit in *pollination (as in the *orchids).

**pollutant** A substance that enters the environment or becomes concentrated within it, and that will, or may, harmfully affect human life or that of desirable species. Pollutants are by-products of human activity (*compare* ALLELOPATHY) and the term embraces noise and the release of substances at temperatures markedly higher than those of the receiving media.

**pollution** The defiling of the natural environment by a *pollutant. *Compare* ALLELOPATHY.

**polyacrylamide gel** A gel used to separate biological molecules (e.g. *proteins of a given range of sizes). It is prepared by mixing a monomer (acrylamide) with a cross-linking agent (N,N'–methylenebisacrylamide) in the presence of a polymerizing agent, and leads to the formation of an insoluble 3-dimensional network of monomer chains, which become hydrated in water. Gels of different pore sizes may be prepared depending upon the relative proportions of the ingredients.

*Polyalthia* (family *Annonaceae) A genus of small trees in which the crown is *monopodial. The flowers are bisexual, with 3 *sepals and 6 *petals which are often showy. There are a few to many fruits. There are about 120 species, occurring in the Old World tropics, especially in Southeast Asia.

**polyarch** Applied to *primary xylem that comprises many strands.

**polycentric** Applied to a *thallus that forms several or many reproductive centres.

**polycentromic** *See* CENTROMERE.

**polychore** *See* WIDE DISTRIBUTION.

**polychronic** *See* POLYTOPISM.

**polyclimax theory** *See* CLIMAX THEORY.

**polycyclic** Applied to a *stele in which the *vascular tissue forms 2 or more concentric rings.

**polygamous** Applied to a plant species in which combinations of male, female, and hermaphrodite flowers occur on the same or different plants.

**polygene** One of a group of *genes that together control a quantitative character such as height or weight. Individually each gene has very little effect on the resulting *phenotype which instead requires the interaction of many genes. Before genes had been described, a multifactorial hypothesis had been constructed to explain the absence of clear-cut segregation into readily recognizable classes that show typical Mendelian ratios, as occurs with quantitative characters. In this case a group of interacting factors corresponds to a group of interacting genes.

**polygenic** See MULTIFACTORIAL.

**polygenic character** A quantitatively variable character (*phenotype) which is dependent upon the interaction of a number of *genes.

**Polygonaceae (buckwheat, dock, rhubarb, sorrel)** A family of *herbs, with a few shrubs and trees, in which the leaves are *simple, entire, and usually alternate. They have an unusual sheathing *stipule known as the *ochrea (or ocrea) at the leaf base. The flowers are bisexual, regular, and are usually held in a *raceme if they are not solitary. They are white, green, or pinkish. The 3–6 *sepals often form a membrane wing around the fruit. There are no *petals, and 6–9 *stamens. The *ovary is *superior, with 2–4 *carpels which are united to produce a single *locule with free *styles. The fruit is a *trigonous *nut with a seed containing a copious *endosperm. The seeds are often dispersed by wind with the aid of the *calyx wing; other seeds are hooked. The family is closely related to the *Plumbaginaceae and is separated into 3 groups based on the geographical zone in which they are distributed; (a) the tropical, half-hardy, climbing trees and shrubs; (b) the shrubby buckwheats of arid habitats in south-eastern Europe, central Asia, and N. America; and (c) the temperate species, including the edible species (e.g. *Rheum rhaponticum*, rhubarb, and *Fagopyrym*

*esculentum*, common buckwheat), the ornamentals (e.g. *Muehlenbeckia* species and *Erigonum* species), and the weeds (e.g. *Rumex* species, sorrels and docks). There are 51 genera, comprising about 1050 species, distributed throughout the world, but with a concentration in the northern temperate regions.

**polygonatous** Of a stem, having many knots.

**polyhaline 1.** Applied to a highly salt-tolerant species that is not usually found outside very saline environments, e.g. *Limonium vulgare* (common sea lavender). **2.** The second most saline zone of a salt-marsh, according to the Venice system for the classification of *brackish waters.

***Polyides rotundus*** See GOAT TANG.

**polymerase** An *enzyme that catalyses the replication and repair of *nucleic acids.

**polymictic** Applied to lakes (e.g. those in high altitudes in the tropics) whose waters are circulating virtually continuously. If periods of stagnation occur, they are very short.

**polymorphic** Occurring in several different forms. *Compare* MONOMORPHIC.

**polymorphism** In genetics, the existence of 2 or more forms that are genetically distinct from one another but contained within the same interbreeding population. The polymorphism may be transient or it may persist over many generations, when it is said to be balanced. Some visible polymorphism can readily be seen in nature. Other examples are cryptic and require biochemical techniques to identify phenotypic differences. Such techniques include gel electrophoresis of *enzymes and other proteins, and the fragmentation of the *DNA molecule by restriction enzymes (which allows the sequencing of *nucleotides), both of which operate nearer to the level of the *genotype. *See also* POLYTOPISM.

**Polynesian floral region** R. Good's (*The Geography of the flowering Plants*, 1974) Polynesian subkingdom within his *palaeotropical kingdom, which contains a small, derived flora with only 9 or 10 endemic genera (*see* ENDEMISM), scattered through the various islands. The affinities are with the Malaysian and Australian regime, and there is no distinct 'Polynesian' flora as such. The great majority of the floral elements have been derived from adjacent floras. *See also* FLORAL PROVINCE and FLORISTIC REGION.

**polypeptide** A linear polymer that consists of 10 or more *amino acids linked by *peptide bonds.

**polyphyletism** The occurrence in taxa of members that have descended via different ancestral lineages. True polyphyletism has traditionally been distinguished from errors of classification, especially at the higher taxonomic levels, where organisms, as a result of *convergent or *parallel *evolution, have been placed wrongly in the same natural group; but some modern phyletic taxonomists would hold that any taxon found to be polyphyletic is unnatural, and so an 'error', which must be disbanded.

**polyploidy** The condition in which an individual possesses 1 or more sets of *homologous *chromosomes in excess of the normal 2 sets found in *diploid organisms. It is caused by the replication within a *nucleus of complete chromosome sets without subsequent nuclear division. Examples are triploidy ($3n$), tetraploidy ($4n$), hexaploidy ($6n$), and octoploidy ($8n$). For example, modern bread wheat (*Triticum aestivum*) has 42 chromosomes as a consequence of *allopolyploidy, a type of polyploidy. It arose from the interbreeding of emmer wheat (*T. turgidum*) with *T. tauschii*, a wild wheat that is found in Iran. *T. turgidum* has 28 chromosomes and *T. tauschii* has 14, so the *zygote would have had 21 chromosomes. This number doubled to 42, thus producing cells with homologous chromosome pairs and a species (*T. aestivum*) capable of producing fertile offspring.

**Polypodiaceae** A family of *leptosporangiate ferns, in which the *sori are mixed, superficial on the lower surface of *fronds, and usually rounded. They have a cosmopolitan distribution. In the wider sense, some 170 genera and several thousand species were included, but in the present-day classification (J. A. Crabbe and A. C. Jermy, *Fern Gazette* 11, 1975) the family is restricted to 56 genera.

**Polypodium** (family *Polypodiaceae) A genus of ferns with *pinnatifid or 1-pinnate fronds borne on creeping fleshy scaly *rhizomes; the *sori are naked and rounded. The sori vein ends on each side of the midrib of the *pinnae, and leaf veins form loops with the free vein ends inside. There are some 75 species, with a cosmopolitan distribution.

**Polyporaceae** (order *Aphyllophorales) A family of *fungi in which the *hymenium typically lines tubes that open to the exterior by pores. The *fruit body may be crust-like, *bracket-like, or hoof-shaped, and may be *annual or *perennial. They grow on wood and are mainly *saprotrophic, but some are *parasitic. The family includes some important tree *pathogens and agents of timber decay.

**polypore** Any *fungus of the order *Aphyllophorales in which the *fruit body bears its *hymenium in a layer of tubes.

**Polyporus** (family *Polyporaceae) A genus of *fungi in which the *fruit body is typically stipitate, the *stipe being central, eccentric, or lateral. The context is white or pale-coloured. The *spores are smooth and elongated. Species are found on wood; some are *parasitic on trees, causing a *white rot. *See also* DRYAD'S SADDLE.

**polyribosome** *See* RIBOSOME.

**polysaccharide** A linear or branched polymer that consists of 10 or more monosaccharides linked by glycosidic bonds (*see* GLYCOSIDE).

**Polysiphonia** (family Rhodomelaceae, order *Ceramiales) A genus of red seaweeds that grow either on stones or as *epiphytes on other *algae. *P. lanosa* is an *obligate parasite which grows on *Ascophyllum*; the *thallus is filamentous and branched.

**polysome** A group of *ribosomes that are attached to a single strand of *m-RNA.

Apparently, each ribosome is at a different stage in the synthesis of the same *polypeptide chain. Accordingly, the number of ribosomes in the polysome varies with the length of the message to be transcribed.

**polysomy** The condition in which a particular *chromosome is represented by several members. Compare DISOMY, NULLISOMY, TETRASOMY, and TRISOMY.

**polystele** A *dictyostele in which in cross-section several *xylems occur as single strands, each surrounded by a *phloem and each unit being a *haplostele type of *protostele. Compare ATACTOSTELE, EUSTELE, PLECTOSTELE, and SOLENOSTELE.

**Polystichum** (family *Aspleniaceae) A genus of *leptosporangiate ferns with spine-toothed *pinnules, and rounded, superficial *sori with circular peltate indusia; the leaves are of a leathery texture. There are some 135 species, of cosmopolitan distribution.

**Polystictus** An obsolete genus of *fungi; species formerly included in this genus have now been transferred to other genera. (*P. versicolor* is now *Coriolus versicolor*; *P. perennis* is now *Coltricia perennis*). The name is still used occasionally as a *form genus for any fungus that forms *annual, leathery, *bracket-shaped *fruit bodies.

**polythetic** Using several or all possible criteria or attributes as the basis for each subdivision of a sample population or for agglomeration of individuals or groups. Compare MONOTHETIC.

**polytopic evolution** See POLYTOPISM.

**polytopic species** See POLYTOPISM.

**polytopism** (polytopic evolution) A type of *monophyletism in which a new (polytopic) *taxon arises in more than one place from *conspecific parents. The chances of this happening simultaneously, except perhaps in the case of subspecies, in each instance are remote; polytopic species may therefore be regarded also as polychronic species.

**Polytrichales** (subclass *Bryidae) An order of *acrocarpous mosses which may

in some species be quite large (e.g. up to 40 cm); the stems are tough, and show some internal differentiation. The leaves have characteristic longitudinal plates or lamellae on the upper surface along the nerve; the leaf bases often sheath the stem. Species are found on the ground in woods, on *heathland, etc.

**Polytrichum** (order *Polytrichales) A genus of mosses in which the plants grow in patches resembling miniature forests; the shoots arise from an underground, *rhizome-like stem. The leaves have a broad, sheathing basal portion which abruptly narrows into a lanceolate to linear-lanceolate blade with a very wide nerve. The *calyptra is characteristically densely hairy and covers the *capsule. A cosmopolitan genus, it is found on the ground in woods, *moors, *heaths, etc., and is *calcifuge. See also HAIR MOSS.

**polytypic** Of a species: divided into subspecies; varying geographically.

**polytypism** The occurrence of *phenotypic variations between populations or groups of a species that are geographically distinct. It is contrasted with *polymorphism, which is variation within a population or group. A species with systematic geographical variation (*subspecies or *clines) is said to be polytypic.

**Pomaderris** (family *Rhamnaceae) A genus of shrubs and trees that have alternate or *opposite, *simple leaves with *stipules. The flowers are small and inconspicuous, regular, bisexual, and usually borne in *cymes. There are 4 or 5 *sepals. The *petals are not always present, but if present they are small and incurved, often folding over the 4 or 5 *stamens. The *ovary is *superior with 2 or 3 *locules, and a simple *style. The fruits are fleshy and the seeds are usually dispersed by mammals and birds which feed on the fruits. Several species are cultivated as ornamentals. There are about 40 species, confined to Australia and New Zealand.

**pome** A *fruit in which the seeds are protected by a tough *carpel wall and the entire *fruit is embedded in a fleshy *receptacle. In an apple, the carpel wall surrounding the seeds comprises the core, which is the true fruit, the edible fleshy part of the apple being the receptacle.

**pomegranate** See PUNICACEAE.

**Pometia** (family *Sapindaceae) A genus of trees in which the leaves are *pinnate. The fruit is a leathery *capsule with a few large, *arillate seeds. The trees yield valuable timber. There are 12 variable species, occurring from Indo-Malaysia to Melanesia.

**pondweeds** See POTAMOGETONACEAE.

**Pontederiaceae** (order Liliales) A family of monocotyledonous (see MONOCOTYLEDON) freshwater, aquatic plants. The leaves have sheathing bases, often petiolate. The flowers are blue, lilac, yellow or white, with 6 *perianth segments, often borne in a showy *inflorescence enclosed by a *spathelike sheath. The fruit is a *capsule or *nutlet. The free-floating water hyacinth (*Eichhornia crassipes*) is probably the world's most serious aquatic weed. Other species are grown as aquatic ornamentals. There are 7 genera, comprising 29 species, occurring in the tropics and subtropics, especially America.

**poor man's weather-glass** See SEA-BELT.

**Pope's buttons** See BULGARIA.

**poplar** See POPULUS.

**poppy** See PAPAVERACEAE.

**population dynamics** The study of factors that influence the size, form, and fluctuations of individual species or genus populations. Emphasis is placed on change, energy flow, and nutrient cycling, with particular reference to *homoeostatic controls. Key factors for study are those influencing natality, mortality, immigration, and emigration.

**population ecology** (autecology) The study of the interaction of a particular species or genus population (or sometimes one of a higher *taxon) with its environment.

**population eruption** See POPULATION EXPLOSION.

**population explosion** (population eruption) The sudden rapid increase in size of the population of a species or genus. The most violent explosions occur when a species is introduced into a new locality where it finds unexploited resources of suitable food, shelter, etc., and a lack of negative controls such as *predators or *parasites. Examples include the population explosions of the prickly pear and of *Rhododendron ponticum* following the introduction of these species into Australia and Britain respectively.

**population genetics** The study of inherited variation in populations of organisms, and its modulation in time and space. Population genetics relates the heritable changes in populations to the underlying individual processes of inheritance and development. Such studies generally involve the estimation of *gene frequencies and the influences of *selection, *mutation, and *migration upon these frequencies in natural (and experimental) populations.

**Populus** (aspens, cottonwoods, poplars; family *Salicaceae) A genus of deciduous, *dioecious, *catkin-bearing trees, which differs from *Salix in having many *stamens in each male flower, toothed catkin scales, flowers with cup-like *discs, and several outer scales to each bud. *Populus tremula* is aspen. There are 35 species, found mostly in the northern temperate zone. Many, including hybrids, are grown for timber.

**porate** Applied to a *pollen grain which has 1 or more *pores.

**porcelain fungus** See OUDEMANSIELLA.

**pore 1.** (pedol.) A void surrounded completely by soil materials and created by the packing of mineral and organic particles. Pores can be filled by any proportion of air or water. **2.** An aperture on the surface of a *pollen grain, which is circular or slightly elliptical. *Compare* COLPUS.

**Porella** (order *Jungermanniales) A genus of leafy *liverworts, including *P. platyphylla*, which is a robust, conspicuous species which can form fairly extensive patches. The stems are 2- or 3-*pinnately branched and up to 8 cm long. The leaves overlap closely in the incubous arrangement. There are 2 lateral rows of leaves; each leaf is divided into 2 parts: a broad, ovate part, and a much smaller, narrower lobe which looks like a separate leaf. There is also a conspicuous row of leaves on the under-surface, so that when the plant is viewed from below it appears to have 5 ranks of leaves. The leaves on the undersurface are broad and rounded with narrowly recurved margins. It is found, often abundantly, on shaded banks, tree roots, etc., mainly in calcareous districts.

**pore space** The total continuous and interconnecting void space in the bulk volume of soil.

**Poria** (family *Polyporaceae) A genus of *fungi in which the *fruit bodies are *resupinate, bearing tubes lined with *hymenium; the context is white or pale-coloured. The spores are colourless, ovoid, and truncate. *Poria* is found on wood.

**porosity** The proportion, as the percentage volume, of the total bulk volume of a body of rock or soil occupied by *pore space.

**porphin** The parent compound of the *porphyrins: it consists of four *pyrrole-like rings, linked into a ring system by four CH groups.

**Porphyra** See BANGIALES and LAVER.

**porphyreus** Purple in colour.

**Porphyridiales** (division *Rhodophyta) An order of *algae that are either single-celled or consist of rows of cells held together by a thick mucilage. *Porphyridium* is a common unicellular soil alga, forming blood-red mucilaginous layers in the soil surface and on rocks and walls.

**porphyrin** A *heterocyclic derivative of a *porphin, which is composed of a *tetrapyrrole ring structure. As such it is capable of combining with a variety of metals and so forms part of the structure of many important biological molecules, including haemoproteins, *chlorophyll, vitamin $B_{12}$, and *cytochromes.

**Portlandian** The youngest stage of the *Jurassic in Europe, overlain by sediments of the *Cretaceous Berriasian Stage and resting in turn on those of the *Kimmerid-gian. It is characterized in S. England by limestones rich in molluscs.

**Portulacaceae** A family of *herbs that are close to *Caryophyllaceae but have only 2 *sepals and *petals that are often joined at the base. There are 22 genera, comprising about 400 species, in temperate and tropical regions, but centred in America.

**positive feedback** In a system, the mechanism by which a process intensifies or accelerates, as each cycle of operation establishes conditions that favour a repetition. Unless checked, positive feedback may lead to loss of control within the system and its eventual failure. *Compare* NEGATIVE FEEDBACK.

**post-climax** In the monoclimax model (*see* CLIMAX THEORY) of *climax vegetation development, *communities differing from the *climatic climax, owing to the presence of more favourable conditions for vegetation development. For example, forest may occupy river valley sites within temperate grassland *biomes because of the greater availability of water in summer in such locations. *Compare* PRE-CLIMAX.

**Post-glacial** *See* HOLOCENE.

**post-medieval woodland** Woodland that is known to have originated on a site that was not wooded prior to about 1650.

**Potamogeton** (family *Potamogeton-aceae) A genus of water plants with creeping, *sympodial *rhizomes and erect, leafy branches, which are chiefly confined to fresh water. The leaves are alternate, *simple or entire, and either all submerged or with some floating. Those with floating leaves tend to have broader leaves, while those with all the leaves submerged tend to have narrow leaves. This is believed to form a series in evolutionary terms from land-based to truly aquatic plants. The flowers are *axillary or terminal *spikes, and are inconspicuous. The parts are in fours. The fruit is a small, green to brown *drupe or *achene containing seeds. These seeds have no *endosperm but contain air, allowing the fruit to float. The plants overwinter as the complete plant, or die back to the rhizome, or form special branches with *tubers or winter buds. There are about 90 species, found throughout the world in a variety of aquatic habitats. Most species are edible, and provide food for many animals.

**Potamogetonaceae (pondweeds)** A cosmopolitan family of aquatic *herbs, found mostly in fresh water, that have submerged and translucent or sometimes leathery and floating leaves. The leaves may be narrow and linear, or elliptical. The inconspicuous flowers are borne in *axillary or terminal *spikes, are pollinated by water or wind, and have 4 *sepals, 4 *stamens, and 4 free *carpels, each producing a fruit that is a *nutlet. The family contains 2 genera, *Potamogeton (about 90 species) and Groenlandia (1 species).

**potassium (K)** An element that is required for healthy plant growth. Potassium *ions neutralize *anionic *macromolecules and organic acids and thus control the water potential of cells. They pass readily through membranes and are important in the movements of leaves and *guard cells. A potassium deficiency leads to reduced growth and to dark or blue-green coloration in the leaves, which may also develop a purple-brown pigment.

**potato** See SOLANUM.

**potato blight** A term that may refer to either late blight or early blight. Early blight is much the less serious of the two diseases and is caused by the *fungus Alternaria solani; dark, concentrically zoned spots appear on the leaves of infected plants. Both early and late blights can also attack tomato plants. See also LATE BLIGHT OF POTATO.

**potato scab** A term that may refer to either common scab or powdery scab. In common scab, caused by Streptomyces scabies, the scabs on the *tubers are corky and superficial; the disease is commonest on light soils and in dry conditions. Powdery scab is caused by Spongospora subterranea, and is most common on heavy soils and in wet seasons; the scabs differ from those of common scab in that they are powdery.

**potential evapotranspiration (PE)** The amount of water that would evaporate from the surface and be transpired by plants were the supply of water unlimited. It is calculated from the mean monthly temperature, with corrections for day length, and was devised by C. W. Thornthwaite as part of his system of *climate classification (see THORNTHWAITE CLIMATE CLASSIFICATION). From PE minus precipitation, an approximate index can be calculated of the extent to which the water available for plants falls short of the amount they are capable of transpiring. Compare ACTUAL EVAPOTRANSPIRATION.

**potential-natural** See NATURAL.

**potometer** An instrument that is designed to measure water uptake in a plant and, indirectly, to estimate *transpiration rates.

**Pottiales (subclass *Bryidae)** An order of *acrocarpous mosses which are usually small; the leaves may be pointed or rounded, and have a nerve that may extend beyond the tip of the leaf in a 'hair point'. There is one family. Genera include *Barbula, Pottia, and *Tortula. They are found in a range of habitats: on walls, on the ground, on tree bark, etc.

**powdery mildew** Either a *fungus of the order *Erysiphales or a plant disease caused by such a fungus. The leaves, fruit, etc., of an infected plant bear characteristic powdery white patches of *conidium-bearing *mycelium; tiny black *perithecia may also be visible. Many types of plant may be affected, usually each by its own particular species or strain of fungus.

**PPI** See PLAN-POSITION INDICATOR.

**PPLO** See PLEUROPNEUMONIA-LIKE ORGANISMS.

**praemorse (premorse)** Having the end terminated abruptly.

**Praghian** See DEVONIAN.

**prairie** A temperate grassland of northern America, dominated by more or less *xeromorphic grasses, which fall into three groups based on stature (tall, mid, and short) with a progressive decrease in rainfall. Various herbaceous broad-leaved *annuals and *perennials are mixed in with the grasses. *See* PALOUSE PRAIRIE.

**pre-adaptation** An *adaptation evolved in one adaptive zone which, quite by chance, proves especially advantageous in an adjacent zone and so allows the organism to radiate into it. No *selection for a future environment is implied.

**pre-Boreal** The first *Flandrian (*Holocene, or post-glacial) stage, a time of rapid forest spread, from about 10 300–9600 BP. Pre-Boreal refers to climatic conditions and in vegetational terms is equivalent to *Pollen Zone IV of the standard British and European postglacial pollen chronology.

**Precambrian** A name that is now used only informally to describe the longest period of geological time, which began with the consolidation of the Earth's crust and ended with the beginning of the *Cambrian Period 542 Ma ago. The Precambrian lasted approximately 4000 Ma; the rocks of this period of geological time are usually altered, and few fossils with hard parts or skeletons have been found within them, although Precambrian limestone rocks in Australia, Siberia, and parts of the USA contain *stromatolites, which are believed to have been formed by *cyanobacteria; frond-like impressions of a supposed plant (*Charnia*) have also been found. Precambrian rocks outcrop extensively in shield areas such as northern Canada and the Baltic Sea. In modern usage the Precambrian has been replaced by the *Hadean, *Archaean, and *Proterozoic Eons.

**precipitable water** The quantity of rainfall that would result from condensation and precipitation of the total moisture in a column of air in the atmosphere. Most atmospheric moisture is contained in the lower atmosphere, below about 5500 m. On average, an atmospheric column of 1 $m^2$ cross-section contains vapour equivalent to 5–25 mm depth of rainfall. The average residence time of moisture in the atmosphere is about 9 days.

**precipitation-efficiency index** Devised in 1931 by C. W. Thornthwaite, an index based on the ratio of mean monthly rainfall and temperature values to evaporation rates. Summation of monthly values gives an annual precipitation-efficiency index (P–E), which is used to define major climatic regions.

**pre-climax 1.** The *community immediately preceding the *climax, especially where this is forest. For example, on light soils in Britain, birchwood is often the pre-climax to oakwood. **2.** In the monoclimax model (*see* CLIMAX THEORY) of climax vegetation development, the name given to communities differing from the *climatic climax owing to environmental conditions that are less favourable for vegetation development than those of the surrounding region. For example, grassland might develop in rain-shadow areas in the lee of a mountain range, in an area that was otherwise forested. *Compare* POSTCLIMAX.

**predation** The interaction between species populations in which one organism, the predator, obtains energy (as food) by consuming, and usually killing, another, the prey. Most typically, a predator is an animal that catches, kills, and eats its prey, but predation also includes feeding by insectivorous plants and the term is sometimes applied to the grazing of plants by animals. In general, grazing activity removes part of a plant but does not kill it, hence it is closer to *parasitism than predation. The consumption of seeds, however, can be regarded as a form of predation if it involves the destruction of the embryo.

**predator** *See* PREDATION.

**predictive dormancy** *See* DORMANCY.

**preferential species** In *phytosociology, a species that is present with varying abundance in several *communities, but is especially abundant and vigorous in one particular community. It belongs to fidelity class 3 in the *Braun-Blanquet phytosociological scheme. *Compare* ACCIDENTAL SPECIES; EXCLUSIVE SPECIES; INDIFFERENT SPECIES; and SELECTIVE SPECIES.

**preformation** *See* EPIGENESIS.

**premorse** *See* PRAEMORSE.

**Pressler borer** *See* INCREMENT BORER.

**pressure potential** The hydrostatic pressure to which water in a liquid phase is subjected. It was known formerly as wall pressure or turgor pressure. In a *turgid plant cell, pressure potential is usually positive but the pressure potential of *xylem in a transpiring plant, which is under considerable tension, will be negative.

**prevailing climax** The most common undisturbed, stable *community that occurs in a region. In any given area various stable communities are found in response to local differences in *habitat. Gradual transitions from one stable type to another occur as environmental controls change. The steady-state community occupying the largest number of non-extreme habitats in the area is the prevailing climax. This approximates to the climatic climax communities for the area. Unlike monoclimax, prevailing climax does not involve consideration of possible long-term climatic changes, or associated habitat changes resulting from weathering and *erosion. Thus, it is a more practicable and readily applied concept. See also CLIMAX THEORY.

**prevernal** In the early spring. The term is used with reference to the 6-part division of the year used by some ecologists, especially in relation to studies of terrestrial and freshwater communities. Compare AESTIVAL; AUTUMNAL; HIBERNAL; SEROTINAL; and VERNAL.

**prey** See PREDATION.

**pride of India** See MELIA.

**primary growth** Growth that results from cell division and subsequent expansion at the apical *meristem. Compare SECONDARY THICKENING.

**primary phloem** *Phloem tissue that develops from the *procambium, *protophloem forming first, followed by *metaphloem.

**primary pigments** In *photosynthesis, pigments that emit electrons which directly drive the photosynthetic reactions. In green plants there are two primary pigments, both of which are forms of *chlorophyll a: P680 and P700. See PHOTOSYSTEM I and PHOTOSYSTEM II.

**primary pit** A *pit that results from incomplete separation of 2 daughter cells in *mitosis.

**primary productivity (primary production)** The rate at which *biomass is produced by photosynthetic and *chemosynthetic autotrophs (mainly green plants) in the form of organic substances, some of which are used as food materials by the autotrophs themselves. Gross primary productivity (GPP) is the total rate of *photosynthesis and chemosynthesis per unit ground area, including that portion of the organic material produced which is used in *respiration during the measurement period. Net primary productivity (NPP) is the rate of production allowing for the amount lost to plant respiration during the measurement period. Net *ecosystem productivity (NEP) is the rate of accumulation of organic material, allowing for both plant respiration and heterotrophic consumption during the measurement period, i.e. NEP = GPP − (respiration by autotrophs + respiration by *heterotrophs). Rates of storage at higher *trophic levels are termed secondary productivities. Strictly, the primary production of an ecosystem (as distinct from its productivity, which is a rate) is the amount of organic material accumulated per unit ground area. It is usual to indicate a period of time, however, since otherwise such data have limited value. Thus in practice the terms 'production' and 'productivity' are often used interchangeably. It is also vital to indicate whether production figures relate to net or gross, and to primary or ecosystem productivity, or to some portion of these, as with a harvested crop. In the oceans, photosynthesis by *phytoplankton in the upper 100 m (the *euphotic zone) accounts for most primary production. Waters in tropical areas are less productive than waters in temperate regions, because in tropical waters the water column undergoes no seasonal vertical mixing and so becomes *oligotrophic. Areas of upwelling of nutrient-rich deep waters have high productivity.

**primary sexual character** An organ that produces *gametes. Examples are the male *anther and female *carpel of flowering plants.

**primary structure** The basic structure of a *polypeptide chain. It relates solely to the type, number, and sequence of *amino acids or *nucleotides in the chain. *Compare* SECONDARY STRUCTURE; TERTIARY STRUCTURE; and QUATERNARY STRUCTURE.

**primary succession (prisere)** A *succession initated on a newly produced bare area, e.g. following glaciation or major earth movements. Since no living remnants of a previously existing *community are present, successional stages and interactions of plants and the physical environment are usually fairly clear, at least initially. *Compare* SECONDARY SUCCESSION.

**primary woodland** A woodland occupying a site that has been continuously wooded (in Britain since the last ice advance) even though it may have been clear-felled, provided that the clear-felling does not break the woodland continuity (i.e., the woodland regenerated or was replanted).

**primary woodland species 1.** Tree species that are confined to *primary woodland. **2.** Tree species that occur in primary woodland but also in localities where their presence is explained by planting or by their survival from the clearance of what was once primary woodland, e.g. in hedges and on stream banks.

**primary xylem** *Xylem tissue that develops from the *procambium, *protoxylem forming first, followed by *metaxylem.

**primitive (evol.)** Preserving the character states of an ancestral stage. The term may be used of a character (as a synonym of *plesiomorph) or, occasionally, of a whole organism.

**primordium** The early cells that serve as the precursors of an organ to which they later give rise by *mitosis during development.

**Primulaceae** A cosmopolitan family of *herbs, with a few low shrubs, in which there are no *stipules to the leaves. The flowers are regular, mostly with parts in fives, but sometimes up to nines. The *corolla is wheel-, bell-, or funnel-shaped and the *petals are usually joined into a tube below. The *stamens are attached to the corolla and lie opposite its lobes. The *ovary is usually *superior and 1-celled, with a free-central *placenta, and the *style is undivided. The family includes many attractive genera (e.g. *Primula*, *Cyclamen*, and *Lysimachia*). Modern classifications recognize 22 genera with about 800 species, found mainly in northern temperate regions.

***Primula vulgaris*** *See* HETEROSTYLY.

**princess palm** *See* DICTYOSPERMA.

**Priscoan** The name formerly given to the earliest of the three subdivisions of the *Precambrian. The time covered by the Priscoan is now ranked as the *Hadean eon, lasting from the formation of the Earth until 3800 Ma ago.

**prisere** *See* PRIMARY SUCCESSION.

**procambium** In a plant, the groups of elongated cells found at the growing tips of roots and stems: they develop into *vascular bundles as the plant continues to grow.

**procaryote** *See* PROKARYOTE.

**procumbent** Lying along the ground.

**producer** In an ecosystem, an organism that is able to manufacture food from simple inorganic substances, i.e. an autotroph, most typically a green plant.

**production 1.** In energy-flow studies, that part of the assimilated food or energy which is retained and incorporated in the *biomass of the organism, but excluding the reproductive bodies released by the organism. This may be regarded as growth. In energy-flow measurements, production is expressed as energy per unit time, per unit area. **2.** *See* PRIMARY PRODUCTIVITY.

**production ecology** The branch of ecology dealing with energy flow and nutrient cycling within *ecosystems.

**production/respiration ratio (P/R ratio)** The relationship between gross *production and total *community respiration. Where P/R = 1 a steady-state community results. This balance may be an instantaneous daily one (e.g. in tropical rain-forest communities), or over a longer period (e.g. in temperate woodlands). If P/R is persistently greater or less than 1, then organic matter either accumulates or is depleted respectively.

**productivity** The rate at which the *biomass increases per unit area. *See* PRIMARY PRODUCTIVITY.

**proenzyme** *See* ZYMOGEN.

**profile** A vertical section through all the constituent *horizons of soil, from the surface to the relatively unaltered parent material.

**profundal zone** The bottom and deepwater area of freshwater ecosystems that lies beyond the depth of effective light penetration (*see* COMPENSATION LEVEL). In shallow freshwater systems, such as ponds, this zone may be missing. *Compare* BENTHIC ZONE.

**progressive evolution** A steady, long-term improvement of evolutionary grade, which has allowed plants and animals to become ever more independent of the aquatic environment in which they first evolved. For example, the sequence *bryophyte, *pteridophyte, *gymnosperm, *angiosperm represents a progressive evolutionary trend.

**progressive succession** The normal sequential development of *communities, from simple communities with few species and low productivity to the optimum sustainable in a given environment. It is contrasted with retrogressive *successional change (*retrogression).

**Progymnospermopsida** The ancestors of the *gymnosperms, which arose in the *Devonian and dwindled to extinction in the latter part of the *Carboniferous. Probably seeds evolved in various different progymnosperms.

**Prokarya** In the widely used five-kingdom classification of living organisms, a superkingdom containing the kingdom *Bacteria.

**prokaryote** A type of organism that is mainly unicellular and in which the cells lack a true *nucleus; in all known prokaryotes the *DNA is present as a loop in the *cytoplasm. Other prokaryotic features include the lack of *chloroplasts and *mitochondria and the possession of small *ribosomes. The prokaryotes comprise two groups (domains): *Eubacteria (the bacteria, including the *cyanobacteria, formerly known as 'blue-green algae') and *Archaea (*see* EXTREMOPHILES).

**proline** A *heterocyclic, non-polar (*see* NON-POLAR MOLECULE) *imino acid, which is present in all *proteins studied to date.

**promeristem** In apical *meristem, the cells initiating growth and those immediately derived from them.

**promoter** A *nucleotide sequence within an *operon, lying between the operator and the *structural gene or genes, which serves as a recognition site and point of attachment for the *RNA *polymerase. It is the starting-point for transcription of the structural gene or genes in the operon, but is not itself transcribed.

**promycelium** The alternative name for the *basidium of *fungi in the orders *Ustilaginales and *Uredinales. It is formed following the *germination of the resting *spores and is often *multiseptate.

**propagule** Any structure that functions in propagation and dispersal, e.g. a *spore or *seed. *See also* MIGRATION.

**prophage** The *genome of a *bacteriophage whose host bacterium responds to its presence lysogenically (*see* LYSOGENY and LYTIC RESPONSE).

**prophase** The first phase of *mitosis and *meiosis I and meiosis II. In mitosis and the first division of meiosis the *chromosomes become visible within the *nucleus, coiling up to produce a series of compact spirals. In meiosis I they also undergo pairing during this phase, the phase then being divided into 5 successive stages: leptotene, zygotene, pachytene, diplotene, and diakinesis. The prophase of the second division of meiosis (prophase II) is very similar to prophase of mitosis: the nucleoli and nuclear membrane break down and the *chromatids shorten and thicken. Many plants go directly from anaphase I to metaphase II and do not have a prophase II.

**Propionibacterium** A genus of *Gram-positive *bacteria in which the cells are variable in shape, forming branched or unbranched, regular or irregular ovoids, rods, or filaments. They are *chemo-organotrophic and grow only in the absence of air. They are found, for example, in dairy products and on skin.

**proplastid** A colourless, double-membrane-bound *organelle, with little internal structure, that acts as a precursor in the development of all *plastids.

**prop root** A tree root that arises from the main stem and provides mechanical support to the plant. *Buttress roots and *stilt roots are prop roots.

**prosenchyma** In *algae or *fungi, tissue that can be seen to contain *hyphae.

**Prostanthera** (mint-bushes; family *Lamiaceae) A genus of shrubs that are usually glandular and yield aromatic oils. The leaves are *opposite. The flowers are solitary and held in the *axils of *bracts or are borne in terminal *panicles or *racemes. The *calyx is 2-lipped and persistent. The *corolla is a short, wide tube with an erect upper lip which is often hooded and 2-lobed, and a lower lip of 3 spreading lobes. There are 4 *stamens. The *style is simple, the *stigma bilobed. The fruit is a *nutlet. There are about 50 species, all endemic (*see* ENDEMISM) to Australia.

**prostheca** In certain types of *bacteria, a narrow, stalk-like extension of the cell.

**prosthetic group** The non-protein component of a *conjugated protein, or the *cofactor of an *enzyme to which it is bound so tightly that it cannot be removed by *dialysis.

**protamine** A generic term for a group of strongly basic, *globular proteins of relatively low molecular weight, which contain large quantities of the *amino acid arginine, but no sulphur. They are found associated with *nucleic acids.

**protandry (adj. protandrous)** In plants (e.g. dandelion), the maturation of *anthers (i.e. male organs) before *carpels (female organs).

**Proteaceae** One of the most distinctive plant families in the southern hemisphere, comprising trees and shrubs in which the leaves are entire or divided, spiral, without *stipules, leathery, and commonly *xeromorphic. The flowers are usually held in showy, dense *inflorescences. They are usually hermaphrodite and *zygomorphic, with 4 *tepals (often becoming recurved), 4 scales, 4 *stamens (often *epipetalous), and a *superior, *unilocular *ovary, with 1 to many *ovules. The fruits are various. There are 75 genera, comprising about 1350 species, occuring in the southern hemisphere, and developed strongly in S. Africa and Australia.

**Proterozoic** The eon of geologic time that followed the *Archaean and preceded the present *Phanerozoic Eons. It is divided into three eras. The Palaeoproterozoic lasted from 2500 Ma to 1600 Ma, the Mesopalaeoproterozoic from 1600 Ma to 1000 Ma, and the Neoproterozoic from 1000 Ma to 542 Ma. The Neoproterozoic culminated with the first abundantly fossiliferous period, the *Ediacaran.

**Proteus** (family *Enterobacteriaceae) A genus of *Gram-negative *bacteria in which the cells are usually rod-shaped, but may vary from ovoid to filamentous under certain conditions. They are *motile, with many *flagella. They are found chiefly in the intestines and faeces of animals, including humans. Some species can cause disease.

**prothallus** A fairly undifferentiated *gametophyte, like those of *Pteridophyta.

**protist** A single-celled member of the *Protoctista.

**Protista** *See* PROTOCTISTA.

**protoco-operation** An interaction of species populations in which both populations benefit, but neither is dependent on the relationship. *Compare* MUTUALISM.

**protocorm** A *tuber structure that develops from the *embryos of lycopods (*Lycopodiaceae) and orchids (*Orchidaceae).

**Protoctista** In the widely used five-kingdom classification system for living organisms, one of the kingdoms within the superkingdom *Eukarya. In the three-domain classification system, a kingdom within the *domain Eukarya. Protoctists are aquatic *eukaryotes, but they are neither animals, nor fungi, nor plants. The kingdom includes naked and shelled amoebas, foraminiferans, zooflagellates, ciliates, dinoflagellates, diatoms (*see* BACILLARIOPHYTA), *algae (including seaweeds), slime moulds, slime nets, and protozoa. Single-celled organisms were formerly known as 'protists' and the kingdom containing them was the Protista. This ranking was abandoned when it became evident that multicellularity evolved many times and that multicellular organisms are closely related to single-celled forms. The name Protoctista means 'first established' from the Greek *protos* 'first' and *ktistos* 'establish'; 'Protista' is no longer used.

**protoderm** The outermost layer of apical *meristem that gives rise to the *epidermis.

**protogyny (adj. protogynous)** A condition in which the female parts develop first. *Compare* PROTANDRY.

**protohemicryptophyte** *See* HEMICRYPTOPHYTE.

**protokeront** A hair-like structure, similar to a *flagellum but differing in basic structure, that occurs on the *plasma membranes of some bacteria.

**protonema** An early stage in the development of a *moss or *liverwort. A protonema is produced when a moss or liverwort *spore germinates. In mosses it usually consists of green, branching filaments; but it is *thalloid (a flat sheet or disc of cells) in *Sphagnum and *Andreaea*, for example, and in many liverworts. The familiar moss or liverwort plant (the *gametophyte) arises from buds which develop on the protonema.

**protophloem** The first *phloem elements to form from the *procambium.

**protoplasm** A complex, translucent, colourless, colloidal substance within each cell, including the *cell membrane, but excluding the large *vacuoles, masses of secretions, ingested material, etc. In animals and plants it is differentiated into nucleoplasm (protoplasm in the nucleus) and *cytoplasm (protoplasm in the rest of the cell).

**protoplast** That part of a plant cell which is actively engaged in metabolic processes (i.e. the *protoplasm other than the *cell wall).

**Protosphagnales** *See* SPHAGNOPSIDA.

**protostele** One of 2 basic *morphologies of *steles, the other being *siphonostele. Protosteles are the more primitive, being characterized by a single strand of *xylem in the centre of the stem or root, and lacking a *pith. Protosteles can be subdivided into *actinostele, *hypophloic haplostele, *haplostele, and *solenostele. *Compare* DICTYOSTELE and MONOSTELE.

**Protosteliomycetes (division *Myxomycota)** A class of primitive *slime moulds in which the feeding stage consists of simple *myxamoebae or small *plasmodia. The myxamoebae do not aggregate prior to fruiting. Species are distributed widely, found on bark, rotting wood, etc.

**prototroph** In bacteriology, a strain of *bacteria that have the nutritional requirements of the wild type or non-mutant species. *Compare* AUXOTROPH.

**protoxylem** The first *xylem elements to form from the *procambium.

**provascular bundle** A strand of the *procambium.

**provirus** A viral *genome which is incorporated into a host genome.

**proximal** Applied to the part of a structure or organ that is closest to its point of attachment to the main body of the plant.

**P/R ratio** *See* PRODUCTION/RESPIRATION RATIO.

**pruinose** Covered with powdery granules; powdery in appearance.

**Prunus (plums, cherries; family *Rosaceae)** A genus of trees and shrubs whose flowers have 5 free *petals, 5 *sepals, about 20 *stamens, and a single *carpel which sits in a cup formed by the *calyx tube but is free from it. The fruit is a 1-seeded *drupe with a stony inner wall and a fleshy outer one. Many kinds of plums and cherries are cultivated for their fruits, and other species are grown for their flowers. There are about 400 species, most of which occur in the northern temperate zone, but some of which are native to the Andes and to the Asian tropics. *See also* AMYGDALUS.

**Prunus serotina** (American cherry) A fast-growing, light-demanding, gregarious tree of the temperate deciduous forests of N. America.

**Prymnesiophyceae** (Haptophyceae) A class of predominantly unicellular, *flagellated *algae (sometimes regarded as protozoa, class *Phytomastigophora) in which there are two smooth (whiplash-type) flagella, and a structure called a haptonema: a filamentous appendage which may be short, or long and whip-like, and which is believed to function mainly in attaching the organism to a solid surface. Species are predominantly marine, often forming a major component of the *phytoplankton in some regions. Many species have a cell covering of many small scales; in some species these scales are impregnated with calcium carbonate and are known as *coccoliths.

**PSI** *See* PHOTOSYSTEM I.

**PSII** *See* PHOTOSYSTEM II.

**psammo-** Prefix derived from the Greek *psammos* meaning 'sand'.

**psammo-littoral zone** The water's-edge zone of sandy shores, both marine and freshwater, where the microscopic plant and animal *communities forming the *psammon are most prolific.

**psammon** The microscopic flora and fauna of the interstitial spaces between sand grains of sea-shore and lake-shore areas.

**psammosere** The characteristic sequence of changes associated with stages in a plant *succession developed on sand-dunes. *See also* SUCCESSION.

**Psathyrella** (family *Coprinaceae) A genus of *fungi in which the *mushroom-shaped fruit bodies are fragile and have a central brittle *stipe. The *spore print is black or dark brown. There are many species, found on the ground, on tree-stumps, etc., often growing in clumps.

**pseudanthia** Reduced, *cymose *inflorescences that consist of both male and female flowers, all of them reduced to 1 *stamen or 1 *carpel respectively.

**Pseudevernia** (order *Lecanorales) A genus of *lichens in which the *thallus is more or less *fruticose. *P. furfuracea* closely resembles *oakmoss (*Evernia prunastri*) but is black on the underside, and bears *isidia rather than *soredia. *Apothecia are rarely formed. This species is common on a range of substrates in hilly districts of the north and west of Britain.

**pseudocarp** (false fruit) A *fruit in which the ripened *ovary and its contents are combined with another structure, often the *receptacle.

**pseudocyphella** A depression or pore in the surface of the *thallus in certain types of *lichen; pseudocyphellae are visible as small pale or white dots.

**pseudoendosperm** The *haploid, nutritive tissue present in the seeds of *gymnosperms, derived from the female *gametophyte. The true *endosperm of *angiosperms is *triploid.

**pseudo-epithecium** In the *fruiting bodies of certain *ascomycetes, a layer above the *hymenium consisting of the tips of *paraphyses in an amorphous matrix.

**pseudofilament** A line of unicellular organisms held together by a gelatinous sheath.

**pseudogamy** The development of the female *gamete into a new plant on stimulation by the male gamete, but without *fertilization.

**pseudogenes** *Genes that have been 'switched off' in evolution and no longer have any function. They are, therefore, entirely neutral and evolve at a constant rate. By comparing pseudogenes in related organisms, a standard can be inferred against which the rate of change in other genes can be measured, enabling further inferences to be made as to the presence or absence of selection pressure in evolution.

**Pseudolarix** (golden larch; family *Pinaceae) A *monotypic genus (*P. kaempferi*) of deciduous conifers, close to *Larix but with female cones that break up when ripe, pointed cone scales, and male cones borne in bunches on spur shoots. The genus is native to eastern China.

**Pseudomonadaceae** A family of *Gram-negative *bacteria; the cells are straight or curved rods, and are typically *motile with polar *flagella. They are *aerobic chemo-organotrophs. There are 4 genera, found in a wide range of habitats. Some species are free-living, others are *pathogenic in plants or animals.

**Pseudomonas** (family *Pseudomonadaceae) A genus of *Gram-negative rod-shaped *bacteria which are typically *motile with polar *flagella. Species are capable of breaking down a wide range of organic compounds and are important agents of mineralization. There are many species. They are common in soils, and in fresh and marine waters. Some can be *pathogenic; for example, *P. aeruginosa* can cause infection of burns, *P. mallei* causes glanders in equines, and *P. solanacearum* can cause wilt in a variety of plants.

**pseudoparaphyse** One of the sterile, thread-like filaments found in the *perithecia and *pseudothecia of certain *fungi. They arise in the upper portion of the *fruit cavity and grow downwards.

**pseudoparenchyma** In *algae and *fungi, tissue composed of closely woven filaments that resembles *parenchyma.

**pseudoperithecium** *See* PSEUDOTHECIUM.

**pseudoplasmodium** In cellular *slime moulds, a *plasmodium-like structure formed by the aggregation of many separate amoeboid cells.

**pseudopodetium** In certain *lichens, a *fruticose part of the *thallus upon which *ascocarps (when present) are borne; a pseudopodetium may resemble certain types of *podetium but differs in its mode of development.

**pseudopodium** In amoeboid cells, a protrusion, usually temporary, of the cell body, which functions in locomotion and in the ingestion of food particles. In slime moulds, an arm-like projection from the body, by which the organism creeps over the surface.

**pseudostem** A false stem formed of the swollen leaf bases, as in the banana.

**pseudo-steppe** An area with *steppe-like vegetation which occurs outside Eurasia. The term 'steppe' strictly refers to the temperate grassland of Eurasia. However, it has also been applied to vegetation on the southern fringe of the Sahara (the Sahel zone), in parts of Namibia, and in south-western Australia.

**Pseudosycidium** The first of the *charophytes, a distinct, new, evolutionary line of green *algae which arose during the *Silurian and whose method of reproduction was sexual. Well-developed male (*antheridia) and female (*oogonia) organs can be identified, the secretion of calcium carbonate around the female organs assisting preservation.

**pseudothecium** (pseudoperithecium) An *ascocarp which resembles a *perithecium but differs from it in its mode of development, and in that it contains *bitunicate *asci.

**Pseudotsuga** (Douglas fir; family *Pinaceae) A genus of evergreen conifers that have leaves like *Abies (fir), but narrow, scaly buds like *Fagus (beech). Female cones have long, projecting, 3-lobed *bract scales which are distinctive. *P. menziesii* is planted extensively for forestry. There are 4 species, 2 occurring in western N. America, 2 in China, and 1 in Japan.

**pseudowhorl** An arrangement of leaves that arise so close together as to appear to arise at the same level (i.e. a *whorl), although in fact they do not do so.

**Psidium** (family *Myrtaceae) A genus of small trees and shrubs, many of which produce edible fruits, notably *P. guajava* (guava), which is widely naturalized. There are 100 species, occurring in neotropical regions.

**Psilocybe** (family *Strophariaceae) A genus of *fungi in which the *fruit bodies are *mushroom-shaped. The cap is moist and fleshy, and is typically more or less conical. The *spore print is purplish-brown or brown. *Psilocybe* species are found on dung, twigs, lawns, etc. When eaten, the fruit bodies have *hallucinogenic properties.

**Psilophytales** Primitive *pteridophytes, which were the earliest vascular plants, from the *Silurian and *Devonian. They had slender, tapering, leafless, or scale-bearing stems up to 50 cm high, often with cone-shaped *sporangia at the top. Recent classifications also include the living psilophytes in this group. The modern psilophytes (Psilopsida) comprise 4 or 5 plant species of the tropics and subtropics, of which *Psilotum* (whisk fern) is the best known.

**psilophytes** *See* PSILOPHYTALES.

**Psilotaceae (order Psilotales)** A family of very ancient and primitive vascular *cryptogams (*Pteridophyta), characterized by forked stems bearing small, scale-like appendages (*Psilotum*), or flattened leaves with a single vein (*Tmesipteris*), creeping *rhizomes without roots, and *spore capsules fused in pairs (*Tmesipteris*) or in triads (*Psilotum*). The *gametophytes are cylindrical and subterranean, and live by means of a *mycorrhizal fungus as *saprophytes. There are 2 genera, and 4–8 species, the only living representatives of the order, probably the most primitive of living vascular plants, and possibly related to the *Devonian fossil group the *Psilophytales. *Tmesipteris* is confined to the Australasian region.

**Psilotum** (whisk ferns; family *Psilotaceae) A genus of very primitive vascular plants without true roots, and with the *spore capsules borne in fused triads on the upper parts of the forking stems, in the *axils of tiny scales. There are no real leaves, but the ridged stems are green, and act as photosynthetic organs. There are 2 species, one (*P. nudum*) occurring widely in the tropics and subtropics, including Australasia and the Pacific islands, extending into southern Spain and into the southern part of the USA; the other, *P. complanatum*, is found in Mexico, Jamaica, and the Pacific.

**Psychotria** (family *Rubiaceae) A large genus of trees and shrubs, many members of which have colourful *inflorescences. There are about 1400 species, occurring in the tropics and subtropics, in regions with a humid climate.

**psychrophile** An *extremophile (domain *Archaea) that thrives in environments where the temperature is low, usually below 15°C.

**psychrotroph** An organism that can grow at low temperatures (e.g. 15°C or below) but that grows better at higher temperatures.

**Pteridium** (bracken; family *Dennstaedtiaceae) A *monotypic genus (*P. aquilinum*) of ferns with deep, creeping *rhizomes, and much dissected, *pinnate *fronds. The *sori are continuous all round the edges of those *pinnae that produce them, and are protected both by the recurved margin of the pinna, and by a delicate *indusial flap on their inner side. It is one of the world's most common ferns, and a pest in many pasture lands in temperate climates, since it is not usually eaten by animals and is encouraged by burning. The genus is cosmopolitan, except for the Arctic and temperate S. America.

**Pteridophyta** (pteridophytes) A division of the plant kingdom, comprising the vascular *cryptogams. They are flowerless plants exhibiting an alternation of 2 distinct and dissimilar generations. The first is a non-sexual, *spore-bearing, *sporophyte generation. It usually appears as a relatively large plant, with stems containing vascular tissue that conducts water and dissolved solutes through the plant, and usually bears the leaves and roots. Spores are produced in *sporangia that are either attached to the leaves (as in ferns) or are on specialized scales grouped into cones (as in horsetails and clubmosses), or in the *axils of leaves on unspecialized stems (as in Psilotaceae and some clubmosses). The second is a sexual, gametophyte generation, in which the plants generally are relatively small, and without differentiation of stem, leaves, or roots. These plants bear male (*antheridia) and female (*archegonia) sex organs, together or on separate plants.

When the eggs in the archegonia are fertilized by sperms from the antheridia, an embryo results: this can grow into a new sporophyte generation. The cells of the sexual, gametophyte generation each contain a single *chromosome set in their nuclei (the *haploid condition). Those of the sporophyte generation each contain a double chromosome set (the *diploid condition), reduced to a single set in the spores. The Pteridophyta, in the wide sense as it is usually understood, includes the classes *Lycopsida (families *Lycopodiaceae, *Selaginellaceae, and *Isoetaceae), *Sphenopsida (or Arthropsida; family *Equisetaceae), *Psilopsida (family *Psilotaceae), and *Filicopsida (the various families of ferns). They first enter the fossil record in the *Silurian.

**pteridophytes** *See* PTERIDOPHYTA.

**Pteridospermales (seed ferns)** An extinct *gymnosperm order, containing the earliest seed plants, which flourished in the *Carboniferous, before disappearing in the *Cretaceous.

***Pteris* (family *Adiantaceae)** A genus of ferns with creeping *rhizomes and long *pinnae, continuously bordered by *sori, which are protected only by the recurved margins of the pinna (without an inner membranaceous *indusium, as in *Pteridium*). The genus is largely tropical. There are about 250 species.

***Pterocarpus* (family *Fabaceae, subfamily *Papilionatae)** A genus of trees that have *pinnate leaves and rather large leaflets. The flowers are showy *racemes, in lax *panicles. The fruit is a disc-like, flat, *indehiscent pod, obliquely beaked, often membranaceous at the edges, containing 1 seed. The trees yield dark, fine-grained, superb cabinet timber (amboyna, angsana, sena from *P. indicus*, padauk from several continental Asian species). The cut bark yields a resin (kino) which is astringent and medicinal. There are about 20 species, with a pantropical distribution.

***Pterocarya* (wing nuts; family *Juglandaceae)** A genus of deciduous trees that have large, *pinnate leaves. The twigs have *septate pith and buds are usually naked. The flowers are in unisexual *catkins. The fruit is a *nut with 2 leafy wings. The trees

yield useful timber. There are 6 species, occurring from the Caucasus to Japan.

**pterochory** Dispersal of winged seeds by the wind.

***Pterocladia*** *See* GELIDIALES.

***Pterostylis* (family *Orchidaceae)** A genus of terrestrial orchids which have thin, soft, fleshy leaves held in a *radical rosette. The plants produce 1 or more underground *tubers. The flowers are green, white, or brownish, and either solitary or held in a terminal *raceme. The middle *sepal and upper 2 *petals form a hood. The 2 lower sepals form a lower lip and are often extended into long, thin tails. The lower lip is either incurved or *reflexed. The *labellum is usually narrow. The fruit is a 3-celled *capsule containing numerous fine seeds. There are about 60 species in the western Pacific region, nearly all endemic (*see* ENDEMISM) to Australia. Many species are cultivated as ornamentals.

**pteryoglutamic acid** *See* FOLIC ACID.

***Ptilotus* (family *Amaranthaceae)** A genus of *herbs and shrubs which are found in both arid and humid habitats. They have entire, alternate or *opposite leaves without *stipules. The flowers are usually bisexual and regular, with 4 or 5 segments of the *perianth which are often fused. They are *axillary and solitary, or in *cymes. There are 1–5 *stamens and a uniovular, *superior *ovary. In several species the flower head is dry, membranaceous, and colourless; in other species flowers are bright and showy. The fruit is a *nut or *berry and the seed has a tough, shiny *testa. A few species are cultivated as ornamentals. There are about 100 species, endemic (*see* ENDEMISM) to Australia.

***Ptychosperma* (family *Arecaceae)** A genus of palms, a few of which are in cultivation. They are clump-forming, with *pinnate leaves, the leaflets being blunt, and *praemorse, with a crown-shaft. The *inflorescences are sparsely branched, below the crown-shaft; 2 male flowers flank a female. The fruits are fleshy and showy. There are 28 species, occurring in the Kei Islands, New Guinea, Solomon Islands, Micronesia, and Queensland.

**pubescent** Covered with soft hair; downy.

***Puccinia*** **(class \*Urediniomycetes)** A genus of \*rust fungi, which form stalked, 2-celled \*teliospores. They parasitize (*see* PARASITISM) monocotyledonous (*see* MONO-COTYLEDON) and dicotyledonous (*see* DI-COTYLEDON) plants throughout the world. Many species cause diseases of economic importance: for example, *P. graminis* causes \*black-stem rust of wheat, *P. horiana* causes chrysanthemum white rust, *P. menthae* causes mint rust, and *P. striiformis* causes yellow rust of wheat and other cereals. There are many species.

**puddled soil** **(poached soil)** Soil in which the structure has been destroyed by the physical impact of raindrops, by tillage when wet, or by trampling by animals.

**puff-ball** The common name for the \*fruit body formed by some \*fungi of the \*Gasteromycetes. Puff-balls are more or less spherical, and the \*spores that develop inside them are released through pores or cracks in the \*peridium. The giant puff-ball is \**Calvatia gigantea*.

**pugioniform** Shaped like a dagger.

**pulque** *See* AGAVE.

**pulse** **1.** The edible \*seeds of any leguminous plant (\*Fabaceae). **2.** An alternative term for an \*algal bloom. **3.** *See* PULSE LABELLING.

**pulse labelling** A technique in which radioisotopes are used for the measurement of the rates of synthesis of compounds within living cells. A suspension of cells or \*organelles is exposed to a small quantity of an \*isotope for a brief period (seconds or minutes), hence the term 'pulse'. This is achieved through the addition to the suspension of a much larger quantity of the stable (unlabelled) isotope of the same compound following the required period of exposure to the radioisotope. The effect of competition between the two isotopes is to reduce to a negligible level the further uptake of the latter. Measurement of the levels of activity in samples under various experimental conditions can yield useful information regarding the factors influencing the uptake and metabolism of compounds.

**pulverulent** Dusty; powdery; covered with powder.

**pulvinate** Cushion-shaped; swollen; convex.

**pulvinus** **1.** A swelling at the base of a \*petiole or \*pinna, made up of cells that are capable of moving water in and out of their \*vacuoles quickly, resulting in movement of the petiole or pinna. **2.** A thickened region at the \*node of a grass stem, often containing an intercalary \*meristem, which can raise the \*culm after \*lodging.

**punah** *See* TETRAMERISTA.

**punas** Arid alpine vegetation on high plateaux along the western side of the Andes. It comprises sparse, tufted grasses, and large, hardy, cushion plants. The pamirs of Tibet are a similar vegetation type.

**punctate** Applied to any structure that is marked by pores or by very small, point-like depressions.

**punctiform** Dot-like in appearance.

**punctuated equilibrium** The theory that \*evolution is characterized by geologically long periods of stability during which little speciation occurs, punctuated by short periods of rapid change.

***Punica*** *See* PUNICACEAE.

**Punicaceae** A family that comprises 1 genus of plants, related to the \*Lythraceae, in which the leaves are \*simple, spiral, and without \*stipules. The flowers are \*perigynous, with 5–8 \*sepals and \*petals, many \*stamens, and many \*carpels which are \*adnate to the \*receptacle. The fruit is seated on the persistent \*calyx, with a leathery rind, and has numerous seeds with partly fleshy leaves. There are 2 species occurring from south-eastern Europe to the Himalayas (*P. granatum*) and Socotra. *P. granatum* is the pomegranate, which is widely cultivated.

**Purbeckian** A late \*Jurassic stage or formation, the type section of which is found in southern England. The sediments were deposited under intertidal to brackish freshwater conditions.

**purine** A basic, nitrogenous compound that resembles a 6-membered *pyrimidine ring fused to a 5-membered *imidazole ring. The 2 principal purines, adenine and guanine, are major constituents of *nucleic acids.

**purple granadilla** See PASSIFLORACEAE.

**purple membrane** In some strains of *Halobacterium* species, a specialized region of the *cell membrane that contains *bacteriorhodopsin. The purple membrane functions in a unique energy-yielding process in which the energy of sunlight is harnessed in the absence of *photosynthesis.

**purple non-sulphur bacteria** See RHODOSPIRILLACEAE.

**purple sulphur bacteria** See CHROMATIACEAE.

**puszta** Hungarian grassland, similar to the N. American prairie.

**putrefaction** The *anaerobic digestion by *bacteria of proteinaceous material (e.g. meat) with the concomitant production of malodorous substances.

**Puya** (family *Bromeliaceae) A genus of terrestrial, monocotyledonous (see MONOCOTYLEDON), stout, unbranched herbs or treelets; *P. raimondii* grows to a height of 9.5 m. There are 168 species, occurring in the Andes.

**PWP** See PERMANENT WILTING POINT.

**pycnidiospore** A *conidium that is produced inside a *pycnidium.

**pycnidium** (pycnium) A flask-shaped or spherical structure within which *conidia are formed in certain types of *fungi. Conidia are released via a pore in the wall of the pycnidium. Pycnidia contain minute spores called *pycniospores.

**pycniospore** (pycnospore) A minute, *haploid *spore formed in a *pycnidium by certain *fungi of the *Uredinales; a pycniospore functions as a male *gamete.

**pycnium** See PYCNIDIUM.

**pycnospore** See PYCNIOSPORE.

**pyinma** See LAGERSTROEMIA.

**pyragallol** A soluble *phenol that, in alkaline solution, will absorb oxygen and is used in respiration and *photosynthesis experiments to estimate the volume of oxygen in a sample.

**pyramid of biomass** A diagrammatic expression of *biomass at different *trophic levels in an *ecosystem, usually plotted as dry matter or *calorific value per unit area or volume. Typically, this gives a gradually sloping pyramid, except where the sizes of organisms vary dramatically from one trophic level to another. In this case, the higher metabolic rate of the smaller organisms may result in a greater biomass of *consumers than of *producers, giving an inverted pyramid. Aquatic communities in winter typically show inverted biomass pyramids. *Compare* PYRAMID OF ENERGY and PYRAMID OF NUMBERS; *see also* ECOLOGICAL PYRAMID.

**pyramid of energy** A diagrammatic expression of the rates of flow of energy through the different *trophic levels of an *ecosystem. It reflects the rates of *photosynthesis, respiration, etc. (and not the standing crop, as in the pyramid of *biomass), and can never be inverted since energy is dissipated through the ecosystem. It is the most fundamental and most useful of the 3 ecological pyramids. *Compare* PYRAMID OF BIOMASS and PYRAMID OF NUMBERS; *see also* ECOLOGICAL PYRAMID.

**pyramid of numbers** A diagrammatic expression of the numbers of individual organisms present at each *trophic level of an *ecosystem. It is the least useful of the 3 types of ecological pyramid since it makes no allowance for the different sizes and metabolic rates of organisms. Typically, it slopes more steeply than the other pyramids and may be inverted, e.g. when based on studies of temperate woodlands in summer. *Compare* PYRAMID OF BIOMASS and PYRAMID OF ENERGY; *see also* ECOLOGICAL PYRAMID.

**pyrenoid** A region of starch formation found in the *chloroplasts of various *algae (e.g. *Spirogyra* and *Chlamydomonas*).

**Pyrenomycetes** (subdivision *Ascomycotina) A class of *fungi that typically form *ascocarps called *perithecia. *Asci are unitunicate and are arranged in a layer (the *hymenium). The class is no longer recognized in most taxonomic schemes.

**Pyrenulales** (subdivision *Ascomycotina) An order of *saprotrophic and lichenized *fungi in which the *thallus is *crustose and the *ascocarps are perithecioid (*see* PERITHECIUM). The *asci are *bitunicate. *Spores are multiseptate and colourless or brown. There are several genera, including *Pyrenula*. *P. nitida* is common on trees in moist western districts of Britain. The thallus is smooth and green to yellowish-grey, typically surrounded and intersected by black lines. The perithecia appear as black dots up to 1 mm across.

**Pyricularia** (class *Hyphomycetes) A *form-genus of plant-*parasitic *fungi which are responsible for some important diseases of cereals and other grasses in the tropics. *P. oryzae* causes the very destructive *blast disease of rice.

**pyriform** Pear-shaped.

**pyrimidine** A basic, 6-membered *heterocyclic compound. The principal pyrimidines (uracil, thymine, and cytosine) are important constituents of *nucleic acids. Thiamine (vitamin $B_1$) is an important pyrimidine derivative, and other derivatives play major roles in *carbohydrate and *lipid metabolism.

**pyro-** A prefix, derived from the Greek *pur*, meaning 'fire', that means 'associated with fire' (e.g. *pyrophyte, literally 'fire plant').

**pyroclimax** *See* FIRE CLIMAX.

**Pyrodinum** *See* DINOPHYTA.

**Pyrolaceae** A family of dicotyledonous (*see* DICOTYLEDON), evergreen herbs whose creeping *rhizomes bear *simple, stalked leaves. The family is close to *Ericaceae but differ in having *petals that are always free, minute seeds, an *ovary incompletely divided into 5 cells, and the herbaceous habit. There are 4 genera, comprising 42 species, confined to the northern temperate and Arctic zones, where they usually occur in coniferous forests or on *heathland, etc., growing in the vegetable litter layer, which they exploit as partial *saprophytes by means of a *mycorrhizal relationship with a fungus.

**pyrophyte** A plant adapted to withstand or to achieve a competitive advantage from fire, e.g. *Themeda triandra*, a grass widespread in Kenyan savannahs.

**pyrrole** A group that consists of four CH units and one NH unit arranged to form a ring.

**Pyrus** (pears; family *Rosaceae) A genus of deciduous, sometimes thorny trees and shrubs, that have *simple leaves and flowers borne in *umbel-like clusters. There are 5 *sepals, 5 *petals, 20–30 purple *stamens, and 2–5 *carpels fused together and with the *receptacle, though the *styles are separate. The flesh of the pear-shaped fruits contains numerous woody, grit-like cells, unlike that of apples (*Malus*). Many cultivars are grown for their fruits. There are about 20 species, occurring in northern temperate Eurasia.

**pyruvic acid** A 3-carbon *keto acid that occupies a central position in cell metabolism. It represents the final product of *glycolysis in aerobic respiration, and subsequently undergoes oxidation to carbon dioxide and acetyl *coenzyme A. During anaerobic respiration it is irreversibly converted to ethyl alcohol and carbon dioxide in plant cells (and reversibly to lactic acid in animal cells). It may also be variously converted into alanine, malic acid, and oxaloacetic acid.

**Pythium** (order *Peronosporales) A genus of *fungi, most of which are *saprotrophs in soil or water. Under certain conditions they may be responsible for a range of plant diseases, including stem and root rots, and *damping-off in seedlings. There are many species.

**$Q_{10}$**  The ratio of the velocity of a reaction at a given temperature to that of the same reaction at a temperature 10°C lower. This ratio is approximately 2.0 for most biological and inorganic reactions. At high temperatures there is usually a decrease in the $Q_{10}$ of biological reactions as *enzymes become more *denatured.

**$QO_2$**  *See* OXYGEN QUOTIENT.

**Q technique**  A method for analysing data, in which observations (N) form the columns and the variables or attributes (n) form the rows in a table or matrix. *Compare* R TECHNIQUE; *see also* INVERSE ANALYSIS and NORMAL ANALYSIS.

**quadrat**  A basic sampling unit of vegetation surveys. Traditionally, 1 m² quadrats were used to sample short, non-woody *communities such as *grasslands and *heathlands. More recently circular and rectangular *quadrats have been used, as well as squares of all sizes, depending on the purpose of the survey; e.g., 10 m² and 20 m² quadrats are commonly used in woodland studies.

**qualitative inheritance**  An inheritance of a character that differs markedly in its expression among individuals of a species: variation in that species is discontinuous. Such characters are usually under the control of *major genes. A principal example is gender. *Compare* QUANTITATIVE INHERITANCE.

**quantasome**  A particle that is embedded in a paracrystalline array on the surface of *thylakoid discs in *chloroplasts. Quantasomes are composed equally of *lipids and *proteins, which include various photosynthetic pigments and redox carriers. For this reason they are considered to be *photosynthetic units. They occur in 2 sizes: the smaller quantasome is thought to represent the site of *photosystem I, the larger to represent the site of *photosystem II.

**quantitative character**  *See* QUANTITATIVE INHERITANCE.

**quantitative inheritance**  An inheritance of a character that depends upon the cumulative action of many *genes, each of which produces only a small effect. Examples of such quantitative characters include spore production in ferns, height of trees, and nectar production in buttercups. Usually the character shows continuous variation, i.e. a gradation from one extreme to the other. *Compare* QUALITATIVE INHERITANCE.

**quantum evolution**  The sudden spurt of diversification in an evolutionary lineage. Typically it results when a population shifts into another adaptive zone. Initially the number of individuals will be small, and rapid genetic and *phenotypic change is feasible in such circumstances, particularly if the new *niche is unoccupied. The concept of quantum evolution has been given special significance in the theory of *punctuated equilibrium.

**quantum speciation**  Traditionally, the rapid *speciation that can occur in small populations isolated from the large, ancestral population, and that are therefore subject to the *founder effect and to *genetic drift. More recently, the term has been assigned a central role in the theory of *punctuated equilibrium.

**quasi-sympatric speciation**  The separation of 1 species into 2 by *adaptation of different sub-populations into different *niches.

**Quaternary**  The sub-era of the *Cenozoic Era of geologic time that began 1.806 Ma ago and continues to the present day. It coincides with the *Pleistogene Period. The Quaternary comprises the *Pleistocene and *Holocene Epochs, during which there have been numerous ice-sheet advances in the northern hemisphere. By the Pleistocene most of the world's faunas and floras had a modern appearance. Although it continues to be widely used, the name Quaternary has now been abandoned and is used only informally.

**quaternary structure** The structure of a *protein that results from the interaction of two or more individual *polypeptides to give larger functional molecules. *Compare* PRIMARY STRUCTURE; SECONDARY STRUCTURE; and TERTIARY STRUCTURE.

**Queensland arrowroot** *See CANNA.*

**Queensland maple** *See FLINDERSIA.*

**Queensland nut** *See MACADAMIA.*

***Quercus*** (oak; family *Fagaceae) A genus of deciduous and evergreen trees and shrubs of the beech family, whose wood has distinctive, broad, compound rays. The buds are scaly. The leaves have small *stipules and are entire, toothed or rarely *crenate. Male flowers are borne in *catkins, female flowers are solitary. The fruit is a *nut seated in a woody cupule, the acorn. Many species are important for timber, tanning (the bark), cork (*Q. suber*), or as ornamentals. There are about 600 species, found in the northern hemisphere temperate zone, subtropical and tropical Asia, and the Andes.

**quiescent centre** A region in the apical *meristem of a root where cell division proceeds very slowly or not at all, but the cells are capable of resuming meristematic activity should tissue surrounding them be damaged.

**quillworts** *See ISOETACEAE.*

**quinone** p-Dioxybenzene or a derivative thereof; many quinones act as electron carriers in *mitochondria and *chloroplasts.

***Quisqualis*** *See COMBRETACEAE.*

q

**raceme** In flowers, an *inflorescence in which the main *axis continues to grow, producing flowers laterally, such that the youngest ones are *apical, or at the centre. Growth is monopodial. *Compare* CORYMB.

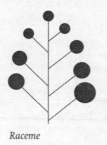

*Raceme*

**racemose** Applied to an *inflorescence in which flowers are developed from the *axillary *meristem, the activity of the meristem continuing at the tips of the main stem and primary laterals.

**rachis** In a plant, the *axis that bears the flower or, if the plant has *compound leaves, the leaflets. *See also* PINNATE.

**racial senescence** *See* PHYLOGER-ONTISM.

***Racomitrium*** (order *Grimmiales) A genus of medium to large mosses, in which the *archegonia (and hence *capsules) are borne terminally on short, lateral branches. Stems are erect or prostrate, often with numerous short branches. It is a cosmopolitan genus, with about 80 species. Most are *saxicolous, mainly on acidic rocks. *R. aciculare* grows on acidic rocks near or in water. *R. aquaticum* grows mainly on wet acidic rocks and on ledges in mountainous regions, often near flowing water. *R. lanuginosum* grows on exposed mountain tops on rock or on *peat, often forming extensive mats and giving its name to the plant community of this habitat: *Racomitrium* *heath (formerly *Rhacomitrium* heath). In this species each leaf has a

long hair point (often as long as the leaf itself), which is rough with minute teeth; the plant has a characteristic grey-green colour, and the shoots appear hoary because of the massed, white hair points.

**radiation ecology** The branch of ecology concerned with the effects of radioactive materials on living systems and on the pathways by which they are dispersed through ecosystems, including their dispersal through the *abiotic environment. The term is used especially with regard to those materials released through human agency.

**radical** Applied to a *leaf that arises from a *rhizome or from the base of a stem.

**radicle** In an *embryo, the rudimentary root.

**radioactive tracer** A radioactive *isotope, whose movement can be monitored, that is used to trace the pathways by which individual substances move through an organism, a living system, the abiotic environment, etc. Non-radioactive chemical analogues of certain substances may be used for the same purpose if their movement can be monitored (e.g. caesium, which can be substituted for potassium).

**radiocarbon dating** ($^{14}$C dating) A dating method for organic material that is applicable to about the last 70 000 years. It relies on the assumed constancy over time of atmospheric $^{14}$C:$^{12}$C ratios (now known not to be valid), and the known rate of decay of radioactive carbon, of which half is lost in a period (the 'half-life') of every 5730 years ± 30 years. The earlier 'Libby standard', 5568 years, is still widely used. In principle, since plants and animals exchange carbon dioxide with the atmosphere constantly, the $^{14}$C content of their bodies when alive is a function of the radiocarbon content of the atmosphere. When an organism dies, this exchange ceases and the radiocarbon fixed in the organism decays at the known half-life rate. Comparison of residual $^{14}$C activity in fossil organic material with modern standards enables the calculation of the age of the samples. Since the method was first devised, it has been realized that the atmospheric $^{14}$C content varies as the cosmic-ray bombardment of the outer atmosphere that generates the $^{14}$C varies. Correction for these fluctuations is possible for about the last 12 000 years by reference to the $^{14}$C contents of long *tree-ring series (e.g. those for *bristlecone pines).

**raffia (bast, bass)** A fibrous tissue from the upper sides of young leaflets of the palm *Raphia vinifera* that is used for tying plants and to make hats, mats, ect. *See* RAPHIA.

**Rafflesia (family *Rafflesiaceae)** A genus of plants that are wholly parasitic on vines of the family *Vitaceae. The flowers are showy and huge (up to 1 m diameter in *R. arnoldii*). They have 4–6 fleshy *tepals, fused at the base, *stamens fused as a ring, and the *ovary *inferior and immersed in a central column, smelling of carrion, and fly-pollinated. It is a prized medicine. There are about 13 species occurring in western Malaysia.

**Rafflesiaceae** A family of plants that are parasites. The vegetative tissues are colourless, *mycelium-like, and immersed in the host. The flowers are often large. There are 8 genera, with about 50 species, occurring in the tropics and subtropics.

**ragwort** *See* SENECIO and ASTERACEAE.

**rain forest** *See* TEMPERATE RAIN FOREST and TROPICAL RAIN FOREST.

**raised bog** An *ombrogenous bog community, typically dome-shaped in section, which develops on former lake sediments, estuarine sites, uniform *clay substrates, and sometimes on the surfaces of *valley bogs. In the British Isles, living (i.e. still growing upward) raised bogs are confined to western areas with a maritime climate, most notably central Ireland. The living bog surface shows a complex of hummocks and pools with different *Sphagnum species occupying different positions in relation to the degree of water-

logging. Hummocks are typically colonized by wet heath species, e.g. heathers (*Calluna vulgaris* and *Erica tetralix*) and sedges (e.g. *Eriophorium* species). The highest point of the dome may be more than 12 m above the *water table of the surrounding land. The dome is sometimes referred to as a cupola. The sloping edges of the dome are termed the rand, and the *rheotrophic mires that surround the dome are the lagg. *See also* OMBROGENOUS PEAT. *Compare* BLANKET BOG.

**Ramalina (order *Lecanorales)** A genus of *lichens in which the *thallus is *fruticose, typically flattened, and erect or pendulous. *Apothecia are typically borne on the edges and/or ends of the *thallus. *Spores are *septate, colourless, and often curved. Species are found on bark and on non-calcareous rocks, in unpolluted regions.

**Ramaria (family Gomphaceae, order *Aphyllophorales)** A genus of *fungi, characterized by their richly branched, coralloid *fruit bodies, which have tough flesh, and by their yellow to brown *spores which are typically rough-surfaced. They are found mostly on the ground; *R. stricta* grows on tree-stumps.

**rambutan** *See* NEPHELIUM.

**ramentum 1.** One of the scales, 1 cell thick, that occur on the young fronds of a fern. As the frond opens most of the ramenta are lost, but some persist on the *rachis. **2.** One of the sterile scales, resembling hairs, that occur among the seeds in cones of the *Mesozoic cycad *Cycadeoidea.*

**ramet** An individual member of a *clone that is capable of separation from other members and then capable of independent life. An individual strawberry plant or clover plant developed from a *stolon are ramets.

**ramie** *See* BOEHMERIA.

**ramiflorous** Borne on the branches.

**ramin** *See* GONYSTYLUS.

**ramón** *See* BROSIMUM.

**rand** The sloping edges of the dome of a *raised bog.

**random assortment** *See* INDEPENDENT ASSORTMENT.

**random genetic drift** *See* GENETIC DRIFT.

**random sample** A sample in which each individual measured or recorded (e.g. organism, site, or *quadrat) is independent of all other individuals and also independent of prominent features of the area or other unit being sampled. In ecological field surveying, random quadrat sampling is most easily achieved by superimposing a grid over the sample area, and identifying a series of random co-ordinates, using random-number tables, at which to locate the quadrats. Alternatively, a *random-walk technique may be used.

**random-walk technique** A method of random sampling in which the direction taken and the distance moved between sample points is determined by random numbers, drawn most usually from random-number tables.

**range 1.** The spread of environmental conditions within which a particular species occurs. **2.** The entire geographical area over which a species occurs. **3.** Extensive, open grazing lands.

**range management** The use of extensive, open grasslands for cattle production or, where appropriate, for the exploitation for human use of semi-domesticated or wild grazing animals (e.g. game cropping on the African *savannah) by methods that maintain the ecosystem. These require the determination of the optimum stocking density that will permit long-term cropping of the grassland without deterioration of the pasture.

**Ranunculaceae** A family of dicotyledons (*see* DICOTYLEDON), mostly *herbs but including a few shrubs, with leaves that are usually alternate, without *stipules, and often palmately lobed or *compound. The flowers are usually regular, with 4 or 5 to many free *sepals and *petals, and numerous free and spirally arranged *stamens and *carpels. These are all features usually regarded as *primitive, but exceptions to most of them occur within the genera of the family. Modern classifications recognize about 58 genera, with about 1750 species

(e.g. *Ranunculus* (buttercups), occurring mostly in the northern temperate, Arctic, and alpine zones. Most are poisonous and very acrid; some are used in medicine; others (e.g. *Delphinium* and *Anemone*) are grown for their beautiful flowers.

**rape** *See* BRASSICA.

**raphe 1.** On some seeds, a ridge caused by the fusion of the funicle to the *integument of the *ovule. **2.** In pennate diatoms (*see* BACILLARIOPHYTA) a split or longitudinal groove in the cell wall.

**Raphia** (family *Arecaceae) A genus of solitary or clump-forming palms, most of which are huge. The leaves are *pinnate and include the longest leaves in the world. The *inflorescences are either terminal, the stem then dying, or *axillary, and form as sinuous, *bract-enclosed cylinders, sometimes branching, with male and female flowers. The fruits are scaly. Palm wine is obtained by tapping the inflorescences of *R. taedigera* and *R. vinifera*. There are about 28 species found in Africa, and one (*R. taedigera*) in tropical America. *See* PIASSAVA and RAFFIA.

**raphide** A needle-shaped crystal, usually of calcium oxalate, found inside a plant cell. Raphides occur in clusters and can be of diagnostic value.

**raspberry** *See* RUBUS.

**raspwort** *See* HALORAGIS.

**rattan** (rotan) *See* CALAMUS; CALOSPATHA; CERATOLOBUS; DAEMONOROPS; KORTHALSIA; and MYRIALEPIS.

**Raunkiaer, Christen** (1876–1960) A Danish ecologist who suggested that the earliest flowering plants grew under tropical conditions and that as their descendants spread to less hospitable regions they evolved a range of mechanisms in order to survive. This led him to develop a system for classifying types of plant according to the positions of their perennating buds in relation to the soil surface (*see* RAUNKIAER'S LIFE-FORM CLASSIFICATION).

**Raunkiaer's life-form classification** A classification of plants, proposed by the Danish botanist C. *Raunkiaer, based on the position of *perennating buds in relation to the soil surface. *See also* CHAMAEPHYTE; EPIPHYTE; GEOPHYTE; HEMICRYPTOPHYTE; PHANEROPHYTE; and THEROPHYTE.

**Rauvolfia** (family *Apocynaceae) A genus of shrubs or small trees, some species of which, especially *R. serpentina*, produce the *alkaloid reserpine, used to counteract high blood pressure. There are 110 species, with a pantropical distribution.

**Ravenala** (family *Strelitziaceae) A *monotypic genus, *R. madagascariensis* of Madagascar, which is an unbranched tree with banana-like leaves held in a single plane like a huge fan. The fruit is a *capsule, the seeds *arillate. It is called 'traveller's palm' because of the potable water trapped by the sheathing leaf bases, and is widely cultivated.

**ray** A line of cells that extends radially across *secondary phloem and *secondary xylem. It is initiated in the vascular *cambium and consists mainly of *parenchyma cells, although in *gymnosperms it may also include *tracheids.

**razor-strop fungus** The common name for the *fruit body of *Piptoporus betulinus*, a *parasite of birch trees. The *fruit body was once used for stropping razors.

**R/B ratio** *See* RESPIRATION–BIOMASS RATIO.

**RDP** *See* RIBULOSE-1,5-DIPHOSPHATE.

**reaction 1. (pedol.)** The degree of acidity or alkalinity of a soil, expressed as a value on the *pH scale. *See also* ACID SOIL and ALKALINE SOIL. **2. (ecol.)** In plant succession, the ability of an individual plant species or vegetation community to modify the physical environment and so favour further successional development. For example, the presence of the first colonizing plants in a *hydrosere will tend to reduce water movement, favouring accelerated silting and so paving the way for plants typical of shallower water to arrive as new colonizers.

**reaction centre** In a *photosynthetic unit, the hypothetical site at which energy derived from the absorption of light is used to transport electrons for *photophosphorylation. The reaction centre contains a *primary pigment (a form of chlorophyll *a*).

**reaction wood** *Wood that forms in response to stresses and has an abnormal structure. In conifers, it is known as compression wood; this is dense and tends to form on the lower side of branches. In *angiosperms, it is known as tension wood; this is often gelatinous and forms on the upper side of branches.

**reading-frame shift** (genetics) In the normal *transcription of a *cistron, *nucleotides are read in threes, the 'reading frame' being determined by the starting-point. Each triplet codes a specific *amino acid; the sequence of *codons therefore dictates the amino acids and their order in a given protein. Certain *mutagens (e.g. acridine dyes), which incorporate themselves between the complementary strands of the *DNA, may cause errors in replication such that the daughter DNA gains or loses a *nucleotide. When this daughter DNA is transcribed (*see* TRANSCRIPTION), the nucleotides will be read in the correct triplets up to the point of mutation, but the extra or missing nucleotide will cause a shift in the reading frame (to left or right) thereafter, such that all subsequent nucleotides will be read in wrong triplets. The result might be a protein with the wrong amino acids (often it is terminated early) but in any case it is dysfunctional.

**reading mistake** In genetics, the placement of an incorrect *amino acid into a *polypeptide chain during protein synthesis (*see* RIBOSOME).

**reafforestation** *See* REFORESTATION.

**recalcitrant seed** A seed that remains viable for only a short time, often 1 year or less, even under conditions that are commonly conducive to longevity. Many tropical genera produce recalcitrant seeds and so must be stored in the growth phase (i.e. as growing plants) rather than as seeds. *See* GENE BANK.

**Recent** *See* HOLOCENE.

**receptacle** **1.** That part of the stem from which all the parts of the flower arise. **2.** In *Asteraceae, the flattened tip of the stem that bears the *bracts and *florets. **3.** In *thallophytes, one of the swellings at the tips of the branches of the *thallus which mark the positions of the chambers (*conceptacles) in which the reproductive organs develop.

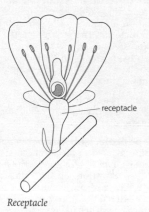

— receptacle

*Receptacle*

**receptor site** A set of reactive chemical groups in the *cell wall of a bacterium that are complementary to a similar set in the tailpiece of a *bacteriophage. It is through the interaction of these that the virus is able to recognize and attach itself to a host bacterium.

**recessive gene** A gene whose *phenotypic effect is expressed in the *homozygous state but masked in the presence of the *dominant *allele (i.e. when the organism is *heterozygous for that gene). Usually the dominant gene produces a functional product whereas the recessive allele does not: both 1 dose and 2 doses per nucleus of the dominant allele therefore lead to expression of its phenotype, whereas the recessive allele is observed only in the complete absence of the dominant allele.

**recipient** In *meromixis, the cell that receives the *chromosome.

**reciprocal averaging** (correspondence analysis) An *ordination method that combines gradient analysis and a system of successive approximation. Its chief merit lies in the clarity with which strong floristic gradients are displayed. Applying the method to a series of samples each containing a number of species begins by assigning an arbitrary (e.g. random) number to each species; these numbers are known as species scores. The scores are weighted according to the abundance of each species in each sample, and the weighted average of the species scores in each sample yields a sample score. The weighted average of all the sample scores yields a new species score. The species and sample scores are then standardized by subtracting the means and dividing by the standard deviation. The procedure is repeated until successive iterations produce little or no changes in the values.

**reciprocal cross** In genetics, one of a pair of crosses in which the two opposite mating types (or sexes) are each coupled with each of two different *genotypes and mated with the reciprocal combination, e.g. male of genotype A × female of genotype B (first cross), and male of genotype B × female of genotype A (the reciprocal cross). Such crosses are used to detect *sex linkage, maternal inheritance, or *cytoplasmic inheritance.

**reciprocal genes** Non-allelic *genes that reciprocate or complement one another. *See* COMPLEMENTARY GENES.

**recombinant** In genetics, an individual or cell with a *genotype produced by recombination, i.e. with combinations of *genes other than those carried in the parents. Recombination results from *independent assortment of *crossing-over. *See also* GENETIC ENGINEERING.

**recombination** The arrangement of *genes in offspring in combinations that differ from those in either parent, and the assortment of *chromosomes into new sets.

**recombination frequency** The number of recombinants divided by the total number of progeny, expressed as a percentage or fraction. Such frequencies indicate relative distances between *loci on a *genetic map.

**recon** The smallest unit of *DNA capable of recombination.

**recreatability (salvageability)** The term used by conservationists to denote a *community or ecosystem that could be readily re-established following disturbance (by motorway building, opencast mining, etc.) without any noticeable change. Certain long-established communities (e.g. in Britain *primary woodlands) are thought not to be recreatable since they would take many hundreds of years to re-establish their present detailed structure and composition and it is virtually impossible to reconstruct the same long-term environmental influences as fashioned their present form.

**rectiflorus** The condition in which the axes of the florets are parallel to the main axis of the *inflorescence.

**recurrence surface** In a peat stratigraphy, the sudden transition from a highly humified peat to a fresh unhumified peat. This change reflects the resumption of wetter conditions and peat growth following a drier period when heath rather than bog plants colonized the older humified peat. Increasingly, these features of peat stratigraphy are coming to be regarded as important indications of past climate change. Objective identification of recurrence surfaces in a stratigraphic profile is best achieved by measuring the degree of *humification of *peat. This involves extracting humic acid and measuring its density by spectrophotometry.

**recuticus** The condition in which there is no apparent *epidermis.

**red algae** The common name for *algae of the group *Rhodophyta. While the majority of marine species are indeed some shade of red, freshwater species may be brown or blue-green. They are known from the *Cambrian and were abundant in the succeeding *Ordovician.

**red cedar** See JUNIPERUS.

**red core of strawberry** A serious disease of strawberries caused by *Phytoph-

thora fragariae (*Oomycota). Infested plants are stunted and their roots become dark brown or black with a characteristic red core (stele).

**red currant** See GROSSULARIACEAE and RIBES.

**red meranti** See SHOREA.

**redox potential** A scale that indicates the reduction (addition of electrons) and oxidation (removal of electrons) for a given material. The position on the scale is expressed as an electric potential in millivolts, normally in the range 0–1300 or 0–1400 mV. The *pH of the sample must be known since this can alter the reading.

**redox reaction** A reaction that involves simultaneous *reduction and *oxidation.

**red podzolic soil (krasnozem)** A soil *profile formed at an advanced stage of weathering and *leaching by the process of *podzolization, being similar in appearance and properties to a *podzol but associated with the greater degree of chemical weathering and higher iron-oxide concentrations of a humid, tropical environment. See also ULTISOLS.

**red rust** An important disease of the tea plant (*Camellia sinensis). Orange-brown, velvety areas appear on the leaves of infected plants. The disease is caused by *algae of the genus *Cephaleuros.

**red snow** See CHLAMYDOMONAS.

**red tide** A phenomenon in which blooms of certain dinoflagellates colour the water of seas or estuaries red or reddish. The dinoflagellates, particularly species of Gonyaulax, are concentrated, along with their toxins, by filter-feeding shellfish, and when the shellfish are subsequently eaten by humans, poisoning results. Fish and shellfish may also die by poisoning.

**reducing sugar** A *sugar that contains an *aldehyde or potential aldehyde group and that has the ability to reduce certain inorganic *ions in solution. These include the cupric ions in Fehling's and Benedict's reagents.

**reduction** A chemical reaction in which atoms or molecules either lose oxygen or gain hydrogen or electrons.

**reduction divisions** In genetics, the two nuclear divisions in *meiosis that produce *daughter nuclei, each of which has half as many *chromosomes as the parental nucleus.

**reduction potential** (electrode potential) A quantitative measure of the ease with which a substance is oxidized or reduced. It represents the voltage necessary to prevent the flow of electrons to or from a half-cell containing aqueous solutions of the oxidized and reduced forms of the test substance when it is connected to a hydrogen half-cell.

**redundant cistron** A *cistron (*DNA sequence) that is present in many copies on one *chromosome, all but one of which are redundant. Examples are the cistrons in the nucleolus organizer coding for the *ribosomal RNA molecules.

**redwood** (coast) See SEQUOIA.

**reed-grass** See PHRAGMITES.

**reedmace** See TYPHA and TYPHACEAE.

**reference species** A species, usually extinct, that occurs as a *fossil of characteristic age and can be used to date geologic deposits or the age of a species lineage.

**reflexed** Bent backwards at a sharp angle.

**reforestation** (reafforestation) **1.** The establishment of a particular type of woodland by planting into an existing, different woodland type. **2.** The replacement of a tree crop by natural or artificial means on land from which a previous wood has been removed.

**refugia** See REFUGIUM.

**refugium** (pl. refugia) A small isolated area where extensive changes, most typically arising from changing climate, have not occurred. Plants and animals formerly characteristic of the region in general now find a refuge from the new unfavourable conditions in these areas. An example might be a mountain summit projecting above a glaciated lowland region. See also RELICT.

**regeneration niche** The specific niche that must exist in order for a plant to become re-established in an area it formerly occupied. A plant that can tolerate a wide range of environmental conditions when mature may nevertheless require very specific conditions for the germination and establishment of its seeds. For example, the Californian redwoods require burned soil and many grassland plants need disturbed, bare microsites for their germination.

**regolith** A general term for the layer of unconsolidated (non-cemented), weathered material, including rock fragments, mineral grains and all other superficial deposits that rest on unaltered, solid bedrock. It reaches its maximum development in the humid tropics, where depths of several hundreds of metres of weathered rock are found. Its lower limit is the weathering front. Soil is regolith that is able to support rooted plants. Compare SAPROLITE.

**Regosols** A reference soil group in the FAO *soil classification that comprises all soils that do not belong to one of the other reference soil groups, i.e. *Acrisols, *Albeluvisols, *Alisols, *Andosols, *Anthrosols, Arenosols, *Calcisols, *Cambisols, *Chernozems, *Cryosols, *Durisols, *Ferralsols, *Fluvisols, *Gleysols, *Gypsisols, *Histosols, *Kastanozems, *Leptosols, *Lixisols, *Luvisols, *Nitisols, *Phaeozems, *Planosols, *Plinthosols, *Podzols, *Solonchaks, *Solonetz, *Umbrisols, and *Vertisols.

**regular distribution** See UNDERDISPERSION.

**regular sample** (systematic sample) One of a number of samples taken at regular intervals, e.g. by the use of regularly spaced *quadrats along some environmental gradient such as a valley side. Though less reliable in certain circumstances than random sampling, systematic sampling may be more practicable and more economical in the time it takes. The chief disadvantages are the possibility that the interval selected may resonate with some unsuspected environmental variable, so giving biased results, and that the form of the sample does not conform with that theoretically assumed for many statistical tests.

**regulator gene** In the *operon theory of *gene regulation, a gene that is involved in switching on or off the *transcription of structural genes. When transcribed, the regulator gene produces a repressor protein, which switches off an *operator gene and hence the operon that this controls. The regulator gene is not part of the operon, and may even be on a different *chromosome. *See also* REPRESSOR.

**reindeer moss** The common name for the *lichen *Cladonia rangiferina* and similar species, which form extensively branched *podetia up to about 8 cm tall; in *C. rangiferina* the apices of the branches are slender and curved, typically all in the same direction. These lichens are found, for example, on *peaty ground in mountains and in Arctic and sub-Arctic regions where, together with other lichens, they form a major component of the winter diet of reindeer, Known in North America as caribou (*Rangifer tarandus*).

**relative-importance value** In ecology, a concept similar in principle to importance value, but based on original, somewhat relative, density–frequency–dominance values. It is the actual basal area, frequency, or density of a particular species, without reference to the density, frequency, and dominance values of other species present in the area.

**relative pollen frequency** (RPF) An expression of *pollen data from sediments for each species, genus, or family, as a percentage of the total pollen count or of the total tree pollen. It is the traditional and most widely used method for preparing pollen diagrams. *Compare* ABSOLUTE POLLEN FREQUENCY; *see also* PALYNOLOGY.

**release factors** In genetics, specific proteins that interact with the growing *polypeptide when a termination *codon is encountered during *transcription, and then mediate the release of the finished polypeptide from the *ribosome.

**relevé** In *phytosociology, the basic field unit recorded. The area should be uniform in floristic composition, and of uniform relief and soil type. The unit size varies with the type of *community, but at least should embrace the *minimal area. Site and soil data, as well as vegetation form, composition, and species, are noted.

**relic** *See* RELICT.

**relict (relic)** Applied to organisms that have survived while other related ones have become extinct. Often the term refers to species that formerly had a much wider distribution and have survived locally through periods of unfavourable conditions (e.g. glacial periods or land submergence) by existing in regions called refugia (*see* REFUGIUM), while becoming extinct elsewhere (e.g. some Arctic-alpine plants). They may be part of a *relict community (e.g. *Dryas octopetala* (mountain avens) in Britain, which was widespread during glacial times but is now restricted to a few mountain tops). It may also refer to a surviving species of a group, the other species of which have become extinct (e.g. *Ginkgo biloba*, the maidenhair tree).

**relict community** A *community that formerly had a much wider distribution but now occurs only very locally. Such contraction can be caused by various factors including climatic change, e.g. glaciation. Communities widespread during glacial times may now be restricted to mountain tops; e.g., Upper Teesdale (England) contains isolated flora characteristic of much of lowland Britain during glacial periods.

**relict coppice** An area of abandoned *coppice woodland.

**relict soil** *See* PALAEOSOL.

**removal time** *See* RESIDENCE TIME.

**rendzina** A *brown earth soil of humid or semi-arid grassland that has developed over calcareous parent material. The term is now obsolete and rendzinas may fall within the orders *Inceptisols or *Mollisols.

**renewable resource** A resource produced as part of the functioning of natural systems at rates comparable with its rate of consumption, e.g. food production by photosynthesis. Limits to renewable resources are determined by flow rate and such resources can provide a sustained yield. Compare FINITE RESOURCE.

**renewal cycle** A *biogeochemical cycle.

**rengas** A group of genera of trees of the family *Anacardiaceae of the Malay archipelago, producing black sap which is highly irritant to the skin and can be fatal. Even seasoned rengas wood can cause a rash.

**reniform** Kidney-shaped.

**repair synthesis** The enzymatic excision and replacement of regions of damaged *DNA. The defective single-stranded segments of DNA are cut out, correct bases are inserted using the complementary strand as a template, and these are interlocked by a DNA polymerase. A polynucleotide *ligase joins the 2 ends of the broken strand to complete the repair. An example is the removal by a nuclease enzyme of a thymine dimer that has been induced by UV radiation of the DNA molecule.

**replicase** Any enzyme capable of catalysing the replication of *DNA or *RNA in either *prokaryotes or *eukaryotes.

**replication** The synthesis of new daughter molecules of nucleic acid from a parent molecule, which acts as a template.

**repressor** In the *operon theory of *gene regulation, a protein produced by a *regulatory gene that inhibits the activity of an *operator gene, and hence switches off an operon. The action of a repressor is determined by some other factor, an effector: either this serves as a co-repressor, and hence is required to switch off the operon, or it inactivates the repressor, and so switches on the operon. See also REGULATOR GENE.

**reptant** Creeping along the ground and rooting.

**resak** See VATICA.

**rescue effect** A species arriving on an island may already be represented there and so may have the effect of reducing the chance of the extinction of that species from the island (i.e. of 'rescuing' it). The rescue effect will be greater on islands that are closer to the mainland source of species than more remote islands because the immigration rate will be higher. Compare EQUILIBRIUM THEORY.

**Resedaceae** A family, mostly of *herbs, with alternate leaves and flowers borne in long *spikes or *racemes. The flowers are irregular, with parts in fours to sevens, but the *stamens are often numerous. The 2–6 *carpels are superior, and often remain open at their tips, a very unusual feature. Reseda odorata (mignonette) has fragrant flowers and is often cultivated. There are 6 genera, with some 75 species, centred in the Mediterranean area, but extending into Africa, western USA, and India.

**Reseda odorata (mignonette)** See RESEDACEAE.

**reserpine** See RAUVOLFIA.

**reservoir pool** A large store of a nutrient at some stage in a *biogeochemical cycle. Reservoir pools are mainly abiotic, but may also be biotic, as in the case of the *biomass of a forest, which represents a considerable store of various elements, especially carbon. Exchanges between the reservoir pool and the *active pool are typically slow by comparison with exchange within the active pool. Human activity, such as the mining of mineral resources and the manufacture and application of fertilizer, may profoundly alter this exchange rate, generally releasing an excess into the active pool which can be accommodated only by establishing a new equilibrium. This may in turn produce unfavourable conditions, manifested as chemical pollution (e.g. excess phosphorus in *eutrophication, excess sulphur in acid precipitation and lake acidification).

**residence time (removal time)** The time that a given substance remains in a particular compartment of a *biogeochemical cycle. In particular, residence time is used to denote the time taken for pollutants to be removed from the atmosphere by natural processes.

**residual (stat.)** A data variability that is not accounted for by a particular statistical test. The residuals of individual data values, i.e. the difference between an observed and a computed value, often give ecologists insight into possible environmental influences on individual data records.

**resilin** A naturally occurring biological rubber, which returns almost to its former shape if deformed, so conserving most of the energy used in its deformation. Resilin contains two unique *amino acids (di- and tri-tyrosine) which give it a characteristic, sapphire-blue fluorescence in ultraviolet light.

**resin** An exudate of tree wood or bark, liquid but becoming solid on exposure, consisting of a complex of *terpenes and similar compounds. It is characteristic of some families, e.g. *Dipterocarpaceae, or groups, e.g. the conifers. In many cases it is of economic value for varnishes, etc. It is produced in specialized cells.

**resource allocation** The way in which the products of *photosynthesis are allocated to different organs within a plant. This changes during the course of a plant's life. Initially, there is a high level of allocation to roots, followed rapidly by shoot and leaf development, and later in the strengthening of support structures. The degree of allocation to reproduction differs between species (see R-SELECTION and K-SELECTION).

**respiration 1.** Oxidative reactions in cellular metabolism involving the sequential degradation of food substances and the use of molecular oxygen as a final hydrogen acceptor; ATP (see ADENOSINE TRIPHOSPHATE), carbon dioxide, and water are the products thus formed. The reactions involved in respiration are the reverse of those in *photosynthesis, i.e. $C_6H_{12}O_6 + 6O_2 \rightarrow 6CO_2 + 6H_2O$. **2.** The physico-chemical processes involved in the transportation of oxygen to and carbon dioxide from the tissues.

**respiration–biomass ratio (R/B ratio)** The relationship between total *community *biomass (i.e. standing crop) and respiration. With larger biomass, respiration will increase but the increase will be less if the individual biomass units or organisms are large (reflecting the inverse relationship between size and metabolic rate). Natural communities tend towards larger organisms and complex structure with low respiration rates per unit biomass.

**respiration quotient (RQ)** The ratio of the amount of carbon dioxide expired to the amount of oxygen consumed during the same period. (RQ = carbon dioxide consumed divided by oxygen utilized.)

**respiratory chain** See ELECTRON-TRANSPORT CHAIN.

**Restio** (family *Restionaceae) A genus of rush-like, monocotyledonous (see MONOCOTYLEDON), *perennial *herbs in which the stems of the plants undertake assimilation. The leaves are reduced to sheaths on the stem. The flowers are small, regular, and *dioecious, arranged in spikelets, often in a nodding *inflorescence. If the *perianth is developed it is composed of 2 series of dry, often *hyaline, segments, and has 3 *stamens. The male flowers often have a rudimentary *ovary, and in the female flowers several *staminodes may be present. The *superior ovary is made up of 1–3 *carpels with a similar number of *locules and *styles. The fruit is dry and *nut-like, or a 3-sided *capsule with many *endospermic seeds. *Restio* species are present as components of grass and sedge communities which often experience seasonal waterlogging. There are 88 species, found in Australia and S. Africa, but with little overlap in the distribution of species between these 2 regions.

**Restionaceae** A family of monocotyledons (see MONOCOTYLEDON) that are rush-like, rhizomatous (see RHIZOME) *herbs. They vary considerably in height, from a few centimetres to 2 m. The leaf-blades are rarely developed, although a dry, withered blade may be present, but at each *node on the tough, wiry shoots there is a sheathing leaf-base. In some species the leaf-bases are well developed, although not photosynthetic. *Ligules are rarely present. The shoots are adapted to become the photosynthetic organs, and are either simple or branched, and in some species are very branched. The stems are circular, semicircular, or almost 4-sided, and sometimes ribbed. Some are solid, others hollow. The flowers are small, regular, and unisexual, and the species are generally *dioecious. Flowers are held in *spikelets in a *panicle. The spikelets are 1- or many-flowered, with a sheathing *spathe. Where the *perianth is developed there are 3–6 thin, dry, similar segments, held in 2 series. In male flowers there are 3 *stamens, often with the rudiments of an *ovary. In female flowers the ovary is *superior with 1–3 *carpels, and 1–3 *locules containing a single *ovule. The fruits are dry and *nut-like, or three-sided *capsules. The seeds contain copious *endosperm. There are 38 genera and about 400 species, mostly found in S. Africa and Australia, often in very dry habitats, or those which experience heavy seasonal rains. The shoots and roots show adaptations to drought and to waterlogging respectively. Some species are used for thatching and matting.

**restorer** A *gene that negates the effects of other genes involved in *male sterility.

**resupinate** Upside-down. Of a fungal *fruit body: flattened against the substrate, with the *spore-bearing layer facing outwards.

**reticulate** Marked with a network pattern (e.g. the veins in a leaf).

**reticulate evolution** The creation of a network of closely related taxa within and at the species level, particularly by *chromosome doubling or *polyploidy. Polyploidy is more common in plants than in animals, and so reticulate evolution is more likely in the former than in the latter.

**reticulate method** A term sometimes used to describe a non-hierarchical clustering technique.

**retrogression** (retrogressive succession) A successional (see SUCCESSION) change, usually from an existing *climax community, leading to a less diverse and less structurally complex *community. The change frequently involves a reduction in *biomass. Retrogression is usually triggered by an environmental factor (e.g. a *pollutant) and disturbance by humans is often in-

volved, e.g. in removing top predators or tree cover and thereby setting in motion a downward development in *ecosystem complexity.

**retrogressive succession** See RETROGRESSION.

**return flow** See INTERFLOW.

**return period** The frequency, based on statistical analysis of past records, with which a particular environmental hazard may be expected.

**reverse mutation** (reversion) The production by further *mutation of a pre-mutation *gene from a mutant gene. This reverse mutation restores the ability of the gene to produce a functional protein. Strictly, reversion is the correction of a mutation, i.e. it occurs at the same site; more loosely, though, the term is applied also to a mutation at another site that masks or suppresses the effect of the first mutation. (In fact, such organisms are not non-mutant, but are double mutants with the same phenotype.) Compare SUPPRESSOR MUTATION.

**reversion** See REVERSE MUTATION.

**revolute** Curved or curled back, most commonly used of the edge of a leaf.

**Rhacomitrium** See RACOMITRIUM.

**rhagadiose** Cracked or fissured.

**Rhamnaceae** A cosmopolitan family of trees and shrubs that have *simple leaves and flowers borne in loose clusters. The flowers have their parts in fours or fives, the *stamens being opposite the *petals, and the 2–4-celled *ovary forms a *berry in the fruit. A dye (sap-green) has been obtained from Rhamnus catharticus (buckthorn); and Frangula alnus (alder buckthorn) has been used for making high-quality charcoal for gunpowder. Other members of the family are used as purgatives. Modern classifications recognize 58 genera, with 875 species.

**Rhamnus catharticus** (buckthorn) See RHAMNACEAE.

**Rhaphidophyceae** (Chloromonadophyceae) A small class of unicellular *algae which are sometimes considered to be members of the *Xanthophyta; they are alternatively regarded as protozoa, class *Phytomastigophora. The cells are typically flagellated with two *flagella, a forward-pointing tinsel type and a backward-pointing whiplash type. They are found in mud at the bottom of freshwater ponds, etc.

**Rhapis** (partridge cane; family *Arecaceae) A genus of delicate, forest-undergrowth fan palms which are common as pot plants. They are clump-forming. The leaflets are blunt. The palms are *dioecious. The flowers have 3 free *carpels, and 1–3 thinly fleshy fruits are produced per flower. There are 9 species, occurring in Japan and southern China.

**rheotaxis** A change in the direction of locomotion in a *motile organism or cell made in response to the stimulus of a current, usually a water current.

**rheotrophic** Applied to a *mire that is fed by flowing water. *Compare* OMBROTROPHIC.

**rheotropism** A tropic response (*see* TROPISM) to a water current.

**Rheum** (family *Polygonaceae) A genus of *herbs in which the leaves are large and *palmately lobed. The *inflorescence is an erect *panicle; the flowers are small and *entomophilous. Rhubarb is the *petiole of *R. rhaponticum*, and medicinal rhubarb (a purge) is *R. officinale*. There are about 50 species, occurring in temperate and subtropical Asia.

**Rhipsalis** (family *Cactaceae) A genus of about 50 species of *epiphytes that have cylindrical, branching stems, and are found mainly in tropical America. They are remarkable in their family in having Old World members in Madagascar and Sri Lanka, but perhaps these were introduced there.

**rhizina** *See* RHIZINE.

**rhizine** (rhizina) A root-like structure in *lichens, made up of *fungal *hyphae and usually functioning in the attachment of the *thallus to the substrate.

**Rhizobiaceae** A family of *Gram-negative *bacteria in which the cells are typically rod-shaped, and *motile with one or more *flagella. They are *chemo-organotrophic, and can grow only in the presence of air. There are 4 genera, found in soil and in association with plants. Some species form mutually beneficial (*symbiotic) associations with plants, while others cause plant diseases.

**Rhizobium** (family *Rhizobiaceae) A genus of *bacteria that occur in soil and in *root nodules on leguminous plants. The bacteria live in the root nodules and benefit the plant by making available to it nitrogen which they fix from the atmosphere. Free-living rhizobia appear to be unable to fix nitrogen.

**Rhizocarpon** (order *Lecanorales) A genus of *crustose *lichens in which the *apothecia are *lecideine and normally black. *Spores are pale to dark brown, 1- or 3-*septate to *muriform. All species are found on rocks. *R. geographicum* occurs on siliceous rocks and walls, mainly in hilly districts; the *thallus is bright yellow, surrounded and intersected by fine black lines, and dotted with small black apothecia.

**Rhizoctonia** (order *Agonomycetales) A *form genus of *fungi which are responsible for a number of important plant diseases, including *damping-off of seedlings and sharp *eyespot of cereals. Some species have *perfect (sexual) states classified in the *Basidiomycotina.

**Rhizoglossum** *See* OPHIOGLOSSACEAE.

**rhizoid** In *mosses and *liverworts: a thread-like structure, often serving to anchor the plant (*gametophyte) to the substrate; rhizoids can absorb water and minerals. In mosses rhizoids are multicellular, in liverworts they are unicellular.

**rhizome** A horizontally creeping underground stem which bears roots and leaves and usually persists from season to season.

**Rhizomnium punctatum** *See* MNIUM.

**rhizomorph** In certain *fungi, a thick, root-like strand of differentiated *hyphae.

**rhizomycelium** A primitive type of *mycelium formed by certain *fungi.

***Rhizophora*** **(family *Rhizophoraceae)** An important genus of 6–9 species of mangrove trees which are pantropical in distribution and *viviparous. Their timber is good for piles, scaffolding, and charcoal; the bark (cutch) is used for tanning.

**Rhizophoraceae** A family of trees and shrubs of tropical rain forest, especially mangroves, in which the leaves are *simple, usually opposite, and have *caducous *stipules. The flowers are hermaphrodite, with 3–16 *perigynous or *epigynous *sepals and *petals, 8 to many *stamens on the edge of the *receptacle, and the *ovary *superior to *inferior. They are related to the *Combretaceae. There are 16 genera, with about 130 species, with a pantropical distribution. Four genera occur as the main trees of mangrove forests, the other occur inland.

**rhizophore** In *Selaginellaceae, a leafless branch that arises from a fork in the stem and grows downward, producing roots at its tip when it reaches the soil.

**rhizoplane** **(root surface)** *See* RHIZO-SPHERE.

***Rhizopus*** **(order *Mucorales)** A genus of *fungi which are *saprotrophic and found on soil, fruit, etc. *R. nigricans* (*R. stolonifer*) forms 'stolons', analogous to the *stolons of higher plants, which grow over the surface of the substratum, attached to it by branching *rhizoids.

**rhizosphere** The area of soil immediately surrounding plant roots, which is altered by their growth, respiration, exchange of nutrients, etc. Within this zone a further zone is sometimes distinguished, called the rhizoplane or root surface.

***Rhododendron*** **(family *Ericaceae)** A genus of shrubs or trees that have big, scaly buds, *simple, alternate, mostly evergreen leaves, and flowers borne in short *racemes or *umbel-like clusters, with 5-lobed, funnel-shaped to bell-shaped, slightly irregular *corollas, and 5 or 10 *stamens. Many species and *hybrids are cultivated for their splendid flowers and sometimes for their foliage. If it is incorporated in quantity by bees, *Rhododendron* nectar may render honey poisonous. There are about 850 species, in 3 major groups. The *Rhododendron* group includes both showy temperate species and the *Vireya* group of delicate, tropical species, some of which have scented flowers. The vast majority of species are confined to the mountains of South-east Asia and the Himalayas, but a few extend westwards into Europe and N. America, and 1 to northern Australia. *Azalea* (with 5 stamens) is sometimes regarded as a separate genus.

**Rhodophyta** **(red algae)** A group of *Eukarya which are mostly red in colour; no *flagellated cells are formed, and the storage product is a type of starch known as *floridean starch. Sexual reproduction tends to be complicated. Red algae may be unicellular, but most are filamentous or membranaceous. The majority occur in the sea, but some are freshwater or terrestrial. The group includes 1 class (Rhodophyceae) with many orders and more than 4000 species. The red seaweeds are more numerous than green and brown seaweeds in temperate and tropical regions, but less numerous in colder regions. Several red seaweeds yield useful products (e.g. *agar, *algin, and *carragheen). The red algae are believed to be one of the oldest groups of *eukaryotic algae.

**Rhodospirillaceae** **(purple non-sulphur bacteria; order *Rhodospirillales)** A family of *anaerobic, anoxygenic, photosynthetic *bacteria which mostly use organic compounds as sources of reduction for *photosynthesis. Most species are *motile, having *flagella. The cells are spherical, spiral, ovoid, or rod-shaped, and range in colour from brownish or greenish to purple. Photosynthetic pigments are attached to membranes which are continuous with the cytoplasmic *membrane. There are several genera. The organisms are found in mud in *eutrophic ponds, ditches, sewage lagoons, etc.

**Rhodospirillales** An order of photosynthetic, mainly aquatic, *Gram-negative *bacteria which contain special *chlorophylls called *bacteriochloro-

r

phylls. *Photosynthesis occurs only in the absence of air, and oxygen is not released. Some species can also grow *chemo-organotrophically in the presence of air. Families are *Chlorobiaceae, *Chloroflexaceae, *Chromatiaceae, and *Rhodospirillaceae.

**Rhodymeniales** (*Rhodophyta, domain *Eukarya) An order of red seaweeds in which the *thallus may be cylindrical or flattened; some (e.g. species of *Lomentaria* and *Chylocladia*) have a segmented or beaded appearance. They are generally found low in the intertidal region, usually attached to rocks. (*Rhodymenia palmata* (*dulse) is now called *Palmaria palmata* and is classified in a separate order, the Palmariales.)

**rhoophilous** Inhabiting creeks.

**rhubarb** *See* RHEUM and POLYGONACEAE.

**Rhus** (family *Anacardiaceae) A genus of shrubs, small trees, or climbers, which produce a resinous, highly irritant exudate. Several species (e.g. sumac) are ornamentals. *R. toxicodendron* is poison ivy of N. America. *R. verniciflua* of China yields lacquer from the trunk and its powdered leaves yield tannin. There are about 200 species, occurring in tropical and warm temperate regions.

**Rhynchostegium** (order *Hypnobryales) A genus of slender to robust mosses, in which the stems are *procumbent and irregularly branched. The leaves are ovate-lanceolate to broadly ovate, with a nerve extending half-way or further up each leaf. The *capsule lid has a long beak; the *seta is smooth and reddish. The genus contains about 195 species, found mostly in the northern hemisphere. *R. riparioides* is aquatic, growing attached to boulders or tree roots by or in streams. The stems may be up to 15 cm long, sparingly branched, and often lacking leaves on their lower parts. *R. confertum* (formerly *Eurhynchium confertum*) is very common in lowland Britain on stone walls, garden rockeries, etc. The leaves are very small (up to 1 mm long) and the plant is inconspicuous except when the *capsules are present.

**Rhynia** *See* TMESIPTERIS.

**rhythm** *See* PERIODICITY and CIRCADIAN RHYTHM.

**Rhytidiadelphus** (order *Hypnobryales) A genus of large, robust mosses in which the shoots are erect to *procumbent. The stems are usually long, and irregularly to *pinnately branched. The leaves are broad at the base, gradually or abruptly narrowing to a sharply pointed tip; each leaf may have a short double nerve, or it may be nerveless. The *seta is deep red and smooth. There are about 8 species, mainly found in northern temperate regions, and including some attractive British species. *R. triquetrus* is often conspicuous, with tall, tough, erect stems from which branches arise irregularly to give a bushy habit; the leaves are large and spreading, and *capsules are rarely formed. It is found mostly in woodland clearings. *R. squarrosus* is a smaller moss with *squarrose leaves; the shoot apices have a characteristic star-like appearance. It is found in moist grassland and is a common weed in lawns.

**rhytidome** *See* BARK.

**Rhytisma** (order *Rhytismatales) A genus of *fungi which are *parasitic on the leaves of trees. *R. acerinum* is common on the leaves of *Acer pseudoplatanus* (sycamore) where it forms conspicuous black blotches (*tar spots) about 1–2 cm across.

**Rhytismatales** (Phacidiales; subdivision *Ascomycotina) An order of *fungi that form apothecioid (*see* APOTHECIUM) *ascocarps immersed in a *stroma or in host tissue; *asci are unitunicate. Species include *saprotrophs and plant *parasites.

**Ribes** (family *Grossulariaceae) A genus of shrubs with simple, alternate, usually *palmately lobed leaves, and flowers in *racemes. These are regular, with 4 or 5 *sepals, *petals, and *stamens, and an *inferior *ovary of 2 *carpels which forms a *berry on top, of which the *calyx persists. Many species are grown for their edible fruits (e.g. *R. rubrum*, redcurrant; *R. nigrum*, blackcurrant; and *R. uva-crispa*, gooseberry), or for their ornamental flowers. There are about 150 species, occurring in the northern temperate zone, and on the mountains of Central and S. America.

**ribitol** A 5-carbon sugar alcohol that forms an integral part of the structure of the *flavins, *riboflavin, *flavin adenine dinucleotide (FAD), and *flavin mononucleotide (FMN).

**riboflavin** Vitamin B$_2$; it consists of an organic base coupled to *ribitol. It occurs widely in nature, being an integral part of the *coenzymes *flavin adenine dinucleotide (FAD) and *flavin mononucleotide (FMN).

**ribonuclease** An enzyme that catalyses the hydrolysis of *RNA, resulting in the formation of mono- and oligonucleotides.

**ribonucleic acid** See RNA.

**ribose** An aldopentose monosaccharide, which comprises the carbohydrate component of *RNA.

**ribosomal RNA** See RNA.

**ribosome** A subcellular granule, 10–20 nm in diameter, composed of *RNA and *protein, which is found in large numbers in all types of cells and in some subcellular *organelles. Ribosomes are the site of protein synthesis: *m-RNA attaches to them and there receives *t-RNA molecules bearing *amino acids. In *eukaryotic cells they are synthesized in the *nucleolus, but are found predominantly in the *cytoplasm, singly or in chains (polysomes or polyribosomes, probably linked by the m-RNA), or attached to the *endoplasmic reticulum, which is then termed 'rough ER'. There are two main types of ribosomes, distinguished by their size: 70S ribosomes occur in *prokaryotes and in the matrix of *chloroplasts and *mitochondria; the slightly larger 80S ribosomes occur in the cytoplasm of eukaryotes. (The 'S' refers to the *Svedberg unit.)

**ribulose-1,5-diphosphate** (RDP) A phosphorylated derivative of the ketopentose monosaccharide ribulose, which acts as an acceptor molecule for carbon dioxide in the *Calvin cycle. The reaction between 1 molecule each of these 2 *substrates results in the formation of 2 molecules of phosphoglyceric acid.

***Riccia*** (order *Marchantiales) A genus of small, *thalloid *liverworts in which the *thallus may be broad or narrow, and usually branches dichotomously (i.e. forks), often resulting in a rosette-like appearance. There are several British species. *R. glauca* (crystalwort) is the commonest, found on damp, cultivated soil from early autumn to late spring; the thallus is light glaucous-green. *R. fluitans* is variable in form and may be free-floating in ponds or ditches or may grow on mud; the free-floating form is narrower and more delicate than the terrestrial form.

**rice** See ORYZA.

**rice stunt** A virus disease which affects rice and other grasses. In addition to stunting, infected rice plants show yellowish streaks on the leaves. The virus is transmitted by sap-sucking leafhoppers.

***Ricinocarpus*** (wedding bush; family *Euphorbiaceae) A genus of erect, *monoecious shrubs with *opposite or alternate leaves which are often linear and revolute (curved back), without *stipules. The flowers are unisexual, but occur in clusters, each of which may comprise both male and female flowers which are held on individual stems. The *calyx is fused and lobed; there are 4–6 white *petals. The male flowers have a column of numerous *stamens fused at the base. The female flowers have a *trilocular *ovary. The fruit is a *capsule. The genus is endemic (see ENDEMISM) to Australia, except for 1 of the 15 species, which is found in New Caledonia.

***Ricinus*** (family *Euphorbiaceae) A *monotypic genus, *R. communis*, the castor-oil plant, a treelet of Africa now widely naturalized elsewhere. The seeds are poisonous but contain a valuable oil.

***Rickettsia*** (family *Rickettsiaceae) A genus of *bacteria which can grow only inside the cells of vertebrate or arthropod hosts. The organisms are rod-shaped. The genus includes the causal agents of typhus fever and various other spotted fevers.

**Rickettsiaceae** (order *Rickettsiales) A family of *bacteria in which the cells

r

are *parasitic in vertebrate and arthropod hosts; the organisms occur in or on the cells of the host, though not in red blood cells. There are several genera. Some species can cause important diseases, including typhus, and are transmitted by lice, fleas, ticks, etc.

**Rickettsiales** An order of *Gram-negative *bacteria, most of which are obligately parasitic (*see* PARASITISM) and can multiply only inside the cells of a living host under natural conditions, although some can be cultivated in cell-free media in the laboratory. Some can cause disease. There are 3 families.

**ridding** An area from which a woodland crop has been taken.

**rill-wash** Eroded material that is concentrated into more or less intermittent trickles and rills on inclined slopes, owing to runoff of water.

**ring-diffuse species** *See* TREE-RING.

**ring-porous species** *See* TREE-RING.

**ring spot** **1.** A ring of pale or yellowish coloration which appears on the leaves of plants infected with certain viruses. **2.** A *fungus disease of *brassicas in which rounded brown spots, about 1 cm across, appear on mature leaves.

**Rinorea** (family *Violaceae) A genus of trees and shrubs in which the flowers are almost regular. There are about 200 species, occurring in the tropics.

**riparian** Pertaining to the bank of a river or shore of a lake.

**ripening** A stage in the development of *fruit, characterized by the softening, colouring, and sweetening of the *tissue, and by a decrease in acidity. Often this is associated with an increase in the rates of respiration, cell expansion, and ethylene production, combined with a loss of *chlorophyll.

**Rivulariaceae** An outdated phycological family of 'blue-green algae' in which genera (such as *Fremyella* and *Rivularia*) were distinguished on the basis of field characteristics which are lost or variable when the organisms are cultured in the laboratory. The organisms are now classified as *cyanobacteria; all those with tapering *trichomes and terminal *heterocysts are included in the genus *Calothrix*.

**RNA (ribonucleic acid)** A *nucleic acid that is characterized by the presence of D-ribose and the *pyrimidine base uracil. It occurs in 3 principal forms, as *messenger-RNA, ribosomal-RNA, and *transfer-RNA, all of which participate in protein synthesis.

**robin's pin-cushion gall** (rose bedeguar) A red and green or crimson, hairy growth on wild (dog) roses (*Rosa canina*), which is formed by the cynipid wasp *Diplolepis rosae*. Its eggs are laid in leaf buds, although the galls appear to grow from the twig or stem. The galls are multilocular, with an average of 30 progeny emerging in June. The average size of the galls is approximately that of a small pea, but the hairs may attain a length of 35 mm. Reproduction is normally *parthenogenetic (offspring arising from unfertilized ova), less than 1 per cent of reared progeny being males.

**Roccella** (order *Opegraphales) A genus of *lichens, in which the *thallus is *fruticose, and attached to its substrate by a basal sheath. The *ascocarps are apothecioid (*see* APOTHECIUM) and round or elongated. *Spores are colourless and usually triseptate (*see* SEPTUM). *Roccella* species are found chiefly on rocks by the sea in warmer regions. Some have been used in the manufacture of certain dyes, including the pH indicator *litmus.

**Rochalimaea** (family *Rickettsiaceae) A genus of *bacteria that live *parasitically in vertebrate and arthropod hosts; they can be cultivated in cell-free media in the laboratory. They can cause disease in humans, and are transmitted from one individual to another by lice.

**rock tripe** The common name for *lichens of the family Umbilicariaceae, particularly *Lasallia pustulata* (*Umbilicaria pustulata*) or *U. esculenta*. *See* IWATAKE.

**roguing** The manual removal of infected or inferior specimens from an otherwise healthy crop of plants.

**Romneya** (family *Papaveraceae) A *monotypic genus (*R. coulteri*) of *glaucous herbs which have big, showy flowers, and are grown as ornamentals, occurring in California and Mexico.

**root** The lower part of a plant, usually underground, by which the plant is anchored and through which water and mineral nutrients enter the plant.

**root cap** The tissue that covers the apex of a root.

**root frequency** In the measurement of vegetation frequency, records based solely on the species rooted in a *quadrat, as distinct from shoot frequency, which includes those species rooted outside the quadrat but with foliage overlapping into the quadrat.

**root hair** An outgrowth (*trichome) from a single epidermal cell in the *piliferous layer of a *root. It is thin-walled and increases the total surface area available for the absorption of water and nutrients.

**root-hair zone** *See* PILIFEROUS LAYER.

**rooting shank** *See* OUDEMANSIELLA.

**root nodule** (actinorrhiza) A small, *gall-like growth on the roots of certain plants, especially leguminous plants (*Fabaceae) but also others including alder (*Alnus* species), bog myrtle (*Myrica gale*), sea buckthorn (*Hippophaë* species), sumach (*Coriaria* species), California lilac (*Ceanothus* species), and the cycads (*Cycadophyta). The nodules develop as a consequence of infection of the roots by bacteria (*Rhizobium* or *Bradyrhizobium* species in the case of legumes, *Actinobacteria in non-legumes). Bacterial colonies then inhabit the root nodules and benefit the plant by fixing atmospheric nitrogen, much of which becomes available to the plant.

**root pressure** The pressure developed in the roots of some species of plants which, when the shoot is cut off, causes fluid to exude from the *vascular system.

**root–shoot ratio** The ratio of the amount of plant *tissues that have supportive functions to the amount of those that have growth functions.

Plants with a higher proportion of roots can compete more effectively for soil nutrients, while those with a higher proportion of shoots can collect more light energy. Large proportions of shoot production are characteristic of vegetation in early *successional phases, while high proportions of root production are characteristic of *climax vegetational phases.

**root surface** (rhizoplane) *See* RHIZO-SPHERE.

***Rosa*** (roses; family *Rosaceae) A genus of prickly, often scrambling shrubs that normally have *pinnate leaves with *stipules. The flowers have usually 5 *sepals and *petals, and numerous free *stamens and free *carpels, but the carpels sit in a deep cup formed by the *calyx tube which becomes fleshy in the fruit, forming the rose hip, a valuable source of vitamin C. It is probably the most widely cultivated ornamental flowering plant. Many species and complex *hybrid forms are cultivated for their showy, often fragrant flowers, which range in colour from white, yellow, and red, to a lilac-blue. The separation of the 7 *chromosome sets in the formation of *gametes is uniquely complex and may be a factor in producing their taxonomic diversity. There are about 100 species, found in the northern temperate zone, with subtropical outliers, but the genus contains many critical taxa (*see* TAXON) whose specific status is uncertain.

**Rosaceae** A large and heterogeneous family of dicotyledonous (*see* DICOTYLE-DON) trees, shrubs, and *herbs that have alternate, *stipulate leaves, and regular flowers, usually with *inferior or partly inferior *ovaries, or with the sometimes free *carpels in a receptacular (*see* RECEPTACLE) cup. There are usually 4 or 5 free *sepals and *petals, and numerous *stamens. Fruits are very varied. They may be *achenes, *drupes, *follicles, or *pomes (as in *Malus). Many are cultivated for their edible fruits (e.g. *Pyrus, Malus, Prunus, Rubus,* and *Fragaria*), or for their flowers. In modern classifications, some 107 genera, with 3100 species are recognized, mostly in temperate zones, but cosmopolitan overall.

**rose** *See ROSA.*

**rose apple** The edible fruits of *Eugenia jambos. See EUGENIA.*

**rose bedeguar** *See* ROBIN'S PIN-CUSHION GALL.

**rosette plant** A plant (e.g. *Bellis perennis*, daisy) whose leaves are spread in a horizontal plane from a short *axis at ground level. Rosette plants are generally found in sparse or low-growing habitats. The advantages include a reduced likelihood of being grazed. *See also* HEMICRYPTOPHYTE.

**rosewood** *See DALBERGIA.*

**rotan** *See* RATTAN.

**rough ER** *See* RIBOSOME.

**rowan** *See SORBUS.*

**royal fern** *See OSMUNDA.*

**royal palm** *See ROYSTONEA.*

***Roystonea* (royal palms; family** *Arecaceae)* A genus of tall, solitary, elegant palms, whose trunks are often bulging. The leaves are *pinnate, the crown-shaft prominent. The *inflorescences are paniculate (*see* PANICLE), and held below the crown-shaft, twin male flowers flanking a female. The *stigma is basal in the fruits. Several are very widely cultivated. There are 6 species, found in southern Florida, the Caribbean region, and northern S. America.

**RPF** *See* RELATIVE POLLEN FREQUENCY.

**RQ** *See* RESPIRATION QUOTIENT.

***r*-selection** A selection for maximizing the intrinsic rate of increase (*r*) of an organism so that when favourable conditions occur, e.g. in a newly formed *habitat, the species concerned can rapidly colonize the area. Such species are opportunists (*see* FUGITIVE SPECIES). An opportunist strategy is advantageous in rapidly changing environments as in the early stages of a *succession. *See also* BET-HEDGING, BIOTIC POTENTIAL and POPULA-TION EXPLOSION; *compare* K-SELECTION.

**R technique** The most usual way to analyse data, in which variables or attributes (*n*) form the data, and the observations for different sample sites or individuals (*N*) form the rows in a table or matrix. *Compare* Q TECHNIQUE; *see also* INVERSE ANALYSIS and NORMAL ANALYSIS TECHNIQUE.

**rubber** *See HEVEA.*

**Rübel, Eduard (1876–1960)** A Swiss phytogeographer who, in collaboration with J. *Braun-Blanquet, helped to develop the classification system of the *Zurich–Montpellier school of phytosociology. This was based originally on the system of A. F. W. *Schimper.

**Rubiaceae** One of the biggest families of flowering plants, comprising trees, shrubs, and a few *herbs, in which the leaves are *simple, *opposite, and usually have interpetiolar *stipules. The flowers are hermaphrodite and *tetra- or *pentamerous, the *corolla tubular, the *stamens *epipetalous, and the *ovary *inferior. Various species are the source of quinine, coffee, and many ornamentals. There are about 637 genera, and 10 700 species, with a cosmopolitan distribution, but strongly centred in the tropics.

***Rubus* (blackberry, raspberry; family** *Rosaceae)* A genus, mostly of shrubs but including a few low *herbs, often with *compound leaves of 3–7 leaflets, and usually with prickly stems. Normally there are 5 *sepals and *petals, and many *stamens and *carpels. The carpels sit on a conical *receptacle and ripen to form a cluster of 1-seeded *drupelets, aggregated into a compound, fleshy fruit. Many species and *hybrids are cultivated for their edible fruits. The genus contains very many species, many of which are critical taxa, often of hybrid origin and frequently apomicts. There are about 250 species, centred in the northern temperate zone, but cosmopolitan.

**ruderal 1. (noun)** A plant that is associated with human dwellings or agriculture, or one that colonizes waste ground. Ruderals are often weeds which have high demands for nutrients and/or are intolerant of competition. In the classification of *plant strategies proposed by J. P. Grime, a plant species found in areas of low stress and high disturbance.

*Compare* COMPETITOR and STRESS-TOLERA-TOR. **2. (adj.)** Applied to such a plant.

**rufous**  Reddish-brown in colour.

**rugose**  Wrinkled; bearing many ridges.

*Rumex*  *See* POLYGONACEAE.

**ruminate**  Of the *endosperm, pale with an irregular pattern of dark lines, looking as though chewed.

**runcinate**  Saw-toothed.

**runner bean**  *See* PHASEOLUS.

**rupestrine**  Growing among rocks. *See also* PETROPHILOUS.

**Ruppiaceae**  An aquatic, monocotyledonous (*see* MONOCOTYLEDON) family of submerged *herbs, usually of brackish or salt water, that have linear or bristle-like leaves, sheathing at their bases, and short, terminal *racemes of minute flowers. These have no *perianths, 2 *stamens, and an *ovary of 4 or more free *carpels which develop long, separate *pedicels in the fruit, forming an *umbel-like structure. *Pollen grains are elongate in shape and rise to the surface in a 'bubble' of air, where pollination occurs. There is 1 genus, *Ruppia*, and about 7 species, occurring throughout temperate and subtropical regions, and in tropical mountain lakes.

**rush**  *See* JUNCUS.

**Russian borer**  *See* PEAT-BORER.

*Russula* (family *Russulaceae) A genus of *fungi, which form fragile *fruit bodies which typically have brightly coloured caps and do not exude a milky latex when damaged. Some species are edible, a few are poisonous. They are found on the ground under trees, with which they often form *mycorrhizal associations.

**Russulaceae** (order *Russulales) A family of *fungi which form *ballistospores on stalked, usually *mushroom-shaped *fruit bodies. *Spores have ornamentations. Species are found on the ground under trees.

**Russulales** (subclass *Holobasidiomycetidae) An order of *fungi characterized by the presence of *sphaerocysts in the tissues of the *fruit body.

**rust**  A plant disease caused by a *fungus of the class *Urediniomycetes. The characteristic symptom is the development of spots or pustules bearing masses of powdery *spores which are usually rust-coloured, yellow, or brown. Infected plants may also show distortions or *gall-like swellings. *Compare* WHITE RUST.

**Rutaceae**  A family of aromatic trees and shrubs which are sometimes thorny. The leaves are mostly *compound, dotted with glands, aromatic when crushed, without *stipules, and sometimes *opposite. The flowers are regular, hermaphrodite, *tetra- or *pentamerous with a *disc, and the *ovary is usually *superior and composed of several fused *carpels. There are 161 genera, with about 1650 species, including the group which provides citrus fruits, occurring in tropical and warm temperate regions, and well represented in Australia and S. Africa.

*R.* **value**  Usually, the multiple correlation coefficient, as distinct from *r*, the simple correlation coefficient.

**rye**  *See* SECALE.

**S** **1.** *See* SULPHUR. **2.** *See* SVEDBERG UNIT.

**s** *See* SELECTION COEFFICIENT.

*Sabal* **(palmetto palm; family \*Arecaceae)** A genus of very hardy fan palms, most of which have a trunk, but some of which occur as stemless clumps. The \*inflorescences are formed among the leaves. The flowers are hermaphrodite. Several are extensively cultivated. There are 14 species, found in southern USA, the Caribbean, and (1 species) in northern S. America.

**saccate** Resembling a bag or sac.

**saccharide** An alternative term for sugar.

*Saccharomyces* **(family \*Saccharomycetaceae)** A genus of \*yeasts that includes some economically important species and strains. In the absence of air, *S. cerevisiae* can ferment sugar to alcohol and carbon dioxide, and strains of this species are used in bread-making, brewing, wine-making, etc. The cells of *S. cerevisiae* are small and elliptical, and reproduce by \*budding. *Saccharomyces* species are found naturally on ripe fruit, soil, etc.

**Saccharomycetaceae (order \*Endomycetales)** A family of \*fungi in which the vegetative stage is composed predominantly of single cells which reproduce by \*budding. A mycelium is formed occasionally in some genera. Most of the \*ascospore-forming \*yeasts belong to this family, including some of economic importance (e.g. *Saccharomyces*). Members of the Saccharomycetaceae are chiefly \*saprotrophic, and found in a range of habitats.

*Saccharum* **(family \*Poaceae)** A genus of large, robust, \*perennial grasses; sugar cane is *S. officinarum*, the only world crop originating from New Guinea. The stem is solid, containing sugar in the soft tissues. There are about 30 species, occurring in the Old World, from the tropics to warm temperate regions.

**saddle fungus** The common name for the saddle-shaped \*fruit bodies of a \*fungus of the genus \**Helvella*; also an alternative name for \*dryad's saddle.

**saffron** Colouring and flavouring obtained from *Crocus sativus* and the dried stigmas that produce them. *See* CROCUS.

**sage** *See* SALVIA.

**sagebrush** *See* ARTEMISIA.

**sagittate** Of a leaf, shaped like an arrow head.

**sago** *See* CYCAS.

**sago palm** *See* METROXYLON.

**Sahel zone** The semi-arid southern border of the Sahara in western Africa, which supports a very dry type of \*savannah including scattered, thorny trees. In recent years the desert has advanced appreciably into the Sahel zone, partly at least as the result of poor land management. *Compare* GUINEA ZONE.

**Sahulland** The name often given to the tropical portion of the combined Australia–New Guinea land mass, as it existed at times of low sea-level during the \*Pleistocene. The faunas of the two present-day components have great similarities, the differences being due mainly to the fact that New Guinea is largely forested and Australia is largely open country. The Sahul shelf, linking New Guinea with Australia, is less than 200 m below the present sea-level.

**sal** *See* SHOREA.

**salac** *See* SALACCA.

*Salacca* **(family \*Arecaceae)** A genus of spiny palms which are usually stemless and occur as big, dense, clumped \*rosettes. The leaves are \*pinnate and large. These palms are \*dioecious, \*inflorescences bursting through the leaf sheaths, sparsely branched into stout, \*flexuous \*spikes, the flowers on them crowded. The fruits are scaly. There are 1–3 seeds embedded in pulp. *S. zalacca* (salac) is widely cultivated, especially in Java, for its sweet, edible fruit. There are 15 species, found from Assam to Sumatra, Malaya, Java, Borneo, and the Philippines.

**Salicaceae** A family of dicotyledonous (*see* DICOTYLEDON) trees and shrubs that have alternate, deciduous, *simple leaves with *stipules and tiny flowers, in unisexual *catkins which are borne on separate male and female plants. The flowers have no *perianth, merely a basal *bract, and have either 2 to many *stamens, or a 1-celled, bicarpellary *ovary with 2 *stigmas. The trees are mainly wind-pollinated. The fruits are *capsules, and the seeds bear long, silky hairs, which help in wind dispersal. Some are important timber trees. There are two genera (*Salix* and *Populus*), comprising 330–50 species, most occurring in the northern temperate zone.

**salic horizon** A soil *horizon, usually below the surface, that contains not less than 2 per cent salt and with a figure of 60 or more for the value calculated as the thickness of the horizon in centimetres multiplied by the percentage of salt. It is a *diagnostic horizon.

**Salicornia** (glassworts; family *Chenopodiaceae) A genus of *annual *herbs and some woody *perennials that have succulent leaves in *opposite pairs, so fused to the stem and to each other that at first sight the plants appear to consist of jointed, cylindrical, fleshy segments, with only the tips of the leaves free. The flowers are solitary, or in groups of 3, in the leaf *axils, and are almost wholly immersed in the fleshy stems. They each consist of 1 or 2 *stamens and a single *carpel. They inhabit salt marshes in coastal or arid regions, and are important colonists of tidal mud-flats. They have extremely high internal suction pressures (up to more than 100 atmospheres) which enable them to take up water from strongly saline solutions. Some species are pickled and eaten as a vegetable. The genus has 13 cosmopolitan species.

**salination** *See* SALINIZATION.

**saline 1.** Pertaining to salt. **2.** An aqueous solution of sodium chloride, with or without other salts, and approximately isotonic with body fluids, that is employed for the temporary maintenance of living cells and tissues.

**saline-sodic soil** Soil that contains more than 15 per cent exchangeable sodium, a saturation extract with a conductivity of more than 0.4 siemens per metre (4 mmhos/cm) (25°C) and in the saturated soil usually has a *pH of 8.5 or less. Either high concentration of salts or high pH, or both, interfere with the growth of most plants.

**saline soil** Soil that contains enough soluble salt to reduce its fertility. The lower limit is usually defined as 0.4 siemens per metre (4 mmhos/cm).

**salinization** (salination in US usage) The process of accumulating soluble salts in soil, usually by an upward capillary movement from a saline *groundwater source.

**Salix** (willows; family *Salicaceae) A genus of trees, some of them large, and shrubs, many of which are dwarf alpine or Arctic species. The *catkin scales are untoothed (they are toothed in *Populus), there are only 2–5 *stamens, and nectaries are present in the flowers which (unlike those of *Populus*) are insect-pollinated. The buds are enveloped by a single outer scale (there are several in *Populus*). *S. alba* var. *caerulea* is the best tree from which to make cricket bats. There are about 300 species, most of them northern temperate or Arctic, but with a few tropical and southern temperate outliers.

**S alleles** (self-sterility genes) *Genes that prevent *self-fertilization by controlling the growth of the *pollen tube, thus causing *male sterility and preventing *inbreeding depression in *monoecious plants. Gametophytic tissue containing the same sterility allele as the *sporophyte is discriminated against.

**Salmonella** (family *Enterobacteriaceae) A genus of *Gram-negative rod-shaped *bacteria. Most species are *motile with many *flagella. The genus includes some important disease-causing bacteria,

S

including the causal agents of typhoid fever, and some types of 'food poisoning'.

**salsify** (*Tragopogon porrifolius*) *See* ASTERACEAE.

**salt marsh** Vegetation often found on mud banks formed at river mouths, showing regular zonation reflecting the length of time different areas are inundated by tides. Sea-water has a high salt content which produces problems of *osmotic pressure for the vegetation, so that only plants that are adapted to this environment (halophytes) can survive.

**salt stress** *Osmotic forces exerted on plants when they are growing in a salt marsh or under other excessively saline conditions.

**salvageability** *See* RECREATABILITY.

*Salvia* (sages; family *Lamiaceae) A genus of *herbs and low shrubs that usually have aromatic foliage and flowers that occur in *whorled *spikes, often with coloured *bracts. The flowers have the *corolla strongly 2-lipped, with the upper lip often strongly arched over the 2 *stamens. These each have only 1 *anther cell developed, on one end of a flexible, see-saw *connective on the tip of the *filament, making the stamen appear branched, with the anther cell on the tip of the longer 'branch'. Varieties of *S. officinalis* are cultivated as pot-herbs, other species for their showy flowers. There are some 900 species, centred in southern Europe and south-western Asia, but occurring widely in tropical and temperate zones.

**Salviniaceae** A family of ferns of specialized, floating, aquatic habit, that have no roots, and leaves in *whorls of 3, 2 of which are oval and float while the third is submerged and divided into root-like, descending lobes. The floating leaves are covered with hairs and waxy *papillae, and are concave above. The slender stems contain a vascular strand surrounding a tiny *pith. The *sporangia are of 2 kinds, *megasporangia and *microsporangia, borne in separate *sporocarps on the tips of the root-like branches of the submerged leaves. The megaspores germinate to form green, female *gametophytes floating at the water surface. There is 1 genus, with 12 species, mostly in tropical Africa, but also present in tropical and temperate America.

**samara** A winged, 1-seeded *nut or *achene, characteristic of *Acer*, but also found in many other plants.

*Sambucus* (elders; family *Caprifoliaceae) A genus of deciduous trees, shrubs, and a few *herbs that have *opposite, *pinnate leaves and a large stem pith. The flowers are in *umbel- or *panicle-like clusters, and usually have a 5-lobed *corolla, and 5 *stamens. The *ovary is 3–5-celled and ripens to a berry-like *drupe in the fruit. The fruits of *S. nigra* are edible, and they and the flowers can be made into wine. There are about 20 species, widespread in temperate and tropical regions, but absent from central and southern Africa.

*Samolus* (family *Primulaceae) A genus of *annual or *perennial *herbs which have erect or trailing stems, and *glabrous, *radical, or alternate leaves. The flowers are bisexual and held in terminal *racemes. The *sepals are fused into a 5-lobed, bell-shaped *calyx, and the *corolla is also fused into a short but broad tube. There are 5 *stamens, and the *ovary and fruit are *half-inferior, the fruit being a *valvate *capsule. There are 15 species, cosmopolitan in distribution.

**sampling** *See* CONTAGIOUS DISTRIBUTION; PLOTLESS SAMPLE; RANDOM SAMPLE; REGULAR SAMPLE; STRATIFIED RANDOM SAMPLE; and TRANSECT.

**samul** (silk cotton) *See* BOMBAX.

**sand** (pedol.) **1.** Mineral particles of diameter 2–0.02 μm in the international system, or 2–0.05 μm in the *USDA system. **2.** A class of soil texture.

**sandalwood** *See* SANTALUM.

**sandwort** *See* ARENARIA.

**Sanger's reagent** A solution of 1-fluoro-2-4-dinitrobenzene, used for the chromatographic detection and quantification of *amino acids, *peptides, and *proteins. Its effectiveness is based on the reaction of the reagent with free alpha- and epsilon-*amino groups to form yellow dinitrophenyl derivatives.

**Sansevieria** (family *Agavaceae) A genus of *perennial herbs, including S. zeylanica which yields bowstring hemp and S. trifasciata (mother-in-law's tongue), a popular house plant with dark green, blotched, strap-like, erect leaves. There are 12 species, found in Arabia, Africa, and Madagascar.

**Santalaceae** A family of *herbs, shrubs, and trees that are semi-parasitic on roots and stems. The flowers are small and inconspicuous, bisexual or unisexual, with 3–6 fused *tepals, the *stamens *adnate to the *perianth, and the *ovary is *inferior or *half-inferior. There are 36 genera, with about 500 species, occurring in tropical and temperate regions.

**Santalum** (sandalwood; family *Santalaceae) A genus of semi-parasitic shrubs and small trees, with *opposite, *simple leaves which are often linear, and roots that attach to the roots of other plants. The small flowers are held in *cymes or *panicles, and are green or white in colour. The *perianth is 4-lobed, and there are 4 *stamens. The fruit is a globular *drupe with an outer, succulent, brightly coloured cover over the hard, woody shell that protects the seed. The succulent part of the fruit is edible. S. spicatum is one of the fragrant sandalwoods used for incense. S. album gives sandalwood oil and is extensively cultivated, especially in India, although it appears not to be native there. The oil is distilled from the wood. There are 8–9 species which are found from south-eastern Asia to the south-western Pacific region and Hawaii.

**sap** The exudate from ruptured tissues emanating from the *vascular system or *parenchyma. See also LATEX and RESIN.

**sapele** See ENTANDROPHRAGMA.

**sap-green** See RHAMNACEAE.

**Sapindaceae** A family mostly of trees and shrubs, with some woody climbers, in which the leaves are *pinnate. The flowers are functionally unisexual, but with both sexes variously developed; they are *tetra- or *pentamerous, with the *sepals

and *petals free, the *disc conspicuous, and the *ovary *superior. There are 145 genera, with 1325 species, occurring in the tropics and subtropics.

**sapling** A young tree, not yet of useful or timber size, arising from a *seed or *sucker.

**sapodilla plum** See ACHRAS.

**Saponaria officinalis** (soapwort) See SAPONIN.

**saponin** Any member of a class of *glycosides that form colloidal (see COLLOID) solutions in water and foam when shaken. Saponins have a bitter taste, hydrolyse (see HYDROLYSIS) red blood cells, and are very toxic to fish. They occur in a wide variety of plants, including Saponaria officinalis (soapwort), which produces a lathery liquid, once widely used for washing wool and still used for delicate textiles, including antique ones.

**Sapotaceae** A family of trees that have white sap. The leaves are spiral and leathery. The flowers are mostly regular with free parts, and the *ovary is *superior. The fruit is a *berry with 1 or a few big, hard seeds. Many species yield valuable timbers, and some produce useful latex (chewing gum, gutta percha). The genera are poorly defined, but there are about 116 of them, with 1100 species, occurring in the tropics and subtropics.

**sappan** (Brazilwood) See CAESALPINIA.

**saprobe** See SAPROTROPH.

**Saprolegnia** (order *Saprolegniales) A genus of *fungi that form a coarse *mycelium in which the many nuclei are not separated by *septa. Species are found in water, mud, and soil. Some species can cause disease in fish.

**Saprolegniales** (class *Oomycota) An order of oomycetes that are found mainly in aquatic habitats. The vegetative state may be unicellular or a well developed *mycelium, depending on species. The order includes many genera. Most species are *saprotrophic but some are *parasitic (e.g. in fish, invertebrates, *algae, and higher plants).

**saprophage** An organism that consumes other, dead, organisms. Saprophages

**S**

form part of the twofold division of the *heterotrophs (organisms that feed on other organisms), and consist mainly of *bacteria and *fungi. They break down complex compounds obtained from dead organisms, absorbing some of the simpler products, but releasing most of the products as inorganic nutrients which can then be used by other organisms. *See also* CONSUMER ORGANISM.

**saprophyte** *See* SAPROTROPH.

**saprotroph (saprobe, saprovore)** Any organism that absorbs soluble organic nutrients from inanimate sources, e.g. from dead plant or animal matter, from dung, etc. If the organism is a plant or is plant-like, it is called a saprophyte. *See also* CONSUMER ORGANISM.

**saprovore** *See* SAPROTROPH.

**sap stain** A stain (often bluish-grey) of the sapwood in freshly cut timber, caused by any of several types of *fungi (e.g. *Penicillium*, *Ceratocystis*, *Cladosporium*). The mechanical strength of the timber is usually little affected.

**sapucaia nut** *See* LECYTHIS.

**sapwood** The active *xylem cells, which are peripheral to the dead, woody centre of the trunks of trees. This part of the xylem is distinguished easily from the central heartwood by its colour.

**SAR** *See* SODIUM ADSORPTION RATIO.

**Saraca (family *Fabaceae, subfamily *Caesalpinoideae)** A genus of small trees in which the young, *pinnate leaves hang as pendulous, coloured tassels. The corymbose (*see* CORYMB) *inflorescences are conspicuous bright yellow or red, borne on trunks or branches. The *sepals form a 4-lobed tube, and are apetalous; there are usually 4–8 *stamens which are long. The pods are big, flat, and *dehiscent. According to tradition, the Buddha was born beneath *S. asoca* (asoka). There are 9 species of *Saraca*, occurring in tropical forests in India and Malesia.

**Sararanga (family *Pandanaceae)** A genus of trees whose leaves are borne in 4 spiral ranks, and open sparsely, producing a falsely *dichotomous branched crown. The *inflorescence is an open, pendulous

*panicle of small, fleshy, kidney-shaped heads in which the tiny flowers are sunken. There are 2 species, occurring in the Philippines, New Guinea, the Bismarcks, and the Solomon Islands.

**Sarcococca (family *Buxaceae)** A genus of about 14 species of evergreen *subshrubs of the Himalayas, and from China to Indo-Malaysia. The inconspicuous but fragrant flowers are borne in *axils. Several winter-flowering varieties are cultivated.

**Sarcostemma (family *Asclepiadaceae)** A genus of climbing, twining, or erect shrubs, herbs, and some trees, whose leaves are reduced or absent, and whose stems are fleshy. The flowers are bisexual, and usually borne in *cymes. The *calyx is a fused tube of 5 segments. The *petals are partially fused along their margins and sometimes folded. The 5 *stamens are attached to the base of the *corolla-tube. There are 10 species, found in tropical and subtropical parts of the Old World, where they are components of *xerophytic communities.

**sarcotesta** In some *fruits (e.g. *Calospatha* and *Daemonorops*), a sheet of edible pulp in which one or more *seeds are embedded.

**Sargassaceae (order *Fucales)** A family of brown seaweeds in which the *thallus shows *monopodial branching; i.e., there is always a distinct main *axis. Air *bladders are present. Species of the genus *Sargassum* are found in warmer regions and occur in huge floating masses, particularly in the Sargasso Sea. *See also* JAPWEED and TRAMP SPECIES.

**sarmentose** Applied to a stoloniferous (*see* STOLON), straggling shrub.

**Sarracenia (family Sarraceniaceae)** A genus of pitcher plants (*compare* NEPENTHACEAE), comprising 8 species, confined to eastern N. America.

**sarsaparilla** *See* SMILAX.

**sassafras** *See* DORYPHORA.

**satinwood** *See* CHLOROXYLON.

**saturated flow** The movement of water through a soil that is temporarily saturated. Most of the loosely held water moves downward, and some moves more slowly laterally.

**saurochory** Dispersal of spores or seeds by snakes or lizards.

**savannah** An extensive tropical vegetation dominated by grasses with varying admixtures of tall bushes and/or trees in open formation. Savannah occurs in diverse tropical environments, although most experience a dry season. Much savannah is no doubt *climatic climax, although extensive tracts are anthropogenic *fire climaxes, and it is generally difficult to distinguish one type from the other.

**savannah woodland** A savannah in which trees and shrubs form a generally light canopy. The trees and bushes are generally *deciduous, yet *evergreens are usually also well represented. Some tall trees occur, but most are stunted and gnarled. They frequently have thick, corky, fire-resistant bark.

**saw sedge** *See* CLADIUM.

*Saxegothaea* (family *Podocarpaceae) A genus of 1 species, *S. conspicua*, an evergreen, *coniferous tree confined to Chile, which has 2-ranked, yew-like foliage. Female cones have fleshy, overlapping, prickly scales. It is just hardy in Britain.

**saxicolous** Growing on stones, rocks, walls, etc.

**Saxifragaceae** **1.** A family of herbs that consists of plants with usually alternate leaves, and regular, *tetra- or *pentamerous flowers, with free *petals and usually 8 or 10 *stamens. There are 2 *carpels, normally united at the base and free above, seated in a *receptacle cup. *Saxifraga* (saxifrage) contains more than 300 species, mostly small, often mat-forming herbs, mostly of alpine or Arctic distribution, many of which have attractive flowers and are much grown in rock gardens. As usually understood, the family contains some 37 genera, with 475 species, of cosmopolitan but largely temperate distribution. **2.** In some modern classifications Saxifragaceae in-cludes the previously separate families *Grossulariaceae (1 genus), Parnassiaceae (1 genus, with 45 species), Hydrangeaceae (15 genera, with 200 species), and *Escalloniaceae (20 genera, with 150 species).

**scab** A general term for any plant disease in which the symptoms include the formation of dry, corky scabs. Examples include apple scab and *potato scab. *See also* STREPTOMYCES.

**scabrous** Rough-surfaced; bearing short, stiff hairs, scales, or points.

*Scaevola* (family *Goodeniaceae) A genus mostly of shrubby trees, with some *herbs, in which the flowers are *zygomorphic, with the *corolla tube split to the base on the top side. The fruit is a *berry. There are 130 species, occurring in the tropics and subtropics, and especially in Australia.

**scalariform** Ladder-like.

**scale** **1.** In vascular plants, a plate-like *trichome. **2.** In some algae, an external covering to the cell.

**scandent** Climbing.

*Scapania* (order *Jungermanniales) A large genus of leafy *liverworts, in which the *thallus is small to large and robust. The leaves occur in 2 ranks; each leaf is divided into 2 lobes, the smaller lobe being folded back over the larger, posterior one and appearing at first sight to be a separate leaf. They are found on rocks, on peat, on tree roots, etc., generally in moist, shady, non-calcareous habitats.

**scape** A leafless flower stalk, e.g. in the plantain.

**scaphoid** Boat-shaped.

**scarious** Dry and membranaceous.

**scarlet elf-cup** The common name for the *apothecium of *Sarcoscypha coccinea* (order *Pezizales), which is cup-shaped, up to 4 cm across, scarlet inside, and whitish outside. It is found on fallen sticks or branches in winter.

**SCE** *See* SISTER CHROMATID EXCHANGE.

*Scenedesmus* (division *Chlorophyta) A genus of microscopic *green algae in which the (non-*motile) cells occur adjacent to one another in *coenobia, the number of cells per coenobium being a multiple of 2, typically 4 or 8. They are very common in all types of aquatic habitat.

**Schefflera** (family *Araliaceae) A genus of trees, shrubs, and woody *epiphytes that have big, *palmately lobed or divided leaves and flowers in *umbels. Several are cultivated as ornamentals. There are about 200 species, occurring in tropical and warm regions.

**Schiff's reagent** A reagent consisting of fuchsin bleached by sulphurous acid that produces a red colour upon reaction with an *aldehyde.

**Schimper, Andreas Franz Wilhelm (1856–1901)** A German botanist and ecologist who was a professor at the Bonn Botanical Institute (1886–98) and at Basle (1891–1901). He showed (1881) that starch grains are not formed in *cytoplasm, and (1883) that *plastids originate from the division of pre-existing plastids. His major work, *Pflanzengeographie auf physiologischer Grundlage* (first edition 1898, first English edition, *Plant-Geography upon a Physiological Basis*, 1903), includes his classification and a systematic account of world vegetation based on physiological adaptation, and did much to establish a sound method for ecological investigation.

**Schistidium** *See* GRIMMIALES.

**Schistostegales (subclass *Bryidae)** An order of mosses, in which the *protonema is persistent and unique in structure; it has light-refracting properties and appears almost luminous. The leaves of the sterile shoots are arranged in 2 ranks and are nerveless. The *capsule lacks a *peristome. There is 1 genus, and 1 species, *Schistostega pennata*. Not common in Britain, it is found mainly in habitats with low light levels (e.g. in caves, rabbit burrows, fissures among siliceous rocks, mine-shafts, etc.) particularly in sandstone areas.

**Schizaeaceae** A family of ferns in which the *sporangia are borne naked and singly on the margins of the leaves, sometimes protected by the incurved leaf margin. The sporangia are relatively mas-

sive, somewhat intermediate in character between those of *eusporangiate and *leptosporangiate ferns, with very short, thick stalks, and a terminal ring of thickened cells; they may contain more than 200 spores. All sporangia mature at the same time (a simple type of development). The *gametophytes are either green and filamentous, like moss *protonemata, or (in *Actinostachys*) may be fleshy, subterranean, cylindrical structures with *mycorrhizal fungi. There are 4 genera and 150 species, mostly tropical, but the family extends into temperate parts of America.

**schizocarp** A dry *fruit, intermediate between *dehiscent and *indehiscent types, that is derived from 2 or more *carpels, each of which matures as a single-seeded unit.

**schizogony** A type of *asexual reproduction in which a new generation of feeding cells results from multiple fission of the parent cell.

**Schizomeria** (family *Cunoniaceae) A genus of 15 species of timber trees, which occur from the Moluccas to Queensland and the Solomon Islands.

**Schizomycetes** An out-dated taxon for the *bacteria.

**Schizophyllaceae (order *Aphyllophorales)** A family of *fungi in which the *fruit bodies bear characteristic longitudinally divided *gills. *Schizophyllum commune* forms fan-shaped fruit bodies that are attached on one side to the substrate. It is found on wood and is common in Europe, especially in the south.

**Schizothrix** *See* LPP GROUP.

**Schleiden, Matthias Jakob (1804–81)** A German botanist who, in collaboration with T. *Schwann, proposed the *cell theory. Schleiden practised law before studying medicine and botany. His studies led him to conclude that all parts of a plant consist of cells or their derivatives, an idea he called 'phytogenesis', publishing an account of it in 1838. For some years he was professor of botany at the Universities of Jena and Dorpat but later he worked as a freelance lecturer and writer.

**schlerenchyma** Structural tissue made up of fibres or stone cells.

***Schoenus*** (family *Cyperaceae) A genus of *perennial *herbs with smooth stems, usually found in damp and marshy habitats. Most are tufted. The leaves are basal or restricted to the lower part of the stem and sheathing; the blades of the leaves are narrow, like those of grass. The flowers are small, inconspicuous, and bisexual, and arranged in 1–4-flowered *spikelets held in a terminal *panicle. The *bract of the lowest spikelet encircles the whole base of the *inflorescence. The *perianth has up to 6 bristles, and there are 3 *stamens and 3 *styles. The fruit is a 3-sided *nut. There are about 100 species, found mainly in Australia and New Zealand.

**Schwann, Theodor** (1810–82) A German physiologist who, in collaboration with M. J. *Schleiden, proposed the *cell theory (and coined the term), according to which all plant and animal tissues are composed of cells, and within an individual organism all the cells are identical (*see also* VIRCHOW, RUDOLF). This was based on a botanical discovery by Schleiden, which he described in 1838, and in 1839 Schwann showed that it also applied to animals. Schwann graduated in medicine at Berlin in 1834, was appointed professor of anatomy at the University of Louvain in 1838, and became a professor at Liège in 1847. The cell theory arose from his studies of yeasts.

**scia-** A prefix, derived from the Greek *skia*, meaning 'shadow', that means 'pertaining to shade or darkness'.

***Sciadopitys*** (Japanese umbrella pine, parasol pine; family *Pinaceae) A *monotypic genus (*S. verticillata*) of conifers with the needle-like leaves fused together in pairs throughout their length, and arranged in *whorls on the shoots, with scale-like leaves at the bases of the whorls. The genus is native to Japan and rare in the wild.

**sciaphilic** (skiaphilic) Shade-loving; applied to plants (called sciophytes) that grow only in shady *habitats.

**scion** *See* GRAFT.

**sciophilous** Applied to the response shown by organisms that have adapted to live entirely in the shade.

**sciophyte** *See* SCIAPHILIC.

***Scirpus*** (club rush; family *Cyperaceae) A genus of rhizomatous (*see* RHIZOME), *perennial, grass-like *herbs which have solid, often 3-angled stems. The leaves usually arise from the base with a closed stem sheath and grass-like blade with no *ligule. The inconspicuous flowers are bisexual and borne in *spikelets of 2 to many. They are held in a terminal, branched *inflorescence. The *perianth is a series of bristles or scales, and there are 3 *stigmas. The *ovary is *superior. There are about 200 species, found throughout the temperate regions of the world. Most species occur in marshland and several are used in medicine. *S. lacustris*, the true bulrush of the northern hemisphere, is used for basketwork.

**scler-** A prefix, derived from the Greek *skleros*, 'hard', that means 'hard' or 'tough'.

**sclereid** *See* SCLERENCHYMA.

**sclerenchyma** The fibrous or woody tissue in a plant that provides mechanical support for it. The tissue is formed from cells whose walls are thickened with *cellulose or *lignin, and when mature the cells usually contain no living protoplasm. The cells may occur singly or in groups and may be long and fibrous or stony. Stone cells (sclereids) are found in *seed coats and some *fruits (e.g. pears).

***Scleroderma*** (order *Sclerodermatales) A genus of *fungi, in which the more or less spherical *fruit bodies consist of spore-forming tissue enclosed by a tough, single-layered *peridium. The spores are spiny or reticulate and are released only on decay of the peridium. There are many species, found in autumn growing under trees in heathland or woodland. *See also* EARTH BALL.

**Sclerodermatales** (class *Gasteromycetes) An order of *fungi, in which the *fruit body occurs on or below ground and is typically more or less spherical (but may resemble an *earth star). *Spores are released via splits in the *peridium.

**sclerophyllous vegetation** Typically scrub, but also woodland, in which the leaves of the trees and shrubs are *evergreen, small, hard, thick, and leathery. These adaptations allow the plants to survive the pronounced hot, dry season of the *Mediterranean-type climate in which sclerophyllous vegetation is best developed.

*Sclerotinia* (order *Helotiales) A genus of *fungi which live *parasitically on a range of higher plants. Stalked *apothecia develop from *sclerotia. *S. fructigena* causes *brown rot in apples and certain other fruits.

*Sclerotium* (order *Agonomycetales) A *form genus of *fungi which have *perfect states classified in either the *Basidiomycotina or the *Ascomycotina. *S. cepivorum* causes the disease *white rot of onions; this species produces hard, resistant sclerotia which can remain in the soil for long periods of time.

**sclerotium** A fungal resting body that is resistant to unfavourable environmental conditions and can remain dormant for long periods of time.

*Scorodocarpus* (family *Olacaceae) A *monotypic genus, comprising *S. borneensis*, a rain-forest tree which produces a hard timber suitable for construction. It is notable for the powerful reek of garlic from all its parts, sometimes at 20 m distance without the tissues having been damaged. It occurs in Sumatra, Malaya, and Borneo.

**scorpioid cyme** A type of *cyme in which the *axis is curved or coiled (like the tail of a scorpion) and the flowers open successively downwards from the apex, always on the convex side of the axis.

**screw pines** *See* PANDANUS.

**Scrophulariaceae** The former name of the family *Plantaginaceae.

**scrub vegetation** A general term for vegetation dominated by *shrubs, i.e. low woody plants, which typically forms an intermediate *community between grass or heath and high forest. *Successional change is not necessarily implied, though the term is often used for the transitional stage in succession to *climax woodland when shrubby plants predominate.

**scutellum** In the fruit of grasses, an organ lying between the *embryo and *endosperm. During germination, the scutellum secretes enzymes involved in the digestion of endosperm.

**Scyphostegiaceae** A *monotypic family (*Scyphostegia borneensis*), which is a small, *dioecious tree in which the leaves are alternate, with inconspicuous *stipules. The flowers are borne in *spikes, and are unisexual, with 6 *tepals, the *stamens forming a column and the *carpels tiny, within an *urceolate structure. The fruit is a fleshy *capsule. It is a curious family, perhaps related to the *Flacourtiaceae, and is endemic (*see* ENDEMISM) to Borneo.

**scyphus** A cup-shaped or funnel-shaped structure (e.g. the cup-shaped *podetia of some species of *Cladonia).

*Scytonema* A genus of filamentous *cyanobacteria (section IV), in which the filaments do not taper, and show frequent false branching owing to a loop of *trichome breaking through the sheath and then rupturing. The filaments have distinct basal and apical regions and form erect tufts, on wet rocks, soil, etc. The genus now includes organisms formerly included in the genus *Tolypothrix*.

**Scytosiphonales** (division *Phaeophyta) An order of *algae in which the *thallus is *parenchymatous and growth is mostly diffuse. *Scytosiphon lomentaria* is a common British species, consisting of loosely tufted, tubular *fronds. *Petalonia* has flat, leaf-like fronds and is found mainly on rocks on the lower shore. *Colpomenia* has a thallus consisting of thin-walled hollow spheres. *See also* OYSTER THIEF.

**sea-belt** (sugar kelp) The common name for the brown seaweed *Laminaria saccharina*. The *thallus consists of a branched, root-like *holdfast, a smooth, slender *stipe, and a yellowish-olive *frond which is single and ribbon-like with an undulating central region and frilly margins. When dry, a whitish, sweet-tasting deposit sometimes appears on the surface of the fronds, hence the alternative name of sugar kelp. Another alternative name is poor man's weather-glass, since under humid conditions the fronds remain soft and limp, whereas under dry conditions they become hard and brittle.

**sea heaths** *See* FRANKENIACEAE.

**sea holly** *See* ERYNGIUM.

**sea lavender (***Limonium vulgare***)** *See* POLYHALINE.

**sea lettuce** The common name for the green, sheet-like seaweed *Ulva lactuca.

**sealing-wax palm** *See* CYRTOSTACHYS.

**sea oak** The common name for the brown seaweed *Halidrys siliquosa* (family *Cystoseiraceae), a *perennial species with regular, alternate branching, and a characteristic zigzag *axis. Stalked air *bladders resembling seed pods are present; these may be up to 5 cm long and are divided internally by *septa into 10 or 12 compartments. They are common and widely distributed at or below low-water mark, or in deep rock pools on sheltered coasts.

**sea sawdust** *See* TRICHODESMIUM.

**sea thong** The common name for the brown seaweed *Himanthalia elongata* (*Himanthaliaceae), found in deep rock pools or at the low-water mark.

**seaweed** The common name for a macroscopic marine *alga. Seaweeds belong to the groups *Rhaeophyta (brown seaweeds), *Rhodophyta (red seaweeds), and *Chlorophyta (green seaweeds).

*Secale* **(rye; family *Poaceae)** A genus of grasses that were cultivated in ancient times and that are now grown on land that is unsuitable for growing wheat, the preferred crop. Domesticated rye (*S. cereale*) is probably descended from *S. montanum*, a weed of wheat and barley, which is native to eastern Turkey. There are 3 species, native to Eurasia.

**secchi disc** A disc used in a simple method for measuring the transparency of water. The disc is 20 cm across, and divided into alternate black and white quadrants. It is lowered into water on a line until the difference between the black and white areas just ceases to be visible, at which point the depth is recorded. The secchi disc provides a convenient method for comparing the transparencies of water at different sites.

**secondary dormancy** *See* DORMANCY.

**secondary growth** *See* SECONDARY THICKENING.

**secondary metabolite** *See* SECONDARY PLANT COMPOUND.

**secondary phloem** In plants in which *secondary thickening occurs, *phloem derived from the vascular *cambium.

**secondary pit** A *pit that develops between cells that were not originally connected, or after separation by mitotic division. *See also* PRIMARY PIT.

**secondary plant compound (secondary metabolite)** A chemical compound produced by a plant that serves no primary function in plant metabolism (e.g. an *alkaloid).

**secondary productivity** *See* PRIMARY PRODUCTIVITY.

**secondary structure** Applied to *proteins, the folding of a *polypeptide or polynucleotide chain along one axis of the molecule; it is stabilized by the formation of intramolecular *hydrogen bonds along the length of the chain.

**secondary succession** *Succession initiated by the disruption of a previously existing seral or *climax community by some major environmental disturbance and leading to a marked change in the stable vegetation *community. Secondary successions occur, for example following fire, after removal of grazing pressure, or when previously cultivated areas are abandoned (as in shifting cultivation). Interactions between plants and the physical environment tend to be less clear in secondary than in *primary successions.

**secondary thickening (secondary growth)** The formation of new *tissue by the repeated lateral division of cells in the *cambium of a woody plant, adding successive layers of new growth. This

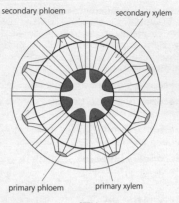

STEM

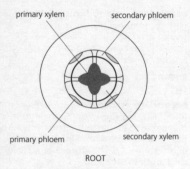

ROOT

*Secondary thickening*

increases the girth of the stem or root, and the growth can be seen as annual rings (*tree-rings). Much of the woody part of a woody plant is the result of secondary thickening.

**secondary woodland** A woodland occupying a site that has not been wooded continuously throughout history (in Britain since the last ice advance). It may be the product of natural *succession or of planting on formerly unwooded land.

**secondary-woodland species** Tree species that are frequent and often abundant in secondary woodland. Usually they are good colonizers, or are abundant in many *habitats, or both.

**secondary xylem** (wood) In plants in which *secondary thickening occurs, *xylem derived from the vascular *cambium.

**secretion** **1.** The act of discharging materials from cells. **2.** Any material discharged from cells.

**secund** Arranged on one side, or curved to one side.

*Securinega* (family *Euphorbiaceae) A genus of small trees which produce very hard wood. There are 20 species, occurring in temperate and tropical regions.

**sedges** *See* CAREX and CLADIUM.

*Sedum* (stonecrops) *See* CRASSULACEAE.

**seed** In the *sexual reproduction of *Spermatophyta (seed plants), the discrete body from which a new plant develops. Formed from a fertilized *ovule, the seed comprises an outer coat (testa) that encloses a food store and an *embryo plant. The food may be stored in the *cotyledons (seed leaves) of the embryo itself or around the embryo in the *endosperm.

**seed leaf** *See* COTYLEDON.

**seedling** A juvenile plant that has developed from a seed.

**seed plants** The group that comprises the *gymnosperms and *angiosperms, plants distinguished by their production of seeds rather than spores. They arose in the *Devonian from *pteridophyte forebears, probably of a *heterosporous character. These early seed plants were gymnosperms. Angiosperms are not certainly represented in the fossil record until the early *Cretaceous.

**segmentation** (cleavage) A process by which a dividing egg-cell, following *fertilization, gives rise to all the cells of the organism.

**segregation, Mendel's law of** *See* MENDEL'S LAWS.

**seismonasty** A nastic movement (*see* NASTY) in plants in response to a sudden stimulation by touch.

*Selaginella* *See* SELAGINELLACEAE.

**Selaginellaceae** A family of the *Pteridophyta in the wide sense, or of the *Lycopsida in the narrower sense, of plants similar to the clubmosses (*Lycopodiaceae), but showing *heterospory, with *megasporangia, each containing usually only 4 large *megaspores (which germinate into female *gametophytes), and *microsporangia, each containing many tiny *microspores (which germinate into male gametophytes). The plants have tiny leaves, like the true clubmosses, but these may be either spirally arranged and all of similar form, or in 4 ranks along the stems, 2 ranks of leaves being much larger than the others. All *Selaginella* leaves have a tiny, tongue-like structure, the *ligule, on the upper surface near the base. All the species have their *sporangia in cones. Probable relatives of the living Selaginellaceae were very important components of the Upper *Carboniferous coal-measure forests (*Lepidodendrales), and many were large trees. There is 1 genus (*Selaginella*), with some 700 species, of world-wide range.

**selangan batu** *See* SHOREA.

**selection** A process that results from the differential reproduction of one *phenotype as compared with other phenotypes in the same population. This determines the relative share of different *genotypes which individuals possess and propagate in a population. The relative probability of survival and reproduction of a phenotype is termed 'fitness' or 'Darwinian fitness'. *See* ADAPTIVE VALUE.

**selection coefficient (s)** A measure of the relative excess of deficiency of fitness of a *genotype compared with another genotype in the population. If $s = 100$, then 1 out of a 100 individuals of a given genotype fails to reproduce.

**selection differential** The difference between the average value of a *quantitative character in the whole population and the average value of those selected to reproduce the next generation.

**selection pressure** The pressure exerted by the environment, through *natural selection, on *evolution. Thus, weak *selection pressures result in little evolutionary change and vice versa.

**selective species** In *phytosociology, a plant species found most frequently in a particular *community, but also present occasionally in others. It is fidelity class 4 of the *Braun-Blanquet scheme. *Compare* ACCIDENTAL SPECIES; EXCLUSIVE SPECIES; INDIFFERENT SPECIES; and PREFERENTIAL SPECIES.

**selective value** *See* ADAPTIVE VALUE.

**selenotropism** A tropic response (*see* TROPISM) to the Moon.

**self-fertilization** The fusion of male and female *gametes from the same individual (as opposed to cross-fertilization, in which the gametes come from different individuals).

**selfish DNA** One of a number of hypotheses advanced in an attempt to explain the presence of surplus *DNA in the *genome which is not *translated into protein. Three hypotheses have been put forward to account for the adaptive advantage of this apparently redundant DNA: (*a*) that extra DNA separates the *genes so as to increase the *cross-over frequency; (*b*) that the possibility of varying the total amount of DNA per cell allows the control of cell volume and cell growth rate; and (*c*) the selfish DNA hypothesis) that *selection acts within the genome favouring any method by which DNA may more rapidly replicate itself, and that this can be better achieved if *phenotypic expression can be bypassed. This is achieved, it is proposed, by DNA spreading laterally so as to be duplicated at new *loci elsewhere in the genome. In this way the DNA may be viewed as acting 'selfishly', since the apparently surplus DNA confers no advantage on the organism bearing it and therefore supplying the materials from which it is made. *See also* SELFISH GENES.

**selfish genes** A term used by some authors (most notably Richard Dawkins) to reinforce their notion that organisms function as agents for the replication of *genes, as opposed to genes functioning as servants of organisms (i.e. that *natural selection operates at the level of the gene). Opponents argue that natural selection operates at the level of the individual, not the gene, since it is the *genome that survives or dies, not individual genes. *See also* SELFISH DNA.

**self-mulching soil** A soil that mixes itself: its surface layers shrink and swell, forming deep cracks into which soil falls.

**self-pollination** The transfer of *pollen from *anther to *stigma of the same plant (either of the same flower or of a different flower but always of the same individual). *See also* POLLINATION.

**self-sterility genes** *See* S ALLELES.

**self-thinning** A progressive decline in the density of a population of growing plants. During the course of the self-thinning process, individuals become larger as the population density declines. If the logarithm of density is plotted against the logarithm of individual weight, then a straight line of negative slope is obtained, and the slope is generally −3/2. This relationship holds for a wide range of plants, from annual weeds to forest trees, and has become known as the 'self-thinning rule'.

**selva** A tropical rain forest. The term is applied specifically to the Amazon Basin, and is often used to denote similar vegetation in other areas.

***Semecarpus*** (family *Anacardiaceae) A genus of trees and treelets in which the leaves are *simple, and often crowded. It is a *rengas genus, producing an irritant black sap. Some species have attractive timber. There are 60 species, occurring in lowland tropical rain forest from Indo-Malaysia to the Solomon Islands.

**semi-desert scrub** A transitional formation type situated between true *desert and more thickly vegetated areas (e.g. between thorn forest and desert or between *savannah and desert). The vegetation is sparser than that of the thorn forest and succulents are more common, as a consequence of the drier climate. Most of the plants are shallow-rooted, and so able to exploit before it evaporates any precipitation that *percolates into the surface layer of the soil.

**semi-evergreen seasonal tropical forest** A distinct formation type, dominated by both *evergreen and *deciduous broad-leaved trees, flanking the rain forest in areas that have a marked dry season. Examples are found on all the continents with tropical territory, but are especially extensive in central and S. America, southeastern Asia, and northern Australia. In Africa, *savannah and related types of vegetation seem to have replaced much of this kind of forest.

**semi-natural community** Vegetation altered by human influence or management in the past, which has taken on a natural aspect owing to the length of time over which the influences have persisted. For example, *heathland and *chalk grassland in Great Britain have long been subject to management and members of each *community have adapted to it. In chalk grassland many plant species are low-growing *rosette plants which avoid being grazed. *Compare* NEAR-NATURAL COMMUNITY.

**semi-natural woodland** On an ancient woodland site, stands of mainly *native trees that have not been obviously planted, whose presence appears to be *natural, and that have been managed over a period, resulting in some change in woodland structure and species composition. On a recent woodland site, the term includes all stands that have originated mainly by natural regeneration.

**semi-permeable** Applied to a membrane whose structure allows the passage only of solvent molecules. A membrane that allows the passage of small molecules but prevents the passage of larger ones is called 'differentially permeable'. The Institute of Biology recommends that biological membranes be described as 'partially permeable'.

**semi-species** A group of organisms that are taxonomically intermediate between a race and a species, with reduced *outbreeding and *gene flow, i.e. with incomplete reproductive isolating mechanisms. Semi-species are thought to represent advanced stages of *speciation.

***Sempervivum*** (house leeks) *See* CRASSULACEAE.

**sempilor** *See* DACRYDIUM ELATUM.

**sena** Timber from *Pterocarpus indica*. *See* PTEROCARPUS.

***Senecio*** (family *Asteraceae) A genus of *annual or *perennial *herbs, shrubs, and trees, many of which are succulent. The leaves are spirally arranged and alternate, without *stipules. The flower heads are solitary or in *corymbs, and are *heterogamous. The involucral (*see* INVOLUCRE) *bracts are in 1 row, sometimes with a few, outer, short bracts. The ray florets are flattened and female, or absent, the disc florets tubular and bisexual. The *achene is ribbed and cylindrical, usually with a simple *pappus. Most species are *xerophytic, with either fleshy leaves and stems, or hairy with inrolled leaves. Several tree species are found at very high altitudes on east African mountains. Several species are ornamentals. Many are now widespread and noxious weeds, e.g. *S. jacobaea* (ragwort). There are about 1500 species, with a cosmopolitan distribution, mainly centred in S. Africa, the Mediterranean, Asia, and America.

**senescence** The complex deteriorative processes that terminate naturally the functional life of an organ or organism. In plants, this may be associated with flowering and fruiting, and an important contributing factor is the decrease in *chlorophyll content and in the ability of the plant to photosynthesize. Plant senescence is controlled by *hormones.

***Senftenbergia sturi*** One of the earliest species referred to the extant fern family, the *Schizaeaceae. Fertile leaf fragments have been described from the Lower *Carboniferous of Scotland.

**senna** *See* CASSIA.

**sensitive plant** *See* MIMOSIA.

**sensitivity analysis** The consideration of a number of factors involved in the mathematical modelling of an *ecosystem and its components. These include feedback and control, and the stability and sensitivity of the system as a whole to changes in some parts of the system. From the analysis, predictions can be made.

**sentry palm** *See* HOWEIA.

**sepal** In a *flower, one of the outer floral leaves, usually greenish, which are borne in a tight spiral or *whorled.

**sepaloid** *Sepal-like.

**sepetir** *See* SINDORA.

**septa** The plural of *septum.

**septate** Having cross-walls (septa).

***Septoria*** **(order** *Sphaeropsidales) A *form genus of *fungi that produce thread-like, *septate *conidia. There are many species, some of which are important plant *pathogens (e.g. *S. nodorum* (*Leptosphaeria nodorum*) causes glume blotch in cereals, *S. oxyspora* (*Selenophoma donacis*) causes halo spot in cereals, and *S. apiicola* causes leaf spot of celery).

**septum** (pl. septa) A cross-wall or partition.

***Sequoia*** **(family** *Taxodiaceae) A *monotypic genus of evergreen conifers, *S. sempervirens* (coast redwood) of the N. American coastal fog belt from Oregon to California. *Sequoia* is a giant, columnar tree, probably the tallest in the world. The bark is soft and spongy, and it yields valuable timber. The leaves are spreading and yew-like.

***Sequoiadendron*** **(family** *Taxodiaceae) A *monotypic genus of evergreen conifers, *S. giganteum* (big tree or wellingtonia), native to the Pacific slopes of the Sierra Nevada in California. These are giant trees, with soft, spongy bark. The leaves are awl-like, being short, and clasping the twig.

**seral stage** A phase in the sequential development of a *climax community. *See* SUCCESSION.

**sere** The characteristic sequence of developmental stages occurring in plant *succession. *See* COMPETITION; ECESIS; MIGRATION; NUDATION; REACTION; STABILIZATION; and SUCCESSION.

***Serianthes*** **(family** *Fabaceae, subfamily *Mimosoideae) A genus of trees that have big, *pinnate leaves, and numerous small leaflets. The pod is flat, and *indehiscent. Several species grow fast. There are 10 species, occurring from eastern Malesia to the Pacific islands.

**serine** An aliphatic, polar (*see* POLAR MOLECULE) *alpha-amino acid. It is often associated with the active site of an *enzyme.

**serological test** A test of similarity between organisms. Plant proteins may be injected into a vertebrate (usually mammalian) bloodstream, which will then raise antibodies specific to those proteins. The blood serum can then be mixed with proteins from another plant species and the extent of the immunological response will be proportional to be similarity of the proteins and, by inference, the relatedness of the plant species.

**serotaxonomy** The classification of very similar plants by means of differences in the *proteins they contain. The technique is based on the highly specific relationship between *antigens and the *antibodies produced in response to them. Protein extracted from a plant is injected into the blood of an animal, where it behaves as an antigen. After an interval for the production of antibodies, a blood sample is taken. This can be used to compare the first plant protein (antigen) with extracts taken from other plants.

**serotinal** In late summer. The term is used with reference to the 6-part division of the year used by some ecologists, especially in relation to studies of terrestrial or freshwater communities. *Compare* AESTIVAL; AUTUMNAL; HIBERNAL; PREVERNAL; and VERNAL.

**serotinous** *See* SEROTINY.

**serotiny** (*adj.* **serotinous**) In certain plants, especially trees (e.g. jack pine (*Pinus banksiana*), lodgepole pine (*P. contorta*), and many species of *Eucalyptus*), the retention of seeds in pods or cones on the tree, often for many years, until a disaster, most commonly the heat of a fire, causes their release. After fire, the seeds fall on ground fertilized by ash in a site cleared of competitors.

**serotonin** 5-hydroxytryptamine, a derivative of *tryptophan. It has been found in nettles, bananas, and tomatoes. In animals, it is a powerful vasoconstrictor found especially in the brain, intestinal fluid, mast cells, and blood platelets, and also in snake and toad venoms; it has

been suggested as a cause of migraines in humans.

**serpentine barrens** Impoverished, often *scrubby or *heathy vegetation associated with serpentine rocks. On weathering these rocks release an excess of *magnesium into the soil, and this often inhibits the development of the natural *climax in the areas concerned.

***Serpula*** (family *Coniophoraceae) A genus of *fungi which includes the *dry-rot fungus *S. lacrymans* (formerly *Merulius lacrymans*), a serious agent of timber decay in buildings. The fungus can spread from one timber to another across brickwork and mortar. Rust-red *spores are produced in shallow pores which form a honeycomb-like layer. Dry rot is rare outside buildings.

**serrate** With tooth-like edges. *Compare* SERRULATE.

**serrated wrack (toothed wrack)** The common name for the brown seaweed *Fucus serratus*. The *thallus is flattened and branched, and the edges are serrated with forward-pointing teeth. The wrack is found near low-water mark and is very common. Small white coiled tubes of the worm *Spirorbis* are commonly found on the surface of the thallus.

**serrulate** With edges bearing very fine tooth-like projections. *Compare* SERRATE.

**sesame** *See* PEDALIACEAE.

***Sesamum indicum*** (sesame) *See* PEDALIACEAE.

**sesquioxides** A general term for the hydrated oxides and hydroxides of iron and aluminium.

**sessile** **1.** Lacking a stalk. **2.** Attached to a substrate; non-*motile.

**seta** **1.** A stiff, hair-like or bristle-like structure. **2.** The stalk of a *moss or *liverwort *capsule.

**setose** Bearing *setae.

**sewage fungus** A slimy growth found in sewage and sewage-polluted waters. It consists of filamentous *bacteria, associated with *fungi and protozoa.

**sex chromosome** A *chromosome whose presence or absence is linked with the sex of the bearer, and plays a role in sex determination. It is present in some sexually reproducing plants (*see* SEX DETERMINATION). The sex that has a *homologous pair of sex chromosomes in the nucleus is said to be *homogametic (referred to as XX), whereas the sex with a dissimilar pair or with an unpaired chromosome is said to be *heterogametic (referred to as X or XY). The homogametic sex produces gametes that are identical in their chromosome sets, all containing one *X-chromosome. The heterogametic sex produces equal numbers of 2 different types of gametes, one with one X-chromosome, and one without (possessing either a *Y-chromosome or none at all). The union of gametes of the 2 sexes thus results in equal numbers of offspring of the 2 sexes. Most plants and a minority of animals are hermaphrodite (i.e. they contain the sexual organs of both male and female). But many higher plants, fungi, and bacteria are dioecious (i.e. have separate sexes), and in these cases sex is most commonly determined by sex chromosomes. Many of the higher plants have the male gender as the heterogametic sex (e.g. *Melandrium*, *Humulus*, and *Rumex*), but, as with the animal kingdom, sometimes the female can be the heterogametic sex (e.g. *Fragaria*).

**sex determination** The mechanism by which sex is determined. In many species, sex is determined at fertilization by the nature of the male *gamete that fertilizes the egg, a Y-bearing male gamete producing the male *zygotes, and an X-bearing male gamete, female zygotes. In plants, this question has received considerable attention. Many plants are *monoecious. In some of these, sex is influenced and perhaps determined by hormones. Only some *diploid, *dioecious plants (e.g. willows) have an XX/XY pattern of sex determination; some others have a single *gene with 2 *alleles that are concerned with sex determination. In most plants it is not clear whether primary sexual determination is under direct genetic control or is the result of a disposition (itself genetically determined) to the appropriate hormone balance for maleness or femaleness. *See also* SEX CHROMOSOME.

**sexine (ectexine, ektexine)** In a *pollen grain, the outer layer of the *exine.

**sexual reproduction** Reproduction that involves the fusion of *haploid nuclei, usually *gametes. In most plants, gametes are formed in the (haploid) *gametophyte by *mitosis; *meiosis occurs in the *sporophyte.

**shaggy ink cap** The common name for the *fruit body of *Coprinus comatus*. The cap is initially white, with white or brownish scales, and initially oval, later becoming bell-shaped. It undergoes autodigestion (*autolysis), dissolving from the edges into an inky-black fluid. The *stipe bears a membranaceous, movable ring. It is common in fields, on grass banks, etc., often growing in clusters. It is edible when young.

**shaggy parasol** *See* PARASOL MUSHROOM.

**Shannon–Wiener index of diversity (information index)** A measure derived from information theories that were developed by Claude Elwood Shannon (1916–2001) and Norbert Wiener (1894–1964) and published in 1949 by Shannon and Warren Weaver (1894–1978). It is widely used in ecology because it is an index that combines richness with evenness. The index is usually represented by the symbol $H$, and is given by $H = -\sum_i^s p_i \log p_i$, where $s$ is the total number of species in the community and $p_i$ is the proportion of species $i$ in the community. Looking at it another way, this is the probability that any random individual taken from the community belongs to species $i$. The higher the $H$ value, the more diverse the community. The evenness component ($J$) can then be calculated by dividing $H$ by $H_{max}$, i.e. the maximum possible diversity for a given number of species and individuals. $H_{max}$ is given by $\ln s$, where $s$ is the total number of species. The greatest value possible for $J$ is 1.0, which represents a perfectly even distribution of the individuals among the species.

**sharka disease** *See* PLUM POX.

**sharp eyespot** *See* EYESPOT.

**shea butter** *See* VITELLARIA.

**sheath** **1.** Of leaves, the base of a blade or stalk that encloses the stem. **2. (of** *\*cyanobacteria) See* CAPSULE (of prokaryotes). **3.** *See* FUNGAL SHEATH.

**shelf fungus** The common name for any \*fungus with a shelf-like \*fruit body that juts out horizontally from its substrate, usually a tree or log. *Compare* BRACKET FUNGUS.

**Shelford's law of tolerance** A law, proposed by V. E. Shelford, that states that the presence and success of an organism depend upon the extent to which a complex of conditions are satisfied. The absence or failure of an organism can be controlled by the qualitative or quantitative deficiency or excess or any one of several factors which may approach the \*limits of tolerance for that organism.

**shelf zone** *See* SUBLITTORAL ZONE.

**shepherd's purse** (*Capsella bursapastoris*) *See* ALBUGO.

**shield fern** *See* DRYOPTERIS.

**shifting cultivation** (slash-and-burn agriculture) The traditional agricultural system of semi-nomadic people, in which a small area of forest is cleared by burning, cultivated for 1–5 years, and then abandoned as soil fertility and crop yields fall and weeds encroach. Ideally vegetation \*succession subsequently returns the plot to \*climax woodland, and soil fertility is gradually restored. Shifting cultivation of this type was once practised world-wide but in modern times it has been primarily associated with tropical rain-forest areas. The system is best suited to low population densities. With increasing population pressure, abandoned plots are often cleared again before a full climax community has been restored, leading eventually to nutrient depletions of the system and degradation of forest to open \*savannah-type woodland, or \*scrub.

**Shigella** (family \*Enterobacteriaceae) A genus of \*Gram-negative, rod-shaped, non-\*motile \*bacteria. They are found in the intestines of humans and higher monkeys and are the causal agents of dysentery.

**shiitake** A mushroom-like \*fungus, *Lentinula edodes* (formerly *Lentinus edodes*) cultivated for food, e.g. in Japan.

**shola** *See* AESCHYNOMENE.

**shoot** A \*stem that is mainly above ground.

**shoot frequency** *See* ROOT FREQUENCY.

**Shorea** (balau, red meranti, white meranti (Malaya), lauan (Philippines), selangan batu (Borneo); family \*Dipterocarpaceae) A genus of trees, many of which are huge, whose fruits have 3 enlarged, wing-like \*sepals. Many species yield valuable timber, and some yield \*dammar. *S. robusta* (sal) is native to the seasonal moist forests of India, and is much planted as \*coppice. According to tradition, the Buddha died beneath one. There are 357 species, in several rather distinct groups. They occur in seasonal and evergreen tropical rain forest, and are native from Sri Lanka to southern China and the Moluccas. They are abundant and have many sympatric species in western Malesia.

**Shorea albida** (alan; family \*Dipterocarpaceae) A huge tree found mainly in the peat-swamp rain forests of northern Borneo, which is remarkable for its occurrence in extensive pure stands, its large range, and the low density of its timber, which is used for light construction, furniture, or packing-cases.

**shoreface** The subtidal coastal zone between the low-water mark and a depth of about 10–20 m. The lower limit of the shoreface corresponds to the position at which waves begin to affect the sea-bed (wave base), and therefore wave action governs the processes active in this area.

**short-day plant** A plant in which flowering is favoured by short days and correspondingly long nights. There are 2 groups of such plants, some species in which there is an absolute requirement for these conditions for the onset of flowering, and other species in which flowering is merely hastened by them. In fact, however, the term is somewhat misleading, in that the critical factor is the period of darkness and short-day plants require a night of more than a minimum duration. Horticulturists exploit this by exposing chrysanthemums to a brief flash of light during the night to delay flowering until Christmas.

**shred** In forestry, to remove the branches from a *standard tree.

**shrub** A *perennial woody plant, less than 10 m tall, which branches below or near ground level into several main stems, although it has no clear trunk. It may be *deciduous (e.g. hawthorn) or *evergreen (e.g. holly). At the end of each growing season there is no die-back of the aerial parts, apart from the loss of foliage. *Compare* HERB; SUBSHRUB; and TREE.

**Si** *See* SILICON.

**siamweed** *See* EUPATORIUM.

**Sida** (family *Malvaceae) A genus of small *shrubs and *herbs, mostly covered in hairs. The leaves are *simple, toothed or lobed, alternate, with *stipules. The flowers are solitary, or held in terminal *spikes or heads, and are usually yellow, orange, or white. They are regular and bisexual. The *calyx is 5-toothed, the *corolla of 5 free *petals. The many *stamens are united into a tube at the base and are joined to the corolla. The *ovary is *superior, with 5 unilocular *carpels. The mature carpels form the *schizocarps, often with a beak or bristle. Several species are grown as ornamentals. There are about 150 species, found in all temperate regions, but centred in America.

**Siderocapsa** A genus of *bacteria in which iron and/or manganese oxides are deposited in material surrounding the spherical or ovoid cells. They are found in aquatic environments rich in iron.

**Siegennian** *See* DEVONIAN.

**sieve cells** Long, slender, tapering cells which form part of the *sieve tube. They lack nuclei but retain *cytoplasm. Each sieve cell ends in a *sieve plate.

**sieve plate** A region of pores that perforate the end of a *sieve cell. Cytoplasmic strands (*see* CYTOPLASM) may pass through the pores, interconnecting adjacent sieve cells and facilitating *translocation. A sieve area is occasionally present in the lateral wall of a sieve cell.

**sieve tube** A series of *sieve cells that lie end to end, forming a tube, in the *phloem.

sieve cell

*Sieve cell*

**sigmoid growth curve** *See* S-SHAPED GROWTH CURVE.

**silage (ensilage)** A type of foodstuff for livestock, prepared from green crops (e.g. grass); the crops are stored in a pit or silo and the *bacteria present on the plants carry out *fermentation, the products of which preserve the plant material from further decay and loss of nutritional value.

**silcrete** *See* DURICRUST.

**silent allele (null allele)** In genetics, an *allele that has no detectable product, and so is not expressed in a *phenotype.

**siliceous** Containing silica. Some kinds of wood are siliceous, and therefore hard to work.

**silicon (Si)** An element that is helpful to or required by certain plants in small quantities. It does not seem to play a role in metabolism but some plants accumulate a large amount in the walls of epidermal and *vascular tissues, and thereby reduce water loss and retard fungal infection. Silicon dioxide (silica) forms an important part of the *cell walls of *diatoms.

**silk cotton (samul)** *See* BOMBAX.

**silky oak** *See* GREVILLEA.

**silt** 1. Mineral soil particles that range in diameter from 0.02–0.002 μm in the

S

international system or 0.05–0.002 μm in the *USDA system. **2.** A class of soil texture.

**Silurian** The third of 6 periods of the *Palaeozoic Era, approximately 443.7–416 Ma ago, whose end is marked by the climax of the Caledonian *orogeny and the filling of several Palaeozoic basins of deposition. It is the period during which land plants first appeared.

**silver leaf** A disease affecting a number of ornamental and fruit trees of the *Rosaceae, especially plum. The characteristic symptom is the silvery sheen that appears on the leaves of infected trees. Branches show a dark brown discoloration and die back. The causal agent is *Chondrostereum purpureum.

**silviculture** The cultivation of forest trees, or woodland management for timber and other wood products. Sometimes the term is used interchangeably with 'forestry', but usually it implies a more *holistic approach.

**Simaroubaceae** A family of trees and shrubs, many of which have a bitter taste and contain compounds of medicinal use. The leaves are *simple or *pinnate, the stalks often jointed. The flowers are small, regular, and *tri- or *pentamerous, with an intrastamenal *disc. The fruit is often *drupe-like. There are 32 genera, with about 170 species, occurring mainly throughout the tropics and extending into temperate Asia.

**similarity coefficient** Any measure of the similarity of two samples. In ecological work, the similarity index devised in 1913 by J. Czekanowki has been widely used. This measures similarity as $C = 2W/(a + b)$, where $a$ and $b$ are the quantities of all the plants (or another commodity) found in the 2 stands, or other units to be compared, and $W$ is the sum of the lesser values for those species common to both units. Complete similarity thus scores 1, complete dissimilarity 0. See also AFFINITY INDEX.

**simple** Of a *leaf, not lobed or divided. Compare COMPOUND.

**simple pit** See BORDERED PIT.

**simple sorus** See GRADATE SORUS and SORUS.

**Sindora** (family *Fabaceae, subfamily *Caesalpinoideae) A genus of big trees which have *pinnate leaves. There are 4 *sepals, 1 *petal, 9 fused *stamens, and 1 free stamen. The fruit is a flat, rounded, woody, *indehiscent, spiny pod with 1 big, flat seed (or several such) seated on a big, hard *aril. The timber is valuable (sepetir). There are 20 species, occurring in south-eastern Asia, and 1 species in western Africa.

**single** (store) To allow a single *pole from a *coppice stool to grow on to form a *standard tree.

**Sino-Japanese floral region** Part of R. Good's (*The Geography of the Flowering Plants*, 1974) boreal kingdom, which corresponds geographically with the Sino-Himalayan-Tibetan mountains, northern and central China, and Japan. The region is differentiated by more than 300 endemic genera (see ENDEMISM), and contains in fact the richest flora in the northern temperate zone. A vast number of garden plants are native to the region, as are a number of important economic plants. See also FLORAL PROVINCE and FLORISTIC REGION.

**sinuate** Curved; having a wavy or indented margin.

**sinus** A space, found for example between the lobes of a plant *leaf.

**siotropism** A tropic response (see TROPISM) to shaking.

**siphonaceous** (siphoneous) Applied to *algae in which the *thallus is not divided up by *septa, i.e. the many nuclei are not compartmentalized into cells. The typical siphonaceous alga has a large *central vacuole surrounded by a layer of *protoplasm, containing *nuclei and *chloroplasts, which lines the *cell wall.

**siphoneous** See SIPHONACEOUS.

**Siphonocladus** (division *Chlorophyta) A genus of filamentous *green algae, in which the *thallus is multicellular, with multinucleate cells and reticulate *chloroplasts. They are found in marine environments, mainly in tropical regions.

**siphonostele**  One of 2 basic *morphologies of *stele, the other being *protostele. Siphonosteles are the more highly developed form, being characterized by a cylinder of *xylem and *phloem surrounding a *pith. Siphonosteles can be subdivided into *amphiphloic siphonostele and *ectophloic siphonostele. *Compare* DICTYOSTELE and MONOSTELE.

**sisal**  *See* AGAVE.

**sister chromatid exchange**  (SCE) An event, similar to *crossing-over, that can occur between sister *chromatids at *mitosis and *meiosis. It may be detected in harlequin chromosomes (sister chromatids that stain differently so that one appears dark and the other light).

**site**  In genetics, the position within a *cistron occupied by a *mutation.

**Sitka spruce**  *See* PICEA.

**skiaphilic**  *See* SCIAPHILIC.

**slash-and-burn agriculture** *See* SHIFTING CULTIVATION.

**sleep movements**  *See* NICTONASTY.

**sleepy grass**  *See* STIPA.

**slickensides**  *See* VERTIC HORIZON.

**slime mould**  A type of *eukaryotic micro-organism in which either a *plasmodium or *pseudoplasmodium is formed. Slime moulds are often included in mycological classification schemes, although probably they are not related to fungi; they are also included in zoological classification schemes, either as protozoa or as a separate phylum (Gymnomyxa, equivalent to the botanical *Myxomycota).

**Sloanea**  (family *Elaeocarpaceae) A genus of trees with pinnately nerved leaves, with *stipules. Some provide useful timber. The flowers are regular, bisexual, axillary or terminal, and usually borne in *racemes or *panicles. There are 4 or 5 *sepals, which are usually free but are occasionally united into a short tube; the *petals are absent, or rarely up to 4 in number and smaller than the *sepals. The

numerous *stamens are free. The *ovary is *superior with 4 or 5 *locules, each with many *ovules. The *style is simple. The fruit is a *capsule, covered in stiff bristles, with *endospermic seeds. The genus was formerly included in the family *Tiliaceae. There are about 100 species, found in Madagascar, tropical Asia, and eastern Australia.

**slop**  *Poles produced by *coppice.

**smaragdine**  Emerald green.

**Smilax**  (family Smilacaceae) A genus of slender, often spiny, woody climbers, which are net-veined *monocotyledons, the leaves having a pair of basal, spiralling *tendrils. The flowers are held in *umbels, and are *dioecious. Sarsaparilla is the dried roots of several of the American species. There are about 200 species, with a tropical and temperate distribution.

**smoke bush**  *See* COTINUS.

**smooth-barked apple**  *See* ANGOPHORA.

**smudge**  A disease that can affect onions, shallots, and leeks. Dark patches, often in the form of concentric rings, appear on the bulbs. The causal agent is the *fungus *Colletotrichum circinans*.

**smudging**  The burning of materials (e.g. oil) to produce a smoke layer that reduces the effect of radiation cooling of the air above the ground surface. It is used as a protective measure (e.g. in fruit-growing areas, especially in frost hollows).

**smut**  A plant disease caused by a *fungus of the order *Ustilaginales. Many types of plant can be affected, but smuts are particularly important in cereals and other grasses. The symptoms include the formation of masses of black soot-like *spores, and infected plants often show some degree of distortion.

**snag**  The standing part of a tree trunk that has snapped above ground level.

**snake's tongue**  *See* OPHIOGLOSSUM.

**snapdragon**  *See* ANTIRRHINUM.

**snow-line**  The lower limit of permanent snow cover. The height of the line varies with latitude; locally it also varies with aspect, because of the relationship to prevailing winds and the quantity of snow deposited, and to summer temperatures, etc.

**snow-patch vegetation** In tundra, and to a lesser extent in alpine environments, late-melting snow patches exert a marked influence on the vegetation. Vegetation that occurs beneath the small snow patches and the peripheral parts of large ones is protected from the rigours of winter, but as the snow melts comparatively early in the summer months, the vegetation is also able to capitalize on most of the *growing season: in these situations, therefore, it is rather luxuriant. Conversely, the larger snow patches, which melt well into or towards the end of the short growing season, tend greatly to restrict the development of vegetation.

**soapwort** (*Saponaria officinalis*) See SAPONIN.

**soboliferous** Forming clumps.

**sociability scale** A 5-point scale used in vegetation analysis to indicate the degree of clumping or gregariousness of an individual plant species, obtained as a visual impression. 1 on the scale indicates a shoot growing singly; 5 indicates shoots growing in large mats or pure populations.

**sodication** In soils, an increase in the percentage of exchangeable sodium. Sodium adsorbs on to soil *cation-exchange sites, causing soil aggregates to disperse, which closes soil *pores and renders the soil impermeable to water. *See also* SODIUM ADSORPTION RATIO.

**sodic soil** **1.** Soil with a sodium content sufficiently high to interfere with the growth of most crop plants. **2.** Soil with more than 15 per cent exchangeable sodium.

**sodium** (Na) An element that is found in all terrestrial plants. It is not essential except in certain $C_4$ salt-tolerant plants, but it has a role in *crassulacean acid metabolism.

**sodium adsorption ratio** (SAR) The tendency for sodium *cations to be adsorbed at *cation-exchange sites in soil at the expense of other cations, calculated as the ratio of sodium to calcium and magnesium in the soil (as the amount of sodium divided by the square root of half the sum of the amounts of calcium and magnesium, where ion concentrations are given in moles per gram). A low sodium content gives a low SAR value. In practice, allowance must be made for other reactions within the soil that do not involve sodium but do affect concentrations of calcium and magnesium. The SAR value is most likely to be changed by irrigation water.

**sodium-coupled transport** The entry of a metabolite into a cell against a concentration gradient, where this entry is coupled to the movement of sodium across the *cell membrane. Sodium readily enters cells because of the concentration of its ions outside the cell membrane which is high compared with that inside. It is thought that carriers exist that can bind to sodium, so exploiting this feature, and also to a metabolite (such as *glucose), which is thereby carried across the membrane.

**soft rot** Any of a range of plant diseases in which the *tissues of the infected plant are softened or even liquefied, typically becoming slimy and malodorous. A common cause of soft rot in carrots and other vegetables is bacterial infection by *Erwinia carotovora*.

**softwood** The wood of coniferous trees. The designation is arbitrary and some softwood is harder than many *hardwoods.

**soil** The natural, unconsolidated, mineral and organic material occurring on the surface of the Earth; it is a medium for the growth of plants.

**soil air** The soil atmosphere, comprising the same gases as in the atmosphere above ground, but in different proportions: it occupies the *pore space of the soil.

**soil classification** A system, analogous to those used in biological *taxonomy, that arranges soil types according to their distinguishing characteristics. Russian scientists were the first to attempt to classify soils in the latter part of the 19th century and many Russian soil names are still used (e.g. *chernozem, *solonetz, and *podzol). By 1975 American scientists at the US Department of Agriculture had devised a classification they called Soil Taxonomy. This system is widely used outside the United States. There are also many national classifications. In 1961 representatives from the Food and Agriculture Organization (FAO) of the United Nations, the United Nations Educational, Scientific, and Cultural Organization (UNESCO), and the International Society of Soil Science (ISS) met to discuss preparing an international classification. This was completed in 1974, updated in 1988, and has been amended several times since. Based on *diagnostic horizons, it divides soils into 30 reference groups and 170 possible subunits.

**soil conservation** The protection of the soil by careful management to prevent physical loss by erosion and to avoid chemical deterioration (i.e. to maintain soil fertility).

**soil formation** The action of the combined primary (weathering and humification) and secondary processes to alter and to rearrange mineral and organic material to form soil, involving the differentiation of soil *profiles and the formation of loose soil from consolidated rock material. *See also* PEDOGENESIS.

**soil horizon** A layer of soil, more or less parallel to the soil surface, that is physically, chemically, and/or biologically distinguishable from the layers above and below it.

**soil management** A variety of practices and operations with respect to soil that aid the production of plants; normally they are planned to allow for sustained yield in the future.

**soil-moisture index** *See* MOISTURE INDEX.

**soil structure** The grouping of individual soil particles into secondary units of aggregates and peds; this grouping is like an internal scaffolding of the soil.

**sola** *See* SOLUM.

**Solanaceae** A family of dicotyledonous (*see* DICOTYLEDON) plants, mostly herbs but some shrubs and trees, having normally alternate leaves, and regular flowers which are mostly *pentamerous. The *petals form a *corolla tube, at least at the base. There are 5 *stamens, alternating with the corolla lobes. The *ovary is of 2 cells, sometimes 4, with many *ovules on axile placentas (*see* PLACENTATION), and a terminal *style. The fruits may be *capsules or *berries. The Solanaceae are an important family, containing many plants of major economic importance, such as *Solanum tuberosum* (potato), and *Lycopersicon esculentum* (tomato); also *Nicotiana tabacum* (tobacco), ornamentals (e.g. *Petunia hybrida* with showy flowers), and many poisonous genera and species rich in *alkaloids (e.g. *Atropa belladonna*, deadly nightshade), some of which are used in medicine. There are some 90 genera, with 2600 species, cosmopolitan in distribution but centred in the tropics.

**Solanum** (family *Solanaceae) A genus of plants, with flowers in which the *stamens are prominent, and converging into a cone, and the *corolla lobes are spreading or reflexed. The genus includes *S. tuberosum* (potato), which is of complex *hybrid origin, from the Andes, and is one of the world's major food plants. Some species have fruits with ornamental *berries. There are some 1400 species, many of them tropical.

**solarization** (heliosis) The inhibition of *photosynthesis at very high light intensities, owing mainly to the photooxidation of certain of the compounds involved.

**solenostele** The most highly developed type of *protostele; it is found in *Selaginella*, often arranged as a *polystele. In ferns, and occasionally in *Selaginella*, the solenostele is amphiphloic, differing from *amphiphloic siphonostele in the absence of a *pith, instead encompassing endodermal cells (*see* ENDODERMIS). In ferns, *leaf gaps may occur, in which case the solenostele may become a *reticulate network and, therefore, be termed a *dictyostele. *Compare* ACTINOSTELE, HAPLOSTELE, and HYPOPHLOIC HAPLOSTELE.

**Solidago** (golden rod; family *Asteraceae) A genus of *perennial *herbs, with stalkless, *simple leaves, and small, yellow flower heads in *panicles. The *bracts of the *involucre of the flower heads overlap in many rows. The heads have both yellow (female or neuter) ray-florets and bisexual *disc-florets. The *achenes are many-ribbed, and have a *pappus of ciliate hairs. *S. canadensis* and *S. gigantea* are much grown in gardens for their showy, golden *panicles of flower heads, but *Solidago* species are regarded as weeds in parts of the USA. There are some 100 species, mostly American, but some European and Asian.

**soligenous mire** A *mire that receives water from rain and slope run-off. *See* AAPA MIRES.

**solodic soil** Leached, formerly saline soil, associated with semi-arid tropical environments, in which the A *horizon has become slightly acid, and the B horizon is enriched with sodium-saturated *clay. The term was used in soil classification systems derived from early Russian systems based on the work of V. V. Dokuchaev, but it is now obsolete.

**Solonchaks** Soils with a layer enriched in soluble salts that is more than 15 cm thick and lies at the surface or only a little way below it. Solonchaks are a reference soil group in the FAO *soil classification.

**solonetz** *Mineral soil at a transitional stage of *leaching or solodization (*see* SOLODIC SOIL) of saline soils, in semi-arid, tropical environments, which has a *sandy, acid A *horizon and a B horizon partially enriched with sodium *clay. The term was used in early systems of *soil classification, but in the *USDA Soil Taxonomy solonetz soils are included in the order *Aridisols and the name is not used. Solonetz are a reference soil group in the FAO soil classification, however.

**soluble RNA (s-RNA)** *Transfer-RNAs that are relatively small molecules and are more soluble in acid than other RNAs.

**solum (pl. sola)** The upper part of a soil *profile, above the parent material, in which processes of soil formation occur, and within which most plant roots and soil animals are found.

**solute potential** *See* OSMOTIC POTENTIAL.

**somatic cell** A body cell (i.e. a cell that is not destined to become a *gamete, and whose *genes will not be passed on to future generations).

**somatic cell hybrid** A hybrid cell resulting from the fusion of two *somatic cells.

**somatic crossing-over** In genetics, *crossing-over during *mitosis of *somatic cells such that parent cells *heterozygous for a given *allele, instead of giving rise to 2 identical heterozygous *daughter cells, give rise to daughter cells one of which is homozygous for one of these alleles, the other being homozygous for the other allele. *Phenotypically differing cell-lines may result. Studies of somatic crossing-over, somatic assort-ment, and cell fusion make up somatic-cell genetics, a modern asexual genetic technique that allows a wide range of *in vitro* manipulation of higher cells, including human cells as well as those of other organisms.

**somatic mutation** A mutation occur-ring in a *somatic cell. If the mutated cell continues to divide, the individual will develop a patch of tissue with a *genotype different from the cells of the rest of the body.

**sombric horizon** A subsurface soil *horizon of well drained, *mineral, tropical and subtropical soils into which *humus has leached downward. Base saturation is less than 50 per cent. It is a *diagnostic horizon.

***Sonneratia*** (family *Sonneratiaceae) A genus of mangrove trees with breathing roots, in which the flowers have many *stamens, inserted on the *calyx tube, and the *ovary is *superior. The flowers are sour-smelling, nocturnal, and bat-pollinated. The fruit is a leathery *berry seated on the persistent calyx. There are 5 species, occurring from eastern Africa and Madagascar to the western Pacific islands.

**sooty mould** A dark, soot-like, *fungal growth which appears on plants infested with sap-sucking insects such as aphids or scale insects; these insects produce a sugary fluid ('honeydew'), and the sooty mould *fungi grow on this. Sooty moulds may belong to various genera of the *Dothideales.

**Sophora** (family *Fabaceae, subfamily *Papilionatae) A genus of trees and shrubs, and a few *perennial *herbs, in which the leaves are alternate, usually *pinnate with numerous leaflets, and with *stipules. The flowers are held in terminal *racemes or leafy *panicles. They are irregular, with the lateral *petals enclosed by the upright petal when in bud, and white, yellow, or blue-violet in colour. The *calyx is composed of 5 fused *sepals with short teeth. The 5 petals are held with the 2 lower ones partly fused, the 2 lateral and 1 upright standard. There are 10 free *stamens. There is a single *carpel, and the *ovary is *superior, with numerous *ovules. The fruit is a fleshy or woody *pod, which is often *indehiscent. The seeds have thick *cotyledons. The timber of the tree species is especially hard and strong (e.g. *S. tetraptera* (kowhai) of New Zealand and Chile is used for bearings in machines, as well as for ornamental purposes). Dye is obtained from *S. japonica*. There are 52 species, found in all tropical and temperate regions, particularly in Asia and America.

**soralium** A mass or cluster of soredia (*see* SOREDIUM).

**Sorbus** (family *Rosaceae) A genus of trees and shrubs with *simple or *pinnate leaves, and no thorns. They are similar to *Pyrus* (pear) in flower structure, with 5 *petals, and many *stamens, but the cartilaginous *carpel walls are not completely united to the apex in the fruit, and the flowers are borne in compound *corymbs, not *simple, *umbel-like *inflorescences. The genus includes *S. aucuparia* (rowan or mountain ash), *S. aria* (whitebeam), and *S. torminalis* (wild service tree). There are about 85 species, some taxonomically very critical, occurring in northern temperate regions.

**Sordariales** (subdivision *Ascomycotina) An order of *fungi in which the *ascocarp is a *perithecium that is not im-

mersed in a *stroma; *asci are unitunicate and cylindrical to club-shaped, and the *ascospores are often dark-coloured when mature. There are many genera. Most are *saprotrophic, and found on dung, wood, or soil.

**soredium** A microscopic structure formed by certain *lichens as a means of *vegetative propagation: it consists of a few fungal *hyphae among which are enmeshed a few cells of the *phycobiont. To the naked eye, soredia in masses appear as a granular or fine powder.

**sori** Plural of *sorus.

**sorocarp** In certain *slime moulds, a fruiting structure consisting of an unenclosed mass of *spores borne at the tip of a stalk.

**sorrel** *See* POLYGONACEAE.

**sorus** (pl. **sori**) In plants, a fruiting structure consisting of a mass of *spores or *sporangia.

**soursop** *See* ANNONA.

**Southampton series classificatory programme** *See* ASSOCIATION ANALYSIS; INFORMATION ANALYSIS; and INVERSE ANALYSIS.

**south Brazilian floral region** Part of R. Good's (*The Geography of the Flowering Plants*, 1974) *neotropical kingdom, which contains a large flora with about 400 endemic genera (*see* ENDEMISM); a high proportion of its several thousand species is also endemic. Many important economic and ornamental plants derive from this region. *See also* FLORAL PROVINCE and FLORISTIC REGION.

**southern bacterial wilt** *See* GRANVILLE WILT.

**southern beech** *See* NOTHOFAGUS and BICENTRIC DISTRIBUTION.

**south temperate oceanic-island floral region** Part of R. Good's (*The Geography of the Flowering Plants*, 1974) Antarctic kingdom which contains a very small flora scattered among the islands of the oceans surrounding Antarctica. There are only 2 endemic genera (*see*

**S**

ENDEMISM), 1 of which is disjunct between several islands. Despite the great distance separating the islands, and their varying latitudes, there is a notable floristic constancy. A characteristic vegetation is tussocky grassland, dominated by *vicarious species of *Poa. See also* FLORAL PROVINCE and FLORISTIC REGION.

### south-west Australian floral region

Part of R. Good's (*The Geography of the Flowering Plants*, 1974) Australian kingdom, which is a very rich floral region with a high degree of *endemism, in many respects rivalling that of the Cape region of S. Africa. The same families are prominent in both floras and they have many growth forms in common. *See also* FLORAL PROVINCE and FLORISTIC REGION.

### soya bean   *See* GLYCINE.

### spadix   a spike of flowers on a swollen *axis.

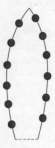

*Spadix*

### Spanish chestnut   *See* CASTANEA.

### Spanish grass   *See* STIPA.

### *Sparassis* (family Sparassidaceae, order *Aphyllophorales)   A genus of fungi in which the *hymenium is borne on the underside of 1 or more flattened, petal-like lobes. *S. crispa* (brain fungus, cauliflower fungus) forms *fruit bodies that are large (20–50 cm across), pale yellow or buff, and densely branched with numerous flattened, crinkled lobes. It is found at the base of conifer trees or tree stumps.

### spathe   In *monocotyledons, a large *bract subtending an *inflorescence.

### *Spathodea* (African tulip tree; family *Bignoniaceae)   A *monotypic genus (*S. campanulata*, formerly *S. nilotica*) of evergreen trees which have *pinnate leaves and large, red flowers in *racemes. The *pod splits into 2 parts. The seeds are winged. They are planted for their attractive flowers and occur in tropical Africa.

### spatulate   Having an end that is broad and flattened, like a spatula.

### spawn   A fungal *mycelium. Among *mushroom growers, a block of manure or other suitable substrate bearing a growth of mycelium of *Agaricus bisporus*, used to start a new culture of mushrooms.

### special adaptation   *See* GENERAL ADAPTATION.

### special creation   The belief that the origin of life and the diversity of life result from acts of God whereby each species was created separately. *Evolution is implicitly rejected as the explanation of these phenomena.

### specialization   A degree of *adaptation of an organism to its environment. A high degree of specialization suggests both a narrow *habitat or *niche and significant *interspecific competition.

### speciation   The separation of populations of plants and animals, originally able to interbreed, into independent evolutionary units which can interbreed no longer, owing to accumulated genetic differences. In *cladistics, the origin of one or more new species occurs inferentially by cladogenesis.

### species (sing. and pl.)   Literally, a group of organisms that resemble one another closely: the term derives from the Latin *speculare*, 'to look'. In taxonomy, it is applied to one or more groups (populations) of individuals that can interbreed within the group but cannot exchange *genes with other groups (populations), or, in other words an interbreeding group of biological organisms that is isolated reproductively from all other organisms (*see* BIOSPECIES). A species can be made up of groups in which members do not actually exchange genes with members of other groups (though in principle they

could do so), as, for example, at the two extremes of a continuous geographical range. However, if some *gene flow occurs along a continuum, the formation of another species is unlikely to occur. Where barriers to gene flow arise (e.g. physical barriers, such as sea, or areas of unfavourable habitat) this reproductive isolation may lead by either local selection or random *genetic drift to the formation of morphologically distinct forms termed races or subspecies. These could interbreed with other races of the same species if they were introduced to one another. Once this potential is lost, through some further evolutionary divergence, the races may be recognized as species, although this concept is not a rigid one. Most species cannot interbreed with others; a few can, but produce infertile offspring; a smaller number may actually produce fertile offspring. The term cannot be applied precisely to organisms whose breeding behaviour is unknown. *See* MORPHOSPECIES and PALAEOSPECIES.

**species–area curve**   *See* MINIMAL AREA.

**species group (superspecies)** A complex of related species that exist *allopatrically (in different geographical areas from one another). They are grouped together because of their morphological similarities, and this grouping can often be supported by experimental crosses in which only certain pairs of species will produce *hybrids.

**species longevity** The persistence of species for long periods of time.

**speedwell**   *See* VERONICA.

**spermatangium** In *Rhodophyta, the organ that produces *spermatia.

**spermatia** In plants, non-*motile sperm, found in *lichens and the *Rhodophyta. *Compare* ANTHEROZOID.

**spermatiophore** A *hypha that bears a *spermatium.

**spermatium** In certain types of *fungi, a microscopic, single-celled, male reproductive cell.

**spermatocyte**   *See* ANTHEROCYTE.

**Spermatophyta** The division of the plant kingdom (Plantae) that includes all of the seed-producing plants. These form two groups, the *gymnosperms and the *angiosperms (flowering plants).

**spermatozoid**   *See* ANTHEROZOID.

**Sphacelariales (division *Phaeophyta)** An order of brown *algae in which growth occurs from a large, dense, brown, apical, meristematic cell. This cell divides transversely, and then the daughter cells divide longitudinally. The *thallus is generally filamentous and branched. *Sphacelaria cirrhosa* is a common British species; it forms small, dense tufts, 0.5–2.5 cm tall, on other seaweeds. It is found throughout the year on the lower shore and below low-water mark.

**Sphaeriales (subdivision *Ascomycotina)** An order of *fungi in which the *ascocarps typically are spherical, hemispherical, or flask-shaped, carbonaceous *perithecia which are immersed in a *stroma. Members include *saprotrophs and plant *parasites.

**sphaerocyst** A spherical cell, clusters of which are present in the flesh of the *fruit bodies in fungi of the *Russulaceae.

**Sphaerophoraceae (order *Caliciales)** A family of *lichens in which the *thallus is *fruticose (branched and solid) or *foliose. The *ascocarps are embedded within the thallus at the edges or tips. *Spores are brown and *aseptate or 1-septate. There are several genera, found chiefly in cool moist regions of the southern hemisphere.

**Sphaeropsidales (class *Coelomycetes)** An order of *imperfect, microscopic *fungi which produce their *conidia in *pycnidia or in *stromata. There are many genera. They are *saprotrophic or *parasitic on plants.

**sphaeroraphide**   *See* DRUSE.

*Sphaerotilus* A genus of *Gram-negative *bacteria which are not assigned to any taxonomic family. Cells are rod-shaped, occurring mostly in chains which are enclosed by thin sheaths; single cells also occur, and these are *motile by means of a tuft of *flagella. *Sphaerotilus* species are *aerobic and *chemo-organotrophic. There is 1 species, *S. natans*, found in

S

sewage treatment plants and in slow-moving freshwater streams polluted with sewage and other effluents. It occurs attached to plants, stones, etc., forming a major component of the slimy growth known as *sewage fungus.

**Sphagnopsida (kingdom *Plantae, division *Bryophyta)** A class of bryophytes that includes the subclass Sphagnidae, with the single order Sphagnales, comprising up to 300 extant species of *Sphagnum*, and the Protosphagnales, mosses known only from the *Permian.

*Sphagnum* **(order Sphagnales, subclass Sphagnidae)** A genus of mosses that differ from other mosses (*Musci) in several respects and are sometimes placed in a separate class: *Sphagnopsida. The plants are characteristically branched with branches in *fascicles of 2–8. The leaves are nerveless and composed mainly of narrow, green, living cells and inflated, colourless, dead cells. The dead cells readily fill with water and *Sphagnum* can absorb water up to at least 20 times its own dry weight. The *capsules of *Sphagnum* are also unique, being roughly spherical with no *peristome. When the capsule is ripe the wall shrinks on drying and the pressure inside builds until the lid is blown off and the spores are ejected. There are up to 300 species, often difficult to distinguish, found, often abundantly, in wet, acidic *habitats: *bogs, *marshes, pools, *moors, wet woodland, damp grassland, etc. The genus is distributed world-wide and covers more of the Earth's surface than any other plant; *Sphagnum* probably accounts for more plant *biomass than any other genus.

**Sphenopsida** A subdivision of the *Pteridophyta, represented today by 1 genus only, *Equisetum*. Sphenopsids flourished in the *Carboniferous coal swamps and one of the fossil types of this period, *Calamites*, included tree-like forms that grew to 30 m.

**spicate** In spikes.

**spicule** A small needle or spine.

**spider plant (*Tradescantia*)** *See* COMMELINACEAE.

**spike** A *raceme in which the flowers are all *sessile.

*Spike*

**spikelet** A small *spike, typical of *Poaceae but also occurring in some reeds (*see* PHRAGMITES) and sedges (*see* CAREX and CLADIUM), that is the fundamental unit of the inflorescence and consists of an *axis, 2 *bracts or *glumes, and 1 or more *florets.

*Spinachia* **(spinach)** *See* CHENOPODIACEAE.

**spindle** The set of *microtubular fibres that appear to move the *chromosomes of *eukaryotes during cell division. The spindle is formed only at *mitosis or *meiosis, appearing at *metaphase and first arranging the chromosomes at its equator. Movement apart of *chromatids then occurs during *anaphase, probably as a result of contraction of the fibres that run from the *centromere (spindle attachment) to the spindle pole, and from pole to pole.

**spindle attachment (centromere)** The region of attachment on a *chromosome that links it to the *spindle at *mitosis or *meiosis. Its position determines its shape at *anaphase (a rod, or J, or V). In a few species the centromeric properties are distributed along the entire length of the chromosome; such species are said to be polycentromic (possessing a diffuse centromere).

*spindle tree* *See* CELASTRACEAE.

**spine** A sharply pointed projection formed from all or part of a modified leaf.

*Spinifex* **(family *Poaceae)** A genus of coastal grasses that are *dioecious. The *spikelets are *sessile, break into small pieces when ripe, and are distributed by the wind. The male spikelets are 2-flowered and arranged in spikes composed of a cluster of 4–6, with long *bracts. The female spikelets are very numerous and held in a large, dense head, each with a spine-like *rachis which is longer than the spikelet and the bract. The plants are *perennial, with thick *rhizomes and *stolon which can spread several metres over the sand. The plant is rough or hairy, with silvery leaves, and a *ligule of hairs. The genus is used to stabilize coastal sand-dunes. There are 3 species, found in east and south-eastern Asia, Australia, and the Pacific region.

**spinney** A wood that consists, or consisted formerly, of thorns (*Crataegus*).

**spinose** Bearing *spines.

**spiral wrack (twisted wrack)** The common name for the brown seaweed *Fucus spiralis* (see FUCUS). The *thallus differs from that of *bladder wrack (*F. vesiculosus*) in lacking air *bladders, and from that of *serrated wrack (*F. serratus*) in that the margins are not serrated; the branches are usually somewhat twisted spirally. It is found attached to rocks rather higher on the shore than other *Fucus* species.

**spire** A young timber tree, the lowest branch of which is at a considerable height.

**Spirillum** A genus of rigid, spirally shaped *bacteria in which the cells usually have a tuft of *flagella at each end. They are *micro-aerophilic and are found in freshwater habitats. An exceptional species, *S. minor*, is *parasitic in rats and can cause rat-bite fever in humans.

**Spirochaeta (family Spirochaetaceae, order *Spirochaetales)** A genus of free-living, spirally shaped *bacteria which are obligately or *facultatively *anaerobic. There are several species, found in sulphide-rich aquatic habitats, including sewage and polluted waters.

**Spirochaetales** An order of *bacteria in which the cells are slender, spiral in shape, and flexible, and 5–250 µm long, depending on species. They are *motile with rotating and flexing movements;

their motility is due to periplasmic *flagella (i.e. flagella that are sandwiched between the *peptidoglycan and outer membrane layers of the *cell wall). Species are *aerobic, *anaerobic, or *facultatively anaerobic. They are *chemo-organotrophs. There are 2 families, with several genera, including some important *pathogens (e.g. *Treponema pallidum*), found in a wide range of habitats. Also, spirochaetes are involved in the *symbiosis between termites and *flagellate protozoa.

**spirochaete** A *bacterium belonging to the order *Spirochaetales.

**Spirogyra (division *Chlorophyta)** A genus of green freshwater *algae, in which the *thallus consists of unbranched filaments. It is capable of a spiral gliding motility when illuminated. The *chloroplast forms a spiral, ribbon-like band which extends the length of the cell. Conjugation occurs by the formation of a tube connecting 2 cells; the contents of 1 cell pass through the tube and fuse with those of the other cell. *Spirogyra* is very common in standing fresh water, especially in spring.

**Spiroplasma (family Spiroplasmat-aceae, order *Mycoplasmatales)** A genus of *bacteria in which the cells are somewhat variable in shape, but all can form helical filaments; the cells lack cell walls. They are *motile by an unknown mechanism; there are no *flagella. They are *chemo-organotrophic, and can grow in the presence or absence of air. They are epiphytic (see EPIPHYTE) or *parasitic in plants and arthropods, and can cause disease (e.g. stubborn disease of citrus trees).

**Spirulina** A genus of filamentous *cyanobacteria (section III) in which the filaments are characteristically spiral in shape. They are capable of gliding motility. They are found in freshwater and marine habitats, mostly in warmer regions; and are also found in hot springs. *Spirulina* is collected and used as food in the region of Lake Chad, Africa.

**Splachnum (order *Funariales)** A genus of mosses that are characteristically found growing on or associated with animal dung.

**S**

The leaves are soft, lanceolate to obovate, with a thin nerve. The *capsule is held erect on a long *seta. Splachnum is a small genus found mainly in the northern hemisphere. S. ampullaceum and S. sphaericum occur in Britain. S. ampullaceum forms dense, light-green tufts on cattle dung. The capsules, borne on red setae, are often numerous and are characteristic in shape, resembling miniature Greek amphorae. They are found mainly in the upland regions of the west and north of Britain, particularly on wet *heaths and *moors, and in *bogs.

**spleenwort** See ASPLENIUM.

**spodic horizon** A subsurface soil *horizon in which organic matter together with aluminium and often iron compounds have accumulated amorphously. It is a *diagnostic horizon. The name is from the Greek spodos meaning 'wood'.

**Spodosols** An order describing soils in which subsurface *horizons contain amorphous materials comprising organic matter and compounds of aluminium and often iron that have accumulated illuvially (see ILLUVIATION). Such soils form in acid material, mainly coarse in texture, in humid cool to temperate climates, often beneath *coniferous forests.

**Spondias** (family *Anacardiaceae) A genus of trees, most of which have *pinnate leaves. The fruit is a slightly angled *drupe with a big round or angled stone, edible in several species (hog plum). There are 10 species occurring in Indo-Malaysia and tropical America.

**spongy mesophyll** In a leaf, *mesophyll tissue comprising cells of irregular shape, some of them lobed, separated by large spaces in which the atmosphere is humid. Spongy mesophyll is the site of gaseous exchange for *photosynthesis and *respiration. See also PALISADE MESOPHYLL.

**spontaneous mutation** In genetics, naturally occurring *mutation, as opposed to one artificially induced by chemicals or irradiation. Usually such mutations are due to errors in the normal functioning of cellular *enzymes.

**sporangiolum** A *sporangium within which there is only a single *spore or a small number of spores.

**sporangiophore** A specialized *hypha which bears a *sporangium.

**sporangiospore** A *spore formed within a *sporangium.

**sporangium** A sac-like structure within which fungal *spores are formed; spores are liberated on rupture of the sporangium wall.

**spore** **1.** A reproductive unit, usually consisting of a single *haploid cell, that is capable of developing into a new organism without fusing with another cell. The release of spores is the main method of dispersal in *Fungi, *algae, *Bryophyta, and *Pteridophyta. **2.** A differentiated bacterial cell which may function as a *propagule or as a resistant structure that allows the organism to survive adverse environmental conditions, often for protracted periods.

**sporeling** A juvenile plant that has arisen from a *spore.

**spore mother cell** (sporocyte) A *diploid plant cell that gives rise to 4 *haploid *spores during *meiosis.

**spore print** The pattern of *spores obtained when the *cap of a fungal *fruit body is placed, *gills or pores down, on a sheet of paper and left for a period of time.

**Sporobolomycetaceae** (class *Blastomycetes) A family of *imperfect *yeasts characterized by the production of *ballistospores. These yeasts are *saprotrophic, and found on plant material.

**sporocarp** (cystocarp) See CARPOSPOROPHYTE.

**sporocyte** See SPORE MOTHER CELL.

**sporodochium** A cushion-shaped mass of fungal *tissue densely covered with *conidiophores.

**sporogenesis** The production of *spores in plants. In *mosses and *liverworts the process occurs in a *sporogonium; in *fungi it occurs in a *sporophore; in other spore-forming plants it occurs in a *sporangium.

**sporogonium** In mosses (*Musci) and liverworts (*Hepaticae), the *sporophyte generation that develops after sexual reproduction and produces *spores.

**sporogony** A process of reproduction in which a *zygote undergoes multiple fission.

**sporophore** A structure upon which *spores are borne directly.

**sporophyll** A leaf in the *Pteridophyta that bears *sporangia. The term is most often applied to the small, scale-like leaves in the cones of *Lycopodium* and *Selaginella*, which bear *sporangia on their upper sides.

**sporophyte** The *spore-producing *diploid generation in the life cycle of plants. In higher plants, such as *angiosperms and *gymnosperms, the sporophyte is the dominant generation, forming the conspicuous plant. In lower plants, such as *mosses, *liverworts, and *hornworts, the *gametophyte is the dominant and conspicuous generation. *See also* ALTERNATION OF GENERATIONS.

**sport** A sudden deviation from type; a *mutation.

**sporulation** The process of *spore formation.

**spraing** A disease of potatoes, which may be caused by either of 2 different viruses, or by adverse growth conditions. *Tubers from infected plants show characteristic crescent-shaped brown marks in the flesh when cut. The viruses are transmitted by eelworms or by soil *fungi.

**spray** Wood trimmings, sold in faggots for kindling.

**spring** *See* SPRINGWOOD.

**springwood** (spring) 1. High forest that has grown from shoots from tree stumps. 2. The ground growth or shoots of new coppice that emerge from the stools of a felled coppice. *See also* TREE-RING.

**spruce** *See* PICEA.

**spur** 1. A short side branch that bears flowers and fruits. 2. In *conifers, a shoot that bears leaves. 3. A tubular projection from a *flower.

**spurge olive** *See* CNEORACEAE.

**spurges** *See* EUPHORBIA.

**spurred valerian** (*Centranthus*) *See* VALERIANACEAE.

**squamulose** Bearing or consisting of small scales (squamules).

**squarrose** Rough, with outstanding processes. Of a *moss, with leaves in which the upper part is curved back, at 90° or more to the lower part.

**s-RNA** *see* SOLUBLE RNA.

**S-shaped growth curve** (sigmoid growth curve) A pattern of growth in which, in a new environment, the population density of an organism increases slowly initially, in a positive acceleration phase; then increases rapidly, approaching an exponential growth rate as in the J-shaped curve; but then declines in a negative acceleration phase until at zero growth rate the population stabilizes. This decline reflects increasing environmental resistance which becomes proportionately more important at higher population densities. This type of population growth is termed density-dependent, since growth rate depends on the numbers present in the population. The point of stabilization, or zero growth rate, is termed the saturation value (symbolized by $K$) or *carrying capacity of the environment for that organism. $K$ represents the upper asymptote of the sigmoidal or S-shaped curve produced when changing population numbers are plotted over time. It is usually summarized mathematically by the *logistic equation. *See* DENSITY-DEPENDENCE. *Compare* J-SHAPED GROWTH CURVE.

**stabilizing selection** *See* MAINTENANCE EVOLUTION.

**Stachybotrys** (class *Hyphomycetes) A *form genus of *fungi in which the vegetative stage is a *septate *mycelium; chains of dark-coloured *conidia are produced from *phialids. Some species produce toxins (trichothecenes) which can lead to poisoning if ingested by humans or other animals.

**Stackhousia** (family *Stackhousiaceae) A genus of *annual or *perennial *herbs with *simple, alternate leaves, which often show *xerophytic features (e.g. succulent leaves). The regular, bisexual flowers are held in clusters, or *racemes, or are solitary, and are composed of a united, 5-lobed *calyx and a partly fused *corolla with the narrow or broad *petals united only in the centre. The fruit is a *schizocarp which splits into 1-seeded segments along the *locule partitions. The large seeds contain a fleshy *endosperm. There are 26 species (25 endemic (*see* ENDEMISM) to Australia and New Zealand); the other is found in Malesia and Micronesia.

**Stackhousiaceae** A family of *xerophytic, *annual or *perennial herbs which have a branched *rhizome system. The leaves are *simple and alternate, with *stipules, and are often succulent or leathery. The flowers are regular and bisexual, and are held in *racemes or clusters spread along the stem . There are 5 fused *sepals forming a lobed *calyx tube, with 5 *petals, either free or partly fused, and 5 *stamens which usually alternate in length. The *ovary is superior with 2–5 uniovular *locules. The *styles are partially fused. The fruit is a *schizocarp which splits along the locule partitions into single-seeded segments. The seeds contain a large embryo and fleshy *endosperm. There are 3 genera, with 28 species, restricted in distribution to Malaysia, Australia, and New Zealand.

**stag's-horn fern** See *PLATYCERIUM*.

**stag's-horn fungus** The common name for the abnormal, antler-like, *fruit bodies of *Lentinus lepideus* found on timber in damp, dark places such as coal mines.

**stamen** The male organ of a flower, comprising a stalk (the filament) and the *anther which is commonly 2-lobed, the lobes united by the *connective.

**staminate** Applied to a flower that produces *stamens and, in some cases, applied to flowers that produce stamens but no *pistils (i.e. male flowers).

**staminode** A rudimentary *stamen which produces no *pollen but which may function as a nectary or *petal.

**stand** (ecol.) **1.** The standing growth of plants, e.g. trees. **2.** A term used in vegetation classification to describe a distinctive plant association that may be recognized elsewhere. The composition

may vary slightly but the recognition of stands enables comparisons between different vegetation *communities to be made. Sometimes the suffix -etum is added to the stem of the generic name of the dominant species.

**standard 1.** A tree that is allowed to grow to its full height. **2.** A single-trunked tree that is large enough to be converted to sawn timber. **3.** A cultivated plant that stands without support because it has been grafted on to a robust, upright stem (e.g. a standard rose).

**standing crop** See BIOMASS.

**Staphyleaceae** A family of trees and shrubs in which the leaves are sometimes *opposite, *trifoliate or *pinnate, and the leaflets serrate, with *stipules. The flowers are hermaphrodite or unisexual. They have 5 *sepals and *petals, 5 *stamens, a *disc, and the *ovary *superior and usually *trilocular. There are 5 genera, with 27 species, occurring in tropical Asia and America, and in the northern temperate zone.

**Staphylococcus** (family *Micrococcaceae) A genus of *bacteria in which the spherical, *Gram-positive cells occur singly, in pairs, and in irregular clusters. They are non-*motile. They are *chemoorganotrophic, and capable of growth in the presence or absence of air. They are found mainly in or on the bodies of mammals. Many can be *pathogenic, causing a variety of conditions, including boils, abscesses, food poisoning, etc.

**star apple** See *CHRYSOPHYLLUM*.

**starch** A homopolysaccharide, consisting of D-*glucose molecules, which is the major storage *carbohydrate of plants. It occurs in 2 forms, the straight-chained amylase and the branched-chain amylopectin.

**starch sheath** An *endodermis that contains *starch grains.

**stasigenesis** The situation in which an evolutionary lineage persists through time without splitting or otherwise changing. So-called 'living fossils' provide examples of stasigenesis.

**stasipatric speciation** A rapid *speciation that may occur among small breeding populations that are not completely isolated genetically or spatially. The term is more or less synonymous with *parapatric speciation.

**stasis (evol.)** A period of little or no evolutionary change; the 'equilibrium' that alternates with 'punctuations' in the theory of *punctuated equilibrium.

**statismospore** A fungal *spore that is not violently discharged at maturity.

**statistical method** In modern usage, a method for analysing data based on probability theory. A statistical method permits the calculation of a value based on observations about some problem that may be tested for significance by comparison with the values that might be expected to arise by chance. Two main categories of statistical method have been developed: classical or parametric tests, and the more recent non-parametric or distribution-free tests. Parametric statistical methods may be applied only to data on an interval scale, and typically they make assumptions about the background population from which the sample is taken, most often that it is normally distributed. Where data are in nominal or ordinal form, or where assumptions about the distribution of data on which a parametric test is based cannot be justified, then non-parametric (distribution-free) methods can be used. In general, parametric tests are more rigorous than non-parametric tests. Formerly, and more colloquially, statistical methods embraced any form of data gathering and analysis. Compare NUMERICAL METHOD.

**stele** The *vascular tissue of a root or stem, consisting of a *xylem, *phloem *pericycle, and sometimes having *pith and *medullary rays. Morphologically, steles can be divided into 2 main categories: *protostele and *siphonostele, the latter possessing a pith. Either type may be a single vascular vessel (*monostele) or divided into several strands (*dictyostele).

**stellate** Star-shaped; radiating in arrangement.

**stem** In a *vascular plant, the part of the plant that bears *buds, leaves, and flowers. It forms the central *axis of the plant and often provides mechanical support. Most commonly it is above ground, but it may lie below ground, and is then termed a *rhizome.

***Stemonitis* (order Stemonitales, class *Myxomycetes)** A genus of *slime moulds in which the *fruit bodies are elongated, narrow, often brownish *sporangia borne on black, glossy stalks. There are many species, found on dead wood, *humus, etc.

**steno-** In ecology, a prefix, derived from the Greek *stenos*, meaning 'narrow', that is used with adjectives describing environmental factors, denoting a limited tolerance by an organism of those factors. Compare EURY-.

**stenoecious** Applied to an organism that can live only in a restricted range of *habitats. See STENO-.

**stenohalic** See STENOHALINE.

**stenohaline (stenohalic)** Very sensitive to changes in *salinity; unable to tolerate a wide range of *osmotic pressures. See STENO-.

**stenospermocarpy** Fruit growth that is stimulated by the *fertilization of *embryo sacs followed by a failure of seed development.

**stenothermal** See STENOTHERMOUS.

**stenothermous (stenothermal)** Unable to tolerate a wide temperature range. See STENO-.

**stephanokont** A *spore or *gamete characterized by a crown of *cilia around the anterior end, giving the appearance of a monk's tonsure; it is typical of *green algae.

**steppe (Eurasian steppe)** A vast, temperate *grassland *biome that stretches from the River Danube in Romania to Dunbey in China. Generally, the dominant vegetation comprises drought-resistant *perennial grasses, but the actual species composition varies from east to west and north to south according to the rate of precipitation. Near the Black Sea, large feather grasses (*Stipa*) and sheep's fescue (*Festuca ovina*) prevail. Here the climate is warm and humid in spring, supporting *ephemeral species (e.g. *Tulipa*) which soon die as the long, hot, dry summer follows. In the central steppe region the spring is cold, supporting few ephemeral species, but in wetter years large numbers of *vegetatively reproducing plants survive (e.g. *Artemisia*). There are 4 discernible belts of latitude, with the highest rainfall in the north: *meadow steppe; dry herbage/turf grass steppe, in which steppe herbage dominates; arid turf grass steppe, which has less steppe herbage; and desert/scrub/turf grass steppe.

**steppe meadow** *See* MEADOW STEPPE.

**Sterculiaceae** A family of trees, shrubs, climbers, and *herbs that have stellate hairs or scales. The leaves are simple or *palmately compound, with *stipules, and the petiole is often long and kneed. The flowers are regular, small, and pentamerous, with a stamenal tube *ovary which has 2–5 *locules. The fruit is commonly dry and *dehiscent. There are about 60 genera, with about 700 species, occurring mainly in the tropics and subtropics.

**Stereaceae** (order *Aphyllophorales) A family of *fungi in which the *fruit bodies are flattened or *resupinate. The *spores are smooth and colourless. Species are found mostly on wood. Some are important *parasites of plants.

**Stereocaulon** (order *Lecanorales) A genus of *lichens in which *pseudopodetia are formed; the pseudopodetia are solid and may bear scale-like, finger-like, or coralloid outgrowths (phyllocladia). *Apothecia are borne at the tips or on the sides of the pseudopodetia. *Spores are multiseptate (*see* SEPTUM) or *muriform. Species are found on acidic substrates, either on rocks or on the ground, chiefly in cool, humid regions or in mountainous areas of warmer regions.

**stereotaxis** *See* THIGMOTAXIS.

**Stereum** (family *Stereaceae) A genus of *fungi which includes both *saprotrophic and *parasitic species. Some species are common on decaying timber in woodland. *See also* CHONDROSTEREUM.

**sterigma** (pl. sterigmata) On a *basidium, a projection that bears a *basidiospore. In most *basidiomycetes the basidium bears 4 sterigmata.

**sterigmata** *See* STERIGMA.

**sterile 1.** Of an organism, unable to produce reproductive structures, i.e. unable to reproduce. **2.** Of land, unable to support the growth of plants, especially cultivated crops. **3.** Of an environment, object, or substance, completely free of all living organisms, including all microorganisms of any type or form.

**steroid** One of a group of derivatives of the fused, reduced, ring compound perhydrocyclo-pentanophenanthrene. As a group, steroids have a wide range of physiological functions in animals. *Compare* PHYTOSTEROL.

**sterrophilous** Growing on moorland.

**Sticta** (order *Peltigerales) A genus of *lichens in which the *thallus is *foliose; the lower surface is downy and is perforated by *cyphellae. The *phycobiont may be a *green alga or a *cyanobacterium. Some species have a strong, characteristically fishy smell. The *apothecia are *lecanorine. *Spores are colourless or brown, with 1–3 *septa. In Britain, *Sticta* species are found chiefly on mossy trees or rocks in moist western districts.

**Stigeoclonium** (division *Chlorophyta) A genus of filamentous *green algae in which the filaments are branched, and the *chloroplast forms a band around the middle of the cell (as in *Ulothrix*). They are common in still or running water. They are tolerant of organic pollution and are sometimes used as an indicator of heavily polluted water.

**stigma** The part of the female reproductive organs on which *pollen grains germinate. *See* CARPEL.

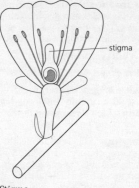

*Stigma*

**Stigonematales** An outdated phycological order of 'blue-green algae' which corresponds with section V of the *cyanobacteria.

**stilt root** A tree root that arises from the lower trunk and runs obliquely to the ground, providing additional support for the tree. Mangroves and certain palm trees (Arecaceae) have stilt roots. *Compare* BUTTRESS ROOT.

**stinkhorn** The common name for the *fungal *fruit body in some members of the *Phallales. The common stinkhorn, *Phallus impudicus*, has a white, spongy, hollow stem 10–20 cm tall. The cap is initially covered with an olive-brown slime containing the *spores, but the slime is eventually removed by flies. This fungus has a strong, fetid odour detectable from a distance of several metres. It is common in woods, gardens, etc. in summer and autumn. The dog stinkhorn, *Mutinus caninus*, is similar but smells less strongly.

**stinking smut** *See* BUNT.

***Stipa*** (needle grasses; family *Poaceae) A genus of tough, sharp, mostly *perennial grasses, whose *spikelets are single flowers and held in *panicles with conspicuous *awns. The spikelets are needle-like. The leaves are tough and narrow, sometimes with a small row of hairs forming a *ligule. There are about 150 species found in temperate and tropical regions, where the grasses are common in arid plains and form the characteristic genus

of the steppes throughout the world. *S. pennata* (feather grass) is commonly found in the Russian steppes. Several species have narcotic properties and are known as sleepy grasses (e.g. *S. inebrians*). *S. tenacissima* (esparto grass or Spanish grass), native to Spain and northern Africa, is used to make ropes, wickerwork, and paper.

**stipe** The stalk or stem of a fungal *fruit body.

**stipitate** Having a *stipe or stalk.

**stipulate** Having a *stipule.

**stipule** An outgrowth, usually occurring in pairs, at or near the base of a leaf *petiole. Stipules may be leaf-like, hard and sharply pointed, sheath-like and protecting the young leaf, or adpressed to the petiole or twig. Occasionally stipules are amplexicaul (i.e. surrounding the clasping the twig). These uncommon positions are valuable aids to plant identification.

**stock** *See* GRAFT.

**stoggal** A tree that has been cut short or has been broken down by the wind, and therefore is not suitable for use as timber.

**stolon** A stem that grows horizontally, a runner (e.g., as in the strawberry).

**stoma (pl. stomata)** A small opening, many of which are found in the epidermal layers of plants, allowing access for carbon dioxide and egress for water. Stomata are surrounded by guard cells which control the pore size.

**stomata** Plural of *stoma.

**stomium** An area of thin-walled cells in a *sporangium or *pollen sac that breaks to release the spores or pollen grains when the surrounding tissue dries.

**stone cell** *See* SCLERENCHYMA.

**stonecrop** (*Sedum*) *See* CRASSULACEAE.

**stonewort** The common name for the brittle, calcified plants of the *Charophyta; *Chara* and *Nitella* are the commonest British genera. The plant consists of a 'stem' or *axis which bears *whorls of branches at intervals (*nodes) along its length. The axis is attached at its base to the substratum by means of branched *rhizoids.

**S**

**stool** **1.** A tree stump that is capable of producing new shoots. **2.** The permanent base of a *coppiced tree.

**store** *See* SINGLE.

**storeyed** *See* STORIED.

**storied** **(storeyed)** Applied to *rays in wood, where these are arranged in a horizontal series as seen in a tangential longitudinal section.

**straits rhododendron** *See* MELASTOMA.

**stramineous** Straw-coloured.

**strangler** A plant that depends on another for physical support and ultimately suppresses the support plant by twining branches or aerial roots, e.g. strangler fig. Stranglers are most typical of tropical rainforest *communities.

**strangleweed** *See* JAPWEED and TRAMP SPECIES.

**strangling fig** A fig tree that germinates high in a host tree and sends roots to the ground which eventually *anastomose and, as they grow, envelop and kill the supporting tree. The *banyans differ in having roots descending from the limbs as well so that a grove of closely growing, stout roots develops.

**stratification** **1.** The arrangement of sediments, sedimentary rocks, soils, etc., in layers (strata). **2.** The placing of seeds between layers of moist peat or sand and exposing them to low temperatures (e.g. by leaving them outdoors through the winter) in order to encourage germination. *Compare* VERNALIZATION.

**stratified random sample** **(partial random sample)** In statistics, a modification of the random sample that is particularly useful when obvious heterogeneity exists in the *community, area, etc. to be investigated. In such instances a simple random sample may fail to record sufficient replicates of a particular subcategory, or may do so only very inefficiently, so preventing a proper statistical monitoring of variability. In a stratified random scheme sample data points are divided into classes (strata) before taking a random sample within each stratum.

**stratocoenosis** The *community of a particular vegetation or physical *habitat layer, e.g. the canopy layer of a woodland or the *hypolimnion of a stratified lake.

**strawberry** *See* FRAGARIA.

**strawberry tree** *See* ARBUTUS.

**streak** A general term for a virus disease in a (usually monocotyledonous) plant in which streaks of yellow colour or necrosis occur on the leaves.

**streaming** The metabolically active movement of the particulate and fluid constituents of the *protoplasm. This can occur within a cell, laterally through protoplasmic material connecting cells, and also longitudinally through files of connected cells.

***Strelitzia*** **(family *Strelitziaceae)** A genus of large herbs (a few with woody trunks) which have *distichous, banana-like leaves. The flowers are held in erect, harshly coloured, orange and blue inflorescences, subtended by vividly coloured *bracts, and some species are cultivated as bird-of-paradise flowers. There are 5 species occurring in southern Africa.

**Strelitziaceae** A small family of large *herbs or plants that often have a woody trunk and banana-like, *petiolate leaves. The flowers are bisexual and *trimerous, with an *inferior *ovary. The fruit is a *capsule, the seeds sometimes *arillate. There are 3 genera, with 7 species, occurring in Africa and America.

***Streptocarpus*** **(Cape primrose; family *Gesneriaceae)** A genus of *annual or *perennial *herbs and subshrubs, several of which are cultivated as ornamentals for their decorative leaves or flowers. There are about 120 species, occurring in tropical and southern Africa and Madagascar.

**Streptococcaceae** A family of *Gram-positive *bacteria in which the cells are spherical or ovoid and occur in pairs, in groups of 4, or in chains. Usually they are non-*motile. They are *chemo-organotrophic, and can grow in the presence or absence of air. They are found in a wide range of habitats.

**Streptococcus** (family *Streptococcaceae) A genus of *Gram-positive, non-*motile *bacteria in which the cells are spherical to ovoid, and often occur in pairs or chains. There are many species, found chiefly as *parasites and *pathogens in warm-blooded animals, including humans. The genus includes the causal agents of tonsillitis and scarlet fever (*S. pyogenes*), pneumonia (*S. pneumoniae*), and dental caries (e.g. *S. mutans*). Not all species are harmful; some are used in the manufacture of certain dairy products, e.g. yoghurt and butter. (These latter organisms are now regarded as belonging to a separate genus, *Lactocossus*.)

**Streptomyces** (order Actinomycetales) A genus of *bacteria in which the spore-bearing *mycelium is variously coloured. Many *antibiotics are obtained from strains of *Streptomyces*: e.g. streptomycin, erythromycin, kanamycin, chloramphenicol, and tetracyclines. There are many species. They are mostly *saprotrophic, and found in soils, for example. A few can be *pathogenic (e.g. *S. scabies* causes common scab of potato).

**stress-tolerator** In the classification of *plant strategies proposed by J. P. Grime, a plant species found in areas of high stress and low disturbance. *Compare* COMPETITOR and RUDERAL.

**striate** Marked with fine lines, ridges, or furrows.

**stringy-bark** A group of *Eucalyptus* species, named from their distinctive bark.

**strobilus** (cone) A group of closely packed *sporophylls arranged around a central axis. In *angiosperms, any cone-like structure.

**stroma** **1.** A cushion-like mass of fungal *tissue in or upon which *spore-bearing structures may develop. **2.** The matrix of *chloroplasts, within which the grana (*see* GRANUM) are embedded. In addition to *starch granules and oil droplets, it contains the *enzymes responsible for the mediation of the *dark reactions of *photosynthesis.

**stromatolite** A rock-like or firmly gelatinous structure, built up over long periods of time from many layers or mats of *cyanobacteria together with trapped sedimentary material. Stromatolites are found mainly in shallow marine waters in warmer regions. Some are still in the process of being formed, e.g. those in Shark Bay, Western Australia. Fossil stromatolites dating from the late *Archaean are also known, although it is not certain that these were formed by cyanobacteria. Stromatolites that formed about 2 billion years ago produced the earliest known reefs, quite different from present-day coral reefs.

**Stropharia** (family *Strophariaceae) A genus of *fungi, with *mushroom-shaped *fruit bodies in which the fleshy cap bears a separable *pellicle and is often slimy. The *stipe is not readily broken from the cap and has an *annulus. The *gills become brown with maturity. *Stropharia* species are found in grassland, on dung, etc.

**Strophariaceae** (order *Agaricales) A family of *fungi in which the *spores in masses appear yellowish-brownish or dusky brown. The *stipe and *pileus of the *fruit body are confluent. Species are found on soil, dung, wood, etc.

**structural gene** A gene that codes for the *amino-acid sequence of a protein. *Compare* REGULATORY GENE.

**structuring method** Any technique for sorting data to reveal important patterns. Classification and ordination methods are examples of structuring methods.

**strychnine** $C_{21}H_{22}N_2O_2$, an *alkaloid that is produced by *Strychnos nux-vomica*. In animals, it is a stimulant of the nervous system in general and in large doses causes convulsions. It can be lethal.

**Strychnos** (family *Loganiaceae) A genus of trees and woody climbers that have hook-like *tendrils. The fruit is a *berry, some fruits containing seeds that are exceedingly poisonous, owing to the presence of *alkaloids. *Curare is obtained from *S. toxifera* of tropical America, *strychnine from *S. nux-vomica* of tropical Asia. There are about 190 species occurring in the tropics.

**stub** A tree that is intermediate between a *stool and a *pollard.

**Sturt's desert pea** *See* CLIANTHUS.

**style** An extension of the *carpel which supports the *stigma.

**Stylidiaceae (trigger plants)** A family of *annual or *perennial *herbs and some shrubs, many of which are *xerophytes. The leaves are alternate, or often in a basal *rosette, *simple, often linear and almost grass-like, and without *stipules. The flowers are held on a *scape, in *racemes or *cymes. They are irregular and bisexual or unisexual, with 5 persistent *sepals, usually united to form 2 lips. The *petals are partially fused, forming a deeply lobed *corolla with a small lower lobe and 2 pairs of upright lobes. There are 2 *stamens, often fused with the *style to form a column which curves partially over the lower lip. When an insect lands on the lower lobe of the flower the column springs up and down, which assists the transfer of *pollen to the insect (this mechanism gives the plants their common name). The *ovary is *inferior, with 2 fused *carpels and 1 or 2 multi-*ovular *locules. The fruit is a *capsule containing small seeds. The family is divided into 2 subfamilies and 2 *tribes separated by their petal and stamen arrangement. A few species are cultivated as ornamentals. There are 5 genera, with about 170 species, found in tropical and temperate regions of Australia and New Zealand, and in restricted regions of Asia and S. America.

*Stylidium* **(trigger plants; family *Stylidiaceae, subfamily Stylidioideae, *tribe Stylideae)** A genus of *herbs and evergreen shrubs in which the flowers are held in *racemes and are irregular, with the 5 *sepals united into a 2-lipped, persistent *calyx. Several species have an irritable *gynostemium which, when released by touch, springs from side to side, assisting in the distribution of the *pollen. Several shrub species of Western Australia are cultivated. There are about 150 species, found in Australasia and south-eastern Asia.

*Stylites* *See* ISOETACEAE.

*Styphelia* **(family *Epacridaceae)** A genus of erect or spreading shrubs with *sessile or short-stemmed, often *lanceolate leaves which are rigid and pointed. The flowers are *axillary, usually solitary, with a cylindrical *corolla with narrow lobes which are hairy on the inside and recurved, exposing the *stamens. The *ovary is composed of 5 uniovular *locules. The *style is composed of many *filaments which are longer than the corolla. The fruit is a *drupe. There are about 130 species, occurring mainly in Australia but also scattered through Malesia and the Pacific islands, often in coastal and montane *heathlands.

**Styraceae** A tropical and warm-temperate family of trees and *shrubs with bark that yields resin used medically and in incense. The leaves are *simple, alternate, usually *entire, and without *stipules. Flowers are regular, usually bisexual, and usually borne in *panicles or *racemes, rarely solitary. The tubular *calyx has 4 or 5 persistent lobes. The *corolla is tubular at the base with 4–7 *valvate lobes; the number of *stamens is either equal to or double the number of corolla valves. The stamens are usually *adnate to the corolla tube or united as a tube. The *ovary is *superior or *inferior, of 3–5 fused *carpels with 3–5 *locules. The fruit is usually a *capsule, sometimes a *drupe, with a persistent calyx, containing one or a few seeds with copious *endosperm. The fruit is eaten by animals, which disperse the seeds. There are 12 genera with about 170 species, found in eastern Asia, western Malesia, south-eastern North America, Central and South America, and around the Mediterranean.

*Styrax* **(family *Styracaceae)** A genus of trees and shrubs, several of which yield a useful resin, *benzoin or gum benjamin, produced by wounding the wood. There are about 120 species, occurring in warm temperate and tropical regions.

**sub-alpine forest** A conifer-dominated forest which occurs mainly in the subalpine zone of temperate latitudes, but with a few extensions south of the Tropic of Cancer. The elevation of the zone increases with decrease in latitude. Generically this forest is closely related to the boreal forest, and indeed the two types have a minority of species in common.

**sub-Atlantic** A colder, wetter climatic phase which followed the more continental climate of *sub-Boreal times. The change from sub-Boreal to sub-Atlantic conditions in Britain began about 2850 BP and roughly coincided with the transition from Bronze to Iron Age cultures. The sub-Atlantic marks a period of renewed *peat growth on *bog surfaces that in late sub-Boreal times were sufficiently dry and humified (*see* HUMIFICATION) to support heath vegetation (e.g. *Calluna vulgaris*, ling or heather). This renewed peat growth gives a major *recurrence surface, the Grenz horizon, which at one time was considered to define the boundaries between Godwin pollen zones VIIb and VIII. *Radiocarbon dating has shown that there are many such recurrence horizons, however, and this use of Godwin pollen zones is now obsolete. *See* POLLEN ZONE.

**sub-Boreal** From Scandinavian evidence, a period during which the climate was cooler than during the preceding *Atlantic *climatic optimum, but not so cold and wet as during the *sub-Atlantic phase that followed. Nowadays the term sub-Boreal is used only loosely.

**subboscus** *See* BOSCUS.

**subclimax** Strictly, the penultimate stage in a *succession to a climatically controlled *climax community (as in *monoclimax theory). Typically, a subclimax community persists for a long time; for example, the forests of *Boreal times may be considered as subclimax to the early Atlantic forests of the post-glacial *climatic optimum.

**suberin** A fatty substance found in or on the surface of *cell walls in *cork and *endodermis (*see* CASPARIAN STRIP). It renders tissues waterproof and protects them from decay.

**suberization** The deposition of *suberin in *cell walls.

**sub-formation** In *phytosociology, a vegetation grouping used by the *Uppsala school of phytosociology, and denoting a geographically distinctive unit of a major formation.

**subfossil** *See* FOSSIL.

**subhymenium** In certain *fungi of the *Basidiomycotina or *Ascomycotina,

a layer of *tissue lying beneath the *hymenium.

**sublittoral zone 1.** In freshwater ecosystems, an alternative name for the *limnetic zone. **2.** (infralittoral zone, shelf zone, subtidal zone) The sea-shore zone lying immediately below the *littoral (intertidal) zone and extending to about 200 m depth or to the edge of the continental shelf. Red and brown *algae are characteristic of this area. Typical animals include sea anemones and corals on rocky shores, and shrimps, crabs, and flounders on sandy shores. It is approximately equivalent to the *circalittoral zone.

**submerged forest** *See* PETRIFIED FOREST.

**submergence marsh** The lower zone of a *salt marsh, from the mean high-water level of neap tides to the general level of mean high water. Typically, this zone experiences more than 360 submergences per annum, usually with more than 1 hour of submergence during daylight each day. Continuous exposure never exceeds 9 days. *Compare* EMERGENCE MARSH.

**sub-nodum** *See* NODUM.

**subsessile** Nearly *sessile.

**subshrub** A plant, smaller than a *shrub, which produces wood only at its base and has abundant growth branching upwards from the base, the upper stems dying back at the end of each growing season. *Compare* HERB and TREE.

**subsidiary cell (accessory cell)** One of the epidermal cells surrounding the *guard cells and forming part of the overall structure of a *stoma. *See also* ANOMOCYTIC.

**subspecies** Technically, a *race of a *species that is allocated a Latin name. The number of races recognized within a species and the allocation of names to them is something of an arbitrary procedure. Systematic and *phenotypic variations do occur within species, but there are no clear rules for identifying them as races or subspecies except that they must be (*a*) geographically distinct, (*b*) populations, not merely morphs, and (*c*) different to some degree from other geographic populations.

**substitutional load** In genetics, the cost in genetic deaths to the population of replacing one *allele by another (a *mutation) in the course of evolutionary change.

**substrate** **1.** (biochem.) The reactant acted upon by an *enzyme. **2.** (substratum) Any object or material upon which an organism grows or to which an organism is attached; an underlying layer or substance.

**substrate mycelium** A part of a *mycelium that remains in contact with the substrate.

**substratum** See SUBSTRATE.

**subtidal zone** See SUBLITTORAL ZONE (2).

**succession** The sequential change in vegetation either in response to an environmental change or induced by the intrinsic properties of the plants themselves. Classically, the term refers to the colonization of a new physical environment by a series of vegetation communities until a final equilibrium state, the *climax, is achieved. The presence of the colonizers, the pioneer plants, modifies the environment so that new species can join or replace the initial colonizers. Changes are rapid at first but slow to a more or less imperceptible rate at the climax stage. The characteristic sequence of developmental stages, i.e. nudation, migration, ecesis, competition, reaction, and stabilization, is termed a sere. During succession, the *ecosystem grows in *biomass, reaching the maximum biomass at climax. This means that during the course of succession the gross *primary production exceeds the total ecosystem respiration, the excess accumulating as biomass. When equilibrium is eventually attained, the ecosystem (community) respiration has risen to a point where it equals gross primary production.

**succinyl coenzyme A** An intermediate compound in the conversion of alpha-ketoglutaric acid to succinic acid during the *citric acid cycle, a process linked to the formation of the energy-rich guanosine triphosphate (GTP). Succinate is not always the final product, as the succinyl *coenzyme A may be employed for acylation reactions or to initiate *porphyrin synthesis.

**succubous (of *liverwort leaves)** See JUNGERMANNIALES.

**succulent** Fleshy.

**sucker** An underground shoot arising *adventitiously from the roots or lower stem of a tree or shrub and emerging from the soil to form a new plant, initially nourished by the parent plant. In cultivated species where grafting (see GRAFT) is practised (e.g. roses and fruit trees), production of suckers from the stock may seriously detract from the vigour of the grafted scion. The term may also be applied to the modified root of a parasite that enables it to extract nutrients from the host.

**sucrase (invertase)** An *enzyme that is responsible for the catalytic *hydrolysis of *sucrose to *fructose and *glucose.

**sucrose** A disaccharide, composed of *fructose and *glucose, which is a common storage and transport sugar in plants. It is known commercially as cane or beet sugar.

**Sudanese park-steppe floral region** Part of R. Good's (*The Geography of the Flowering Plants*, 1974) African subkingdom within his *palaeotropical kingdom, which has a not very rich flora in which species of *Acacia*, grasses, and palms tend to dominate the vegetation. See also FLORAL PROVINCE and FLORISTIC REGION.

**sudden oak death** A disease caused by the fungus-like organism *Phytophthora ramorum* that has resulted in extensive tree death in the United States and was reported in a few localities in Britain in 2003. *P. ramorum* infects a wide range of trees, including oaks, but is also found in some shrubs, such as *Rhododendron* and *Viburnum*.

**suffocation disease** A disease of rice in which reddish-brown patches spread from the tips of the leaves downwards; the roots of the plants blacken and rot. The condition is encouraged by poorly drained soils and appears to be associated with the production of toxic levels of hydrogen sulphide by *bacteria in the soil.

**suffruticose     chamaephyte**   *See*
CHAMAEPHYTE.

**sugar**   A member of a group of
water-soluble *carbohydrates that have
a low molecular weight and are composed
of one or more simple compounds
(monosaccharides).

**sugar apple**   *See* ANNONA.

**sugar beet** (*Beta*)   *See* CHENOPODI-
ACEAE.

**sugar cane**   *See* SACCHARUM.

**sugar kelp**   *See* SEA-BELT.

**sugar palm**   *See* ARENGA.

*Suillus* (order *Boletales)   A genus of
*boletes in which the *spore print is
olive, cinnamon, or brownish in colour.
In most species the surface of the cap is
glutinous. There are many species, found
on the ground in association with conifer
trees.

**sulcal**   *See* SULCATE.

**sulcate** (sulcal)   Marked with ridges,
grooves, or furrows.

**sulphur** (S)   An element that is
needed for plant life. It is found covalently
bound, especially in *proteins, where
it stabilizes their structures. It is also
involved in *oxidation and *reduction
reactions. Sulphur-deficient plants become chlorotic (*see* CHLOROSIS) and
etiolated (*see* ETIOLATION).

**sulphur fungus** (sulphur polypore)
The common name for the *fruit bodies of
*Laetiporus sulphureus*, which are *bracket-
like, typically growing in tiers. The upper
surface is irregular, suede-like, and
initially a bright sulphur-yellow or orange-
yellow, later fading to a dull whitish
colour. The tubes on the underside are sulphur-yellow with small, more or less circular pores. The *spore print is white. They
are found on various broad-leaved (living
or dead) trees, particularly oak.

**sulphur polypore**   *See* SULPHUR FUNGUS.

**sulphur-reducing organism** (domains
*Archaea and *Bacteria)   The *pheno-
type that utilizes sulphur-containing
compounds, rather than oxygen, in

its respiratory pathways, often releasing
hydrogen sulphide. Both *archaebacteria
and bacteria include sulphur-reducing
phenotypes.

**sulphur tuft**   The common name for
the dense clusters of *fruit bodies of
*Hypholoma fasciculare*, in which the cap is
convex, 2–7 cm across, and bright yellow,
becoming brownish towards the centre.
The stem, flesh, and *gills are yellow,
although the gills later become dark
brown. They are very common throughout
the year on tree stumps.

**sulphydryl** (thiol)   The radical group—SH.

**sumac**   *See* RHUS.

**sundew**   *See* DROSERACEAE.

**sunflower** (*Helianthus annuus*)   *See*
ASTERACEAE.

**superdominance**   *See* OVERDOMINANCE.

**supergene**   A segment of *chromosome that is protected from *crossing-over
and so is transmitted intact from generation to generation, like a *recon. The term
is used to refer to several closely linked
*gene *loci that affect a single *trait or a
series of inter-related traits, such as the
genes coding for the expression of pin
and thrum flower phenotypes in the
primrose and other species of the genus
*Primula*. Genes involved in the same biochemical function, e.g. those controlling
the synthesis of *tryptophan, are often
clustered together in close *linkage
within the *genome to form a supergene. The bacterial *operon is one such
example.

**superior**   Applied to an *ovary when
the other organs of the flower are inserted
below it. *Compare* INFERIOR.

**superspecies**   *See* SPECIES GROUP.

**supersuppressor**   *See* SUPPRESSOR MUT-
ATION.

**suppression**   *See* SUPPRESSOR MUTATION.

**suppressor mutation** (suppressor)   In
genetics, a second *mutation that masks
the *phenotypic effects of an earlier
mutation. This second mutation occurs at

a different site in the *genome (i.e. it is not a strict reversion). Intragenic suppression results from a second mutation that corrects the functioning of the mutant gene (e.g. a mutation of a different nucleotide in the same triplet, such that the *codon then encodes the original *amino acid). Intergenic suppression results from mutation of a different *gene, the product of which compensates for the dysfunction in the first (e.g. a mutation that produces a mutant *transfer-RNA molecule that inserts an amino acid in response to a nonsense codon, thus continuing a protein that would otherwise have been terminated). If a single suppressor mutation can suppress more than one existing mutation, it is said to be a supersuppressor. Compare REVERSE MUTATION.

**supralittoral zone**  The sea-shore zone immediately above the *littoral fringe and beyond the reach of tidal submergence, though affected by sea spray.

**'survival of the fittest'**  See NATURAL SELECTION.

**suspensor**  **1.** In seed plants and some vascular *cryptogams, a filamentous structure that differentiates from the *embryo, then elongates, pushing the embryo into the *endosperm or *gametophyte tissue, where it can obtain nourishment. **2.** In *Mucorales, a cell that supports the *gametangium.

**Svedberg unit**  **(S)** The unit of measurement in which sedimentation coefficients are expressed. It is equal to $10^{-13}$ seconds and is usually given for the solvent water at 20°C. It is named after the Swedish physical chemist Theodor Svedberg (1884–1971).

***Swainsona***  **(family *Fabaceae, subfamily *Papilionatae)** A genus of *herbs and small shrubs that are either *glabrous or with close-lying hairs. The leaves are alternate and narrowly *pinnate with many leaflets, often with *stipules, which are usually broad at the base. The flowers are in *axillary *racemes, and are usually violet, purple, or red. The *calyx tube has teeth of unequal length, and the *petals are held in a butterfly shape. The upper *stamen is

free; the remaining 9 are fused together. The *superior *ovary contains numerous *ovules. The pod is ovoid or oblong, and often turgid. There are about 50 species, all endemic (see ENDEMISM) to Australia and New Zealand.

**swamp**  A wet area, dominated by emergent aquatic vegetation, that is normally covered by water all year and is not subject to drying out during the summer. In European usage, the term is usually applied to herbaceous wetland ecosystems, such as reed beds, but in American usage 'swamp' is used only of forested wetlands. The American equivalent of the European swamp is a 'marsh'. Compare FEN.

**swamp cypress**  **1.** See ACTINOSTROBUS. **2.** **(Chinese swamp cypress)** See GLYPTOSTROBUS.

**swamp sepetir**  See SINDORA.

**swarm cell (swarmer)**  In *Myxomycetes, a flagellated, amoeba-like cell, lacking a *cell wall, produced when a *spore germinates. Swarm cells are usually produced singly, but spores sometimes release up to 4. They ingest food, reproduce by division, and mature as *gametes which fuse to form *zygotes. These develop into a *plasmodium.

**swarmer**  See SWARM CELL.

**sweet bay**  See LAURUS.

**sweet chestnut**  See CASTANEA.

**sweet flag**  See ACORUS.

**sweet gum**  See LIQUIDAMBAR.

**sweet marjoram**  See ORIGANUM.

**sweet potato**  See IPOMOEA.

**sweetsop**  See ANNONA.

***Swietenia***  **(mahogany; family *Meliaceae)** A genus containing the true mahoganies, which are deciduous trees with *pinnate leaves, and flowers that are functionally unisexual. The fruit is a *capsule. They yield valuable timber. There are 3 species, occurring from Mexico to Brazil, and in the Caribbean region.

***Swintonia***  **(family *Anacardiaceae)** A genus of trees that have *simple leaves and in which the *petals enlarge to form wings around the fruit, which is a *nut. There are 12 species, occurring from south-eastern Asia to western Malesia.

**sword sedge**   See LEPIDOSPERMA.

**sycamore   1. (UK)** See ACER. **2. (Near East)** See FICUS. **3. (N. America)** See PLATANUS.

**Sydney red gum**   See ANGOPHORA.

**symbiont**   A symbiotic organism. See also SYMBIOSIS.

**symbiosis**   A general term describing the situation in which dissimilar organisms live together in close association. As originally defined, the term embraces all types of mutualistic and parasitic relationships. In modern use it is often restricted to mutually beneficial species interactions, i.e. *mutualism. Compare COMMENSALISM and PARASITISM. See also CONJUNCTIVE SYMBIOSIS and DISJUNCTIVE SYMBIOSIS.

**Symingtonia** (family *Hamamelidaceae) A genus of trees that have trinerved leaves, and large, adpressed *stipules covering the bud. The flowers are held in heads and become woody in the fruit. There are many winged seeds. There are 2 species, occurring in the Himalayas, and from southern China to western Malesia.

**sympatric**   Applied to species that occupy similar *habitats or whose habitats invariably overlap. Compare ALLOPATRIC.

**sympatric evolution**   The development of new taxa from the ancestral *taxon, within the same geographic range. (It is geographically possible for interbreeding to occur between the potential new taxa, but for some reason this does not happen.) Because of the difficulty of envisaging what the reasons might be, few authorities until recently accepted the reality of sympatric evolution, except for certain special kinds of organisms; but recent studies have shown that *chromosomal *mutation can set up a partial barrier to interbreeding, sufficient to permit sympatric speciation. This requires the development of some form of reproductive isolating mecha-

nism which has arisen by selection within a geographically confined area; it may be structural, physiological, or genetic.

**sympatry**   The occurrence of species together in the same area. The differences between closely related species usually increase (*diverge) when they occur together, in a process called character displacement, which may be morphological or ecological. Compare ALLOPATRY.

**Sympieza**   See FYNBOS.

**symplast**   The continuum of *cytoplasm through a plant that results from the cytoplasm in different cells being connected by *plasmodesmata, forming an important pathway, e.g. for *auxins.

**symplesiomorph**   Applied to a character state, primitive (*plesiomorph) and shared between 2 or more taxa (see TAXON). Shared possession of a symplesiomorph character state is not evidence that the taxa in question are related.

**sympodial**   Applied to a type of branching in which an apparent main *axis is made up of many lateral branches, each arising from the one before; i.e., each is an extension growth from lateral axes, not from the original tip.

**synapomorphic**   Applied to *apomorphic features possessed by 2 or more taxa (see TAXON) in common. If the two groups share a character state that is not the primitive one, it is plausible that they are related in an evolutionary sense, and only synapomorph character states can be used as evidence that taxa are related. Phylogenetic trees are built up by discovering groups united by synapomorphies.

**synapsis**   The side-by-side pairing of *homologous chromosomes during the zygotene stage of *meiotic *prophase.

**syncarp**   A compound *fruit of 2 or more fused *carpels.

**Syncarpia** (Australian turpentine; family *Myrtaceae) A genus of large forest trees, which often reach massive size, up to 50 m in height. The bark is persistent and fibrous. The leaves are of 2 forms, like those of Eucalyptus. The juvenile leaf is elliptical, *pubescent, and short-stalked;

S

the adult leaves are *opposite, *peduncu-late, ovate or elliptical, often grouped to appear as though they are *whorled. They are glossy and green on the upper surface, and hairy or pubescent below. The flowers are held in globular heads on a thick stalk with several flowers in the head. The *sepals are short and rounded, the 4 or 5 *petals free. The *stamens are long, numerous, and free, often held in 2 rows. The single *style is slender. The *capsules are 3-celled, hard and woody, with a persistent *calyx, and contain many small seeds. The timber is valuable, being heavy, hard, and durable. It is used in construction work in Australia. The trees are sometimes planted for shade and as wind-breaks. There are 5 species, occur-ring in rain forests from the Moluccas to Queensland.

**syncarpous** With the *carpels *con-crescent.

***Synchytrium*** **(order** *Chytridiales**)** A genus of *fungi that are plant *parasites. *S. endobioticum* is an important *pathogen of potatoes, causing black wart disease.

**syndetocheilic** In some *gym-nosperms, applied to a type of *stoma in which the 2 *guard cells and the subsidiary cells are all derived from a single mother cell. *Compare* HAPLOCHEILIC. *See also* MESOGENOUS.

***Synechocystis*** A genus of *cyano-bacteria (section I) which tentatively includes the former phycological genera *Aphanocapsa*, *Eucapsis*, *Merismopedia*, and *Microcystis*.

**synecology** The study of whole plant and animal *communities, including the study of terrestrial ecosystems, biological aspects of oceanography, and applied prob-lems of human management and alter-ation of ecosystems. *Compare* AUTECOLOGY.

**synergid (pl. synergidae)** One of 2 *haploid cells that lie inside the *embryo sac, beside the *ovum. Synergidae nourish the *ovum.

**synergism** The result of combined factors, each of which influences a process

in the same direction but which, when combined, give a greater effect than they would acting separately.

**syngamy** The union of nuclei of 2 *gametes to produce a *zygote nucleus, following *fertilization.

**syngeneic** *See* ISOGENEIC.

**synonym** In taxonomy, a plant name that differs from the official name. Usually, it is an older name that does not conform to the rules governing priority in the application of names.

**synthetic theory** A modern theory of *evolution, incorporating Darwinian thinking, Mendelian genetics (*see* MENDEL'S LAWS), and an understanding of *genes and genetic change at the molecular level.

**syntype** All specimens in a type series in which no *type specimen was designated.

***Synura*** *See* CHRYSOPHYTA.

**synusia** A distinct layer in an area of vegetation that is composed of plants of a similar life-form, e.g. *Polytrichum* moss on the ground layer of an open oak wood. Different synusiae may appear in the same layer at various seasons of the year.

**synzoochory** *See* ZOOCHORY.

***Syringa*** **(lilac; family** *Oleaceae**)** A genus of shrubs and small trees with *simple, *oval, *opposite leaves, 4-lobed *corollas, *capsular fruits, and winged seeds. The *panicles of fragrant, lilac-coloured, white, or purple flowers make this a favourite garden shrub. The name 'syringa' is sometimes misleadingly applied to *Philadelphus* (mock orange), which is not related to lilac. There are 25 species, occurring from south-eastern Europe to eastern Asia.

**syringa** A name used misleadingly for *Philadelphus*, especially *P. coronarius*.

**systematic sample** *See* REGULAR SAMPLE.

***Syzygium*** *See* EUGENIA.

**Tacca** (family *Taccaceae) A genus of *rhizomatous herbs in which the leaves are radical, broad, and usually divided. The flowers are subtended by *bracts, the inner ones being long and thread-like. East Indian arrowroot comes from the tubers of *T. leontopetaloides*. There are 10 species, occurring in the tropics, especially in South-east Asia.

**Taccaceae** A family of tropical, *perennial, *rhizomatous herbs in which the large, simple or lobed leaves are radical and often on long *petioles. The regular, bisexual flowers are held in *umbels subtended by long and broad *bracts, which often form an *involucre below the *inflorescence. The *sepals and *petals are similar, and are held in 2 series of 3 segments fused to form a broad, bulbous tube to which the 6 *stamens are attached (*epipetalous). The *ovary is *inferior, of 3 fused *carpels forming a single *locule with numerous *ovules. The *style is short, with 3 *stigmas. The fruit is a *berry with many seeds, which have much *endosperm. There are 2 genera, 1 *monotypic, separated by their leaf and fruit characteristics. There are 31 species, found throughout tropical regions.

**tachytely** A rate of *evolution within a group that is much faster than the average, or *horotelic, rate. Such accelerated evolution typically occurs when an organism enters a new adaptive zone and initiates an *adaptive radiation to fill the available *niches.

**tactic movement** *See* TAXIS.

**Tahitian chestnut** *See* INOCARPUS.

**taiga** The name applied by many authorities to the whole of the *boreal forest, but by some only to the more open, park-like tracts along the northern fringe of the boreal forest, otherwise known as *lichen woodland.

**take-all** A disease of wheat, barley, and other grasses, caused by the *fungus *Gaeumannomyces graminis*. The *pathogen infects the roots of the plant, causing root rot and blackening of the stem base; the ears are bleached and empty (whiteheads).

**tal** *See* BORASSUS.

**Talauma** (family *Magnoliaceae) A genus of trees and shrubs, like *Magnolia* except that the fruiting *carpels form a woody or cartilaginous mass, separate from the *axis, and indehiscent (i.e. the upper portion of the carpels breaking away). There are 50 species, occurring in tropical and subtropical Asia and America.

**talipot palm** *See* CORYPHA.

**Tamaricaceae** A family of dicotyledonous (*see* DICOTYLEDON) trees and shrubs with minute, scale- or needle-like leaves, resembling those of conifers. The tiny, regular flowers, which are *tetra- or *pentamerous, are normally grouped in dense spikes or *racemes. The *stamens are usually free, and the *ovary is 1-celled and *superior, ripening to a *capsule. Several *Tamarix* species (tamarisks), particularly *T. gallica* are frequently planted for ornament or shelter, or are naturalized, on the coasts of north-western Europe and elsewhere, as they survive salt spray and strong winds very well. There are 5 genera and 87 species, occurring from south-western Europe to central Asia, and southern Africa.

**tamarind 1.** *See* TAMARINDUS. **2. (native tamarind)** *See* DIPLOGLOTTIS.

**Tamarindus** (tamarind; family *Fabaceae, subfamily *Caesalpinoideae) A genus of 1 species, *T. indica*, which is native to tropical Africa and western Asia, and is now widely cultivated. It is a big, deciduous tree. The leaves are *pinnate. The *pod is *dehiscent, with several seeds set in a pulp which is much valued as a foodstuff and enters world commerce as a basis for sauces.

**tamarisk** (*Tamarix*)  *See* TAMARICACEAE.

**Tamarix** (tamarisk)  *See* TAMARICACEAE.

**Tamus communis** (black bryony) *See* DIOSCOREACEAE.

**tangle**  The common name for the large brown seaweed *Laminaria digitata*. The *holdfast is much branched and root-like, the *stipe is thick, flexible, smooth, and cylindrical, 'and the blade is oar-shaped and divided into many strap-like ribbons. It is found at and below low-water mark.

**tannia**  *See* XANTHOSOMA.

**tannin**  A generic term for complex, non-nitrogenous compounds containing *phenols, *glycosides, or hydroxy acids, which occur widely in plants. They are toxic substances with astringent properties, whose principal function appears to be to render plant tissues unpalatable to herbivores. Tannins often build up during the individual development of leaves, rendering them less palatable to herbivores as they age.

**Tansley, Sir Arthur George (1871–1955)**  A British ecologist and conservationist who emphasized *ecology as an 'approach to botany through the direct study of plants in their natural conditions' (*Practical Plant Ecology*, 1923). He also pointed out the fact that since plants exist in communities the ecologist should be concerned with the structure of communities, or 'plant sociology'. This view became central to most British and American ecological theory. Tansley coined the term *ecosystem in 1935 (in 'The use and abuse of vegetational terms and concepts', *Ecology* 16: 284–307), although it may have been used earlier by his colleague Roy Clapham. Tansley's book *Types of British Vegetation* (1911) paved the way for vegetation description in Britain. Tansley was a lecturer at the University of Cambridge (1907–23), where much of his ecological work was done, and professor of botany at the University of Oxford (1927–37). He was instrumental in founding the British Ecological Society (1913) and was its first president. His many books include *The British Islands and their Vegetation* (1939) and *Britain's Green Mantle* (1949). *See also* CLEMENTS, FREDERIC EDWARD.

**tapetum**  **1.** A layer of cells, rich in food, which surrounds the *spore mother cells in *Tracheophyta (e.g. in the *pollen sacs of the *anther in a fern *sporangium). This layer may either break down to form the periplasmodium, a fluid that is absorbed by the developing *microspores, or it may last longer, breaking down shortly before anther *dehiscence and secreting substances into the *locule. **2.** A layer of nutritive cells within the *sporangium of a fern.

**Taphrinales** (subdivision *Ascomycetales) An order of *fungi that are *parasitic on a range of flowering plants. *Ascocarps are not formed, the *asci developing from *hyphae or from cells produced by hyphal fragmentation. Genera include *Taphrina*; *T. deformans* is responsible for *peach leaf curl and other diseases involving leaf distortion. *See also* WITCHES' BROOM.

**tapioca**  *See* MANIHOT.

**tap root**  A large, descending, central root.

**tapu cloth**  *See* BROUSSONETIA.

**Taraxacum** (family *Asteraceae) A genus of herbs, many of which contain latex. *T. officinale* (dandelion) is a cosmopolitan weed. Some species are cultivated as ornamentals or salad plants, and in Russia rubber is obtained from the latex of *T. bicorne*. There are 60 species but they are *apomictic, so there are hundreds of microspecies. They occur in northern temperate regions and temperate S. America.

**tare**  *See* VICIA.

**taro**  *See* COLOCASIA; *compare* GIANT TARO and GIANT SWAMP TARO.

**tar spot**  A disease of sycamore (*Acer pseudoplatanus*) in which black spots appear on the leaves as a result of infection by *Rhytisma acerinum*. The name may also refer to the *fungus itself.

**taurgya**  A Burmese word that is widely used to describe the practice, in many tropical countries, of establishing plantations by planting and tending tree seedlings with food crops. Food cropping is ended after 1–2 years as the trees grow up.

**tautonym** A taxonomic name in which the names for genus and species are the same.

**Taxaceae** A family of mostly *dioecious conifers with the *ovules solitary or in pairs and not borne in cones, but surrounded when ripe by a fleshy sheath or *aril, which is often coloured and resembles a *berry. This is always open at the tip, exposing the *pollination pore, or *micropyle, of the *ovule. The male flowers are borne in tiny globular cones; the *pollen grains are unwinged; and no resin cells or canals occur in the wood. There are 6 genera and 18 species, including *Taxus* (yew), in the northern temperate zone and in New Caledonia.

**Taxales** An order of *gymnosperms, and the youngest in the subdivision *Coniferopsida, dating from late *Cretaceous times.

**taxis (tactic movement)** A change in direction of locomotion in a *motile micro-organism or cell, made in response to certain types of external stimulus, e.g. the presence of particular chemicals (chemotaxis), changes in light intensity (phototaxis), or changes in temperature (thermotaxis), etc.

**Taxodiaceae** A family of conifers (or, according to some authors, a *tribe of the family *Pinaceae). The leaves are spirally arranged (not in opposite pairs as in *Cupressaceae) and the leaves are evergreen, hard, and scale-, spine-, or awl-like in most genera, though in a few (e.g. *Metasequoia* and *Taxodium*) they are soft and deciduous. The cones are globular and hard, with the *bracts completely fused to the *ovule-bearing scales (as in Cupressaceae but unlike Pinaceae). The family consists of 10 genera and only 13 species. It includes many ancient *relict *monotypic genera, like *Sequoiadendron*, *Sequoia*, *Metasequoia*, and *Cryptomeria*, all of which have an extremely limited distribution as natives today though formerly they were mostly very widespread (*epibiontics). They are widely cultivated in many temperate regions;

for example, *Sequoiadendron giganteum* (wellingtonia or big tree), once widespread in the world in *Pliocene times, is now confined as a native to the Sierra Nevada of California.

**Taxodium** (swamp cypress; family *Taxodiaceae) A genus of conifers with alternate, strap-shaped deciduous leaves, and remarkable 'knees' on its roots in wet soil, which come to the surface like the *pneumatophores of mangroves, and evidently help to provide oxygen for the root system. There are 2 species, *T. distichum*, confined to the USA, and *T. mucronatum* in Mexico.

**taxon (pl. taxa)** A group of organisms of any taxonomic rank, e.g. family, genus, or species.

**taxonomy** The scientific classification of organisms.

**Taxus** (yew; family *Taxaceae) A genus of conifers with erect, solitary *ovules surrounded by fleshy cup-like red *arils with a sweet taste. The whole plants are highly poisonous (containing the *alkaloid taxine) except for the red arils of the seeds. *T. baccata* is native in western Europe and is notable for its longevity; some specimens planted in churchyards in Britain and in France are believed to be about 1000 years old. It is the only genus of the family in Europe. There are 7 species, occurring mainly in the northern temperate zone but also in Mexican and Malesian mountains.

**tea** See CAMELLIA and THEACEAE.

**teak** See TECTONA and VERBENACEAE.

**tea-tree** See LEPTOSPERMUM.

**Tecoma** (family *Bignoniaceae) A genus of woody climbers that have *pinnate leaves and colourful trumpet-shaped flowers. Several are grown as ornamentals. There are 12 species, occurring from Florida to Argentina.

**Tectona** (family *Verbenaceae) A genus of deciduous trees that have big, *opposite leaves. The flowers are small and held in big terminal *panicles. The fruit is surrounded by the inflated *calyx. *T. grandis* (teak) is the commonest and most wide-ranging species, much cultivated for its durable, attractive timber. There are 3 species, occurring in seasonal south-eastern Asia.

**tegulicolous** Growing on tiles, e.g. on a roof.

**teleutosorus** *See* TELIUM.

**teleutospore** *See* TELIOSPORE.

**Teliomycetes (subdivision *Basidiomycotina)** A class of *fungi that do not form *basidiocarps. The class is no longer recognized in most taxonomic schemes.

**teliospore (teleutospore)** A resting *spore formed by *fungi of the classes Urediniomycetes (*rust fungi) and Ustilaginomycetes, which includes many parasitic species.

**telium (teleutosorus)** A region of *spore-bearing fungal *tissue in a host plant infected with a *rust fungus. Telia vary in structure among the various species of rust. The spores produced in a telium are called *teliospores.

**telocentric chromosome** A *chromosome that has the *centromere at one end.

***Telopea* (family *Proteaceae)** A genus of tall shrubs, with alternate, often leathery leaves which are entire or toothed, with or without *stipules. The flowers are irregular, bisexual, and held in pairs. They are showy, red, and composed of 4 *perianth lobes with reflexed tips. Two to 4 scales, like *petals, alternate with the lobes. The 4 *stamens are attached to the lobes. The *ovary is *superior, with a single *carpel and 4 *ovules in the single *locule. The fruit is a hard, *sessile *follicle with single-winged seeds. There are 4 species, endemic (*see* ENDEMISM) to eastern Australia and Tasmania.

**telophase** The fourth and final phase of *mitosis and the 2 divisions of *meiosis, during which the *spindle disappears, *nucleoli reappear, the *nuclear membranes start to develop around the 2 groups of *daughter chromosomes/chromatids, and the *chromosomes return to their extended state, in which they are no longer visible. The *nuclei then enter a resting stage as they were before division occurred. In many higher plants, telophase of the first meiotic division is omitted.

**Teloschistales (subdivision *Ascomycotina)** An order of lichenized and *lichenicolous *fungi; in lichenized members the *thallus may be *crustose, *foliose, or *fruticose, and is typically yellow, orange, or red. There are several genera, found mostly in nitrogen-rich habitats, particularly in warm, dry regions.

**temperate deciduous forest** *Deciduous summer forest dominated by broad-leaved hardwoods, which occurs over large tracts in the mid-latitudes of Europe, N. America, and eastern Asia, but which is restricted in the southern hemisphere to Chilean Patagonia. Much of the original forest has been cleared for cultivation and pasture.

**temperate grassland** A type of vegetation which includes in the northern hemisphere the N. American *prairies and the Eurasian *steppe, and in the southern hemisphere the *veld of S. Africa, the *pampas of Argentina, the Canterbury Plains of New Zealand, as well as extensive areas in Australia. Much of this grassland has been ploughed up or converted into pasture for domesticated animals. Although most of this grassland occurs in parts of the temperate zone where there is a seasonal deficit of moisture, there is often enough to permit tree growth. Presumably fires, both natural and anthropogenic, have been important in the development of this grassland.

**temperate phage** *See* PLAQUE and VIRULENT PHAGE.

**temperate rain forest** Forest that develops in temperate regions where rainfall is high (typically 1500–3000 mm) or fog is frequent (e.g. on the Pacific coast of N. America). Broadleaved evergreen trees are common, conifers are often present, and there are many *epiphytes and climbers. Such forests occur in coastal areas of the south-eastern United States and north-western N. America, in southern Chile, in parts of Australia and New Zealand, and in southern Japan and China.

**temple tree** (frangipani) *See* PLUMERIA.

**temporary wilting** *Wilting that occurs in hot weather when the rate of *transpiration exceeds the rate at which a plant can absorb moisture from the soil. The plant recovers when the temperature falls. *Compare* PERMANENT WILTING POINT.

**tendril** Part of a stem, leaf, or *petiole that is modified as a delicate, commonly twisted, thread-like appendage. It is an aid to climbing, as in *Cucurbitaceae.

**tendrillate** With *tendrils.

**tension wood** *See* REACTION WOOD.

**tepal** One of the *perianth members in those flowers where there is no distinction between *calyx and *corolla.

*Tephrosia* (family *Fabaceae, subfamily *Papilionatae) A genus of shrubs and herbs, which have *pinnate leaves, usually with several leaflets but sometimes with very few. The leaves are often silky below. The *inflorescence is a terminal *raceme held opposite the leaf, or an *axillary raceme. The irregular flowers are red, purple, or white, with the *calyx teeth about equal in length. The *petals are clawed, with the upright *petal hairy on its outer surface, and the wing petals held obliquely, often slightly adherent to the incurved keel. The upper *stamen is usually free, the remaining 9 being fused in a tube. The *ovary contains many *ovules. There are more than 400 species, found in tropical and subtropical regions of the world, but especially in Africa.

**terentang** The pale, attractively figured timber of *Campnosperma*.

**terete** Circular in cross-section, used, for example, of a plant stem.

*Terminalia* (family *Combretaceae) A genus of large, often deciduous or semideciduous trees, which have distinctive, tiered branching, and the leaves held in terminal clusters. The flowers are held in *panicles or *spikes, and are apetalous, with 10 *stamens, and a *disc present. The fruits are woody, and sometimes winged. The trees yield useful timber, and some are used for dyeing and tanning (*see* MYROBALAN). A few have edible fruits.

There are about 150 species, occurring throughout the tropics.

**ternary fission** A type of cell division that results in the formation of 3 daughter cells from a single parent cell.

**ternate** Compound, and divided into 3 more or less equal parts.

**Ternstroemiaceae** An old name for the family *Theaceae.

**terpene** A hydrocarbon that is composed of two or more *isoprene units. Terpenes may be linear or cyclic molecules or combinations of both, and include important biological compounds such as vitamins A, E, and K, *carotenoids, *phytol, gibberellic acid, natural rubber, and some *lipids.

**terra rossa** European soils, red in colour, that developed over limestone. They are deep and ancient, some of them being pre- *Pleistocene. They are now classified as *Inceptisols or *Mollisols.

**terricolous** Growing on soil or on the ground.

**Tertiary** The first sub-era of the *Cenozoic Era, which began about 65.5 Ma ago and ended 1.806 Ma ago. The Tertiary followed the *Mesozoic and comprises two periods, *Palaeogene and *Neogene. During the Tertiary *angiosperms superseded the *gymnosperms as the dominant plants. Although the name Tertiary is widely used, in years to come it is likely to become obsolete in formal use.

**testa** The seed coat. *See* INTEGUMENT.

**Tethys Sea** (Neotethys) The sea that more or less separated the two great *Mesozoic supercontinents of *Laurasia (in the north) and *Gondwana (in the south). That land bridges between the two supercontinents existed for much of the Mesozoic is attested to by the cosmopolitan character of dinosaur faunas and of some plants and by the existence of 2 distinct floral assemblages in Laurasia and Gondwana.

**tetrad** **1.** Four homologous *chromatids in a bundle during the first meiotic prophase and metaphase. **2.** The

4 haploid cells resulting from a single *meiosis.

**tetrad analysis** The use of tetrads (4 haploid cells resulting from a single *meiosis) to study the behaviour of *chromosomes and *genes in *crossing-over during meiosis. Such analyses require organisms in which the products of meiosis are held together and so can be counted as units, as is true of the *meiospores contained in the ascus sac of various fungi such as *Aspergillus, Neurospora*, and *Saccharomyces*.

**Tetragonia** (family *Aizoaceae) A genus of *annual and *perennial, sometimes scrambling *herbs or low shrubs in which the leaves are *simple, alternate and *opposite, and often succulent. The flowers are regular, bisexual, and usually yellow or green in colour. The 5–8 *sepals are fused and often attached to the *ovary. There are no *petals, but sometimes the numerous *stamens are *petal-like. The ovary is *superior, *inferior, or *half-inferior. The fruit is a juicy or leathery *capsule enclosed in the hardened *calyx tube. There are 50–60 species, all restricted to the southern hemisphere.

**tetrameric** *See* DIMER.

**Tetramerista** (family *Tetrameristaceae) A *monotypic genus, *T. glabra*, consisting of a rain-forest tree, occurring in Sumatra, Malaya, and Borneo. It yields a useful timber, punah.

**Tetrameristaceae** A small family of 2 genera of plants, each with 1 species, occurring in Guyana and western Malesia, and related to the *Theaceae.

**tetramerous** Having parts in fours.

**Tetraphidales** (subclass *Bryidae) An order of small, *acrocarpous mosses; the leaves are usually crowded at the tops of the sterile stems, forming cup-shaped 'gemma cups' within which discoid *gemmae are formed. The *capsule is unique in having a *peristome with only 4 teeth. They are found on *peaty banks, rotten logs, etc.

**tetraploidy** *See* POLYPLOIDY.

**tetrapyrroles** A group of pigments that are present in most living organisms.

Most of those found in plants are light-sensitive. (Those in animals serve a variety of functions.) The tetrapyrrole molecule consists of 4 joined *pyrrole rings, a pyrrole ring consisting of 4 CH units and 1 NH unit. Tetrapyrroles may be linear (i.e. the pyrrole rings are joined side by side) or cyclic.

**tetrarch** *Primary xylem that has 4 strands.

**tetrasomy** The condition in which a particular *chromosome is represented by 4 members (i.e. the tetraploid condition). *Compare* DISOMY, NULLISOMY, POLYSOMY, and TRISOMY.

**Tetraspora** (division *Chlorophyta) A genus of microscopic *green algae which appear to be closely related to *Chlamydomonas. The vegetative cells are non-*motile and typically occur in gelatinous colonies (e.g. *Tetraspora gelatinosa* forms amorphous green masses which may be visible to the naked eye). They are found in still freshwater habitats.

**tetrasporangium** A *sporangium in which 4 *spores (tetraspores) form.

**tetraspore** *See* TETRASPORANGIUM.

**tetro-allelic** Applied to a *polyploid in which 4 different *alleles exist at a given *gene *locus.

**thalline margin** *See* LECANORINE.

**thallophyte** A plant that is not differentiated into root, stem, and leaves; the plant body is known as a *thallus. In some classifications, thallophytes are ranked as a subkingdom, Thallophyta. *Compare* EMBRYOPHYTE.

**thallose** Flattened, and showing no differentiation into stem and leaves. *See* HEPATICAE.

**thallospore** A fungal *spore that develops from part of a fungal *hypha.

**thallus** A primitive type of vegetative plant body that is not differentiated into stems, leaves, and roots, although *analogous structures may be present. The term is used mainly of non-vascular plants, e.g. *algae, *fungi, *lichens, and *liverworts. *See also* THALLOPHYTE.

**thalweg** A line joining the lowest points of successive cross-sections, either along a river channel or, more generally, along the valley that it occupies.

**Thamnidiaceae** (order *Mucorales) A family of *fungi in which *sporangiola are formed. In many genera columellate *sporangia are also formed. Species are found on dung, soil, decaying vegetation, etc.

**Theaceae** A family of tropical and subtropical trees and shrubs, and some scrambling climbers, in which the leaves are alternate, often evergreen, and leathery, without *stipules. The flowers are regular, often solitary, showy, and usually bisexual. There are 4–7 *sepals and *petals and numerous *stamens which are either free, in bundles, or in a tube. The *ovary is generally *superior, with 3–5 fused *carpels with separate *locules and free *styles. The fruit is a *capsule, *berry, or *achene, with a persistent *calyx. The seeds have little *endosperm. The family is often split into 8 or more smaller families, separated by the characteristics of the reproductive system. Many species are ornamentals, including the well-known *Camellia japonica*, but the best-known and economically most valuable member of the family is *C. sinensis*, the tea plant. This species has been cultivated for many centuries and contains caffeine, polyphenols, and essential oils. There are 28 genera, with about 520 species, most of which are restricted to tropical and subtropical regions, centred in Asia and America.

**theca** The shell-like structure surrounding the cell in certain *algae.

**Thelephora** (family *Thelephoraceae) A genus of *fungi in which the *fruit bodies are variously shaped, according to species; they may be more or less branched, are typically leathery in texture, and bear a well-developed, wrinkled or smooth *hymenium. *T. terrestris* (earth fan) forms fan-shaped *fruit bodies 3–8 cm across, erect or horizontal, reddish to brown, often in clusters. They are common in conifer woods and heaths.

**Thelephoraceae** (order *Aphyllophorales) A family of *fungi in which the flesh of the *fruit body usually darkens or turns greenish when treated with potassium hydroxide. Typical spores are ornamented. Most species are found on the ground, but some grow on wood.

**Theligonaceae** (order *Haloragales) A small family of *annual and *perennial dicotyledonous (*see* DICOTYLEDON) *herbs in which the leaves are *simple and fleshy, the lower ones *opposite the upper ones alternate, with fused pairs of membranaceous *stipules. The flowers are in unisexual clusters of 1–3, the males globose, the females tubular. The *ovary is a single *carpel, developing into a nut-like *drupe with a fleshy *endosperm, which is dispersed by ants. The family comprises 1 genus, with 3 species, occurring from the Mediterranean region to China, and Japan.

**Thelypteridaceae** A family of *leptosporangiate ferns with the mixed type of *sorus. The sori are superficial, small, near the *pinnule margins, and have either no *indusium or a kidney-shaped one that is soon *evanescent. The leaves are normally *bipinnate and hairy; scales on the *frond stalk or stipe are few. *Thelypteris thelypteroides* (marsh fern) is the most familiar European and temperate N. American species; it has creeping *rhizomes and grows in *fen peat. The family is now regarded as containing some 30 genera and 900 species, mostly tropical.

**thelytoky** Obligatory *parthenogenesis, such that populations consist entirely of females, with occasional functionless males. It is the only genetic system in which *fertilization (the union of egg and sperm) is eliminated completely. Evolution in thelytokous species can therefore take place only by favourable *mutations occurring in a single individual, and persisting in the line descending from that individual.

**Themeda** (kangaroo grasses; family *Poaceae) A genus of coarse, tussock-forming, *perennial grasses whose *spikelets are borne on *racemes in pairs, usually one stalked and sterile, the other *sessile and fertile. The racemes are single or paired, or held in a *panicle with many branches. There are usually subtending leaves with the *inflorescence. There are 19 species, found in southern and eastern Africa, Asia, and Australia. The genus includes *T. triandra*, the dominant grass of the savannah areas of tropical Africa, and *T. australis*, which is important in Australia.

**Theobroma** (family *Sterculiaceae) A genus of understorey trees, all of restricted natural range, in which the flowers are *cauliflorous and *ramiflorous, and *pentamerous. The fruit is a large *drupe, the seeds numerous, and surrounded by pulp derived from the *integuments. Cocoa comes from *T. cacao*. There are 20 species, occurring in the rain forests of western S. America.

**theory of differentiation** See DIFFERENTIATION, THEORY OF.

**theory of generic cycles** See GENERIC CYCLES, THEORY OF.

**thermic** See PERGELIC.

**thermocline** Generally, a gradient of temperature change, but applied more particularly to the zone of rapid temperature change between the warm surface waters (*epilimnion) and cooler deep waters (*hypolimnion) in a thermally stratified lake in summer.

**thermoduric** Applied to an organism (usually a micro-organism) that can tolerate relatively high temperatures; often used specifically for micro-organisms that survive the process of pasteurization.

**thermophile** An *extremophile (domain *Archaea) that thrives in environments where the temperature is high, typically up to 60°C. Compare HYPERTHERMOPHILE.

**thermophilous species** A warmth-loving species. In pollen analysis (*see* PALYNOLOGY) the term refers in particular to a species, genus, or family characteristic of warmer environments than those otherwise indicated by the pollen record.

**therophyte** One of *Raunkiaer's life-form categories, of plants that complete their life cycle rapidly during periods when conditions are favourable and survive unfavourable conditions (e.g. heat or competition) as seeds;

they are thus annual or ephemeral plants. Therophytes are very typical of desert environments and cultivated land. *Compare* CHAMAEPHYTE; CRYPTOPHYTE; HEMICRYPTOPHYTE; and PHANEROPHYTE.

**thiamine** Vitamin B$_1$. It contributes to the formation of the important *coenzyme thiamine pyrophosphate which is involved in the oxidative decarboxylation of alpha-keto acids and transketolase reactions.

**Thiessen polygons** A diagram used in studies of plant competition in which the position of each plant is plotted and joined by a straight line to each of its neighbours. The joining lines are then bisected by lines drawn at right angles to them, forming polygons. The size of polygons indicates the density of plants.

**thigmotaxis (stereotaxis)** A change in direction of locomotion in a *motile organism or cell made in response to a tactile stimulus (touch), where the direction of movement is determined by the direction from which the stimulus is received. It may inhibit movement, causing the organism to come into close contact with a surface. Thigmotaxis is commonly observed in insects.

**thigmotropism (haptotropism)** The tropic response of a plant organ to the stimulus of touch.

**thin** In a cultivated crop, to remove some plants in order to increase the area available to others. In forestry, to remove some of the trees in an area to enable those that remain to grow larger, especially in girth.

**Thiobacillus** A genus of *bacteria not assigned to any taxonomic family. The cells are *Gram-negative, rod-shaped, and often motile with a single *flagellum at one end. Energy is obtained by the oxidation of sulphur or sulphur-containing compounds, e.g. sulphide and thiosulphate. Most species require the presence of air for growth. There are several species, found in soil, marine and freshwater environments, acidic mine waters, hot sulphur springs, etc.

**thiol** See SULPHYDRYL.

***Thismea***   *See* BURMANNIACEAE.

**thongweed**   *See* SEA-THONG.

**thorn**   A woody, projecting structure with a sharp point that is derived from the leaf, stem, or branch of the plant and is connected to its vascular system.

**thorn-apple**   *See* DATURA.

**thorn forest (thorn scrub, thorn woodland)** A tropical type of vegetation with thorny shrubs and bushy trees, perhaps with a few taller trees, set in a sparse ground flora in which grasses are often lacking. It is in this last respect that thorn forest differs principally from *savannah woodland, the paucity of grasses reflecting the increased aridity of the climate. Thorn forest merges in even drier regions into semi-desert scrub.

**thorn scrub**   *See* THORN FOREST.

**thorn woodland**   *See* THORN FOREST.

**threonine**   An aliphatic, polar (*see* POLAR MOLECULE) *alpha-amino acid.

**throughflow**   *See* INTERFLOW.

**Thuidiales (subclass *Bryidae)** An order of *pleurocarpous mosses which range in size from minute to large. The primary stems are creeping, with secondary stems prostrate to ascending, irregularly branched to regularly 1- to 3-*pinnate. The leaves are typically ovoid, nerveless or with a short double nerve or a single nerve ending below the leaf tip. There are several genera, found in a wide range of habitats. *Thuidium tamariscinum* is a very attractive moss, bright green in colour, and intricately, tripinnately branched to give a *fern-like appearance. It is a common and conspicuous moss in woods and on shady banks.

***Thuja* (family *Cupressaceae)** A genus of evergreen, coniferous trees in which the leaves are small, and closely adpressed to the flattened twigs. The cones have 3 or 4 pairs of scales, only the middle ones being fertile and bearing paired seeds. *T. plicata* is western red cedar, with weather-resistant timber that is used for shingles, greenhouses, etc. All members of the genus are known as arbor vitae. There are 5 species, occurring in eastern Asia and N. America.

**thylakoid**   One of the membranaceous discs or sacs that form the principal sub-units of a *granum in *chloroplasts.

**Thymelaeaceae**   A family of shrubs, with a few trees and climbers in which the bark is fibrous. The wood often has included *phloem. The leaves are *simple and sometimes *opposite. The flowers are hermaphrodite and regular. The *calyx is tubular, commonly coloured, and lobed; *petals are often absent or *episepalous; the *stamens are episepalous; the *ovary is *superior. The seeds have a *caruncle or *aril. There are 50 genera, with about 720 species, occurring in the tropics, especially in Africa, and in temperate regions.

**thymine**   The *pyrimidine base that occurs in *DNA.

**thyrse**   Applied to an *inflorescence that is densely branched, comprising a *racemose central axis with *cymose lateral branches.

**Thyrsopteridaceae**   A family of tree-ferns of ancient origin, included by some authors with other tree-ferns in the Cyatheaceae. They have marginal *gradate sori, which in *Thyrsopteris* are produced on specialized fertile *pinnae, and they also have cup-like *indusia. In *Cibotum*, the sori are on the margins of ordinary pinnae and have a 2-lipped *indusium. There are 3 genera. *Thyrsopteris*, with only one species, *T. elegans*, is endemic (*see* ENDEMISM) to Juan Fernandez and seems to be a very ancient relict. *Cibotum* has 12 species and *Culcita* 7 species. Both are widespread in tropical and subtropical forests.

**tidal flat**   An area of intertidal sand flat, mud flat, and marsh developed in some lagoons in *mesotidal areas, and in protected bays and estuarine areas along *macrotidal coasts. Extensive tidal flats occur in the Wash (eastern England), Waddenzee (Netherlands), and along the German coast of the North Sea. Tidal flats also occur in warmer climates, as in the Persian Gulf, where carbonate and evaporite deposits develop. In tropical areas tidal flats tend to be colonized by mangrove swamps.

**tidal inlet** A narrow channel that connects the open sea with a lagoon. Tidal inlets often occur in barrier island systems and are typified by small-scale deltas at each end of the inlets, resulting from the high-velocity, tidal currents that flow through the channels.

**tidal range** The difference in height between consecutive high and low waters. The tidal range varies from a maximum during spring tides to a minimum during neap tides. In tide tables daily high- and low-water heights are given for each geographical locality mentioned.

***Tieghemopanax*** (family *Araliaceae) A genus of trees and shrubs with alternate, *pinnate leaves which are almost or completely *sessile, and often toothed. The flowers are regular, small, and unisexual, held in *umbels. The *calyx is very small, often without distinct teeth. There are 5 *petals, occasionally only 4, and the same number of *stamens. The *ovary is of 5 fused *carpels. The fruit is a *drupe with 5 seeds. There are 35 species, found only in Australia and Vanuatu.

***Tilia*** (lime, linden, basswood family *Tiliaceae) A genus of trees with large, blunt buds, and alternate, usually heart-shaped, leaves on long stalks. The flowers are in small stalked *cymes with a large oblong *bracteole fused to the *inflorescence stalk for some distance. Flowers are yellow or white, fragrant, and insect-pollinated (unlike most northern temperate forest trees); and have 5 free *sepals and *petals, many *stamens, and a 5-celled *ovary which forms a globular or ovoid 1-celled, 1–3-seeded *nut. The nectar is eagerly sought by bees, who produce an excellent honey where it is plentiful. The timber was formerly much used for wood-carving; Grinling Gibbons (1648–1721) often used it. The young leafy shoots were once used to feed stalled cattle where herbage was scarce, and a kind of bread was made from the starch-rich *phloem tissues. There are some 45 species in the northern temperate zone. *Tilia* species, especially *T. cordata*, were dominant in pre-Neolithic times in many English and European forests, but they have become much scarcer as a result of human activities, though remaining locally important in less modified forests in Eurasia and N. America. The *hybrid *T. × europea* (a natural hybrid between *T. cordata* and *T. platyphyllos*) is much planted in parks and avenues in Europe.

**Tiliaceae** A family of dicotyledonous (*see* DICOTYLEDON) trees, shrubs, and a few herbs. The leaves are usually alternate, and the inflorescences *cymose. Flowers are regular, bisexual, with *superior *ovaries; the (usually) 5 *sepals are valvate in bud and there are normally 5 *petals and many free *stamens. The fruit is a *capsule, a *drupe, or a *nut. A number of *Tilia* species are used for timber; *Corchorus capsularis* is the source of *jute; and other species are grown for ornament. There are about 48 genera, with 725 species, occurring mainly in the tropics, but with some temperate species.

**tiller** In grasses, a lateral shoot arising at ground level.

***Tilletia*** (family *Tilletiaceae) A genus of *smut *fungi which form unicellular *teliospores and which typically attack the *ovaries of a host plant (a member of the *Poaceae). There are many species. *See also* BUNT.

**Tilletiaceae** (order *Ustilaginales) A family of *smut *fungi in which *basidiospores are formed apically on *septate or non-septate *basidia. All members are *parasitic on plants, some causing important diseases of cereals.

**tilth** The physical condition of the soil that determines its suitability for plant cultivation.

**timber 1.** Wood in the form of unsquared logs. **2.** Tree trunks that are suitable for beams or for sawing into planks.

**timber-line** (waldgrenze) A line that marks the altitudinal limit of trees that are in a close canopy and that grow erect and tall. It occurs below the *tree line, and below the *kampfzone (in which the trees often show the *krummholz condition).

**time series** A data set in which the intervals are of equal time and arranged in order of occurrence. The series may be for individual or averaged values which can be analysed by statistical techniques, including spectrum or harmonic analyses.

**timothy-grass bacillus** *See* MYCOBACTERIUM.

**tinder fungus (hoof fungus)** The common name for the *fruit bodies of *Fomes fomentarius*. These are hoof-shaped; the upper surface is pale brown or greyish-brown with concentric furrows, while the tubes on the underside are pale becoming brown. The *spore print is white. In Britain, tinder fungus is found usually on birch trees and occurs mostly in Scotland and northern England; in northern Europe and the USA it is common on beech and other trees.

**tissue** A group of cells of similar type working in a co-ordinated manner towards a common function. In plants, they are normally bound together by their *cell walls. Some fluids are also considered to be tissues.

**tissue culture** Individual cells of an organism, that are grown in a sterile medium containing the nutrients the cells require. The technique can yield information about the nutritional requirements of individual cells, and whole plants can often be grown from tissue samples or even from single cells. In *gene banks, tissue culture methods may be used to store plants, in a non-bulky growth phase, if their seeds do not remain viable for long periods of time. It is also possible to save crop plant varieties from extinction through an attack by *pathogens by isolating and culturing meristematic tissue (*see* MERISTEM), which is rarely attacked. Crosses can also be made between otherwise incompatible plants by removing *embryos before they are aborted and growing them by this method.

***Tmesipteris*** (family *Psilotaceae) A genus of primitive *cryptogamic vascular plants which (unlike *Psilotum*, the other genus of the family) have strap-shaped, pointed, one-veined leaves, and *sporangia fused in pairs on to a pair of *bracts. Like *Psilotum*, it has no roots, only

*rhizoids on its *rhizomes. The genus is of great interest for its probable relationship to the most ancient vascular plants known, the *Psilophytales of *Devonian times, such as *Rhynia* and *Hornea*. There are probably only 2 species, confined to Australasia and some Pacific islands.

**toadstool** A loose term for any umbrella-shaped fungal *fruit body, or for any such *fruit body that is inedible or poisonous. *Compare* MUSHROOM.

**tobacco** *See* NICOTIANA.

**tobacco mosaic** A disease of tobacco, tomato, and related plants, in which the leaves become distorted and are marked with a characteristic mosaic of light and dark green. It is caused by an *RNA-containing virus, the tobacco mosaic virus.

**tobacco necrosis virus** An *RNA-containing virus which can infect a wide range of plants; symptoms of infection range from small leaf lesions to *necrosis of the entire plant. It can be transmitted by the soil fungus *Olpidium brassicae*.

**tobacco rattle** A disease that occurs in *tobacco, tomato, *potato, and other plants. Symptoms vary with the nature of the plant infected. The causal agent is a virus which consists of 2 entirely separate particles, both of which must be present in the plant for replication of the complete virus to occur.

**tocopherol** Vitamin E; one of a group of fat-soluble *terpene compounds that function as anti-oxidants in cells.

**toddy 1.** Palm wine, prepared by fermenting the sugary exudate of young, bruised *inflorescences of several palms, especially *Borassus* and *Cocos*. *See also* NYPA. **2.** A mixture of spirits and water, sweetened and spiced.

***Todea*** **(family *Osmundaceae)** A *monotypic genus (*T. barbata*) of primitive ferns, which, like *Osmunda*, has certain features that are intermediate between the ancient *eusporangiate and the modern *leptosporangiate ferns: more than 1 initial cell may contribute to the formation of each *sporangium and more than 500 *spores may be produced in each one, but the sporangia are stalked. *Todea* produces its thick-stalked sporangia on the under-surface of the *fronds only, and not aggregated into *sori, while in *Osmunda* the sporangia are borne in marginal clusters on certain *pinnae. *T. barbata* is found in the temperate regions of the Old World southern hemisphere.

**tolerance, limits of** *See* LIMITS OF TOLERANCE.

**Tolypothrix** *See* SCYTONEMA.

**tomato** *See* LYCOPERSICON.

**tomentose** Woolly; covered with a fine mesh of hairs.

**tomentum** A mat of fine hairs which gives a woolly appearance to the undersides of some rhododendron leaves, certain fungal structures, lichen *thalli, etc.

**tonoplast** The membrane around a *vacuole.

**Toona** (family *Meliaceae) A genus of deciduous or semi-deciduous trees with *pinnate leaves which are similar to *Cedrela* but whose flowers have a flattish *disc, and the *petals and *stamens partly free. Some yield valuable timber (e.g. Australia cedar and Burma cedar). There are 6 poorly defined species, occurring from India to Australia.

**toothed wrack** *See* SERRATED WRACK.

**topogenous mire** (topogenous peat) A type of bog that forms under climatic conditions of reduced rainfall, with consequent lower humidity and summer drought, which restrict the growth of wetland vegetation to areas where precipitation is concentrated (e.g. valley bottoms).

**topogenous peat** *See* TOPOGENOUS MIRE.

**toposequence** A sequence of soils in which distinctive soil characteristics are related to topographic situation.

**topsoil 1.** The superficial layer of soil moved in cultivation. **2.** The A *horizon of a soil *profile. **3.** Any surface layer of soil.

**toral** *See* TORUS.

**Torrey, John** (1796–1873) An American botanist and chemist, who held professorships in chemistry, mineralogy, geology, and natural history before being appointed, in 1836, New York State botanist. While still a student he was one of the founders, in 1817, of the Lyceum of Natural History which later became the New York Academy of Sciences. He originated the *Flora of North America* and between 1838 and 1843 collaborated on its early sections with his former pupil, Asa *Gray (who completed it).

**Torreya** (family *Taxaceae) A genus of conifers which differs from *Taxus* in its *ovules, which develop into plum-like, hard seeds with a green, purple-striped *aril, and its stiff, spine-tipped leaves. *T. californica* (nutmeg tree) is cultivated as an ornamental. There are 5 species, occurring in the USA, China, and Japan.

**Tortula** (order *Pottiales) A genus of mosses which form greenish, sometimes hoary patches or cushions. The leaf nerve is usually excurrent. *Capsules are erect, ellipsoid or cylindrical, straight or slightly curved, with a beaked lid. There are about 280 species, found mainly in temperate regions. *T. muralis* is common on brick and stone walls throughout Britain, even in towns. The leaves are broadly tongue-shaped, and the nerve extends well beyond the leaf tip to form a long hair point. *T. ruraliformis* (*T. ruralis* ssp. *ruraliformis*) often forms extensive carpets on sand-dunes; when moist the plants are orange-brown or golden-green in colour.

**torus** (adj. **toral**) Ring-shaped, like a doughnut.

**totara** *See* PODOCARPUS.

**touch-me-not** (*Impatiens*) *See* BALSAMINACEAE.

**Tournefort, Joseph Pitton de** (1656–1708) A French botanist who became a professor at the Jardin du Roi in Paris and is remembered for producing a system of plant classification and nomenclature in the 1690s. His *Institutiones Rei Herbariae* (1700) helped to bridge the gap between the work of *Bauhin and *Linnaeus. He also wrote *A Voyage into the Levant* (1718) in which he was the first to describe the common azalea and rhododendron, as well as other plants he collected during this journey. The tribe Tournefortia of the *Boraginaceae was named for him.

**toxin** A microbial product which is poisonous to plants or animals.

**trabecula** A projection of a *cell wall that extends as a bar across a cell *lacuna or *lumen.

**trace element** *See* ESSENTIAL ELEMENT.

**trachea** A vessel; a tube-like series of non-living cells in the *xylem of a plant, supplying mechanical support and involved in the transport of water and salts.

**tracheary element** A *xylem element that is involved in the transport of water, i.e. a *tracheid or *vessel element.

**tracheid** One of the long, cylindrical, tapered cells in a *trachea. The cell is dead and the wall has bands of *lignin in it which add structural strength. Its function is to form part of the trachea and so to conduct *xylem fluids.

**Tracheophyta** (vascular plants; king-dom *Plantae) A division comprising plants that have vascular tissues (*xylem and *phloem) through which water and nutrients are transported. In many modern classifications this division embraces the divisions *Pteridophyta and *Spermatophyta.

**Trachycarpus** (family *Arecaceae) A genus of palms several of which are in cultivation and are very hardy. The trunk is usually single. The leaves are fan-like, the flowers hermaphrodite, with 3 free *carpels. *T. fortunei* (chusan or windmill palm) is hardy outdoors in Britain. There are 4 species, occurring in subtropical eastern Asia from the Himalayas to Japan.

**Tradescantia** (spiderplant, wandering Jew) *See* COMMELINACEAE.

**Tragopogon porrifolius** (salsify) *See* ASTERACEAE.

**trait** Any detectable *phenotypic property of an organism; a character.

**trama** A fungal *tissue that supports a *hymenium, e.g. the sterile tissue of the *gill of an *agaric.

**Trametes** (family *Polyporaceae) A genus of *fungi which form *bracket-like *fruit bodies. There are many species. A number of *fungi formerly included in this genus (e.g. *Coriolus versicolor* and *Lenzites betulina*) have been transferred to other genera.

**tramp species** Species that have been spread around the world inadvertently by human commerce. For example, the brown weed *Sargassum muticum* (*japweed or strangleweed) was inadvertently introduced into British coastal waters, probably with the importation of the Japanese oyster (*Crassostrea gigas*); its common name refers to its effects on pipes and outboard engines. *Eichhornia crassipes* (water hyacinth), taken to the USA from Venezuela by Japanese exhibitors at the 1884 Cotton Exposition in New Orleans, escaped and is now a serious weed of many rivers in low latitudes throughout the world.

**transaminase** (aminotransferase) An *enzyme that catalyses a *transamination reaction.

**transamination** The transfer of an *amino group from an *amino acid to a *keto acid in a reaction catalysed by a transaminase. This is the principal method in cells for the synthesis of non-essential amino acids, and requires a pyridosal or pyrido-oxamine phosphate as a *coenzyme.

**transcellular streaming** The movement of constituents of the *protoplasm through linear files of longitudinally orientated plant cells.

**transcription** The polymerization of ribonucleotides into a strand of *RNA in a sequence complementary to that of a single strand of *DNA. By this means the genetic information contained in the latter is faithfully matched in the former. The process is mediated by a DNA-dependent RNA polymerase.

**transduction** The transfer of bacterial genetic material from one *bacterium to another via a *phage. The genetic material of one bacterial cell becomes incorporated into phage particles which, after release from the dead host cell, then act as vectors carrying it to other bacterial cells.

**transect (isonome)** A linear vegetation sampling method most commonly used to investigate an environmental gradient, e.g. of salinity in a *salt marsh. Generally a line is drawn across the vegetation zoning and recordings are made at intervals along its length.

**transferase** An *enzyme that catalyses the transfer of a functional group from one substance to another.

**transfer cell** A type of cell found in higher plants, with a great abundance of protuberances in its wall and many *plasmodesmata, which enhance solute movements. These cells are found in association with cells of the *phloem and *xylem.

**transfer ribonucleic acid** See TRANSFER-RNA.

**transfer-RNA (transfer ribonucleic acid, t-RNA)** A generic term for a group of small *RNA molecules, each composed of 70–80 *nucleotides arranged in a clover-leaf pattern stabilized by hydrogen bonding. They are responsible for binding *amino acids and transferring these to the *ribosomes during the synthesis of a polypeptide (i.e. during translation). At the *ribosomes, which are attached to the *messenger-RNA (m-RNA), the 'reading frame' indicates the 3 m-RNA nucleotides that form the next triplet codon in the sequence: whichever t-RNA molecule carries the complementary anticodon can associate with the ribosome such that the amino acid that it bears can be joined on to the end of the growing polypeptide.

**transgenic** Applied to an organism that contains genetic material from another organism, usually supplied by molecular biological techniques.

**transglycosylation** A mechanism for glycosidic (see GLYCOSIDE) bond formation, particularly during *polysaccharide synthesis. *Nucleoside phosphate derivatives act as activated donor compounds in which the energies of their glycosidic bonds are partially conserved in the reaction products. Glycosides cannot be synthesized spontaneously from free monosaccharides owing to the high negative free energy ($-\Delta G$) of the *hydrolysis reaction.

**transient polymorphism** The presence in a population at a particular *gene *locus of alternative *alleles in which one is progressively replaced by another, in contrast to a balanced polymorphism where the alleles are in equilibrium with each other.

**transition** In genetics, a type of *mutation (a nucleotide-pair substitution) that involves the replacement in *DNA or *RNA of one *purine with another, or of one *pyrimidine with another. An example is the change of GC (guanine-cytosine) to AT (adenine-thymine).

**translation** The polymerization of *amino acids into a *polypeptide chain whose structure is determined genetically. This process occurs on a *ribosome and involves several small proteins, *m-RNA, and *t-RNA. The sequence of amino acids in the chain is specified by that of the *nucleotides in the m-RNA (these being read as *codons, i.e. in groups of three), and this in turn follows the sequence of nucleotides in the *DNA.

**translocation 1.** The movement of dissolved substances within a plant, usually from the site of synthesis or uptake to centres of growth or storage. **2.** A change in the arrangement of genetic material, altering the location of a *chromosome segment. The most common forms of translocation are reciprocal, involving the exchange of chromosome segments between 2 non-*homologous chromosomes. A chromosomal segment may also move to a new location within the same chromosome or in a different chromosome, without reciprocal exchange; these kinds of translocation are sometimes called transposition. **3.** The movement of soil materials in solution or in suspension from one *horizon to another.

**transmutation** The transformation of one element into another by radioactive decay. The term is also sometimes used for the change of one species or type to another.

**transpiration**   The loss of water vapour from a plant to the outside atmosphere. It takes place mainly through the *stomata of leaves and the *lenticels of stems. Its function is disputed. It may reduce leaf temperature, but its absence from some tropical plants would suggest that this is not essential. It may also be important in mineral absorption and *translocation. However, it may be merely an inevitable concomitant of gaseous exchange which, to be efficient, requires open stomata; as gases are exchanged, water is lost.

**transposable elements**   A chromosomal *locus that may be transposed from one spot to another within and among the *chromosomes of the complement. The process occurs through breakage on either side of these loci and their subsequent insertion into a new point either on the same or a different chromosome.

**traveller's palm**   See RAVENALA.

***Trebouxia***   (division *Chlorophyta) A genus of unicellular, non-*motile green *algae which are the commonest *phycobionts in *lichens; free-living *Trebouxia* cells are found on trees, fences, etc. Phycobiont cells show some differences from their free-living counterparts, e.g. in the proportion of photosynthetic products formed, and in the amount of such products released from the cells (which is greater in the phycobiont).

**tree**   A woody plant that may grow more than 10 m tall. Characteristically it has 1 main stem, although many trees (e.g. oak and ash) may grow multi-stemmed forms. At the end of each growing season there is no die-back of aerial parts, apart from the loss of foliage. *Compare* HERB; SHRUB; and SUBSHRUB.

**tree-borer**   See INCREMENT BORER.

**tree-fern**   See CYATHEACEAE and THYRSO-PTERIDACEAE.

**tree-line**   **(baumgrenze)**   A line through the last of the stunted trees, forming the latitudinal or altitudinal limit beyond which the climate is too cold for trees to grow. This is normally the region where the mean summer temperature is lower than 10 °C. The tree line marks the boundary between *tundra vegetation and the bare rock, snow, and ice of the high Arctic and Antarctic. On mountainsides the height of the tree line varies with latitude (i.e. it is higher in the tropics than in higher latitudes) and climate type. Trees below the tree line are erect and grow in dense stands. Trees close to the tree line are sparsely scattered and prostrate or nearly so. *See* KRUMMHOLZ.

**tree lungwort**   The common name for the *lichen *Lobaria pulmonaria*. The *thallus is broadly lobed; the upper surface is bright green when wet but pale greenish-grey when dry, while the lower surface is brown, and marked with a network of short, dark brown hairs. The main habitat is ancient woodland, so it is now rare in most parts of Britain, although it is still fairly common in parts of the west and in Scotland. This lichen was used extensively in folk medicine for treating various lung disorders, owing to its fancied resemblance to a lung.

**tree ring**   **(annual ring, growth ring)** A sheath of cells appearing as one of a series of concentric rings in the cross-section of a woody stem. Each ring is usually the result of a single yearly growth flush starting in spring and ceasing in the late summer. The new wood (*xylem) cells arise from renewed activity of the vascular *cambium. A sharp boundary usually occurs between the rings since cells formed in spring are typically large, thin-walled, and appear pale in colour when compared with the small, thick-walled cells of later summer. In some dicotyledonous species (*see* DICOTYLEDON) vessel distribution also varies, with most vessels occuring in the early (spring) wood. These ring-porous species contrast with ring-diffuse species in which the vessels are distributed evenly throughout the growth ring. The growth check that causes variable cambial activity may be temperature or water stress or both. In some extreme environments growth may not always be renewed on an annual basis, leading to absent rings ('missing years') and false series (multiple rings in the same year). Dating by tree rings is difficult in these circumstances. *See also* DENDROCHRONOLOGY; DENDROECOLOGY; FIRE

SCAR; VASCULAR CAMBIUM; VESSEL ELEMENT; and TRACHEID.

**tree-ring analysis** *See* DENDRO-CHRONOLOGY; DENDROCLIMATOLOGY; and DENDROECOLOGY.

**tree-ring index** An annual tree-ring width that has been standardized for age. Most dendrochronological work is based on tree-ring indices or standardized series, rather than on the original ring widths.

**tree veld** In S. Africa, grassland with an open or light cover of trees which gives it a parkland aspect. It is thought that in the absence of fire much, if not all, of the veld would develop into forest or *scrub.

**Tremadocian** The oldest of 7 stages that comprise the *Ordovician Period lasting from 488.3–478.6 Ma ago. Mudstones and sandstones of the Tremadoc series occur in N. America, Ireland, Wales, England, and Scandinavia. They mark the shelf and slope areas that occur on the southern edge of the northern Iapetus Ocean.

**Tremandraceae** A family of small shrubs in which the leaves are *simple, often narrow, *opposite or alternate, and without *stipules. Occasionally the shrubs are glandular and have winged stems. The flowers are often very brightly coloured, usually red or purple. They are regular, bisexual, solitary, and *axillary, with 3–5 *sepals and *petals. There are twice as many *stamens as petals. The *ovary is *superior, of 2 fused *carpels with a single *style. The fruit is a *capsule which splits open along the *locule partitions. The seeds contain copious *endosperm and are occasionally hairy. The family is related to the *Pittosporaceae and *Polygonaceae. Some species are cultivated as ornamentals. There are 3 genera, with 43 species, within the family, all of them confined to Australia, especially the south-west.

**Tremella** (order *Tremellales) A large genus of *fungi, in which the *fruit body is gelatinous (or absent in some *parasitic species) and is typically convoluted or cushion-like. Most species are *saprotrophic, for example, on wood, but some are *parasitic on the fruit bodies of other fungi.

**Tremellales** (subclass *Phragmobasidiomycetidae) An order of *fungi in which the basidia are divided longitudinally by *septa. *Fruit bodies are typically gelatinous when wet, cartilaginous when dry. They are mostly *saprotrophic, usually growing on wood. There are many genera.

**trend surface analysis** A special case of multiple regression analysis, in which the independent variables are spatial coordinates. This enables assessment of spatial trend in data values. The technique is used, in particular, in geographical studies.

**Trentepohlia** (division *Chlorophyta) A genus of *algae in which the *thallus is typically bright orange and minutely filamentous, resembling orange velvet to the naked eye. It is found on rocks, leaves, tree bark, etc.; also as the *phycobiont in certain *lichens.

**Treponema** (family Spirochaetaceae, order *Spirochaetales) A genus of flexible, spirally shaped *bacteria which are *anaerobic or *micro-aerophilic *chemoorganotrophs. There are many species, found in the alimentary and reproductive tracts in humans and other animals. Some species are *pathogenic, e.g. *T. pallidum* is the causal agent of syphilis.

**triallelic** Applied to a *polyploid in which 3 different *alleles exist at a given *locus.

**triarch** *Primary xylem that has 3 strands.

**Triassic** The earliest (251–199.6 Ma ago) of the three periods of the *Mesozoic Era. As a result of the mass extinctions of the late *Palaeozoic, the Triassic contained many new faunal and floral elements, including certain *gymnosperms.

**tribe** In plant *taxonomy, a rank between family and genus, comprising genera whose shared features serve to distinguish them from other genera within the family. The names of tribes bear the suffix -eae. Tribes may be grouped to form subfamilies and divided to form subtribes.

**tricarboxylic acid cycle** *See* CITRIC ACID CYCLE.

**trichoblast 1.** In root *epidermis, a specialized cell that develops into a root hair. **2.** In some red algae (*Rhodophyta), a hair-like branch.

***Trichoderma*** (class *Hyphomycetes) A *form genus of *fungi in which *conidia are produced in *phialids; the walls of the conidia are slimy, and masses of the *spores stick together at the tip of the phialid. Some species have *perfect states in the *Hypocreales.

***Trichodesmium*** A genus of filamentous *cyanobacteria which are closely related to *Oscillatoria and perhaps should be included in that genus. The filaments occur in masses embedded in a flocculent mucilage. Species are unusual among cyanobacteria in that they are planktonic, often forming a major component of the plankton in tropical seas. They are capable of *nitrogen fixation. *Heterocysts are not formed. Gas vacuoles allow the colonies to float, and when seas are calm they form conspicuous surface blooms (sea sawdust) measuring up to several kilometres across.

**trichogyne** An extension, often hair-like, from the female *gametangium that receives the male *gamete or nucleus prior to *fertilization. It is found in certain green and red algae (*Chlorophyta and *Rhodophyta), *ascomycetes, and *lichens.

***Tricholoma*** (family *Tricholomataceae) A genus of *fungi that form *mushroom-shaped *fruit bodies; the *stipe is fleshy and central, and the *gills are only narrowly attached to the stipe. The *spore print is white or pale pink; the *spores are smooth and do not stain dark violet when treated with iodine. There are many species, found on the ground, mostly in woodland. *See also* BLEWIT.

**Tricholomataceae** (order *Agaricales) A family of *fungi that form white or pink *spores in *stipitate *fruit bodies which typically bear *gills. There are many genera. They are mainly *saprotrophic, but some are *parasitic in woody plants.

***Trichomanes*** (bristle-ferns; family *Hymenophyllaceae) A genus of filmy ferns that differs from *Hymenophyllum* in the cylindrical or bell-shaped *indusium surrounding each marginal, *gradate sorus, and in its filamentous (not flat as in *Hymenophyllum*) *gametophytes. The tip of the soral placenta eventually projects as a bristle from the *indusium. In the wide sense there are some 330 species, but some authors divide the genus into many smaller genera. The plants occur throughout most tropical zones, with a few *outliers in mild, moist, temperate zones (e.g. *T. radicans*, Killarney bristle-fern, found in western parts of the British Isles).

**trichome 1.** An outgrowth from an epidermal cell (e.g. a root hair). **2.** In certain *bacteria and *cyanobacteria, a chain of vegetative cells; a cyanobacterial trichome is often surrounded by a slimy sheath.

***Trichophyton*** (class *Hyphomycetes) A *form genus of *fungi which occur either as *saprotrophs (e.g. in soil), or as *parasites of animals and humans, causing diseases of the skin, hair, or nails. Some species have *perfect states classified in the *Gymnoascales.

***Trichospermum*** (family *Tiliaceae) A genus of small trees that have flattish sprays of leaves, fibrous bark (used for cordage), and twigs with *stipule scars. The flowers are held in *cymes, and are *pentamerous. The fruit is *capsular. There are 36 species, occurring from the Nicobar Islands to the western Pacific and tropical America.

**trichothallic growth** Growth that occurs only in particular, well-defined regions, e.g. at or near the base of the filament in certain brown algae (*Phaeophyta).

**tricolpate** Of a *pollen grain, having 3 colpi (*see* COLPUS).

**trifoliate** Applied to leaves that are composed of 3 leaflets.

***Trifolium*** (clovers; family *Fabaceae) A genus of *annual or *perennial herbs, with leaves that are trifoliate, stalked, and with *stipules *adnate to their stalks. The flowers are usually in dense, *racemose heads, with short stalks or none. The *petals are persistent, and the *pods very small, short, and *indehiscent, remaining enclosed in the *calyx, and often also by the standard petal. The genus includes many important fodder plants (especially *T. pratense* and *T. repens*) and, like other Fabaceae, fixes nitrogen by means of *root-nodule bacteria, so increasing soil fertility. There are 238 species, centred in southern Europe and western Asia but extending to other temperate and tropical regions.

**trifurcate** Forming 3 branches.

**trigger plants** See *STYLIDIUM* and STYLIDIACEAE.

**Triglochin** (arrow-grass) See JUNCA-GINACEAE.

**Trigonella** (fenugreek, foenugreek; family *Fabaceae, subfamily *Papilion-atae) A genus of *annual herbs with toothed, trifoliate leaves. The flowers are blue, yellow, or whitish, solitary or in clusters, and small and pea-like (*see PISUM*); they develop into slender, curved *pods up to 15 cm long. *T. foenum-graecum* is widely cultivated. The fresh plant is eaten as a vegetable, and the seeds are used in curry, chutney, and flavourings. Fenugreek is grown as a fodder crop in northern Africa. Most species are from the Mediterranean region and Asia, but they also occur in Macronesia, S. Africa, and Australia.

**Trigoniaceae** A family of forest trees and shrubs, closely related to Polygalaceae, in which the leaves are *simple with small *stipules. The flowers are *zygomorphic, with 5 *sepals, 3 or 5 *petals, 3–12 *stamens, on 1 side, and the *ovary *superior with 3 *locules. The fruit is a *capsule. There are 4 genera, with 35 species, occurring in Madagascar, western Malesia, and tropical America.

**trigonous** Triangular in cross-section.

**trilete** Applied to a spore bearing a *triradiate scar, owing to the possession of tetrahedral symmetry by the *tetrad.

**Trilliaceae** A family of monocotyledonous (*see MONOCOTYLEDON*) *herbs, close to *Liliaceae, but with erect, simple stems bearing leaves in an *opposite pair, or in a single *whorl near the summit of the stem, and having *reticulate veining, un-usual in monocotyledons. The flowers are terminal, usually solitary, or sometimes in a small *umbel, with free *sepals and *petals each in 1 whorl, and with the parts in either threes or fours. The number of *stamens is equal to that of the *perianth segments; the *ovary is *superior, and usually 1-celled. The fruit is a *berry or a fleshy *capsule. There are 4 genera, with 53 species, in the northern temperate zone. *Paris*, found in Europe and temperate Asia, has flower parts in fours; *Trillium*, of N. America, has them in threes.

**trilocular** Having three *locules.

**trimeric** See DIMER.

**trimerous** Having parts in threes, as in the flowers of most monocotyledons (*see* MONOCOTYLEDON).

**tri-nerved** Of a leaf, with 3 main nerves.

**Triodia** (family *Poaceae) A genus of coarse grasses that are similar in superficial appearance to *Spinifex and are commonly known as such. They form tussocks, usually 50 cm in height and 0.5–1 m wide. The leaves are softly hairy on the sheath and edges, but hairless at the apex. They are strongly nerved, with needle-like, rigid blades. The flower *panicle is usually loose and nodding. There are 35 species, all endemic (*see* ENDEMISM) to Australia, with little agricultural value. *T. basedowii* (lobed spinifex) is the most common grass of the central Australian grassland communities on sand plains.

**Triplochiton scleroxylon** (family *Sterculiaceae) A fast-growing western African tree, nowadays extensively planted for its valuable timber (obeche), which is often exported.

**triploid** Applied to a cell with 3 sets (3*n*) of *chromosomes in its *nucleus, or to an organism composed of such cells (as opposed for example to *haploid (*n*) or *diploid (2*n*) cells or organisms). *See also* POLYPLOIDY.

**triquetrous** Triangular in cross-section with a sharp angle at each corner.

**triradiate** Branching in 3 directions, making a Y-shape.

**trisomy** The condition in which a particular *chromosome is represented by 3 members (i.e. the triploid condition). *Compare* DISOMY, NULLISOMY, POLYSOMY, and TETRASOMY.

**Trithurieae** *See* CENTROLEPIDACEAE.

**Triticum** (wheat; family *Poaceae) A genus of domesticated grasses that yields wheat, the most important cereal crop of temperate regions. The ancestry of the cultivated wheats is uncertain but they are believed to be derived from T. *monococcum* (einkorn wheat), native to the eastern Mediterranean region. The widely cultivated species include T. *aestivum* (bread wheat) and T. *durum* (durum wheat, used to make pasta).

**Triuridaceae** A family of monocotyledonous (*see* MONOCOTYLEDON) small tropical *herbs which are always *saprophytic. The plants lack photosynthetic pigments and their leaves are reduced to scales distributed along the colourless or reddish stem. The regular flowers are either *dioecious or *monoecious, and unisexual or bisexual, and are borne in *racemes. There are up to 10 *perianth segments held in series, which are equal or unequal; in some species a tuft of hair is present in the tip of the segment. The 2–6 *stamens are free, sometimes with some infertile stamens in the separate male or female flowers. In the male flowers there may be 3 fertile stamens and 3 *staminodes. After flowering the segments reflex. The female flowers often have *staminodes. The *ovary is *superior with free, unfused, uniovular *carpels which each have a single *style. The fruit is a dense mass containing the seeds, which have an oily, white *endosperm. There are 6 genera, with 42 species, present in all the tropical and warm regions of the world.

**t-RNA** *See* TRANSFER-RNA.

**Trochodendraceae (order Trochodendrales)** A family of evergreen forest trees containing a single species, *Trochodendron aralioides*, of Korea and Japan. The tree has leaves that are alternate, apparently *whorled, diamond-shaped, and glossy, with long stalks. The tiny green flowers have no *perianth, and are borne in *inflorescences at the ends of branches.

The fruits are *follicles. The family has the 'primitive' feature of having no vessels. Birdlime is made from the bark.

**Tropaeolum** (family Tropaeolaceae) A genus of dicotyledonous (*see* DICOTYLEDON) *herbs, often succulent. Many species climb by means of twining leaf stalks. The leaves may be entire or *palmately lobed, with long stalks. The flowers are showy, with 5 irregular *petals, and 5 *sepals forming a spur below. T. *majus* is the nasturtium, a common garden ornamental from S. America, once cultivated as a salad plant. There are 86 species, occurring from Mexico to Patagonia.

**trophic** Pertaining to nutrition, food, or feeding.

**trophic level** A step in the transfer of food or energy within a chain. There may be several trophic levels within a system, for example *producers (autotrophs), primary *consumers (herbivores), and secondary consumers (carnivores); further carnivores may form fourth and fifth levels. There are rarely more than 5 levels since usually by this stage the amount of food or energy is greatly reduced. *See also* ECOSYSTEM.

**tropical forest** A category of vegetation comprising a variety of formation types including rain forest, *semi-evergreen seasonal forest including *monsoon forest, and *deciduous seasonal forest.

**tropical montane forest** Typically, a forest of short trees 10–12 m high in a single stratum, their trunks and boughs being mishapen and covered in *mosses, *lichens, and *liverworts. This is the 'mossy forest' which coincides with the belts of mist and cloud at heights of 2000–3000 m. If the montane forest persists above the cloud zone, as it does in New Guinea, for example, then the trees are taller and the 'mosses' less in evidence.

**tropical rain forest** The most prolific, productive, and the oldest of all the world vegetation types, which occurs in the equatorial lowlands and is characterized by up to 3 tree strata, and a profusion of *lianes, *epiphytes, and

**t**

*saprophytes. The trees are *evergreen and flowering, and fruiting and leafing continue all year. Tropical rain forest occurs in lowlands where rainfall is heavy and there is little or no dry season. Floristically, 4 main regions of tropical rain forest are recognized: the African, American, Indo-Malaysian, and Australsian. These share few woody species and genera in common.

**tropical seasonal forest** A tropical forest that grows in regions with a marked dry season. There is some defoliation during the dry season, the degree depending on the severity of the moisture deficit. The structure of the forest is also simpler than that of the rain forest, with fewer tree strata, and less luxuriant growths of climbing and herbaceous plants.

**tropic movement** See TROPISM.

**tropism** (tropic movement) A directional response by a plant to a stimulus. It may be positive or negative (i.e. towards or away from the source of the stimulus). The plant may respond by growth or *turgor changes so that parts of it bend towards, away from, or at right angles to the direction of the stimulus. The suffix -tropism is used in relation to responses to particular stimuli; e.g., phototropism is a response to light, geotropism to gravity, and chemotropism to a chemical substance.

**'true' fungi** See EUMYCOTA.

**'true' mosses** See BRYIDAE.

**truffle** The underground *fruit body of certain species of *Tuber. Truffles are edible and highly esteemed. T. aestivum (summer truffle) is 2–8 cm in diameter and has a dark brown, warty surface. The inside is greyish with a network of white veins. It is occasionally found in England in the soil of beech woods. T. melanosporum (the black or Périgord truffle) occurs in S. Europe and resembles T. aestivum but is rather smaller (1–3 cm).

**trumpet vine** See CAMPSIS.

**tryptophan** A *heterocyclic, *non-polar, *alpha-amino acid.

**Tsuga** (hemlock firs; family *Pinaceae) A genus of evergreen conifers, with narrow, oblong, flat, blunt, stalked leaves arranged spirally on the stems, but spreading laterally and usually unequal in length, with 2 white bands lengthwise beneath. The cones are small, rounded, and *pendulous, with persistent scales and inconspicuous *bracts. There are about 10 species, some native to temperate N. America, others to eastern Asia. The genus includes several valuable timber trees, often grown commercially.

**tube-nucleus** The vegetative nucleus (one of 3 nuclei in the male *gametophyte) in the growing *pollen tube of a flowering plant; it degenerates after *double fertilization.

**Tuber** (order *Pezizales) A genus of *fungi in which the *fruit bodies occur underground. The *hymenium is not exposed to the exterior, and spores appear to be dispersed by animals that eat the fruit bodies. The fruit bodies of Tuber species are edible and highly prized as *truffles (T. aestivum summer truffle; T. melanosporum, Périgord or black truffle; T. uncinatum, Burgundy truffle).

**tuber** A swollen stem or root that functions as an underground storage organ. Stem tubers (in *potatoes, for example) often produce *buds along the stem from which aerial stems arise the following season. Root tubers produce no buds, or produce buds only at the point where the tuber is attached to the stem of the plant.

**tubercle bacillus** Any species of *Mycobacterium that can cause tuberculosis, but particularly M. tuberculosis.

**tubule** A small tube.

**tulip** See TULIPA.

**Tulipa** (tulip; family *Liliaceae) A genus of bulbous *herbs with large, erect, usually solitary, terminal flowers of cup or bell shape, with 6 similar *perianth segments, and without nectaries. The *stigma is 3-lobed. There are some 100 species, in Europe, northern Africa, and central Asia. Many species and garden hybrids are cultivated for their fine flowers of varied colours. The garden tulip (T. gesneriana) was introduced to Europe from Turkey in the sixteenth century. There was a craze in the seventeenth century, reaching a peak between 1630 and 1640, particularly in the Netherlands, for breeding and growing new forms.

**tulip tree (tulip poplar)** See LIRIODENDRON and SPATHODEA; see also BICENTRIC DISTRIBUTION.

**Tumbunan division** The New Guinean biota above about 1200 m altitude, including Australian rain forests between Cooktown and Mackay, and between Gympie and Illawarra. Compare IRIAN DIVISION.

**tundra** A treeless plain of the Arctic and Antarctic characterized by a low, 'grassy' sward. Actually, although grasses are rarely absent, sedges (*Carex* species), rushes (*Juncus* species), and wood rushes (*Luzula* species) are the dominant plants, together with *perennial herbs, dwarf woody plants, and various *bryophytes and *lichens.

**Tundra Soil** One of the Great Soil Groups, within suborder 1 of the order Zonal Soils of the 1949 *USDA system of soil classification, based originally on the work of V. V. Dokuchaev, but now superseded by the USDA Soil Taxonomy in which Tundra Soils are classified as *Inceptisols. They occur on ground that drains poorly (mainly because of permafrost), and are acid, 30–60 cm deep, have a high content of organic matter at the surface, and a microrelief formed by freezing and thawing. Their formation, and the decomposition of organic matter, is inhibited by the low temperature.

**tung oil** See ALEURITES.

**tunica** In *angiosperms, the cap of cells at the apical *meristem, overlying the *corpus.

**turbinate** Shaped like a spinning top and attached at the point.

**turgor** The rigidity of a plant and its cells and organs resulting from hydrostatic pressure exerted on the cell walls.

**turgor pressure** See PRESSURE POTENTIAL.

**turmeric** See CURCUMA and ZINGIBERACEAE.

**turnip** See BRASSICA.

**turnover number** A measure of *enzyme activity, normally expressed as the number of moles of *substrate transformed into 1 or more products per mole of enzyme.

**turnover rate** A measure of the movement of an element in a *biogeochemical cycle. Turnover rate is calculated as the rate of flow into or out of a particular nutrient pool, divided by the quantity of nutrient in that pool. Thus, it measures the importance of a particular nutrient flux in relation to the pool size. Compare TURNOVER TIME.

**turnover time** The measure of the movement of an element in a *biogeochemical cycle; the reciprocal of *turnover rate. Turnover time is calculated by dividing the quantity of nutrient present in a particular nutrient pool or reservoir by the flux rate for that nutrient element into or out of the pool. Turnover time thus describes the time it takes to fill or empty that particular nutrient reservoir.

**turpentine** See PISTACIA and SYNCARPIA.

**twisted wrack** See SPIRAL WRACK.

**two-way table** See CONTINGENCY TABLE.

**tyloses** In older wood, and sometimes in the vessels of *herbs, hollow ingrowths developed from adjacent *parenchyma cells that eventually cause blockage, Tyloses often fill with resins, gums, tannins, or other pigmented materials (some of which are used as dyes), giving the wood a characteristic colour.

**type specimen (holotype)** An individual plant or animal chosen by taxonomists to serve as the basis for naming and describing a new species or variety. Compare LECTOTYPE and SYNTYPE.

***Typha*** (reedmace; family *Typhaceae) A genus of tall, stout, *rhizomatous herbs which grow in shallow water. They are usually more than 2 m in height, with long, simple stems. The leaves arise from the base and are linear and elongated with a thick, spongy inner tissue, sheathing at the base. The plant relies on wind pollination and wind dispersal. The leaves of *T. latifolia* (reedmace) are used as weaving material for baskets and mats. There are 10–12 species, found in shallow, freshwater habitats of temperate and tropical regions throughout the world.

**Typhaceae** (reedmaces, cat's-tails, sometimes erroneously called bulrushes) A family of monocotyledonous (*see* MONO-COTYLEDON) *herbs, that have erect, un-branched stems. The flowers are tiny, but numerous, in a terminal, cylindrical *spike. The lower part of this spike consists of close-packed female flowers, densely surrounded by hairs (which may represent the *perianth), the upper part consists of male flowers with 2–5, often united, *stamens. Each female floret has a stalked, unilocular *ovary. The fruits are *achenes which dry and split. Species of *Typha* are important colonists of shallow, fresh water, where they help to consolidate muds. There is 1 genus, with 10–12 species.

**tyrosine** An aromatic, polar (*see* POLAR MOLECULE), *alpha-amino acid.

**ubiquinone** *Coenzyme Q; a generic term for a group of compounds structurally related to vitamin K that function as electron carriers in the electron-transport chain of *mitochondria.

**ullucu** *See* ULLUCUS TUBEROSUS.

***Ullucus tuberosus*** (ullucu; family *Basellaceae) A *herb with twining stems, endemic (*see* ENDEMISM) to the high Andes, and frost-resistant, which is cultivated for its edible, potato-like tubers. The *anthers are porous and the fruit is a *berry.

**Ulmaceae** A family of trees with *simple, alternate leaves, often asymmetrical at the base, and with small, usually hermaphrodite flowers in dense clusters on the twigs. The flowers have a 4–8-lobed *perianth with imbricate lobes, 4–8 *stamens opposite the perianth lobes, and an *ovary of 2 fused *carpels. They are wind-pollinated. The tiny, dry or fleshy fruits are often winged. Modern classifications recognize some 16 genera, with about 140 species, mostly in the northern temperate zone.

***Ulmus*** (elms; family *Ulmaceae) A genus of mostly large trees, in which the leaves are asymmetrical at the base, and usually hairy, at least beneath. The hermaphrodite flowers appear in clusters before the leaves. The tiny *nutlets have a broad, oval wing that aids wind dispersal, and is notched at the tip. The virulent strain of Dutch elm disease that was introduced about 1968 from N. America into Europe has destroyed most European elm trees, but the roots are not normally killed and many elms are re-establishing themselves from sucker shoots. The disease is caused by a fungus (*Ophiostoma ulmi*) and is apparently always transmitted by the elm-bark beetle (*Scolytus*). Some eastern Asian elms appear to be immune to this disease. Elm wood is used sometimes for making coffins, and hollowed-out elm stems were formerly used for water pipes,

as the wood does not decay readily when waterlogged. There are 18 species, occurring in the northern temperate zone, and in the mountains of the Asian tropics.

***Ulota*** *See* ORTHOTRICHIALES.

***Ulothrix*** (division *Chlorophyta) A genus of unbranched, filamentous, green *algae, each cell of which contains a single *chloroplast which forms an incomplete band around the middle of the cell. The filaments are attached at their bases to rocks, etc., in marine or fresh water.

**ULR** *See* UNIT LEAF RATE.

**Ultisols** *Mineral soils, an order identified by an *argillic B *horizon with a base saturation of less than 35 per cent, and red in colour from iron oxide concentration. Ultisols are leached, acid soils, associated with humid subtropical forest environments. *See also* RED PODZOLIC SOILS.

**ultracentrifugation** Centrifugation carried out at high rotor speeds (up to 75 000 rpm) and therefore under high centrifugal forces (up to 750 000 g). Analytical ultracentrifuges are employed to determine the mass and to some extent the shape of molecules; preparative ultracentrifuges are used to separate the components of mixtures on the bases of these parameters.

***Ulva*** (division *Chlorophyta) A genus of green seaweeds in which the *thallus is flat and sheet-like, and is 2 cells thick. *U. lactuca* (sea lettuce, green *laver) is a common British species; the *holdfast is *perennial, producing new *fronds each year. It is found attached to rocks in the intertidal zone, and in brackish water in estuaries.

**umbel** An *inflorescence in which all the *pedicels arise at the apex of an *axis. It is commonly compound, usually umbrella-shaped, and is characteristic of *Apiaceae.

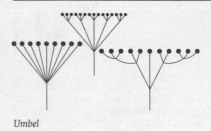

*Umbel*

**Umbelliferae**   *See* APIACEAE.

***Umbilicaria*** **(order \*Lecanorales)** A genus of \*lichens in which the \*foliose \*thallus is roughly circular in outline, either entire or lobed, and is attached to the substrate at a point in the centre of its lower surface. The \*apothecia are \*lecideine and black. Species are found mainly on non-calcareous rocks, chiefly in hilly and mountainous districts. *See also* IWATAKE.

**umbilicate   1.** Having a central pit or depression (e.g. of a fungal \*pileus). **2.** Of a \*lichen, having a more or less circular \*thallus that is attached to a substrate by a central point on its lower surface (e.g. \*Umbilicaria).

**umbo** A convex protuberance or swelling in the centre, as in the \*pileus of certain fungal \*fruit bodies.

**umbonate** Having or resembling an \*umbo.

**umbric epipedon** A surface soil \*horizon similar to a \*mollic epipedon but with a base saturation of less than 50%. It is a \*diagnostic horizon. The name is from the Latin *umbra* meaning 'shade'.

**Umbrisols** Soils with an \*umbric epipedon. Umbrisols are a reference soil group in the FAO \*soil classification.

**unavailable water** Water that is present in the soil but that cannot be absorbed rapidly enough by plants for their needs because it is held so strongly to the surface of soil particles.

***Uncaria*** **(family \*Rubiaceae)** A genus of stout, woody climbers which have woody, recurved hooks. Several are medicinal (e.g. *U. gambir* provides gambier, used for both medicine and tanning). There are 34 species, with a pantropical distribution.

**unconfined aquifer**   *See* AQUIFER.

**underdispersion (regular distribution)** In plant ecology, the situation in which the pattern of individuals of a given plant species within a \*community is not random, but regular, with similar numbers recorded in all \*quadrats. *See* PATTERN ANALYSIS.

**underdominance**   *See* BALANCED POLYMORPHISM.

**understorey species** The trees of the lower canopy levels in a woodland ecosystem, as distinct from emergent, crown, or upperstorey species. Some species (e.g. \*hazel and \*banana) are characteristic of the understorey, but others may be younger specimens of emergent species.

**underwood   1.** A wood, either growing or cut, that consists of \*coppice or \*pollard regrowth, or small shrubs and saplings grown from \*seed or \*suckers. **2.** The lower storey of a forest crop.

**unequal crossing-over** In genetics, a \*cross-over after improper pairing between \*chromosome \*homologues that are not perfectly aligned. The result is 1 cross-over \*chromatid with 1 copy of the segment and another with 3 copies. The phenomenon was first described at the *Bar* \*locus in *Drosophila* (fruit fly).

**uniaxial** The condition in which a plant has a single central axis without branches.

**unicentric distribution** The occurrence of endemic species (*see* ENDEMISM) in very local areas. In Scandinavia, for example, about 40 endemics are found in just 1 or 2 areas; these areas are thought by some authorities to have been refuges from the glaciation of the late \*Pleistocene (i.e. they were nunataks above the ice sheets).

**unicorn plant**   *See* MARTYNIACEAE.

**unilocular** With one \*locule.

**uniramous** Unbranched.

**uniseriate** Arranged in a single row.

u

**unisexual flower** A flower that possesses either *stamens or *carpels but not both. A plant may be unisexual (*dioecious), possessing only male flowers or female flowers; or it may be *monoecious with male and female reproductive organs borne in the same flower or in different unisexual flowers but on the same plant.

**unit leaf rate (ULR)** The rate of *photosynthesis per unit area of leaf. *Primary productivity is thus the unit leaf rate multiplied by the *leaf-area index.

**unitunicate** Applied to an *ascus in which layers of the ascus wall do not separate during *ascospore release.

**univalent** A single *chromosome observed during *meiosis when *bivalents are also present. A univalent has no pairing (synaptic) mate. An example is the *sex chromosome of an XO male.

**universal veil** A membranaceous structure which encloses the young *fruit body in certain types of *agaric; the universal veil is ruptured as the fruit body expands.

**univoltine** Applied to species in which 1 generation reaches maturity each year.

**upas tree (ipoh)** *See ANTIARIS.*

**Uppsala school of phytosociology** A set of floristic methods intended mainly for the classification of vegetation *communities, developed by G. E. *Du Rietz and others in Uppsala, Sweden, starting in 1921, at first quite independently of the work of the *Zurich–Montpellier (Z–M) team. More recently the approaches have converged as many ideas have been modified in the light of Z–M work in an attempt to link the two. The main distinction between them lies in the emphasis placed by the Uppsala scheme on *exclusive or *preferential species of high *constancy or dominance, rather than *fidelity, to define the basic community types.

**uracil** A *pyrimidine base that occurs in *RNA.

**urceolate** Flask-shaped.

**Urediniomycetes** (*rust fungi; subdivision *Basidiomycotina) A class of *fungi that are obligate parasites (*see* PARASITISM) of a wide variety of plants including *angiosperms, *gymnosperms, and *pteridophytes. Many have complex life cycles that involve the sequential formation of up to 5 different types of *spore. Some rusts can complete their *life cycle in a single host plant, while others require two different host species.

**urediniospore** *See UREDOSPORE.*

**uredinium** *See UREDIUM.*

**urediospore** *See UREDOSPORE.*

**uredium (uredinium, uredosorus)** A region of fungal *tissue within which *spores (called *uredospores) develop in a plant infected with certain rust fungi (class *Urediniomycetes).

**uredosorus** *See UREDIUM.*

**uredospore (urediospore, urediniospore)** A type of fungal *spore that is borne on a thin stalk (*pedicel) in a structure known as a *uredium.

**uridine** The *nucleotide formed when *uracil is linked to ribose sugar.

**uridylic acid** The *nucleotide formed from *uracil.

***Urtica*** (nettle; family *Urticaceae) A genus of *annual or *perennial *herbs with ridged stems, *opposite, *simple leaves, and numerous stinging hairs. The flowers are small, green, and unisexual, in clusters along *spike-like stems in leaf *axils, and both sexes of flowers may occur on the same plant, or the plants may be *dioecious. There are 45 species, mostly found in temperate regions.

**Urticaceae** A family of small trees, shrubs, or dicotyledonous (*see* DICOTYLEDON) *herbs, often with stinging hairs, with *simple leaves, and only small, usually unisexual, clustered flowers. The flowers have a *perianth of 4 or 5 segments, 4 or 5 *stamens in the male flowers, and a 1-called *ovary in the female. The fruits are either dry *nutlets or *drupes. Modern classifications recognize about 52 genera and 1050 species, distributed widely in most regions of the world.

**USDA** United States Department of Agriculture.

**Usnea** (order *Lecanorales) A genus of *fruticose *lichens, commonly known as beard lichens owing to their tufted, filamentous appearance and frequently pendulous habit. The numerous species are often variable and are notoriously difficult to distinguish from one another. *Usnea* species are very sensitive to atmospheric pollution and are found only in unpolluted, humid areas, particularly mountain woodlands.

**usnic acid** A substance, with some *antibiotic activity, that is found in certain species of lichens. Usnic acid is active against some Gram-positive (*see* GRAM REACTION) bacteria (and has been used medicinally, in ointments for the treatment of local skin infections).

**Ustilaginales** (*smut fungi; class *Ustilaginomycetes) An order of *fungi that are mostly *parasitic on *angiosperms, often causing diseases of economic importance, especially in cereals.

The fungus forms masses of dark, powdery *spores in the leaves, stems, flowers, or fruits of the host plant.

**Ustilaginomycetes** (subdivision *Basidiomycotina) A class of *fungi that do not form sexual organs and do not produce macroscopic *basidiocarps. Members include plant *parasites and *saprotrophs.

**Ustilago** (family Ustilaginaceae, order *Ustilaginales) A genus of *smut fungi in which *teliospore germination gives rise to a transversely *septate *promycelium which bears lateral and apical *basidiospores. Species are *parasitic on a range of plants, causing some important diseases (e.g. *U. nuda* is responsible for loose *smut of barley).

**utile** *See* ENTANDROPHRAGMA.

**Utricularia** (bladderwort) *See* LENTIBULARIACEAE.

**Vaccinium** (bilberry; family *Ericaceae)
A genus of mostly low shrubs, often evergreen, with alternate, *simple leaves, *tetra-
or *pentamerous flowers with mostly bell-
or urn-shaped *corollas and *inferior
*ovaries, and *berry-like fruits. There are
some 450 species, found mostly in the northern (temperate zone and the Arctic, with
some tropical mountain outliers. Several
species are cultivated, particularly the American cranberry (*V. macrocarpon*), with reflexed
corolla lobes, from whose berries cranberry
sauce is made; European cranberry (*V. oxycoc-
cos*) is a similar but smaller species, also with
edible berries, found in bogs throughout the
northern temperate zone.

**vacuole**  A membrane-bound sac that
is found in many cells, normally acting
as a storage organ of various types. A large
central vacuole is a particular feature of
many plant cells, where it can occupy
80–90 per cent of the total cell volume.

**vadose zone**  See WATER TABLE.

**vagile**  Applied to a plant that is free to
move about. *Compare* SESSILE.

**vagility**  The inherent power of movement possessed by individuals or *dias-
pores. Vagility in plants is often greater
than commonly realized; *spores and
seeds may float in the air for several miles
and are very efficient means of *dispersal.

**valency**  A measure of the number of
other *ions of a chemical element that can
be combined with a particular atom.

**valerian**  See VALERIANACEAE.

**Valerianaceae**  A family of dicotyledonous (*see* DICOTYLEDON) *herbs with opposite or *wholly radical leaves, which are
often *pinnatifid and have no *stipules.
The *inflorescence is a *cymose, often
dense or sometimes *umbellate head of
small flowers, which may be irregular. The
*calyx is either merely a ring, or forms a
feathery *pappus in fruit. The *corolla is
funnel-shaped, and sometimes spurred

below. There are 1–4 *stamens, attached to
the usually 5-lobed *corolla. The *ovary is
*inferior, 3-celled, but with only 1 cell
fertile. The fruit is dry and *indehiscent.
Modern classifications recognize about
17 genera, with 400 species, distributed
widely except in Australia. *Valerianella* (corn
salad) has been used as a vegetable. *Centranthus* (spurred valerian) is grown in gardens.
*Valeriana* (valerian) has medicinal uses.

**Valerianella** (corn salad)  See VALERI-
ANACEAE.

**valine**  An  aliphatic,  *non-polar,
*alpha-amino acid.

**vallecular canal (cortical canal)**  In
*Equisetum* and some of its fossil relatives,
one of a number of large, air-filled, intercellular channels running the length of each
*internode and positioned approximately
between the *vascular bundles. They lie outside and are larger than the *carinal canals.

**valley bog**  A *mire community that
develops in wet valley bottoms, valleys
with some downstream impedance, or
badly drained hollows. Many European
valley mires have layers of charcoal in
their stratigraphy below the *peat, suggesting that fire may originally have initiated peat formation by creating a charcoal
deposit that effectively seals off basal soils
from water penetration and causes waterlogging. The *ground water is base-poor
and conditions are acidic. This type of
mire is flow-fed (rheotrophic), so technically it is a poor *fen type of community
rather than a true *bog. The supply of nutrients is determined by the concentration
of elements in the drainage water (which
is usually low) and the rate of water flow
through the system. The central part of
the mire often has the fastest water flow
and is hence less *oligotrophic than the
lateral parts of the mire. In extensive valley mires the acidic, lateral mire expanses
may become elevated and *ombrotrophic
if rainfall is adequate. *Compare* BLANKET
BOG; OMBROGENOUS PEAT; and RAISED BOG.

**valvate** Applied to the arrangement (*aestivation) of *sepals or *petals in a flower *bud such that these parts meet at their edges and do not overlap. Where they overlap, the aestivation is described as *imbricate.

**valve** 1. One of the portions into which a *fruit splits. 2. In a diatom (*Bacillariophyta), a silica *theca, either the *epithecium or the *hypothecium.

**Van Valen's 'law'** When plotted as cumulative curves on a logarithmic scale, the duration frequencies of many species tend to show a more or less straight-line relationship with time. In effect, taxa generally become extinct regardless of their age. The 'law' is named after its discoverer, L. Van Valen.

**variance** *See* MEAN SQUARE.

**variation** Differences displayed by individuals within a species, and which may be favoured or eliminated by *natural selection. In sexual reproduction, reshuffling of *genes in each generation ensures the maintenance of variation. The ultimate source of the variation is *mutation, which produces fresh genetic material.

**variegation** The phenomenon in some plants in which patches of 2 or more different colours occur on the leaves or flowers; variegation may be an inherited characteristic or may be due to virus infection. The persistence of variegation in some plants suggests that it has adaptive significance. In plants such as *Sansevieria*, variegation may serve a camouflage function, especially in dappled light. A similar avoidance of predation may be served by the 'V' shaped variegation pattern on clover leaves that makes them appear smaller than they actually are.

**vascular bundle** A discrete, longitudinal strand that consists principally of *vascular tissue (*xylem and *phloem). Groups of these form a continuous conductive system throughout the plant along which water and soluble nutrients pass. This vascular system also contributes to the structure support of the plant.

**vascular cambium** *Cambium that produces *secondary phloem on its outer side and *secondary xylem on its inner side.

**vascular cryptogam** A vascular plant that reproduces by spores rather than seeds.

**vascular cylinder** *See* STELE.

**vascular plants** *See* TRACHEOPHYTA.

**vascular system** *See* VASCULAR BUNDLE and VASCULAR TISSUE.

**vascular tissue** *Tissue through which water and nutrients move. *See also* VASCULAR BUNDLE; PHLOEM; and XYLEM.

*Vatica* (family *Dipterocarpaceae) A genus of medium to small, evergreen trees which yield a siliceous timber (resak). There are 65 species, occurring in southern India, Sri Lanka, and from Hainan to New Guinea.

*Vaucheria* (division *Xanthophyta) A genus of *algae which have little in common with the other members of the Xanthophyta apart from the presence of forward-directed tinsel and backward-directed whiplash *flagella on the *spermatozoids. The algae are bright green, filamentous, *coenocytic, and irregularly branched. They are common in still or slow-moving fresh water and on damp soil.

*Vavaea* (family *Meliaceae) A genus of trees or treelets which have *simple leaves held in terminal clusters, and tiered branches. The fruit is a *berry, with the *pericarp rather woody. There are 4 species, occurring from Sumatra to Polynesia.

**Vavilov, Nikolai Ivanovich (1887–1943)** A Russian plant geneticist whose extensive field studies, in Iran, Afghanistan, Ethiopia, China, and Central and S. America, led him to the view that the greatest variation in species occurs in certain restricted areas (centres of diversity) which he believed identified the regions in which those species originated (centres of origin). He returned from his travels with a large collection of specimens he intended to use for study and to breed new varieties. He was professor of botany at the University of Saratov (1917–21) and later became head of the Lenin All-Union Academy of Agricultural Sciences. He was elected a member of the Academy of Sciences of the USSR in 1929, and a foreign member of the Royal Society of London in 1942. He was opposed by T. D. *Lysenko, who had him removed from his

V

positions in 1940. Vavilov is believed to have spent the years from 1940 in prison and to have died at Magadan, Siberia.

**vegetable ivory** The bony *endosperm of certain palms, mainly *Phytelephas* (New World) and *Metroxylon* (eastern tropics). See also CORYPHA.

**vegetation** A collection of plants of diverse or the same species. It is classified in many different ways, according to origin, appearance, and functional characteristics. Geographically, much emphasis is given to *climax vegetation, which reflects the influence of climate, soils, and humans.

**vegetation mosaic** The pattern of different plant *communities, or stages of the same community. The term is applied particularly to communities that show cyclical change (e.g. the *Calluna* cycle on heathlands), with examples of all stages being present together in a typically extensive and well-developed community.

**vegetative** Applied to a stage or structure that is concerned with feeding and growth rather than with *sexual reproduction. Vegetative reproduction is *asexual reproduction.

**vegetative cell** An actively growing cell, as opposed to a cell that forms *spores (*spore mother cell).

**vegetative nucleus** The *tube-nucleus of a *pollen grain in a flowering plant.

**vegetative propagation** (**vegetative reproduction**) A reproductive process that is asexual and so does not involved a recombination of genetic material. It involves unspecialized plant parts which may become reproductive structures, e.g. roots, stems, or leaves. Compared with *sexual reproduction, it represents a saving of material and energy for the plant. It is especially common among grasses.

**vegetative reproduction** See VEGETATIVE PROPAGATION.

**vegetative state** 1. A stage in the *life cycle of a plant when reproduction proceeds *asexually by detachment of some part of the plant body and its subsequent development into a complete plant. 2. The noninfective state in a phage during which the *genome multiplies actively

and controls the synthesis by the host (a *bacterial cell) of the materials necessary for the production of infective particles (i.e. more phages and phage *DNA). These are released by lysis of the host cell.

**Veillonellaceae** A family of *Gram-negative *bacteria that grow only in the absence of air. Cells are spherical and characteristically occur in pairs. They are non-*motile, and *chemo-organotrophic. They are found as *parasites in animals, including humans, occurring chiefly in the intestine.

**vein** A *vascular bundle, or group of vascular bundles lying parallel to one another and very close together, in a leaf.

**vein-banding** A symptom of some virus diseases of plants in which bands of lighter or darker colour occur along the main veins of a leaf.

**vein-clearing** A symptom of some virus diseases of plants in which the veins become unnaturally clear or translucent.

**velamen** Several layers of densely packed, dead cells on the *epidermis of the aerial roots of epiphytic (see EPIPHYTE) orchids and other plants that absorb water as it flows over them.

**veld** Extensive grasslands in the east of the interior of S. Africa, often with a scattering of trees or bushes. In these instances the names *tree veld and bush veld respectively are applied. True grass veld is confined to the high terrain.

**Vellozia** (family Velloziaceae) A genus of monocotyledonous (see MONOCOTYLEDON) plants that have woody, branching stems, and crowded, narrow, *xeromorphic leaves with persistent bases. The flowers are regular and trimerous, and the *ovary is *inferior. The fruit is a woody *capsule. There are 124 species, occurring mostly in seasonal tropical America, but also in Africa and Madagascar.

**velum** 1. A flap of membranous tissue that protects a *sporangium in *Isoetaceae. 2. In certain toadstools and mushrooms, the remains of the *partial veil after it has ruptured.

**velvet shank** See FLAMMULINA.

**venation** The arrangement of the veins or nerves of a leaf.

**Venezuela and Guiana floral region**
Part of R. Good's (*The Geography of the Flowering Plants*, 1974) *Neotropical kingdom, a region that has affinities with the adjacent Andean and Amazon floral region, but nevertheless constitutes an independent unit with about 100 endemic genera (*see* ENDEMISM). The flora is poorly known and has yielded no plants of any great value. *See also* FLORAL PROVINCE and FLORISTIC REGION.

**Venice system** A system for the classification of *brackish water based on the percentage of chlorine contained in the water.

**venter** The swollen base of an *archegonium that contains the *megaspore.

***Ventilago*** (family *Rhamnaceae) A genus of trees and shrubs of which many are climbers using hooks. They have *opposite or alternate, *simple leaves without *stipules. The flowers are small, inconspicuous, bisexual, regular, and usually borne in *cymes. There are 4 or 5 *sepals and 4 or 5 incurved *petals, often enclosing the *stamens. The *ovary is *superior with 2 or 3 *locules with a single *ovule. The fruit has a single wing on the upper end. There are 35 species, found mainly in Asia, but there is 1 species in Africa and another in Madagascar.

**ventral** The surface nearest the substrate or facing the axis.

***Venturia*** (order *Dothideales) A genus of *fungi that occur as *parasites and *pathogens of higher plants. The *ascospores are typically greenish-yellow, elongated and *septate, and are formed in *pseudoperithecia which are immersed in the substrate. *V. inaequalis* is the cause of *apple scab.

**Verbenaceae** A family of *herbs, shrubs, trees, and woody climbers, mostly with *opposite leaves that are variously arranged, but are always irregular and hermaphrodite, with *superior *ovaries. The *calyx and *corolla are often 2-lipped, and the *petals are joined into a tube below. The ovary has a terminal *style, and is usually 4-celled. The fruit is usually a *drupe, sometimes a *capsule, or rarely consists

of 4 *nutlets. The family contains many useful plants, e.g. *Tectona grandis* (teak), valuable for its timber in south-eastern Asia. Species of *Verbena* and *Clerodendrum* are grown for their attractive flowers. *Avicennia* is a mangrove of coastal swamps. Modern classifications recognize some 91 genera and 1900 species. Most are tropical and subtropical, but some extend into temperate zones.

**vernal** In the late spring. The term is used with reference to the 6-part division of the year used by some ecologists, especially in relation to studies of terrestrial or freshwater communities. *Compare* AESTIVAL; AUTUMNAL; HIBERNAL; PREVERNAL; and SEROTINAL.

**vernalization** The treatment of *germinating seeds with low temperatures to induce flowering at a particular preferred time. For example, winter varieties of wheat can be sown in the spring and then be exposed to a temperature just above 0°C for a few weeks. The result of this is that they behave like spring varieties and flower in the same year (otherwise they would continue to grow *vegetatively and would not flower until the following year). The stimulus is perceived by the *apical meristem (either in the *embryo or as an apical bud), and some plant hormones such as *gibberellin can be used to achieve the same effect. *Compare* STRATIFICATION.

**vernation** The arrangement of *bud scales or young leaves in a shoot bud. They

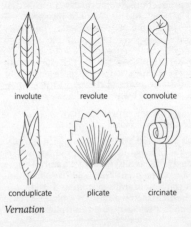

involute    revolute    convolute

conduplicate    plicate    circinate

*Vernation*

may be rolled lengthwise (circinate); pleated lengthwise (plicate); rolled from the sides to the centre of the underside (involute); rolled from the sides to the centre of the upper side (revolute); rolled from the sides with one side folded around the other (convolute); or folded so that each leaf clasps those next to it inside the fold (conduplicate).

**Veronica** (speedwell; family *Plantaginaceae) A genus of *annual or *perennial *herbs or shrubs that have simple or lobed, *opposite leaves without *stipules. The flowers are usually blue, but sometimes pink or white. They are bisexual and are held in *axillary or terminal *racemes, or, if they are 5-lobed, solitarily in leaf axils. The *calyx is fused and 4-lobed, or 5-lobed but with a very small upper lobe. The *corolla is fused into a very short tube with 4 lobes; the upper, composed of 2 fused components, being the largest. There are 2 *stamens and a *superior *ovary of 2 fused *carpels. The fruit is an elongated, flattened *capsule. There are about 250 species, present throughout temperate regions in a wide variety of habitats. Many species are cultivated as ornamentals.

**Verrucariales** (subdivision *Ascomycotina) An order of (mainly lichenized) *fungi in which the *thallus is usually *crustose, sometimes *foliose or *squamulose, and grey to brown in colour. The *ascocarps are perithecioid (*see* PERITHECIUM) and may be *sessile or immersed in the thallus or substrate. There are several genera, found mainly on rocks, walls, etc.

**verrucose** Warty in appearance.

**vertic horizon** A soil *horizon composed of clayey material with polished *ped surfaces, often with slickensides (crack surfaces caused by swelling and shrinking), or wedge-shaped soil *aggregates. The name is from the Latin *vertere*, to turn.

**verticillate** Arranged in a *whorl or whorls.

**Verticillium** (class *Hyphomycetes) A *form genus of *fungi that produce colourless, *aseptate *conidia. Species are responsible for a number of important plant diseases, including a type of *wilt.

**verticillium wilt** A plant disease caused by a *fungus of the genus *Verticillium*. The infected plant may show *wilting, with bending downwards and yellowing of the lower leaves. When the stem is cut well above ground level, characteristic brown streaks can be seen inside the stem following the conducting elements.

**Vertisols** *Mineral soils, an order of soils that contain more than 30 per cent by weight swelling *clay (e.g. montmorillonite), and that expand when wet and contract when dry to produce a self-inverting soil and an undulating (gilgai) microrelief. Vertisols are associated with seasonally wet and dry environments, and are extensive in the tropics, forming beneath grassland. Vertisols are a reference soil group in the FAO *soil classification.

**vesicle 1.** Generally, any small *bladderlike structure containing a fluid. **2.** A small, membrane-bound, fluid-filled sphere that occurs, often in large numbers, in the cytoplasm of many eukaryotic cells. Vesicles may be variously associated with the uptake and discharge of materials in cells and also with their transport and storage.

**vessel** *See* TRACHEA and VESSEL ELEMENT.

**vessel element** A *tracheary cell with thickened walls and *perforation plates at either end, many vessel elements being joined end to end in *xylem to form a vessel. *See* TRACHEA and TRACHEID.

**vetch** *See* VICIA.

**viability** The probability that a fertilized egg will survive and develop into an adult organism, often applied to plant *germination experiments with comparisons across phenotypic classes under standard specific environmental conditions.

**Vibrio** (family *Vibrionaceae) A genus of *bacteria in which the cells are straight or curved rods, and typically have 1 or more *flagella at 1 end. They are *chemo-organotrophic, and can grow in the presence or absence of air. They are found primarily in aquatic environments and in association with aquatic animals (e.g. copepods). The genus includes some important *pathogens of humans (e.g. the causal agent of cholera), fish, and shellfish.

V

**Vibrionaceae** A family of *Gram-negative *bacteria in which the cells are straight or curved rods, and are typically *motile with 1 or more *flagella at 1 end. There are several genera, found as *saprotrophs in freshwater or marine environments, and in association with aquatic animals; the family also includes some important *pathogens of humans and other animals.

**vicariad** See VICARIANCE.

**vicariance (vicariad, vicarious species)** The geographical separation of a species so that 2 closely related species or a species pair result, 1 species being the geographical counterpart of the other. There are many examples in the floras of N. America and Eurasia.

**vicariance biogeography** A school of biogeographical thought, derived from Croizat's *panbiogeography. Supporters of this school maintain that the distribution of organisms depends on their normal means of dispersal; e.g., disjunctions are explicable in terms of new barriers (rivers, rises in sea-level, etc.) having split formerly continuous ranges, rather than in terms of the organisms hopping over already existing barriers. Thus, they reject sweepstake routes and similar concepts, postulating instead former land bridges and even vanished continents where there is sufficient coincident plant and animal distribution.

**vicarious distribution** Closely related species pairs that are derived from a common ancestor are said to be vicarious. When 1 species replaces the other geographically (as opposed to ecologically) their distribution is vicarious. There are many herbaceous plants with a vicarious distribution between N. America and Europe.

**vicarious species** See VICARIANCE.

**Vicia (vetches; family *Fabaceae, subfamily *Papilionatae)** A genus of herbs, most of which climb by means of leaf-*tendrils. The leaves are usually *pinnate, in several pairs, with no terminal leaflets. The flowers are in *axillary *racemes, and have an obliquely truncate *stamen-tube. The *style is normally cylindrical, and either hairless or downy all round, and bearded below the *stigma (in the similar genus Lathyrus the style is flattened, and bearded on its upper side). The *pods are 2-valved, flattened, and contain several seeds in each. Vicia faba (broad bean) is an important vegetable, V. sativa (common vetch or tare) an important fodder plant. There are some 140 species, mostly northern temperate, but also occurring in S. America and tropical E. Africa.

**Victoria (family *Nymphaeaceae)** A genus of giant water-lilies (V. amazonica, formerly V. regia), famous for its enormous, almost circular leaves, up to 2 m across with radial rims, and V. cruziana, which occur in rivers and lakes in the Amazon region.

**Vigna (family *Fabaceae, subfamily *Papilionatae)** A genus of vetch-like (see VICIA), twining *herbs, usually with 3 leaflets, and with showy flowers grouped on long *peduncles. Nine of the *stamens are joined below, 1 is free. The *style has a line of hairs along 1 side only. The *pods are not flattened. There are about 150 species, occurring in the topics, especially in the Old World.

**Villafranchian** An age that is dated at base at approximately 3 Ma ago. It lasted approximately 2 Ma and therefore crosses the Late *Pliocene/Early *Pleistocene boundary.

**Vinca (periwinkle; family *Apocynaceae)** A genus of 7 species of creeping shrubs or *perennial *herbs, found in Europe and western Asia.

**Viola (pansy, violet; family *Violaceae)** A genus of dicotyledonous (see DICOTYLEDON) herbs with *simple alternate or basal leaves. The flowers are *zygomorphic, with 5 *sepals and 5 *petals, the lower 2 often fused, with prominent guide-lines and a contrasting centre. Many species have been domesticated and *hybridized to produce the showy pansies grown in garden borders. There are about 500 species, of widespread temperate range.

**Violaceae** A family of dicotyledonous (see DICOTYLEDON) *herbs, and some shrubs and trees, in which the leaves are *simple and *stipulate. The flowers are usually solitary, though sometimes in *racemes, mostly irregular, with 5 free *sepals, *petals, and *stamens, and the *corolla is frequently spurred. The *superior *ovary is 1-celled, with usually 3 *parietal placentae, with a single, often curved and thickened *style. Modern classifications recognize some 23 genera, with 930 species, found throughout much of the world. *Viola* (violets, pansies), containing herbs with alternate leaves and spurred corollas, is the largest genus and is mainly temperate. Various species are cultivated for their attractive flowers.

**violet** *See* Viola *and* Violaceae.

**Virchow, Rudolf (1821–1902)** A German pathologist and anthropologist (and also a politician of liberal views) whose studies of pathogenic organisms led him to the discovery that cells are derived from other cells, which he summarized as *omnis cellula e cellula* ('every cell from a cell').

*Vireya* *See* Rhododendron.

**Virginia creeper** *See* Parthenocissus.

**virion** An individual virus particle.

**viroid** A piece of infectious *nucleic acid. Viroids appear to resemble viruses in some respects, but consist only of small, closed circles of *RNA: there is no capsid. They can cause disease in plants (e.g. potato spindle tuber disease and hop stunt).

*Virola* **(family *Myristicaceae)** A genus of plants some of which are big trees and an important source of timber (banak). *V. suri-namensis* is one of the major timber species of the Amazon rain forest. There are about 60 species occurring in central and southern regions of tropical America.

**virulent phage** A *bacteriophage that causes the destruction of the host *bacterium by *lysis. *Temperate phages, on the other hand, rarely cause lysis.

**virus** A type of non-cellular 'organism' which has no metabolism of its own. It consists mainly or solely of a *nucleic acid *genome (*RNA or *DNA) enclosed by *protein; in some cases there is also a *lipoprotein envelope. In order to replicate (multiply), a virus must infect a cell of a suitable host organism where it redirects the host-cell metabolism to manufacture more virus particles. The progeny viruses are released, with or without concomitant destruction of the host cell, and then can infect other cells. All types of organism, from bacteria to humans, are susceptible to infection by viruses; virus infections may be asymptomatic or may lead to more or less severe disease.

**viscotaxis** A change in direction of locomotion in a *motile organism or cell, made in response to a change in the viscosity of the surrounding medium.

*Viscum* **(mistletoe)** *See* Loranthaceae.

**visible** In genetics, applied to a *mutant whose *phenotype may actually be observed, as opposed to *lethals, whose occurrence may be inferred only from the absence of an expected class of individuals in the progeny of a cross designed to detect induced mutants.

**vital stain** A stain that is capable of entering and staining a living cell without causing it injury.

*Vitellaria* **(formerly *Butyrospermum*; family *Sapotaceae)** A *monotypic genus (*V. paradoxa*) which is a north-tropical African tree whose seeds yield shea butter when pressed.

**vitric horizon** A soil *horizon that contains more than 10% volcanic glass or other volcanic material. The name is from the Latin *vitrum* meaning 'glass'.

**vitta** *See* Apiaceae.

**viviparous** Applied to: (*a*) a plant (e.g. mangroves) whose seeds *germinate within and obtain nourishment from the *fruit; or (*b*) a plant (e.g. some grasses) that reproduces vegetatively from shoots rather than an inflorescence.

**volunteer plant** A domesticated plant that has resulted from natural propagation, as opposed to having been deliberately planted by humans.

**volva** A cup-like sheath surrounding the base of the *stipe in some *agaric *fruit bodies; the volva is a remnant of the *universal veil after the rupture of the latter.

*Volvox* **(division *Chlorophyta)** A genus of *green algae (sometimes alternatively regarded as protozoa, class *Phytomastigophora) in which the cells occur in spherical, *motile colonies (*coenobia) of 500 to many thousands of cells, according to species. The individual cells resemble those of *Chlamydomonas* species. The genus shows considerable complexity in morphology and sexual reproduction. Species are found in various aquatic habitats: puddles, ponds, ditches, lakes, etc.

**Wahlenbergia** (Australian bluebell; family *Campanulaceae) A genus of small herbs with alternate or *opposite leaves which are sometimes toothed. The blue, solitary flowers are held on long stalks with 5 *sepals, and a 5-lobed, bell-shaped *corolla. There are 5 *stamens, and a 3-lobed *stigma. The *capsule is urn-shaped, and splits into 3 to release the numerous, very fine seeds. Many species are *annuals, and many are cultivated. There are about 150 species found especially in southern temperate regions.

**waldgrenze** *See* TIMBER-LINE.

**wallaba** *See* EPERUA.

**Wallace, Alfred Russel (1823–1913)** A British naturalist who worked in the E. Indies and as a result of his observations there arrived independently at a theory of evolution by natural selection. This was communicated to Charles *Darwin and their views were combined in the paper they presented jointly to the Linnean Society of London on 1 July 1858.

**wall pressure** *See* PRESSURE POTENTIAL.

**walnut** *See* JUGLANS.

**wandering Jew** (*Tradescantia*) *See* COMMELINACEAE.

**Warming, Johannes Eugenius Bülow (1841–1924)** A Danish botanist whose work laid the foundations for and greatly stimulated the study of modern plant ecology. He maintained that plant communities should be studied in relation to their surroundings and developed a basis for their classification. This view was expounded in his book *Plantesamfund* (1895, published in English in 1925 as *Oecology of Plants*). From 1882 to 1885 he was professor of botany at the Royal Institute of Technology in Stockholm, and from 1885 to 1911, professor of botany and director of the botanical gardens at Copenhagen.

**warping** A traditional farming practice in which a river is permitted to flood low-lying ground temporarily in order to supply silt in which crops can be grown without irrigation.

**warp soil** Soil consisting principally of *silt deposited on land by deliberate flooding (warping).

**Washingtonia** (family *Arecaceae) A genus of massive palms which are hardy and cultivated. The leaves are fan-like and persistent, forming a shaggy, brown petticoat investing the trunk. The flowers are hermaphrodite, with a single trilocular *ovary. The fruit contains 1 seed. There are 2 species, 1 occurring in California, the other in Mexico.

**waste** **1.** An area of unenclosed land used for common pasture. **2.** Unlicensed felling of trees. **3.** Over-exploitation of *woodland or wood-pasture or the failure to enclose *coppice after felling, resulting in stock being allowed to graze to the detriment of the undergrowth.

**water chestnut** *See* ELEOCHARIS.

**watercress** *See* NASTURTIUM.

**water hyacinth** *See* EICHHORNIA and PONTEDERIACEAE.

**water lettuce** *See* PISTIA.

**water-lily** *See* NYMPHAEA and NYMPHAEACEAE.

**water milfoil** *See* MYRIOPHYLLUM.

**water mould** An aquatic *fungus; sometimes specifically a fungus of the order *Saprolegniales.

**water potential** **1.** A measure of the energy that causes water to enter plant cells. There is a net movement of water from a region of high water potential to one of low water potential. **2.** Surface tension. *See* CAPILLARY MOISTURE.

**water table** The upper surface of *groundwater or the level below which the material is permanently saturated with water. The region below the water table is the phreatic or vadose zone.

**Watson, James Dewey** (b. 1928) The American geneticist who, with F. *Crick and M. Wilkins, won the 1962 Nobel Prize for Physiology or Medicine for their modelling of the *DNA molecule. Watson and Crick worked at the Cavendish Laboratory, Cambridge; since 1968 Watson has been director of the Cold Spring Harbor Laboratory.

**Watson–Crick model** The currently accepted model for the structure of *DNA, as proposed by J. D. Watson and F. H. C. Crick (1953). It is suggested that DNA is composed of two right-handed, anti-parallel polynucleotide chains coiled around a common axis to form a double helix. This structure is maintained by *hydrogen bonds formed between the chains through the base pairing of adenine to thymine and cytosine to guanine.

**waverer** A young tree that is left standing after *coppice wood has been cut.

**wax cap** The common name for the waxy, typically brightly coloured, *fruit bodies of *fungi of the family *Hygrophoraceae.

**wax flower** *See* ERIOSTEMON.

**wax palm** *See* CARNAUBA WAX and *COPERNICIA*.

**wedding bush** *See* RICINOCARPUS.

**weed** A plant that occurs opportunistically on land that has been disturbed by human activity (*see* RUDERAL) or on cultivated land where it competes for nutrients, water, sunlight, or other resources with cultivated plants. The presence of weeds among crop plants can result in reduced yield because of competition; in gardens weeds may be unsightly; in waterways they can become a hindrance to navigation (e.g. the water hyacinth, *Eichhornia crassipes*).

**Weichselian** *See* DEVENSIAN.

**Weinmannia** (family *Cunoniaceae) A genus of trees in which the leaves are *opposite, *pinnate, with big *stipules, and the leaflets toothed. The flowers are tiny, with 4 *petals, 8 *stamens, and the *ovary *superior and bilocular. The fruit

is a *capsule. There are about 190 species, occurring in Madagascar, the Mascarenes, from Malaysia to New Zealand and the Pacific islands, and from Mexico to Chile.

**Weismann, August (1834–1914)** A German biologist who established the improbability, if not impossibility, of the inheritance of acquired characteristics, as required by *Lamarck's theory of *adaptation.

**wellingtonia** *See* SEQUOIADENDRON and TAXODIACEAE.

**Welwitschia** The sole genus, with only one species (*W. mirabilis*), of the remarkable *gymnosperm family Welwitschiaceae (related to Ephedraceae (*see* EPHEDRA) and Gnetaceae (*see* GNETUM) of the deserts of south-western Africa). It has a barrel-like stem with a deep tap root, bearing on its rim big, tattering, strap-like leaves, and flowers that are small, *dioecious, and covered by *bracts, with the single, erect *ovule within the *perianth tube of the female flower.

**Wenlock (Wenlockian)** The stratigraphic name given to the Middle *Silurian succession throughout Europe. Reef-dwelling organisms abound on the bedding planes of the Wenlock limestone of the Welsh borderlands.

**West African rain-forest floral region** Part of R. Good's (*The Geography of the Flowering Plants*, 1974) African subkingdom, which contains a rich but poorly documented flora which has provided some valuable economic plants, e.g. *Coffea liberica*, and others of more local importance, including timber trees. *See also* FLORAL PROVINCE and FLORISTIC REGION.

**West and Central Asiatic floral region** Part of R. Good's (*The Geography of the Flowering Plants*, 1974) Boreal kingdom, which contains a limited, specialized flora, because much of the region comprises dry mountains and deserts. There are about 150 endemic genera (*see* ENDEMISM), virtually all small. Barley and some types of wheat probably originated in this region. *See also* FLORAL PROVINCE and FLORISTIC REGION.

**western red cedar** *See* THUJA.

**W**

**wet rot** A type of timber decay found only in wood with a high moisture content. It is caused by the cellar *fungus, *Coniophora puteana.*

**wheat** *See* TRITICUM.

**whiplash** An alternative, little-used word for *flagellum.

**whisk fern** *See* PSILOTUM.

**whitebeam** *See* SORBUS.

**white blister** *See* WHITE RUST and ALBUGINACEAE.

**white coral fungus** *See* CLAVULINA.

**white ipecacuanha** *See* HYBANTHUS.

**white lauan** *See* PARASHOREA.

**white meranti** *See* SHOREA.

**white rot 1.** A disease of onions, shallots, and leeks, caused by the *fungus *Sclerotium cepivorum.* Leaves of infected plants turn yellow, and a fluffy, white *mycelium appears on the bulb. Small, black *sclerotia can be seen in the mycelium and these can survive in the soil for many years. **2.** A type of timber decay in which the cellulose, hemicellulose, and lignin components of the wood are decomposed, leaving the wood soft, white, and fibrous.

**white rust 1. (white blister)** A disease affecting brassicas and other plants, caused by species of the genus *Albugo.* White, blister-like spots are formed on the leaves, and infected plants may show some distortion. **2.** A disease of chrysanthemums caused by *Puccinia horiana.* Buff or white pustules appear on the undersides of the leaves and may become brown and necrotic (*see* NECROSIS). This disease first appeared in Britain in 1963 and is now a notifiable disease.

**white seraya** *See* PARASHOREA.

**whorl** The arrangement in which leaves, *petals, etc. all arise at the same point on an *axis, encircling it.

***Widdringtonia*** (family *Cupressaceae) A genus of small to medium-sized, evergreen, cypress-like conifers with small, scale leaves arranged in pairs. The cones have only 4 scales. The timber of *W.*

*cedarbergensis* (Clanwilliam cedar) is very valuable and is used for high-quality furniture. Young trees of *W. nodiflora* (Mlanji cedar) are used as a substitute for Christmas trees. There are 5 species, occurring in central and S. Africa.

**wide distribution (polychore distribution)** A situation in which taxonomic groups of plants have a very extensive distributional range, spanning several *floral kingdoms or regions, e.g. 'wides' may be cosmopolitan, subcosmopolitan, tropical, or temperate.

**wig tree** *See* COTINUS.

**wilderness** An extensive area of land that has never been permanently occupied by humans or subjected to their intensive use (e.g. for mineral extraction or cultivation) and that exists in a *natural, or nearly natural state. Wilderness areas are selected for their ecological wholeness, rather than for the presence of any particular *biota, landscape, or recreational attraction. In the USA, where wilderness areas have been formally designated, no economic use is allowed except by presidential decree in extreme emergency. The areas are free from traffic, and the number and activities of visitors are carefully controlled. Elsewhere, the concept merges with that of national parks, wilderness areas often being zones of more restricted public access within the park areas.

**wildlife** Any undomesticated organisms, although the term is sometimes restricted to wild animals, excluding plants.

**wild oat (*Avena fatua*)** *See* HERBICIDE.

**wild service tree** *See* SORBUS.

**wild-type gene** The *allele most frequently observed at a given *gene *locus.

**wilga** *See* GEIJERA.

**willow** *See* SALIX.

**willow moss** The common name for the moss *Fontinalis antipyretica. See* FONTINALIS.

**wilt** A type of plant disease in which wilting (i.e. loss of plant turgidity) is a principal symptom. Wilts are frequently due to infection of the plant by a fungus belonging to the *Deuteromycotina. As an example, *see* VERTICILLIUM WILT.

**wilting** The limpness found when plant *tissues contain insufficient water to hold the cells rigid. This may occur when the rate of *transpiration exceeds the rate at which water is able to enter the root system from a soil containing ample water, causing temporary wilting from which the plant recovers when the transpiration rate falls. It may also be due to a deficiency of water in the soil. When water shortage is prolonged or acute, the plant may reach a point from which recovery is impossible even if abundant water is subsequently supplied. *See* PERMANENT WILTING POINT.

**wilting coefficient** *See* PERMANENT WILTING POINT.

**wilting point** *See* PERMANENT WILTING POINT.

**Windermere Interstadial (Late-Devensian Interstadial)** A relatively warm period that occurred towards the end of the last (*Devensian) glaciation in Britain. The event took place about 13 000–11 000 radiocarbon (*see* RADIOCARBON DATING) years BP. It includes the *Bølling, Older *Dryas, and *Allerød chronozones of Scandinavia. The pollen sequence shows a sharp rise followed by a marked fall in birch and juniper, marking the beginning and end respectively of the interstadial.

**windmill palm** *See* TRACHYCARPUS.

**wing nut** *See* PTEROCARYA.

**Winkler method** The standard chemical procedure for measuring oxygen in water.

**Winteraceae** A family of trees and shrubs in which the leaves are *simple, alternate, dotted with glands, and without *stipules. The flowers have many free parts. The wood is without vessels. Winteraceae are the southern hemisphere counterpart to the Magnoliaceae. There are 5 genera, with about 60 species, occurring from Borneo to the Pacific islands and Australasia, and also in S. America.

**Wisconsin School** A group of ecologists, led by J. T. Curtis and his associates, who developed a range of simple *ordination methods in 1950 and later, while studying the vegetation of Wisconsin. In the most widely applied scheme, density–frequency-dominance (DFD) values are estimated for all species recorded in the field. Importance values derived from these enable identification of *leading dominants. Stands with the same leading dominant are grouped together and importance values for all species recalculated within the groups. These values are used to assign climax adaptation numbers and subsequently a continuum index is derived. Later developments, notably by Bray and Curtis (1957), improved the way in which stand similarities were expressed and portrayed graphically, by using a similarity index calculated from stand data rather than using the subjectively assessed continuum index.

**witches' broom** A dense, broom-like cluster of proliferating branches or twigs at a particular point in a tree or shrub; from a distance it may resemble a bird's nest. Witches' brooms may be caused by infection with any of a number of organisms, including certain *rusts; *Taphrina deformans* and *T. turgida* cause witches' brooms in birch trees (*Betula pubescens* and *B. pendula* respectively).

**witches' butter** *See* EXIDIA.

**witch hazel** *See* HAMAMELIDACEAE.

**wobble hypothesis** A theory proposed to explain the partial degeneracy of the *genetic code in that some *t-RNA molecules can recognize more than one *codon. It is proposed that the first 2 bases in the codon and *anticodon will form complementary pairs in the normal antiparallel fashion. However, a degree of steric freedom or 'wobble' is allowed in the base pairing at the third position. Thus, for serine, 6 *m-RNA codons may be paired with only 3 t-RNA anticodons.

**Wolffia** (family *Lemnaceae) A genus of minute floating herbs, which have a tiny *thallus with no roots or leaves. *W. arrhiza* is the smallest known flowering plant and is scarcely visible to the naked eye. There are 7 species, with a cosmopolitan distribution.

**wood** **1.** *The secondary xylem of *dicotyledons and conifers, which forms a dense growth during *secondary thickening, providing the mechanical support which allows *trees to grow to a considerable height. **2.** An area of trees, often associated with a particular name (e.g. Hayley Wood) that denotes a district area. **3.** The produce of *coppice or underwood that is not of timber size.

**wood blewit (Lepista nuda)** See BLEWIT.

**woodland** **1.** A vegetation *community that includes mature trees, which are usually spaced more widely and so are more spreading in form that *forest trees (i.e. their crowns are not touching and they do not form a closed canopy). Woodland is often defined as having 40 per cent canopy closure or less. Between the trees, grass, *heath, or *scrub communities typically develop, giving a park-like landscape. **2.** A general term for a wooded landscape, often used generally, e.g. broadleaved woodland, or to describe a number of separate wooded areas, e.g. the Estate Woodlands. Colloquially, the terms 'forest' and 'woodland' are often used interchangeably in Britain.

**wood mushroom** The common name for the *fruit body of *Agaricus silvicola. The *fruit body resembles that of the cultivated *mushroom but has an odour of aniseed. It is found on the ground in woods during late summer and autumn and is edible.

**wood-pasture** Land on which trees grow and where farm livestock or deer are grazed systematically. See also POLLARDING.

**Woodsia** **(family *Aspleniaceae)** A genus of small ferns, characterized by a short *rhizome, very scaly frond undersides, and *orbicular *sori on the undersides of the leaves, with an *indusium surrounding the base of the sorus, which is either split into hair-like segments from the beginning, or starts as a cup, splitting into hair-like segments later. There are 21 species, occurring in the Arctic and in northern temperate mountain regions.

**wood sorrel** See OXALIS.

**wood sugar** See XYLOSE.

**woody** Applied to plants that contain *wood.

**wormwood** See ARTEMISIA.

**wound response** The metabolic activity that follows the disruption of plant structure by external forces, e.g. by wind or freezing. This activity is controlled by *hormones in higher plants and typically includes an increased rate of synthesis of *callose and gummy substances, the formation of more *endoplasmic reticulum, and more *mitotic divisions, followed by the differentiation of roots and *buds.

**wound-tumour virus** An *RNA-containing, plant-infecting virus which can cause the formation of *galls in a wide range of plants. The virus can be transmitted from plant to plant by leafhoppers.

**wrack** The common name for any seaweed of the family *Fucaceae. See also BLADDER WRACK; EGG WRACK; SERRATED WRACK; and SPIRAL WRACK.

**Wright's inbreeding coefficient** See COEFFICIENT OF INBREEDING.

**Würm** See DEVENSIAN.

W

**Xanthomonas** (family *Pseudomonadaceae*) A genus of *Gram-negative *bacteria in which cells are straight rods, each with a single polar *flagellum. They are *aerobic *chemo-organotrophs. All species are plant *pathogens, causing various types of disease in a wide range of plants.

**xanthophyll** *See* CAROTENOID.

**Xanthophyllum** (family *Polygalaceae*) A genus of rain-forest trees whose leaves often dry yellowish. The flowers are *pentamerous, usually with one hood-like *petal. The fruit is *indehiscent, woody or somewhat fleshy, containing 1 seed or many. The trees yield useful timber. There are 93 species, occurring from India and Hainan to the Solomon Islands.

**Xanthophyta** (yellow-green algae) A division of algae in which the *chloroplasts are yellow-green and which form *motile cells with 1 long, forward-directed tinsel *flagellum and 1 much shorter, backward-directed whiplash flagellum. These cells typically function as motile spores (*zoospores). 'Mature' organisms may be unicellular, colonial, filamentous, or multinucleate and *coenocytic (*siphonaceous). Immotile unicellular forms can usually convert readily to the flagellated motile form. Xanthophytes are found mainly in freshwater habitats. Genera include *Botrydium and *Vaucheria.

**Xanthoria** (order *Teloschistales) A genus of *lichens in which the *thallus is *foliose, usually bright yellow or orange in colour, and is attached to the substrate by *rhizines on the lower surface. *X. parietina* is perhaps the best-known of all lichens; its bright yellow rosette-like thalli, each bearing crowded *apothecia in the centre, are a familiar sight on farm buildings, on rocks by the sea, etc. This species is moderately tolerant of air pollution.

**Xanthorrhoea** (family *Xanthorrhoeaceae*) A genus of grass-trees, which are woody, *perennial plants with narrow, sheathing, and linear leaves, and dry, papery flowers. The outer parts of the flower are like *glumes, and thicker than the inner parts which are membranaceous and dry. The plants are often very tall, and are found in dry, open habitats and in *Eucalyptus* woodland, often on soils of low fertility. *X. hastilis* (black-boy) is found in the Australian bush, where it forms a tussock with a long, bulrush-like *spike of flowers. It yields a resin from the leaf bases which is used in varnish and sealing wax. There are 15 species, found only in Australia.

**Xanthorrhoeaceae** (grass-trees) A family of monocotyledonous (*see* MONOCOTYLEDON), stout, woody, perennial plants with *rhizomes, which are often *xerophytic. They are tall, with few branches, and *simple, linear, sheathing leaves. The regular flowers are bisexual or unisexual, on *dioecious plants, and are held in *panicles, *spikes, or clusters. The *perianth comprises 2 *whorls of 3 segments, and is usually dry and papery. There are 2 whorls of 3 *stamens, the inner whorl usually attached to the flower segments. The *superior *ovary is of 3 fused *carpels, with 3 *locules containing 1 or more *ovules. The fruit is a *capsule, or a single-seeded *nut. The plants are often components of dry, open *Eucalyptus* scrub or woodland. There are 9 genera, with 60 species, of which 7 genera are endemic (*see* ENDEMISM) to Australia.

**Xanthosoma** (family *Araceae) A genus of plants, several species of which are cultivated for their starchy, edible, lateral tubers, known as tannia or yautia. There are 45 species, occurring in the W. Indies, and in central and S. America.

**X-chromosome** The *sex chromosome found in a double dose in the *homogametic sex and in a single dose in the *heterogametic sex. Unlike the

*Y-chromosome, it contains numerous *genes (which therefore show *sex-linkage; *see* SEX DETERMINATION).

**xenia** A situation in which the *genotype of the *pollen influences the developing embryo or the maternal tissue (*endosperm) of the *fruit to produce an observable effect on the *seed. For example, a variety of maize with white endosperm may be *pollinated by one possessing dark yellow endosperm to give seeds with pale yellow endosperm.

**xenogamy** *Fertilization involving *pollen and *ovules from different flowers on genetically distinct plants.

**xeric** A dry, as opposed to a wet (hydric) or intermediate (mesic), environment.

**xeromorphic (zeromorphic)** Applied to organisms that show morphological adaptations that enable them to withstand drought.

**xerophile (zerophile)** A plant of warm, dry *desert environments in Alphonse de *Candolle's (1874) classic temperature-based scheme of world vegetation zones. *Compare* MEGATHERM.

**xerophyte (zerophyte)** A plant that can grow in very dry conditions and is able to withstand periods of drought. The adaptations include an ability to store water, waxy leaves and leaves reduced to spines to avoid water loss through *transpiration, and short life cycles (ephemeral) that can be completed when sufficient water is available.

**xerosere** The characteristic sequence of *communities reflecting the developmental stages of a plant *succession that begins in an arid environment.

**xylan** A *polysaccharide containing *xylose that is commonly found in the *cell walls of higher plants. The molecules consist of a backbone of beta-xylose units linked between the first and fourth carbon atom of neighbouring sugar units, with side chains of various degrees of complexity and containing a variety of other sugars. *See also* HEMICELLULOSE.

*Xylaria* (order *Sphaeriales) A genus of

wood-rotting *fungi. *X. hypoxylon* forms branching, antler-like *fruit bodies (*stromata), the upper parts of which are white with powdery *conidia. *Perithecia are formed on the lower parts of the stromata. *X. polymorpha* forms clusters of finger-like stromata on old tree stumps.

**xylary** Pertaining to *xylem.

**xylem** A plant *tissue consisting of various types of cells, which transports water and dissolved substances towards the leaves. It can be distinguished from the *phloem by the presence of vertical systems of dead cells with thick, *lignified walls. *See also* TRACHEA and TRACHEID.

*Xylocarpus* (family *Meliaceae) A genus of mangrove forest trees which have big, spreading buttresses. The fruit is a large, spherical *capsule. The seed has a corky *testa, adapted thereby to water dispersal. The trees yield a valuable cabinet timber. There are 3 species, occurring in the Old World tropics from E. Africa to the Pacific.

**xylophilous** Preferring to grow on wood.

**xylose (wood sugar)** An aldopentose sugar that is commonly found in plants and especially in woody tissues.

**Xyridaceae** A family of monocotyledonous (*see* MONOCOTYLEDON), herbaceous plants, which are mostly *perennial. They are rush-like marsh plants with radical, sheathing leaves, and dense *spikes or heads of flowers. The leaves are linear and cylindrical, or with narrow, flat blades. The flowers are bisexual and held in the *axil of a stiff *bract. The inner of the 3 *sepals forms a hood over the *petals, while the others are keeled. The *corolla is a tube with 3 lobes, and is usually yellow in colour. The *stamens are held opposite the *corolla lobes and are attached to the tube. The *ovary is *superior, with the 3 *carpels fused to form a single *locule with many *ovules, and a 3-lobed *style. The corolla remains on the seed *capsule. There are 5 genera, with about 260 species, distributed mainly throughout tropical and subtropical regions, and centred in America.

**X**

**yam** *Tubers of about 10 cultivated species of *Dioscorea*, most of which are tropical, but 2 of which (*D. opposita* and *D. japonica*) are cultivated in temperate regions in Asia. In the USA the name is also applied to the sweet potato (*Ipomoea batatas*), a somewhat similar but unrelated edible tuber. The yam bean is a name given to the seed of several species of *Fabaceae (e.g. *Pachyrhizus erosus* and *Sphenostylis stenocarpa*) which also produce edible tubers.

**yautia** *See* XANTHOSOMA.

**Y-chromosome** The *sex chromosome found only in the *heterogametic sex. It usually differs in size from the *X-chromosome and contains few or no *major genes. Often only a short part of it pairs with the X-chromosome at *meiosis. *See* SEX DETERMINATION.

**yeast** A general term for a *fungus that can exist in the form of single cells, reproducing by fission (*see* binary fission) or by *budding. Sometimes the name refers more specifically to *Saccharomyces cerevisiae*.

**yellow-green algae** *See* XANTHOPHYTA.

**yellows** A general term for any plant disease in which yellowing or *chlorosis is a characteristic symptom. Yellows diseases are often caused by *viruses but may be caused by *bacteria or *fungi.

**yerba maté** *See* ILEX.

**Yersinia** (family *Enterobacteriaceae) A genus of *Gram-negative *bacteria in which the cells are ovoid or rod-shaped, and *motile or non-motile. They occur as *parasites and *pathogens of humans, other mammals, birds, and fish. *Y. pestis* (formerly called *Pasteurella pestis*) is the causal agent of plague.

**ylang ylang** (macassar oil) *See* CANANGA.

**Yoda's power law** A numerical description of the process of self-thinning among plant seedlings. Beyond a certain density of sowing, the number of surviving plants is not related to the initial *seed density; instead, a constant relationship is evident between the density of survivors and their total *biomass. K. Yoda et al. (1963) summarized this relationship as $W = C\varrho^{-3/2}$ where $W$ is the dry weight of surviving plants, $\varrho$ is the density of the surviving plants, and $C$ is a constant reflecting the growth characteristic of the species concerned.

**yom hin** *See* CHUKRASIA.

**Zamiaceae** A family of *gymnosperms which forms the bulk of the cycads (*Cycadaceae). The *pinnules are straight in *vernation and the male and female *sporophylls are in determinate cones, the female being scale-like with a thickened base, usually with 2 *ovules. There are 8 genera, with 85 species, distributed in tropical and temperate Australia, America, and Africa.

**Zannichelliaceae** A family of submerged, *perennial, aquatic, monocotyledonous (*see* MONOCOTYLEDON) *herbs of fresh or salt water, comprising plants that have slender, creeping *rhizomes, and stems bearing narrow, linear, *opposite, *whorled, or alternate leaves, with sheathing bases, and tiny flowers, solitary in leaf *axils or in small clusters. The flowers are unisexual. There is either no *perianth, or one composed of 3 tiny scales. Male flowers have 1–3 *stamens, with 1 or 2 *anther cells, while in female flowers up to 9 *carpels are free from each other and terminate in a long beak bearing the *stigma. Pollination occurs under water, and the fruits are *achenes. There are 4 genera, with some 7 cosmopolitan species.

**Zantedeschia** (family *Araceae) A genus of 6 species of tropical African plants, which includes *Z. aethiopica*, the florists' arum lily.

**zapote** *See* MANILKARA.

**Z-chromosome** The *sex chromosome found in both *heterogametic females and *homogametic males.

**Zea** (family *Poaceae) A genus of stout grasses that have separate male and female *inflorescences. *Z. mays* (corn, maize), an *annual, is now believed to have arisen by prolonged selection from wild forms; it has male flowers in terminal plumes (tassels) and female flowers on lower leaf *axils, forming in the fruit a cob with large, starchy grains on a stout axis. It is the most important cereal in the world,

after rice and wheat. There are 4 species, native to Central America. *See also* GENETIC EROSION.

**Zelkova** (family *Ulmaceae) A genus of deciduous trees and shrubs, which have smooth bark. The leaves are spiral and toothed. The fruit is a *nut. They yield good cabinet wood. There are 5 species occurring from the Mediterranean region to eastern Asia.

**zeromorphic** *See* XEROMORPHIC.

**zerophile** *See* XEROPHILE.

**zerophyte** *See* XEROPHYTE.

**zinc** (Zn) An element that is required by plants. It is found bound to a variety of *enzymes, stabilizing them and also being involved in catalysis. Deficiency in plants prevents the expansion of leaves and internodes, giving a rosette style of plant.

**Zingiberaceae** (ginger, cardamom, turmeric) The ginger family, comprising *rhizomatous *herbs, many of which are huge, and spicy in all their parts. The leaves have *pinnate nervation and a sheathing base with a 2-ranked *ligule. The aerial stems are oblique, but in *Costus* and its related genera spiral. *Inflorescences are borne either on leafy stems or separately from the rhizome. There are 53 genera, with about 1200 species, occurring, mostly in rain forest, throughout the tropics but chiefly in Indo-Malaysia.

**Zizyphus** (family *Rhamnaceae) A genus of small trees in which the *stipules are often modified as thorns. The leaves have 3 main nerves. The flowers are *pentamerous. The fruit is a *drupe, edible in many species. (Jujube is *Z. jujuba*, occurring from India to southern China; *Z. lotus* is believed to be the lotus fruit of ancient times.) There are 86 species, occurring in the tropics and subtropics.

**Zn** *See* ZINC.

**zonation** 1. The broad distribution of vegetation according to latitude and altitude. The control is primarily climatic, and similar vegetation zones are encountered on the flanks of high tropical mountains to those found at sea-level between the tropics and the poles. **2.** The division of an *ecosystem into distinct vertical layers that experience particular *abiotic conditions. This is particularly clear in the distribution of plants and animals on a rocky seashore, where different species inhabit a series of horizontal strips or belts of the shore, approximately parallel to the water's edge. In many places the strips (zones) are sharply bounded by the differently coloured seaweeds that populate them.

**zoochory (synzoochory)** Dispersal of spores or seeds by animals.

***Zoogloea* (family *Pseudomonadaceae)** A genus of *bacteria in which young cells are *motile with a single polar *flagellum, but later clump together to form flocs which may be visible to the naked eye. They are *aerobic *chemoorganotrophs, found in organically polluted, freshwater habitats and in aerobically treated sewage.

**Zoopagales (class *Zygomycetes)** An order of *fungi that form *zygospores and asexual *spores which are not forcibly discharged. Species are found in soil and water and are *parasitic (e.g. on certain amoebae, nematodes, and other fungi).

**zoosporangium** A *sporangium within which *motile *spores (*zoospores) are formed.

**zoospore** A *motile (flagellated) cell which may function as a disseminative unit and/or as a *gamete.

**zooxanthellae** Unicellular dinoflagellates that live symbiotically (*see* SYMBIOSIS) with certain corals.

**Zosteraceae** A family of *perennial, submerged, marine, monocotyledonous (*see* MONOCOTYLEDON) *herbs, with creeping or *tuberous *rhizomes, flattened stems, linear sheathing leaves, and unisexual flowers in vertical rows, borne within a leaf sheath on 1 side of the flattened stem. There is no *perianth. Male flowers each consist of a single *stamen with thread-like *pollen grains. Female flowers consist of an *ovary with 2 *stigmas. *Pollination occurs in the sea. Modern classifications recognize 3 genera, with some 17 species, found in most temperate seas.

**Z-scheme** A diagrammatic representation of the electron flow in *cyclic phosphorylation and *non-cyclic phosphorylation, showing the change in energy potential of the electrons.

**Zurich–Montpellier school of phytosociology (Montpellier school of phytosociology)** A group led by J. *Braun-Blanquet and his associates, who developed a set of floristic methods for vegetation classification (in 1927 and later) at Zurich and Montpellier. These have been widely adopted in Europe although they are less accepted elsewhere. The aim was to provide a framework for the classification of the vegetation of the world, but in practice the scheme is most useful in regional and national surveys. The approach depends on detailed field surveying to identify vegetation associations, which can then be grouped hierarchically into alliances, orders, classes, etc., with the vegetation circle (global scale) being the most complex hierarchical level. Suffixes added to the genitive stem of the generic names of the plants label the *communities so identified and indicate the hierarchical status of the community:

| RANK | ENDING |
| --- | --- |
| class | -etea |
| order | -etalia |
| alliance | -ion |
| association | -etum |
| sub-association | -etosum |
| variant | (specific name used) |

An extensive ecological literature discusses the system and introduces many modifications. The most often quoted objections relate to the use of *homogeneous stands only in the description of vegetation; the concept of *minimal area as it is used to define homogeneity; and in particular to the use of *fidelity, and the associated problem of defining faithful species in order to characterize the associations. *Compare* UPPSALA SCHOOL.

Z

**zwitterion** A dipolar *ion (i.e. one with both negative and positive charges and therefore no net charge). *Amino acids in solution at their isoelectric point (i.e. the *pH at which they are electrically neutral) usually exist in this form, when the *amino group is protonated ($-NH_3^+$) and the carbonyl (CO) group dissociated ($-COO^-$).

**zygomorphic (irregular)** Bilaterally symmetrical, as a *snapdragon flower, and therefore divisible into equal halves in only 1 plane.

**Zygomycetes (subdivision Zygomycotina)** A class of *fungi in which the vegetative stage is typically a well-developed, branched, *coenocytic *mycelium. Members are *saprotrophs or *parasites on other fungi, plants, or invertebrates.

**Zygomycotina (division *Eumycota)** A subdivision of *fungi characterized by the formation of non-*motile asexual *spores (aplanospores). Sexual reproduction involves the fusion of *gametangia followed by zygospore formation. There are 2 classes. The subdivision includes *saprotrophic and *parasitic species.

**zygophore** In fungi of the *Mucorales, a specialized *hypha which bears a *zygospore.

**Zygophyllaceae** A family of *xerophytic or *halophytic, mostly woody, *perennial shrubs, with some *herbs and trees, in which the leaves are *opposite, usually fleshy, leathery, or hairy, with *stipules which often become spiny. The branches are sometimes joined at the *nodes. The flowers are regular, bisexual, and are held in *cymes, paired or solitary. They have 4 or 5 overlapping *sepals and the same number of *petals, usually also overlapping, although in some species the petals are absent. The *stamens are in *whorls of 5 with up to 3 whorls. The *ovary is *superior, usually of 5 fused *carpels, and often winged. There are usually 5 *locules with numerous *ovules. The *stigmas are lobed and held in a short *style. The fruit is a *dehiscent *capsule, or *berry- or *drupe-like, containing *endospermic seeds. There are 6 subfamilies, divided mainly by the structure of their fruit. The family contains some valuable timber trees, e.g. of *Guaiacum* species (lignum vitae) which give a very durable timber. Other species produce edible fruit, and some are used medicinally. There are 27 genera, with about 250 species, found in tropical and subtropical regions of both hemispheres.

**Zygophyllum (family *Zygophyllaceae)** A genus of low-lying, small, bushy shrubs found in deserts and arid regions. They have fleshy, *opposite leaves, mostly *bifoliate, or occasionally with several pairs of leaflets, and often with spiny *stipules. The stems are fleshy. The flowers are solitary or paired, terminal, and yellow or white in colour, with a red or purple spot at the base of the *petal. There are 4 or 5 overlapping *sepals and the same number of petals, which are usually overlapping and clawed. There are twice the number of *stamens, held in *whorls. The *ovary is *superior, with 4 or 5 *locules, and a simple *style with a very small *stigma. The fruit is a *capsule, angled along the locule partitions, or winged. There are 80 species, found in tropical and subtropical regions of the Old World, and in Australia.

**zygospore** A thick-walled, usually pigmented, resting *spore formed by fungi of the *Zygomycotina.

**zygote** The fertilized *ovum of an animal or plant formed from the fusion of male and female *gametes, when, under normal circumstances, the *diploid *chromosome number is restored, in the stage before it undergoes division.

**zygotene** *See* MEIOSIS and PROPHASE.

**zymase** A heat-labile, non-dialysable (*see* DIALYSIS) *enzyme fraction, derived from yeast cells, which catalyses alcoholic fermentation.

**zymogen (pro-enzyme)** The inactive precursor of an *enzyme, subsequently activated by specific partial proteolysis.

**zymogenous** Applied to organisms whose presence in a given *habitat is transient; the numbers of such organisms fluctuate greatly in response to the availability of particular nutrients, for example.

# Endangered Plants

A taxon is classified as extinct (Ex) if it is no longer known to exist in the wild despite repeated searches of the type localities and other known or likely places where it might be found.

It is classed as extinct/endangered (Ex/E) if it is considered possibly to be extinct in the wild.

It is classed as endangered (E) if it is in danger of extinction and unlikely to survive in the wild if the causal factors threatening its survival continue operating. This category includes taxa whose numbers have been reduced to a critical level or whose habitats have been so drastically reduced that they are deemed to be in immediate danger of extinction.

A taxon is classed as critically endangered (CR) if it faces a high risk of extinction in the wild in the immediate future. Such risks are defined according to criteria A to E, below.

A The population has been reduced, either
  1 as an observed, estimated, inferred, or suspected reduction of at least 80% over the last 10 years or three generations, whichever is the longer, based on (and specifying) any of the following:
    (*a*) direct observation
    (*b*) an index of abunance of the taxon
    (*c*) a decline in the area of occupancy, extent of occurrence, and/or quality of habitat
    (*d*) actual or potential levels of exploitation
    (*e*) the effects of introduced taxa, hybridization, pathogens, pollutants, competitors, or parasites;
  2 as a reduction of at least 80%, projected or suspected to occur within the next 10 years or three generations, whichever is the longer, based on (and specifying) any of the conditions stipulated in 1*b*, *c*, *d*, or *e* (above).
B The extent of the occurrence of a taxon is estimated to be less than 100 km² or the area it occupies is estimated to be less than 10 km² and estimates indicate any two of the following:
  1 the distribution is severely fragmented or the taxon is known to exist at only a single location
  2 a continuing decline has been observed, inferred, or projected in:
    (*a*) the extent of occurrence
    (*b*) the area of occupancy
    (*c*) the area, extent, and/or quality of habitat
    (*d*) the number of locations or subpopulations
    (*e*) the number of mature individuals
  3 extreme fluctuations in any of the following:
    (*a*) the extent of occurrence
    (*b*) the area of occupancy
    (*c*) the number of locations or subpopulations
    (*d*) the number of mature individuals.
C The population is estimated to number less than 250 mature individuals and either:
  1 an estimated continuing decline of at least 25% within three years or one generation, whichever is the longer, or
  2 a continuing decline, observed, projected, or inferred, in the number of mature individuals and the structure of the population in the form of either

    (*a*) severe fragmentation (i.e. no subpopulation is estimated to contain more than 50 mature individuals) or

    (*b*) all individuals are in a single subpopulation.

D  The population is estimated to number less than 50 mature individuals.

E  Quantitative analysis shows the probability of extinction in the wild is at least 50% within 10 years or three generations, whichever is the longer.

*Note:* at present the list has been completed only for Europe, and only for the endangered (E) and extinct/endangered (Ex/E) categories. Lists have not yet been released of plants in the other categories.

| Species | Common name | Family | Country of Origin | Threat status |
|---------|-------------|--------|-------------------|---------------|
| *Abies nebrodensis* | abete dei nebrodi; Sicilian fir | Pinaceae | Italy | E |
| *Achillea horanszkyi* | horánszky-cickafark | Compositae | Hungary | E |
| *Adenocarpus ombriosus* | | Leguminosae | Spain | E |
| *Adonis cyllenea* | Mount Killini pheasant's-eye | Ranunculaceae | Greece | E |
| *Aeonium mascaense* | | Crassulaceae | Spain | E |
| *Aichryson bethencourtianum* | | Crassulaceae | Spain | E |
| *Aichryson pachycaulon* | | Crassulaceae | Spain | E |
| *Allium rouyi* | | Alliaceae | Spain | E |
| *Alyssum fastigiatum* | | Cruciferae | Spain | E |
| *Anagyris latifolia* | | Leguminosae | Spain | E |
| *Androcymbium europaeum* | | Colchicaceae | Spain | E |
| *Androcymbium rechingeri* | Rechinger's lily | Colchicaceae | Greece | E |
| *Andryala levitomentosa* | | Compositae | Romania | E |
| *Anthemis glaberrima* | | Compositae | Greece | E |
| *Anthyllis lemanniana* | | Leguminosae | Portugal | E |
| *Antirrhinum lopesianum* | | Scrophulariaceae | Portugal | E |
| *Apium bermejoi* | | Umbelliferae | Spain | E |
| *Aquilegia litardierei* | ancolie de Litardière | Ranunculaceae | France | E |
| *Arenaria grandiflora* | | Caryophyllaceae | Spain | E |

| Species | Common name | Family | Country of Origin | Threat status |
|---|---|---|---|---|
| *Arenaria nevadensis* | Sierra Nevada sandwort | Caryophyllaceae | Spain | E |
| *Arenaria norvegica* | | Caryophyllaceae | United Kingdom | E |
| *Argyranthemum adauctum* | | Compositae | Spain | E |
| *Argyranthemum lemsii* | | Compositae | Spain | E |
| *Argyranthemum lidii* | Tirma daisy | Compositae | Spain | E |
| *Argyranthemum pinnatifidum* | | Compositae | Portugal | E |
| *Argyranthemum sundingii* | | Compositae | Spain | E |
| *Argyranthemum thalassophytum* | | Compositae | Portugal | E |
| *Argyranthemum vincentii* | | Compositae | Spain | E |
| *Armeria euscadiensis* | | Plumbaginaceae | Spain | E |
| *Armeria helodes* | | Plumbaginaceae | Italy | E |
| *Armeria humilis* | | Plumbaginaceae | Portugal | E |
| *Armeria maritima* | | Plumbaginaceae | Romania | E |
| *Armeria pseudarmeria* | | Plumbaginaceae | Portugal | E |
| *Artemisia granatensis* | royal manzanilla | Compositae | Spain | E |
| *Artemisia molinieri* | armoise de Molinier | Compositae | France | E |
| *Asparagus fallax* | | Asparagaceae | | |
| *Asphodelus bento-rainhae* | abrotea | Asphodelaceae | Portugal | E |
| *Asplenium petrarchae* | | Aspleniaceae | Spain | E |

| Species | Common name | Family | Country | Status |
|---|---|---|---|---|
| *Asteriscus schultzii* | | Compositae | Spain | E |
| *Astragalus idaeus* | | Leguminosae | Greece | Ex/E |
| *Astragalus peterfii* | | Leguminosae | Romania | E |
| *Astragalus physocalyx* | swalen-calyx milk vetch | Leguminosae | Bulgaria | Ex/E |
| *Astragalus raphaelis* | | Leguminosae | Italy | E |
| *Atractylis arbuscula* | | Compositae | Spain | E |
| *Atractylis preauxiana* | piña de mar | Compositae | Spain | E |
| *Barlia metlesicsiana* | | Orchidaceae | Spain | E |
| *Bassia saxicola* | | Chenopodiaceae | Italy | E |
| *Bellevalia hackelii* | | Hyacinthaceae | Portugal | E |
| *Bencomia brachystachya* | Bencomo's burnet | Rosaceae | Spain | E |
| *Bencomia exstipulata* | | Rosaceae | Spain | E |
| *Bencomia sphaerocarpa* | Hierro bencomo's burnet | Rosaceae | Spain | E |
| *Biscutella divionensis* | lunetière de Dijon | Cruciferae | France | E |
| *Borderea chouardii* | Chouard's flowering yam | Dioscoreaceae | Spain | E |
| *Brassica bourgeaui* | | Cruciferae | Spain | E |
| *Brassica macrocarpa* | egadi cabbage; cavolo delle egadi | Cruciferae | Italy | E |
| *Bromus grossus* | brome à fleurs nombreuses; brome épais; zware drefs; schwere trespe | Graminae | Belgium, Luxembourg, Switzerland | E |
| *Bromus interruptus* | interrupted brome | Graminae | United Kingdom | E |

| Species | Common name | Family | Country of Origin | Threat status |
|---|---|---|---|---|
| *Bupleurum capillare* | | Umbelliferae | Greece | E |
| *Bupleurum kakiskalae* | | Umbelliferae | Greece | E |
| *Bunium brevifolium* | | Umbelliferae | Portugal | E |
| *Bystropogon wildpretii* | | Labiatae | Spain | E |
| *Campanula gelida* | | Campanulaceae | Czechoslovakia | E |
| *Campanula secundiflora* | | Campanulaceae | Yugoslavia | E |
| *Carex perraudieriana* | | Cyperaceae | Spain | E |
| *Centaurea alba* | | Compositae | Greece | E |
| *Centaurea borjae* | | Compositae | Spain | E |
| *Centaurea jankae* | | Compositae | Romania | E |
| *Centaurea kalambakensis* | | Compositae | Greece | E |
| *Centaurea lactiflora* | | Compositae | Greece | E |
| *Centaurea leucophaea* | centaurée gris cendré, à deuxformes | Compositae | France | Ex/E |
| *Centaurea linaresii* | | Compositae | Spain | E |
| *Centaurea maculosa* | centaurée tachetée, blanchâtre | Compositae | France | E |
| *Centaurea niederi* | | Compositae | Greece | E |
| *Centaurea peucedanifolia* | | Compositae | Greece | E |
| *Centaurea pontica* | | Compositae | Romania | E |

| Species | Common name | Family | Country | Status |
|---|---|---|---|---|
| *Centaurium quadrifolium* | | Gentianaceae | Spain | E |
| *Centaurium rigualii* | | Gentianaceae | Spain | E |
| *Centranthus trinervis* | camarezza sardo-corsa; centranthe à trois nervures | Valerianaceae | France, Italy | E |
| *Cerastium alsinifolium* | | Caryophyllaceae | Czechoslovakia | E |
| *Ceropegia ceratophora* | | Asclepiadaceae | Spain | E |
| *Chaenorhinum tenellum* | | Scrophulariaceae | Spain | E |
| *Chaenorhinum minus* | chaenorrhinum à feuilles rougeâtres | Scrophulariaceae | France | E |
| *Chamaecytisus nejceffii* | | Leguminosae | Bulgaria | E |
| *Chamaemeles coriacea* | buxa da rocha | Rosaceae | Portugal | E |
| *Cheirolophus anagensis* | | Compositae | Spain | E |
| *Cheirolophus arboreus* | | Compositae | Spain | E |
| *Cheirolophus duranii* | Hierro knapweed | Compositae | Spain | E |
| *Cheirolophus falcisectus* | | Compositae | Spain | E |
| *Cheirolophus ghomerythus* | Gomera knapweed | Compositae | Spain | E |
| *Cheirolophus junonianus* | La Palma knapweed | Compositae | Spain | E |
| *Cheirolophus lagunae* | | Compositae | Spain | E |
| *Cheirolophus massonianus* | | Compositae | Portugal | E |
| *Cheirolophus metlesicsii* | | Compositae | Spain | E |
| *Cheirolophus puntallanensis* | | Compositae | Spain | E |
| *Cheirolophus santosabreui* | | Compositae | Spain | E |

| Species | Common name | Family | Country of Origin | Threat status |
|---|---|---|---|---|
| *Cheirolophus satarataensis* | | Compositae | Spain | E |
| *Cheirolophus tagananensis* | | Compositae | Spain | E |
| *Cistus heterophyllus* | | Cistaceae | Spain | E |
| *Clematis elisabethae-carolae* | | Ranunculaceae | Greece | E |
| *Cochlearia polonica* | warzucha polska | Cruciferae | Poland | E |
| *Coincya rupestris* | Coincy's rock cabbage | Cruciferae | Spain | E |
| *Consolida samia* | | Ranunculaceae | Greece | E |
| *Convolvulus fernandesii* | | Convulvaceae | Portugal | E |
| *Convolvulus lopezsocasii* | Lanzarote bind-weed | Convulvaceae | Spain | E |
| *Convolvulus massonii* | | Convulvaceae | Portugal | E |
| *Coronopus navasii* | | Cruciferae | Spain | E |
| *Crambe arborea* | Guimar cliff cabbage | Cruciferae | Spain | E |
| *Crambe sventenii* | Fuerteventura cabbage; Fuerteventura cliff cabbage | Cruciferae | Spain | E |
| *Crataegus monogyna* | aubépine de mer | Rosaceae | France | E |
| *Crepis crocifolia* | | Compositae | Greece | E |
| *Daphne rodriguezii* | | Thymelaeaceae | Spain | E |
| *Dendriopoterium pulidoi* | Pulido's burnet | Rosaceae | Spain | E |
| *Deschampsia maderensis* | | Graminae | Portugal | E |

| | | | | |
|---|---|---|---|---|
| *Dianthus morisianum* | | Caryophyllaceae | Italy | E |
| *Dianthus plumarius* | | Caryophyllaceae | Hungary | E |
| *Dorycnium spectabile* | | Leguminosae | Spain | E |
| *Draba dorneri* | | Cruciferae | Romania | E |
| *Draba dubia* | | Cruciferae | Spain | E |
| *Drosera rotundifolia* | | Droseraceae | France | E |
| *Echinospartum algibicum* | | Leguminosae | Spain | Ex/E |
| *Echium auberianum* | | Boraginaceae | Spain | E |
| *Echium saetabense* | | Boraginaceae | Spain | E |
| *Erica andevalensis* | | Ericaceae | Spain | E |
| *Erodium cazorlanum* | | Geraniaceae | Spain | E |
| *Erodium rupicola* | | Geraniaceae | Spain | E |
| *Erucastrum palustre* | erucastro friulano | Cruciferae | Italy | E |
| *Euphorbia anachoreta* | | Euphorbiaceae | Portugal | E |
| *Euphorbia bourgeauana* | | Euphorbiaceae | Spain | E |
| *Euphorbia gaditana* | | Euphorbiaceae | Spain | E |
| *Euphorbia handiensis* | | Euphorbiaceae | Spain | E |
| *Euphorbia margalidiana* | lleterassa | Euphorbiaceae | Spain | E |
| *Femeniasia balearica* | | Compositae | Spain | E |
| *Ferula latipinna* | | Umbelliferae | Spain | E |
| *Festuca brigantina* | | Graminae | Portugal | E |

| Species | Common name | Family | Country of Origin | Threat status |
|---|---|---|---|---|
| Festuca henriquesii | | Graminae | Portugal | E |
| Festuca lahonderei | fétuque de Lahondère | Graminae | France | E |
| Fritillaria euboeica | | Liliaceae | Greece | E |
| Fumaria caroliana | fumeterre de charles | Papaveraceae | France | E |
| Geocaryum bornmuelleri | | Umbelliferae | Greece | E |
| Geranium cazorlense | | Geraniaceae | Spain | E |
| Geranium maderense | | Geraniaceae | Portugal | E |
| Globularia ascanii | agaete bird's tongue | Globulariaceae | Spain | E |
| Globularia sarcophylla | Tirajana bird's tongue | Globulariaceae | Spain | E |
| Globularia stygia | styx globularia | Globulariaceae | Greece | E |
| Goodyera macrophylla | | Orchidaceae | Portugal | E |
| Gyrocaryum oppositifolium | | Boraginaceae | Spain | E |
| Halimium verticillatum | | Cistaceae | Portugal | E |
| Helianthemum bramwelliorum | | Cistaceae | Spain | E |
| Helianthemum bystropogophyllum | Gran Canarian hairy rock-rose | Cistaceae | Spain | E |
| Helianthemum cirae | | Cistaceae | Spain | E |
| Helianthemum gonzalezferrari | | Cistaceae | Spain | E |
| Helianthemum inaguae | | Cistaceae | Spain | E |
| Helianthemum juliae | | Cistaceae | Spain | E |

| | | | | |
|---|---|---|---|---|
| Helianthemum lini | | Cistaceae | Spain | E |
| Helianthemum teneriffae | | Cistaceae | Spain | E |
| Helichrysum alucense | | Compositae | Spain | E |
| Helichrysum sonogynum | | Compositae | Spain | E |
| Herniaria fontanesii | | Illecebraceae | Italy | E |
| Hieracium chaunotrichum | | Compositae | Czechoslovakia | E |
| Hieracium texedense | | Compositae | Spain | E |
| Hypericum aciferum | | Guttiferae | Greece | E |
| Hypericum reflexum | | Guttiferae | Spain | E |
| Hypochoeris oligocephala | Teno cat's ear | Compositae | Spain | E |
| Iberis runemarkii | | Cruciferae | Greece | E |
| Ilex perado | | Aquifoliaceae | Spain | E |
| Isoetes azorica | | Isoetaceae | Portugal | E |
| Isoetes malinverniana | calamaria malinverniana | Isoetaceae | Italy | E |
| Isoplexis chalcantha | cockscomb foxglove | Scrophulariaceae | Spain | E |
| Isoplexis isabelliana | Tamadaba foxglove | Scrophulariaceae | Spain | E |
| Juniperus brevifolia | | Cupressaceae | Portugal | E |
| Juniperus cedrus | | Cupressaceae | Portugal, Spain | E |
| Knautia lebrunii | knautia de le brun | Dipsacaceae | France | E |
| Kunkeliella psilotoclada | | Santalaceae | Spain | E |
| Lamyropsis microcephala | Sardinian thistle; cardo | Compositae | Italy | E |

| Species | Common name | Family | Country of Origin | Threat status |
|---|---|---|---|---|
| *Laurentia canariensis* | | Campanulaceae | Spain | E |
| *Laserpitium longiradium* | long-rayed laser | Umbelliferae | Spain | E |
| *Lavandula buchii* | | Labiatae | Spain | E |
| *Lavatera phoenicea* | | Malvaceae | Spain | E |
| *Lepidium cardamines* | | Cruciferae | Spain | E |
| *Leucojum fabrei* | nivéole de fabre | Amaryllidaceae | France | E |
| *Leuzea longifolia* | | Compositae | Portugal | E |
| *Ligusticum albanicum* | vratik i shqipërisë | Umbelliferae | Albania | Ex/E |
| *Ligusticum lucidum* | | Umbelliferae | Spain | E |
| *Limonium arborescens* | | Plumbaginaceae | Spain | E |
| *Limonium brassicifolium* | | Plumbaginaceae | Spain | E |
| *Limonium cavanillesii* | | Plumbaginaceae | Spain | E |
| *Limonium companyonis* | | Plumbaginaceae | France | E |
| *Limonium dendroides* | | Plumbaginaceae | Spain | E |
| *Limonium dufourei* | | Plumbaginaceae | Spain | E |
| *Limonium estevei* | | Plumbaginaceae | Spain | E |
| *Limonium fruticans* | | Plumbaginaceae | Spain | E |
| *Limonium imbricatum* | | Plumbaginaceae | Spain | E |
| *Limonium laetum* | | Plumbaginaceae | Italy | E |

| | | | | |
|---|---|---|---|---|
| *Limonium lausianum* | | Plumbaginaceae | Italy | E |
| *Limonium macrophyllum* | | Plumbaginaceae | Spain | E |
| *Limonium macropterum* | | Plumbaginaceae | Spain | E |
| *Limonium magallufianum* | | Plumbaginaceae | Spain | E |
| *Limonium majoricum* | | Plumbaginaceae | Spain | E |
| *Limonium malacitanum* | | Plumbaginaceae | Spain | E |
| *Limonium neo-castellonense* | | Plumbaginaceae | Spain | E |
| *Limonium optimae* | | Plumbaginaceae | Italy | E |
| *Limonium pavonianum* | | Plumbaginaceae | Italy | E |
| *Limonium pseudodictyocladon* | | Plumbaginaceae | Spain | E |
| *Limonium pseudolaetum* | | Plumbaginaceae | Italy | E |
| *Limonium pulviniforme* | | Plumbaginaceae | Italy | E |
| *Limonium redivivum* | | Plumbaginaceae | Spain | E |
| *Limonium rigualii* | | Plumbaginaceae | Spain | E |
| *Limonium spectabile* | | Plumbaginaceae | Spain | E |
| *Limonium sventenii* | | Plumbaginaceae | Spain | E |
| *Limonium tauromenitanum* | | Plumbaginaceae | Italy | E |
| *Limonium todaroanum* | | Plumbaginaceae | Italy | E |
| *Linaria hellenica* | malea toadflax | Scrophulariaceae | Greece | E |
| *Linum dolomiticum* | pilisi len | Linaceae | Hungary | E |
| *Lithodora nitida* | | Boraginaceae | Spain | E |

| Species | Common name | Family | Country of Origin | Threat status |
|---|---|---|---|---|
| *Logfia neglecta* | gnaphale négligé; cottonière negligée; vergeten viltkruid; verkanntes filzkraut | Compositae | Belgium, France | Ex/E |
| *Lotus berthelotii* | pico paloma; dove's beak | Leguminosae | Spain | E |
| *Lotus eremiticus* | | Leguminosae | Spain | E |
| *Lotus genistoides* | | Leguminosae | Spain | E |
| *Lotus kunkelii* | succulent birdsfoot-trefoil; jinamar bird's foot trefoil | Leguminosae | Spain | E |
| *Lotus leptophyllus* | | Leguminosae | Spain | E |
| *Lotus maculatus* | | Leguminosae | Spain | E |
| *Lotus pyranthus* | | Leguminosae | Spain | E |
| *Lysimachia minoricensis* | | Primulaceae | Spain | E |
| *Marcetella maderensis* | | Rosaceae | Portugal | E |
| *Marsilea azorica* | | Marsileaceae | Portugal | E |
| *Micromeria glomerata* | | Labiatae | Spain | E |
| *Micromeria leucantha* | | Labiatae | Spain | E |
| *Micromeria pineolens* | | Labiatae | Spain | E |
| *Micromeria rivas-martinezii* | | Labiatae | Spain | E |
| *Micromeria taygetea* | | Labiatae | Greece | E |
| *Minuartia glaucina* | | Caryophyllaceae | Czechoslovakia | E |

| Species | Common name | Family | Country | Status |
|---|---|---|---|---|
| *Minuartia graminifolia* | | Caryophyllaceae | Romania | E |
| *Minuartia hirsuta* | | Caryophyllaceae | Romania | E |
| *Minuartia wettsteinii* | | Caryophyllaceae | Greece | E |
| *Monanthes adenoscepes* | | Crassulaceae | Spain | E |
| *Monanthes dasyphylla* | | Crassulaceae | Spain | E |
| *Monanthes niphophila* | | Crassulaceae | Spain | E |
| *Monanthes wildpretii* | | Crassulaceae | Spain | E |
| *Monizia edulis* | | Umbelliferae | Portugal | E |
| *Murbeckiella sousae* | | Cruciferae | Portugal | E |
| *Musschia wollastonii* | | Campanulaceae | Portugal | E |
| *Myosotis rehsteineri* | Lake Constance forget-me-not; Rehsteiners Vergissmeinnicht | Boraginaceae | Austria, Germany, Hungary, Italy, Liechtenstein, Switzerland | E |
| *Myosotis retusifolia* | | Boraginaceae | Portugal | E |
| *Myosotis ruscinonensis* | dune forget-me-not; myosotis du roussillon | Boraginaceae | France | E |
| *Myosotis solange* | | Boraginaceae | Greece | E |
| *Narcissus pseudonarcissus* | | Amaryllidaceae | Spain | E |
| *Narcissus scaberulus* | | Amaryllidaceae | Portugal | E |
| *Narcissus tortifolius* | | Amaryllidaceae | Spain | E |
| *Naufraga balearica* | Balearic castaway; naufragée des Baléares | Umbelliferae | France, Spain | E |

| Species | Common name | Family | Country of Origin | Threat status |
|---|---|---|---|---|
| *Nepeta sphaciotica* | Cretan mint | Labiatae | Greece | E |
| *Normania nava* | | Solanaceae | Spain | E |
| *Normania triphylla* | | Solanaceae | Portugal | E |
| *Odontites granatensis* | | Scrophulariaceae | Spain | E |
| *Oenanthe conioides* | water dropwort; Elbe water dropwort | Umbelliferae | Germany | E |
| *Omphalodes littoralis* | | Boraginaceae | Spain | E |
| *Ononis christii* | | Leguminosae | Spain | E |
| *Onopordum nogalesii* | Fuerteventura thistle | Compositae | Spain | E |
| *Onosma fastigiata* | orcanette atlantique | Boraginaceae | France | E |
| *Onosma pseudarenaria* | | Boraginaceae | Romania | E |
| *Orchis scopulorum* | | Orchidaceae | Portugal | E |
| *Orobanche berthelotii* | | Scrophulariaceae | Spain | Ex/E |
| *Parolinia aridanae* | | Cruciferae | Spain | E |
| *Parolinia schizogynoides* | Gomera shrubby stock | Cruciferae | Spain | E |
| *Phalaris maderensis* | | Graminae | Portugal | E |
| *Phleum crypsoides* | | Graminae | Italy | E |
| *Phlomis margaritae* | | Labiatae | Spain | E |
| *Pimpinella rupicola* | | Umbelliferae | Spain | E |
| *Pinguicula vallisneriifolia* | | Lentibulariaceae | Spain | E |

| | | | | |
|---|---|---|---|---|
| *Pinguicula vulgaris* | | Lentibulariaceae | Poland | E |
| *Plantago atrata* | | Plantaginaceae | Czechoslovakia | E |
| *Plantago malato-belizii* | | Plantaginaceae | Portugal | E |
| *Polygala sinisca* | | Polygalaceae | Italy | E |
| *Polygala supina* | amareala | Polygalaceae | Romania | E |
| *Polygonum albanicum* | nejcë shqiptare | Polygonaceae | Albania | E |
| *Polystichum drepanum* | | Dryopteridae | Portugal | E |
| *Polystichum falcinellum* | | Dryopteridae | Portugal | E |
| *Primula wulfeniana* | | Primulaceae | Romania | E |
| *Psilotum nudum* | | Psilotaceae | Spain | E |
| *Pterocephalus virens* | | Dipsaceae | Spain | E |
| *Pulicaria burchardii* | | Compositae | Spain | E |
| *Pulicaria canariensis* | | Compositae | Spain | E |
| *Pyrus magyarica* | magyar vadkörte | Rosaceae | Hungary | E |
| *Ranunculus cabrerensis* | | Ranunculaceae | Spain | E |
| *Ranunculus degenii* | zhabinë e degenit | Ranunculaceae | Albania | Ex/E |
| *Ranunculus radinotrichus* | | Ranunculaceae | Greece | E |
| *Ranunculus weyleri* | | Ranunculaceae | Spain | E |
| *Rhaponticum canariensis* | | Compositae | Spain | E |
| *Rhinanthus halophilus* | | Scrophulariaceae | Germany | E |
| *Rhynchosinapis johnstonii* | | Cruciferae | Portugal | E |

| Species | Common name | Family | Country of Origin | Threat status |
|---|---|---|---|---|
| *Ribes sardoum* | Sardinian gooseberry; ribes di sardegna | Grossulariaceae | Italy | E |
| *Rivasgodaya nervosa* | | Leguminosae | Spain | E |
| *Rothmaleria granatensis* | | Compositae | Spain | E |
| *Ruta microcarpa* | Gomeran rue | Rutaceae | Spain | E |
| *Salicornia veneta* | Venetian glasswort; salicornia veneta | Chenopodiaceae | Italy | E |
| *Salvia broussonetii* | | Labiatae | Spain | E |
| *Salvia herbanica* | | Labiatae | Spain | E |
| *Sambucus palmensis* | | Caprifoliaceae | Spain | E |
| *Saxifraga cintrana* | | Saxifragaceae | Portugal | E |
| *Scrophularia smithii* | | Scrophulariaceae | Spain | E |
| *Sedum nudum* | | Crassulaceae | Spain | E |
| *Senecio caespitosus* | | Compositae | Portugal | E |
| *Senecio coincyi* | | Compositae | Spain | E |
| *Senecio elodes* | | Compositae | Spain | E |
| *Senecio hadrosomus* | | Compositae | Spain | E |
| *Senecio hansenii* | | Compositae | Spain | E |

| | | | | |
|---|---|---|---|---|
| *Senecio hermosae* | | Compositae | Spain | E |
| *Senecio lagascanus* | Gomera knapweed | Compositae | Portugal | E |
| *Seseli intricatum* | | Umbelliferae | Spain | E |
| *Sideritis cystosiphon* | | Labiatae | Spain | E |
| *Sideritis discolor* | | Labiatae | Spain | E |
| *Sideritis infernalis* | | Labiatae | Spain | E |
| *Sideritis marmorea* | | Labiatae | Spain | E |
| *Sideritis nervosa* | | Labiatae | Spain | E |
| *Silene cintrana* | | Caryophyllaceae | Portugal | E |
| *Silene hicesiae* | | Caryophyllaceae | Italy | E |
| *Silene nocteolens* | | Caryophyllaceae | Spain | E |
| *Silene orphanidis* | | Caryophyllaceae | Greece | E |
| *Silene rothmaleri* | | Caryophyllaceae | Portugal | E |
| *Silene stockenii* | | Caryophyllaceae | Spain | E |
| *Solenanthus reverchonii* | | Boraginaceae | Spain | E |
| *Sonchus arboreus* | | Compositae | Spain | E |
| *Sonchus gandogeri* | | Compositae | Spain | E |
| *Sonchus pinnatus* | | Compositae | Spain | E |
| *Sonchus radicatus* | | Compositae | Spain | E |

| Species | Common name | Family | Country of Origin | Threat status |
|---|---|---|---|---|
| *Sonchus wildpretii* | | Compositae | Spain | E |
| *Sorbus maderensis* | | Rosaceae | Portugal | E |
| *Stemmacantha cynaroides* | | Compositae | Spain | E |
| *Stipa austroitalica* | italian stipa; lino delle fate | Graminae | Italy | E |
| *Stipa veneta* | | Graminae | Italy | E |
| *Symphytum cycladense* | Sikinos comfrey | Boraginaceae | Greece | E |
| *Taeckholmia heterophylla* | | Compositae | Spain | E |
| *Teline benehoavensis* | | Leguminosae | Spain | E |
| *Teline linifolia* | | Leguminosae | Spain | E |
| *Teline nervosa* | | Leguminosae | Spain | E |
| *Teline osyroides* | | Leguminosae | Spain | E |
| *Teline rosmarinifolia* | | Leguminosae | Spain | E |
| *Teline salsoloides* | | Leguminosae | Spain | E |
| *Thymelaea broterana* | | Thymelaeaceae | Portugal | E |
| *Thymelaea tartonraira* | passerine de thomas | Thymelaeaceae | France | E |

| | | | | |
|---|---|---|---|---|
| *Thymelaea thomasii* | passerine de thomas | Thymelaeaceae | France | E |
| *Tolpis glabrescens* | | Compositae | Spain | E |
| *Trisetum burnoufii* | | Graminae | France | Ex/E |
| *Tuberaria major* | | Cistaceae | Portugal | E |
| *Tulipa didieri* | tulipe de Didier | Liliaceae | France, Switzerland | E |
| *Tulipa grengiolensis* | | Liliaceae | Switzerland | E |
| *Tulipa lortetii* | tulipe de Lortet | Liliaceae | France | E |
| *Tulipa mauriana* | tulipe de Maurienne | Liliaceae | France | E |
| *Tulipa montisandrei* | tulipe du Mont-André | Liliaceae | France | E |
| *Tulipa platystigma* | tulipe à stigmate aplati | Liliaceae | France | E |
| *Verbascum cylleneum* | | Scrophulariaceae | Greece | E |
| *Veronica oetaea* | Mount Iti speedwell | Scrophulariaceae | Greece | E |
| *Vicia bifoliolata* | | Leguminosae | Spain | E |
| *Viola cheiranthifolia* | | Violaceae | Spain | E |
| *Viola plantaginea* | | Violaceae | Spain | Ex/E |
| *Viola stolonifera* | | Violaceae | Spain | E |
| *Wulfenia baldaccii* | vulfenia e baldaçit | Scrophulariaceae | Albania | Ex/E |

# The Universal Genetic Code

| amino acid | abbreviation | codons |
| --- | --- | --- |
| alanine | Ala | GCA, GCC, GCG, GCU |
| arginine | Arg | AGA, AGG, CGA, CGG, CGC, CGU |
| asparaginine | Asn | AAC, AAU |
| aspartic acid | Asp | GAC, GAU |
| cysteine | Cys | UGC, UGU |
| glutamic acid | Glu | GAA, GAG |
| glutamine | Gln | CAA, CAG |
| glycine | Gly | GGA, GGC, GGG, GGU |
| histidine | His | CAC, CAU |
| isoleucine | Ile | AUA, AUC, AUU |
| leucine | Leu | CUA, CUC, CUG, CUU, UUA, UUG |
| lysine | Lys | AAA, AAG |
| methionine | Met | AUG |
| phenylalanine | Phe | UUC, UUU |
| proline | Pro | CCA, CCC, CCG, CCU |
| serine | Ser | AGC, AGU, UCA, UCC, UCG, UCU |
| threonine | Thr | ACA, ACC, ACG, ACU |
| tryptophan | Trp | UGG |
| tyrosine | Tyr | UAC, UAU |
| valine | Val | GUA, GUC, GUG, GUU |
| stop codon | | UAA, UAG, UGA |

# The Geologic Time-Scale

| Eon/ Eonothem | Era/ Erathem | Sub-era | Period/ System | Epoch/ Series | Began Ma |
|---|---|---|---|---|---|
| P H A N E R O Z I C | | Quaternary | Pleistogene | Holocene | 0.11 |
| | | | | Pleistocene | 1.81 |
| | Cenozoic | Tertiary | Neogene | Pliocene | 5.3 |
| | | | | Miocene | 23.03 |
| | | | Palaeogene | Oligocene | 33.9 |
| | | | | Eocene | 55.8 |
| | | | | Palaeocene | 65.5 |
| | Mesozoic | | Cretaceous | Late | 99.6 |
| | | | | Early | 145.5 |
| | | | Jurassic | Late | 161.2 |
| | | | | Middle | 175.6 |
| | | | | Early | 199.6 |
| | | | Triassic | Late | 228 |
| | | | | Middle | 245 |
| | | | | Early | 251 |
| | Palaeozoic | Upper | Permian | Late | 260.4 |
| | | | | Middle | 270.6 |
| | | | | Early | 299 |
| | | | Carboniferous | Pennsylvanian | 318.1 |
| | | | | Mississipian | 359.2 |
| | | | Devonian | Late | 385.3 |
| | | | | Middle | 397.5 |
| | | | | Early | 416 |
| | | Lower | Silurian | Late | 422.9 |
| | | | | Early | 443.7 |
| | | | Ordovician | Late | 460.9 |
| | | | | Middle | 471.8 |
| | | | | Early | 488.3 |
| | | | Cambrian | Late | 501 |
| | | | | Middle | 513 |
| | | | | Early | 542 |
| P R O T E R O Z O I C | Neoproterozoic | | Ediacaran | | 630 |
| | | | Cryogenian | | 850 |
| | | | Tonian | | 1000 |
| | Mesoproterozoic | | Stenian | | 1200 |
| | | | Ectasian | | 1400 |
| | | | Calymmian | | 1600 |
| | Palaeoproterozoic | | Statherian | | 1800 |
| | | | Orosirian | | 2050 |
| | | | Rhyacian | | 2300 |
| | | | Siderian | | 2500 |

# The Geologic Time-Scale—cont'd

| Eon/ Eonothem | Era/ Erathem | Sub-era | Period/ System | Epoch/ Series | Began Ma |
|---|---|---|---|---|---|
| A R C H A E A N | Neoarchaean | | | | 2800 |
| | Mesoarchaean | | | | 3200 |
| | Palaeoarchaean | | | | 3600 |
| | Eoarchaean | | | | 3800 |
| H A D E A N | Swazian | | | | 3900 |
| | Basin Groups | | | | 4000 |
| | Cryptic | | | | 4567.17 |

(Source: International Union of Geological Sciences, 2004.
Note: Hadean is an informal name. The Hadean, Archaean, and Proterozoic Eons cover the time formerly known as the Precambrian. Tertiary has been abandoned as a formal name and Quaternary is likely to be abandoned in the next few years, although both names are still widely used.)

# SI Units (Système International d'Unités)

| Quantity | Name of unit | Symbol | Equivalent | Reciprocal |
|---|---|---|---|---|
| length | metre | m | 3.281 feet | 1 ft = 0.3048 m |
| mass | kilogram | kg | 2.2 pounds | 1 lb = 0.454 kg |
| time | second | s | | |
| electric current | ampere | A | | |
| thermodynamic temperature | kelvin | K | 1°C = 1.8°F | 1°C = 1 K |
| luminous intensity | candela | cd | | |
| amount of substance | mole | mol | | |

## Supplementary units

| Quantity | Unit | Symbol |
|----------|------|--------|
| plane angle | radian | rad |
| solid angle | steradian | sr |

## Derived SI units

| Quantity | Name of unit | Symbol | Equivalent | Reciprocal |
|----------|-------------|--------|------------|------------|
| frequency | hertz | Hz | | |
| energy | joule | J | 0.2388 calories | 1 cal = 4.1868 J |
| force | newton | N | 0.225 pounds force | 1 lbf = 4.448 N |
| power | watt | W | 0.00134 horse power | 1 hp = 745.7 W |
| pressure | pascal | Pa | 0.00689 pounds force/sq. inch | 1 lbf/sq.in = 145 Pa |
| electric charge | coulomb | C | | |
| electric potential difference | volt | V | | |
| electric resistance | ohm | Ω | | |
| electric conductance | siemens | S | | |
| electric capacitance | farad | F | | |
| magnetic flux | weber | Wb | | |
| inductance | henry | H | | |
| magnetic flux density | tesla | T | | |
| luminous flux | lumen | lm | | |
| illuminance | lux | lx | | |
| absorbed dose | gray | Gy | | |
| activity | becquerel | Bq | | |
| dose equivalent | sievert | Sv | | |

# Multiples used with SI units

| Name of multiple | Symbol | Value (multiply by) |
|---|---|---|
| atto | a | $10^{-18}$ |
| femto | f | $10^{-15}$ |
| pico | p | $10^{-12}$ |
| nano | n | $10^{-9}$ |
| micro | $\mu$ | $10^{-6}$ |
| milli | m | $10^{-3}$ |
| centi | c | $10^{-2}$ |
| deci | d | $10^{-1}$ |
| deca | da | 10 |
| hecto | h | $10^{2}$ |
| kilo | k | $10^{3}$ |
| mega | M | $10^{6}$ |
| giga | G | $10^{9}$ |
| tera | T | $10^{12}$ |
| peta | P | $10^{15}$ |
| exa | E | $10^{18}$ |

# Oxford Paperback Reference

**A Dictionary of Chemistry**

Over 4,200 entries covering all aspects of chemistry, including physical chemistry and biochemistry.

'It should be in every classroom and library ... the reader is drawn inevitably from one entry to the next merely to satisfy curiosity.'

*School Science Review*

**A Dictionary of Physics**

Ranging from crystal defects to the solar system, 3,500 clear and concise entries cover all commonly encountered terms and concepts of physics.

**A Dictionary of Biology**

The perfect guide for those studying biology – with over 4,700 entries on key terms from biology, biochemistry, medicine, and palaeontology.

'lives up to its expectations; the entries are concise, but explanatory'

*Biologist*

'ideally suited to students of biology, at either secondary or university level, or as a general reference source for anyone with an interest in the life sciences'

*Journal of Anatomy*

# OXFORD

# Oxford Paperback Reference

**A Dictionary of Psychology**
Andrew M. Colman

Over 10,500 authoritative entries make up the most wide-ranging
dictionary of psychology available.

'impressive ... certainly to be recommended'
*Times Higher Educational Supplement*

'Comprehensive, sound, readable, and up-to-date, this is probably the
best single-volume dictionary of its kind.'
*Library Journal*

**A Dictionary of Economics**
John Black

Fully up-to-date and jargon-free coverage of economics. Over 2,500
terms on all aspects of economic theory and practice.

**A Dictionary of Law**

An ideal source of legal terminology for systems based on English law.
Over 4,000 clear and concise entries.

'The entries are clearly drafted and succinctly written ... Precision for the
professional is combined with a layman's enlightenment.'
*Times Literary Supplement*

# OXFORD

# Oxford Paperback Reference

## Concise Medical Dictionary

Over 10,000 clear entries covering all the major medical and surgical specialities make this one of our best-selling dictionaries.

'"No home should be without one" certainly applies to this splendid medical dictionary'

*Journal of the Institute of Health Education*

'An extraordinary bargain'

*New Scientist*

'Excellent layout and jargon-free style'

*Nursing Times*

## A Dictionary of Nursing

Comprehensive coverage of the ever-expanding vocabulary of the nursing professions. Features over 10,000 entries written by medical and nursing specialists.

## An A-Z of Medicinal Drugs

Over 4,000 entries cover the full range of over-the-counter and prescription medicines available today. An ideal reference source for both the patient and the medical professional.

**OXFORD**

# More Social Science titles from OUP

**The Globalization of World Politics**
John Baylis and Steve Smith

The essential introduction for all students of international relations.

'The best introduction to the subject by far. A classic of its kind.'
Dr David Baker, University of Warwick

**Macroeconomics**
**A European Text**
Michael Burda and Charles Wyplosz

'Burda and Wyplosz's best-selling text stands out for the breadth of its coverage, the clarity of its exposition, and the topicality of its examples. Students seeking a comprehensive guide to modern macroeconomics need look no further.'
Charles Bean, Chief Economist, Bank of England

**Economics**
Richard Lipsey and Alec Chrystal

The classic introduction to economics, revised every few years to include the latest topical issues and examples.

**VISIT THE COMPANION WEB SITES FOR THESE CLASSIC TEXTBOOKS AT:**

**www.oup.com/uk/booksites**

OXFORD

# Oxford Companions

'Opening such books is like sitting down with a knowledgeable friend. Not a bore or a know-all, but a genuinely well-informed chum ... So far so splendid.'

*Sunday Times* [of *The Oxford Companion to Shakespeare*]

For well over 60 years Oxford University Press has been publishing Companions that are of lasting value and interest, each one not only a comprehensive source of reference, but also a stimulating guide, mentor, and friend. There are between 40 and 60 Oxford Companions available at any one time, ranging from music, art, and literature to history, warfare, religion, and wine.

Titles include:

**The Oxford Companion to English Literature**
Edited by Margaret Drabble
'No guide could come more classic.'

Malcolm Bradbury, *The Times*

**The Oxford Companion to Music**
Edited by Alison Latham
'probably the best one-volume music reference book going'

*Times Educational Supplement*

**The Oxford Companion to Western Art**
Edited by Hugh Brigstocke
'more than meets the high standard set by the growing number of Oxford Companions'

*Contemporary Review*

**The Oxford Companion to Food**
Alan Davidson
'the best food reference work ever to appear in the English language'

*New Statesman*

**The Oxford Companion to Wine**
Edited by Jancis Robinson
'the greatest wine book ever published'

*Washington Post*

OXFORD

# Oxford Paperback Reference

**The Kings of Queens of Britain**
John Cannon and Anne Hargreaves

A detailed, fully-illustrated history ranging from mythical and pre-conquest rulers to the present House of Windsor, featuring regional maps and genealogies.

**A Dictionary of Dates**
Cyril Leslie Beeching

Births and deaths of the famous, significant and unusual dates in history – this is an entertaining guide to each day of the year.

'a dipper's blissful paradise ... Every single day of the year, plus an index of birthdays and chronologies of scientific developments and world events.'

*Observer*

**A Dictionary of British History**
Edited by John Cannon

An invaluable source of information covering the history of Britain over the past two millennia. Over 3,600 entries written by more than 100 specialist contributors.

Review of the parent volume
'the range is impressive ... truly (almost) all of human life is here'
Kenneth Morgan, *Observer*

# OXFORD

# Oxford Paperback Reference

**The Concise Oxford Dictionary of English Etymology**
T. F. Hoad

A wealth of information about our language and its history, this reference source provides over 17,000 entries on word origins.

'A model of its kind'

*Daily Telegraph*

**A Dictionary of Euphemisms**
R. W. Holder

This hugely entertaining collection draws together euphemisms from all aspects of life: work, sexuality, age, money, and politics.

Review of the previous edition
'This ingenious collection is not only very funny but extremely instructive too'

Iris Murdoch

**The Oxford Dictionary of Slang**
John Ayto

Containing over 10,000 words and phrases, this is the ideal reference for those interested in the more quirky and unofficial words used in the English language.

'hours of happy browsing for language lovers'

*Observer*

# OXFORD

# Oxford Paperback Reference

**The Concise Oxford Dictionary of Quotations**
Edited by Elizabeth Knowles

Based on the highly acclaimed *Oxford Dictionary of Quotations*, this paperback edition maintains its extensive coverage of literary and historical quotations, and contains completely up-to-date material. A fascinating read and an essential reference tool.

**The Oxford Dictionary of Humorous Quotations**
Edited by Ned Sherrin

From the sharply witty to the downright hilarious, this sparkling collection will appeal to all senses of humour.

**Quotations by Subject**
Edited by Susan Ratcliffe

A collection of over 7,000 quotations, arranged thematically for easy look-up. Covers an enormous range of nearly 600 themes from 'The Internet' to 'Parliament'.

**The Concise Oxford Dictionary of Phrase and Fable**
Edited by Elizabeth Knowles

Provides a wealth of fascinating and informative detail for over 10,000 phrases and allusions used in English today. Find out about anything from the 'Trojan horse' to 'ground zero'.

OXFORD

# Oxford Paperback Reference

**The Concise Oxford Dictionary of World Religions**
Edited by John Bowker

Over 8,200 entries containing unrivalled coverage of all the major world religions, past and present.

'covers a vast range of topics ... is both comprehensive and reliable'
*The Times*

**The Oxford Dictionary of Saints**
David Farmer

From the famous to the obscure, over 1,400 saints are covered in this acclaimed dictionary.

'an essential reference work'
*Daily Telegraph*

**The Concise Oxford Dictionary of the Christian Church**
E. A. Livingstone

This indispensable guide contains over 5,000 entries and provides full coverage of theology, denominations, the church calendar, and the Bible.

'opens up the whole of Christian history, now with a wider vision than ever'
Robert Runcie, former Archbishop of Canterbury

# OXFORD

# More Art Reference from Oxford

**The Grove Dictionary of Art**

The 34 volumes of *The Grove Dictionary of Art* provide unrivalled coverage of the visual arts from Asia, Africa, the Americas, Europe, and the Pacific, from prehistory to the present day.

'succeeds in performing the most difficult of balancing acts, satisfying specialists while ... remaining accessible to the general reader'

*The Times*

**The Grove Dictionary of Art – Online**
**www.groveart.com**

This immense cultural resource is now available online. Updated regularly, it includes recent developments in the art world as well as the latest art scholarship.

'a mammoth one-stop site for art-related information'

*Antiques Magazine*

**The Oxford History of Western Art**
Edited by Martin Kemp

From Classical Greece to postmodernism, *The Oxford History of Western Art* is an authoritative and stimulating overview of the development of visual culture in the West over the last 2,700 years.

'here is a work that will permanently alter the face of art history ... a hugely ambitious project successfully achieved'

*The Times*

**The Oxford Dictionary of Art**
Edited by Ian Chilvers

*The Oxford Dictionary of Art* is an authoritative guide to the art of the western world, ranging across painting, sculpture, drawing, and the applied arts.

'the best and most inclusive single-volume available'

Marina Vaizey, *Sunday Times*